T0181110

International Handbooks on Information Systems

Series Editors

Peter Bernus, Jacek Błażewicz, Günter Schmidt, Michael Shaw

Titles in the Series

M. Shaw, R. Blanning, T. Strader and A. Whinston (Eds.)
Handbook on Electronic Commerce
ISBN 978-3-540-65882-1

J. Błażewicz, K. Ecker, B. Plateau and D. Trystram (Eds.)
Handbook on Parallel and Distributed Processing
ISBN 978-3-540-66441-3

H.H. Adelsberger, Kinshuk, J.M. Pawlowski and D. Sampson (Eds.)
Handbook on Information Technologies for Education and Training
ISBN 978-3-540-74154-1, 2nd Edition

C.W. Holsapple (Ed.)
Handbook on Knowledge Management 1 Knowledge Matters
ISBN 978-3-540-43527-3
Handbook on Knowledge Management 2 Knowledge Directions
ISBN 978-3-540-43848-9

J. Błażewicz, W. Kubiak, I. Morzy and M. Rusinkiewicz (Eds.)
Handbook on Data Management in Information Systems
ISBN 978-3-540-43893-9

P. Bernus, P. Nemes and G. Schmidt (Eds.)
Handbook on Enterprise Architecture
ISBN 978-3-540-00343-4

S. Staab and R. Studer (Eds.)
Handbook on Ontologies
ISBN 978-3-540-40834-5

S.O. Kimbrough and D.J. Wu (Eds.)
Formal Modelling in Electronic Commerce
ISBN 978-3-540-21431-1

P. Bernus, K. Merlins and G. Schmidt (Eds.)
Handbook on Architectures of Information Systems
ISBN 978-3-540-25472-0, 2nd Edition

S. Kirn, O. Herzog, P. Lockemann and O. Spaniol (Eds.)
Multiagent Engineering
ISBN 978-3-540-31406-6

J. Błażewicz, K. Ecker, E. Pesch, G. Schmidt and J. Węglarz (Eds.)
Handbook on Scheduling
ISBN 978-3-54028046-0

F. Burstein and C.W. Holsapple (Eds.)
Handbook on Decision Support Systems 1
ISBN 978-3-540-48712-8

F. Burstein and C.W. Holsapple (Eds.)
Handbook on Decision Support Systems 2
ISBN 978-3-540-48715-9

D. Seese, Ch. Weinhardt and F. Schlottmann (Eds.)
Handbook on Information Technology in Finance
ISBN 978-3-540-49486-7

Steffen Staab • Rudi Studer (Eds.)

Handbook on Ontologies

Second Edition

Editors

Prof. Dr. Steffen Staab
University of Koblenz-Landau
Faculty of Informatics
Universitätsstr. 1
56016 Koblenz
Germany
staab@uni-koblenz.de

Prof. Dr. Rudi Studer
University of Karlsruhe
Institute AIFB
Englerstr. 11
76131 Karlsruhe
Germany
studer@aifb.uni-karlsruhe.de

ISBN 978-3-662-49995-5 ISBN 978-3-540-92673-3 (eBook)
DOI 10.1007/978-3-540-92673-3
Springer Dordrecht Heidelberg London New York

Cover design: WMX Design

Printed on acid-free paper

Springer is part of Springer Science+Business Media (www.springer.com)

To Angela & Irene

Preface

The second edition of the Handbook on Ontologies provides an up-to-date comprehensive overview of the field of ontologies that is evolving rather fast. Since the first edition of the handbook that was finished in 2003 and published in 2004, ontologies have achieved an even more important role with respect to the advancement of established information systems, of systems for data and knowledge management, or of systems for collaboration and information sharing as well as for the development of revolutionary fields such as semantic technologies and, more specifically, the Semantic Web. By covering a broad range of aspects that are related to ontologies, i.e. language and engineering aspects, infrastructures and applications, the reader of this handbook may either get a broad, comprehensive picture of the field of ontologies or he may investigate specific aspects of ontologies that are most relevant for her or his work.

Major Changes with Respect to the 1st Edition

Between the time we wrote the preface to the 1st edition of the ontology handbook, 5 years ago, and today, a large amount of research work, development and use of ontologies have happened. Therefore, the handbook has undergone major changes from the first to the second revision.

At the level of the coarsest granularity, the reader may discover one completely new part (Part III) on – *Ontologies*. This part now covers the description of some very intriguing ontologies that were not around in 2003. Thereby, this part mirrors the fact that – unlike in 2003 – finding ontologies on the Web is now easy, selecting the right one may be the hard and learning by example what is a good ontology or what is a promising field of application for ontologies is absolutely necessary.

Furthermore, the reader may discover that the part on Ontology Infrastructure has been divided into two parts, one on *Infrastructure for Ontologies* and one on *Ontology-Based Infrastructures and Methods*. The first extends the scope of ontologies by providing a larger extent of scalability for dealing with ontologies. The second extends the scope of ontologies by presenting refined and new approaches for putting ontologies into different types of software infrastructures and methods. The latter kind does in fact constitute a generic type of application of ontologies, one that is independent of a particular target domain of application.

Finally, one may find again core parts of the first edition, such as *Part I: Ontology Representation Languages*, *Part II: Ontology Engineering* and *Part VI: Ontology-Based Applications.*

However, within these parts 21 (sic!) of the 36 overall chapters had to be written from scratch. Nearly all the remaining chapters have undergone substantial changes to make them up-to-date. We will not describe these changes in detail, but we want to present to the reader the flow he now finds in the second edition of the ontology handbook.

Overview of the 2nd Edition of the Handbook

'Ontology' is still a rather overloaded term, which is used with several different meanings. This handbook does not consider the philosophical notion of an 'ontology' as addressed in philosophy for more than two thousand years by investigating questions like "what exists?". It rather approaches the notion of ontologies from a Computer Science point of view. In Computer Science, ontologies started to become a relevant notion in the 1990s, related mostly to work in Knowledge Acquisition at that time. In this context the basic definition of an ontology was coined as follows: "An ontology is an explicit specification of a conceptualization" (cf. [7]). I.e. an ontology provides a specification of a conceptualization of generic notions like time and space or of an application domain like knowledge management or life sciences. Starting from this initial definition, various characterizations of ontologies have been developed resulting in the following and nowadays most frequently seen definition: "An ontology is a formal, explicit specification of a shared conceptualization" (cf. [6]). 'Explicit' refers to the fact that all elements of an ontology are explicitly defined, whereas 'formal' means that the ontology specification is given in a language that comes with a formal syntax and semantics, thus resulting in machine executable and machine interpretable ontology descriptions. Finally, 'shared' captures the aspect that an ontology is representing consensual knowledge that has been agreed on by a group of people, typically as a result of a social process.

What Is an Ontology? preceeding the main body of the handbook will further elaborate the definition of what an ontology is:

Nicola Guarino, Daniel Oberle, and Steffen Staab:
What Is an Ontology?

In order not to appeal to intuition and to fix the terminology precisely, the introduction will give a very formal approach to the definition of ontologies.

The formally well-versed reader may enjoy the formal accuracy of the definition of ontology – at a level of precision that has been sorely lacking so far. However, the more application-oriented reader may focus only on the run-through example given in the introduction and rather warm up with the applications he may find in *Part VI: Ontology-Based Applications.*

Part I: Ontology Representation Languages

The main body of the handbook starts with a part on current representation languages for ontologies in correlation with other aspects, such as data, the Web and rules. The first chapter describes the family of ontology languages described as *Description Logics* that constitutes the foundation for the majority of ontology work found nowadays. Description Logics is a subset of first-order predicate logics and combines expressiveness with a well-understood logical framework:

U. Sattler, F. Baader, and I. Horrocks: *Description Logics*

The second chapter describes an approach to ontologies that is derived from work in the area of logics-based databases. *Ontologies in F-Logic* ties in neatly with existing database frameworks and has found many up-and-running commercial applications.

J. Angele, G. Lausen, and M. Kifer: *Ontologies in F-Logic*

As explained in *What is an Ontology?* the aspect of sharing is central to the notion of ontologies. The Web is the current, almost pervasive means to share information and also knowledge. A lightweight representation for data and knowledge on the Web is the *Resource Description Framework (RDF).* Its core objectives and its connection to logical frameworks are explained in the following chapter.

J.Z. Pan: *Resource Description Framework*

Very soon after the definition of RDF and RDF Schema, ontology engineers and users recognized the need for a more expressive ontology representation language. The outcome of a joint development process is explained in *Web Ontology Language: OWL.* OWL and particularly its flavor of OWL-DL now constitute the language of choice for representing an ontology in the Semantic Web.

G. Antoniou and F. van Harmelen: *Web Ontology Language: OWL*

When ontologies were developed to improve the expressiveness of knowledge bases, they were typically used in connection with production rule-based systems. While production rule-based systems have severe disadvantages with regard to manageability because of a lack of declarativeness, new developments since the first edition of the handbook have shown how logical rules may be included into ontology languages such as description logics. Some corresponding results are presented in Chapter "Ontologies and Rules".

B. Parsia and P. Hitzler: *Ontologies and Rules*

This part of the book will be very helpful to the reader if he wants to understand the representational underpinnings of ontologies. Many examples in the chapters will help him to get an intuitive understanding of the representational issues. Advanced issues and more complete descriptions are pointed out in the many references given in these chapters.

Part II: Ontology Engineering

The second part of the handbook deals with the practical development of ontologies. The first chapter presents a generic model of ontology development that is now state-of-the-art and found in many variations in textbooks on the topic of ontology engineering. In fact, it also introduces the different aspects of ontology engineering found in the remainder of this part of the ontology handbook.

Y. Sure, S. Staab, and R. Studer: *Ontology Engineering Methodology*

While the general methodological blueprint reflects concerns that one would also find in a software engineering process, the next chapter focuses on an issue that becomes almost unavoidable for large, realistic ontologies. It shows how to develop ontologies in a distributed setting where most experts cannot afford to assemble often – if at all.

H.S. Pinto, C. Tempich, and S. Staab:
Ontology Engineering and Evolution in a Distributed World Using DILIGENT

The sound engineering of ontologies needs sophisticated tools. The following two chapters describe *Formal Concept Analysis* and *OntoClean*, tools that both aim at improving the inheritance relationships of specified concepts. The first does so by analysing the correlation between intensions and extensions of concepts, while the second investigates how the variability of a concept is constrained by the intended conceptualization.

G. Stumme: *Formal Concept Analysis*

N. Guarino and C.A. Welty: *An Overview of OntoClean*

Beyond conceptual relationships, ontology engineers need to express specific concerns: knowledge about knowledge, part-whole-relationships, etc. The chapter on *Ontology Design Patterns* explains how the idea of software design patterns can be adopted in ontologies to provide an understandable and expressive model.

A. Gangemi and V. Presutti: *Ontology Design Patterns*

Such ontology design patterns may be filled by manual work, but the use of machine learning mechanisms as a tool for suggesting ontological constructs is an increasingly important means.

P. Cimiano, A. Mädche, S. Staab, and J. Völker: *Ontology Learning*

When learning an ontology, the induction mechanisms need to distinguish between the – possibly multiple – names of a concept or relation and the concept or relation itself. Thus, it constructs a lexicon. Investigating existing lexica, one finds that these actually contain many more specific hints useful for reuse during ontology construction.

G. Hirst: *Ontology and the Lexicon*

At the end of the ontology engineering process, the resulting ontology needs to be matched against the requirements. Such requirements may be task or domain specific (e.g. high precision when retrieving knowledge); however, there are also general criteria of soundness of ontologies that are analysed by Vrandečić.

D. Vrandečić: *Ontology Evaluation*

The whole process of ontology engineering may be supported by specific tools that allow the management of specification and design documents as well as ontology-specific concerns such as traceability information, patterns, lexica, etc. Though the full support of all these aspects has not been realized by any environment, the current state-of-the-art is elaborated on by Mizoguchi and Kozaki.

R. Mizoguchi and K. Kozaki: *Ontology Engineering Environments*

The part on ontology engineering closes the ultimate issue of concern about ontologies in *any* kind of application: ontologies are supposed to improve the total cost of operating a system by improving system aspects such as efficiency or quality. However, with regard to the total cost of ownership, one also needs to consider the amount of time and money to be invested in the construction and the maintenance of the ontology.

E. Simperl and C. Tempich:
Exploring the Economical Aspects of Ontology Engineering

Only if the overall balance between investment and return yields a sufficiently large margin, the employment of an ontology can be taken into consideration. As will be shown in the parts on Ontology-Based Infrastructures and Applications, the improvement of ontology engineering, an increased amount of experience and sound and scalable infrastructure, now contribute successfully to the uptake of ontologies and ontology technologies.

Part III: Ontologies

Some important experiences of ontology engineering are captured in *structures* of existing ontologies and in the *use* of existing ontologies. Authors have distinguished different types of ontologies at different levels of generality and for different types of purposes (cf. [1,3,7]): 'Top-level ontologies', sometimes also called 'foundational ontologies', capture general concepts that are domain-independent, like an event. Or they specify the conceptualization of common sense knowledge, e.g. about space and time. 'Domain ontologies' model concepts and relations that are relevant for a specific domain, like 'gene' in a life science domain. Similarly, 'Task ontologies' describe concepts that are specific for a task at hand, like 'symptom' for a diagnostic task. Finally, at the lowest level of abstraction, so-called 'Application Ontologies' are specified that combine domain and task ontologies and extend them with more refined domain and task specific concepts and relations, like 'fever' as a symptom for diagnosis in the medical domain.

A widely used foundational ontology has been defined with DOLCE:

S. Borgo and C. Masolo: *Foundational Choices in DOLCE*

DOLCE is re-used in some domain and application ontologies in order to provide a sound and comprehensively specific, yet extensible framework. Extensibility is a core concern, because it is impossible to formalize domains like *Software* or *Multimedia* completely.

D. Oberle, S. Grimm, and S. Staab: *An Ontology for Software*

R. Arndt, R. Troncy, S. Staab, and L. Hardman:
COMM: A Core Ontology for Multimedia Annotation

The next chapter targets concerns about processes and tasks. It reflects the fact that the integration of procedural aspects into static constraints specified in ontologies is gaining importance because of application domains such as Web Services (Chapter "Semantic Web Services").

M. Grüninger: *Using the PSL Ontology*

The final two chapters of this part consider the re-use of knowledge structures toward a more comprehensive formalization in an ontology. Both in biomedicine and in the domain of managing cultural objects, the need for rich structuring of the complex domains have led to a long tradition in defining knowledge structures and to a fruitful field of application for ontologies.

N. Shah and M. Musen:
Ontologies for Formal Representation of Biological Systems

M. Doerr: *Ontologies for Cultural Heritage*

Part IV: Infrastructures for Ontologies

The scalability of ontology technology is crucial to its uptake in industrial settings. Thereby, one can find an overwhelming development. As recently as 1994, Benchmarking of systems (cf. [5]) surveyed complete ontology reasoning that would work on rather weakly expressive languages with a couple of hundred entities at the most (i.e. concepts and/or instances). The situation has changed completely with current infrastructures for ontologies that target high to very high scalability of dealing with ontological structures.

Current work targets simple ontological structures in RDF with billions of triples[1]. The first chapter of this part explains state-of-the-art systems for *RDF Storage and Retrieval.*

A. Hertel, J. Broekstra, and H. Stuckenschmidt:
RDF Storage and Retrieval Systems

With regard to description logic languages like OWL-DL, one nowadays finds ontology reasoning systems that can also handle 10^5 concepts in ontological concept definitions occurring in practice (cf. also [4]).

R. Möller and V. Haarslev: *Tableau-Based Reasoning*

Furthermore, developments in recent years have joined means of optimization used in the fields of logic programming and logic databases (also cf. Chapter "Ontologies in F-Logic") with the field of description logics (Chapter "Description Logics") in order to achieve higher scalability for ontological reasoning with databases (more specifically: A-Boxes).

B. Motik: *Resolution-Based Reasoning for Ontologies*

Beyond infrastructures for single ontologies, this part also presents approaches towards managing *multiple* ontologies. In order to search for and re-use ontologies it is necessary to index them. *Ontology Repositories* provide such means for supporting the process for search and re-use.

J. Hartmann, R. Palma, and A. Gómez-Pérez: *Ontology Repositories*

On the Semantic Web ontologies, and the entities they contain are supposed to cross-reference to each other to allow for a seamless use of ontologies. If such cross-references do not yet exist, infrastructures for establishing ontology mappings allow for their definition.

N.F. Noy: *Ontology Mapping*

[1] Cf. the billion triple challenge at `http://challenge.semanticweb.org/`

Part V: Ontology-Based Infrastructure and Methods

Infrastructures and methods may be extended with ontology technologies to benefit from the agreement on a shared vocabulary and the underlying reasoning technology. These characteristics are frequently given when the domain of applications of the infrastructures and methods is *inherently complex*, must be *kept extensible*, and requires the interaction of some *user*.

The first two chapters on ontology-based infrastructures and methods fall into the domain of software. The first contribution identifies different opportunities in the Software Engineering lifecycle where ontologies may play a role – the 'user' here is the software developer.

D. Gašević, N. Kaviani, and M. Milanović:
Ontologies and Software Engineering

The second contribution of this kind presents an approach for describing dynamic access to software provided through the means of Web Services. The approach targets a software environment where software building block may be composed ad hoc, the user being either a software developer, a person configuring software, or even an end user.

J. de Bruijn, M. Kerrigan, M. Zaremba, and D. Fensel:
Semantic Web Services

The next two chapters present the use of ontologies in data analysis task. *Ontologies for Machine Learning* show how domain complexity may be used by machine learning algorithms in order to enhance effectiveness and/or explainability of machine learning and data mining results.

S. Blöhdorn and A. Hotho: *Ontologies for Machine Learning*

The second chapter of this kind shows how ontologies are used for capturing the complexities of entities and relationships that may be discovered from texts using automated information extraction.

C. Nédellec, A. Nazarenko, and R. Bossy: *Information Extraction*

In the final chapter of this part Dzbor, Motta and Gridinoc consider an ontology-enhanced infrastructure for user interaction, more specifically an ontology-extended browser.

M. Dzbor, E. Motta, and L. Gridinoc:
Browsing and Navigation in Semantically Rich Spaces:
Experiences with Magpie Applications

While such infrastructures and methods as described in this part of the handbook can be considered as constituting some kind of application, they are generic and may be used in very different domains, e.g. from various types of business processes (e.g. customer relationship management or knowledge management) to health and to information repositories (recommender systems, portals). In the following part, we target some of these specific areas.

Part VI: Ontology-Based Applications

Over the last 15 years, ontology-based applications have been spreading and maturing. One now finds ontology-based applications in areas as diverse as customer support and engineering of cars.

Coming from knowledge acquisition and knowledge-based systems, a core area of application for ontologies has been knowledge management. But, rather than fully capturing knowledge about a particular domain, the idea of using ontologies in knowledge was that one would agree on a common vocabulary and use it for knowledge shared by formal as well as by informal means, e.g. texts, while making best use of the reasoning technology in the background.

A. Abecker and L. van Elst: *Ontologies for Knowledge Management*

As mentioned earlier (Chapter "Ontologies for Formal Representation of Biological Systems"), biomedical applications have been using ontological structuring for a long time – albeit tending towards less formal structures. There, the most prominent domain of application was data integration for human use. Now, however, the target of ontologies in this domain is the support for computational support of biological data, e.g. in experiments and simulations.

R. Stevens and P. Lord: *Application of Ontologies in Bioinformatics*

Two more chapters target the use of ontologies in portals. The domains of art and cultural heritage have a long interest in comprehensive classifications to manage the many artifacts available, e.g. in museums, and present them to the public.

E. Hyvönen: *Semantic Portals for Cultural Heritage*

The final chapter of the handbook targets the domain of digital libraries and shows how ontology-based recommender systems can facilitate access to such libraries – especially in situations where traditional recommender systems stall because of the so-called "cold start problem", i.e. the disadvantage that it is difficult to provide good recommendations when only little is known about the user of the system.

S.E. Middleton, D. De Roure, and N.R. Shadbolt:
Ontology-Based Recommender Systems

Conclusion

As the size of the handbook is physically limited, this last chapter will conclude the handbook. Many more chapters would be necessary to fully capture the range of ontology technologies and applications. New ontology representation languages have been researched and become input for standardization processes, such as the extension of OWL-DL into OWL-2. New tools are being developed to provide infrastructure for ontologies via Web frontends. New applications pop up almost daily and in unforeseen domains. Hence, this handbook cannot be complete and will not be in the next couple of years. However, we see this as a vital sign for the area of research and development of ontology and ontology technologies and not as a detriment to the intended usefulness of this handbook. We hope that you, the reader, will enjoy its content and make productive use of it.

Koblenz & Karlsruhe, October 2008

Steffen Staab
Rudi Studer

Acknowledgments

We gratefully acknowledge efforts by all the different authors who also acted as peer reviewers. We have been greatly dependent on their dedication to provide high-quality papers as well as comprehensive reviews of other papers now in the handbook. We sincerely thank Thomas Franz and Gerd Gröner, University of Koblenz, for organizational support of the paper handling and reviewing process. We acknowledge funding by EU projects NeOn, IST-2005-027595, `http://www.neon-project.org`, and X-media, FP6-26978, `http://www.x-media-project.org`, that has enabled this support.

References

1. Stephan Grimm, Pascal Hitzler, and Andreas Abecker. Knowledge representation and ontologies. In Rudi Studer, Stephan Grimm, and Andreas Abecker, editors, *Semantic Web Services*, pages 51–105. Springer Verlag, 2007.
2. Thomas R. Gruber. A translation approach to portable ontology specifications. *Knowledge Acquisition*, 5(2):199–220, 1993.
3. Nicola Guarino. Semantic matching: Formal ontological distinctions for information organization, extraction, and integration. In M. T. Pazienza, editor, *Information Extraction: A Multidisciplinary Approach to an Emerging Information Technology*, pages 139–170. Springer Verlag, 1997.
4. Volker Haarslev and Ralf Möller. High performance reasoning with very large knowledge bases: A practical case study. In Bernhard Nebel, editor, *IJCAI*, pages 161–168. Morgan Kaufmann, 2001.
5. Jochen Heinsohn, Daniel Kudenko, Bernhard Nebel, and Hans-Jürgen Profitlich. An empirical analysis of terminological representation systems. *Artif. Intell.*, 68(2):367–397, 1994.
6. Rudi Studer, V. Richard Benjamins, and Dieter Fensel. Knowledge engineering: Principles and methods. *Data & Knowledge Engineering*, 25(1-2):161–197, 1998.
7. Gert-Jan van Heijst, Guus Schreiber, and Bob Wielinga. Using explicit ontologies in kbs development. *Intl. J. Human-Computer Studies*, 46(2/3):183–292, 1997.

Contents

Part V Ontology-Based Infrastructure and Methods

Part VI Ontology-Based Applications

What Is an *Ontology*?

Nicola Guarino[1], Daniel Oberle[2], and Steffen Staab[3]

[1] ITSC-CNR, Laboratory for Applied Ontology, 38100 Trento, Italy,
nicola.guarino@cnr.it
[2] SAP Research, CEC Karlsruhe, 76131 Karlsruhe, Germany,
d.oberle@sap.com
[3] University of Koblenz-Landau, ISWeb, 56016 Koblenz, Germany,
staab@uni-koblenz.de

Summary. The word "ontology" is used with different senses in different communities. The most radical difference is perhaps between the philosophical sense, which has of course a well-established tradition, and the computational sense, which emerged in the recent years in the knowledge engineering community, starting from an early informal definition of (computational) ontologies as "explicit specifications of conceptualizations". In this paper we shall revisit the previous attempts to clarify and formalize such original definition, providing a detailed account of the notions of *conceptualization* and *explicit specification*, while discussing at the same time the importance of *shared* explicit specifications.

1 Introduction

The word "ontology" is used with different meanings in different communities. Following [9], we distinguish between the use as an uncountable noun ("Ontology," with uppercase initial) and the use as a countable noun ("an ontology," with lowercase initial) in the remainder of this chapter. In the first case, we refer to a philosophical discipline, namely the branch of philosophy which deals with the *nature* and *structure* of "reality." Aristotle dealt with this subject in his Metaphysics[1] and defined Ontology[2] as the science of "being *qua* being," i.e., the study of attributes that belong to things because of their very nature. Unlike the experimental sciences, which aim at discovering and modeling reality under a certain perspective, Ontology focuses on the

[1] The first books of Aristotle's treatises, known collectively as "Organon," deal with the nature of the world, i.e., physics. Metaphysics denotes the subjects dealt with in the rest of the books – among them Ontology. Philosophers sometimes equate Metaphysics and Ontology.

[2] Note, that the term "Ontology" itself was coined only in the early seventeenth century [13].

S. Staab and R. Studer (eds.), *Handbook on Ontologies,* International Handbooks on Information Systems, DOI 10.1007/978-3-540-92673-3,

nature and structure of things per se, independently of any further considerations, and even independently of their actual existence. For example, it makes perfect sense to study the Ontology of unicorns and other fictitious entities: although they do not have actual existence, their nature and structure can be described in terms of general categories and relations.

In the second case, which reflects the most prevalent use in Computer Science, we refer to *an* ontology as a special kind of information object or computational artifact. According to [7, 8], the account of existence in this case is a pragmatic one: "For AI systems, what 'exists' is that which can be represented."

Computational ontologies are a means to formally model the structure of a system, i.e., the relevant entities and relations that emerge from its observation, and which are useful to our purposes. An example of such a system can be a company with all its employees and their interrelationships. The ontology engineer analyzes relevant entities[3] and organizes them into *concepts* and *relations*, being represented, respectively, by unary and binary predicates.[4] The backbone of an ontology consists of a generalization/specialization hierarchy of concepts, i.e., a taxonomy. Supposing we are interested in aspects related to human resources, then *Person*, *Manager*, and *Researcher* might be relevant concepts, where the first is a superconcept of the latter two. *Cooperates-with* can be considered a relevant relation holding between persons. A concrete person working in a company would then be an instance of its corresponding concept.

In 1993, Gruber originally defined the notion of an ontology as an "explicit specification of a conceptualization" [7].[5] In 1997, Borst defined an ontology as a "formal specification of a shared conceptualization" [1]. This definition additionally required that the conceptualization should express a *shared* view between several parties, a consensus rather than an individual view. Also, such conceptualization should be expressed in a (formal) machine readable format. In 1998, Studer et al. [15] merged these two definitions stating that: "An ontology is a formal, explicit specification of a shared conceptualization."

[3] Entity denotes the most general being, and, thus, subsumes subjects, objects, processes, ideas, etc.

[4] Unfortunately, the terminology used in Computer Science is problematic here. What we call "concepts" in this chapter may be better called "properties" or "categories." Regrettably, "property" is used to denote a binary relation in RDF(S), so we shall avoid using it. Also, Smith made us aware that the notion of "concept" is quite ambiguous [14]. A way to solve the terminological conflict is to adopt the philosophical term "universal," which roughly denotes those entities that can have instances; particulars are entities that do not have instances. What we call "concepts" correspond to unary universals, while "relations" correspond to binary universals.

[5] Other definitions of an ontology have surfaced in the literature, e.g., [16] or [11], which are similar to Gruber's. However, the one from Gruber seems to be the most prevalent and most cited.

All these definitions were assuming an informal notion of "conceptualization," which was discussed in detail in [9]. In the following, we shall revisit such discussion, by focusing on the three major aspects of the definition by Studer et al.:

- What is a *conceptualization*?
- What is a proper *formal, explicit specification*?
- Why is '*shared*' of importance?

It is the task of this chapter to provide a concise view of these aspects in the following sections. It lies in the nature of such a chapter that we have tried to make it more precise and formal than many other useful definitions of ontologies that do exist – but that do not clarify terms to the degree of accuracy that we target here.

Accordingly, the reader new to the subject of ontologies may prefer to learn first about applications and examples of ontologies in the latter parts of this book and may decide to return to this opening chapter once he wants to see the common *raison d'être* behind the different approaches.

2 What is a *Conceptualization*?

Gruber [7, 8] refers to the notion of a conceptualization according to Genesereth and Nilsson [5], who claim: "A body of formally represented knowledge is based on a conceptualization: the objects, concepts, and other entities that are assumed to exist in some area of interest and the relationships that hold among them. A conceptualization is an abstract, simplified view of the world that we wish to represent for some purpose. Every knowledge base, knowledge-based system, or knowledge-level agent is committed to some conceptualization, explicitly or implicitly."

Despite the complex mental nature of the notion of "conceptualization," Genesereth and Nilsson choose to explain it by using a very simple mathematical representation: an extensional relational structure.

Definition 2.1 (Extensional relational structure) An extensional relational structure, *(or a* conceptualization according to Genesereth and Nilsson*), is a tuple* $(D, \mathbf{R})$ *where*

- D *is a set called the universe of discourse*
- $\mathbf{R}$ *is a set of relations on* D

Note that, in the above definition, the members of the set $\mathbf{R}$ are ordinary mathematical relations on D, i.e., sets of ordered tuples of elements of D. So each element of $\mathbf{R}$ is an *extensional* relation, reflecting a *specific* world state involving the elements of D, such as the one depicted in Fig. 1, concerning the following example.

Example 2.1 *Let us consider human resources management in a large software company with 50,000 people, each one identified by a number (e.g., the social security number, or a similar code) preceded by the letter I. Let us assume that our universe of discourse D contains all these people, and that we are only interested in relations involving people. Our* **R** *will contain some unary relations, such as Person, Manager, and Researcher, as well as the binary relations reports-to and cooperates-with.*[6] *The corresponding extensional relation structure* $(D, \mathbf{R})$ *looks as follows:*

- $D = \{I000001, ..., I050000, ...\}$
- $\mathbf{R} = \{Person, Manager, Researcher, cooperates\text{-}with, reports\text{-}to\}$

Relation extensions reflect a specific world. Here, we assume that Person comprises the whole universe D and that Manager and Researcher are strict subsets of D. The binary relations reports-to and cooperates-with are sets of tuples that specify every hierarchical relationship and every collaboration in our company. Some managers and researchers are depicted in Fig. 1. Here, I046758, a researcher, reports to his manager I034820, and cooperates with another researcher, namely I044443.

- $Person = D$
- $Manager = \{..., I034820, ...\}$
- $Researcher = \{..., I044443, ..., I046758, ...\}$
- $reports\text{-}to = \{..., (I046758, I034820), (I044443, I034820), ...\}$
- $cooperates\text{-}with = \{..., (I046758, I044443), ...\}$

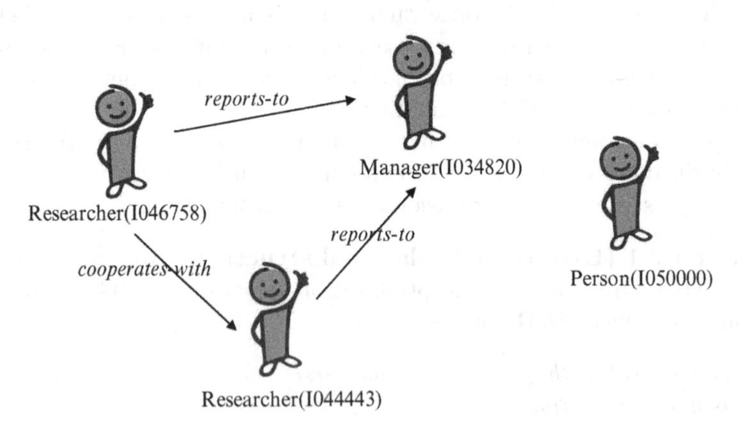

Fig. 1. A tiny part of a specific world with persons, managers, researchers, and their relationships in the running example of human resources in a large software company

[6] The name of a person could also be assigned via relations, e.g., *firstname(I046758,'Daniel')* and *lastname(I046758,'Oberle')*.

Despite its simplicity, this extensional notion of a conceptualization does not really fit our needs and our intuition, mainly because it depends too much on a specific state of the world. Arguably, a conceptualization is about concepts. Now, should our concept of *reports-to* change when the hierarchical structure of our company changes? Indeed, as discussed in [9], a conceptualization should not change when the world changes. Otherwise, according to the Genesereth and Nilsson's view given in Definition 2.1, every specific people interaction graph, such as the one depicted in Fig. 1, would correspond to a different conceptualization, as shown in Example 2.2.

Example 2.2 *Let us consider the following alteration of Example 2.1 with* $D' = D$ *and* $\mathbf{R}' = \{Person, Manager, Researcher, reports\text{-}to', cooperates\text{-}with\}$ *where* $reports\text{-}to' = reports\text{-}to \cup \{(I034820, I050000)\}$.

Although we only added one new reporting relationship, it is obvious that $(D, \mathbf{R}) \neq (D', \mathbf{R}')$ and, thus, we have two different conceptualizations according to Genesereth and Nilsson.

The problem is that the extensional relations belonging to $\mathbf{R}$ reflect a specific world state. However, we need to focus on the meaning of the underlying concepts, which are independent of a single world state: for instance, the meaning of *cooperates-with* lies in the particular way two persons act in the company.

In practice, understanding such meaning implies having a rule to decide, observing different behavior patterns, whether or not two persons are cooperating. Suppose that, in our case, for two persons *I046758* and *I044443* to cooperate means that (1) both declare to have the same goal; (2) both do something to achieve this goal. Then, the meaning of "cooperating" can be defined as a *function* that, for each global behavioral context involving all our universe, gives us the list of couples who are actually cooperating in that context. The reverse of this function *grounds* the meaning of a concept in a specific world state. Generalizing this approach, and abstracting from time for the sake of simplicity, we shall say that an *intensional* relation[7] (as opposed to an *extensional* relation) is a function from a set of maximal world states (the global behavioral contexts in our case) into extensional relations. This is the common way of expressing intensions, which goes back to Carnap [2] and is adopted and extended in Montague's semantics [4].

To formalize this notion of intensional relation, we first have to clarify what a "world" and a "world state" is. We shall define them with reference to the notion of "system," which will be given for granted: since we are dealing with computer representations of real phenomena, a system is simply the given piece of reality we want to model, which, at a given degree of granularity, is

[7] To underly their link with conceptualizations, Guarino has proposed to call such intensional relations "conceptual relations" in [10].

"perceived" by an observing agent (typically external to the system itself) by means of an array of "observed variables."[8]

In our case, this system will be an actual group of people interacting in certain ways. For the sake of simplicity, we shall assume to observe this system at a granularity where single persons can be considered as atoms, so we shall abstract, e.g., from body parts. Moreover, we shall assume that the only observed variables are those which tell us whether a person has a certain goal (belonging to a pre-determined list), and whether such person is actually acting to achieve such goal. Supposing there is just one goal, we have $50{,}000 + 50{,}000 = 100{,}000$ variables. Each combination of such variables is a world state. Two different agents (outside the observed system) will share the same meaning of "cooperating" if, in presence of the same world states, will pick up the same couples as instances of the *cooperates-with* relation. If not, they will have different conceptualizations, i.e., *different ways of interpreting their sensory data.* For instance, an agent may assume that sharing a goal is enough for cooperating, while the other may require in addition some actual work aimed at achieving the goal.

Definition 2.2 (World) *With respect to a specific system S we want to model, a* world state *for S is a maximal observable state of affairs, i.e., a unique assignment of values to all the observable variables that characterize the system. A* world *is a totally ordered set of world states, corresponding to the system's evolution in time. If we abstract from time for the sake of simplicity, a world state coincides with a world.*

At this point, we are ready to define the notion of an intensional relation in more formal terms, building on [9], as follows:

Definition 2.3 (Intensional relation, or *conceptual relation*) *Let S be an arbitrary system, D an arbitrary set of distinguished elements of S, and W the set of world states for S (also called worlds, or possible worlds). The tuple $<D, W>$ is called a* domain space for S, *as it intuitively fixes the space of variability of the universe of discourse D with respect to the possible states of S. An* intensional relation *(or conceptual relation) ρ^n of arity n on $<D, W>$ is a total function $\rho^n : W \rightarrow 2^{D^n}$ from the set W into the set of all n-ary (extensional) relations on D.*

Once we have clarified what a conceptual relation is, we give a representation of a conceptualization in Definition 2.4. Below, we also show how the conceptualization of our human resources system looks like in Example 2.3.

Definition 2.4 (Intensional relational structure, or *conceptualization*) *An* intensional relational structure *(or a* conceptualization *according to Guarino) is a triple $\mathbf{C} = (D, W, \Re)$ with*

[8] It is important to note that, if we want to provide a well-founded, grounded account of meaning, this system needs to be first of all a physical system, and not an abstract entity.

- D *a universe of discourse*
- W *a set of possible worlds*
- $\Re$ *a set of conceptual relations on the domain space* $<D, W>$

Example 2.3 *Coming back to the Examples 2.1 and 2.2, we can see them as describing two different worlds compatible with the following conceptualization* **C***:*

- $D = \{I000001, ..., I050000, ...\}$ *the universe of discourse*
- $W = \{w_1, w_2, ...\}$ *the set of possible worlds*
- $\Re = \{Person^1, Manager^1, Researcher^1, cooperates\text{-}with^2, reports\text{-}to^2\}$ *the set of conceptual relations*

For the sake of simplicity, we assume that the unary conceptual relations, viz., $Person^1$*,* $Manager^1$*, and* $Researcher^1$*, are rigid, and, thus, map to the same extensions in every possible world. We do not make this specific assumption here for the binary* $reports\text{-}to^2$ *and* $cooperates\text{-}with^2$*:*

- *for all worlds w in W:* $Person^1(w) = D$
- *for all worlds w in W:* $Manager^1(w) = \{..., I034820, ...\}$
- *for all worlds w in W:* $Researcher^1(w) = \{..., I044443, ..., I046758, ...\}$
- $reports\text{-}to^2(w_1) = \{..., (I046758, I034820), (I044443, I034820), , ...\}$
- $reports\text{-}to^2(w_2) = \{..., (I046758, I034820), (I044443, I034820), (I034820, I050000), ...\}$
- $reports\text{-}to^2(w_3) = ...$
- $cooperates\text{-}with^2(w_1) = \{..., (I046758, I044443), ...\}$
- $cooperates\text{-}with^2(w_2) = ...$

3 What is a Proper *Formal, Explicit Specification*?

In practical applications, as well as in human communication, we need to use a *language* to refer to the elements of a conceptualization: for instance, to express the fact that *I046758* cooperates with *I044443*, we have to introduce a specific symbol (formally, a predicate symbol, say *cooperates-with*, which, in the user's intention, is intended to represent a certain conceptual relation. We say in this case that our language (let us call it **L**) *commits* to a conceptualization.[9] Suppose now that **L** is a first-order logical language, whose nonlogical symbols (i.e., its *signature*, or its *vocabulary*) are the elements of the set {*I046758, I044443, cooperates-with, reports-to*}. How can we make sure

[9] Of course, properly speaking, it is an *agent* who commits to a conceptualization while using a certain language: what we call the *language commitment* is an account of the competent use of the language by an agent who adopts a certain conceptualization.

that such symbols are *interpreted* according to the conceptualization we commit to? For instance, how can we make sure that, for somebody who does not understand English, *cooperates-with* is not interpreted as corresponding to our conceptualization of *reports-to*, and vice versa? Technically, the problem is that a logical signature can, of course, be interpreted in arbitrarily many different ways. Even if we fix *a priori* our interpretation domain (the *domain of discourse*) to be a subset of our *cognitive domain*, the possible interpretation functions mapping predicate symbols into proper subsets of the domain of discourse are still unconstrained. In other words, once we commit to a certain conceptualization, we have to make sure to only admit those *models* which are *intended* according to the conceptualization. For instance, the intended models of the *cooperates-with* predicate will be those such that the interpretation of the predicate returns one of the various possible extensions (one for each possible world) of the conceptual relation denoted by the predicate. The problem however is that, to specify what such possible extensions are, we need to *explicitly* specify our conceptualization, while conceptualizations are typically in the mind of people, i.e., *implicit*.

Here emerges the role of ontologies as "explicit specifications of conceptualizations." In principle, we can explicitly specify a conceptualization in two ways: *extensionally* and *intensionally*. In our example, an *extensional* specification of our conceptualization would require listing the extensions of every (conceptual) relation for all possible worlds. However, this is impossible in most cases (e.g., if the universe of discourse D or the set of possible worlds W are infinite) or at least very impractical. In our running example, we are dealing with thousands of employees and their possible cooperations can probably not be fully enumerated. Still, in some cases it makes sense to partially specify a conceptualization in an extensional way, by means of examples, listing the extensions of conceptual relations in correspondence of selected, stereotypical world states. In general, however, a more effective way to specify a conceptualization is to fix a language we want to use to talk of it, and to constrain the interpretations of such a language in an *intensional* way, by means of suitable axioms (called *meaning postulates* [2]). For example, we can write simple axioms stating that *reports-to* is asymmetric and intransitive, while *cooperates-with* is symmetric, irreflexive, and intransitive. In short, an ontology is just a set of such axioms, i.e., a logical theory designed in order to capture the intended models corresponding to a certain conceptualization and to exclude the unintended ones. The result will be an *approximate* specification of a conceptualization: the better intended models will be captured and non-intended models will be excluded (cf. Fig. 2).

The axioms for intensionally and explicitly specifying the conceptualization can be given in an *informal* or *formal* language **L**. As explained in the introduction, [15] requires that the explicit specification must be formal in addition to what proposed in [1,7]. 'Formal' refers to the fact that the expressions must be machine readable, hence natural language is excluded. Let us now discuss all the notions above in a more formal way.

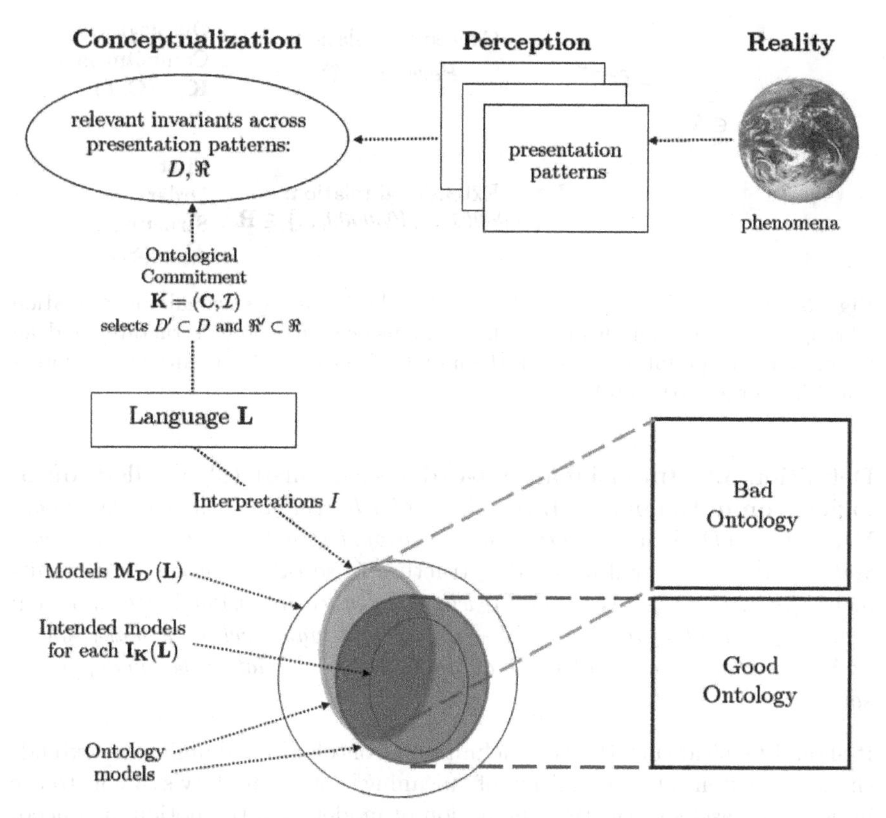

Fig. 2. The relationships between phenomena occurring in reality, their perception (at different times), their abstracted conceptualization, the language used to talk about such conceptualization, its intended models, and an ontology

3.1 Committing to a Conceptualization

Let us assume that our language $\mathbf{L}$ is (a variant of) a first-order logical language, with a vocabulary $\mathbf{V}$ consisting of a set of constant and predicate symbols (we shall not consider function symbols here). We shall introduce the notion of ontological commitment by extending the standard notion of a (extensional) first order structure to that of an *intensional* first order structure.

Definition 3.1 (Extensional first-order structure) *Let* $\mathbf{L}$ *be a first-order logical language with vocabulary* $\mathbf{V}$ *and* $S = (D, \mathbf{R})$ *an extensional relational structure. An* extensional first order structure *(also called* model *for* $\mathbf{L}$*) is a tuple* $M = (S, I)$*, where* I *(called* extensional interpretation function*) is a total function* $I : \mathbf{V} \rightarrow D \cup \mathbf{R}$ *that maps each vocabulary symbol of* $\mathbf{V}$ *to either an element of* D *or an extensional relation belonging to the set* $\mathbf{R}$.

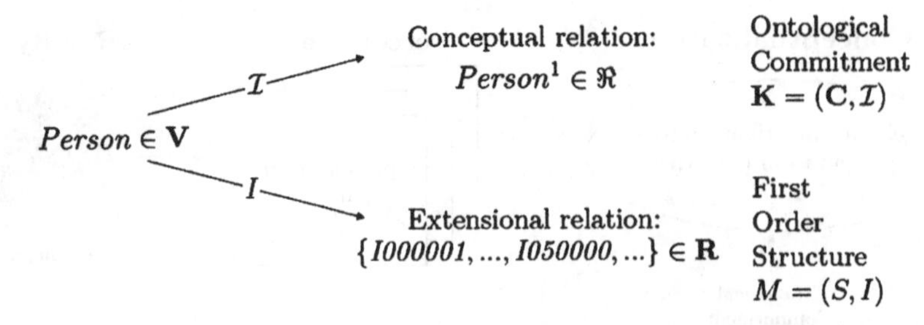

Fig. 3. The predicate symbol *Person* has both an extensional interpretation (through the usual notion of model, or extensional first-order structure) and an intensional interpretation (through the notion of ontological commitment, or intensional first order structure)

Definition 3.2 (Intensional first-order structure) (also called: ontological commitment) *Let* **L** *be a first-order logical language with vocabulary* **V** *and* $\mathbf{C} = (D, W, \Re)$ *an intensional relational structure (i.e., a conceptualization). An* intensional first order structure *(also called* ontological commitment*) for* **L** *is a tuple* $\mathbf{K} = (\mathbf{C}, \mathcal{I})$*, where* $\mathcal{I}$ *(called* intensional interpretation function*) is a total function* $\mathcal{I} : \mathbf{V} \rightarrow D \cup \Re$ *that maps each vocabulary symbol of* **V** *to either an element of* D *or an intensional relation belonging to the set* $\Re$.

It should be clear now that the definition of ontological commitment extends the usual (extensional) definition of "meaning" for vocabulary symbols to the intensional case, substituting the notion of model with the notion of conceptualization. Figure 3 captures this idea.

Example 3.1 *Coming back to our Example 2.1, the vocabulary* **V** *coincides with the relation symbols, i.e.,* **V** $= \{$*Person, Manager, Researcher, reports-to, cooperates-with*$\}$*. Our ontological commitment consists of mapping the relation symbol Person to the conceptual relation* $Person^1$ *and proceeding alike with Manager, Researcher, reports-to, and cooperates-with.*

3.2 Specifying a Conceptualization

As we have seen, the notion of ontological commitment is an extension of the standard notion of model. The latter is an extensional account of meaning, the former is an intensional account of meaning. But what is the relationship between the two? Of course, once we specify the intensional meaning of a vocabulary through its ontological commitment, somehow we also constrain its models. Let us introduce the notion of *intended model* with respect to a certain ontological commitment for this purpose.

Definition 3.3 (Intended models) *Let* $\mathbf{C} = (D, W, \Re)$ *be a conceptualization,* **L** *a first-order logical language with vocabulary* **V** *and ontological* *commitment* $\mathbf{K} = (\mathbf{C}, \mathcal{I})$. *A model* $M = (S, I)$, *with* $S = (D, \mathbf{R})$, *is called an* intended model *of* $\mathbf{L}$ *according to* $\mathbf{K}$ *iff*

1. *For all constant symbols* $c \in \mathbf{V}$ *we have* $I(c) = \mathcal{I}(c)$
2. *There exists a world* $w \in W$ *such that, for each predicate symbol* $v \in \mathbf{V}$ *there exists an intensional relation* $\rho \in \Re$ *such that* $\mathcal{I}(v) = \rho$ *and* $I(v) = \rho(w)$

The set $\mathbf{I_K}(\mathbf{L})$ *of all models of* $\mathbf{L}$ *that are compatible with* $\mathbf{K}$ *is called the set of* intended models *of* $\mathbf{L}$ *according to* $\mathbf{K}$.

Condition 1 above just requires that the mapping of constant symbols to elements of the universe of discourse is identical. Example 2.1 does not introduce any constant symbols. Condition 2 states that there must exist a world such that every predicate symbol is mapped into an intensional relation whose value, for that world, coincides with the extensional interpretation of such symbol. This means that our intended model will be – so to speak – a description of that world. In Example 2.1, for instance, we have that, for w_1, $I(\mathit{Person}) = \{\mathit{I000001}, ..., \mathit{I050000}, ...\} = \mathit{Person}^1(w_1)$ and $I(\mathit{reports\text{-}to}) = \{..., (\mathit{I046758}, \mathit{I034820}), (\mathit{I044443}, \mathit{I034820}), (\mathit{I034820}, \mathit{I050000}), ...\} = \mathit{reports\text{-}to}^2(w_1)$.

With the notion of intended models at hand, we can now clarify the role of an ontology, considered as a logical theory designed to account for the intended meaning of the vocabulary used by a logical language. In the following, we also provide an ontology for our running example.

Definition 3.4 (Ontology) *Let* $\mathbf{C}$ *be a conceptualization, and* $\mathbf{L}$ *a logical language with vocabulary* $\mathbf{V}$ *and ontological commitment* $\mathbf{K}$. *An* ontology $\mathbf{O_K}$ *for* $\mathbf{C}$ *with vocabulary* $\mathbf{V}$ *and ontological commitment* $\mathbf{K}$ *is a logical theory consisting of a set of formulas of* $\mathbf{L}$, *designed so that the set of its models approximates as well as possible the set of intended models of* $\mathbf{L}$ *according to* $\mathbf{K}$ *(cf. also Fig. 2).*

Example 3.2 *In the following we build an ontology* O *consisting of a set of logical formulae. Through* O_1 *to* O_6 *we specify our human resources domain with increasing precision.*

Taxonomic Information. We start our formalization by specifying that Researcher and Manager are sub-concepts of Person:
$O_1 = \{\mathit{Researcher}(x) \rightarrow \mathit{Person}(x), \mathit{Manager}(x) \rightarrow \mathit{Person}(x)\}$

Domains and Ranges. We continue by adding formulae to O_1 *which specify the domains and ranges of the binary relations:*
$O_2 = O_1 \cup \{\mathit{cooperates\text{-}with}(x, y) \rightarrow \mathit{Person}(x) \wedge \mathit{Person}(y), \mathit{reports\text{-}to}(x, y) \rightarrow \mathit{Person}(x) \wedge \mathit{Person}(y)\}$

Symmetry. cooperates-with can be considered a symmetric relation:
$O_3 = O_2 \cup \{\mathit{cooperates\text{-}with}(x, y) \leftrightarrow \mathit{cooperates\text{-}with}(y, x)\}$

Transitivity. Although arguable, we specify reports-to as a transitive relation:
$O_4 = O_3 \cup \{\mathit{reports\text{-}to}(x, z) \leftarrow \mathit{reports\text{-}to}(x, y) \wedge \mathit{reports\text{-}to}(y, z)\}$

Disjointness There is no Person who is both a Researcher and a Manager:
$O_5 = O_4 \cup \{Manager(x) \rightarrow \neg Researcher(x)\}$

3.3 Choosing the Right Domain and Vocabulary

On the basis of the discussion above, we might conclude that an ideal ontology is one whose models exactly coincide (modulo isomorphisms) with the intended ones. Things are not so simple, however: even a "perfect" ontology like that may fail to exactly specify its target conceptualization, if its *vocabulary* and its *domain of discourse* are not suitably chosen. The reason for that lies in the distinction between the *logical* notion of *model* and the *ontological* notion of *possible world*. The former is basically a combination of assignments of abstract relational structures (built over the domain of discourse) to vocabulary elements; the latter is a combination of actual (observed) states of affairs of a certain system. Of course, the number of possible models depends both on the size of the vocabulary and the extension of the domain of discourse, which are chosen more or less arbitrarily, on the basis of what appears to be relevant to talk of. On the contrary, the number of world states depends on the observed variables, even those which – at a first sight – are considered as irrelevant to talk of. With reference to our example, consider the two models where the predicates of our language (whose signature is reported above) are interpreted in such a way that their extensions are those described respectively in Examples 2.1 and 2.2. Each model corresponds to a different pattern of relationships among the people in our company, but, looking at the model itself, nothing tells us what are the world states where a certain pattern of relationships holds. So, for example, it is impossible to discriminate between a conceptualization where *cooperates-with* means that two persons cooperate when they are just sharing a goal, and another where they need also do something to achieve that goal. In other words, each model, in this example, will "collapse" many different world states. The reason of this is in the very simple vocabulary we have adopted: with just two predicates, we have not enough expressiveness to discriminate between different world states. So, to really capture our conceptualization, we need to extend the vocabulary in order to be able to talk of sharing a goal or achieving a goal, and we have to introduce goals (besides persons) in our domain of discourse. In conclusion, the degree to which an ontology specifies a conceptualization depends (1) on the richness of the domain of discourse; (2) on the richness of the vocabulary chosen; (3) on the axiomatization. In turn, the axiomatization depends on language expressiveness issues as discussed in Sect. 3.4.

3.4 Language Expressiveness Issues

At one extreme, we have rather informal approaches for the language **L** that may allow the definitions of terms only, with little or no specification of the

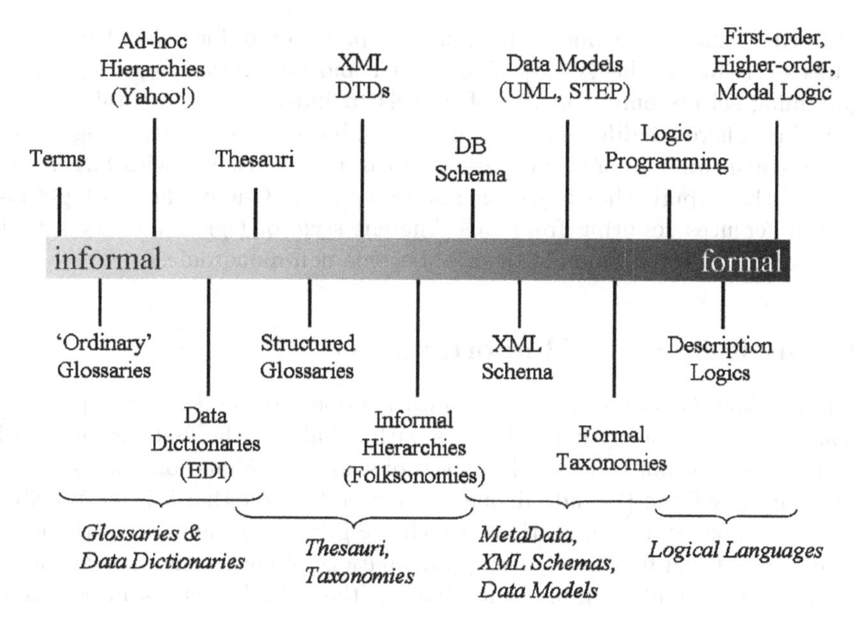

Fig. 4. Different approaches to the language **L** according to [17]. Typically, logical languages are eligible for the formal, explicit specification, and, thus, ontologies

meaning of the term. At the other end of the spectrum, we have formal approaches, i.e., logical languages that allow specifying rigorously formalized logical theories. This gives rise to the continuum introduced by [17] and depicted in Fig. 4. As we move along the continuum, the amount of meaning specified and the degree of formality increases (thus reducing ambiguity); there is also increasing support for automated reasoning (cf. Chapters "Tableau-Based Reasoning" and "Resolution-Based Reasoning for Ontologies").

It is difficult to draw a strict line of where the criterion of formal starts on this continuum. In practice, the rightmost category of logical languages is usually considered as formal. Within this rightmost category one typically encounters the trade-off between *expressiveness* and *efficiency* when choosing the language **L**. On the one end, we find higher-order logic, full first-order logic, or modal logic. They are very expressive, but do often not allow for sound and complete reasoning and if they do, reasoning sometimes remains untractable. At the other end, we find less stringent subsets of first-order logic, which typically feature decidable and more efficient reasoners. They can be split in two major paradigms. First, languages from the family of *description logics (DL)* (cf. chapter "Description Logics"), e.g., OWL-DL (cf. chapter "Web Ontology Language: OWL"), are strict subsets of first-order logic. The second major paradigm comes from the tradition of *logic programming (LP)* [3] with one prominent representor being F-Logic (cf. chapter "Ontologies in F-Logic"). Though logic programming often uses a syntax comparable to first-order logics, it assumes a different interpretation of formulae. Unlike the Tarski-style model theory [18] of first-order and description logic, logic programming selects only a subset of models to judge semantic entailment of formulae. There are different ways to select subsets of models resulting in different semantics – all of them geared to deal more efficiently with larger sets of data than approaches based on first-order logic. One of the most prominent differences resulting from this different style of logical models is that expressive logic programming theories become non-monotonic.

4 Why is *Shared* of Importance?

A formal specification of a conceptualization does not need to be a specification of a *shared* conceptualization. As outlined above, the first definitions of "ontologies" did not consider the aspect of sharing [6,8] and only later it was introduced by Borst [1]. Indeed, one may correctly argue that it is not possible to share whole conceptualizations, which are private to the mind of the individual. What can be shared, are approximations of conceptualizations based on a limited set of examples and showing the actual circumstances where a certain conceptual relation holds (for instance, actual situations showing cases where the *cooperates-with* relationship occurs). Beyond mere examples it is also possible to share meaning postulates, i.e., explicit formal constraints (e.g., the relationship *cooperates-with* is symmetric). Such definitions, however, presuppose a mutual agreement on the primitive terms used in these definitions. Since however meaning postulates cannot fully characterize the ontological commitment of primitive terms, one may recognize that sharing of conceptualizations is at best partial.

For practical usage of ontologies, it turned out very quickly that without at least such minimal shared ontological commitment from ontology stakeholders, the benefits of having an ontology are limited. The reason is that an ontology formally specifies a domain structure *under the limitation* that its stakeholder understand the primitive terms in the appropriate way. In other words, the ontology may turn out useless if it is used in a way that runs counter to the shared ontological commitment. In conclusion, any ontology will always be less complete and less formal than it would be desirable in theory. This is why it is important, for those ontologies intended to support large-scale interoperability, to be *well-founded*, in the sense that the basic primitives they are built on are sufficiently well-chosen and axiomatized to be generally understood.

4.1 Reference and Meaning

For appropriate usage, ontologies need to fulfill a further function, namely facilitating the communication between the human and the machine – referring to terminology specified in the ontology – or even for facilitating inter-machine and inter-human communication. The communication situation can

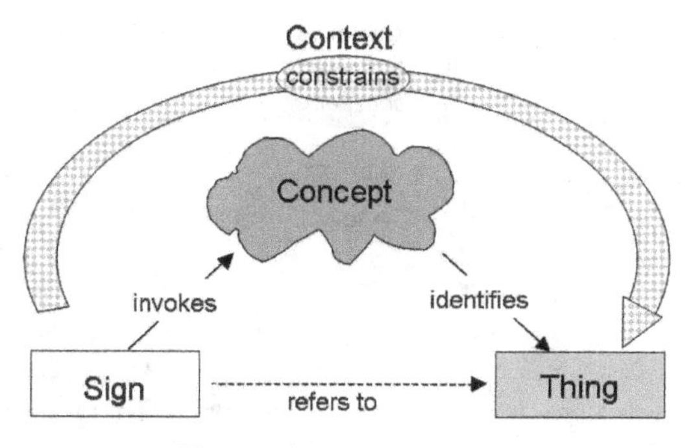

Fig. 5. Semiotic triangle

be illustrated using the semiotic triangle of Ogden and Richard [12], following thoughts by Peirce, Saussure, and Frege (cf. Fig. 5).

All agents, whatever their commitment to an ontology is, find themselves in a communication situation illustrated using the semiotic triangle: The sender of a message may use a word or – more generally – a sign like the string "Person" to stand for a concept the sender has in his own "mind." He uses the sign in order to refer to abstract or concrete things in the world, which may, but need not be, physical objects. The sender also invokes a concept in the mind of an actor receiving this sign. The receiver uses the concept in order to point out the individual or the class of individuals the sign was intended to refer to. Thereby, the interpretation of the sign as a concept as well as its use in a given situation depends heavily on the receiver as well as the overall communication context. Therefore, the meaning triangle is sometimes supplemented with further nodes in order to represent the receiver or the context of communication. We have illustrated the context by an instable arrow from sign to thing that constrains possible acts of reference. Note that the act of reference remains indirect, as it is mediated by the mental concept. Once the concept is invoked, it behaves (so to speak) as a function that, given a particular context (i.e., the world state mentioned in previous sections), returns the things we want to refer to. Moreover, the correspondences between sign, concept, and thing are weak and ambiguous. In many communication circumstances, the usage of signs can erroneously invoke the wrong concepts and represent different entities than intended to.

This problem is further aggravated when a multitude of agents exchanges messages in which terms do not have a prescribed meaning. Unavoidably, different agents will arrive at different conclusions about the semantics and the intention of the message.

When agents commit to a common ontology they can limit the conclusions possibly associated with the communications of specific signs, because not all relations between existing signs may hold and logical consequences from the usage of signs are implied by the logical theory specifying the ontology. Therefore the set of possible correspondences between signs, concepts and

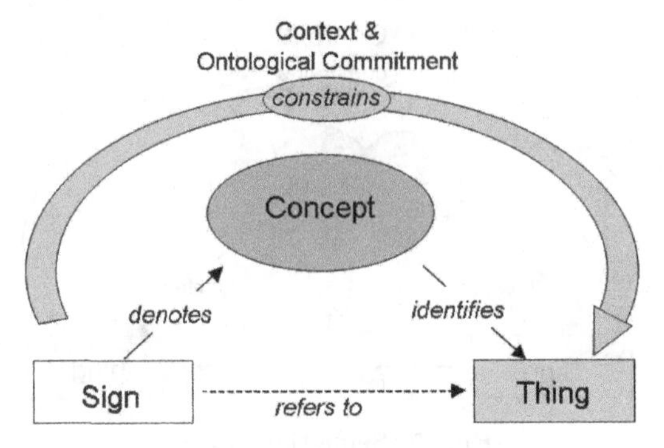

Fig. 6. Semiotic triangle revisited

real-world entities is strongly reduced – ideally up to a situation where the message becomes completely unambiguous (cf. Fig. 6). Thereby, not only the act of reference becomes clearer, but also the connection between sign and concept changes from a weakly defined relationship of "invokes" into a logically precise meaning of "denotes." Likewise, the meaning of a concept is now determined by a precise logical theory (contrast Figs. 5 and 6).

5 Discussion

In this chapter we have introduced three core aspects of computational ontologies: conceptualizations, specifications of conceptualizations, and shared ontological commitments. These are very broad categories suitable to investigate many different formalisms and fields of applications.

In fact, they are not even the only aspects of ontologies, which may be classified into different types, depending on the way they are used. For instance, the primary purpose of *top-level ontologies* lies in providing a broad view of the world suitable for many different target domains. *Reference ontologies* target the structuring of ontologies that are derived from them. The primary purpose of *core ontologies* derives from the definition of a super domain. *Application ontologies* are suitable for direct use in reasoning engines or software packages – and this list is not yet complete and will require many more experiences yet to be made.

Acknowledgments

We would like to thank our colleagues Aldo Gangemi, Susan Marie Thomas, Marta Sabou, as well as Pascal Hitzler for their fruitful reviews and discussions that helped to shape this contribution.

References

1. W. Borst. *Construction of Engineering Ontologies.* PhD thesis, Institute for Telematica and Information Technology, University of Twente, Enschede, The Netherlands, 1997.
2. R. Carnap. *Meaning and Necessity – A Study in Semantics and Modal Logic.* The University of Chicago Press, second edition, 1956.
3. S. K. Das. *Deductive Databases and Logic Programming.* Addison Wesley, 1992.
4. D. R. Dowty, R. Wall, and S. Peters. *Introduction to Montague Semantics*, volume 11 of *Studies in Linguistics and Philosophy.* Springer, Heidelberg, 1980.
5. M. R. Genesereth and N. J. Nilsson. *Logical Foundations of Artificial Intelligence.* Morgan Kaufmann, Los Altos, CA, 1987.
6. T. R. Gruber. Towards principles for the design of ontologies used for knowledge sharing. In N. Guarino and R. Poli, editors, *Formal Ontology in Conceptual Analysis and Knowledge Representation.* Kluwer Academic Publishers, Deventer, The Netherlands, 1993.
7. T. R. Gruber. A Translation Approach to Portable Ontologies. *Knowledge Acquisition*, 5(2):199–220, 1993.
8. T. R. Gruber. Toward Principles for the Design of Ontologies Used for Knowledge Sharing. *International Journal of Human Computer Studies*, 43(5–6): 907–928, 1995.
9. N. Guarino and P. Giaretta. Ontologies and Knowledge Bases: Towards a Terminological Clarification. In N. Mars, editor, *Towards Very Large Knowledge Bases: Knowledge Building and Knowledge Sharing*, pages 25–32. IOS Press, Amsterdam, 1995.
10. N. Guarino. Formal Ontology in Information Systems. In N. Guarino, editor, *Formal Ontology in Information Systems. Proceedings of FOIS'98, Trento, Italy, June 6-8, 1998*, pages 3–15. IOS Press, Amsterdam, 1998.
11. J. Hendler. Agents and the Semantic Web. *IEEE Intelligent Systems*, 16(2): 30–37, 2001.
12. C. K. Ogden and I. A. Richards. *The Meaning of Meaning: A Study of the Influence of Language upon Thought and of the Science of Symbolism.* Routledge & Kegan Paul Ltd., London, tenth edition, 1923.
13. P. Øhrstrøm, J. Andersen, and H. Schärfe. What has happened to ontology. In F. Dau, M.-L. Mugnier, and G. Stumme, editors, *Conceptual Structures: Common Semantics for Sharing Knowledge, 13th International Conference on Conceptual Structures, ICCS 2005, Kassel, Germany, July 17–22, 2005, Proceedings*, volume 3596 of *Lecture Notes in Computer Science*, pages 425–438. Springer, Heidelberg, 2005.
14. B. Smith. Beyond Concepts: Ontology as Reality Representation. In A. C. Varzi and L. Vieu, editors, *Formal Ontology in Information Systems – Proceedings of the Third International Conference (FOIS 2004)*, pages 73–85. IOS Press, Amsterdam, 2004.
15. R. Studer, R. Benjamins, and D. Fensel. Knowledge engineering: Principles and methods. *Data & Knowledge Engineering*, 25(1–2):161–198, 1998.
16. W. Swartout and A. Tate. Ontologies. *IEEE Intelligent Systems*, 14(1):18–19, 1999.
17. M. Uschold. Ontologies and Semantics for Seamless Connectivity. *SIGMOD Record*, 33(4):58–64, 2004.
18. R. L. Vaught. Alfred Tarski's Work in Model Theory. *The Journal of Symbolic Logic*, 51(4):869–882, 1986.

Part I

Ontology Representation Languages

Description Logics

Franz Baader[1], Ian Horrocks[2], and Ulrike Sattler[3]

[1] Institut für Theoretische Informatik, TU Dresden, Germany,
baader@tcs.inf.tu-dresden.de
[2] Computing Laboratory, Oxford University, Oxford, UK,
ian.horrocks@comlab.ox.ac.uk
[3] Department of Computer Science, University of Manchester, Manchester, UK,
sattler@cs.man.ac.uk

Summary. In this chapter, we explain what description logics are and why they make good ontology languages. In particular, we introduce the description logic $\mathcal{SHIQ}$, which has formed the basis of several well-known ontology languages, including OWL. We argue that, without the last decade of basic research in description logics, this family of knowledge representation languages could not have played such an important rôle in this context.

Description logic reasoning can be used both during the design phase, in order to improve the quality of ontologies, and in the deployment phase, in order to exploit the rich structure of ontologies and ontology based information. We discuss the extensions to $\mathcal{SHIQ}$ that are required for languages such as OWL and, finally, we sketch how novel reasoning services can support building ontologies.

1 Introduction

The aim of this section is to give a brief introduction to description logics, and to argue why they are well-suited as ontology languages. In the remainder of the chapter we will put some flesh on this skeleton by providing more technical details with respect to the theory of description logics, and their relationship to state of the art ontology languages. More detail on these and other matters related to description logics can be found in [6].

1.1 Ontologies

There have been many attempts to define what constitutes an ontology, perhaps the best known (at least amongst computer scientists) being due to Gruber: "an ontology is an explicit specification of a conceptualisation" [49].[1] In this context, a conceptualisation means an abstract model of some aspect

[1] This was later elaborated to "a formal specification of a shared conceptualisation" [21].

S. Staab and R. Studer (eds.), *Handbook on Ontologies*, International Handbooks on Information Systems, DOI 10.1007/978-3-540-92673-3,

of the world, taking the form of a definition of the properties of important concepts and relationships. An explicit specification means that the model should be specified in some unambiguous language, making it amenable to processing by machines as well as by humans.

Ontologies are becoming increasingly important in fields such as knowledge management, information integration, cooperative information systems, information retrieval and electronic commerce. One application area which has recently seen an explosion of interest is the so called *Semantic Web* [18], where ontologies are set to play a key rôle in establishing a common terminology between agents, thus ensuring that different agents have a shared understanding of terms used in semantic markup.

The effective use of ontologies requires not only a well-designed and well-defined ontology language, but also support from reasoning tools. Reasoning is important both to ensure the quality of an ontology, and in order to exploit the rich structure of ontologies and ontology based information. It can be employed in different phases of the ontology life cycle. During ontology design, it can be used to test whether concepts are non-contradictory and to derive implied relations. In particular, one usually wants to compute the concept hierarchy, i.e. the partial ordering of named concepts based on the subsumption relationship. Information on which concept is a specialization of another, and which concepts are synonyms, can be used in the design phase to test whether the concept definitions in the ontology have the intended consequences or not. This information is also very useful when the ontology is deployed.

Since it is not reasonable to assume that all applications will use the same ontology, interoperability and integration of different ontologies is also an important issue. Integration can, for example, be supported as follows: after the knowledge engineer has asserted some inter-ontology relationships, the integrated concept hierarchy is computed and the concepts are checked for consistency. Inconsistent concepts as well as unintended or missing subsumption relationships are thus signs of incorrect or incomplete inter-ontology assertions, which can then be corrected or completed by the knowledge engineer.

Finally, reasoning may also be used when the ontology is deployed. As well as using the pre-computed concept hierarchy, one could, for example, use the ontology to determine the consistency of facts stated in annotations, or infer relationships between annotation instances and ontology classes. More precisely, when searching web pages annotated with terms from the ontology, it may be useful to consider not only exact matches, but also matches with respect to more general or more specific terms – where the latter choice depends on the context. However, in the deployment phase, the requirements on the efficiency of reasoning are much more stringent than in the design and integration phases.

Before arguing why description logics are good candidates for such an ontology language, we provide a brief introduction to and history of description logics.

1.2 Description Logics

Description logics (DLs) [6, 16, 30] are a family of knowledge representation languages that can be used to represent the knowledge of an application domain in a structured and formally well-understood way. The name *description logics* is motivated by the fact that, on the one hand, the important notions of the domain are described by concept *descriptions*, i.e. expressions that are built from atomic concepts (unary predicates) and atomic roles (binary predicates) using the concept and role constructors provided by the particular DL. On the other hand, DLs differ from their predecessors, such as semantic networks and frames, in that they are equipped with a formal, *logic*-based semantics.

In this introduction, we only illustrate some typical constructors by an example. Formal definitions are given in Sect. 2. Assume that we want to define the concept of "A man that is married to a doctor and has at least five children, all of whom are professors." This concept can be described with the following concept description:

$$\mathsf{Human} \sqcap \neg\mathsf{Female} \sqcap \exists\mathsf{married}.\mathsf{Doctor} \sqcap (\geq 5\,\mathsf{hasChild}) \sqcap \forall\mathsf{hasChild}.\mathsf{Professor}$$

This description employs the Boolean constructors *conjunction* ($\sqcap$), which is interpreted as set intersection, and *negation* ($\neg$), which is interpreted as set complement, as well as the *existential restriction* constructor ($\exists R.C$), the *value restriction* constructor ($\forall R.C$), and the *number restriction* constructor ($\geq n\, R$). An individual, say Bob, belongs to $\exists\mathsf{married}.\mathsf{Doctor}$ if there exists an individual that is married to Bob (i.e. is related to Bob via the $\mathsf{married}$ role) and is a doctor (i.e. belongs to the concept Doctor). Similarly, Bob belongs to $(\geq 5\,\mathsf{hasChild})$ iff he has at least five children, and he belongs to $\forall\mathsf{hasChild}.\mathsf{Professor}$ iff all his children (i.e. all individuals related to Bob via the $\mathsf{hasChild}$ role) are professors.

In addition to this description formalism, DLs are usually equipped with a terminological and an assertional formalism. In its simplest form, *terminological axioms* can be used to introduce names (abbreviations) for complex descriptions. For example, we could introduce the abbreviation $\mathsf{HappyMan}$ for the concept description from above. More expressive terminological formalisms allow the statement of constraints such as

$$\exists\mathsf{hasChild}.\mathsf{Human} \sqsubseteq \mathsf{Human},$$

which says that only humans can have human children. A set of terminological axioms is called a TBox. The *assertional formalism* can be used to state properties of individuals. For example, the assertions

$$\mathsf{HappyMan}(\mathsf{BOB}), \quad \mathsf{hasChild}(\mathsf{BOB}, \mathsf{MARY})$$

state that Bob belongs to the concept $\mathsf{HappyMan}$ and that Mary is one of his children. A set of such assertions is called an ABox, and the named individuals that occur in ABox assertions are called ABox individuals.

Description logic systems provide their users with various inference capabilities that deduce implicit knowledge from the explicitly represented knowledge. The *subsumption* algorithm determines subconcept–superconcept relationships: C is subsumed by D iff all instances of C are necessarily instances of D, i.e. the first description is always interpreted as a subset of the second description. For example, given the definition of HappyMan from above, HappyMan is subsumed by $\exists$hasChild.Professor – since instances of HappyMan have at least five children, all of whom are professors, they also have a child that is a professor. The *instance* algorithm determines instance relationships: the individual i is an instance of the concept description C iff i is always interpreted as an element of C. For example, given the assertions from above and the definition of HappyMan, MARY is an instance of Professor. The *consistency* algorithm determines whether a knowledge base (consisting of a set of assertions and a set of terminological axioms) is non-contradictory. For example, if we add $\neg$Professor(MARY) to the two assertions from above, then the knowledge base containing these assertions together with the definition of HappyMan from above is inconsistent.

In order to ensure reasonable and predictable behavior of a DL system, these inference problems should at least be decidable for the DL employed by the system, and preferably of low complexity. Consequently, the expressive power of the DL in question must be restricted in an appropriate way. If the imposed restrictions are too severe, however, then the important notions of the application domain can no longer be expressed. Investigating this trade-off between the expressivity of DLs and the complexity of their inference problems has been one of the most important issues in DL research. Roughly, the research related to this issue can be classified into the following four phases.

Phase 1 (1980–1990) was mainly concerned with implementation of systems, such as KLONE, K-REP, BACK, and LOOM [24, 70, 71, 80]. These systems employed so-called *structural subsumption algorithms*, which first normalize the concept descriptions, and then recursively compare the syntactic structure of the normalized descriptions [73]. These algorithms are usually relatively efficient (polynomial), but they have the disadvantage that they are complete only for very inexpressive DLs, i.e. for more expressive DLs they cannot detect all the existing subsumption/instance relationships. At the end of this phase, early formal investigations into the complexity of reasoning in DLs showed that most DLs do not have polynomial-time inference problems [23, 74]. As a reaction, the implementors of the CLASSIC system (the first industrial-strength DL system) carefully restricted the expressive power of their DL [22, 79].

Phase 2 (1990–1995) started with the introduction of a new algorithmic paradigm into DLs, so-called *tableau-based algorithms* [40, 54, 88]. They work on propositionally closed DLs (i.e. DLs with full Boolean operators) and are complete also for expressive DLs. To decide the consistency of a knowledge base, a tableau-based algorithm tries to construct a model of it by breaking down the concepts in the knowledge base, thus inferring new constraints on the elements of this model. The algorithm either stops because all attempts to build a model failed with obvious contradictions, or it stops with a "canonical" model. Since in propositionally closed DLs, subsumption and satisfiability can be reduced to consistency, a consistency algorithm can solve all the inference problems mentioned above. The first systems employing such algorithms (KRIS and CRACK) demonstrated that optimized implementations of these algorithm lead to an acceptable behavior of the system, even though the worst-case complexity of the corresponding reasoning problems is no longer in polynomial time [9, 27]. This phase also saw a thorough analysis of the complexity of reasoning in various DLs [39–41]. Another important observation was that DLs are very closely related to modal logics [85].

Phase 3 (1995–2000) is characterized by the development of inference procedures for very expressive DLs, either based on the tableau-approach [58,60] or on a translation into modal logics [35–38]. Highly optimized systems (FaCT, RACE, and DLP [50, 55, 78]) showed that tableau-based algorithms for expressive DLs lead to a good practical behavior of the system even on (some) large knowledge bases. In this phase, the relationship to modal logics [36, 86] and to decidable fragments of first-order logic was also studied in more detail [19, 45–47, 76], and applications in databases (like schema reasoning, query optimization, and integration of databases) were investigated [28, 29, 31].

We are now at the beginning of *Phase 4*, where industrial strength DL systems employing very expressive DLs and tableau-based algorithms are being developed, with applications like the Semantic Web or knowledge representation and integration in bio-informatics in mind.

1.3 Description Logics as Ontology Languages

As already mentioned above, high quality ontologies are crucial for many applications, and their construction, integration, and evolution greatly depends on the availability of a well-defined semantics and powerful reasoning tools. Since DLs provide for both, they should be ideal candidates for ontology languages. That much was already clear ten years ago, but at that time there was a fundamental mismatch between the expressive power and the efficiency of reasoning that DL systems provided, and the expressivity and the large knowledge bases that ontologists needed [42]. Through the basic research in DLs of the last 10–15 years that we have summarized above, this gap between the needs of ontologist and the systems that DL researchers provide has finally become narrow enough to build stable bridges.

The suitability of DLs as ontology languages has been highlighted by their role as the foundation for several web ontology languages, including OWL, an ontology language standard developed by the W3C Web-Ontology Working Group[2] (see chapter "Web Ontology Language: OWL"). OWL has a syntax

[2] `http://www.w3.org/2001/sw/WebOnt/`

based on RDF Schema, but the basis for its design is the expressive DL $\mathcal{SHIQ}$ [60],[3] and the developers have tried to find a good compromise between expressiveness and the complexity of reasoning. Although reasoning in $\mathcal{SHIQ}$ is decidable, it has a rather high worst-case complexity (ExpTime). Nevertheless, highly optimized $\mathcal{SHIQ}$ reasoners such as FaCT++ [95], Racer [52] and Pellet [91] behave quite well in practice.

Let us point out some of the features of $\mathcal{SHIQ}$ that make this DL expressive enough to be used as an ontology language. Firstly, $\mathcal{SHIQ}$ provides number restrictions that are more expressive than the ones introduced above (and employed by earlier DL systems). With the *qualified number restrictions* available in $\mathcal{SHIQ}$, as well as being able to say that a person has at most two children (without mentioning the properties of these children):

$$(\leq 2\,\mathsf{hasChild}),$$

one can also specify that there is at most one son and at most one daughter:

$$(\leq 1\,\mathsf{hasChild}.\neg\mathsf{Female}) \sqcap (\leq 1\,\mathsf{hasChild}.\mathsf{Female}).$$

Secondly, $\mathcal{SHIQ}$ allows the formulation of complex terminological axioms like "humans have human parents":

$$\mathsf{Human} \sqsubseteq \exists\mathsf{hasParent}.\mathsf{Human}.$$

Thirdly, $\mathcal{SHIQ}$ also allows for *inverse roles*, *transitive roles*, and *subroles*. For example, in addition to hasChild one can also use its inverse hasParent, one can specify that hasAncestor is transitive, and that hasParent is a subrole of hasAncestor.

It has been argued in the DL and the ontology community that these features play a central role when describing properties of aggregated objects and when building ontologies [43, 83, 93]. The actual use of a DL providing these features as the underlying logical formalism of the web ontology language OWL [57] substantiates this claim [93].[4]

Finally, we would like to briefly mention three extensions to $\mathcal{SHIQ}$ that are often used in ontology languages (we will discuss them in more detail in Sect. 4).

Complex roles are often required in ontologies. For example, when describing complex physically composed structures it may be desirable to express the fact that damage to a part of the structure implies damage to the structure as a whole. This feature is particularly important in medical ontologies: it is supported in the Grail DL [81], which was specifically designed for use with medical terminology, and in another medical terminology application using

[3] To be exact, it is based on $\mathcal{SHOIN}$.

[4] The more expressive qualified number restrictions are not supported by OWL, but are featured in the proposed OWL 2 extension (see Sect. 4).

the comparatively inexpressive DL $\mathcal{ALC}$, a rather complex "work around" is performed in order to capture this kind of information [89].[5]

It is quite straightforward to extend $\mathcal{SHIQ}$ so that this kind of propagation can be expressed: simply allow for the use of complex roles in role inclusion axioms. E.g. $\mathsf{hasLocation} \circ \mathsf{partOf} \sqsubseteq \mathsf{hasLocation}$ expresses the fact that things located in part of something are also located in the thing as a whole. Although this leads to undecidability in general, syntactic restrictions can be devised that lead to a decidable logic [59].

Concrete domains [7,69] integrate DLs with concrete sets such as the real numbers, integers, or strings, and built-in predicates such as comparisons $\leq$, comparisons with constants ≤ 17, or isPrefixOf. This supports the modelling of concrete properties of abstract objects such as the age, the weight, or the name of a person, and the comparison of these concrete properties. Unfortunately, in their unrestricted form, concrete domains can have dramatic effects on the decidability and computational complexity of the underlying DL [69].

Nominals are special concept names that are to be interpreted as singleton sets. Using a nominal Turing, we can describe all those computer scientists that have met Turing by $\mathsf{CSientist} \sqcap \exists \mathsf{hasMet}.\mathsf{Turing}$. Again, nominals can have dramatic effects on the complexity of a logic [94]. The extension of $\mathcal{SHIQ}$ with nominals is usually called $\mathcal{SHOIQ}$.

2 The Expressive Description Logic $\mathcal{SHIQ}$

In this section, we present syntax and semantics of the expressive DL $\mathcal{SHIQ}$ [60] (although, as can be seen in chapter "Web Ontology Language: OWL", the DL underlying OWL is, in some respects, slightly more expressive).

In contrast to most of the DLs considered in the literature, which concentrate on constructors for defining concepts, the DL $\mathcal{SHIQ}$ also allows for rather expressive roles. Of course, these roles can then be used in the definition of concepts.

Definition 1 (Syntax and semantics of $\mathcal{SHIQ}$-roles and concepts). *Let* $\mathbf{R}$ *be a set of* role names, *which is partitioned into a set* $\mathbf{R}_+$ *of transitive roles and a set* $\mathbf{R}_P$ *of normal roles. The set of all* $\mathcal{SHIQ}$-roles *is* $\mathbf{R} \cup \{r^- \mid r \in \mathbf{R}\}$, *where* r^- *is called the* inverse *of the role* r.

Let $\mathbf{C}$ *be a set of* concept names. *The set of* $\mathcal{SHIQ}$-concepts *is the smallest set such that*

1. *every concept name* $A \in \mathbf{C}$ *is a* $\mathcal{SHIQ}$*-concept,*
2. *if* C *and* D *are* $\mathcal{SHIQ}$*-concepts and* r *is a* $\mathcal{SHIQ}$*-role, then* $C \sqcap D$, $C \sqcup D$, $\neg C$, $\forall r.C$, *and* $\exists r.C$ *are* $\mathcal{SHIQ}$*-concepts,*

[5] In this approach, so-called *SEP-triplets* are used both to compensate for the absence of transitive roles in $\mathcal{ALC}$, and to express the propagation of properties across a distinguished "part-of" role.

3. if C is a $\mathcal{SHIQ}$-concept, r is a simple[6] *$\mathcal{SHIQ}$-role, and $n \in \mathbb{N}$, then $(\leqslant n\, r.C)$ and $(\geqslant n\, r.C)$ are $\mathcal{SHIQ}$-concepts.*

An interpretation *$\mathcal{I} = (\Delta^{\mathcal{I}}, \cdot^{\mathcal{I}})$ consists of a set $\Delta^{\mathcal{I}}$, called the* domain *of $\mathcal{I}$, and a function $\cdot^{\mathcal{I}}$ that maps every role to a subset of $\Delta^{\mathcal{I}} \times \Delta^{\mathcal{I}}$ such that, for all $p \in \mathbf{R}$ and $r \in \mathbf{R}_+$,*

$$\langle x, y\rangle \in p^{\mathcal{I}} \;\; \textit{iff} \;\; \langle y, x\rangle \in (p^-)^{\mathcal{I}},$$
$$\textit{if } \langle x, y\rangle \in r^{\mathcal{I}} \textit{ and } \langle y, z\rangle \in r^{\mathcal{I}} \textit{ then } \langle x, z\rangle \in r^{\mathcal{I}}.$$

The interpretation function $\cdot^{\mathcal{I}}$ of an interpretation $\mathcal{I} = (\Delta^{\mathcal{I}}, \cdot^{\mathcal{I}})$ maps, additionally, every concept to a subset of $\Delta^{\mathcal{I}}$ such that

$$(C \sqcap D)^{\mathcal{I}} = C^{\mathcal{I}} \cap D^{\mathcal{I}}, \qquad (C \sqcup D)^{\mathcal{I}} = C^{\mathcal{I}} \cup D^{\mathcal{I}}, \qquad \neg C^{\mathcal{I}} = \Delta^{\mathcal{I}} \setminus C^{\mathcal{I}},$$
$$(\exists r.C)^{\mathcal{I}} = \{x \in \Delta^{\mathcal{I}} \mid \textit{There is some } y \in \Delta^{\mathcal{I}} \textit{ with } \langle x, y\rangle \in r^{\mathcal{I}} \textit{ and } y \in C^{\mathcal{I}}\},$$
$$(\forall r.C)^{\mathcal{I}} = \{x \in \Delta^{\mathcal{I}} \mid \textit{For all } y \in \Delta^{\mathcal{I}}, \textit{ if } \langle x, y\rangle \in r^{\mathcal{I}}, \textit{then } y \in C^{\mathcal{I}}\},$$
$$(\leqslant n\, r.C)^{\mathcal{I}} = \{x \in \Delta^{\mathcal{I}} \mid \sharp r^{\mathcal{I}}(x, C) \leqslant n\},$$
$$(\geqslant n\, r.C)^{\mathcal{I}} = \{x \in \Delta^{\mathcal{I}} \mid \sharp r^{\mathcal{I}}(x, C) \geqslant n\},$$

where $\sharp M$ denotes the cardinality of the set M, and $r^{\mathcal{I}}(x, C) := \{y \mid \langle x, y\rangle \in r^{\mathcal{I}}$ and $y \in C^{\mathcal{I}}\}$. If $x \in C^{\mathcal{I}}$, then we say that x is an instance *of C in $\mathcal{I}$, and if $\langle x, y\rangle \in r^{\mathcal{I}}$, then y is called an* r-successor *of x in $\mathcal{I}$.*

So far, we have fixed the syntax and semantics of concepts and roles. Next, we define how they can be used in a $\mathcal{SHIQ}$ TBox. Please note that authors sometimes distinguish between a role hierarchy or RBox and a TBox – we do not make this distinction here.

Definition 2 (TBox). *A* role inclusion axiom *is of the form $r \sqsubseteq s$, where r, s are $\mathcal{SHIQ}$-roles. A* general concept inclusion *(GCI) is of the form $C \sqsubseteq D$, where C, D are $\mathcal{SHIQ}$-concepts.*

A finite set of role inclusion axioms and GCIs is called a TBox.

An interpretation $\mathcal{I}$ is a model *of a TBox $\mathcal{T}$ if it satisfies all axioms in $\mathcal{T}$, i.e. $C^{\mathcal{I}} \subseteq D^{\mathcal{I}}$ holds for each $C \sqsubseteq D \in \mathcal{T}$ and $r^{\mathcal{I}} \subseteq s^{\mathcal{I}}$ holds for each $r \sqsubseteq s \in \mathcal{T}$.*

A concept definition is of the form $A \equiv C$, where A is a concept name; it can be seen as an abbreviation for the two GCIs $A \sqsubseteq C$ and $C \sqsubseteq A$.

In addition to describing the relevant notions of an application domain, a DL knowledge base may also contain knowledge about the properties of specific individuals (or objects) existing in this domain. This done in the assertional part of the knowledge base (ABox).

[6] We refer the interested reader to [60] for a definition of simple roles: roughly speaking, a role is simple if it is neither transitive nor has a transitive sub-role. Only simple roles are allowed in number restrictions to ensure decidability [60].

Definition 3. *Let* $\mathbf{I}$ *be a set of* individual names *disjoint from* $\mathbf{R}$ *and* $\mathbf{C}$. *For* $a, b \in \mathbf{I}$ *individual names,* C *a possibly complex* $\mathcal{SHIQ}$ *concept, and* r *a* $\mathcal{SHIQ}$ *role, an expression of the form*

- $C(a)$ *is called a* concept assertion, *and*
- $r(a, b)$ *is called a* role assertion.

A finite set of concept and role assertions is called an ABox.

An interpretation function $\cdot^{\mathcal{I}}$, *additionally, is required to map every individual name* $a \in \mathbf{I}$ *to an element* $a^{\mathcal{I}} \in \Delta^{\mathcal{I}}$. *An interpretation* $\mathcal{I}$ satisfies

- *a concept assertion* $C(a)$ *if* $a^{\mathcal{I}} \in C^{\mathcal{I}}$, *and*
- *a role assertion* $r(a, b)$ *if* $\langle a^{\mathcal{I}}, b^{\mathcal{I}} \rangle \in r^{\mathcal{I}}$.

An interpretation that satisfies each concept assertion and each role assertion in an ABox $\mathcal{A}$ *is called a* model *of* $\mathcal{A}$.

Inference problems for concepts are defined w.r.t. a TBox. Inference problems for individuals additionally involve an ABox.

Definition 4. *The concept* C *is called* satisfiable with respect to the TBox $\mathcal{T}$ *iff there is a model* $\mathcal{I}$ *of* $\mathcal{T}$ *with* $C^{\mathcal{I}} \neq \emptyset$. *Such an interpretation is called a* model of C *w.r.t.* $\mathcal{T}$. *The concept* D subsumes *the concept* C *w.r.t.* $\mathcal{T}$ *(written* $C \sqsubseteq_{\mathcal{T}} D$*) if* $C^{\mathcal{I}} \subseteq D^{\mathcal{I}}$ *holds for all models* $\mathcal{I}$ *of* $\mathcal{T}$. *Two concepts* C, D *are* equivalent *w.r.t.* $\mathcal{T}$ *(written* $C \equiv_{\mathcal{T}} D$*) if they subsume each other.*

The ABox $\mathcal{A}$ *is called* consistent with respect to the TBox $\mathcal{T}$ *iff there exists a model of* $\mathcal{T}$ *and* $\mathcal{A}$. *The individual* a *is called an* instance of the concept C with respect to the TBox $\mathcal{T}$ and the ABox $\mathcal{A}$ *iff* $a^{\mathcal{I}} \in C^{\mathcal{I}}$ *holds for all models of* $\mathcal{I}$ *of* $\mathcal{T}$ *and* $\mathcal{A}$.

By definition, equivalence can be reduced to subsumption. In addition, subsumption can be reduced to satisfiability since $C \sqsubseteq_{\mathcal{T}} D$ iff $C \sqcap \neg D$ is unsatisfiable w.r.t. $\mathcal{T}$. Satisfiability and the instance problem can be reduced to the consistency problem since C is satisfiable w.r.t. $\mathcal{T}$ if the ABox $\{C(a)\}$ is consistent w.r.t. $\mathcal{T}$, and a is an instance of C w.r.t. $\mathcal{T}$ and $\mathcal{A}$ if the ABox $\mathcal{A} \cup \{\neg C(a)\}$ is inconsistent w.r.t. $\mathcal{T}$.

As mentioned above, most DLs are (decidable) fragments of (first-order) predicate logic [5, 19]. Viewing role names as binary relations, concept names as unary relations, and individual names as constants, for example, the role inclusion axiom $r \sqsubseteq s^-$ translates into $\forall x \forall y. r(x, y) \Rightarrow s(y, x)$, the concept assertion $(A \sqcap B)(a)$ translates into $A(a) \wedge B(a)$, and the GCI $A \sqcap \exists r.C \sqsubseteq D \sqcup \forall s^-.E$ translates into

$$\forall x.(A(x) \wedge \exists y.r(x, y) \wedge C(y)) \Rightarrow (D(x) \vee \forall y.s(y, x) \Rightarrow E(y)).$$

This translation preserves the semantics: we can easily view DL interpretations as predicate logic interpretations, and then prove, e.g. that each model of a concept C w.r.t. a TBox $\mathcal{T}$ is a model of the translation of C conjoined with the (universally quantified) translations of $\mathcal{T}$.

The reasoning services that can decide the inference problems introduced above can be implemented using various algorithmic techniques, including tableaux-based techniques (see chapter "Tableau-Based Reasoning") and resolution-based techniques (see "Resolution-Based Reasoning for Ontologies").

3 Describing Ontologies in $\mathcal{SHIQ}$

In general, an ontology can be formalised in a DL knowledge base as follows. Firstly, we restrict the possible worlds by introducing restrictions on the allowed interpretations. For example, to express that, in our world, we want to consider humans, which are either muggles or sorcerers, we can use the GCIs

$$\mathsf{Human} \sqsubseteq \mathsf{Muggle} \sqcup \mathsf{Sorcerer} \quad \text{and} \quad \mathsf{Muggle} \sqsubseteq \neg\mathsf{Sorcerer}.$$

Next, to express that humans have exactly two parents and that all parents and children of humans are human, we can use the following GCI:

$$\begin{aligned}\mathsf{Human} \sqsubseteq\ & \forall \mathsf{hasParent}.\mathsf{Human} \sqcap (\leqslant 2\ \mathsf{hasParent}.\top) \sqcap (\geqslant 2\ \mathsf{hasParent}.\top) \sqcap \\ & \forall \mathsf{hasParent}^{-}.\mathsf{Human},\end{aligned}$$

where $\top$ is an abbreviation for the top concept $A \sqcup \neg A$.[7]

In addition, we consider the *transitive* role hasAncestor, and the role inclusion

$$\mathsf{hasParent} \sqsubseteq \mathsf{hasAncestor}.$$

The next GCI expresses that humans having an ancestor that is a sorcerer are themselves sorcerers:

$$\mathsf{Human} \sqcap \exists \mathsf{hasAncestor}.\mathsf{Sorcerer} \sqsubseteq \mathsf{Sorcerer}.$$

Secondly, we can define the relevant notions of our application domain using concept definitions. Recall that the concept definition $A \equiv C$ stands for the two GCIs $A \sqsubseteq C$ and $C \sqsubseteq A$. A concept name is called *defined* if it occurs on the left-hand side of a definition, and *primitive* otherwise.

We want our concept definitions to have definitional impact, i.e. the interpretation of the primitive concept and role names should uniquely determine the interpretation of the defined concept names. For this, the set of concept definitions together with the additional GCIs must satisfy three conditions:

1. There are no multiple definitions, i.e. each defined concept name must occur at most once as the left-hand side of a concept definition.

---

[7] When the qualifying concept is $\top$, this is equivalent to an unqualified restriction, and it will often be written as such, e.g. ($\leqslant 2$ hasParent).

2. There are no cyclic definitions, i.e. no cyclic dependencies between the defined names in the set of concept definitions.[8]
3. The defined names do not occur in any of the additional GCIs.

In contrast to concept definitions, the GCIs in $\mathcal{SHIQ}$ may well have cyclic dependencies between concept names. An example are the above GCIs describing humans.

As a simple example of a set of concept definitions satisfying the restrictions from above, we define the concepts grandparent and parent[9]:

$$\mathsf{Parent} \equiv \mathsf{Human} \sqcap \exists \mathsf{hasParent}^{-}.\top,$$
$$\mathsf{Grandparent} \equiv \exists \mathsf{hasParent}^{-}.\mathsf{Parent}.$$

The TBox consisting of the above concept definitions and GCIs, together with the fact that hasAncestor is a transitive superrole of hasParent, implies the following subsumption relationship:

$$\mathsf{Grandparent} \sqcap \mathsf{Sorcerer} \sqsubseteq \exists \mathsf{hasParent}^{-}.\exists \mathsf{hasParent}^{-}.\mathsf{Sorcerer},$$

i.e. grandparents who are sorcerers have a grandchild who is a sorcerer. Though this conclusion may sound reasonable given the assumptions, it requires quite some reasoning to obtain it. In particular, one must use the fact that hasAncestor (and thus also $\mathsf{hasAncestor}^{-}$) is transitive, that $\mathsf{hasParent}^{-}$ is the inverse of hasParent, and that we have a GCI that says that children of humans are again humans.

To sum up, a $\mathcal{SHIQ}$-TBox can, on the one hand, axiomatize the basic notions of an application domain (the primitive concepts) by GCIs, transitivity statements, and role inclusions, in the sense that these statements restrict the possible interpretations of the basic notions. On the other hand, more complex notions (the defined concepts) can be introduced by concept definitions. Given an interpretation of the basic notions, the concept definitions uniquely determine the interpretation of the defined notions.

The *taxonomy* of such a TBox is then given by the subsumption hierarchy of the defined concepts. It can be computed using a subsumption algorithm for $\mathcal{SHIQ}$ (see chapters "Tableau-Based Reasoning" and "Resolution-Based Reasoning for Ontologies"). The knowledge engineer can test whether the TBox captures her intuition by checking the satisfiability of the defined concepts (since it does not make sense to give a complex definition for the empty concept), and by checking whether their place in the taxonomy corresponds to their intuitive place. The taxonomy of our example TBox would contain, for example, the fact that Grandparent is subsumed by Parent which is, in turn, subsumed by Human – if this is not intended, then the knowledge engineer

[8] In order to give cyclic definitions definitional impact, one would need to use fixpoint semantics for them [1, 75].

[9] In addition to the role hasParent, which relates children to their parents, we use the concept Parent, which describes all humans having children.

would need to go back and modify the TBox. The expressive power of $\mathcal{SHIQ}$ together with the fact that one can "verify" the TBox in the sense mentioned above is the main reason for $\mathcal{SHIQ}$ being well-suited as an ontology language [43, 83, 93].

In case we have, in addition to our TBox $\mathcal{T}$, also an ABox $\mathcal{A}$, we can first ask a DL reasoner to check whether $\mathcal{A}$ is consistent w.r.t. $\mathcal{T}$ to make sure that our assertions in $\mathcal{A}$ conform with the axioms expressed in $\mathcal{T}$. Consider the following ABox:

$$\mathcal{A} = \{\,\mathsf{Human}(\mathsf{Harry}),\ \mathsf{Sorcerer}(\mathsf{Bob}) \\ \mathsf{hasParent}(\mathsf{Harry}, \mathsf{Bob})\},$$

and let $\mathcal{T}$ consist of all axioms in this section. We can first use a DL reasoner to prove that $\mathcal{A}$ is consistent w.r.t. $\mathcal{T}$. Next, we can *query* $\mathcal{A}$ through $\mathcal{T}$. For example, we can ask a DL reasoner to retrieve all instances of Human w.r.t. $\mathcal{T}$ and $\mathcal{A}$. This would result in Harry and Bob being returned: for the former, this information is explicit in $\mathcal{A}$, for the latter, this is implied by the GCI which states that parents of humans are humans. Similarly, both Harry and Bob are instances of Sorcerer w.r.t. $\mathcal{T}$ and $\mathcal{A}$: for Harry, this is a consequence of the GCI which states that offsprings of sorcerers are sorcerers. As a final example, let us point out that our ABox contains no instance of $\forall\mathsf{hasParent}.\mathsf{Sorcerer}$: even though all *explicitly known* parents of Harry are sorcerers, Harry could have other parents (and indeed must have another parent) who may or may not be a sorcerer – this feature of DL semantics is known as the "open world assumption" [5].

4 Extensions and Variants of $\mathcal{SHIQ}$

The ontology language OWL extends $\mathcal{SHIQ}$ with nominals and concrete datatypes; see chapter "Web Ontology Language: OWL." In this section, we discuss the consequences of these extensions on the reasoning problems in $\mathcal{SHIQ}$.

Concrete datatypes, as available in OWL, are a very restricted form of concrete domains [7]. For example, using the concrete domain of all nonnegative integers equipped with the $<$ predicate, a (functional) role age relating (abstract) individuals to their (concrete) age, and a (functional) subrole father of hasParent, the following axiom states that children are younger than their fathers:

$$\mathsf{Animal} \sqsubseteq (\mathsf{age} < (\mathsf{father} \circ \mathsf{age})).$$

Extending expressive DLs with concrete domains may easily lead to undecidability [8, 68]. In OWL, however, no datatype predicates are supported – only XML schema datatypes (such as integer and string) and enumerations (such as $\{1, 2, 5, 7\}$) can be used in descriptions. These restrictions are enough to ensure that decidability is not compromised (in fact in [77], decidability of $\mathcal{SHIQ}$ extended with a more general type of concrete domains is shown).

Concerning nominals, things become a bit more complicated: nominals are individual names used as concepts, as in $\mathsf{Catholic} \sqcap \exists \mathsf{hasSeen}.\{\mathsf{Pope}\}$ and thus allow the use of individuals not only in ABoxes, but also in concept expressions and TBoxes. Firstly, we can use the same (relativised axiomatization) technique as used for $\mathcal{SHIQ}$ in [94] to translate $\mathcal{SHIQ}$ extended with nominals into a fragment of C2, the two-variable fragment of first order logic with counting quantifiers [48, 76]. Since this translation is polynomial, satisfiability and subsumption are decidable in NExpTime. This is optimal since the problem is also NExpTime-hard [94]. Roughly speaking, the combination of GCIs (or transitive roles and role inclusions), inverse roles, and number restrictions with nominals is responsible for this leap in complexity (from ExpTime for $\mathcal{SHIQ}$ to NExpTime). Until recently, no "practical" decision procedure for $\mathcal{SHOIQ}$, i.e. the extension of $\mathcal{SHIQ}$ with nominals, had been described, where by "practical" we mean a decision procedure that works in some "goal-directed" way, in contrast to "blindly" guessing a model $\mathcal{I}$ of at most exponential size and then checking whether $\mathcal{I}$ is indeed a model of the input. An extension of the tableaux algorithm for $\mathcal{SHIQ}$ has, however, now been developed [59], has been successfully implemented in the FaCT++ and Pellet systems, and seems to work well on realistic ontologies [90].

Finally, as mentioned above, it is quite straightforward to extend $\mathcal{SHIQ}$, or even $\mathcal{SHOIQ}$, with complex role inclusion axioms. The resulting DL, $\mathcal{SROIQ}$ [56], is the basis for a recent proposal to extend the OWL language, the extended language being called OWL 2.[10] In addition to complex role inclusion axioms, OWL 2 also supports qualified number restrictions, and more expressive datatypes than OWL.

5 Reasoning Beyond the Standard Inference Problems

As argued in the introduction, standard reasoning services for concepts (such as satisfiability and subsumption algorithms) can be used in different phases of the ontology life cycle. In the design phase, they can test whether concepts are non-contradictory and can derive implied relations between concepts. However, for these services to be applied, one already needs a sufficiently developed TBox. The result of reasoning can then be used to develop the TBox further. Until recently, however, DL systems provided no reasoning support for writing this initial TBox. The development of so-called non-standard inferences in DLs (like computing least common subsumers [13,32,65,67], most specific concepts [10, 66], rewriting [14], approximation [26], and matching [3, 11, 12, 20]) tries to overcome this deficit. These kinds of inferences are sketched in the first subsection.

---

[10] `http://www.w3.org/TR/2008/WD-owl2-syntax-20081202/`

In the presence of ABoxes, one often wants to ask queries that are more complex than simple instance queries involving only one individual and one concept. So-called conjunctive queries, which are treated in the second subsection, overcome this deficit.

5.1 Non-standard Inferences

In this subsection, we will sketch how non-standard inferences can support building a DL knowledge base.

Assume that the knowledge engineer wants to introduce the definition of a new concept into the TBox. In many cases, she will not develop this new definition from scratch, but rather try to re-use things that are already present in some knowledge base (either the one she is currently building or a previous one). In a chemical process engineering application [72,82], we have observed two ways in which this is realized in practice:

1. The knowledge engineer decides on the basic structure of the newly defined concept, and then tries to find already defined concepts that have a similar structure. These concepts can then be modified to obtain the new concept.
2. Instead of directly defining the new concept, the knowledge engineer first gives examples of objects that belong to the concept to be defined, and then tries to generalize these examples into a concept definition.

Both approaches can be supported by the non-standard inferences mentioned above, though this kind of support is not yet provided by any of the existing DL systems.

The first approach can be supported by matching concept patterns against concept descriptions. A *concept pattern* is a concept description that may contain variables that stand for descriptions. A *matcher* σ of a pattern D onto the description C replaces the variables by concept descriptions such that the resulting concept $\sigma(D)$ is equivalent to C. For example, assume that the knowledge engineer is looking for concepts concerned with individuals having a son and a daughter sharing some characteristic. This can be expressed by the pattern

$$\exists \mathsf{hasChild}.(\mathsf{Male} \sqcap X) \sqcap \exists \mathsf{hasChild}.(\mathsf{Female} \sqcap X).$$

The substitution $\sigma = \{X \mapsto \mathsf{Tall}\}$ shows that this pattern matches the description $\exists \mathsf{hasChild}.(\mathsf{Male} \sqcap \mathsf{Tall}) \sqcap \exists \mathsf{hasChild}.(\mathsf{Female} \sqcap \mathsf{Tall})$. Note, however, that in some cases the existence of a matcher is not so obvious.

The second approach can be supported by algorithms that compute most specific concepts and least common subsumers. Assume that the examples are given as ABox individuals $i_1, \ldots, i_k$. In a first step, these individuals are generalized into concepts by respectively computing the most specific (w.r.t. subsumption) concepts $C_1, \ldots, C_k$ in the available DL that have these individuals as instances. In a second step, these concepts are generalized into one concept by computing the least common subsumer of $C_1, \ldots, C_k$, i.e. the least concept description (in the available DL) that subsumes $C_1, \ldots, C_k$. In this context, rewriting of concepts comes into play since the concept descriptions produced by the algorithms for computing least common subsumers may be rather large (and thus not easy to comprehend and modify for the knowledge engineer). Rewriting minimizes the size of these description without changing their meaning by introducing names defined in the TBox.

Until now, the results on such non-standard inferences are restricted to DLs that are considerably less expressive than $\mathcal{SHIQ}$. For some of them, they only make sense if used for inexpressive DLs. For example, in DLs that contain the disjunction constructor, the least common subsumer of $C_1, \ldots, C_k$ is simply their disjunction, and computing this is of no help to the knowledge engineer. What one would like to obtain as a result of the least common subsumer computation are the structural similarities between the input concepts.

Thus, support by non-standard inferences can only be given if one uses DLs of restricted expressive power. However, this also makes sense in the context of ontology engineering. In fact, the users that will require the most support are the naive ones, and it is reasonable to assume that they will not use (or even be offered) the full expressive power of the underlying DL. This two-level approach is already present in tools like Protégé [64], which offer a frame-like user interface. Using this simple interface, one gets only a fragment of the expressive power of OWL. To use the full expressive power, one must type in DL expressions.

Another way to overcome the gap between DLs of different expressive power is to use the approximation inference [26]. Here, one tries to approximate a given concept description C in an expressive DL $\mathcal{L}_1$ by a description D in a less expressive DL $\mathcal{L}_2$. When approximating from above, D should be the least description in $\mathcal{L}_2$ subsuming C, and when approximating from below, D should be the greatest description $\mathcal{L}_2$ subsumed by C.

5.2 Queries

As we have seen in Sect. 3, given an ontology consisting of an ABox and possibly a TBox, we can retrieve instances of concepts from it, thereby explicating knowledge about concept instances in the given ontology. In this sense, we can use concepts as a query language. It has turned out, however, that this is a rather weak query language which does not allow one to query, for example, for humans whose parents are married. Continuing the example from Sect. 3, could express this query as a *conjunctive query*:

$$q(x) :- \mathsf{Human}(x), \mathsf{hasParent}(x, y), \mathsf{hasParent}(x, z), \mathsf{married}(y, z)$$

Conjunctive queries are well-known in the database community and have been suggested as an expressive query language for DLs [29]. Their answers can be sets of individual names from the ABox as in the example query above or, more generally, sets of tuples (if we have more than one answer variable). Roughly speaking, individual names from the ABox are in the answer set if, for each model of the ontology, we can find a match from the variables into the model's domain such that all conjuncts in the query are satisfied. Hence, as in instance retrieval, all axioms in the ontology are taken fully into account when answering queries. However, in contrast to the standard reasoning problems, and especially instance retrieval, answering conjunctive queries cannot be reduced to consistency. This problem is, however, decidable for a variety of logics [29] and it turned out to remain decidable even if transitive roles are used in the query [44].

6 Conclusion

The emphasis in DL research on a well-defined, logic-based semantics and a thorough investigation of the basic reasoning problems, together with the availability of highly optimized systems for very expressive DLs, makes this family of knowledge representation formalisms an ideal starting point for defining ontology languages. The standard reasoning services such as consistency checking, computation of the taxonomy, testing for unsatisfiable concepts, and instance retrieval, are provided by highly optimised, state-of-the-art DL systems for very expressive DLs. Optimizations of these systems for large ABoxes and the implementation of conjunctive query answering algorithms are active research areas.

To be used in practice, the domain expert also needs tools that further support knowledge acquisition (i.e. building ontologies), maintenance (i.e. evolution of ontologies), and integration and inter-operation of ontologies. First steps in this direction have already been taken. For example, Protégé [64] and SWOOP [63] are tools that support the development of OWL ontologies. On a more fundamental level, non-standard inferences that support building and maintaining knowledge bases are now important topics of DL research. These include the inference problems discussed in Sect. 5.1 but also others that we have not discussed there due to space limitations: for example, tool support has been developed to explain subsumption and unsatisfiability and to repair unsatisfiable concepts (for example, see [62,87]) and to support modular design and re-use of ontologies (for example, see [33]).

In this chapter we have concentrated on very expressive Description Logics that are the formal basis for the web ontology language OWL. For the sake of completeness, we mention here some recent results on inexpressive DLs that are relevant in the context of ontology applications. Several biomedical ontologies, such as SNOMED [92] and the Gene Ontology [34], are based on rather inexpressive DLs, whose main distinguishing feature is that they disallow value restrictions ($\forall r.C$), but provide for existential restrictions ($\exists r.C$). Recently, it has turned out that such inexpressive DLs with existential restrictions behave much better w.r.t. computational complexity than the corresponding DLs with value restrictions. For example, the subsumption problem in $\mathcal{EL}$, which allows for conjunction, existential restrictions, and the top concept, stays polynomial in the presence of (cyclic or acyclic) concept definitions [2] and even arbitrary GCIs [25]. In [4] it is shown that these polynomiality results also hold for extensions of $\mathcal{EL}$ by constructors that are of interest for ontology applications, such as the bottom concept (which allows disjointness statements to be formulated), nominals, a restricted form of concrete domains, and a restricted form of so-called role-value maps. A first implementation of the polynomial-time subsumption algorithm for such an extension of $\mathcal{EL}$ behaves well on very large bio-medical ontologies [15].

References

1. Franz Baader. Using automata theory for characterizing the semantics of terminological cycles. *Annals of Mathematics and Artificial Intelligence*, 18(2–4): 175–219, 1996.
2. Franz Baader. Terminological cycles in a description logic with existential restrictions. In Georg Gottlob and Toby Walsh, editors, *Proc. of the 18th Int. Joint Conf. on Artificial Intelligence (IJCAI 2003)*, pages 325–330. Morgan Kaufmann, Los Altos, 2003.
3. Franz Baader, Sebastian Brandt, and Ralf Küsters. Matching under side conditions in description logics. In Bernhard Nebel, editor, *Proc. of the 17th Int. Joint Conf. on Artificial Intelligence (IJCAI 2001)*, pages 213–218. Morgan Kaufmann, Seattle, WA, 2001.
4. Franz Baader, Sebastian Brandt, and Carsten Lutz. Pushing the $\mathcal{EL}$ envelope. In *Proc. of the 19th Int. Joint Conf. on Artificial Intelligence (IJCAI 2005)*, 2005.
5. Franz Baader, Diego Calvanese, Deborah McGuinness, Daniele Nardi, and Peter Patel-Schneider, editors. *The Description Logic Handbook.* Cambridge University Press, 2003.
6. Franz Baader, Diego Calvanese, Deborah McGuinness, Daniele Nardi, and Peter F. Patel-Schneider, editors. *The Description Logic Handbook: Theory, Implementation and Applications.* Cambridge University Press, 2003.
7. Franz Baader and Philipp Hanschke. A schema for integrating concrete domains into concept languages. In *Proc. of the 12th Int. Joint Conf. on Artificial Intelligence (IJCAI'91)*, pages 452–457, 1991.
8. Franz Baader and Philipp Hanschke. Extensions of concept languages for a mechanical engineering application. In *Proc. of the 16th German Workshop on Artificial Intelligence (GWAI'92)*, volume 671 of *Lecture Notes in Computer Science*, pages 132–143. Springer, Berlin, 1992.
9. Franz Baader and Bernhard Hollunder. A terminological knowledge representation system with complete inference algorithm. In *Proc. of the Workshop on Processing Declarative Knowledge (PDK'91)*, volume 567 of *Lecture Notes in Artificial Intelligence*, pages 67–86. Springer, Berlin, 1991.
10. Franz Baader and Ralf Küsters. Computing the least common subsumer and the most specific concept in the presence of cyclic $\mathcal{ALN}$-concept descriptions.

In *Proc. of the 22nd German Annual Conf. on Artificial Intelligence (KI'98)*, volume 1504 of *Lecture Notes in Computer Science*, pages 129–140. Springer, Berlin, 1998.
11. Franz Baader and Ralf Küsters. Matching in description logics with existential restrictions. In *Proc. of the 7th Int. Conf. on Principles of Knowledge Representation and Reasoning (KR 2000)*, pages 261–272, 2000.
12. Franz Baader, Ralf Küsters, Alex Borgida, and Deborah L. McGuinness. Matching in description logics. *Journal of Logic and Computation*, 9(3):411–447, 1999.
13. Franz Baader, Ralf Küsters, and Ralf Molitor. Computing least common subsumers in description logics with existential restrictions. In *Proc. of the 16th Int. Joint Conf. on Artificial Intelligence (IJCAI'99)*, pages 96–101, 1999.
14. Franz Baader, Ralf Küsters, and Ralf Molitor. Rewriting concepts using terminologies. In *Proc. of the 7th Int. Conf. on Principles of Knowledge Representation and Reasoning (KR 2000)*, pages 297–308, 2000.
15. Franz Baader, Carsten Lutz, and Bontawee Suntisrivaraporn. CEL – a polynomial-time reasoner for life science ontologies. In Ulrich Furbach and Natarajan Shankar, editors, *Proc. of the Int. Joint Conf. on Automated Reasoning (IJCAR 2006)*, volume 4130 of *Lecture Notes in Artificial Intelligence*, pages 287–291. Springer, Berlin, 2006.
16. Franz Baader and Ulrike Sattler. An overview of tableau algorithms for description logics. *Studia Logica*, 69(1):5–40, 2001.
17. Sean Bechhofer, Ian Horrocks, Carole Goble, and Robert Stevens. OilEd: a reason-able ontology editor for the semantic web. In *Proc. of the 2001 Description Logic Workshop (DL 2001)*, pages 1–9. CEUR (`http://ceur-ws.org/`), 2001.
18. Tim Berners-Lee. *Weaving the Web*. Harpur, San Francisco, 1999.
19. Alexander Borgida. On the relative expressiveness of description logics and predicate logics. *Artificial Intelligence*, 82(1–2):353–367, 1996.
20. Alexander Borgida and Deborah L. McGuinness. Asking queries about frames. In *Proc. of the 5th Int. Conf. on the Principles of Knowledge Representation and Reasoning (KR'96)*, pages 340–349, 1996.
21. Pim Borst, Hans Akkermans, and Jan Top. Engineering ontologies. *International Journal of Human-Computer Studies*, 46:365–406, 1997.
22. Ronald J. Brachman. "Reducing" CLASSIC to practice: knowledge representation meets reality. In *Proc. of the 3rd Int. Conf. on the Principles of Knowledge Representation and Reasoning (KR'92)*, pages 247–258. Morgan Kaufmann, Los Altos, 1992.
23. Ronald J. Brachman and Hector J. Levesque. The tractability of subsumption in frame-based description languages. In *Proc. of the 4th Nat. Conf. on Artificial Intelligence (AAAI'84)*, pages 34–37, 1984.
24. Ronald J. Brachman and James G. Schmolze. An overview of the KL-ONE knowledge representation system. *Cognitive Science*, 9(2):171–216, 1985.
25. Sebastian Brandt. Polynomial time reasoning in a description logic with existential restrictions, GCI axioms, and – what else? In Ramon López de Mántaras and Lorenza Saitta, editors, *Proc. of the 16th Eur. Conf. on Artificial Intelligence (ECAI 2004)*, pages 298–302, 2004.
26. Sebastian Brandt, Ralf Küsters, and Anni-Yasmin Turhan. Approximation and difference in description logics. In D. Fensel, F. Giunchiglia, D. McGuiness, and

M.-A. Williams, editors, *Proc. of the 8th Int. Conf. on Principles of Knowledge Representation and Reasoning (KR 2002)*, pages 203–214. Morgan Kaufmann, Los Altos, 2002.

27. Paolo Bresciani, Enrico Franconi, and Sergio Tessaris. Implementing and testing expressive description logics: preliminary report. In *Proc. of the 1995 Description Logic Workshop (DL'95)*, pages 131–139, 1995.
28. Martin Buchheit, Francesco M. Donini, Werner Nutt, and Andrea Schaerf. A refined architecture for terminological systems: terminology = schema + views. *Artificial Intelligence*, 99(2):209–260, 1998.
29. Diego Calvanese, Giuseppe De Giacomo, and Maurizio Lenzerini. On the decidability of query containment under constraints. In *Proc. of the 17th ACM SIGACT SIGMOD SIGART Symp. on Principles of Database Systems (PODS'98)*, pages 149–158, 1998.
30. Diego Calvanese, Giuseppe De Giacomo, Maurizio Lenzerini, and Daniele Nardi. Reasoning in expressive description logics. In Alan Robinson and Andrei Voronkov, editors, *Handbook of Automated Reasoning*, chapter 23, pages 1581–1634. Elsevier, Amsterdam, 2001.
31. Diego Calvanese, Giuseppe De Giacomo, Maurizio Lenzerini, Daniele Nardi, and Riccardo Rosati. Description logic framework for information integration. In *Proc. of the 6th Int. Conf. on Principles of Knowledge Representation and Reasoning (KR'98)*, pages 2–13, 1998.
32. William W. Cohen, Alex Borgida, and Haym Hirsh. Computing least common subsumers in description logics. In William Swartout, editor, *Proc. of the 10th Nat. Conf. on Artificial Intelligence (AAAI'92)*, pages 754–760. AAAI Press/The MIT Press, Austin, TX, 1992.
33. Bernardo Cuenca Grau, Ian Horrocks, Yevgeny Kazakov, and Ulrike Sattler. Modular reuse of ontologies: theory and practice. *Journal of Artificial Intelligence Research*, 31:273–318, 2008.
34. The Gene Ontology Consortium. Gene ontology: tool for the unification of biology. *Nature Genetics*, 25:25–29, 2000.
35. Giuseppe De Giacomo. *Decidability of Class-Based Knowledge Representation Formalisms*. PhD thesis, Dipartimento di Informatica e Sistemistica, Università di Roma "La Sapienza", 1995.
36. Giuseppe De Giacomo and Maurizio Lenzerini. Boosting the correspondence between description logics and propositional dynamic logics. In *Proc. of the 12th Nat. Conf. on Artificial Intelligence (AAAI'94)*, pages 205–212, 1994.
37. Giuseppe De Giacomo and Maurizio Lenzerini. Concept language with number restrictions and fixpoints, and its relationship with μ-calculus. In *Proc. of the 11th Eur. Conf. on Artificial Intelligence (ECAI'94)*, pages 411–415, 1994.
38. Giuseppe De Giacomo and Maurizio Lenzerini. TBox and ABox reasoning in expressive description logics. In *Proc. of the 5th Int. Conf. on the Principles of Knowledge Representation and Reasoning (KR'96)*, pages 316–327, 1996.
39. Francesco M. Donini, Bernhard Hollunder, Maurizio Lenzerini, Alberto Marchetti Spaccamela, Daniele Nardi, and Werner Nutt. The complexity of existential quantification in concept languages. *Artificial Intelligence*, 2–3:309–327, 1992.
40. Francesco M. Donini, Maurizio Lenzerini, Daniele Nardi, and Werner Nutt. The complexity of concept languages. In *Proc. of the 2nd Int. Conf. on the Principles of Knowledge Representation and Reasoning (KR'91)*, pages 151–162, 1991.

41. Francesco M. Donini, Maurizio Lenzerini, Daniele Nardi, and Werner Nutt. Tractable concept languages. In *Proc. of the 12th Int. Joint Conf. on Artificial Intelligence (IJCAI'91)*, pages 458–463, 1991.
42. Jon Doyle and Ramesh S. Patil. Two theses of knowledge representation: language restrictions, taxonomic classification, and the utility of representation services. *Artificial Intelligence*, 48:261–297, 1991.
43. Dieter Fensel, Frank van Harmelen, Michel Klein, Hans Akkermans, Jeen Broekstra, Christiaan Fluit, Jos van der Meer, Hans-Peter Schnurr, Rudi Studer, John Hughes, Uwe Krohn, John Davies, Robert Engels, Bernt Bremdal, Fredrik Ygge, Thorsten Lau, Bernd Novotny, Ulrich Reimer, and Ian Horrocks. On-to-knowledge: ontology-based tools for knowledge management. In *Proceedings of the eBusiness and eWork 2000 (eBeW'00) Conference*, October 2000.
44. Birte Glimm, Ian Horrocks, Carsten Lutz, and Ulrike Sattler. Conjunctive query answering for the description logic $\mathcal{SHIQ}$. *Journal of Artificial Intelligence Research*, 31:157–204, 2008.
45. Erich Grädel. Guarded fragments of first-order logic: a perspective for new description logics? In *Proc. of the 1998 Description Logic Workshop (DL'98)*. CEUR Electronic Workshop Proceedings, `http://ceur-ws.org/Vol-11/`, 1998.
46. Erich Grädel. On the restraining power of guards. *Journal of Symbolic Logic*, 64:1719–1742, 1999.
47. Erich Grädel, Phokion G. Kolaitis, and Moshe Y. Vardi. On the decision problem for two-variable first-order logic. *Bulletin of Symbolic Logic*, 3(1):53–69, 1997.
48. Erich Grädel, Martin Otto, and Eric Rosen. Two-variable logic with counting is decidable. In *Proc. of the 12th IEEE Symp. on Logic in Computer Science (LICS'97)*, 1997.
49. Thomas Gruber. A translation approach to portable ontology specifications. *Knowledge Acquisition*, 5(2):199–220, 1993.
50. Volker Haarslev and Ralf Möller. RACE system description. In *Proc. of the 1999 Description Logic Workshop (DL'99)*, pages 130–132. CEUR Electronic Workshop Proceedings, `http://ceur-ws.org/Vol-22/`, 1999.
51. Volker Haarslev and Ralf Möller. Expressive ABox reasoning with number restrictions, role hierarchies, and transitively closed roles. In *Proc. of the 7th Int. Conf. on Principles of Knowledge Representation and Reasoning (KR 2000)*, pages 273–284, 2000.
52. Volker Haarslev and Ralf Möller. RACER system description. In *Proc. of the Int. Joint Conf. on Automated Reasoning (IJCAR 2001)*, volume 2083 of *Lecture Notes in Artificial Intelligence*, pages 701–705. Springer, Berlin, 2001.
53. Patrick Hayes. RDF model theory. W3C Working Draft, April 2002. `http://www.w3.org/TR/rdf-mt/`.
54. Bernhard Hollunder, Werner Nutt, and Manfred Schmidt-Schauß. Subsumption algorithms for concept description languages. In *Proc. of the 9th Eur. Conf. on Artificial Intelligence (ECAI'90)*, pages 348–353. Pitman, London, 1990.
55. Ian Horrocks. Using an expressive description logic: FaCT or fiction? In *Proc. of the 6th Int. Conf. on Principles of Knowledge Representation and Reasoning (KR'98)*, pages 636–647, 1998.
56. Ian Horrocks, Oliver Kutz, and Ulrike Sattler. The even more irresistible $\mathcal{SROIQ}$. In *Proc. of the 10th Int. Conf. on Principles of Knowledge Representation and Reasoning (KR 2006)*, pages 57–67. AAAI Press, New York, 2006.

57. Ian Horrocks, Peter F. Patel-Schneider, and Frank van Harmelen. From $\mathcal{SHIQ}$ and RDF to OWL: the making of a web ontology language. *Journal of Web Semantics*, 1(1):7–26, 2003.
58. Ian Horrocks and Ulrike Sattler. A description logic with transitive and inverse roles and role hierarchies. *Journal of Logic and Computation*, 9(3):385–410, 1999.
59. Ian Horrocks and Ulrike Sattler. A tableaux decision procedure for $\mathcal{SHOIQ}$. In *Proc. of the 19th Int. Joint Conf. on Artificial Intelligence (IJCAI 2005)*, pages 448–453, 2005.
60. Ian Horrocks, Ulrike Sattler, and Stephan Tobies. Practical reasoning for very expressive description logics. *Journal of the Interest Group in Pure and Applied Logic*, 8(3):239–264, 2000.
61. Ian Horrocks, Ulrike Sattler, and Stephan Tobies. Reasoning with individuals for the description logic $\mathcal{SHIQ}$. In David McAllester, editor, *Proc. of the 17th Int. Conf. on Automated Deduction (CADE 2000)*, volume 1831 of *Lecture Notes in Computer Science*, pages 482–496. Springer, Berlin, 2000.
62. Aditya Kalyanpur, Bijan Parsia, Evren Sirin, and Bernardo Cuenca-Grau. Hendler. Repairing Unsatisfiable Concepts in OWL Ontologies. In *Proc. of 3rd Europ. Semantic Web Conf. (ESWC 2006)*, number 4011 of LNCS, Springer, Berlin, 2006.
63. Aditya Kalyanpur, Bijan Parsia, Evren Sirin, Bernardo Cuenca-Grau, and James Hendler. SWOOP: a web ontology editing browser. *Journal of Web Semantics*, 4(2), 2005.
64. Holger Knublauch, Ray Fergerson, Natalya Noy, and Mark Musen. The Protégé OWL Plugin: an open development environment for semantic web applications. In Sheila A. McIlraith, Dimitris Plexousakis, and Frank van Harmelen, editors, *Proc. of the 2004 International Semantic Web Conference (ISWC 2004)*, number 3298 in LNCS, pages 229–243. Springer, Berlin, 2004.
65. Ralf Küsters and Alex Borgida. What's in an attribute? Consequences for the least common subsumer. *Journal of Artificial Intelligence Research*, 14:167–203, 2001.
66. Ralf Küsters and Ralf Molitor. Approximating most specific concepts in description logics with existential restrictions. In F. Baader, editor, *Proc. of the Joint German/Austrian Conference on Artificial Intelligence, 24th German / 9th Austrian Conference on Artificial Intelligence (KI 2001)*, volume 2174 of *Lecture Notes in Artificial Intelligence*. Springer, Berlin, 2001.
67. Ralf Küsters and Ralf Molitor. Computing least common subsumers in ALEN. In Bernard Nebel, editor, *Proc. of the 17th Int. Joint Conf. on Artificial Intelligence (IJCAI 2001)*, pages 219–224. Morgan Kaufmann, Los Altos, 2001.
68. Carsten Lutz. NEXPTIME-complete description logics with concrete domains. In *Proc. of the Int. Joint Conf. on Automated Reasoning (IJCAR 2001)*, volume 2083 of *Lecture Notes in Artificial Intelligence*, pages 45–60. Springer, Berlin, 2001.
69. Carsten Lutz. Description logics with concrete domains – a survey. In *Advances in Modal Logics Volume 4*. World Scientific Publishing Co. Pte. Ltd., Singapore, 2003.
70. Robert MacGregor. The evolving technology of classification-based knowledge representation systems. In John F. Sowa, editor, *Principles of Semantic Networks*, pages 385–400. Morgan Kaufmann, Los Altos, 1991.

71. Eric Mays, Robert Dionne, and Robert Weida. K-Rep system overview. *SIGART Bulletin*, 2(3):93–97, 1991.
72. Ralf Molitor. *Unterstützung der Modellierung verfahrenstechnischer Prozesse durch Nicht-Standardinferenzen in Beschreibungslogiken.* PhD thesis, LuFG Theoretical Computer Science, RWTH-Aachen, Germany, 2000. In German.
73. Bernhard Nebel. *Reasoning and Revision in Hybrid Representation Systems*, volume 422 of *Lecture Notes in Artificial Intelligence.* Springer, Berlin, 1990.
74. Bernhard Nebel. Terminological reasoning is inherently intractable. *Artificial Intelligence*, 43:235–249, 1990.
75. Bernhard Nebel. Terminological cycles: semantics and computational properties. In John F. Sowa, editor, *Principles of Semantic Networks*, pages 331–361. Morgan Kaufmann, Los Altos, 1991.
76. Leszek Pacholski, Wieslaw Szwast, and Lidia Tendera. Complexity of two-variable logic with counting. In *Proc. of the 12th IEEE Symp. on Logic in Computer Science (LICS'97)*, pages 318–327. IEEE Computer Society Press, Los Alamitos, CA, 1997.
77. Jeff Z. Pan and Ian Horrocks. Semantic web ontology reasoning in the $\mathcal{SHOQ}(\mathbf{D_n})$ description logic. In *Proc. of the 2002 Description Logic Workshop (DL 2002)*, 2002.
78. Peter F. Patel-Schneider. DLP. In *Proc. of the 1999 Description Logic Workshop (DL'99)*, pages 9–13. CEUR Electronic Workshop Proceedings, `http://ceur-ws.org/Vol-22/`, 1999.
79. Peter F. Patel-Schneider, Deborah L. McGuiness, Ronald J. Brachman, Lori Alperin Resnick, and Alexander Borgida. The CLASSIC knowledge representation system: guiding principles and implementation rational. *SIGART Bulletin*, 2(3):108–113, 1991.
80. Christof Peltason. The BACK system – an overview. *SIGART Bulletin*, 2(3):114–119, 1991.
81. A. Rector, S. Bechhofer, C. A. Goble, I. Horrocks, W. A. Nowlan, and W. D. Solomon. The GRAIL concept modelling language for medical terminology. *Artificial Intelligence in Medicine*, 9:139–171, 1997.
82. Ulrike Sattler. *Terminological knowledge representation systems in a process engineering application.* PhD thesis, RWTH Aachen, 1998.
83. Ulrike Sattler. Description logics for the representation of aggregated objects. In *Proc. of the 14th Eur. Conf. on Artificial Intelligence (ECAI 2000)*, 2000.
84. Andrea Schaerf. Reasoning with individuals in concept languages. *Data and Knowledge Engineering*, 13(2):141–176, 1994.
85. Klaus Schild. A correspondence theory for terminological logics: preliminary report. In *Proc. of the 12th Int. Joint Conf. on Artificial Intelligence (IJCAI'91)*, pages 466–471, 1991.
86. Klaus Schild. *Querying Knowledge and Data Bases by a Universal Description Logic with Recursion.* PhD thesis, Universität des Saarlandes, Germany, 1995.
87. Stefan Schlobach, Zhisheng Huang, Ronald Cornet, and Frank van Harmelen. Debugging Incoherent Terminologies. *Journal of Automated Reasoning*, 39(3):317–349, 2007.
88. Manfred Schmidt-Schauß and Gert Smolka. Attributive concept descriptions with complements. *Artificial Intelligence*, 48(1):1–26, 1991.
89. Stefan Schulz and Udo Hahn. Parts, locations, and holes – formal reasoning about anatomical structures. In *Proc. of AIME 2001*, volume 2101 of *Lecture Notes in Artificial Intelligence.* Springer, Berlin, 2001.

90. Evren Sirin, Bernardo Cuenca Grau, and Bijan Parsia. From wine to water: optimizing description logic reasoning for nominals. In *Proc. of the 10th Int. Conf. on Principles of Knowledge Representation and Reasoning (KR 2006)*, pages 90–99. AAAI Press, New York, 2006.
91. Evren Sirin and Bijan Parsia. Pellet: an OWL DL reasoner. In *Proc. of the 2004 Description Logic Workshop (DL 2004)*, 2004.
92. K.A. Spackman, K.E. Campbell, and R.A. Cote. SNOMED RT: a reference terminology for health care. *Journal of the American Medical Informatics Association*, pages 640–644, 1997. Fall Symposium Supplement.
93. Robert Stevens, Ian Horrocks, Carole Goble, and Sean Bechhofer. Building a reasonable bioinformatics ontology using OIL. In *Proceedings of the IJCAI-2001 Workshop on Ontologies and Information Sharing*, pages 81–90. CEUR (`http://ceur-ws.org/`), 2001.
94. Stephan Tobies. *Complexity Results and Practical Algorithms for Logics in Knowledge Representation*. PhD thesis, RWTH Aachen, 2001.
95. Dmitry Tsarkov and Ian Horrocks. FaCT++ description logic reasoner: system description. In *Proc. of the Int. Joint Conf. on Automated Reasoning (IJCAR 2006)*, volume 4130 of *Lecture Notes in Artificial Intelligence*, pages 292–297. Springer, Berlin, 2006.

Ontologies in F-Logic

Jürgen Angele[1], Michael Kifer[2], and Georg Lausen[3]

[1] Ontoprise GmbH, Amalienbadstrasse 36, D-76227 Karlsruhe, Germany, angele@ontoprise.de
[2] State University of New York at Stony Brook, Stony Brook, NY 11794-4400, USA, kifer@cs.stonybrook.edu
[3] Albert-Ludwigs-Universität Freiburg, Georges-Koehler-Allee, Gebäude 51, D-79110 Freiburg, Germany, lausen@informatik.uni-freiburg.de

Summary. *Frame Logic* (*F-logic*) combines the advantages of conceptual modeling that come from object-oriented frame-based languages with the declarative style, compact and simple syntax, and the well defined semantics of logic-based languages. F-logic supports typing, meta-reasoning, complex objects, methods, classes, inheritance, rules, queries, modularization, and scoped inference. In this paper we describe the capabilities of knowledge representation systems based on F-logic and illustrate the use of this logic for ontology specification. We give an overview of the syntax and semantics of the language and discuss the main ideas behind the various implementations. Finally, we present a concrete application deployed in the automotive industry.

1 Introduction

A conceptual model (or an ontology) is an abstract, declarative description of the information for an application domain. It includes the relevant vocabulary, constraints on the valid states of the information, and the ways to draw inferences from that information.

Conceptual modeling has a long history in the area of database systems. It began with the seminal work on the *Entity-Relationship* (ER) *model* [8], which divided the world into *entity types* and *relationship types*. An entity type is a homogeneous set of *entities* specified via their attributes and ranges. An entity represents a concrete object that belongs to one or more entity types and whose structure conforms to the types it belong to. A relationship type is a homogeneous set of *relationships*. It is specified by the entity types that are involved in the relationship and the *roles* these types play in that relationship. The ER model is thus a rudimentary language for specifying ontologies for the kinds of information that is natural to store in relational databases. ER was later extended to *extended entity relationship* (EER) model [20,29] by adding specialization, generalization, grouping, and other features. Subsequent modeling languages, like UML, were greatly influenced by ER and EER.

S. Staab and R. Studer (eds.), *Handbook on Ontologies,* International Handbooks on Information Systems, DOI 10.1007/978-3-540-92673-3,

Declarative query languages, like SQL, were central to relational database systems from their inception. The most attractive aspect of database query languages is the fact that their queries say *which* things to find rather than how to find them. Clearly, such languages are much harder to implement than the traditional imperative languages, like C. In the early days, database query languages had to be severely restricted in order to enable reasonably efficient implementations. However, as applications grew in sophistication, computing power increased, and our knowledge of algorithms for query processing expanded, the use of rule-based languages for processing information became more and more attractive. Further push came from the Semantic Web, which increased the awareness of the need for logic-based languages for processing ontologies and other distributed knowledge on the Web. This awareness led W3C to create a working group that was chartered with creation of a *rule interchange format* (RIF) – a family of standardized languages intended to facilitate the exchange of rule-based applications over the Web.[1]

Datalog [2] is the basis of all database rule languages. It has a model-theoretic semantics, can be efficiently implemented, and is reasonably expressive. However, it does not support function symbols, which are important for representing objects, and it is a poor choice as a modeling language. To improve the modeling power of logic-based languages, a number of extensions were proposed. These include more powerful kinds of negation, function symbols, high-level modeling constructs, and frame-based syntax. *F-logic* (or *Frame Logic*) [23] has emerged as a popular extension that provides all these features. As conceptual modeling goes, it faces little competition among rule-based logic languages. It accounts in a clean and declarative manner for most of the structural aspects of frame-based and object-oriented languages and, at the same time, is as powerful as any rule-based language for knowledge representation. An overview of logic-based languages can be found in [24]. Related, but limited languages have also been proposed for semi-structured and XML databases (e.g., [17,26]).

There are several major implementations of F-logic, including *FLORID* [25], *OntoBroker*™ [12], and *FLORA-2* [36]. Each implementation introduces a number of extensions to F-logic as well as restrictions to make their particular implementation methods more effective. FLORID and FLORA-2 were developed in the academia. Their main goal is to provide free platforms for experimenting with innovative features in the design of an F-logic based rule language. OntoBroker™ is a commercial system. Its main emphasis is on efficiency and integration with external tools and systems.

This paper takes a view of F-logic as an ontology modeling language as well as a language for building applications that use these ontologies. The ability to span both sides of the engineering process, ontologies and applications, is a particularly strong aspect of F-logic. This should be contrasted with the current state of semantic Web applications, which specify ontologies

[1] http://www.w3.org/2005/rules/wg.html

declaratively, using the OWL language [28], but then work with these ontologies using imperative languages, like Java, and in this mismatch loose many of the advantages of OWL's logic-based modeling.

In the following sections we give an overview of the syntax and the semantics of F-logic by illustrating the main features through a number of simple examples. Then we describe the ways in which F-logic can be implemented. Towards the end we present a real-life ontology-based application of F-logic in the automotive industry.

2 F-Logic by Example

In this paper we use the *new syntax* for F-logic – a simplified and extended version of the original syntax introduced in [23]. It was developed by the F-logic Forum group[2] and incorporates experience gained in the course of a decade of using F-logic in real life applications. The main points of the new syntax are summarized in [13]. All major F-logic based systems are in the process of migrating to the new syntax and some (e.g., FLORA-2) have already done so.

2.1 A Simple Ontology-Based Application

F-logic is an object-oriented language and ontologies are modeled in this language in an object-oriented style. One starts with class hierarchies, proceeds with type specification, defines the relationships among classes and objects using rules, and finally populates the classes with concrete objects.

The first part of the example presented in Fig. 1 is a small ontology. It states that every `woman` and `man` is a `person`. Objects have attributes. In the example, `person`-objects have the attributes `father, mother, daughter, son`, and these attributes have ranges. For example, the range of the attribute `son` is `man` and of `mother` is `woman`. The statements that specify the types of the attributes (which use the `*=>` sign) are called *signatures*. We will explain the significance of the symbol "*" in `*=>` in due time.

Then follows a set of rules, which say what else can be derived from the ontology. The first rule says that if `?X` is the `father` of `?Y` and `?Y` is a `man`, then `?Y` is a `son` of `?X`. A similar relationship holds for sons and mothers, and for daughters, fathers, and mothers. All variables in a rule are implicitly quantified outside of the rule (and for that reason the quantifiers are dropped). For instance, from the logical point of view the first rule in Fig. 1 is just an abbreviation for

```
∀?X ∀?Y (?X[son -> ?Y] ← ?Y:man[father -> ?X])
```

[2] http://projects.semwebcentral.org/projects/forum/

```
/* ontology consisting of a class hierarchy and signatures */
woman::person.
man::person.
person[father{0:1} *=> man].
person[mother{0:1} *=> woman].
person[daughter *=> woman].
person[son *=> man].
/* rules consisting of a rule head and a rule body */
?X[son -> ?Y] :- ?Y:man[father -> ?X].
?X[son -> ?Y] :- ?Y:man[mother -> ?X].
?X[daughter -> ?Y] :- ?Y:woman[father -> ?X].
?X[daughter -> ?Y] :- ?Y:woman[mother -> ?X].
/* facts */
Abraham:man.
Sarah:woman.
Isaac:man[father -> Abraham, mother -> Sarah].
Ishmael:man[father -> Abraham, mother -> Hagar:woman].
Jacob:man[father -> Isaac, mother -> Rebekah:woman].
Esau:man[father -> Isaac, mother -> Rebekah].
/* query */
?- X:woman[son -> ?Y[father -> Abraham]].
```

Fig. 1. A simple ontology-based application

Once the ontology is ready, we populate it with facts, some of which are specified at the end of the example. These facts tell us that Abraham is a man and Sarah is a woman. We also learn that Isaac is a man and his parents are Abraham and Sarah. Similar information is supplied on a number of other well-known individuals. These facts are part of the object base in our example. Other facts can be derived via deductive rules or other inference rules, such as inheritance. For instance, we can derive that Isaac is a son of Abraham even though this is not stated explicitly.

The last statement in Fig. 1 is a query to the object base. It asks to find all the women who have sons by Abraham. The answers are `?X = Sarah, ?Y = Isaac` and `?X = Hagar, ?Y = Ishmael`. The query illustrates how object descriptions can be nested within other object descriptions and yield concise and natural specification. Nesting is mostly a syntactic sugar, however. The same query can be written as a conjunction of non-nested expressions:

```
?- ?X:woman and ?X[son -> ?Y] and ?Y[father -> Abraham].
```

Note that all methods and attributes are multi-valued. For instance, from the rules we can derive `Abraham[son -> {Isaac,Ishmael}]` or

```
Abraham[son -> Isaac].
Abraham[son -> Ishmael].
```

Sometimes the nature of the application domain calls suggests cardinality constraints on some attributes in an ontology. In our example, the `father` and the `mother` attributes should have the cardinality constraint {0:1}, since there can be no more than one mother or father. Our ontology allows for the possibility that there can be no mother or father because the parents' identity may not be known or because there are none (notably, Adam and Eve).

The original F-logic explicitly distinguished between functional methods (whose cardinality constraint is {0:1}) and set-valued methods (whose cardinality is {0:∗}). It did not use the cardinality constraints syntax, but used instead => to denote functional methods in the schema and =>> for set-valued methods. To specify facts, the symbols -> and ->> were used, respectively. However, experience has shown that this notation is too error prone (it is easy to forget the extra ">") and not sufficiently flexible. For instance, it was hard to specify that some attribute is mandatory, i.e., must always have at least one value. Using the new syntax this is easy enough. For example, the following states that the name of a person must always be known:

```
person[name{1:*} *=> _string].
```

Here `_string` is a new built-in data type for strings – see Sect. 2.4.

2.2 Objects and Their Properties

Figure 1 shows the main components of an F-logic knowledge base: class hierarchies, signatures, rules, and objects. In this section we will take a closer look at each of these components in turn.

Object Names and Variable Names

Object names and variable names, also called *id-terms*, are the basic syntactic elements of F-logic. To distinguish object names from variable names, the later are prefixed with the ?-sign. Examples of object names are `Abraham`, `man`, `daughter`, and of variable names are `?X`, or `?method`. Object names can take several different forms:

- *Symbol.* A symbol is a sequence of characters enclosed in quotation marks (e.g., `"ab*-@c"`). Alphanumeric strings do not require quotes (e.g., `abc123`, `parent`).
- *A primitive data type.* An object that belongs to a primitive data type has the form `"..."^^typename` (e.g., `"12:22:33"^^_time`, `"123"^^_integer`). Primitive data types are discussed in Sect. 2.4.
- *A numeric shorthand.* Integers, decimals, and floating point numbers have shorthand notation. For instance, `"123"^^_integer` can be written simply as `123`, `"123.45"^^_decimal` as `123.45`, and `"123.45E-1"^^_float` as `123.45E-1`.

Complex id-terms are created from function symbols and other id-terms as usual in predicate logic: `couple(Abraham, Sarah)`, `f(?X)`. An id-term that contains no variable is called a *ground id-term*.

Methods

Application of a method to an object is specified using *data-F-atoms*. A remarkable feature of F-logic is that methods are also represented as objects and can be handled like regular objects without any special language support. For instance, in Fig. 1 the method names `father` and `son` are also object names just like `Isaac` and `Abraham`. Variables may appear anywhere, which enables queries about method names like

```
?- Abraham[?X->?].
?X = son
```

This query returns only one answer because only one attribute has values for the object `Abraham`. However, if we ask a query about the attributes defined in the *schema* of the object `Abraham` then more answers are returned:

```
?- Abraham[?X=>?].                                   (1)
?X = son, daughter, father, mother
```

This is because Abraham is a man, who is a person, and the class `person` has four attributes in its signature (the first group of statements in Fig. 1). We will explain later why we use `=>` in (1), while Fig. 1 uses `*=>` in the signatures for the class `person`. Note that both of the above queries use a special variable `?`, the *don't-care variable.* Each occurrence of this variable is treated as a completely new variable and answer substitutions are never returned for that variable.

Sometimes a method may take arguments. For example, Jacob's sons are born by different women. To express this, we introduce a version of the method `son`, which takes a parameter that denotes the mother. Parameters are also objects and are represented by id-terms.

```
Jacob[son(Leah)->{Reuben, Simeon, Levi, Judah, Issachar, Zebulun},
      son(Rachel)->{Joseph, Benjamin},
      son(Zilpah)->{Gad, Asher},                               (2)
      son(Bilhah)->{Dan, Naphtali}].
```

We could add one more parameter to indicate the order in which the sons were born:

```
Jacob[son(Leah,1)->Reuben, son(Leah,2)->Simeon,
      son(Leah,3)->Levi, son(Leah,4)->Judah,
      son(Bilhah,5)->Dan, son(Bilhah,6)->Naphtali,
      son(Zilpah,7)->Gad, son(Zilpah,8)->Asher,                (3)
      son(Leah,9)->Issachar, son(Leah,10)->Zebulun,
      son(Rachel,11)->Joseph, son(Rachel,12)->Benjamin].
```

Note that in (1), (2), and (3) above, the same method `son` is used with different numbers of parameters. This is one of the forms of *overloading* supported by F-logic. Given the object base described in Fig. 1, the query

```
?- Jacob[son -> ?X].
```

yields all twelve sons of Jacob. In contrast, the query

```
?- Jacob[son(Rachel) -> ?X].
```

returns only `?X = Joseph` and `?X = Benjamin`. Note that variables in a query can be bound only to individual objects and never to sets of objects. Thus, the above query does not return one answer, the set `?X = {Joseph, Benjamin}`, but two answers. In each of these answers `?X` is bound to exactly one element. However, all of the following queries return the answer true:

```
?-Jacob[son -> {Joseph, Benjamin}].
?-Jacob[son -> Joseph].
?-Jacob[son -> Benjamin].
```

If we wanted to know whether `{Joseph, Benjamin}` is *precisely* the set of all Jacob's sons by Rachel then negation must be used (see [23], Sect. 12.4.1.2).

F-logic also supports Boolean methods. These return no values, but only state whether a property is true or not. For instance, `Jacob[married]`.

Class Hierarchies

Class hierarchies are defined with the help of isa-F-atoms and subclass-F-atoms. An *isa-F-atom* of the form `o:c` states that an object `o` is a member of class `c`. The members of a class typically are called the *instances* of the class. A *subclass-F-atom* of the form `sc::cl` says that the class `sc` is a subclass of the class `cl`. In the following example the first three isa-F-atoms say that `Abraham` and `Isaac` are instances of the class `man`, whereas `Sarah` is an instance of the class `woman`. The next two subclass-F-atoms state that both `man` and `woman` are subclasses of the class `person`:

```
Abraham:man.
Isaac:man.
Sarah:woman.
woman::person.
man::person.
```

In F-logic, classes are also objects and thus are represented as id-terms. Hence, classes can have methods defined on them, and they can be instances of other classes. As mentioned earlier, methods are also objects and, as such, can belong to classes (or can themselves serve as classes). This can be helpful when one needs to attach meta-information to ontology elements – for example, to annotate them with provenance information:

```
son:Relation[authoredby -> Hans].
```

Furthermore, variables are permitted at all positions in isa- and subclass-F-atoms, so objects, methods, and classes are represented and queried uniformly using the same language facilities. In this way, F-logic naturally supports the meta-information facility. In contrast to other object-oriented languages where an object can be an instance of exactly one most specific class (e.g., ROL [24]), F-logic permits to be an instance in several, possibly incomparable, most specific classes. Likewise, a class can have several incomparable most specific superclasses. Thus, the class hierarchy is a directed acyclic graph.

Expressing Information About an Object: F-Molecules

F-molecules are used to make several different assertions about the same object in a compact way. For example, the following F-molecule says that Isaac is a man, his father is Abraham, and Jacob and Esau are amongst his sons.

```
Isaac:man[father -> Abraham, son -> {Jacob,Esau}].        (4)
```

This F-molecule is equivalent to a conjunction of the following statements:

```
Isaac:man.
Isaac[father -> Abraham].
Isaac[son -> Jacob].
Isaac[son -> Esau].
```

Note that sets can equivalently be split into subsets. Thus, for Isaac's sons in (4) we have an equivalent representation:

```
Isaac[son -> {Jacob}].
Isaac[son -> {Esau}].
```

For singleton sets, the braces around set elements can be omitted:

```
Isaac[son -> Jacob].
Isaac[son -> Esau].
```

An important feature of F-molecules is that they can be nested. The following molecules are examples of nesting of data-F-atoms and isa-F-atoms.

```
Isaac[father -> Abraham:man[son(Hagar:woman) -> Ishmael],
      mother -> Sarah:woman].
Jacob:(man::person).                                        (5)
Jacob[(father:method) -> Isaac].
```

Nesting allows one to specify the properties of an object locally without the need to split a complex statement into verbose conjunctions where the different assertions may appear far apart. However, nesting is just a syntactic sugar, which does not increase the expressive power. The statements in (5) can be "unnested" and represented as the following set of facts:

```
Isaac[father -> Abraham].
Abraham:man.
Abraham[son(Hagar) -> Ishmael].
Hagar:woman.
Isaac[mother -> Sarah].
Sarah:woman.
man::person.
Jacob:man.
Jacob[father -> Isaac].
father:method.
```

Formally speaking, a nested molecule is equivalent to a conjunction of its atomic components.

As usual with nesting of expressions, there are precedence rules. In F-logic, molecules are processed left-to-right, and if this is not what one expects then parentheses must be used. For instance, the first clause below says that Isaac is a man and he believes in God, whereas the second clause says that Isaac is a man and that the class `man`, when treated as an *object*, believes in God.

```
Isaac:man[believesin -> God].
Isaac:(man[believesin -> God]).
```

Signatures

Signature-F-atoms specify the schema of a class; they declare the methods that apply to the various classes, the types of the arguments used by those methods, and the methods' ranges. In addition, cardinality constraints can be specified, as we saw previously. Syntactically, signature atoms are similar to data-F-atoms, but the arrow `*=>` is used instead of `->`. Here are some examples of signatures:

```
person[father{0:1} *=> man].
person[daughter *=> woman].
man[son(woman) *=> man].
```

The first states that the single-valued method `father` is defined for instances of the class `person` and the range of that method is the class `man`. We know that this is a single-valued method because of the cardinality constraint that says that a person can have at most one father. The second signature defines the multi-valued method `daughter` for the class `person`. It says that the objects returned by this method belong to the class `woman`. How do we know that `daughter` is a multi-valued method? By its cardinality constraint (more precisely, the lack of it). Unless stated otherwise, a method is multi-valued and its default cardinality constraint is `{0:*}`, which does not really constrain anything. The third signature-F-atom declares the multi-valued method `son`, which applies to objects of the class `man`. The method takes arguments of type `woman`. The result of the method must be an object of class `man`. Boolean combinations of ranges can also be used. For instance (momentarily changing the theme of our example), we could define teaching assistants to be both students and employees:

```
course[teachingAssistant *=> (student and employee)].        (6)
```

Intersection of ranges can actually be specified without the and-operator:

```
course[teachingAssistant *=> student].
course[teachingAssistant *=> employee].                        (7)
```

Union of ranges is harder to specify without an operator that works on classes, but the or-operator saves the day:

```
course[instructor *=> (professor or lecturer)].
```

F-logic supports method overloading. This means that methods denoted by the same object name may be declared for different classes. Methods may also be overloaded and used with different numbers of parameters, as we saw in examples (2) and (3), which use the method `son` of class `man` with both one and two parameters. The corresponding signature-F-atoms look like this:

```
man[son(woman) *=> man].
man[son(woman,integer) *=> man].
```

As with data-F-atoms, signatures can be combined and nested. For example:

```
person[father{0:1} *=> man[son(woman) *=> man, son(woman,integer) *=> man]].
```

Inheritable and Non-inheritable Methods

Now we are ready to explain the difference between the symbols `=>` and `*=>` in signatures. F-logic distinguishes between *inheritable* and *non-inheritable* methods. Inheritable methods are roughly like instance methods in Java. One defines them for a class, they are inherited by subclasses and instances, and they are invoked by applying them to instances of a class. Non-inheritable methods are like static methods in Java. They are defined for classes and make sense only when applied to classes.[3]

In F-logic, to declare a signature of an inheritable method, use `*=>`. When a method is non-inheritable, use `=>`. Inheritable methods are inherited to subclasses also as inheritable methods. For instance, in Fig. 1, we declared several signatures for the class `person`. Since `man` is a subclass of `person`, the

[3] Java permits application of static methods to instances of the classes for which those methods are defined. However, this is a (perhaps unfortunate) syntactic sugar. The result of such an application is the same as applying the static method to the corresponding class.

same signatures are inherited by that class, so the following is implied by the semantics of F-logic:

```
man[father{0:1}*=>man, mother{0:1}*=>woman, daughter *=> woman, son *=> man].
```

However, when inheritable methods are inherited to the instances of the classes, they become non-inheritable. For instance, `Isaac` is an instance of the class `man`. When the methods `father`, `mother`, etc., are inherited by `Isaac` from `man`, these methods become non-inheritable, so the following is implied by the semantics of F-logic:

```
Isaac[father{0:1}=>man, mother{0:1} => woman, daughter => woman, son => man].
```

The stars are gone and `=>` is used instead of `*=>`. This is why the query in Fig. 1 uses `=>` instead of `*=>`. Note that even if the object `Isaac` were also treated as a class, e.g., the class of Isaac's camels, it would not necessarily make sense to further inherit the methods `father`, `mother`, etc., down to the instances of that class. This is why inheritable methods loose their inheritability when they are inherited by class instances.

F-Molecules Without Properties

If we want to represent an object without giving any properties, we have to attach an empty method-specification list to the object name:

```
Thing[].
```

This statement asserts that the object `Thing` exists, but does not state any properties. The query that asks about existence of this object

```
?- Thing[].
```

will return true, but

```
?- Thing[foo -> ?].
```

will return false, if the property `foo` had not been defined. The fact `Thing[ ]` is different from the fact `Thing`. This latter says that `Thing` is a proposition, which implies that the query

```
?- Thing.
```

(which is different from the query `?- Thing[ ]`) would succeed.

Predicate Symbols

Experience shows that it is convenient to be able to use predicates alongside objects. In F-logic, predicate symbols are used in the same way as in other deductive languages, such as Datalog. A predicate formula is constructed out of a predicate symbol followed by one or more arguments included in parentheses. Such a formula is called a *P-atom*. The following are examples of P-atoms.

```
married(Isaac,Rebekah).
male(Jacob).
sonof(Isaac,Rebekah,Jacob).
Thing.
```

The last P-atom above is a 0-ary predicate symbol, a proposition. Information expressed by P-atoms can usually be represented by F-atoms, as shown below:

```
Isaac[marriedto -> Rebekah].
Jacob:man.
Isaac[son(Rebekah) -> Jacob].
```

Nesting of F-molecules inside P-molecules is permitted and handled similarly to nesting F-molecules. For instance, the P-molecule

```
married(Isaac[father -> Abraham], Rebekah:woman).
```

is equivalent to the following set of P- and F-atoms:

```
married(Isaac,Rebekah).
Isaac[father -> Abraham].
Rebekah:woman.
```

F-logic also supports predicate signatures, but we do not discuss this here.[4]

2.3 Path Expressions

Path expressions are a standard fixture in most object-oriented languages. In F-logic, a path expression of the form `obj.expr` denotes the set of objects {`a1,a2,...`}, such that `obj[expr -> {a1,a2,...}]` is true. The expression in the path expression can be a simple attribute or a method application. Method expressions can be further applied to the results of path expressions, and in this way longer path expressions can be constructed: `obj.expr1.expr2.expr3`. Here are some path expressions and the sets of objects they refer to in the example of Fig. 1.

```
Isaac.son                      {Jacob, Esau}
Jacob.son(Rachel,11)           {Joseph}              (8)
Esau.father.father.son         {Isaac, Ishmael}
```

The second line illustrates an application of a method in a path expression, and the third line shows a path expression where expressions are applied

[4] Details can be found at http://projects.semwebcentral.org/projects/forum/forum-syntax.html.

repeatedly. Since a method can yield a set of results, one might wonder about the meaning of an expression such as this:

```
Abraham.son.son
```

Since `Abraham.son` is a set, what does the second attribute, `son`, apply to? In F-logic, the answer is that it applies to every object denoted by `Abraham.son`, and the results of all such applications are unioned. Thus, this path expression denotes the set of all Abraham's grandchildren.

Path expressions in F-logic were first introduced [23] and subsequently refined and extended in [14]. As a result of this enhancement, path expressions can appear anywhere an object can.

In imperative object-oriented languages, path expressions provide the only way to navigate object relationships. In F-logic, most general way to navigate through objects is to use F-molecules and combine them with logical connectives `and` and `or`. However, the use of path expressions can simplify formulas in many cases. Since a path expression, such as `obj.expr`, denotes all the objects that can bind to the variable `?X` in `obj[expr -> ?X]`, path expressions can help eliminate variables. For instance, Abraham's grandsons can be represented in either of the following ways:

```
?- Abraham.son.son=?X .
?- Abraham.son[son -> ?X].
?- Abraham[son -> ?Y] and ?Y[son -> ?X].
```

These queries have two answers, one per each of Abraham's grandsons. The first query is most concise. The second query combines path expression notation with frame-based notation. It is slightly longer, but both queries use just one variable. The third query does not use path expressions; it is bulkier than the first two, and requires two variables instead of one. It should be clear that by stacking more method applications in one path expression one can eliminate many variables and thus simplify some queries and rules.

2.4 Built-in Data Types and Methods

The new syntax for F-logic supports a large number of XML Schema data types and the corresponding built-ins. The built-ins are largely the same as in XQuery, but F-logic follows more elegant, object-based conventions.

The most important data types are `_string`, `_integer`, `_decimal`, `_iri`, `_time`, `_dateTime`, and `_duration`. By convention, the built-in types and methods start with an underscore. The constants that belong to these types are denoted as `"..."^^_type`. The first three of the built-in types above are self-explanatory. The type `_iri` is for representing *International Resource Identifiers* (e.g., `"http://foo.bar.com/a/b/c#fgh?id=7"^^_iri`), which generalize URLs and are used to denote objects on the Web. This type provides methods such as `_schema` (`http` in our case), `_host` (`foo.bar.com`), `_path`

(`/a/b/c`), and more. The type `_time` represents a time point within one day (for instance, `"12:33:56"^^_time`), and `_dateTime` is a type of arbitrary time points (for instance `"2007-06-22T10:23:55+03:00"^^_dateTime` represents the time point of 10:23:55 on June 22, 2007 with time zone three hours ahead of Greenwich). The type `_duration` represents temporal durations (for instance, `"P2Y3DT1H3.4S"^^_duration` represents the duration of two years, three days, one hour, and 3.4 seconds). The duration-objects can be added to or subtracted from the time and dateTime objects. The following examples illustrate these types and their corresponding built-ins.

```
?- "file:///abc/cde/efg"^^_iri[_scheme->?P].              // ?P = file
?- "mailto:me@foo.com"^^_iri[_user->?U,_host->?H].
                                            // ?U = me, ?H = foo.com
?- "2007-11-22T23:33:55.234"^^_dateTime[_hour->?Hr].    // ?Hr = 23
?- "P21Y11M12DT11M55S"^^_duration[_year->?Yr1].         // ?Yr = 21
?- "21:22:55"^^_time[_add("PT2H1M1S"^^_duration)->?X].
                                          // ?X = "23:23:56"^^_time
```

Along with the primitive data types come variables that are restricted to that data type only. For instance, `?L^^_iri` can unify only with the objects of the primitive data type `_iri`, which is described above. Thus, if we had

```
Obj[location ->"file:///abc/cde/efg"^^_iri].
Obj[location ->NewYork].
?- Obj[location->?L^^_iri].
```

then the only answer will be `?L = "file:///abc/cde/efg"^^_iri`. The object `NewYork` will not be returned because it is not an IRI.

2.5 Rules

Rules are perhaps the best candidates for building applications around ontologies. We have already seen examples of rules in Fig. 1. In general, a rule is an expression of the form *head :- body*, where the head of the rule is an F-molecule and the body is a Boolean combination of F-molecules or negated F-molecules. Conjunctions are represented using the `and` keyword and disjunctions using `or`. Commas can also be used in place of `and`, and the semicolon is a shorthand for `or`. Molecules may contain variables, and all variables are implicitly ∀-quantified outside of the rule. Rules in F-logic have a logical semantics as developed in the fields of logic programming and deductive databases. Consider the following rule from Fig. 1:

```
?X[son->?Y] :- ?Y:man[father->?X].
```

Its semantics can be informally explained as follows. Whenever we can find id-terms for the variables `?X` and `?Y` so that all molecules in the body become either existing or derived facts, then the head of the rule is derived after applying the same substitutions to `?X` and `?Y` in the head. In case of our rule above, its body is true for `?X=Abraham` and `?Y=Isaac` or `?Y=Ismael;` or for `?X=Isaac` and either `?Y=Jacob` or `?Y=Esau`. This is due to the facts that `Isaac:man[father -> Abraham]`, `Ishmael:man[father -> Abraham]`, `Jacob: man[father -> Isaac]`, `Esau:man[father -> Isaac]` are explicitly given in Fig. 1. From these facts and the above rule we can derive

```
Abraham[son -> Isaac], Abraham[son -> Ishmael]
Isaac[son -> Jacob], Isaac[son -> Esau].
```

We can write these facts in a more concise way as `Abraham[son -> {Isaac, Ishmael}]` and `Isaac[son -> {Jacob,Esau}]`. Similarly, with the rule

```
?X[grandson -> ?Y] :- ?Y:man[father -> ?Z], ?Z:man[father -> ?X]].
```

we can derive `Abraham[grandson -> {Jacob, Esau}]`. When molecules in a rule body are negated, a form of the well-founded semantics is typically used in F-logic systems [18,35]. Note that, since F-logic allows variables over method and class names, formulas like `?X:?Y[?X -> ?Y]` are legal and might, in fact, match some facts in a particular ontology. For instance, if both `abc:cde` and `abc[abc -> cde]` are true). There are many interesting applications for F-logic's meta-information capabilities. For instance, one can write rules for checking type correctness:

```
?O[typeError -> ?M] :- ?O[?M => ?T], ?O[?M -> ?V], not ?V:?T.
```

This rule says that a type error exists if there is a data molecule `?O[?M -> ?V]` that does not conform to the signature `?O[?M => ?T]` for the method `?M`. If a type error is found, `?O` is bound to the object where the type error exists, and `?M` to the offending method.

Note that here both the signature and the data molecule can be derived rather than explicitly given. Also, due to signature inheritance, it is enough to use `?O[?M => ?T]` instead of `?C[?M *=> ?T]`, for some superclass `?C` of `?O`. This is because if `c[m *=> t]` and `o:c` are true for some `o`, `c`, `m`, `t`, then `o[m => t]` is derived by inheritance.

Rules can be recursive. For example, given a genealogy (a parenthood relationship), we may want to specify ancestry information as follows:

```
?X[ancestor -> ?Y] :- ?X[parent -> ?Y].
?X[ancestor -> ?Y] :- ?X[ancestor -> ?Z], ?Z[parent -> ?Y].
```

A more complex case is when we want to combine ancestry information with the information about the number of generations by which an ancestor is removed from the subject person.

```
?X[generation(?Y) -> 1] :- ?X[parent -> ?Y].
?X[generation(?Y) -> ?N] :-
                    ?X[generation(?Z) -> ?N1], ?Z[parent -> ?Y], ?N is ?N1+1.
```

Note that the generation method is, in general, multi-valued, since there can be multiple genealogical lines connecting a pair of individuals.

The genealogy example gives us an opportunity to illustrate the powerful facility of aggregate functions supported by most F-logic based systems:

```
?X[shortestAncestryLine(?Y) -> ?N] :- ?N = min{?L|?X[generation(?Y) -> ?L]}.
```

This rule says that the length of the shortest ancestry line between any pair of individuals, `?X` and `?Y`, is some number `?N` that is computed as the smallest `?L` such that `?X[generation(?Y) -> ?L]` is true.

2.6 Scoped Inference: Modularization and Integration

The concept of scoped inference [19, 21] is central to modularization and integration of knowledge. It was first proposed in TRIPLE [30] and FLORA-2 [36], and the F-logic Forum group has adopted this concept as the main vehicle for modularization of F-logic ontologies.[5]

The concept of a module is well known in software engineering, and it is equally important in knowledge engineering. It is especially important for representing distributed knowledge, such as ontologies scattered over the Web, since rules and concepts that belong to different ontologies may interact in subtle and unintended ways.

The basic idea is that a knowledge base is a collection of *scopes of inference* or *modules*. Each module is a collection of rules and facts. The notion of a rule is extended as follows. As before, it is a statement of the form ***Head* :- *Body***. The head literal is a predicate or an F-molecule – still no change here. The notion of a body of a rule is also unchanged – a Boolean combination of predicates and F-molecules. However, now these predicates and molecules can optionally be labeled with module references like this: *pred-or-molecule@module-name*. Note that only the formulas that occur in a rule body can have references to modules. The formulas in the head cannot.

A rule of the form *Head :- Body* that belongs in a particular module, *M*, defines *Head* for that module. A subformula of the form *L@N* inside *Body* is a query to module *N*, asking whether *L* is implied by the knowledge base that resides in module `N`. For instance, some data source, `gendata`, may provide information about parents of various individuals. One may not be able to (or may not want to) insert new rules into that data source in order to preserve the integrity of the data. However, it is possible to create a different module, say `mygenealogy`, put rules there, and reference the information in the data source `gendata`:

```
?X[ancestor -> ?Y] :- ?X[parent -> ?Y]@gendata.
?X[ancestor -> ?Y] :- ?X[parent -> ?Z]@gendata, ?Z[ancestor -> ?Y].
```

[5] http://projects.semwebcentral.org/projects/forum/forum-syntax.html

Here, the molecules of the form ...[ancestor -> ...] are defined in the same module where our two rules belong, i.e., mygenealogy. The literals of the form ...[parent -> ...]@gendata are queries to the module gendata where the parenthood information resides. Some other module might query both of these modules, but the answers might be different. For instance, the queries

```
?- ?X[parent -> ?Y]@gendata.
?- ?X[parent -> ?Y]@mygenealogy.
```

will likely return different answers, since the parent attribute is not defined in mygenealogy. Thus, the first query will return all it knows about the parenthood relationship among individuals, while the second query will return nothing. Likewise,

```
?- ?X[ancestor -> ?Y]@gendata.
?- ?X[ancestor -> ?Y]@mygenealogy.
```

will return different answers: the ancestor attribute is not defined in module gendata so the first query will return nothing, while the second will return the transitive closure of the parent attribute (which mygenealogy imports from gendata using the above rules).

Module names can be arbitrary strings. Some systems even allow module names to be arbitrary terms. However, in case of public ontologies, module names most often are URIs.

Modules can be created in different ways. First, a new module can be created on-the-fly and rules can be added to it at run time. For instance, in FLORA-2, the following query will create the module mygenealogy and drop the above rules into it:

```
?-newmodule{mygenealogy},
  insertrule{((?X[ancestor -> ?Y] :- ?X[parent -> ?Y]@gendata),
             (?X[ancestor -> ?Y] :-
                          ?X[parent -> ?Z]@gendata, ?Z[ancestor -> ?Y])
             )@mygenealogy }.
```

Another method is to create a file containing rules and facts, and then load it into a module. For example, in the FLORA-2 system this is done as follows:

```
?- _load(myfile>>mygenealogy).
```

Finally, a file can be designated to specifically contain rules for a particular module using the declaration of the form

```
:- module mygenealogy.
```

at the beginning of the file.

Besides modularization, the concept of a module is a potent vehicle for integration of and reasoning about ontologies that reside at different sources.

If one just unions the rules and the facts found at the sources of interest, as implied by the import mechanism of the OWL language, the rules may contradict each other or have subtle and unintended interactions. In contrast, if different sources are treated as separate modules, one can differentiate among the information residing at these sources and specify the appropriate integration rules. These rules may give preference to some sources, partially or completely disregard information supplied by others, or clearly flag conflicting information.

F-logic modules can be imported into other modules. This allows one to construct ontologies in a hierarchical way. For instance, an upper level ontology may be defined in one module and a domain-specific ontology, defined in another module, might inherit concept definitions from the upper-level ontology. This can be conveniently specified by the following statement included at the top of the domain ontology:

```
:- importmodule myupperlevelontology.
```

The effect of this statement is that every concept and method defined in the upper ontology can be used in the domain ontology without the need for the `@myupperlevelontology` designator.

The reader is referred to [16] for further details on the powerful mechanism of modules in F-logic.

2.7 Inheritance

In frame-based systems, inheritance comes in two forms: *structural* and *behavioral.* Structural inheritance means that declarations of structure in a class are also inherited by subclasses. Behavioral inheritance deals with default method definitions. The F-logic Forum group has decided to include structural inheritance as a core feature, but made behavioral inheritance optional.

Structural inheritance is simpler, and we look at it first. Consider the signature declarations such as

```
person[father{0:1} *=> man].
person[daughter *=> woman].
```

Since `man::person` holds, it should follow that a man's father is also a man:

```
man[father{0:1} *=> man].
man[daughter *=> woman].
```

should logically follow from the above. Structural inheritance is monotonic in the sense that types inherited from superclasses are never overwritten, but are accumulated instead. In the following example, class `workingStudent` is defined as subclass of the classes `worker` and `student`. A worker is paid by at least one company and a student may be paid by research institutions.

```
worker[paidBy{1:*} *=> company].
student[paidBy *=> researchInstitution].
workingStudent::worker.
workingStudent::student.
```

As structural inheritance is monotonic, it follows that whenever a working student is paid, the money must come from an institution which is a company and a research institution as well.

Behavioral inheritance is much more complex because it can be overwritten by explicit or derived information specified for subclasses. For instance, most humans have their heart on the left and so it is reasonable to specify this *as a default*. In F-logic, defaults for classes are specified using the `*->` arrow style:

```
person[heartPosition *-> left].
```

But dextrocardiacs have their hearts on the right, which is expressed as

```
dextrocardiac[heartPosition *-> right].
dextrocardiac :: person.
```

Suppose John Doe is a dextrocardiac: `JohnDoe:dextrocardiac`. He inherits `heartPosition *-> left` from the class `person` and `heartPosition*-> right` from the class `dextrocardiac`. Although this is a contradiction, the class `dextrocardiac` is more specific to John Doe than the class `person`, so inheritance from `dextrocardiac` *overrides* inheritance from `person`.

Unlike structural inheritance, behavioral inheritance is *nonmonotonic*. This means that inferences that were made from a set of facts and rules, S, may no longer hold from a bigger set $S' \supseteq S$. To see this, let us return to our biblical genealogy. We know `Abraham:person`, and there has been no indication that he was a dextrocardiac. According to the common understanding of inference by inheritance, it should follow that

```
Abraham[heartPosition -> left].
```

It also follows that, for example, `man[heartPosition* -> left]`. (Properties are inherited to individual class instances as non-inheritable methods and to subclasses as inheritable ones. This is why we use the `->` style arrow for `Abraham` and `*->` for `man`.)

Suppose now that new information says that `Abraham` was a dextrocardiac. The rules of inheritance then tell us that inheritance from `dextrocardiac` overrides inheritance from `person` so `Abraham[heartPosition -> left]` must become false while `Abraham[heartPosition -> right]` must become true. In other words, a larger set of premises no longer entails the old conclusion that Abraham's heart is on the left.

In F-logic, class hierarchies can be defined by rules, so it is not possible to "eyeball" an ontology and tell which classes are subclasses of other classes and which inheritance overrides what. This is further complicated by the fact that class hierarchies may depend on negative facts derived through default negation. The complete treatment of the semantics for inheritance in F-logic is beyond the scope of this paper. Several semantics have been proposed [3, 23,27,35], but the F-logic Forum group adopted the semantics defined in [35].

3 Implementations of F-Logic

Implementations of F-logic based systems can be roughly divided into two categories: those that are based on native object-oriented deductive engines and those that use relational deductive engines. The engines can be further divided into bottom-up and top-down engines.

FLORID implements F-logic using a dedicated bottom-up deductive engine, which handles objects directly through an object manager. In that sense, it is similar to object-oriented databases. In contrast, FLORA-2 and OntoBroker^{TM} use relational engines, which do not support objects directly. Instead, both systems translate F-logic formulas into statements that use predicates (relations) instead of F-logic molecules, and then execute them using relational deductive engines. FLORA-2's target engine is XSB – a Prolog-like inference engine with numerous enhancements, which make XSB into a more declarative and logically complete system than the usual Prolog implementations. XSB's inference mode is top-down with a number of bottom-up-like extensions. OntoBroker^{TM} uses its own relational deductive engine. Its main inference mode is bottom-up, but it includes several enhancements inspired by top-down inference, such as dynamic filtering [22] and Magic Sets [10].

The translation from F-logic into the relational syntax used by FLORA-2 and OntoBroker^{TM} was defined in [23]. The main ideas are as follows:

(1) First, molecular expressions are replaced by equivalent conjunctions of atomic molecules. We illustrated this process in earlier sections.
(2) Next, these atomic expressions are represented by first-order predicates.
(3) The resulting set of rules is augmented with additional "closure rules" to capture the specific semantics of F-logic. Some rules are needed to express statements such as the transitivity of the subclass relationship; other rules implement the semantics of inheritance and other features.

Table 1 shows the second stage in the translation process. Whenever an F-logic specification is split into modules, the predicates `type`, `sub`, `isa`, etc., in the table are disambiguated for different modules by either adding an additional argument or by specializing predicate names for each module.

The following are examples of some of the closure rules added in stage (3) of the translation process:

```
// closure rules for ?X :: ?Y
sub(?X, ?Z) :- sub(?X, ?Y) and sub(?Y, ?Z).
```

F-atom	Predicate
C[A(B1,...,Bn) => R]	type(C,A(B1,...,Bn),R)
C[A(B1,...,Bn) *=> R]	defaulttype(C,A(B1,...,Bn),R)
A::B	sub(A,B)
o:C	isa(o,C)
o[A -> b]	data(o,A,b)
o[A* -> b]	defaultdata(o,A,b)
p(b1,...,bn)	p(b1,...,bn)

Table 1. Transformation of F-logic atoms into predicate notation.

```
// closure rules for ?X : ?C
isa(?O, ?C) :- sub(?C1, ?C) and isa(?O, ?C1).

// structural inheritance of signatures
defaulttype(?C1, ?A, ?T) :- sub(?C1, ?C2) and defaulttype(?C2, ?A, ?T).
type(?O, ?A, ?T) :- isa(?O, ?C) and defaulttype(?C, ?A, ?T).
```

In systems that support non-monotonic inheritance, additional rules are included [35]. The resulting rule sets then processed using the well-founded semantics for rule-based languages [18]. As an optimization, Ontobroker^{TM} recognizes special cases where a simpler, stratified semantics can be used, while FLORID allows the user to explicitly define the stratification.

4 An Industrial Application: Configuration of Test Cars

We will now describe a project in the automotive industry where ontologies have two main purposes: (1) representing and sharing knowledge to optimize business processes for testing cars and (2) integration of live data into this optimization process. A car manufacturer has a fleet of test cars. The cars are frequently reconfigured and tested with the new configuration. Reconfiguration could mean changing the engine, the gear, the electric system, etc. When parts are replaced, many of their interdependencies must be taken into account. In many cases these dependencies are known only by certain human experts and require significant amount of communication between different departments in the manufacturer's plant, between the manufacturer and suppliers, and between different suppliers. Often test cars are misconfigured and do not work. Thus, if dependencies can be checked by a computer without building misconfigured cars, manufacturer's costs can be significantly lowered. The same advantage applies to the development of new cars, collaboration between the manufacturer and suppliers, and the built-to-order process.

Besides describing human knowledge about various domains, ontologies serve as mediators between data sources [6]. This enables retrieving up-to-date data about parts from the legacy systems of the manufacturers. This integration aspect is handled in more detail elsewhere [5].

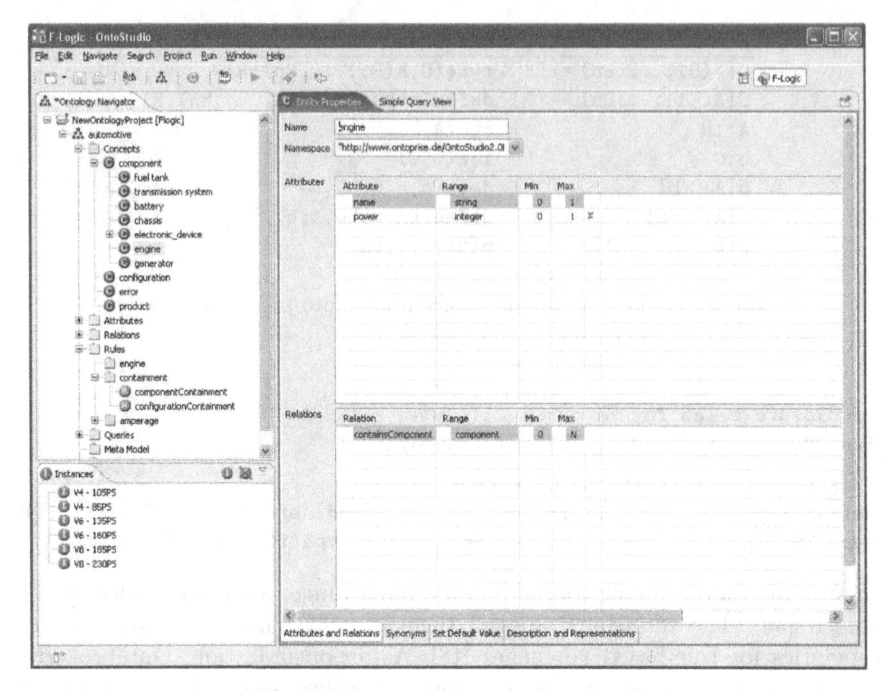

Fig. 2. Part of the automotive ontology

The ontology created in this project is used in two different ways. It is integrated into a software assistant which helps the engineer in configuring test cars. The engineer asks the assistant about a reconfigured system and the system responds with dependencies that have to be taken into account and also supplies the contact information for the relevant experts. In addition, the assistant provides explanations to help the engineer understand and validate the decisions made by the software assistant.

Creation of the ontology for the auto manufacturer is supported by OntoBroker™ and OntoStudio™ [1] (cf. Fig. 2). The former is the already mentioned inference engine for F-logic and the latter is a complete graphical environment for ontology and rule engineering. The basic part of the ontology describes parts and the part-of hierarchy. Much of the ontology structure was gleaned from legacy systems. Figure 4 shows an excerpt of that ontology. It indicates that, for example, a gear is part of a car and switching lever is a part of the gear. For motors, attributes like maximum power and type, are shown.

Without the rules, an ontology can describe only simple relationships between concepts, like a part being a component of another part, or being connected to another part. More complex relationships require the use of rules and constraints. The following are examples of such rules and constraints:

Rule 1: If a part is assigned to a sub-system then sub-parts are also assigned to the same sub-system.

Rule 2: The sub-part relation is transitive.
Rule 3: If the fuel injector is damaged then the motor is damaged
Rule 4: Cars that do not have a 12 cylinder motor, fulfill the DIN norm.
Constraint 1: The maximum power of the motor must not exceed the maximum power of the brakes.
Constraint 2: The filter in the catalyst must suit the motor fuel.

In F-logic, these rules are represented as follows:

Rule 1:
```
?Z[sub_system -> ?Y] :- ?X[has_part -> ?Z, sub_system -> ?Y].
```
Rule 2:
```
?X[has_part -> ?Z] :- ?X.has_part.has_part = ?Z.
```
Rule 3:
```
?X[damaged] :- ?Y:fuel_inj[damaged], ?X:motor, ?Y.part_of = ?X.part_of.
```
Rule 4:
```
?X[fulfills -> DIN] :- ?X:Car[has_part -> ?Y:motor], not ?Y[cyl -> 12].
```
Constraint 1:
```
!- ?X:motor[max_power -> ?Z1], (?Y:brake).max_power =< abs(?Z3).
```
Constraint 2:
```
!- ?X:motor[fuel_type -> ?Z], ?Y:filter[fuel_type -> ?Z].
```

Taken separately, each of the above rules or constraints is simple. However, it is not the complexity of individual rules that determines the difficulty of the problem, but the overwhelming amount of such rules and constraints. The rules can interfere with each other and make the task of configuring a correct test car complex and error prone. On the other hand the simplicity of the individual rules indicates that such ontologies can be created and maintained by domain experts, like mechanical engineers in our case.

5 F-Logic and the Semantic Web

F-logic has strong following in the Semantic Web area. Although it was not chosen as a standard for ontology representation, the most frequent ontological constructs map straightforwardly into F-logic. For example, Bruijn and Heymans [7] show how to embed RDF(S) in F-logic. The advantages of F-logic for building ontology-based applications were recognized early on in the development of the Semantic Web. For instance, Decker et al. [11] and Staab et al. [31] advocate F-logic as an inference service for RDF. Two W3C member submissions, SWSL-Rules[6] and WRL,[7] are also based on F-logic.

[6] http://www.w3.org/Submission/2005/SUBM-SWSF-SWSL-20050909/
[7] http://www.w3.org/Submission/WRL/

Now that the activities around adding rules to the semantic Web are in full swing, the W3C's RIF working group[8] is adding F-logic style frames to the core part of the RIF specification. Frames are also the main mechanism for combining RDF data with RIF-compliant languages.

F-logic plays even more important role in the area of Semantic Web Services, where the need to go beyond ontologies is most evident. It is one of the core ideas underlying the Web Services Modeling Ontology (WSMO)[9] and Semantic Web Services Framework (SWSF).[10]

6 Conclusions

The focus of this paper is the use of F-logic as a language for representing ontologies and building intelligent applications on top. F-logic is a frame-based logic language, which supports object-oriented style of application development. A remarkable feature of F-logic is the ability to reason about objects and their schema naturally and without the need for special language features.

A number of implementations of F-logic exist – both commercial and open-source academic systems (OntoBroker™, FLORID, FLORA-2). There is now vast experience with using F-logic for building intelligent information systems that rely on ontologies for extensibility and interoperability. Many practical applications, such as [4,33], have shown that F-logic is not only an ideal tool for declarative modeling of complex real-life application domains, but also that the inference engines for F-logic have matured to the extent that enables commercial use.

References

1. J. Angele, M. Erdmann, H.P. Schnurr, D. Wenke: Ontology based knowledge management in automotive engineering scenarios. In: Proceedings of ESTC 2007, 1st European Semantic Technology Conference, 31.05.–01.06.2007, Vienna, Austria.
2. S. Abiteboul, R. Hull, V. Vianu: Foundations of Databases. Addison-Wesley, Reading, MA, 1995.
3. S. Abiteboul, G. Lausen, H. Uphoff, E. Waller: Methods and Rules. In: Proceedings SIGMOD Conference, 1993, pp. 32–41.
4. J. Angele, H.-P. Schnurr, S. Staab, R. Studer: The Times they are A-Changin' – The corporate history analyzer. In: D. Mahling and U. Reimer (eds.). Proceedings of the Third International Conference on Practical Aspects of Knowledge Management. Basel, Switzerland, October 30–31, 2000.

[8] http://www.w3.org/2005/rules/wg.html
[9] http://www.wsmo.org/index.html
[10] http://www.w3.org/Submission/SWSF/

5. J. Angele, H.-P. Schnurr: Do not use this gear with a switching lever! Automotive industry. In: Proceedings of the Industrial Track of the Fourth International Semantic Web Conference (ISWC2005), Galway, Ireland, November 6–10, 2005.
6. J. Angele, E. Mnch, H. Oppermann, H. Rudat, H. Schnurr: Customer service accelerated by semantics. In: Proceedings of the Industrial Track of the Fifth International Semantic Web Conference (ISWC2006), Athens, USA, November 7–9, 2005.
7. J. de Bruijn, S. Heymans: RDF and logic: Reasoning and extension. In: Proceedings of the 6th International Workshop on Web Semantics (WebS 2007), Regensburg, Germany. IEEE Computer Society Press, Los Alamitos, CA, 2007.
8. P.P. Chen: The entity relationship model. Toward a unified view of data. ACM Transactions on Database Systems, 1, 9–36, 1976.
9. F. Baader, D. Calvanese, D. McGuinness, D. Nardi, P. Patel-Schneider (eds.): The Description Logic Handbook. Cambridge University Press, 2003.
10. C. Beeri, R. Ramakrishnan: On the power of magic. Journal of Logic Programming, 10, 255–300, 1991.
11. S. Decker, D. Brickley, J. Saarela, J. Angele: A query and inference service for RDF. In: Proceedings of the W3C Query Language Workshop (QL-98), Boston, MA, 3–4 December, 1998.
12. S. Decker, M. Erdmann, D. Fensel, R. Studer: OntoBroker: Ontology based access to distributed and semi-structured information. In: R. Meersman et al. (ed.). Database Semantics: Semantic Issues in Multimedia Systems. Kluwer Academic, Boston, MA, 1999.
13. F-logic Forum Group. http://projects.semwebcentral.org/projects/forum/forum-syntax.html.
14. J. Frohn, G. Lausen, H. Uphoff: Access to objects by path expressions and rules. In: Proceedings VLDB, 1994, pp. 273–284.
15. J. Frohn, R. Himmeröder, P.-T. Kandzia, C. Schlepphorst: How to write F-logic programs in FLORID. Available from ftp://ftp.informatik.uni-freiburg.de/pub/florid/tutorial.ps.gz, 1997.
16. G. Yang, M. Kifer, C, Zhao, H. Wan: The FLORA-2 manual. Available from http://flora.sourceforge.net/documentation.php.
17. H. Garcia-Molina, Y. Papakonstantinou, D. Quass, A. Rajaraman, Y. Sagiv, J.D. Ullman, V. Vassalos, J. Widom: The TSIMMIS approach to mediation: Data models and languages. JIIS 8(2), 117–132, 1997.
18. A. Van Gelder, K.A. Ross, J.S. Schlipf. The well-founded semantics for general logic programs. Journal of the ACM, 38(3), 620-650, 1991.
19. M. Kifer: Nonmonotonic reasoning in FLORA-2. In: Proceedings of Logic Programming and Nonmonotonic Reasoning. Lecture Notes in Computer Science 3662. Springer, Berlin, 2005, pp. 1–12.
20. M. Kifer, A. Bernstein, P.M. Lewis: Database systems: An application oriented approach, 2nd edition. Addison-Wesley, Reading, MA, 2005.
21. M. Kifer, J. de Bruijn, H. Boley, D. Fensel: A realistic architecture for the semantic Web. In: Proceedings of Rules and Rule Markup Languages for the Semantic Web, Lecture Notes in Computer Science 3791. Springer, Berlin, 2005, pp. 17–29.
22. M. Kifer, E.L. Lozinskii: A framework for an efficient implementation of deductive databases. In: Proc. of the 6th Advanced Database Symposium, Aug. 1986, Tokyo Japan, pp. 109–116.

23. M. Kifer, G. Lausen, J. Wu: Logical foundations of object-oriented and frame-based languages. Journal of the ACM, 42, 741–843, 1995.
24. M. Liu: Deductive database languages: Problems and solutions. ACM Computing Surveys, 31(1), 27–62, 1999.
25. B. Ludäscher, R. Himmeröder, G. Lausen, W. May: Christian Schlepphorst. Managing semistructured data with FLORID: A deductive object-oriented perspective. Information Systems 23(8), 589–613, 1998.
26. W. May: A rule-based querying and updating language for XML. In: Proceedings of DBPL 2001, LNCS 2397, pp. 165–181.
27. W. May, P.-T. Kandzia: Nonmonotonic inheritance in object-oriented deductive database languages. Journal of Logic and Computation, 11(4), 2001.
28. M. Dean, D. Connolly, F. van Harmelen, J. Hendler, I. Horrocks, D.L. McGuinness, P.F. Patel-Schneider, L.A. Stein: OWL Web Ontology Language 1.0 Reference. WWW Consortium, November, 2002.
29. R. Elmasri, S.B. Navathe: Fundamentals of database systems, 5th edition. Addison-Wesley, Reading, MA, 2006.
30. M. Sintek, S. Decker: TRIPLE – A query, inference, and transformation language for the semantic Web. International Semantic Web Conference (ISWC), June 2002.
31. S. Staab, M. Erdmann, A. Mädche, S. Decker: An extensible approach for modeling ontologies in RDF(S). In: Rolf Grütter (ed.). Knowledge Media in Healthcare: Opportunities and Challenges. Idea Group Publishing, Hershey, 2001.
32. Y. Sure, S. Staab, J. Angele: OntoEdit: Guiding ontology development by methodology and inferencing. In: R. Meersman, Z. Tari et al. (eds.). Proceedings of the Confederated International Conferences CoopIS, DOA and ODBASE 2002, 28th October–1st November, 2002, University of California, Irvine, USA, Springer, Berlin, LNCS 2519, pp. 1205–1222.
33. S. Staab, A. Maedche: Knowledge portals – Ontologies at work. AI Magazine, 21(2), Summer 2001.
34. G. Yang, M. Kifer: Reasoning about anonymous resources and meta statements on the semantic Web. Journal on Data Semantics, LNCS 2800, 69–97, 2003.
35. G. Yang, M. Kifer: Inheritance in rule-based frame systems: Semantics and inference. Journal on Data Semantics, VII,79–135, 2006.
36. G. Yang, M. Kifer, C. Zhao: FLORA-2: A rule-based knowledge representation and inference infrastructure for the semantic Web. In: Proceedings of International Conference on Ontologies, Databases and Applications of Semantics (ODBASE-2003). Springer, Berlin, LNCS 2888, pp. 671–688.

Resource Description Framework

Jeff Z. Pan

University of Aberdeen, Aberdeen, UK, jpan@csd.abdn.ac.uk

Summary. This chapter introduces Resource Description Framework (RDF), the W3C recommendation for semantic annotations in the Semantic Web. It will cover the syntax and semantics of RDF, as well as its relation with the W3C OWL Web Ontology Language. To address the mismatch between RDF and OWL-DL, the most expressive decidable fragment of the OWL standard, we introduce a novel variant of RDF(S), called RDFS-FA, which provides a solid semantic foundation for many of the latest Description Logic-based SW ontology languages, such as OWL-DL and OWL2-DL.

1 Introduction: Heading for the Semantic Web

In *Realising the Full Potential of the Web* [2], Tim Berners-Lee identifies two major objectives that the Web should fulfil. The first goal is to enable people to work together by allowing them to share knowledge. The second goal is to incorporate tools that can help people analyse and manage the information they share in a meaningful way. This vision has become known as the *Semantic Web* (SW) [3].

The Web's provision to allow people to write online content for other people is an appeal that has changed the computer world. This same feature that is responsible for fostering the first goal of the Semantic Web, however, hinders the second objective. Much of the content on the existing Web, the so-called *syntactic Web*, is human but not machine readable. Furthermore, there is great variance in the quality, timeliness and relevance [2] of Web resources (i.e. Web pages as well as a wide range of Web accessible data and services) that makes it difficult for programs to evaluate the worth of a resource.

The vision of the Semantic Web is to augment the syntactic Web so that resources are more easily interpreted by programs (or 'intelligent agents'). The enhancements will be achieved through the *semantic markups* which are machine-understandable annotations associated with Web resources.

Encoding semantic markups will necessitate the Semantic Web adopting an annotation language. To this end, the W3C (World Wide Web Consortium)

S. Staab and R. Studer (eds.), *Handbook on Ontologies*, International Handbooks on Information Systems, DOI 10.1007/978-3-540-92673-3,

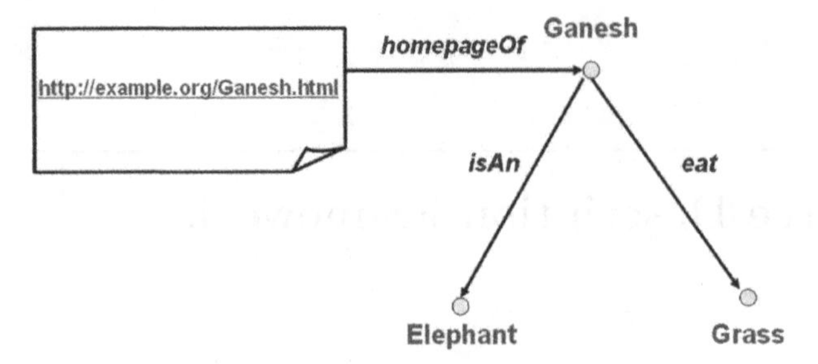

Fig. 1. RDF annotations in a directed labeled graph

community has developed a recommendation called resource description framework (RDF) [13]. The development of RDF is an attempt to support effective creation, exchange and use of annotations on the Web.

Example 1. Annotating Web Resources in RDF

As shown in Fig. 1, we can associate an RDF annotation[1] to `http://example.org/Ganesh.html` and state that it is the homepage of the resource Ganesh, which is an elephant and eats grasses.

We invite the reader to note that the above RDF annotations are different from HTML [27] mark-ups in that they describe the contents of Web resources, instead of the presentation of Web pages.

Annotations alone do not establish the semantics of what is being marked-up. For example, the annotations presented in Fig. 1 do not explain what elephants mean. The rest of the chapter is organised as follows. Section 2 presents RDF and two ways of providing semantics to RDF annotations. Section 3 introduces RDF Schema (or RDFS for short) and its semantics. Section 4 explains the semantic mismatch between RDF(S) and OWL-DL, while Sect. 5 introduces a sub-language of RDF, called RDFS-FA, which on the one hand has a semantics that is compatible with OWL-DL and on the other hand still allows meta-classes and meta-properties. Section 6 concludes the chapter.

2 Annotation and Meaning

The vision of the Semantic Web is to make Web resources (not just HTML pages, but a wide range of Web accessible data and services) more understandable to machines. Machine-understandable annotations are, therefore, introduced to describe the content and functions of Web resources.

[1] See Sect. 2 for precise definitions of RDF syntax.

2.1 RDF

RDF [13] as a W3C recommendation provides a *data model* for annotations in the Semantic Web. It is built upon earlier developments such as the Dublin Core (see Sect. 2.2) and the platform for Internet content selectivity (PICS) [26] content rating initiative.

An RDF statement (or RDF triple) is of the form:

$$\text{subject property object.} \tag{1}$$

RDF annotates Web resources in terms of named properties. Values of named properties (i.e. objects) can be URIrefs of Web resources or literals, viz. representations of data values (such as integers and strings). A set of RDF statements is call an *RDF graph.*

To represent RDF statements in a machine-processable way, RDF defines a specific extensible markup language (XML) syntax, referred to as RDF/XML [14]. RDF-annotated resources (i.e. subjects) are usually named by Uniform Resource Identifier references. Uniform resource identifiers (URIs) are strings that identify Web resources [7]. Uniform resource locators (URLs) are a particular type of URIs, i.e. those have network locations. A *URI reference* (or URIref) is a URI, together with an optional fragment identifier at the end. For example, the URI reference `http://www.example.org/Elephant#Ganesh` consists of the URI `http://www.example.org/Elephant` and (separated by the `#` character) the fragment identifier `Ganesh`. As a convention, *name spaces*, which are sources where multiple resources are from, are (usually) URIs with the `#` character. For example, `http://www.example.org/Elephant#` is a name space. Resources without URIrefs are called *blank nodes*; a blank node indicates the existence of a resource, without explicitly mentioning the URIref of that resource. A *blank node identifier*, which is a local identifier, can be used to allow several RDF statements to reference the same blank node. As RDF/XML is verbose, in this chapter, we use the Notation 3 (or N3) syntax of RDF, where each RDF statement is of the form (1). Figure 2 shows an RDF graph in N3 syntax, where the '@prefix' introduces shorthand identifications (such as 'ex:') of XML namespaces and a semicolon ';' introduces another property of the same subject. In these statements, the

```
@prefix rdf: <http://www.w3.org/1999/02/22-rdf-syntax-ns#>
@prefix ex:  <http://example.org/#>
@prefix elp: <http://example.org/Animal#>

elp:Ganesh ex:mytitle "A resource called Ganesh" ;
             ex:mycreator "Pat Gregory" ;
             ex:mypublisher _:b1 .
_:b1 elp:name "Elephant United" .
```

Fig. 2. RDF statements

annotated resource is elp:Ganesh, which is annotated with three properties ex:mytitle, ex:mycreator and ex:mypublisher. Note that `_:b1` is a blank node identifier.

Given that RDF alone does not specify the intended meaning for Web resources, how do we provide meaning to Web resources through annotations? The meaning comes either from pre-agreed informal semantics, e.g. from Dublin Core, or from ontologies.

2.2 Dublin Core

One way of giving meaning to annotations is to provide some pre-agreed informal semantics for a set of information properties. For example, the Dublin Core Metadata Element Set [5] provides 15 'core' information properties, such as 'Title', 'Creator', 'Date', with descriptive semantic definitions (in natural language). One can use these information properties in, e.g. RDF or META tags of HTML.

If we replace the properties ex:mytitle, ex:mycreator and ex:mypublisher used in Fig. 2 with dc:title, dc:creator and dc:publisher as shown in Fig. 3, Dublin Core compatible intelligent agents can then understand that the title of the Web resource is 'A resource called Ganesh', and the creator is Pat Gregory. This is not possible for the RDF statements in Fig. 2 because, in general, users may use arbitrary names for the title, creator and publisher properties, etc.

The limitation of the 'pre-agreed informal semantics' approach is its inflexibility, i.e. only a limited range of pre-agreed information properties can be expressed.

2.3 Ontology

An alternative approach is to use ontologies to specify the meaning of Web resources. *Ontology* is a term borrowed from philosophy that refers to the science of describing the kinds of entities in the world and how they are

```
@prefix rdf: <http://www.w3.org/1999/02/22-rdf-syntax-ns#>
@prefix dc:  <http://purl.org/dc/elements/1.1/>
@prefix elp: <http://example.org/Animal#>

elp:Ganesh dc:title "A resource called Ganesh" ;
           dc:creator "Pat Gregory" ;
           dc:publisher _:b1 .
_:b1 elp:name "Elephant United" .
```

Fig. 3. Dublin core properties in RDF statements

related. In computer science, ontology is, in general, a 'representation of a shared conceptualisation' of a specific domain [8, 30]. It provides a shared and common *vocabulary*, including important concepts, properties and their definitions, and *constraints*, sometimes referred to as background assumptions regarding the intended meaning of the vocabulary, used in a domain that can be communicated between people and heterogeneous, distributed application systems.

The ontology approach is more flexible than the pre-agreed informal semantics approach because users can customise vocabulary and constraints in ontologies. For example, applications in different domains can use different ontologies. Typically, ontologies can be used to specify the meaning of Web resources (through annotations) by asserting resources as instances of some important concepts and/or asserting resources relating to resources by some important properties defined in ontologies.

Ontologies can be expressed in Description Logics. An ontology usually corresponds to a TBox in Description Logics (see chapter "Description Logics"). Vocabulary in an ontology can be expressed by named concepts and roles, and concept definitions can be expressed by equivalence introductions. Background assumptions can be represented by general concept and role axioms. Sometimes, an ontology corresponds to a DL knowledge base. For example, in the OWL Web ontology language to be introduced in chapter "Web Ontology Language: OWL," an ontology also contains instances of important concepts and relationships among these instances, which can be represented by DL assertions. In the rest of the chapter, we will introduce RDF Schema (RDFS), an ontological schema language, and a novel modification of RDF(S) as a semantic foundation for many of the latest Description Logics-based SW ontology languages, including OWL-DL and OWL 1.1.

3 RDFS: A Web Ontological Schema Language

Following W3C's 'one small step at a time' strategy, RDFS can be seen as a first try to support expressing simple ontologies with RDF syntax. In RDFS, predefined Web resources rdfs:Class, rdfs:Resource and rdf:Property can be used to define classes (concepts), resources and properties (roles), respectively.

Unlike Dublin Core, RDFS does not predefine information properties but a set of meta-properties that can be used to represent background assumptions in ontologies:

- rdf:type: the instance-of relationship
- rdfs:subClassOf: the property that models the subsumption hierarchy between classes
- rdfs:subPropertyOf: the property that models the subsumption hierarchy between properties

```
@prefix rdf:  <http://www.w3.org/1999/02/22-rdf-syntax-ns#>
@prefix rdfs: <http://www.w3.org/2000/01/rdf-schema#>
@prefix elp:  <http://example.org/Animal#>

elp:Animal rdf:type rdfs:Class .
elp:Habitat rdf:type rdfs:Class .
elp:Elephant rdf:type rdfs:Class ; rdfs:subClassOf elp:Animal .
elp:liveIn rdf:type rdf:Property ;
            rdfs:domain elp:Animal ; rdfs:range elp:Habitat .
elp:south-sahara rdf:type elp:Habitat .
elp:Ganesh rdf:type elp:Elephant ; elp:liveIn elp:south-sahara .
```

Fig. 4. An RDFS ontology

- rdfs:domain: the property that constrains all instances of a particular property to describe instances of a particular class
- rdfs:range: the property that constrains all instances of a particular property to have values that are instances of a particular class

RDFS statements are simply RDF triples; viz. RDFS provides no syntactic restrictions on RDF triples. Figure 4 shows an animal ontology in RDFS; it has three classes, i.e. elp:Animal, elp:Habitat and elp:Elephant (which is rdfs:subClassOf elp:Animal), and a property elp:*liveIn*, the rdfs:domain and rdfs:range of which are elp:Animal and elp:Habitat, respectively. In addition, it states that the resource elp:Ganesh is an instance of elp:Elephant, and that it elp:*liveIns* an elp:Habitat called elp:south-sahara.

At a glance, RDFS is a simple ontological schema langauge that supports only class and property hierarchies, as well as domain and range constraints for properties. According to the RDF Model Theory (RDF MT) to be explained in Sect. 3.2, however, it is more complicated than that (see Proposition 1 on page 79).

3.1 RDF(S) Datatyping

RDF(S) provides a specification of datatypes and data values; accordingly, it allows the use of datatypes defined by any external type systems, e.g. the XML Schema type system, which conform to this specification.

Definition 1. (Datatype) *A* datatype d *is characterised by a lexical space,* $L(d)$*, which is a non-empty set of Unicode strings; a value space,* $V(d)$*, which is a non-empty set, and a total mapping* $L2V(d)$ *from the lexical space to the value space.*

For example, *boolean* is a datatype with value space $\{true, false\}$, lexical space {"T", "F","1","0"} and lexical-to-value mapping {"T"$\mapsto$ *true*, "F"$\mapsto$ *false*, "1"$\mapsto$ *true*, "0"$\mapsto$ *false*}.

Definition 2. (Typed and Plain Literals) Typed literals *are of the form "v" ^^u, where v is a Unicode string, called the* lexical form *of the typed literal, and u is a URI reference of a datatype.* Plain literals *have a lexical form and optionally a* language tag *as defined by [1], normalised to lowercase.*

The denotation of a typed literal is the value mapped from its enclosed Unicode string by the lexical-to-value mapping of the datatype associated with its enclosed datatype URIref. For example, "1"^^xsd:boolean is a typed literal that represents the boolean value *true*, while "1"^^xsd:integer represents the integer 1. Plain literals, e.g. "1", are considered to denote themselves [9].

The associations between datatype URI references (e.g. xsd:boolean) and datatypes (e.g. *boolean*) can be provided by datatype maps defined as follows.

Definition 3. (Datatype Map) *We consider a datatype map* $\mathbf{M}_d$ *that is a partial mapping from datatype URI references to datatypes.*

Example 2. Datatype Map $\mathbf{M}_{d1} = \{\langle \text{xsd:string}, \mathit{string} \rangle, \langle \text{xsd:integer}, \mathit{integer} \rangle\}$ is a datatype map, where xsd:string and xsd:integer are datatype URI references, and *string* and *integer* are datatypes. ◇

A datatype map may include some built-in XML Schema datatypes (as seen in Example 2), while other built-in XML Schema datatypes are problematic and thus unsuitable for various reasons. For example, xsd:ENTITIES is a list-value datatype that does not fit the RDF datatype model.[2] Please note that derived XML Schema datatypes are not RDF(S) datatypes, because there is no standard way to access a derived XML Schema datatype through a URI reference. Therefore, there is no way to include a derived XML Schema datatype in a datatype map.

3.2 RDF Model Theory

RDF MT provides semantics not only for RDFS ontologies, but also for RDF triples. RDF MT is built on *simple interpretations*. To simplify presentations, in this chapter we do not cover blank nodes, which are identified by local identifiers instead of URIrefs.

Definition 4. (Simple Interpretation) *Given a set of URI references* $\mathbf{V}$, *a* simple interpretation I *of* $\mathbf{V}$ *in the RDF model theory is defined by:*

- *A non-empty set* $\mathbf{IR}$ *of resources, called the* domain *(or* universe*) of* I
- *A set* $\mathbf{IP}$*, called the* set of properties *in* I
- *A mapping* $IEXT$*, called the* extension function*, from* $\mathbf{IP}$ *to the powerset of* $\mathbf{IR} \times \mathbf{IR}$
- *A mapping* IS *from URIrefs in* $\mathbf{V}$ *to* $\mathbf{IR} \cup \mathbf{IP}$

[2] See the RDF semantics document [9] for the complete list of RDF(S) built-in datatypes.

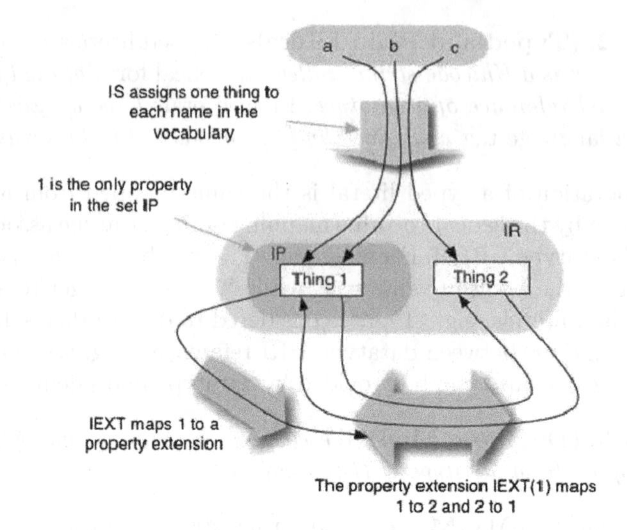

Fig. 5. A simple interpretation of **V** = {a,b,c} (from [9])

Given a triple [s p o .], *I*([s p o .]) = *true if s,p,o* ∈ **V**, IS(p) ∈ **IP**, *and* ⟨IS(s), IS(o)⟩ ∈ $IEXT$(IS(p)); *otherwise, I*([s p o .]) = *false*.

Given a set of triples S, I(S) = *false if I(*[s p o .]*)* = *false for some triple* [s p o .] *in S, otherwise I(S)* = *true. I* satisfies *S, written as I* ⊨ *S if I(S)* = *true; in this case, we say I is a simple interpretation of S.*

Note that Definition 4 does not specify the relationship between **IP** and **IR**, i.e. **IP** may or may not be disjoint with **IR**. Figure 5 presents a simple interpretation I of **V** = {a,b,c}, where the URIref b is simply interpreted as a property because IS(b) = 1 ∈ **IP**, and $IEXT$(IS(b)), the extension of IS(b), is a set of pairs of resources that are in **IR**, i.e. {⟨1, 2⟩,⟨2, 1⟩}. Since ⟨IS(a), IS(c)⟩ ∈ $IEXT$(IS(b)), I([a b c .]) = *true*; hence, we can conclude that I satisfies [a b c .].

The semantics of RDF triples is given in terms of RDF-Interpretations.

Definition 5. *(***RDF-Interpretation***) Given a set of URI references* **V** *and the set* **rdfV**, *called the* RDF vocabulary, *of URI references in the rdf: namespace, an RDF-interpretation of* **V** *is a simple interpretation I of* **V** ∪ **rdfV** *that satisfies:*

1. *For p* ∈ **V** ∪ **rdfV**, IS(p) ∈ **IP** *iff* ⟨IS(p), IS(rdf:Property)⟩ ∈ $IEXT$(IS(rdf:type))
2. *All the RDF axiomatic statements*[3]

Condition 1 of Definition 5 implies that each member of **IP** is a resource in **IR**, due to the definition of $IEXT$ in Definition 4; in other words, RDF-interpretations require **IP** to be a subset of **IR**. RDF axiomatic statements

[3] Readers are referred to [9] for the list of the RDF axiomatic statements.

mentioned in Condition 2 are RDF statements about RDF built-in vocabularies in **rdfV**; e.g. [rdf:type rdf:type rdf:Property.] is an RDF axiomatic statement. According to Definition 5, any RDF-interpretation I should satisfy [rdf:type rdf:type rdf:Property.], viz. IS(rdf:type) should be in **IP**.

Finally, the semantics of RDFS statements written in RDF triples is given in terms of RDFS-Interpretations.

Definition 6. *(***RDFS-Interpretation***) Given* **rdfV***, a set of URI references* **V** *and the set* **rdfsV***, called the* RDFS vocabulary*, of URI references in the rdfs: namespace, an RDFS-interpretation I of* **V** *is an RDF-interpretation of* **V** ∪ **rdfV** ∪ **rdfsV** *which introduces:*

- *A set* **IC***, called the set of classes in I*
- *A mapping ICEXT (called the* class extension function*) from* **IC** *to the set of subsets of* **IR**

and satisfies the following conditions (let x,y,u,v be URIrefs in **V** ∪ **rdfV** ∪ **rdfsV***)*[4]*:*

1. IS(x) $\in ICEXT(IS$(y)$)$ *iff* $\langle IS$(x)$, IS$(y)$\rangle \in IEXT(IS$(rdf:type)$)$
2. **IC** $= ICEXT(IS$(rdfs:Class)$)$ *and* **IR** $= ICEXT(IS$(rdfs:Resource)$)$,
3. *If* $\langle IS$(x)$, IS$(y)$\rangle \in IEXT(IS$(rdfs:domain)$)$ *and* $\langle IS$(u)$, IS$(v)$\rangle \in IEXT(IS$(x)$)$, *then* IS(u) $\in ICEXT(IS$(y)$)$
4. *If* $\langle IS$(x)$, IS$(y)$\rangle \in IEXT(IS$(rdfs:range)$)$ *and* $\langle IS$(u)$, IS$(v)$\rangle \in IEXT(IS$(x)$)$, *then* IS(v) $\in ICEXT(IS$(y)$)$
5. $IEXT(IS$(rdfs:subPropertyOf)$)$ *is transitive and reflexive on* **IP**
6. *If* $\langle IS$(x)$, IS$(y)$\rangle \in IEXT(IS$(rdfs:subPropertyOf)$)$, *then* IS(x),IS(y) $\in$ **IP** *and* $IEXT(IS$(x)$) \subseteq IEXT(IS$(y)$)$
7. $IEXT(IS$(rdfs:subClassOf)$)$ *is transitive and reflexive on* **IC**
8. *If* $\langle IS$(x)$, IS$(y)$\rangle \in IEXT(IS$(rdfs:subClassOf)$)$, *then* IS(x),IS(y) $\in$ **IC** *and* $ICEXT(IS$(x)$) \subseteq ICEXT(IS$(y)$)$
9. *If* IS(x) $\in$ **IC**, *then* $\langle IS$(x)$, IS$(rdfs:Resource)$\rangle \in IEXT(IS$(rdfs:subClassOf)$)$

and satisfies all the RDFS axiomatic statements.[5]

Condition 1 indicates that a 'class' is not a strictly necessary but convenient semantic construct [9] because the class extension function $ICEXT$ is simply 'syntactic sugar' and is defined in terms of $IEXT$. Handling classes in this way can be counter-intuitive (cf. Proposition 1). Condition 2 to 8 are about RDFS meta-properties rdfs:domain, rdfs:range, rdfs:subPropertyOf and rdfs:subClassOf. Condition 9 ensures that all classes are sub-classes of rdfs:Resource.

Proposition 1. *The RDFS statements* [rdfs:Resource rdf:type rdfs:Class .] *and* [rdfs:Class rdfs:subClassOf rdfs:Resource.] *are always true in all RDFS-interpretations.*

[4] We only focus on the core RDFS primitives, i.e. the RDFS predefined meta-properties introduced on page 75.

[5] Again, readers are referred to [9] for a list of the RDFS axiomatic statements, which includes, e.g. [rdf:type rdfs:range rdfs:Class.].

Proof. For [rdfs:Resource rdf:type rdfs:Class.]:

1. According to the definition of IS and Definition 5, for any resource x, we have $IS(\mathrm{x}) \in \mathbf{IR}$. Due to $\mathbf{IR} = ICEXT(IS(\text{rdfs:Resource}))$ and Condition 1 in Definition 6, $\langle IS(\mathrm{x}), IS(\text{rdfs:Resource})\rangle \in IEXT(IS(\text{rdf:type}))$. Since rdf:Property is a built-in resource, we have $\langle IS(\text{rdf:Property}), IS(\text{rdfs:Resource})\ \rangle \in IEXT(IS(\text{rdf:type}))$.
2. Due to [rdf:type rdfs:range rdfs:Class.] (an RDFS axiomatic statement), $\langle IS(\text{rdf:Property}), IS(\text{rdfs:Resource})\rangle \in IEXT(IS(\text{rdf:type}))$ and Condition 4 in Definition 6, we have $IS(\text{rdfs:Resource}) \in ICEXT(IS())$ rdfs:Class. Therefore, for any RDFS-interpretation I, we have I $\models$ [rdfs:Resource rdf:type rdfs: Class.].

For [rdfs:Class rdfs:subClassOf rdfs:Resource .]: According to the definition of **IC**, every class is its member, including $IS(\text{rdfs:Class})$, viz.$IS()$ rdfs:Class $\in$ **IC**. Due to Condition 9 of Definition 6, $\langle IS(\text{rdfs:Class}), IS(\text{rdfs:Resource})\rangle \in IEXT\ (IS(\text{rdfs:subClassOf}))$; hence, for any RDFS-interpretation I, we have I $\models$ [rdfs:Class rdfs:subClassOf rdfs:Resource.] □

The two RDFS statements in Proposition 1 suggest a strange situation for rdfs:Class and rdfs:Resource as discussed in [18]: On the one hand, rdfs:Resource is an instance of rdfs:Class; on the other hand, rdfs:Class is a sub-class of rdfs:Resource. Hence is rdfs:Resource an instance of its sub-class? Users may find this counter-intuitive and thus hard to understand – this is why we say that RDF(S) is more complicated than it appears. We will address this issue in Sect. 5.

Now we define RDFS-interpretations w.r.t. a datatype map $\mathbf{M}_d$.

Definition 7. (RDFS $\mathbf{M}_d$-Interpretation) *Given a datatype map* $\mathbf{M}_d$, *an* RDFS $\mathbf{M}_d$-interpretation *I of a vocabulary* **V** *is any RDFS-interpretation of* $\mathbf{V} \cup \{u \mid \exists\ d.\langle u, d\rangle \in \mathbf{M}_d\}$ *which introduces*

- *A distinguished subset* **LV** *of* **IR**, *called the* set of literal values, *which contains all the plain literals in* **V**
- *A mapping IL from typed literals in* **V** *into* **IR**

and satisfies the following extra conditions:

1. $\mathbf{LV} = ICEXT(IS(\text{rdfs:Literal}))$
2. *For each pair* $\langle u, d\rangle \in \mathbf{M}_d$
   (a) $ICEXT(d) = V(d) \subseteq \mathbf{LV}$
   (b) *There exist* $d \in \mathbf{IR}$ *s.t.* $IS(\mathrm{u}) = d$
   (c) $IS(\mathrm{u}) \in ICEXT(IS(\text{rdfs:Datatype}))$
   (d) *For* "s"^^u' $\in \mathbf{V}$, $IS(\mathrm{u}') = d$, *if* $s \in L(d)$, *then* $IL(\text{"s"\^{}\^{}u'}) = L2S(d)(s)$, *otherwise,* $IL(\text{"s"\^{}\^{}u'}) \notin \mathbf{LV}$,
3. *If* $d \in ICEXT(IS(\text{rdfs:Datatype}))$, *then* $\langle d, IS(\text{rdfs:Literal})\rangle \in IEXT(\mathit{rdfs{:}subClassOf})$.

According to Definition 7, **LV** is a subset of **IR**; i.e. literal values are resources. Condition 1 ensures that the class extension of rdfs:Literal is **LV**. Condition 2) asserts that RDF(S) datatypes are classes, condition 2) ensures that there is a resource d for datatype d in $\mathbf{M}_d$, condition 2) ensures that the class rdfs:Datatype contains the datatypes used in any satisfying $\mathbf{M}_d$-interpretation, and condition 2) explains why the range of IL is **IR** rather than **LV** (because, for "s"^^u, if $s \notin L(IS(\text{u}))$, then $IL(\text{"}s\text{"\^{}\^{}}u) \notin \mathbf{LV}$). Condition 3 requires that RDF(S) datatypes are sub-classes of rdfs:Literal.

If the datatypes in the datatype map $\mathbf{M}_d$ impose disjointness conditions on their value spaces, it is possible for an RDF graph to have no RDFS $\mathbf{M}_d$-interpretation which satisfies it, i.e. there exists a *datatype clash*. For example,

```
_:x rdf:type xsd:string.
_:x rdf:type xsd:decimal.
```

would constitute a datatype clash because the value spaces of xsd:string and xsd:decimal are disjoint. In RDF(S), an ill-typed literal does not in itself constitute a datatype clash, cf. Condition 2) in Definition 7, but a graph which entails that an ill-typed literal has rdf:type rdfs:Literal would be inconsistent.

Having described the semantics, we now briefly discuss reasoning in RDF(S). Entailment is the key inference problem in RDF(S), which can be defined on the basis of interpretations. Indeed, cRDF is impossible to express contradictions if we do not consider datatypes.

Definition 8. *(***RDF Entailments***) Given two sets of RDF statements S_1 and S_2, and a datamap $\mathbf{M}_d$, S_1* simply entails *(*RDF-entails, RDFS-entails, RDFS-$\mathbf{M}_d$-entails*) S_2 if all the simple interpretations (RDF-interpretations, RDFS-interpretations, RDFS $\mathbf{M}_d$-interpretation, resp.) of S_1 also satisfy S_2.*

4 Mismatch between RDF(S) and OWL-DL

This section describes the relation between RDF(S) and OWL-DL, which is a key sub-language of the standard (W3C recommendation) Web Ontology Langauge. One key question is whether it is possible to use an RDF(S) inference engine to do OWL-DL reasoning, or vice versa. The short answer is *no*, and this section explains why.

The OWL recommendation actually consists of three languages of increasing expressive power: OWL-Lite, OWL-DL and OWL-Full. *OWL-Lite* and *OWL-DL* are basically very expressive description logics (DLs). OWL-Full provides the same set of constructors as OWL-DL, but allows them to be used in an unconstrained way (in the style of RDF). OWL-Full is undecidable, because it combines the OWL expressivity with the meta-modelling architecture of RDF(S) [15].[6] Accordingly, OWL-DL is the most expressive decidable

[6] Another reason that OWL-Full is undecidable is that it does not impose restrictions on the use of transitive properties [12].

sub-language of OWL. More details of the OWL language can be found in chapter "Web Ontology Language: OWL."

This section discusses *both* the syntactic and semantic mismatches between RDF(S) and OWL-DL. From the syntax aspect, OWL-DL heavily restricts the syntax of RDF(S), viz. some RDF(S) annotations are not recognisable by OWL-DL agents, since they are syntactically ill formed. The RDF/XML syntax form of an OWL-DL ontology is *valid*, iff it can be translated (according to the mapping rules provided in [25]) from the abstract syntax form of the ontology. Actually, it is far from an easy task to check if an RDF graph is an OWL-DL ontology [11], since no inverse mapping is defined in the OWL specification.

From the semantics aspect, OWL-DL has an RDF MT-style semantics, in which (including built-in) classes and properties are treated as objects (or resources) in the domain. In order to make it equivalent to the direct semantics of OWL-DL [25], the domain of discourse is divided into several disjoint parts. In particular, the interpretations of classes, properties, individuals and OWL/RDF vocabulary are strictly separated. Therefore, classes and properties, unsurprisingly, *cannot* be treated as ordinary resources as they are in RDF MT. Strictly speaking, even those RDF(S) statements which are valid OWL-DL statements do not share the same meaning in an RDF(S) ontology and an OWL-DL ontology.

OWL-Full seems to be a bridge between RDF(S) and OWL-DL; however, there exist at least three known issues that the RDF-style semantics for OWL-Full needs to solve, and a proven solution has yet to be given. The first issue is about entailment [23]. Consider the following question: does the following individual axiom

```
Individual(ex:John
  type(intersectionOf(ex:Student ex:Employee ex:European)))
```

entail the individual axiom

```
Individual(ex:John
  type(intersectionOf(ex:Student ex:European)))?
```

In OWL-DL, the answer is simply 'yes', since intersectionOf(ex:Student ex:Employee ex:European) is a sub-class of intersectionOf(ex:Student ex:European). Since in RDF(S) every class is a resource, OWL-Full needs to make sure of the existence of the resource intersectionOf(ex:Student ex:European) in every possible interpretation; otherwise, the answer will be 'no' which leads to a disagreement between OWL-DL and OWL-Full. In general, OWL-Full introduces so called *comprehension principles* to add all the missing resources into the domain for all the OWL class descriptions. It has yet to be proved that the proper resources are all added into the universe, no more and no less, and that the added resources will not bring any side-effects.

The second issue is about contradiction classes [11, 23, 24]. In OWL-Full, it is possible to construct a class the instances of which have no rdf:type relationship linked to:

```
_:c owl:onProperty rdf:type; owl:allValuesFrom _:d.
_:d owl:complementOf _:e.
_:e owl:oneOf _:l
_:l rdf:first _:c; rdf:rest rdf:nil.
```

The above triples require that rdf:type relates members of the class _ : c to anything but _ : c. It is impossible for one to determine the membership of _ : c. If an object is an instance of _ : c, then it is not; but if it is not then it is – this is a contradiction class. Note that it is not a valid OWL-DL class, as OWL-DL disallows using rdf:type as an object property. With naive comprehension principles, resources of contradiction classes would be added to all possible OWL-Full interpretations, which thus have ill-defined class memberships. To avoid the issue, the comprehension principles must also consider avoiding contradiction classes. Unsurprisingly, devising such comprehension principles took a considerable amount of effort [11], and no proof has ever shown that all possible contradiction classes are excluded in the comprehension principles of OWL-Full.

The third issue is about the size of the universe [10]. Consider the following question: is it possible that there is only one object in an interpretation of the following OWL ontology?

```
Individual(elp:Ganesh type(elp:Elephant))
DisjointClasses(elp:Elephant elp:Plant)
```

In OWL-DL, classes are not objects, so the answer is 'yes': The only object in the domain is the interpretation of **elp:Ganesh**, the elp:Elephant class thus has one instance, i.e. the interpretation of **elp:Ganesh**, and the elp:Plant class has no instances. In OWL-Full, since classes are also objects, besides **elp:Ganesh**, the classes elp:Elephant and elp:Plant should both be mapped to the only one object in the universe. This is not possible because the interpretation of **elp:Ganesh** is an instance of elp:Elephant, but not an instance of elp:Plant; hence, elp:Elephant and elp:Plant should be different, i.e. there should be at least two objects in the universe. As the above axioms are valid OWL-DL axioms, this example shows that OWL-Full disagrees with OWL-DL on valid OWL-DL ontologies. To partially address this issue, the OWL specification weakens the relations between OWL-DL and OWL-Full by claiming (with a sketched proof) that, given two OWL-DL ontologies O1 and O2, O1 entails O2 w.r.t. the OWL-DL semantics implies that O1 entails O2 w.r.t. the OWL-Full semantics. Furthermore, this example shows that the interpretation of OWL-Full has different features than the interpretation of standard first order logic (FOL) model theoretic semantics. This raises the question as to whether it is possible to layer FOL languages on top of RDF(S).

It should be noted that *for some* the above presentation of the three issues might be a little too negative about the situation w.r.t. OWL-Full and OWL-DL: the first two issues are difficulties that have, in theory, been claimed to be solved by the use of comprehension principles and restrictions on the syntactic form of OWL-DL's RDF serialisation. From this perspective, the main side effect of comprehension principles is that all OWL-Full models have infinite domains; hence, any OWL-DL ontologies that have only finite models are necessarily inconsistent when treated as OWL-Full ontologies. This leads to the third issue and demonstrates why, in the OWL specification, the relations between OWL-Full and OWL-DL is weakened.

5 RDFS-FA: Connecting RDF(S) and OWL-DL

In this section, we introduce *RDFS-FA* (RDFS with Fixed layered metamodelling architecture), as a sub-language of RDF(S), to restore the desired connection between RDF(S) and OWL-DL. RDFS-FA addresses the following characteristics of RDF(S):

- RDF triples have built-in semantics.
- Classes and properties, including built-in classes and properties of RDF(S) and its subsequent languages such as OWL, are treated as objects (or resources) in the domain.
- There are no restrictions on the use of built-in vocabularies.

Intuitively, RDFS-FA provides a UML like meta-modelling architecture. Let us recall that RDFS has a non-layered meta-modelling architecture; resources in RDFS can be classes, objects and properties at the same time, viz. classes and their instances (as well as relationships between the instances) are the same layer. RDFS-FA, instead, divides up the universe of discourse into a series of strata (or layers). The built-in modelling primitives of RDFS are separated into different strata of RDFS-FA, and the semantics of modelling primitives depend on the stratum they belong to. Theoretically there can be a large number of strata in the meta-modelling architecture; in practice, four strata (as shown in Fig. 6) are usually enough. The UML-like meta-modelling architecture makes it easier for users who are familiar with UML to understand and use RDFS-FA.

In RDFS-FA, classes cannot be objects and vice versa;[7] in RDFS, Web resources can be classes, properties, objects or even datatypes all at once. We argue that RDFS-FA is more intuitive than RDFS based on the following observation: when users design their ontologies, a common concern is to decide

[7] Classes can be regarded as mega-objects in upper strata of the meta-modelling architecture.

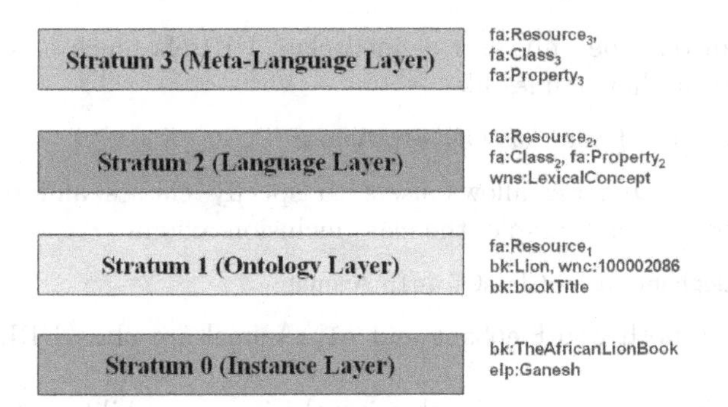

Fig. 6. The UML-like meta-modelling architecture (number of strata = 4) of RDFS-FA

```
@prefix fa:  <http://dl-web.man.ac.uk/rdfsfa/ns#>
@prefix elp:  <http://example.org/Animal#>

elp:Animal fa:type2 fa:Class2 .
elp:Habitat fa:type2 fa:Class2 .
elp:Elephant fa:type2 fa:Class2 ; fa:subClassOf2 elp:Animal .
elp:liveIn fa:type2 fa:AbstractProperty2 ;
          fa:domain2 elp:Animal ; fa:range2 elp:Habitat .

elp:south-sahara fa:type1 elp:Habitat .
elp:Ganesh fa:type1 elp:Elephant ; elp:liveIn elp:south-sahara .
```

Fig. 7. An RDFS-FA ontology

whether to model something in the domain as a class or as an object (see also [17]). This concern suggests that users intuitively tend to assume that classes and objects should be different from each other. Therefore, layered meta-models could be more intuitive than non-layered meta-models.

Readers are referred to [21] for a formal introduction of RDFS-FA ontologies and their semantics. Informally speaking, an RDFS-FA ontology is a set of RDFS-FA axioms, which are basically RDF triples (in N3 syntax)[8] with extra syntactic rules, which (1) disallow arbitrary use of its built-in vocabulary and (2) enable the use of meta-classes and meta-properties in specified layers as well as the use of annotation properties.

Figure 7 shows an example RDFS-FA ontology. Firstly, the layering structure is clear. elp:Animal, elp:Habitat, elp:Elephant and elp:*liveIn* are in stratum 1 (the Ontology layer), while elp:Ganesh and elp:south- sahara are in stratum 0 (the Instance Layer). Secondly, RDFS-FA disallows arbitrary use of its built-in vocabulary. For example, in class inclusion axioms, the

[8] Here we use the N3 syntax, instead of the RDF/XML syntax, as it is more compact.

subjects can only be only user-defined class URIrefs (such as elp:Animal), which could disallow triples like

fa:Resource$_1$ fa:subClassOf$_2$ elp:Animal .

Furthermore, RDFS-FA allows users to specify classes and properties in specified strata. For example, the class inclusion axiom

elp:Elephant fa:subClassOf$_2$ elp:Animal .

requires that both elp:Elephant and elp:Animal are class URIrefs in stratum 1.

We conclude this section by showing the interoperability between RDFS-FA and OWL-DL. It is much easier to layer OWL-DL, syntactically *and* semantically, on top of RDFS-FA than on top of RDF(S). In particular, there is a one-to-one bidirectional mapping (see [21] for details) between the RDFS-FA axioms in strata 0-1 and OWL-DL axioms in OWL abstract syntax. For example, the RDFS-FA class inclusion axiom [C_1 fa:subClassOf$_2$ D_1.] can be mapped to the OWL class axiom (SubClassOf C_1 D_1) and vice versa. In the syntactic level, it is easier to layer OWL-DL on top of RDFS-FA than on top of RDF(S), due to the above bidirectional mapping. Let us recall that, according to the OWL Semantics and Abstract Syntax document [25], the mapping between OWL-DL axioms, or *OWL axioms* for short, and RDF(S) statements is *only* unidirectional, i.e. from OWL axioms to RDF(S) statements. For example, we can map the following OWL axiom SubClassOf (C_1 D_1) to the RDF(S) statement [C_1 rdfs:subClassOf D_1.], with an implicit OWL constraint, viz., C_1 and D_1 can only be class URIrefs, but not URIrefs for properties or individuals, etc. However, the above RDF(S) statement without such (implicit) constraint cannot be correctly mapped to the OWL axiom (SubClassOf C_1 D_1). In the semantic level, it can be shown that the above bidirectional mapping is a semantics-preserving mapping [21].

It has be shown [22] that we can extend OWL DL with the meta-modelling architecture of RDFS-FA into OWL-FA, and that OWL-FA is also decidable.

6 Related Work

As earlier works [4, 16] pointed out, RDFS has a non-standard and non-fixed layer meta-modelling architecture, which makes some elements in the model have multiple roles in the RDFS specification. Therefore, it makes even the RDFS specification itself somehow confusing and difficult to understand for users. To clear up any confusion, Pan and Horrocks [18] proposed a Fixed layer meta-modelling Architecture for RDFS, reducing the multiple roles of RDFS built-in primitives by stratifying them into different layers of the meta-modelling architecture. Subsequently, the RDF Model Theory (RDF MT) [9] gave an official semantics for RDF and RDFS, justifying the dual roles by treating both classes and properties as objects in the universe. Pan and Horrocks [19] suggested that RDFS could have two kinds of semantics, i.e. RDF MT and the stratified semantics of RDFS(FA).

Horst [29] extends RDF MT to cover some OWL constructors and axioms by proposing the so-called pD* semantics. Interestingly, the pD* semantics is in line with the 'if-semantics' of RDFS and weaker than the 'iff-semantics' that is used in the RDF-compatible semantic for OWL DL and OWL Full. One of the motivations of having the iff-semantics in the RDF-compatible semantic for OWL is to solve the 'too few entailment' problem [19]. Note that the iff-semantics is *not* relevant to the direct semantics of OWL DL. Among the 15 OWL URIs, the pD* interprets `owl:FunctionalProperty`, `owl:InverseFunctionalProperty`, `owl: SymmetricProperty` and `owl:Transi- tiveProperty` as the if conditions of the standard mathematical definitions. The `owl:inverseOf` is interpreted as that if two properties are owl:inverseOf-related, then their extensions are each other's inverse as binary relations. The pD* semantics requires that two classes are equivalent if and only if they are both subclasses of each other. `owl:equivelantProperty` is treated in a similar way to `owl:equivalentClass`. The pD* semantics interprets `owl:sameAs` as an equivalence relation.In particular, the pD* semantics includes the iff condition for `owl:hasValue`. But for `owl:someValueFrom` and `owl:allValueFrom`, the pD* semantics still includes half of OWL's iff conditions. If two classes are owl:disjointWith-related the pD* semantics requires their extensions are disjoint. The pD* semantics requires that the extensions of `owl:sameAs` and `owl:differentForm` are disjoint. Based on the pD* semantics discussed above, the corresponding pD* entailment rules are also given in [29]. It consists of 23 rules to illustrate that what conclusion can be deduced from some given premises. These rules are proved to be sound and complete with respect to the pD* semantics.

Patel-Shneider et al. [25] extended RDFS with OWL constructors to OWL Full, which keeps the meta-modelling architecture of RDFS. Motik [15] shows that the meta-modelling architecture of OWL Full contributes to its undecidability. Motik [15] also provides two alternative meta-modelling approaches for OWL DL, i.e. the contextual approach and the HiLog approach.

- In the context approach, the names for classes, properties and individuals are not distinct and are interpreted depending on the context; i.e. they are interpreted by class interpretation functions, property interpretation functions and individual interpretation functions, respectively. Intuitively speaking, this approach provides a 'two-layered' meta-modelling architecture, i.e. the instance layer and class layer. OWL FA provides a 'multi-layered' meta-modelling architecture. At a quick glance, the 'two-layered' and the 'multi-layered' meta-modelling architectures should be similar; however, the example we show later in this section indicates that they are quite different.

- The HiLog approach is closer to the spirit of OWL Full meta-modelling. It has a 'two-step' interpretation function for classes, which first maps symbols to resources in the domain and then maps these resources to a set of resources in the domain. Intuitively speaking, this approach provides a 'one-layered' meta-modelling architecture, in the sense that classes and individuals are both interpreted as resources in the domain. Note that it is dificult/impossible to map classes in the 'one-layered' meta-modelling architecture to the 'multi-layered' meta-modelling architectures such as that of MOF.

We now use an example in [15] to illustrate some of the differences among the above two approaches and our approach. Let us consider the following knowledge base[9] $\Sigma =\{$ $\texttt{Harry} :_1 \mathsf{Eagle}, \texttt{Harry} :_1 \neg\mathsf{Aquila}, \mathsf{Eagle} =_1 \mathsf{Aquila}\}$. In the contextual approach, since Eagle and Aquila as concepts and as individuals are independent, Σ is satisfiable. In the HiLog approach, it is not satisfiable because Eagle and Aquila are interpreted as the same object, let us call it a, and Harry cannot be both in and not in the concept extension of a. In OWL FA, Σ is unsatisfiable because the meta-individual equality axiom $\mathsf{Eagle} =_1 \mathsf{Aquila}$ indicates two concepts Eagle and Aquila are equivalent, and $\texttt{Harry}^{\mathcal{I}}$ cannot be both in and not in $\mathsf{Eagle}^{\mathcal{I}}$. This example indicates the contextual semantics (at least sometimes) is not as intuitive as the Hilog semantics and the FA semantics.

Let us conclude this section by briefly comparing the three approaches. In terms of syntax, the contextual and Hilog approaches seem to be more elegant in that they do not have to change the syntax of OWL DL, while the FA approach introduces strata numbers to facilitate the 'multi-layered' meta-modelling architecture. In terms of semantics, it seems that the FA approach is closer to the Hilog approach (according to the above example). It is an interesting peace of future work to investigate more detailed differences between the Hilog approach and the FA approach. In terms of computability, the FA approach is closer to the contextual approach in that we can reduce the reasoning services (such as knowledge base satisfiability) to existing DL reasoning services. Finally, the contextual approach and the Hilog approach have not covered datatypes yet, while the FA approach covers datatypes. In order to support datatypes in the contextual approach, some extra syntax may be needed for OWL DL, otherwise it is difficult to distinguish the contexts. For example, in $\exists R.E$, E can be either a class or a datatype. It is not clear how to support datatypes in the Hilog approach yet.

Other existing approaches either limit the extension of RDF(S) to only a property-related subset of OWL with a weaker semantics proposed by ter Horst ([28, 29]), or weaken the semantic connection between the individual interpretation and class interpretation of a given URI [6], hence failing to propagate important inferences from meta-classes to classes (see [21]).

[9] In [15], the subscripts are not used.

7 Conclusion

In this chapter, we have presented RDF. RDF is a standard syntax for Semantic Web annotations and languages. RDF Schema is an ontological schema language that supports only class and property hierarchies, as well as domain and range constraints for properties. RDF(S) has a key role in supporting such compatibility by providing a common basis on which more expressive SW languages can be built. Recent research, however, has shown that there exist syntactic and semantic mismatch between RDF(S) and OWL-DL. Accordingly, this chapter includes a novel modification of RDF(S), called RDFS-FA, which provides a solid semantic foundation for many of the latest Description Logic-based SW ontology languages, and imposes no limitation on its extension to more expressive Description Logics (such as OWL-DL, OWL2-DL and OWL-Eu [20]).

In chapter "RDF Storage and Retrieval Systems," we will further describe entailment and querying over RDF(S) ontologies. As for RDFS-FA, reasoning in RDFS-FA and its OWL extension, OWL-FA, is discussed in [22]; such reasoning can be performed by reduction to OWL-DL reasoning.

References

1. H. Alvestrand. Rfc 3066 – tags for the identification of languages. Technical report, IETF, Jan 2001. `http://www.isi.edu/in-notes/rfc3066.txt`.
2. Tim Berners-Lee. Realising the Full Potential of the Web. W3C Document, URL `http://www.w3.org/1998/02/Potential.html`, Dec 1997.
3. T. Berners-lee. Semantic Web Road Map. W3C Design Issues. URL http://www.w3.org/DesignIssues/Semantic.html, Oct. 1998.
4. J. Broekstra, M. Klein, S. Decker, D. Fensel, F. van Harmelen, and I. Horrocks. Enabling knowledge representation on the web by extending rdf schema. In *Proc. of the International World Wide Web Conference*, 2001.
5. DCMI. Dublin Core Metadata Element Set, Version 1.1: Reference Description. DCMI Recommendation, URL `http://dublincore.org/documents/dces/`, June 2003.
6. J. de Bruijn, E. Franconi, and S. Tessaris. Logical reconstruction of normative RDF. In *OWL: Experiences and Directions Workshop*, 2005.
7. Joint W3C/IETF URI Planning Interest Group. URIs, URLs, and URNs: Clarifications and Recommendations 1.0. URL `http://www.w3.org/TR/uri-clarification/`, 2001. W3C Note.
8. T. R. Gruber. *Towards Principles for the Design of Ontologies Used for Knowledge Sharing*, chapter of Formal Ontology in Conceptual Analysis and Knowledge Representation. Kluwer Academic, New York, 1993.
9. P. Hayes. RDF Semantics. Technical report, W3C, Feb 2004. W3C recommendation, `http://www.w3.org/TR/rdf-mt/`.
10. I. Horrocks and P. F. Patel-Schneider. Three Theses of Representation in the Semantic Web. In *Proc. of the 12th International World Wide Web Conference*, pages 39–47. ACM, New York, 2003.

11. I. Horrocks, P. F. Patel-Schneider, and F. van Harmelen. From SHIQ and RDF to OWL: The Making of a Web Ontology Language. *Journal of Web Semantics*, 1(1):7–26, 2003.
12. I. Horrocks, U. Sattler, and S. Tobies. Practical Reasoning for Expressive Description Logics. In *Proc. of the 6th International Conference on Logic for Programming and Automated Reasoning*, pages 161–180, 1999.
13. G. Klyne and J. J. Carroll. Resource Description Framework (RDF): Concepts and Abstract Syntax. URL `http://www.w3.org/TR/rdf-concepts/`, Feb 2004. Series Editor: Brian McBride.
14. F. Manola and E. Miller. RDF Primer, W3C Recommendation. URL `http://www.w3.org/TR/rdf-primer/`, Feb 2004. Series Editor: Brian McBride.
15. B. Motik. On the Properties of Metamodeling in OWL. In *Proc. of the 4th International Semantic Web Conference*, 2005.
16. W. Nejdl, M. Wolpers, and C. Capella. The RDF Schema Specification Revisited. In *Modelle und Modellierungssprachen in Informatik und Wirtschaftsinformatik, Modellierung 2000*, 2000.
17. N. Noy. Representing Classes as Property Values on the Semantic Web, Jul. 2004.
18. J. Z. Pan and I. Horrocks. Metamodeling Architecture of Web Ontology Languages. In *Proceeding of the Semantic Web Working Symposium (SWWS)*, 2001.
19. J. Z. Pan and I. Horrocks. RDFS(FA) and RDF MT: Two Semantics for RDFS. In *Proc. of the 2nd International Semantic Web Conference*, 2003.
20. J. Z. Pan and I. Horrocks. OWL-Eu: Adding Customised Datatypes into OWL. *Journal of Web Semantics*, 4(1):29–48, 2006.
21. J. Z. Pan and I. Horrocks. RDFS(FA): Connecting RDF(S) and OWL DL. In *IEEE Transactions on Knowledge and Data Engineering*, pages 192–206, 2007.
22. J. Z. Pan, I. Horrocks, and G. Schreiber. OWL FA: A Metamodeling Extension of OWL DL. In *Proc. of OWLED2005*, 2005.
23. P. F. Patel-Schneider. Layering the Semantic Web: Problems and Directions. In *Proc. of the 1st International Semantic Web Conference*, 2002.
24. P. F. Patel-Schneider. Two Proposals for a Semantic Web Ontology Language. In *Proc. of the International Description Logic Workshop*, 2002.
25. P. F. Patel-Schneider, P. Hayes, and I. Horrocks. OWL Web Ontology Language Semantics and Abstract Syntax. Technical report, W3C, Feb. 2004. W3C Recommendation.
26. PICS. Platform for Internet Content Selectivity. `http://www.w3.org/PICS/`, 1997.
27. D. Raggett, A. Le Hors, and I. Jacobs. HTML 4.01 Specification. W3C Recommendation, `http://www.w3.org/TR/html4/`, dEC. 1999.
28. H. J. ter Horst. Extending the RDFS Entailment Lemma. In *Proc. of the 3rd International Semantic Web Conference*, pages 77–91, 2004.
29. H. J. ter Horst. Completeness, Decidability and Complexity of Entailment for RDF Schema and a Semantic Extension Involving the OWL Vocabulary. *Journal of Web Semantic*, 3(2), 2005.
30. M. Uschold and M. Gruninger. Ontologies: Principles, Methods and Applications. *The Knowledge Engineering Review*, 1996.

Web Ontology Language: OWL

Grigoris Antoniou[1] and Frank van Harmelen[2]

[1] FORTH-ICS and Department of Computer Science, University of Crete, Crete, Greece, antoniou@ics.forth.gr
[2] Department of AI, Vrije Universiteit Amsterdam, Amsterdam, The Netherlands, frank.van.harmelen@cs.vu.nl

Summary. The expressivity of RDF and RDF Schema that was described in [12] is deliberately very limited: RDF is (roughly) limited to binary ground predicates, and RDF Schema is (again roughly) limited to a subclass hierarchy and a property hierarchy, with domain and range definitions of these properties.

However, the Web Ontology Working Group of W3C [10] identified a number of characteristic use-cases for Ontologies on the Web which would require much more expressiveness than RDF and RDF Schema. It proceeded to define OWL, the language that is aimed to be the standardised and broadly accepted ontology language of the Semantic Web.

In this chapter, we first describe the motivation for OWL in terms of its requirements, and the resulting non-trivial relation with RDF Schema. We then describe the various language elements of OWL in some detail.

1 Requirements for Ontology Languages

Ontology languages allow users to write explicit, formal conceptualizations of domains models. The main requirements are: (a) a well-defined syntax; (b) a well-defined semantics; (c) efficient reasoning support; (d) sufficient expressive power; and (e) convenience of expression.

The importance of a *well-defined syntax* is clear, and known from the area of programming languages; it is a necessary condition for *machine-processing* of information. OWL builds upon RDF and RDFS and has the same kind of syntax.

Formal semantics describes precisely the meaning of knowledge. "Precisely" here means that the semantics does not refer to subjective intuitions, nor is it open to different interpretations by different persons (or machines). The importance of formal semantics is well-established in the domain of mathematical logic, among others.

S. Staab and R. Studer (eds.), *Handbook on Ontologies*, International Handbooks on Information Systems, DOI 10.1007/978-3-540-92673-3,

One use of formal semantics is to allow humans to reason about the knowledge. For ontological knowledge we may reason about:

- *Class membership:* If x is an instance of a class C, and C is a subclass of D, then we can infer that x is an instance of D.
- *Equivalence of classes:* If class A is equivalent to class B, and class B equivalent to class C, then A is equivalent to C, too.
- *Consistency:* Suppose we have declared x to be an instance of the class A. Further suppose that A is a subclass of B, and A and B are declared disjoint. Then we have an inconsistency which points to a probable error in the ontology.
- *Classification:* If we have declared that certain property-value pairs are sufficient conditions for membership of a class A, then if an individual x satisfies such conditions, then x must be an instance of A.

Semantics is a prerequisite for *reasoning support*: Derivations such as the above can be made mechanically, instead of being made by hand. Reasoning support is important because it allows one to check the consistency of the ontology and the knowledge, check for unintended relationships between classes, and automatically classify instances in classes. Checks like the above are valuable for *designing* large ontologies, where multiple authors are involved, and *integrating and sharing* ontologies from various sources.

Formal semantics and reasoning support is usually provided by mapping an ontology language to a known logical formalism, and by using automated reasoners that already exist for those formalisms. OWL is (partially) mapped on a description logic, and makes use of existing reasoners such as FaCT and RACER.

1.1 Limitations of the Expressive Power of RDF Schema

RDF and RDFS allow the representation of *some* ontological knowledge. The main modelling primitives of RDF/RDFS concern the organization of vocabularies in typed hierarchies: subclass and subproperty relationships, domain and range restrictions, and instances of classes. However a number of other features are missing. Here we list a few:

- *Local scope of properties:* `rdfs:range` defines the range of a property, say `eats`, for all classes. Thus in RDF Schema we cannot declare range restrictions that apply to some classes only. For example, we cannot say that cows eat only plants, while other animals may eat meat, too.
- *Disjointness of classes:* Sometimes we wish to say that classes are disjoint. For example, `male` and `female` are disjoint. But in RDF Schema we can only state subclass relationships, e.g. `female` is a subclass of `person`.
- *Boolean combinations of classes:* Sometimes we wish to build new classes by combining other classes using union, intersection and complement. For example, we may wish to define the class `person` to be the disjoint union of the classes `male` and `female`. RDF Schema does not allow such definitions.

- *Cardinality restrictions:* Sometimes we wish to place restrictions on how many distinct values a property may or must take. For example, we would like to say that a person has exactly two parents, and that a course is taught by at least one lecturer. Again such restrictions are impossible to express in RDF Schema.
- *Special characteristics of properties:* Sometimes it is useful to say that a property is *transitive* (like "greater than"), *unique* (like "is mother of"), or the *inverse* of another property (like "eats" and "is eaten by").

So we need an ontology language that is richer than RDF Schema, a language that offers these features and more. In designing such a language one should be aware of the *tradeoff between expressive power and efficient reasoning support.* Generally speaking, the richer the language is, the more inefficient the reasoning support becomes, often crossing the border of non-computability. Thus we need a compromise, a language that can be supported by reasonably efficient reasoners, while being sufficiently expressive to express large classes of ontologies and knowledge.

1.2 Compatibility of OWL with RDF/RDFS

Ideally, OWL would be an extension of RDF Schema, in the sense that OWL would use the RDF meaning of classes and properties (`rdfs:Class`, `rdfs:subClassOf`, etc.), and would add language primitives to support the richer expressiveness identified above.

Unfortunately, the desire to simply extend RDF Schema clashes with the trade-off between expressive power and efficient reasoning mentioned before. RDF Schema has some very powerful modelling primitives, such as the `rdfs:Class` (the class of all classes) and `rdf:Property` (the class of all properties). These primitives are very expressive, and will lead to uncontrollable computational properties if the logic is extended with the expressive primitives identified above.

1.3 Three Species of OWL

All this has lead to a set of requirements that may seem incompatible: efficient reasoning support and convenience of expression for a language as powerful as a combination of RDF Schema with a full logic.

Indeed, these requirements have prompted W3C's Web Ontology Working Group to define OWL as three different sublanguages, each of which is geared towards fulfilling different aspects of these incompatible full set of requirements:

- *OWL Full:* The entire language is called OWL Full, and uses all the OWL languages primitives (which we will discuss later in this chapter). It also allows to combine these primitives in arbitrary ways with RDF and RDF

Schema. This includes the possibility (also present in RDF) to change the meaning of the pre-defined (RDF or OWL) primitives, by applying the language primitives to each other.
The advantage of OWL Full is that it is fully upward compatible with RDF, both syntactically and semantically: any legal RDF document is also a legal OWL Full document, and any valid RDF/RDF Schema conclusion is also a valid OWL Full conclusion. As a disadvantage, the language has become so powerful as to be undecidable, dashing any hope of guarantees on complete and efficient reasoning.

- *OWL DL:* In order to regain computational efficiency, OWL DL (short for: Description Logic) is a sublanguage of OWL Full which restricts the way in which the constructors from OWL and RDF can be used. We will give details later, but roughly this amounts to disallowing application of OWL's constructor's to each other, and thus ensuring that the language corresponds to a well studied description logic.
The advantage of this is that it permits efficient reasoning support.
The disadvantage is that we loose full compatibility with RDF: an RDF document will in general have to be extended in some ways and restricted in others before it is a legal OWL DL document. Conversely, every legal OWL DL document is still a legal RDF document.
- *OWL Lite:* An even further restriction limits OWL DL to a subset of the language constructors. For example, OWL Lite excludes enumerated classes, disjointness statements and arbitrary cardinality (among others).
The advantage of this is a language that is both easier to grasp (for users) and easier to implement (for tool builders).
The disadvantage is of course a restricted expressivity.

Ontology developers adopting OWL should consider which sublanguage best suits their needs. The choice between OWL Lite and OWL DL depends on the extent to which users require the more-expressive constructs provided by OWL DL and OWL Full. The choice between OWL DL and OWL Full mainly depends on the extent to which users require the meta-modeling facilities of RDF Schema (e.g. defining classes of classes, or attaching properties to classes). When using OWL Full as compared to OWL DL, reasoning support is less predictable since complete OWL Full implementations will be impossible.

There are strict notions of upward compatibility between these three sublanguages:

- Every legal OWL Lite ontology is a legal OWL DL ontology
- Every legal OWL DL ontology is a legal OWL Full ontology
- Every valid OWL Lite conclusion is a valid OWL DL conclusion
- Every valid OWL DL conclusion is a valid OWL Full conclusion

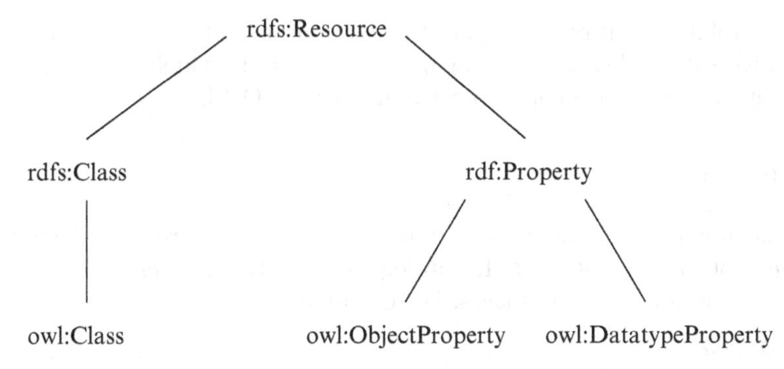

Fig. 1. Subclass relationships between OWL and RDF/RDFS

OWL still uses RDF and RDF Schema to a large extent:

- All varieties of OWL use RDF for their syntax
- Instances are declared as in RDF, using RDF descriptions and typing information
- OWL constructors like `owl:Class`, `owl:DatatypeProperty` and `owl:ObjectProperty` are all specialisations of their RDF counterparts. Figure 1 shows the subclass relationships between some modelling primitives of OWL and RDF/RDFS.

The original hope in the design of OWL was that there would be a downward compatibility with corresponding re-use of software across the various layers. However, the advantage of full downward compatibility for OWL (that any OWL aware processor will also provide correct interpretations of any RDF Schema document) is only achieved for OWL Full, at the cost of computational intractability.

2 The OWL Language

2.1 Syntax

OWL builds on RDF and RDF Schema, and uses RDF's XML syntax. Since this is the primary syntax for OWL, we will use it here, but it will soon become clear that RDF/XML does not provide a very readable syntax. Because of this, other syntactic forms for OWL have also been defined:

- An XML-based syntax which does not follow the RDF conventions. This makes this syntax already significantly easier to read by humans.
- An abstract syntax which is used in the language specification document. This syntax is much more compact and readable then either the XML syntax or the RDF/XML syntax.

- A graphical syntax based on the conventions of the UML language (Universal Modelling Language). Since UML is widely used, this will be an easy way for people to get familiar with OWL.

2.2 Header

OWL documents are usually called *OWL ontologies*, and are RDF documents. So the root element of a OWL ontology is an `rdf:RDF` element which also specifies a number of namespaces. For example:

```
<rdf:RDF
    xmlns:owl ="http://www.w3.org/2002/07/owl#"
    xmlns:rdf ="http://www.w3.org/1999/02/22-rdf-syntax-ns#"
    xmlns:rdfs="http://www.w3.org/2000/01/rdf-schema#"
    xmlns:xsd ="http://www.w3.org/2001/XLMSchema#">
```

An OWL ontology may start with a collection of assertions for housekeeping purposes. These assertions are grouped under an `owl:Ontology` element which contains comments, version control and inclusion of other ontologies. For example:

```
<owl:Ontology rdf:about="">
  <rdfs:comment>An example OWL ontology</rdfs:comment>
  <owl:priorVersion
       rdf:resource="http://www.mydomain.org/uni-ns-old"/>
  <owl:imports rdf:resource="http://www.mydomain.org/persons"/>
  <rdfs:label>University Ontology</rdfs:label>
</owl:Ontology>
```

The only one of these assertions which has any consequences for the logical meaning of the ontology is `owl:imports` : this lists other ontologies whose content is assumed to be part of the current ontology. Notice that while namespaces are used for disambiguation purposes, imported ontologies provide definitions that can be used. Usually there will be an import element for each used namespace, but it is possible to import additional ontologies, for example ontologies that provide definitions without introducing any new names.

Also note that `owl:imports` is a transitive property: if ontology A imports ontology B, and ontology B imports ontology C, then ontology A also imports ontology C.

2.3 Class Elements

Classes are defined using a `owl:Class` element.[1] For example, we can define a class `associateProfessor` as follows:

[1] `owl:Class` is a subclass of `rdfs:Class`.

```
<owl:Class rdf:ID="associateProfessor">
 <rdfs:subClassOf rdf:resource="#academicStaffMember"/>
</owl:Class>
```

We can also say that this class is disjoint from the `professor` and `assistantProfessor` classes using `owl:disjointWith` elements. These elements can be included in the definition above, or can be added by referring to the id using `rdf:about`. This mechanism is inherited from RDF.

```
<owl:Class rdf:about="#associateProfessor">
 <owl:disjointWith rdf:resource="#professor"/>
 <owl:disjointWith rdf:resource="#assistantProfessor"/>
</owl:Class>
```

Equivalence of classes can be defined using a `owl:equivalentClass` element:

```
<owl:Class rdf:ID="faculty">
 <owl:equivalentClass rdf:resource="#academicStaffMember"/>
</owl:Class>
```

Finally, there are two predefined classes, `owl:Thing` and `owl:Nothing` . The former is the most general class which contains everything (everything is a thing), the latter is the empty class. Thus every class is a subclass of `owl:Thing` and a superclass of `owl:Nothing`; in addition, a class may be equivalent to `owl:Thing` or `owl:Nothing`.

2.4 Property Elements

In OWL there are two kinds of properties:

- *Object properties* which relate objects (instances of classes, that is, interesting elements of the domain of discourse) to other objects.
  Examples are `isTaughtBy`, `supervises` etc.
- *Datatype properties* which relate objects to datatype values.
  Examples are `phone`, `title`, `age`, etc. OWL does not have any predefined data types, nor does it provide special definition facilities. Instead it allows one to use XML Schema data types, thus making use of the layered architecture the Semantic Web. Datatype properties are expressed using `owl:DatatypeProperty`

Here is an example of a datatype property.

```
<owl:DatatypeProperty rdf:ID="age">
 <rdfs:range rdf:resource=
   "http://www.w3.org/2001/XLMSchema#nonNegativeInteger"/>
</owl:DatatypeProperty>
```

User-defined data types will usually be collected in an XML schema, and then used in an OWL ontology.

Here is an example of a property, expressed using `owl:ObjectProperty`

```
<owl:ObjectProperty rdf:ID="isTaughtBy">
 <rdfs:domain rdf:resource="#course"/>
 <rdfs:range rdf:resource="#academicStaffMember"/>
 <rdfs:subPropertyOf rdf:resource="#involves"/>
</owl:ObjectProperty>
```

More than one domain and range may be declared. In this case the intersection of the domains, respectively ranges, is taken.

OWL allows us to relate "inverse properties", via `owl:inverseOf` . A typical example is `isTaughtBy` and `teaches`.

```
<owl:ObjectProperty rdf:ID="teaches">
 <rdfs:range rdf:resource="#course"/>
 <rdfs:domain rdf:resource="#academicStaffMember"/>
 <owl:inverseOf rdf:resource="#isTaughtBy"/>
</owl:ObjectProperty>
```

Actually domain and range can be inherited from the inverse property (interchange domain with range).

Equivalence of properties can be defined using `owl:equivalentProperty`.

```
<owl:ObjectProperty rdf:ID="lecturesIn">
 <owl:equivalentProperty rdf:resource="#teaches"/>
</owl:ObjectProperty>
```

2.5 Property Restrictions

With `rdfs:subClassOf` we can specify a class C to be subclass of another class C'; then every instance of C is also an instance of C'.

Now suppose we wish to declare, instead, that the class C satisfies certain conditions, that is, all instances of C satisfy the conditions. Obviously it is equivalent to saying that C is subclass of a class C', where C' collects all objects that satisfy the conditions. That is exactly how it is done in OWL, as we will show. Note that, in general, C' can remain anonymous, as we will explain below.

The following element requires first year courses to be taught by professors only (according to a questionable view, older and more senior academics are better at teaching).

```
<owl:Class rdf:about="#firstYearCourse">
 <rdfs:subClassOf>
  <owl:Restriction>
   <owl:onProperty rdf:resource="#isTaughtBy"/>
   <owl:allValuesFrom rdf:resource="#Professor"/>
```

```
  </owl:Restriction>
  </rdfs:subClassOf>
</owl:Class>
```

owl:allValuesFrom is used to specify the class of possible values the property specified by owl:onProperty can take (in other words: all values of the property must come from this class). In our example, only professors are allowed as values of the property isTaughtBy.

We can declare that mathematics courses are taught by staff member with ID 949358 as follows:

```
<owl:Class rdf:about="#mathCourse">
 <rdfs:subClassOf>
  <owl:Restriction>
   <owl:onProperty rdf:resource="#isTaughtBy"/>
   <owl:hasValue rdf:resource="#949352"/>
  </owl:Restriction>
 </rdfs:subClassOf>
</owl:Class>
```

owl:hasValue states a specific value that the property, specified by owl:onProperty must have. And we can declare that all academic staff members must teach at least one undergraduate course as follows:

```
<owl:Class rdf:about="#academicStaffMember">
 <rdfs:subClassOf>
  <owl:Restriction>
   <owl:onProperty rdf:resource="#teaches"/>
   <owl:someValuesFrom rdf:resource="#undergraduateCourse"/>
  </owl:Restriction>
 </rdfs:subClassOf>
</owl:Class>
```

Let us compare owl:allValuesFrom and owl:someValuesFrom . The example using the former requires *every* person who teaches an instance of the class, a first year subject, to be a professor. In terms of logic we have a *universal quantification.*

The example using the latter requires that *there exists* an undergraduate course that is taught by an instance of the class, an academic staff member. It is still possible that the same academic teaches postgraduate courses, in addition. In terms of logic we have an *existential quantification.*

In general, an owl:Restriction element contains a owl:onProperty element, and one or more restriction declarations. One type of restriction declarations are those that define restrictions on the kinds of values the property can take:
owl:allValuesFrom, owl:hasValue and owl:someValuesFrom. Another type of restrictions are *cardinality restrictions*, expressed by owl:minCardinality or owl:maxCardinality . For example, we can require every course to be taught by at least someone.

```
<owl:Class rdf:about="#course">
 <rdfs:subClassOf>
  <owl:Restriction>
   <owl:onProperty rdf:resource="#isTaughtBy"/>
   <owl:minCardinality rdf:datatype="&xsd;nonNegativeInteger">
     1
   </owl:minCardinality>
  </owl:Restriction>
 </rdfs:subClassOf>
</owl:Class>
```

Notice that we had to specify that the literal “1” is to be interpreted as a nonNegativeInteger (instead of, say, a string), and that we used the `xsd` namespace declaration made in the header element to refer to the XML Schema document.

Similarly, one can declare an upper limit on the number of class elements (for instance, that a department must have at most thirty members) using `owl:maxCardinality`.

It is possible to specify a precise number. For example, a PhD student must have exactly two supervisors. This can be achieved by using the same number in `owl:minCardinality` and `owl:maxCardinality`. For convenience, OWL offers also `owl:cardinality` .

We conclude by noting that `owl:Restriction` defines an anonymous class which has no id, is not defined by `owl:Class` and has only a local scope: it can only be used in the one place where the restriction appears. When we talk about classes, please bare in mind the twofold meaning: classes that are defined by `owl:Class` with an id, and local anonymous classes as collections of objects that satisfy certain restriction conditions, or as combinations of other classes, as we will see shortly. The latter are sometimes called *class expressions*.

2.6 Special Properties

Some properties of property elements can be defined directly:

- `owl:TransitiveProperty` defines a transitive property, such as “has better grade than”, “is taller than”, “is ancestor of”, etc.
- `owl:SymmetricProperty` defines a symmetric property, such as “has same grade as”, “is sibling of”, etc.
- `owl:FunctionalProperty` defines a property that has at most one unique value for each object, such as “age”, “height”, “directSupervisor”, etc.
- `owl:InverseFunctionalProperty` defines a property for which two different objects cannot have the same value, for example the property “isTheSocialSecurityNumberfor” (a social security number is assigned to one person only).

An example of the syntactic form of the above is:

```
<owl:ObjectProperty rdf:ID="hasSameGradeAs">
  <rdf:type rdf:resource="&owl;TransitiveProperty" />
  <rdf:type rdf:resource="&owl;SymmetricProperty" />
  <rdfs:domain rdf:resource="#student" />
  <rdfs:range rdf:resource="#student" />
</owl:ObjectProperty>
```

2.7 Boolean Combinations

It is possible to talk about Boolean combinations (union, intersection, complement) of classes (be it defined by `owl:Class` or by class expressions). For example, we can say that courses and staff members are disjoint as follows:

```
<owl:Class rdf:about="#course">
 <rdfs:subClassOf>
  <owl:Restriction>
   <owl:complementOf rdf:resource="#staffMember"/>
  </owl:Restriction>
 </rdfs:subClassOf>
</owl:Class>
```

The `owl:complementOf` says that every course is an instance of the complement of staff members, that is, no course is a staff member. Note that this statement could also have been expressed using `owl:disjointWith`.

The union of classes is built using `owl:unionOf` .

```
<owl:Class rdf:ID="peopleAtUni">
 <owl:unionOf rdf:parseType="Collection">
  <owl:Class rdf:about="#staffMember"/>
  <owl:Class rdf:about="#student"/>
 </owl:unionOf>
</owl:Class>
```

The `rdf:parseType` attribute is a shorthand for an explicit syntax for building list with `<rdf:first>` and `<rdf:rest>` tags. Such lists are required because the built-in containers of RDF have a serious limitation: there is no way to close them, i.e. to say "these are all the members of the container". This is because, while one graph may describe some of the members, there is no way to exclude the possibility that there is another graph somewhere that describes additional members. The list syntax provides exactly this facility, but is very verbose, which motivates the `rdf:parseType` shorthand notation.

Note that this does not say that the new class is a subclass of the union, but rather that the new class is *equal* to the union. In other words, we have stated an *equivalence of classes*. Also, we did not specify that the two classes must be disjoint: it is possible that a staff member is also a student.

Intersection is stated with `owl:intersectionOf`.

```
<owl:Class rdf:ID="facultyInCS">
 <owl:intersectionOf rdf:parseType="Collection">
  <owl:Class rdf:about="#faculty"/>
  <owl:Restriction>
   <owl:onProperty rdf:resource="#belongsTo"/>
   <owl:hasValue rdf:resource="#CSDepartment"/>
  </owl:Restriction>
 </owl:intersectionOf>
</owl:Class>
```

Note that we have built the intersection of two classes, one of which was defined anonymously: the class of all objects belonging to the CS department. This class is intersected with `faculty` to give us the faculty in the CS department. Further we note that Boolean combinations can be nested arbitrarily. The following example defines administrative staff to be those staff members that are neither faculty nor technical support staff.

```
<owl:Class rdf:ID="adminStaff">
 <owl:intersectionOf rdf:parseType="Collection">
  <owl:Class rdf:about="\#staffMember"/>
   <owl:Class>
    <owl:complementOf>
     <owl:Class>
      <owl:unionOf rdf:parseType="Collection">
       <owl:Class rdf:about="\#faculty"/>
       <owl:Class rdf:about="\#techSupportStaff"/>
      </owl:unionOf>
     </owl:Class>
    </owl:complementOf>
   </owl:Class>
 </owl:intersectionOf>
</owl:Class>
```

2.8 Enumerations

An enumeration is a `owl:oneOf` element, and is used to define a class by listing all its elements.

```
<owl:Class rdf:ID="daysOfTheWeek">
  <owl:oneOf rdf:parseType="Collection">
   <owl:Thing rdf:about="\#Monday"/>
   <owl:Thing rdf:about="\#Tuesday"/>
   <owl:Thing rdf:about="\#Wednesday"/>
   <owl:Thing rdf:about="\#Thursday"/>
   <owl:Thing rdf:about="\#Friday"/>
   <owl:Thing rdf:about="\#Saturday"/>
```

```
  <owl:Thing rdf:about="\#Sunday"/>
 </owl:oneOf>
</owl:Class>
```

2.9 Instances

Instances of classes are declared as in RDF. For example:

```
<rdf:Description rdf:ID="949352">
 <rdf:type rdf:resource="#academicStaffMember"/>
</rdf:Description>
```

or equivalently:

```
<academicStaffMember rdf:ID="949352"/>
```

We can also provide further details, such as:

```
<academicStaffMember rdf:ID="949352">
  <uni:age rdf:datatype="&xsd;integer">
    39
  </uni:age>
</academicStaffMember>
```

Unlike typical database systems, OWL does not adopt the *unique names assumption*, thus: just because two instances have a different name (or: ID), that does not imply that they are indeed different individuals. For example, if we state that each course is taught by at most one one staff member:

```
<owl:ObjectProperty rdf:ID="isTaughtBy">
  <rdf:type rdf:resource="&owl;FunctionalProperty" />
</owl:ObjectProperty>
```

and we subsequently state that a given course is taught by two staff members:

```
<course rdf:about="CIT1111">
  <isTaughtBy rdf:resource="#949318"/>
  <isTaughtBy rdf:resource="#949352"/>
</course>
```

this does *not* cause an OWL reasoner to flag an error. After all, the system could validly infer that the resources "949318" and "949352" are apparently equal. To ensure that different individuals are indeed recognised as such, we must explicitly assert their inequality:

```
<lecturer rdf:ID="949318">
  <owl:differentFrom rdf:resource="#949352">
</lecturer>
```

Because such inequality statements occur frequently, and the required number of such statements would explode if we wanted to state the inequality of a large number of individuals, OWL provides a shorthand notation to assert the pairwise inequality of all individuals in a given list using `owl:AllDifferent`:

```
<owl:AllDifferent>
  <owl:distinctMembers rdf:parseType="Collection">
    <lecturer rdf:about="#949318"/>
    <lecturer rdf:about="#949352"/>
    <lecturer rdf:about="949111"/>
  </owl:distinctMembers>
</owl:AllDifferent>
```

Note that `owl:distinctMembers` can only be used in combination with `owl:AllDifferent`.

2.10 Datatypes

Although XML Schema provides a mechanism to construct user-defined datatypes (e.g. the datatype of `adultAge` as all integers greater than 18, or the datatype of all strings starting with a number), such derived datatypes cannot be used in OWL. In fact, not even all of the many the built-in XML Schema datatypes can be used in OWL. The OWL reference document lists all the XML Schema datatypes that can be used, but these include the most frequently used types such as string, integer, boolean, time and date.

2.11 Versioning Information

We have already seen the *owl:priorVersion* statement as part of the header information to indicate earlier versions of the current ontology. This information has not formal model-theoretic semantics but can be exploited by humans readers and programs alike for the purposes of ontology management.

Besides *owl:priorVersion*, OWL has three more statements to indicate further informal versioning information. None of these carry any formal meaning:

- An `owl:versionInfo` statement generally contains a string giving information about the current version, for example RCS/CVS keywords.
- An `owl:backwardCompatibleWith` statement contains a reference to another ontology. This identifies the specified ontology as a prior version of the containing ontology, and further indicates that it is backward compatible with it. In particular, this indicates that all identifiers from the previous version have the same intended interpretations in the new version. Thus, it is a hint to document authors that they can safely change their documents to commit to the new version (by simply updating namespace declarations and `owl:imports` statements to refer to the URL of the new version).

- An `owl:incompatibleWith` on the other hand indicates that the containing ontology is a later version of the referenced ontology, but is not backward compatible with it. Essentially, this is for use by ontology authors who want to be explicit that documents cannot upgrade to use the new version without checking whether changes are required.

2.12 Layering of OWL

Now that we have discussed all the language constructors of OWL, we can completely specify which features of the language can be used in which sublanguage (OWL Full, DL and Lite):

OWL Full

In OWL Full, all the language constructors can be used in any combination as long as the result is legal RDF.

OWL DL

In order to exploit the formal underpinnings and computational tractability of Description Logics, the following constraints must be obeyed in an OWL DL ontology:

- *Vocabulary partitioning:* Any resource is allowed to be only either a class, a datatype, a datatype property, an object property, an individual, a data value or part of the built-in vocabulary, and not more than one of these. This means that, for example, a class cannot be at the same time an individual, or that a property cannot have some values from a datatype and some values from a class (this would make it both a datatype property and an object property).
- *Explicit typing:* Not only must all resources be partitioned (as prescribed in the previous constraint), but this partitioning must be stated explicitly. For example, if an ontology contains the following:

```
<owl:Class rdf:ID="C1">
    <rdfs:subClassOf rdf:resource="#C2" />
</owl:Class>
```

  this already entails that C2 is a class (by virtue of the range specification of rdfs:subClassOf). Nevertheless, an OWL DL ontology must *explicitly* state this information:

```
<owl:Class rdf:ID="C2"/>
```

- *Property separation:* By virtue of the first constraint, the set of object properties and datatype properties are disjoint. This implies that inverse properties, and functional, inverse functional and symmetric characteristics can never be specified for datatype properties.

- *No transitive cardinality restrictions:* No cardinality restrictions may be placed on transitive properties (or their superproperties, which are of course also transitive, by implication).
- *Restricted anonymous classes:* Anonymous classes are only allowed in the domain and range of `owl:equivalentClass` and `owl:disjointWith`, and in the range (not the domain) of `rdfs:subClassOf`.

OWL Lite

An OWL ontology must be an OWL DL ontology, and must further satisfy the following constraints:

- The constructors `owl:oneOf`, `owl:disjointWith`, `owl:unionOf`, `owl:complementOf` and `owl:hasValue` are not allowed
- Cardinality statements (both minimal, maximal and exact cardinality) can only be made on the values 0 or 1, and no longer on arbitrary non-negative integers.
- `owl:equivalentClass` Statements can no longer be made between anonymous classes, but only between class identifiers (Figs. 2 and 3).

OWL DL has a formal model-theoretic semantics [4] providing a rigorous and provably decidable semantics for the language. As discussed in chapter "Resource Description Framework," DLs are a decidable subset of first order logic (FOL), being restricted to the 2-variable fragment of FOL. The translation of languages constructs from OWL to Description Logic constructors is given in tables 2 and 3. The OWL constructs are given in terms of their abstract syntax [4].

Abstract Syntax	DL Syntax
`owl:Thing`	$\top$
`owl:Nothing`	$\bot$
`intersectionOf(`C_1 C_2 `...)`	$C_1 \sqcap C_2$
`unionOf(`C_1 C_2 `...)`	$C_1 \sqcup C_2$
`complementOf(`C`)`	$\neg C$
`oneOf(`o_1 `...)`	$\{o_1, \ldots\}$
`restriction(`R `someValuesFrom(`C`))`	$\exists R.C$
`restriction(`R `allValuesFrom(`C`))`	$\forall R.C$
`restriction(`R `hasValue(`o`))`	$R : o$
`restriction(`R `minCardinality(`n`))`	$\geq n\, R$
`restriction(`R `minCardinality(`n`))`	$\leq n\, R$
`restriction(`U `someValuesFrom(`D`))`	$\exists U.D$
`restriction(`U `allValuesFrom(`D`))`	$\forall U.D$
`restriction(`U `hasValue(`v`))`	$U : v$
`restriction(`U `minCardinality(`n`))`	$\geq n\, U$
`restriction(`U `maxCardinality(`n`))`	$\leq n\, U$

Fig. 2. OWL DL descriptions

Abstract Syntax	DL Syntax
Class(A partial $C_1 \ldots C_n$)	$A \sqsubseteq C_1 \sqcap \ldots \sqcap C_n$
Class(A complete $C_1 \ldots C_n$)	$A = C_1 \sqcap \ldots \sqcap C_n$
EnumeratedClass(A $o_1 \ldots o_n$)	$A = \{o_1, \ldots, o_n\}$
SubClassOf(C_1 C_2)	$C_1 \sqsubseteq C_2$
EquivalentClasses($C_1 \ldots C_n$)	$C_1 = \ldots = C_n$
DisjointClasses($C_1 \ldots C_n$)	$C_i \sqcap C_j = \bot, i \neq j$
DatatypeProperty(U super(U_1)...super(U_n)	$U \sqsubseteq U_i$
domain(C_1) ...domain(C_m)	$\geq 1\, U \sqsubseteq C_i$
range(D_1) ...range(D_l)	$\top \sqsubseteq \forall U.D_i$
[Functional])	$\top \sqsubseteq \leq 1\, U$
SubPropertyOf(U_1 U_2)	$U_1 \sqsubseteq U_2$
EquivalentProperties($U_1 \ldots U_n$)	$U_1 = \ldots = U_n$
ObjectProperty(R super(R_1)...super(R_n))	$R \sqsubseteq R_i$
domain(C_1) ...domain(C_m)	$\geq 1\, R \sqsubseteq C_i$
range(C_1) ...range(C_l)	$\top \sqsubseteq \forall R.C_i$
[inverseOf(R_0]	$R = (^- R_0)$
[Symmetric]	$R = (^- R)$
[Functional]	$\top \sqsubseteq \leq 1\, R$
[InverseFunctional]	$\top \sqsubseteq \leq 1\, R^-$
[Transitive])	$Tr(R)$
SubPropertyOf(R_1 R_2)	$R_1 \sqsubseteq R_2$
EquivalentProperties($R_1 \ldots R_n$)	$R_1 = \ldots = R_n$

Fig. 3. OWL DL axioms and facts

The decidability of the logic ensures that sound and complete DL reasoners can be built to check the consistency of an OWL ontology, i.e. verify whether there are any logical contradictions in the ontology axioms. Furthermore, reasoners can be used to derive inferences from the asserted information, e.g. infer whether a particular concept in an ontology is a subconcept of another, or whether a particular individual in an ontology belongs to a specific class.

At the time of this writing, an effort is underway to define another sublanguage, sometimes referred to as RDFS+ and other times as OWL Very Lite, which is intended to be a much simpler version that provides only simple reasoning extensions to RDFS to allow for very efficient scalability.

OWL DLP

As explained above, OWL is based on Description Logic. Since this is a fragment of FOL, it inherits many of the properties of that logic. In particular, it inherits the open-world assumption and the non-unique name assumption. This has lead to the definition of an interesting sublanguage of OWL DL, named OWL DLP. OWL DLP is not part of the official W3C OWL-species layering, but is nevertheless sufficiently interesting to deserve some discussion here.

The *open-world assumption* says that we cannot conclude some statement x to be false simply because we cannot show x to be true. After all, our axioms may be simply non-committal on the status of x. In other words, we may not deduce falsity from a the absence of truth. The opposite assumption (closed-world assumption, *CWA*) would allow to derive falsity from the inability to derive truth. The choice between open world semantics and closed world semantics is a recurring issue in the design of web-based KR languages. Although in general the Web would seem more suited to open-world reasoning (and indeed both RDFS and OWL adopt an open-world semantics), there are many use-cases where a closed-world semantics is appropriate: students in a class, customers of a company, cities in a country are all examples of closed sets: if a student is not listed as enrolled, we can safely assume she is not enrolled. Although useful in many cases, there is currently no practical mechanism in RDFS or OWL to state that a given set of individuals (or facts) is "closed". The only (limited) possibility is to empose an `owl:maxCardinality` constraint on a role.

The unique name assumption has been discussed above: if we encounter two individuals with different names, we can safely assume they are indeed different individuals. Typically, database systems assume a single, unique name for each individual. Again, on the web this assumption would be too strong. In a world as large as the web, many individuals are known under multiple names ("Jim Hendler", "James Hendler", "Prof. J. Hendler", "the co-chair of the OWL Working Group", etc.). When encountering two such different names, we should safely assume that they may or may not designate the same individual, until further reasoning decides the issue one way or the other. OWL contains a simple device to state that all individuals in an enumerated set are known to be different (i.e. that they are not just different names for some of the same individuals), but this language construct (`owl :AllDifferent`) requires the explicit enumeration of these names, which can be either impractical, or even impossible in principle.

Hence, for both the CWA and the UNA, ontologies are sometimes in need of one, and sometimes in need of the other. This conundrum was nicely resolved in [13], which identified a fragment of OWL baptised DLP, for Description Logic Programming: this fragment is the largest fragment on which the choice for CWA and UNA does not matter as depicted in Fig. 4. That is to say, OWL DLP is weak enough so that the differences between the choices do not show up. The advantage of this is that people or applications that wish to make different choices on these assumptions can still exchange ontologies in OWL DLP without harm. Of course, as soon as they go outside OWL DLP, they will notice that they draw different conclusions from the same statements. In other words, they will notice that they disagree on the semantics.

Fortunately, DLP is still large enough that it can be used for useful representation and reasoning tasks. It allows the use of such OWL constructors as class and property equivalence, equality and inequality between individuals,

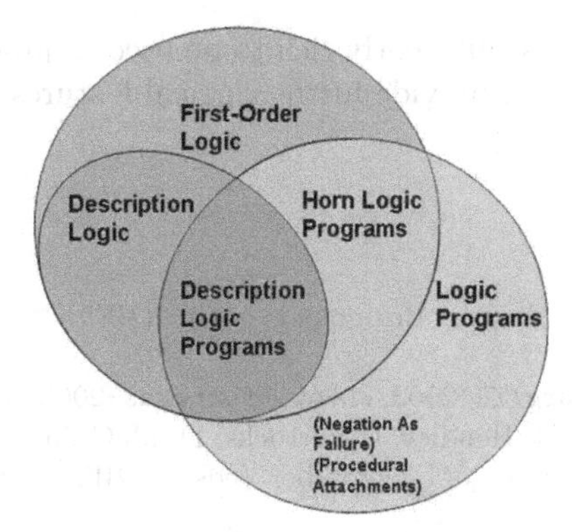

Fig. 4. Relation of OWL-DLP to other KR languages

inverse, transitive, symmetric and functional properties, and the intersection of classes. It excludes however constructors such as intersection and arbitrary cardinality-constraints.

These constructors do not only allow useful expressivity for many practical cases, while guaranteeing correct interchange between OWL reasoners independent of their CWA and UNA, they also allow for a translation into efficiently implementable reasoning techniques based on databases and logic programs.

As is already clear from the above two points, RDFS and OWL do not allow any form default reasoning, even though many years of KR applications have shown this to be a very useful device for dealing with incomplete knowledge. This would be particularly important in a world as large as the Web, where not all properties of all objects will be explicitly known, but must often be inferred by default until shown otherwise. However, a lack of concensus in the KR community on how to best formalise defaults has prevented such features to be included in the Semantic Web standardised representation languages.

Finally, a point often raised is that the large and open world of the Web will almost certainly need some forms of uncertainty and fuzziness. Again, lack of concensus has prevented such language features to be included, although it would seem clear that they will ultimately be needed in some form or other, either in the representation or in the inference mechanisms.

3 Summary

OWL is the proposed standard for Web ontologies. It allows us to describe the semantics of knowledge in a machine-accessible way. OWL builds upon RDF and RDF Schema: (XML-based) RDF syntax is used; instances are defined using RDF descriptions; and most RDFS modelling primitives are used. Formal semantics and reasoning support is provided through the mapping of OWL on logics. Predicate logic and description logics have been used for this purpose. While OWL is sufficiently rich to be used in practice, extensions are in the making. They will provide further logical features, including rules.

References

1. D. McGuinness, F. van Harmelen (eds.). *OWL Web Ontology Language Overview*
`http://www.w3.org/TR/2003/WD-owl-features-20030331/`.
2. F. van Harmelen, J. Hendler, I. Horrocks, D. McGuinness, P. Patel-Schneider, L. Stein. In: M. Dean, G. Schreiber (eds.), *OWL Web Ontology Language Reference*
`http://www.w3.org/TR/2003/WD-owl-ref-20030331/`.
3. M. Smith, C. Welty, D. McGuinness. *OWL Web Ontology Language Guide*
`http://www.w3.org/TR/2003/WD-owl-guide-20030331/`.
4. P. Patel-Schneider, P. Hayes, I. Horrocks. *OWL Web Ontology Language Semantics and Abstract Syntax*
`http://www.w3.org/TR/2003/WD-owl-semantics-20030331/`.
5. J. Hefflin. *Web Ontology Language (OWL) Use Cases and Requirements*
`http://www.w3.org/TR/2003/WD-webont-req-20030331/`.

Further interesting articles related to DAML+OIL and OIL include:

6. J. Broekstra et al. Enabling knowledge representation on the Web by extending RDF schema. In: *Proc. 10th World Wide Web Conference (WWW'10)*, 2001.
7. D. Fensel et al. OIL: an ontology infrastructure for the semantic Web. *IEEE Intelligent Systems* 16, 2, 2001.
8. D. McGuinness. Ontologies come of age. In: D. Fensel et al. (eds), *The Semantic Web: Why, What, and How.* MIT Press, Cambridge, MA, 2002.
9. P. Patel-Schneider, I. Horrocks, F. van Harmelen. Reviewing the design of DAML+OIL: an ontology language for the semantic Web, *Proceedings of AAAI'02.*

The key site on OWL is:

10. `http://www.w3.org/2001/sw/WebOnt/`.

The two most relevant chapters from this Handbook are:

11. B. McBride. The resource description framework (RDF) and its vocabulary description language RDFS. In: S. Staab, R. Studer (eds.), *The Handbook on Ontologies in Information Systems*, Springer, Berlin, 2003.
12. F. Baader, I. Horrocks, U. Sattler. Description Logics. In: S. Staab, R. Studer (eds.), *The Handbook on Ontologies in Information Systems*, Springer, Berlin, 2003.

The work on DLP has been reported in:

13. B. N. Grosof, I. Horrocks, R. Volz, S. Decker. Description logic programs: combining logic programs with description logic. In: *Proceedings of the Twelfth World Wide Web Conference*, pp. 48–57, 2003.

Ontologies and Rules

Pascal Hitzler[1] and Bijan Parsia[2]

[1] AIFB, Universität Karlsruhe (TH), Karlsruhe, Germany,
hitzler@aifb.uni-karlsruhe.de
[2] University of Manchester, Manchester, UK, bparsia@cs.man.ac.uk

Summary. Ontologies and rules are two established paradigms in knowledge modelling, and play an important role for the Semantic Web. In this chapter, we present an introduction to common approaches for combining OWL ontologies and rules. In particular, we cover the Semantic Web Rules Language SWRL and Description Logic Programs DLP, and give pointers to the literature.

1 Introduction

The Web Ontology Language OWL, as introduced in chapter "Web Ontology Language: OWL," is the language recommended by the World Wide Web consortium (W3C) for expressing ontologies for the Semantic Web. OWL is based on Description Logics, see chapter "Description Logics," and as such is based on first-order predicate logic as underlying knowledge representation and reasoning paradigm.

Throughout the advent of the Semantic Web, however, F-Logic as an alternative approach for expressing ontologies has been based on rules, more precisely on the logic programming paradigm, see chapter "Ontologies in F-Logic." Due to this, and also due to the importance of rule-based systems in industrial practice, it is natural to ask how OWL and rule-based knowledge-representation and reasoning can be combined.

Achieving a conceptually clear integration of OWL and rules, however, is not a trivial task. The reason for the difficulties lies in the fact that OWL – based on first-order predicate logic – adheres to the open-world assumption, while rules generally follow the closed-world assumption. Consequently, the semantics of OWL and rules differs considerably, and achieving a meaningful, intuitive, and formally clear combined semantics is not straightforward.

Research on the topic of ontologies and rules has thus spawned into several different directions. In this chapter, we provide an overview of the state of the art, by discussing briefly some of the approaches which we consider to be

S. Staab and R. Studer (eds.), *Handbook on Ontologies*, International Handbooks on Information Systems, DOI 10.1007/978-3-540-92673-3,

most promising or interesting. Since we want to be brief and to the point, our selection is naturally subjective, but we made an effort to include references to many of the recent publications on the topic.

In Sect. 2, we present the Semantic Web Rules Language SWRL, a rule *extension* of OWL DL which adheres to the open-world paradigm, and is thus entirely in the spirit of the OWL DL language. We also introduce the decidable fragment of SWRL known as DL-Safe rules. SWRL adds to the expressive power of OWL by allowing the modelling of certain axioms which lie outside the capability of OWL DL.

In Sect. 3, we change perspective and consider rule *fragments* of OWL DL. In particular, we present its naive Horn fragment DLP – description logic programs – and a more sophisticated fragment called Horn-$\mathcal{SHIQ}$, which encompasses DLP.

In Sect. 4, we briefly address *hybrid* approaches, i.e. theories and systems which combine OWL with some existing rules language, and thus combine the open- and the closed-world assumption.

We conclude in Sect. 5.

Throughout the chapter, we will employ the syntax for Description Logics as introduced in chapter "Description Logics."

2 SWRL and DL-Safe Rules

Attempts to combine some sort of rules with a description logic go back at least as far as the Classic system [2]. Starting in the late 1990s, there were a number of attempts to combine Datalog (as the premier deductive database language) with description logics; notable examples include $\mathcal{AL}$-log [6] and CARIN [22]. While partially motivated by the desire to increase the expressive power of both components, these attempts (esp. $\mathcal{AL}$-log) at hybrid systems were strongly constrained by the desire to retain both the modeling and the computational properties of the respective components, and even, perhaps, the implementation techniques or actual implementations. (For example, the proof procedure for $\mathcal{AL}$-log described in [6] calls an independently developed description logic reasoner as a oracle.)

In general, when considering combination formalisms, the upper bound is the unrestricted union of the two systems. In essence, the Semantic Web Rule Language (SWRL)[1] [13] is the unrestricted union of OWL DL (i.e. roughly the description logic $\mathcal{SHOIN}(D)$) and (binary) function-free Horn logic. The result is a very expressive formalism which is, unsurprisingly, undecidable. Also, it is comparatively not well understood: there are no native reasoners for it, relationships to other formalisms are not precisely mapped out, and there is almost no experience in using SWRL for ontology modeling. However, it does serve as a unifying overarching formalism for various rule and rule like extension to OWL DL. It is, for example, a superset of (binary) $\mathcal{AL}$-log and of CARIN.

[1] http://www.w3.org/Submission/SWRL/

Decidability can be regained with the imposition of a *safety* condition on SWRL rules. Essentially, the possible values of (explicit) variables in SWRL rules are restricted to named individuals only which confines the direct effects of such rules to the ABox. This safety condition is known as "DL-Safety" and such SWRL rules are generally called "DL-Safe SWRL rules" or "DL-Safe rules". Not only are DL-Safe rules (in combination with OWL-DL) decidable, but reasonable implementations are emerging.

SWRL was first proposed under the name OWL Rules Language [13] before gaining built-in atoms and a more RuleML[2] flavored syntax. In that form, it was published as a W3C member submission. The decidability of SWRL rules with the DL-Safety condition was established in [31, 32], and further elaborated in [28], including a discussion of a robust implementation of (most of) OWL plus DL-safe rules, KAON2.

2.1 Definition of SWRL

SWRL contains OWL DL as a proper part, that is, all OWL DL axioms are SWRL axioms. Additionally, a SWRL knowledge base may contain a set of *rules*, which consist of an antecedent (body) and a consequent (head) which themselves are sets of SWRL atoms.

A SWRL atom may be of the following[3] forms:

- Unary atoms:
  $C(arg1)$ where C is an arbitrary OWL DL class expression
  $D(dataArg1)$ where D is a datatype URI or an enumerated value range
- Binary atoms:
  $P(arg1, arg2)$ where P is an object property
  $Q(arg1, dataArg1)$ where Q is a datatype property
  $arg1 = arg1$ equality, or "sameAs"
  $arg2 \neq arg2$ inequality, or "differentFrom"

Where arguments are of the form:

$arg1 \mid arg2$ these are either individuals denoting URIs or individual ranging variables
$dataArg1$ these are data literals or data value ranging variables

A SWRL rule is of the form:

- $Atom_1 \wedge \ldots \wedge Atom_n \rightarrow Atom_{n+1} \wedge \ldots \wedge Atom_m$

where atoms 1 through n form the antecedent (i.e. body) and atoms $n + 1$ through m form the consequent (i.e. the head).

[2] http://www.ruleml.org/

[3] In addition to the forms directly below, SWRL also allows for "built-in" atoms which are discussed in Sect. 2.3.

The semantics of SWRL are traditionally given via an extension of the "direct" model theory for OWL DL.[4] However, for current purposes it is a little more perspicuous to present the semantics as an extension of the standard translation[5] of description logics to first order logic. To avoid various tedious details of the translation of $\mathcal{SHOIN}(D)$ which are irrelevant to understanding SWRL, we give only the translation of the simpler description logic $\mathcal{SH}$, that is, omitting nominals, inverse roles, number restrictions, and datatypes (but retaining (in)equality). We therefore also eliminate SWRL atoms with data valued arguments.

In the table below, $\mathcal{X}$ and $\mathcal{Y}$ are meta-linguistic variables that range over constants (i.e. names of individuals) or object variables (i.e. x, y, and z). A represents an arbitrary atomic concept, C and D represent arbitrary class expressions, a and b represent arbitrary constants, and P represents an arbitrary role:

- Atomics

Term	Translation
$\pi(x)$ or $\pi(y)$	y or x (resp.)
$\pi(A, \mathcal{X})$	$A(\mathcal{X})$
$\pi(P, \mathcal{X}, \mathcal{Y})$	$P(\mathcal{X}, \mathcal{Y})$

- Concept expressions

Concept	Translation
$\pi(A, \mathcal{X})$	$A(\mathcal{X})$
$\pi(\neg C, \mathcal{X})$	$\neg\pi(C, \mathcal{X})$
$\pi(C \sqcap D, \mathcal{X})$	$\pi(C, \mathcal{X}) \wedge \pi(D, \mathcal{X})$
$\pi(C \sqcup D)$	$\pi(C, \mathcal{X}) \vee \pi(D, \mathcal{X})$
$\pi(\exists P.C, \mathcal{X})$	$\exists\pi(\mathcal{X})(\pi(P, \mathcal{X}, \pi(\mathcal{X})) \wedge \pi(C, \pi(\mathcal{X})))$
$\pi(\forall P.C, \mathcal{X})$	$\forall\pi(\mathcal{X})(\pi(P, \mathcal{X}, \pi(\mathcal{X})) \rightarrow \pi(C, \pi(\mathcal{X})))$

- Axioms[6]

Axiom	Translation
$\Pi(C \sqsubseteq D)$	$\forall x(\pi(C, x) \rightarrow \pi(D, x))$
$\Pi(P \sqsubseteq Q)$	$\forall x, y(\pi(P, x, y) \rightarrow \pi(Q, x, y))$
$\Pi(Trans(P))$	$\forall x, y, z(\pi(P, x, y) \wedge \pi(P, y, z) \rightarrow \pi(P, x, z))$
$\Pi(a : C)$	$\pi(C, a)$
$\Pi(< a, b >: P)$	$\pi(P, a, b)$
$\Pi(a = b)$	$a = b$
$\Pi(a \neq b)$	$a \neq b$

[4] As in the SWRL submission.

[5] For a basic discussion of the standard translation in a modal logic context, see chapter "Ontologies in F-Logic" of [1]. For variants of the standard translation for $\mathcal{SHOIN}(D)$ and other description logics see [32] and [16].

[6] Class and property equivalence axioms can be defined as a pair of inclusion axioms.

It is clear that, except for transitivity axioms, $\mathcal{SH}$ can be encoded into first order logic using only two distinct variables (i.e. x and y) and transitivity only needs one additional variable (i.e. z). To extend this translation to a SWRL extension of $\mathcal{SH}$, we need an arbitrary supply of variables, that is, at least enough for each distinct variable in the rule with the largest number of distinct variables in a given rule set. To help distinguish translations of $\mathcal{SH}$ axioms from translations of SWRL rules, we shall use capital letters from $\{X, Y, Z\}$ with subscripts when necessary.

Let $\Psi = \psi_1 \wedge \ldots \wedge \psi_n \rightarrow \psi_{n+1} \wedge \ldots \wedge \psi_m$ be a SWRL rule, such that $Var(\Psi) = V_1 \ldots V_k$ are all the SWRL variables in the atoms $\psi_1 \ldots \psi_m$. Then we can extend the translation function, π, as follows (note that the meta-variables $\mathcal{X}$ and $\mathcal{Y}$ now also range over SWRL variables):

- Atomics

Term	Translation
$\pi(x)$ or $\pi(y)$ or $\pi(\mathcal{V})$	y or x or $\mathcal{V}$(resp.)
$\pi(A, \mathcal{X})$	$A(\mathcal{X})$
$\pi(P, \mathcal{X}, \mathcal{Y})$	$P(\mathcal{X}, \mathcal{Y})$

- Axioms

Axiom	Translation
$\pi(\Psi)$	$\forall V_1 \ldots V_k(\pi(\psi_1) \wedge \ldots \wedge \pi(\psi_n) \rightarrow \pi(\psi_{n+1}) \wedge \ldots \wedge \pi(\psi_m))$
	Where a ψ is of the form $A(\mathcal{X})$ or $P(\mathcal{X}, \mathcal{Y})$

One immediate consequence of the translation is that it is obvious that SWRL rules can entirely replace the role axioms of $\mathcal{SH}$:

Axiom	FOL
$Trans(P)$	$\forall x, y, z(P(x, y) \wedge P(y, z) \rightarrow P(x, z))$
$P(X, Y) \wedge P(Y, Z) \rightarrow P(X, Y)$	$\forall X, Y, Z(P(X, Y) \wedge P(Y, Z) \rightarrow P(X, Z))$
$P \sqsubseteq Q$	$\forall x, y(P(x, y) \rightarrow Q(x, y))$
$P(X, Y) \rightarrow Q(X, Y)$	$\forall X, Y(P(X, Y) \rightarrow Q(X, Y)$

Clearly, replacing the variables in the first order translation of the $\mathcal{SH}$ axioms with their uppercased versions results in the first order translation of the SWRL rule. Thus $\mathcal{SH}$+SWRL is trivially reducible to $\mathcal{ALC}$+SWRL. Of course, concept inclusion axioms may be encoded as SWRL rules analogously to how property inclusions are (at least, if we allow concept expressions as the functor of atoms; otherwise, we need at least some definitions). If, as the SWRL submission does, we permit SWRL rules with empty antecedents, then we can assimilate the ABox axioms (i.e. type, property, and (in)equality assertions) as well.

These reductions, as observed in [13], highlight the fact that Description Logics can be seen as rule languages, given a sufficiently liberal notion of rule. The flip side is that SWRL rules have more in common with description logic axioms than with Datalog rules and certainly more than with production or ECA rules.

While we know of no user study comparing the usability of explicit rules to Description Logic style variable free axioms (and we believe that this would be a very difficult study to conduct), there is no shortage of claims that one style or the other is more usable, transparent, or easy to learn. For example, some researchers working on controlled natural languages (CNLs) for OWL argue that it is better to produce a CNL sentence corresponding to the SWRL rule (including explicit variables!) for transitivity instead of using the more succinct "Property P is transitive". At least one trade off seems clear: Description Logic "variable free" style is less cluttered with (clearly redundant!) variables, whereas SWRL style is probably initially more comfortable for people familiar with rule languages such as Prolog or with first order logic or with query languages like SQL. The lack of "visual noise" for standard Description Logic axioms (i.e. concept and property subsumption, transitivity, inverse, etc.) seems decisive except for explication purposes. Similarly, the operator style syntax of Description Logics makes it easier to manage complex nested expressions. This suggests that there may be user advantages to continuing to adopt more expressive role constructors (such as the expanded role composition operator in OWL 2).

The undecidability of $\mathcal{SHOIN}$+SWRL can easily be seen by the fact that it is possible to encode the transitivity of an otherwise simple role using a SWRL rule. Transitive roles in number restrictions are a well known source of undecidability for $\mathcal{SHIQ}$ (e.g. see [14]), and thus of OWL DL as well. Of course, it is also easy to directly encode undecidable domino tiling problems in SWRL (as described in [13], for which $\mathcal{SH}$+SWRL is sufficient) or other known undecidable extensions (such as role-value-maps).

We have seen that we can see SWRL as a very straightforward generalisation of description logics. The addition of arbitrary variables to conditionals brings in a lot of expressive power. However, SWRL is a somewhat strange fragment of first order logic. It lacks, at least, n-ary predications and function symbols, but it nevertheless is only semi-decidable. Of course, pure Prolog is another case of an undecidable fragment of first order logic, but there the question of "Why not move to full first order logic?" has a well known answer: Prolog has a proof procedure which gives rise to a powerful, reasonably efficient in the common case, and understandable performance model for computations. In fact various compromises (such as omitting the occurs check) have well understood performance (and soundness/completeness) tradeoffs. Unfortunately, there is no such body of knowledge about SWRL yet.

2.2 Definition of DL-Safe SWRL

The free syntax of SWRL (e.g. conjunctions in the head and arbitrary concept expressions as atoms) helps emphasis the similarities of SWRL rules with Description Logic axioms, but it obscures the relation between SWRL rules and rules systems based on Horn clauses, such as Datalog. However, it is easy enough to transform an arbitrary SWRL knowledge base (consisting of a rules part and a Description Logic part) into a form where all the SWRL rules consequents contain only a single conjunct and all the atoms are atomic. (Essentially, definitions are added to the description logic part that replace each complex class atom with a fresh atomic atom, and conjunctive consequences are eliminated via the Lloyd–Topor transformations [24] as mentioned in the SWRL submission.)

SWRL ontologies in this form of a knowledge base resemble hybrid Description logic and Datalog knowledge bases. By adjusting the rules portion to be semantically more like Datalog rules (that is, so that the rules act only on the explicit facts in the knowledge and other rules), the hybrid flavor becomes stronger including regained decidability. In the translation to first order logic, one can impose the safety restriction (of variables to named individuals) by means of a special DL predicate, O, which is true of all named individuals (i.e. there is a fact of the form $O(a)$ for all individual names, a, appearing in the ontology) and does not appear in any concept expression of the ontology (and thus not in the scope of any existential quantifier).[7] Then, for each variable in the DL-safe rule, an O-atom with that variable is added to the body.[8]

That is, we extend the translation to first order in the following way. Let $\Omega(K) = c_1 \ldots c_n$ be the set of individual names appearing in a knowledge base K. $K' = K \cup \{c_1 : O \ldots c_n : O\}$. Π is then applied to the elements of K'. Furthermore, let $\Psi_{dls} = \alpha_1 \wedge \ldots \wedge \alpha_n \rightarrow \alpha_{n+1}$ be a DL-Safe rule such that every α is atomic and $Var(\Psi_{dls}) = v_1 \ldots v_k$ are all the SWRL variables in the atoms $\alpha_1 \ldots \alpha_{n+1}$. Then:

Axiom Translation
$\pi(\Psi_{dls})\ \forall v_1 \ldots v_k(\pi(\alpha_1) \wedge \ldots \wedge \pi(\alpha_n) \wedge O(v_1) \wedge \ldots \wedge O(v_k) \rightarrow \pi(\alpha_{n+1}))$

It is clear from the translation that the DL-Safe version of the SWRL rule is much weaker than the unrestricted version. Consider the following simple example (adapted from [38]):

$$\text{Foot} \sqsubseteq \exists\text{partOf.Leg} \tag{1}$$

$$\text{BurnOnFoot} \sqsubseteq \exists\text{locatedIn.Foot} \tag{2}$$

$$\text{LegInjury} \equiv \exists\text{locatedIn.Leg} \tag{3}$$

$$< \text{burn1, foot1} > : \text{locatedIn} \tag{4}$$

$$< \text{foot1, leg1} > : \text{partOf} \tag{5}$$

$$\text{leg1} : \text{Leg} \tag{6}$$

$$\text{foot1} : \text{Foot} \tag{7}$$

$$\text{burn2} : \text{BurnOnFoot} \tag{8}$$

$$\text{locatedIn(X, Y)} \wedge \text{partOf(Y, Z)} \rightarrow \text{locatedIn(X, Z)} \tag{9}$$

[7] Thanks to an anonymous reviewer for some clarifications on this point.

[8] We presume normal Datalog safety, i.e. that every variable in the head appears in the body.

If we interpret (8) as an unrestricted SWRL rule, then we can conclude:

$$< \text{burn1}, \text{leg1} > : \text{locatedIn} \tag{10}$$
$$\text{burn1} : \text{LegInjury} \tag{11}$$
$$\text{BurnOnFoot} \sqsubseteq \text{LegInjury} \tag{12}$$
$$\text{burn2} : \text{LegInjury} \tag{13}$$

whereas, if we impose DL-Safety on (8), we can only conclude (10) and (11). Clearly, with DL-Safety, rules can still interact with Description Logic axioms, but only through new ground facts. It is because we infer (10) (via (9)) that we can conclude (11) (via the class definition (3)). Thus, DL-Safety constrains rules to work on and through ground facts.

DL-Safe SWRL rules, in addition to being much more computationally reasonable, may be more cognitively adequate, perhaps *due* to their relative expressive weakness. For users coming from a database background, they can be seen as a data manipulation language. It is at least plausible that it is somewhat easier for the typical user to understand why a missing fact blocks an entailment, than to understand why a subsumption fails because of there only being necessary, but not sufficient, conditions defined.

2.3 Built-ins

This more programmatic feel is enhanced by the presence of built-in atoms, that is, atoms with a fixed, predefined interpretation. The SWRL submission includes built-ins for value comparison, mathematics, and string manipulation among others. There are several issues with built-ins with perhaps the most prominent being how they are to be interpreted if their variables are under-instantiated when being evaluated (that is, whether they should be interpreted as arbitrary constraints or more procedurally, e.g. "throwing" an error when an binding is of an inappropriate type or must be drawn arbitrarily from the domain rather than determined by the knowledge base). However, SWRL built-ins seem fairly popular: at least they are reported as desired, as a supplement to OWL's datatype facility. In spite of this clamor, it is unclear what the right semantics of built-ins (or even just data property atoms) should be, in part due to the fact that the most natural reading of data property atoms (and built-ins) is as general constraints, whereas many users talk about built-ins as if they were procedural attachments.

3 Rule Fragments of OWL

3.1 Description Logic Programs

DLP are a naive *Horn fragment* of OWL. They inherit their semantics from OWL, thus adhering to the open world assumption. At the same time, DLPs can be transformed *syntactically* into Logic Programming syntax, and thus provide a kind of basic interoperability between OWL and Logic Programming. Let it be noted, though, that DLP is *not* a *common* fragment of OWL and Logic Programming, because the two semantics normally considered are different. Nevertheless, the two semantics have a clearly understood and strong relationship, about which we will talk later in Sect. 3.2.

Compared to OWL DL, DLP is a rather primitive ontology language. It strictly contains, however, the OWL DL fragment of RDFS (see chapter "Resource Description Framework.") and has the pleasing property of having polynomial data complexity. It is thus one of the tractable OWL fragments[9] discussed within the currently ongoing OWL 2 standardisation effort.[10]

According to an analysis in [40], most of the existing OWL ontologies are almost completely contained in DLP.[11] An example is the Semantic Web Research Community ontology, SWRC [39],[12] which is the most imported ontology on the web.[13]

DLP was originally presented in [11], and a thorough treatment can be found in [40].

Definition of DLP

Originally, DLP was presented as fragment of OWL. We present DLP in an alternative and more constructive way, which exposes the modeling capabilities and the limitations of it. It can be considered a kind of *normal form* for DLP.

We need to fix terminology first. We understand DLP as a *semantic* fragment of OWL, i.e. we abstract (for the time being) from a concrete syntax: Every OWL statement which is *semantically* equivalent – in the sense of first order logic – to a (finite) set of function-free Horn clauses constitutes a valid DLP statement.[14] Allowing integrity constraints, we call the resulting fragment DLP IC (or just IC). Allowing integrity constraints and equality, we call the resulting fragment DLP ICE (or ICE). We write DLP^+ for the (semantic) fragment common to OWL DL and (function-free non-disjunctive) Datalog. Analogously, we write DLP^+ IC, IC^+, etc.

[9] http://www.w3.org/Submission/2006/SUBM-owl11-tractable-20061219/

[10] http://www.w3.org/2007/OWL/

[11] We would like to mention that it has been argued whether this analysis is appropriate.

[12] http://www.aifb.uni-karlsruhe.de/about.html

[13] http://ebiquity.umbc.edu/blogger/2007/06/15/how-owlimport-is-used/

[14] In our terminology, the set of OWL Lite statements $\{C \sqsubseteq D \sqcup E, D \equiv E\}$ would not qualify as a set of DLP statements, although it is semantically equivalent to $\{C \sqsubseteq D, D \equiv E\}$, which is expressible in DLP. We are well aware of this restriction, but will not be concerned with it in the moment, because this more general notion of semantic equivalence is not readily accessible by syntactic means. Note, however, that $C \sqsubseteq D \sqcup D$ qualifies as a DLP statement, since it is semantically equivalent to $C \sqsubseteq D$.

Now to the definition. Allowed are the following, where a, b, a_i stand for individuals, C stands for a concept name and $R, Q, R_i, Q_{i,j}$ stand for role names.

- ABox:
  $C(a)$ (individual assertion)
  $R(a, b)$ (property assertion)
  $a = b$ (ICE) (individual equivalence)
- Property Characteristics:
  $R \equiv Q$ (equivalence)
  $R \sqsubseteq Q$ (subproperty)
  $\top \sqsubseteq \forall R.C$ $(C \neq \bot)$ (domain)
  $\top \sqsubseteq \forall R^{-}.C$ $(C \neq \bot)$ (range)
  $R \equiv Q^{-}$ (inverse)
  $R \equiv R^{-}$ (symmetry)
  $\top \sqsubseteq \leq 1 R$ (ICE) (functionality)
  $\top \sqsubseteq \leq 1 R^{-}$ (ICE) (inverseFunctionality)
- TBox: We allow expressions of the form

$$\exists Q_{1,1}^{(-)} \ldots \exists Q_{1,m_1}^{(-)}.\mathtt{Left}_1 \sqcap \cdots \sqcap \exists Q_{k,1}^{(-)} \ldots \exists Q_{k,m_k}^{(-)}.\mathtt{Left}_k \sqsubseteq \forall R_1^{(-)} \ldots \forall R_n^{(-)}.\mathtt{Right}$$

  where the following apply.
  - For DLP we allow $\mathtt{Left}_j$ to be of the forms C, $\{o_1, \ldots, o_n\}$, $\bot$ or $\top$, and $\mathtt{Right}$ to be of the forms C or $\top$.
  - For DLP IC we allow $\mathtt{Left}_j$ to be of the forms C, $\{o_1, \ldots, o_n\}$, $\bot$, or $\top$, and $\mathtt{Right}$ to be of the form C, $\top$, or $\bot$.
  - For DLP ICE we allow $\mathtt{Left}_j$ to be of the forms C, $\{o_1, \ldots, o_n\}$, $\bot$, or $\top$, and $\mathtt{Right}$ to be of the form C, $\top$, $\bot$, or $\{o\}$.
  - For the DLP$^+$ versions we furthermore allow $\mathtt{Right}$ to be of the form $\exists R^{(-)}.\{a\}$.

  The superscript $^{(-)}$ shall indicate that an inverse symbol may occur in these places. Note that (by a common abuse of notation) we allow any of k, m_i, n to be zero. For $k = 0$ the left hand side becomes $\top$. Note also that we could have disallowed $\bot$ on the left and $\top$ on the right, since in either case the statement becomes void. Likewise, it would suffice to require $n = 0$ in all cases, since universal quantifiers on the right are expressable using existentials on the left. Disallowing the existential quantifiers on the left (while keeping universals on the right) is also possible, but at the expense of the introduction of an abundance of new concept names. As an example, note that $\exists R.C \sqcap \exists Q.D \sqsubseteq E$ would have to be translated into the set of statements $\{C_1 \sqcap D_1 \sqsubseteq E, C \sqsubseteq \forall R^{-}.C_1, D \sqsubseteq \forall Q^{-}.D_1\}$, where C_1 and D_1 are new concept names. Our representation is more compact.

Using any of the established syntaxes of the OWL language, an OWL axiom is said to be in DLP if its translation into Description Logic syntax results in a finite set of statements of the above mentioned form.

An Example

We give a small example ontology which displays the modeling expressivity of DLP:

For the TBox, we model the following sentences

(1) Every man or woman is an adult
(2) A grown-up is a human who is an adult
(3) A woman who has somebody as a child, is a mother
(4) An orphan is the child of humans who are dead
(5) A lonely child has no siblings
(6) AIFB researchers are employed by the University of Karlsruhe

They can be written in DL syntax as follows – the axioms actually constitute an $\mathcal{ALCIO}$ TBox (see chapter "Description Logics").

$$\mathrm{Man} \sqcup \mathrm{Woman} \sqsubseteq \mathrm{Adult} \tag{14}$$

$$\mathrm{GrownUp} \sqsubseteq \mathrm{Human} \sqcap \mathrm{Adult} \tag{15}$$

$$\mathrm{Woman} \sqcap \exists \mathrm{childOf}^{-}.\top \sqsubseteq \mathrm{Mother} \tag{16}$$

$$\mathrm{Orphan} \sqsubseteq \forall \mathrm{childOf}.(\mathrm{Dead} \sqcap \mathrm{Human}) \tag{17}$$

$$\mathrm{LonelyChild} \sqsubseteq \neg \exists \mathrm{siblingOf}.\top \tag{18}$$

$$\mathrm{AIFBResearcher} \sqsubseteq \exists \mathrm{employedBy}.\{\mathrm{UKARL}\} \tag{19}$$

Using the forms of DLP statements which we introduced, the TBox can be written as follows.

$$\mathrm{Man} \sqsubseteq \mathrm{Adult} \qquad (1)$$

$$\mathrm{Woman} \sqsubseteq \mathrm{Adult} \qquad (1)$$

$$\mathrm{GrownUp} \sqsubseteq \mathrm{Human} \qquad (2)$$

$$\mathrm{GrownUp} \sqsubseteq \mathrm{Adult} \qquad (2)$$

$$\mathrm{Woman} \sqcap \exists \mathrm{childOf}^{-}.\top \sqsubseteq \mathrm{Mother} \qquad (3)$$

$$\mathrm{Orphan} \sqsubseteq \forall \mathrm{childOf}.\mathrm{Dead} \qquad (4)$$

$$\mathrm{Orphan} \sqsubseteq \forall \mathrm{childOf}.\mathrm{Human} \qquad (4)$$

$$\mathrm{LonelyChild} \sqsubseteq \forall \mathrm{siblingOf}.\bot \qquad (5)$$

$$\mathrm{AIFBResearcher} \sqsubseteq \exists \mathrm{employedBy}.\{\mathrm{UKARL}\} \qquad (6)$$

We note that for (5) we require DLP IC, while for (6) we require DLP$^+$. As an example for an RBox, we use the following.

$\text{parentOf} \equiv \text{childOf}^-$	parentOf and childOf are inverse roles.
$\text{parentOf} \sqsubseteq \text{ancestorOf}$	parentOf is a subrole of ancestorOf.
$\text{fatherOf} \sqsubseteq \text{parentOf}$	fatherOf is a subrole of parentOf.
$\top \sqsubseteq \forall \text{ancestorOf.Human}$	Human is the domain of ancestorOf.
$\top \sqsubseteq \leq 1 \text{fatherOf}^-$	fatherOf is inverse functional.

We can populate the classes and roles by means of an ABox, e.g. in the following way.

$$\begin{aligned} \{\text{Ian}, \text{Benjamin}, \text{Raphael}, \text{Horrocks}\} &\sqsubseteq \text{Man} \\ \{\text{Yue}, \text{Ulrike}\} &\sqsubseteq \text{Woman} \\ \text{Ian} &= \text{Horrocks} \\ < \text{Ian}, \text{UMAN} > &: \text{employedBy} \ldots \end{aligned}$$

Note that an ABox statement such as

$$\{\text{Yue}, \text{Ulrike}\} \sqsubseteq \text{Woman}$$

is simply syntactic sugar for the two statements

$$\text{Yue} : \text{Woman} \qquad \text{Ulrike} : \text{Woman}.$$

We therefore consider it to be part of the ABox. To be precise, the original statement actually is (syntactically) not in OWL Lite, but the equivalent set of three ABox statements is. The statement Ian = Horrocks requires DLP ICE.

Note also that class inclusions cannot in general be replaced by equivalences. For example, the statement

$$\text{Adult} \sqsubseteq \text{Man} \sqcup \text{Woman}$$

is not in DLP.

Relation to Logic Programming

A DLP knowledge base can be translated into Horn logic, and the latter can be expressed using Logic Programming syntax. To be more precise, DLP translates syntactically into Datalog. Let us first continue our example, giving the knowledge base in Datalog form.

The TBox is as follows.

$$
\begin{aligned}
\text{Adult}(y) &\leftarrow \text{Man}(y) && (1)\\
\text{Adult}(y) &\leftarrow \text{Woman}(y) && (1)\\
\text{Human}(y) &\leftarrow \text{GrownUp}(y) && (2)\\
\text{Adult}(y) &\leftarrow \text{GrownUp}(y) && (2)\\
\text{Mother}(y) &\leftarrow \text{childOf}(x,y) \wedge \text{Woman}(y) && (3)\\
\text{Dead}(y) &\leftarrow \text{Orphan}(x) \wedge \text{childOf}(x,y) && (4)\\
\text{Human}(y) &\leftarrow \text{Orphan}(x) \wedge \text{childOf}(x,y) && (4)\\
&\leftarrow \text{LonelyChild}(x) \wedge \text{siblingOf}(x,y) && (5)\\
y = \text{UKARL} &\leftarrow \text{AIFBResearcher}(x) \wedge \text{employedBy}(x,y) && (6)
\end{aligned}
$$

Translating the RBox yields the following statements.

$$
\begin{aligned}
\text{parentOf}(x,y) &\leftarrow \text{childOf}(y,x)\\
\text{childOf}(x,y) &\leftarrow \text{childOf}(y,x)\\
\text{ancestorOf}(x,y) &\leftarrow \text{parentOf}(x,y)\\
\text{parentOf}(x,y) &\leftarrow \text{fatherOf}(x,y)\\
\text{Human}(y) &\leftarrow \text{ancestorOf}(x,y)\\
y = z &\leftarrow \text{fatherOf}(y,x) \wedge \text{fatherOf}(z,x)
\end{aligned}
$$

Translated as such, DLP knowledge bases can also be evaluated under Datalog semantics, which differs from the OWL semantics. The two semantics, however, coincide on the set of inferred instances of named classes, i.e. ABox reasoning in DLP can be done using Datalog semantics. The formal relationship between the two semantics is as follows.

Theorem 1. *Let K be a DLP knowledge base and let K' be the translation of K into Datalog syntax. Let C be a named class and a be a named individual. Then $K \models_{OWL} C(a)$ under the OWL semantics iff $K \models_{Datalog} C(a)$ under the Datalog semantics.*[15]

Very recently, in the wake of the already mentioned forthcoming revision of the OWL standard, which will be based on the $\mathcal{SROIQ}$ description logic, a

[15] Note that – since we have non-disjunctive rules – the only effect of integrity constraints is that they can render the knowledge base to be inconsistent. Equality also needs some explanations: The theorem assumes that we do not use the unique name assumption, and equality thus has the same meaning as it has under OWL DL, e.g. two different constants can be equal, meaning that they denote the same individual.

naive rule fragment of $\mathcal{SROIQ}$ has been identified which considerably extends DLP, but is not a fragment of $\mathcal{SHOIN}$(D) (i.e. of OWL DL). It is called $\mathcal{SROIQ}$ Rules, and the interested reader shall be pointed to [21] (see also [10,35]).

3.2 Horn-$\mathcal{SHIQ}$

Horn-$\mathcal{SHIQ}$ is another Horn fragment of OWL DL, which encompasses DLP. To be precise, Horn-$\mathcal{SHIQ}$ has a feature which lies outside OWL DL, namely the use of qualified number restrictions.[16] The corresponding fragment lying within OWL DL would be Horn-$\mathcal{SHIN}$, which is obtained from Horn-$\mathcal{SHIQ}$ by simply disallowing the use of qualified number restrictions.

Horn-$\mathcal{SHIQ}$ has the pleasing property that it is of tractable (i.e. polynomial) data complexity [15,28], while its combined complexity is ExpTime [20]. It thus provides striking balance between expressivity and ABox reasoning scalability. While Horn-$\mathcal{SHIQ}$ can be dealt with by any of the standard OWL reasoners, it is specifically supported by KAON2[17] [28] – see chapter "Resolution-Based Reasoning for Ontologies" – and by the new HermiT system[18] [33].

The original definition of Horn-$\mathcal{SHIQ}$, due to [15,28], remained implicit. It was defined as the fragment of $\mathcal{SHIQ}$ which, after transformation by the KAON2 algorithms, resulted in (non-disjunctive) Datalog. It is outside the scope of this chapter to detail the underlying KAON2 algorithms, and we refer the reader to chapter "Resolution-Based Reasoning for Ontologies" and to [28] for this. Instead, we give an alternative definition by means of a grammar.

Definition of Horn-$\mathcal{SHIQ}$

The following definition is taken from [19], where also a formal proof can be found that this definition coincides with the original notion.

We say that a $\mathcal{SHIQ}$ axiom $C \sqsubseteq D$ is *Horn* if the concept expression $\neg C \sqcup D$ has the form $\mathbf{C}_1^+$ as defined by the context-free grammar in Table 1. A $\mathcal{SHIQ}$ knowledge base with an extensionally reduced ABox is *in Horn-$\mathcal{SHIQ}$* if all of its TBox axioms are Horn.

It is easily seen by referring to the definition on page 120, that DLP IC is indeed contained in Horn-$\mathcal{SHIQ}$. Just note that the use of $\{o_1, \ldots, o_n\}$ on the `Left` is removed by extensionally reducing the ABox. Intuitively speaking, Horn-$\mathcal{SHIQ}$ adds the free use of role restrictions to DLP, as, e.g. existential restriction can be used freely.

---

[16] However, Horn-$\mathcal{SHIQ}$ lies within the proposed OWL 2 language, see http://www.w3.org/Submission/owl11-tractable/

[17] http://kaon2.semanticweb.org/

[18] http://www.cs.man.ac.uk/~bmotik/HermiT/

Table 1. A grammar for defining Horn-$\mathcal{SHIQ}$

$\mathbf{C}_1^+$	$\leftarrow$	$\top \mid \bot \mid \neg\mathbf{C}_1^- \mid \mathbf{C}_1^+ \sqcap \mathbf{C}_1^+ \mid \mathbf{C}_0^+ \sqcup \mathbf{C}_1^+ \mid \exists\mathbf{R}.\mathbf{C}_1^+ \mid \forall\mathbf{S}.\mathbf{C}_1^+ \mid \forall\mathbf{R}.\mathbf{C}_0^+ \mid$ $\mid \geq n\,\mathbf{R}.\mathbf{C}_1^+ \mid \leq 1\,\mathbf{R}.\mathbf{C}_0^- \mid \mathbf{A}$
$\mathbf{C}_1^-$	$\leftarrow$	$\top \mid \bot \mid \neg\mathbf{C}_1^+ \mid \mathbf{C}_0^- \sqcap \mathbf{C}_1^- \mid \mathbf{C}_1^- \sqcup \mathbf{C}_1^- \mid \exists\mathbf{S}.\mathbf{C}_1^- \mid \exists\mathbf{R}.\mathbf{C}_0^- \mid \forall\mathbf{R}.\mathbf{C}_1^- \mid$ $\mid \geq 2\,\mathbf{R}.\mathbf{C}_0^- \mid \leq n\,\mathbf{R}.\mathbf{C}_1^+ \mid \mathbf{A}$
$\mathbf{C}_0^+$	$\leftarrow$	$\top \mid \bot \mid \neg\mathbf{C}_0^- \mid \mathbf{C}_0^+ \sqcap \mathbf{C}_0^+ \mid \mathbf{C}_0^+ \sqcup \mathbf{C}_0^+ \mid \forall\mathbf{R}.\mathbf{C}_0^+$
$\mathbf{C}_0^-$	$\leftarrow$	$\top \mid \bot \mid \neg\mathbf{C}_0^+ \mid \mathbf{C}_0^- \sqcap \mathbf{C}_0^- \mid \mathbf{C}_0^- \sqcup \mathbf{C}_0^- \mid \exists\mathbf{R}.\mathbf{C}_0^- \mid \mathbf{A}$

A, **R**, and **S** denote the sets of all concept names, role names, and simple role names, respectively. The presentation is slightly simplified by exploiting associativity and commutativity of $\sqcap$ and $\sqcup$, and by omitting $\geq 1\,R.C$ if $\exists R.C$ is present. The grammar for Horn-$\mathcal{SHIN}$ is obtained by replacing the qualifying class by $\top$ in all number restrictions

An Example

As an example for a Horn-$\mathcal{SHIQ}$ knowledge base, consider the ontology in Table 2, which exemplifies the expressivity possible in Horn-$\mathcal{SHIQ}$. Note in particular the free use of role restrictions which is not possible in DLP.

Relation to Logic Programming

Horn-$\mathcal{SHIQ}$ can be translated into Datalog syntax, i.e. it can be understood as a rule fragment of $\mathcal{SHIQ}$. We do not have the space to detail the underlying KAON2 translation algorithms, and refer to chapter "Resolution-Based Reasoning for Ontologies" and [28] for this. It shall be noted, though, that this transformation is *not* an equivalence translation in the sense that the original ontology and its translation have the same models. The relationship is more intricate, as given in the following theorem from [28].

Theorem 2. *Let K be a Horn-$\mathcal{SHIQ}$ knowledge base and let $D(K)$ be the transformation of K into Datalog syntax resulting from the application of the KAON2 transformation algorithms. Then the following hold.*

1. *K is unsatisfiable if and only if $D(K)$ is unsatisfiable.*
2. *$K \models \alpha$ if and only if $D(K)$ entails α under Datalog semantics, where α is of the form $A(a)$ or $R(a, b)$, and A is an atomic concept.*
3. *$K \models C(a)$ for a non-atomic concept C if and only if, for Q a new atomic concept, $Q(a)$ is entailed by $D(K \cup \{C \sqsubseteq Q\})$ under the Datalog semantics.*

In order to exemplify the mentioned transformation, we give a translation of the ontology from Table 2 into a logic program which can be executed under Prolog with tabling.[19] This example is due to [36].

[19] Using, e.g. XSB-Prolog, http://xsb.sourceforge.net/

Table 2. An example ontology in Horn-$\mathcal{SHIQ}$

	TBox/RBox		
(1)	Parent	$\equiv$	$\exists$ hasChild.$\top$
(2)	Person	$\sqsubseteq$	$\exists$ childOf.Person
(3)	ManyChildren	$\sqsubseteq$	$\geq$2 hasChild.$\top$
(4)	NoSiblings	$\sqsubseteq$	Person $\sqcap\, \forall$ childOf.($\leq$1 hasChild.$\top$)
(5)	childOf	$\equiv$	hasChild^{-1}

ABox	
$<$ Elaine, Sir Lancelot $>$	: hasChild
Lancelot du Lac	: noSiblings
$<$ Lancelot du Lac, Elaine $>$	: childOf

So consider the ontology given in Table 2. The corresponding translation to logic programming is given in Table 3. Let us first consider the upper part, which shows the rules directly created in the translation. Some of the rules clearly represent (part of) some $\mathcal{SHIQ}$-axiom, as is the case for "person(X) $\leftarrow$ nosiblings(X)." and axiom (4). Other rules are obtained by more complicated reasoning steps, such as, e.g. "parent(X) $\leftarrow$ manychildren(X)." which is obtained from axioms (3) and (1). While such rules are still fairly self-explanatory, there are also a number of axioms that include predicates of the form $S_f(X, X_f)$ which do not appear in the original knowledge base. These predicates are introduced during the transformation process, more precisely during a step for eliminating Skolem-functions. Intuitively, $S_f(X, Y)$ holds if and only if $Y = f(X)$. However, the predicates S_f are only satisfied for a finite number of constants, since arbitrary application of functions is not needed and might even lead to undecidability. The exact number of additional function symbols may vary from case to case. Finally, two of the rules represent integrity constraints by means of the predicate inc, in the sense that inc evaluates to `true` if the integrity constraint is violated.

The rules in the lower part of Table 3 define various auxiliary predicates that are needed for the correctness of the translation. In order to restrict these definitions to a finite number of terms, we introduce a predicate O that specifies the individuals for which the program applies. In our case, these are just the individuals from the ABox. Using O, we define S_f as discussed above. Further, we introduce a predicate HU defining which terms are considered in the program, namely individuals from O and their immediate successors for each function symbol. The remaining rules yield a necessary equality theory, restricted to the terms in HU.

The resulting program now allows us to conclude several ABox statements. For example, we can derive that "parent(Elaine)" and that "Sir Lancelot $\approx$ Lancelot du Lac".

Table 3. Continuation of the example from Table 2: Its translation into Horn-logic, consisting of the translated rules (middle) and auxiliary axioms (bottom)

$$
\begin{aligned}
\mathsf{person}(X) &\leftarrow \mathsf{nosiblings}(X).\\
\mathsf{person}(X_{f3}) &\leftarrow \mathsf{person}(X), S_{f3}(X, X_{f3}).\\
\mathsf{parent}(X) &\leftarrow \mathsf{haschild}(X, Y).\\
\mathsf{parent}(X) &\leftarrow \mathsf{manychildren}(X).\\
\mathsf{haschild}(Y, X) &\leftarrow \mathsf{childof}(X, Y).\\
\mathsf{haschild}(X, X_{f1}) &\leftarrow \mathsf{manychildren}(X), S_{f1}(X, X_{f1}).\\
\mathsf{haschild}(X, X_{f2}) &\leftarrow \mathsf{parent}(X), S_{f2}(X, X_{f2}).\\
\mathsf{haschild}(X, X_{f0}) &\leftarrow \mathsf{manychildren}(X), S_{f0}(X, X_{f0}).\\
\mathsf{childof}(X, X_{f3}) &\leftarrow \mathsf{person}(X), S_{f3}(X, X_{f3}).\\
\mathsf{childof}(Y, X) &\leftarrow \mathsf{haschild}(X, Y).
\end{aligned}
$$

$$
\begin{aligned}
Y_1 \approx Y_2 &\leftarrow \mathsf{nosiblings}(X), \mathsf{childof}(X, Z), \mathsf{haschild}(Z, Y_1), \mathsf{haschild}(Z, Y_2).\\
\mathsf{inc} &\leftarrow \mathsf{manychildren}(X), \mathsf{nosiblings}(X_0), \mathsf{childof}(X_0, X).\\
\mathsf{inc} &\leftarrow X_{f1} \approx X_{f0}, \mathsf{manychildren}(X), S_{f1}(X, X_{f1}), S_{f0}(X, X_{f0}).
\end{aligned}
$$

$$
\begin{aligned}
S_f(X, f(X)) &\leftarrow \mathsf{O}(X).\\
\mathsf{HU}(X) &\leftarrow \mathsf{O}(X).\\
\mathsf{HU}(f(X)) &\leftarrow \mathsf{O}(X). \qquad (\text{for } f \in \{f_0, f_1, f_2, f_3\})\\
X \approx X &\leftarrow \mathsf{HU}(X).\\
X \approx Y &\leftarrow Y \approx X, \mathsf{HU}(X), \mathsf{HU}(Y).\\
X \approx Z &\leftarrow X \approx Y, Y \approx Z, \mathsf{HU}(X), \mathsf{HU}(Y), \mathsf{HU}(Z).\\
C(Y) &\leftarrow C(X), X \approx Y, \mathsf{HU}(X), \mathsf{HU}(Y).\\
&\quad (\text{for } C \in \{\mathsf{person}, \mathsf{parent}, \mathsf{manychildren}, \mathsf{nosiblings}\})\\
R(Y_1, Y_2) &\leftarrow R(X_1, X_2), X_1 \approx Y_1, X_2 \approx Y_2, \mathsf{HU}(X_1), \mathsf{HU}(X_2), \mathsf{HU}(Y_1), \mathsf{HU}(X_2).\\
&\quad (\text{for } R \in \{\mathsf{childof}, \mathsf{haschild}\})\\
&\mathsf{O}(\text{Elaine}). \quad \mathsf{O}(\text{Sir Lancelot}). \quad \mathsf{O}(\text{Lancelot du Lac}).
\end{aligned}
$$

4 Hybrid Approaches

In Sect. 2, we discussed the SWRL rule extension of OWL, which basically follows the design principles of OWL DL by adhering to the open world assumption and by being semantically based on first-order predicate logic. In Sect. 3, we discussed rule languages which are fragments of OWL DL, and thus inherit its semantics.

We have focussed on these perspectives for several reasons:

- They align with OWL DL semantically in a very natural way
- They are supported by some of the most prominent and powerful OWL reasoning engines
- They appear to be least disputed as to their importance for ontology modeling

At the same time, there is a multitude of proposals for *hybrid* systems which comprise both classical OWL reasoning and traditional rule-based approaches like logic programming in different variants. The quest for such hybrid solutions is currently still ongoing, but is not yet close to a conclusion. We briefly discuss the two approaches which we consider to be most mature at this stage.

The first of these is called Hybrid MKNF knowledge bases due to [29, 30], which roughly is an autoepistemic extension of OWL DL with DL-safe rules. The result is a seamless integration of open- and closed-world reasoning within a single framework which encompasses both OWL DL reasoning and prominent forms of non-monotonic reasoning. Recent investigations [17, 18] strive at establishing an implementable semantics for this approach.

The second approach is based on an integration of OWL DL reasoning with Answer Set Programming, which has been realised as the dlvhex system [8,9,37]. The integration is less strong as for Hybrid MKNF knowledge bases, and basically consists of two reasoning engines which interact in a bidirectional way when reasoning over knowledge bases.

Many additional approaches are currently being investigated and proposed, ranging from tightly integrated ones to loosely coupled systems. We list recent references as a starting point for the interested reader [4, 5, 7, 12, 23, 25–27, 34].

5 Conclusions

Our discussion of ontologies and rules is centered on two specific paradigms of "ontology" formalisms and rule formalisms: Description Logics and logic programming (though we have only lightly touched on issues with various forms of non-monotonicity and similar core features of logic programming systems). While arguably these are both very prominent, one might say dominant, formalisms in the ontology engineering communities today – and there is an enormous amount of work on their combination as we have seen – historically, this was not always the case. For example, if we consider expert systems of the 1970s and 1980s such as Mycin [3] we find that production rule languages were quite prominent as rule formalisms for ontologies, and for knowledge representation more generally. Today, that community generally represents itself as dealing with "business rules" almost exclusively.[20] This is reflected in the make up of the currently ongoing rules interchange format (RIF) working group[21] of the W3C.

Integrating *logic programming* and business rules, much less description logics, has proven to be challenging. This is rather surprising given the obvious

[20] See http://www.businessrulesgroup.org/brmanifesto.htm for a taste of business rule concerns.

[21] http://www.w3.org/2005/rules/wg

parallels between Prolog and OPS5 style rules and the natural thought that use of forward chaining inference methods – or more specifically variants of the Rete algorithm – is irrelevant to the semantics of the rules. It seems *very* natural to think that a (simple) relational database with some (simple) event-condition-action (ECA) rules is equivalent to that relational database with some Datalog rules. However, a database administrator may have reason to prefer that the ECA rule evaluation *modified the database*. In fact, that may have been a key feature of those rules: consider a situation where one wished to validate certain inputs but *only at input time*, that is, subsequent operations are licensed to violate the validation criteria. In such contexts, the notion of "assert" and "retract" have representational significance. If you add, in the action language, the ability to execute code fragments written in a programming language, it seems clear that there is a fundamental divergence. Unfortunately, unlike with Description Logics and logic programming, we do not currently have a well established common semantic framework (i.e. modal theory) ready to hand to aid us with integration.

Acknowledgement

The first named author acknowledges support by the Deutsche Forschungsgemeinschaft under the ReaSem project.

References

1. Patrick Blackburn, Johan F. A. K. van Benthem, and Frank Wolter. *Handbook of Modal Logic, volume 3* (*Studies in Logic and Practical Reasoning*). Elsevier Science, New York, 2006.
2. Ronald J. Brachman, Deborah L. McGuiness, Peter F. Patel-Schneider, and Lori A. Resnick. Living with CLASSIC: when and how to use a KL-ONE-like language. In John Sowa, editor, *Principles of Semantic Networks*. Morgan Kaufmann, San Mateo, US, 1990.
3. Bruce Buchanan and Edward Shortliffe. *Rule-Based Expert Systems: The MYCIN Experiments of the Stanford Heuristic Programming Project*. Addison-Wesley, Reading, MA, 1984.
4. Jos de Bruijn, Thomas Eiter, Axel Polleres, and Hans Tompits. Embedding non-ground logic programs into autoepistemic logic for knowledge-base combination. In *Proceedings of the Twentieth International Joint Conference on Artificial Intelligence (IJCAI-07)*, Hyderabad, India, January, 6–12 2007. AAAI, New York.
5. Jos de Bruijn, David Pearce, Axel Polleres, and Agustín Valverde. Quantified equilibrium logic and hybrid rules. In Marchiori et al., editors, *Web Reasoning and Rule Systems, First International Conference, RR 2007, Innsbruck, Austria, June 7–8, 2007, Proceedings*, volume 4524 of *Lecture Notes in Computer Science*. Springer, Berlin, 2007, pages 58–72.

6. Francesco M. Donini, Maurizio Lenzerini, Daniele Nardi, and Andrea Schaerf. AL-log: Integrating datalog and description logics. *Journal of Intelligent Information Systems*, 10(3):227–252, 1998.
7. Wlodzimierz Drabent and Jan Maluszynski. Well-founded semantics for hybrid rules. In Marchiori et al., editors, *Web Reasoning and Rule Systems, First International Conference, RR 2007, Innsbruck, Austria, June 7–8, 2007, Proceedings*, volume 4524 of *Lecture Notes in Computer Science*. Springer, Berlin, 2007, pages 1–15.
8. Thomas Eiter, Giovambattista Ianni, Roman Schindlauer, and Hans Tompits. dlvhex: A prover for semantic-web reasoning under the answer-set semantics. In *2006 IEEE / WIC / ACM International Conference on Web Intelligence (WI 2006), 18-22 December 2006, Hong Kong, China*. IEEE Computer Society, Los Alamitos, CA, 2006, pages 1073–1074.
9. Thomas Eiter, Giovambattista Ianni, Roman Schindlauer, and Hans Tompits. Effective integration of declarative rules with external evaluations for semantic-web reasoning. In York Sure and John Domingue, editors, *The Semantic Web: Research and Applications, 3rd European Semantic Web Conference, ESWC 2006, Budva, Montenegro, June 11–14, 2006, Proceedings*, volume 4011 of *Lecture Notes in Computer Science*. Springer, Berlin, 2006, pages 273–287.
10. Francis Gasse, Ulrike Sattler, and Volker Haarslev. Rewriting rules into SROIQ axioms. In *Proc. 21st Int. Workshop on Description Logics (DL-08), Dresden, Germany*, 2008.
11. Benjamin Grosof, Ian Horroks, Raphael Volz, and Stefan Decker. Description logic programs: Combining logic programs with description logics. In *Proc. of WWW 2003, Budapest, Hungary, May 2003*. ACM, New York, 2003, pages 48–57.
12. Stijn Heymans, Davy Van Nieuwenborgh, and Dirk Vermeir. Nonmonotonic ontological and rule-based reasoning with extended conceptual logic programs. In *Proceedings ESWC2005*, volume 3532 of *Lecture Notes in Computer Science*. Springer, Berlin, 2005, pages 392–407.
13. Ian Horrocks and Peter F. Patel-Schneider. A proposal for an OWL rules language. In *Proc. of the Thirteenth International World Wide Web Conference (WWW 2004)*. ACM, New York, 2004, pages 723–731.
14. Ian Horrocks, Ulrike Sattler, and Stephan Tobies. Practical reasoning for very expressive description logics. *Logic Journal of the IGPL*, 8(3):239–263, 2000.
15. Ullrich Hustadt, Boris Motik, and Ulrike Sattler. Data complexity of reasoning in very expressive description logics. In *Proceedings of the 19th International Joint Conference on Artificial Intelligence (IJCAI)*, 2005, pages 466–471.
16. Ullrich Hustadt, Renate A. Schmidt, and Lilia Georgieva. A survey of decidable first-order fragments and description logics. *Journal of Relational Methods in Computer Science*, 1:251–276, 2004. Invited overview paper.
17. Matthias Knorr, Jose Alferes, and Pascal Hitzler. Towards tractable local closed world reasoning for the semantic web. In J. Neves, M. F. Santos, and J. Machado, editors, *Progress in Artificial Intelligence, 13th Portuguese Conference on Aritficial Intelligence, EPIA 2007, Guimaraes, Portugal, December 3–7, 2007, Proceedings*, volume 4874 of *Lecture Notes in Computer Science*. Springer, Berlin, 2007, pages 3–14.
18. Matthias Knorr, Jose Julio Alferes, and Pascal Hitzler. A Coherent Well-founded model for Hybrid MKNF knowledge bases. In: Malik Ghallab, Constantine

D. Spyropoulos, Nikos Fakotakis, Nikos Avouris (eds.), *Proceedings of the 18th European Conference on Artificial Intelligence, ECAI2008,* Patras, Greece, July 2008. IOS Press, 2008, pp. 99–103.
19. Markus Krötzsch, Pascal Hitzler, Denny Vrandečić, and Michael Sintek. How to reason with OWL in a logic programming system. In *Proc. 2nd Int. Conf. on Rules and Rule Markup Languages for the Semantic Web (RuleML 2006)*, Athens, Georgia, USA. Springer, Berlin, 2006.
20. Markus Krötzsch, Sebastian Rudolph, and Pascal Hitzler. Complexity of Horn description logics. In *Proceedings of the Twenty-Second Conference on Artificial Intelligence, AAAI-07, Vancouver, British Columbia, Canada, July 2007*, pages 452–457.
21. Markus Krötzsch, Sebastian Rudolph, and Pascal Hitzler. Description Logic Rules. In: Malik Ghallab, Constantine D. Spyropoulos, Nikos Fakotakis, Nikos Avouris (eds.), *Proceedings of the 18th European Conference on Artificial Intelligence, ECAI2008,* Patras, Greece, July 2008. IOS Press, 2008, pp. 80–84.
22. Alon Y. Levy and Marie-Christine Rousset. CARIN: A representation language combining Horn rules and description logics. In *European Conference on Artificial Intelligence*, 1996, pages 323–327.
23. Alon Y. Levy and Marie-Christine Rousset. Combining Horn rules and description logics in CARIN. *Artificial Intelligence*, 104:68–78, 1998.
24. John W. Lloyd. Foundations of logic programming (second, extended edition). In *Springer Series in Symbolic Computation.* Springer, New York, 1987.
25. Thomas Lukasiewicz. A novel combination of answer set programming with description logics for the semantic web. In Enrico Franconi, Michael Kifer, and Wolfgang May, editors, *The Semantic Web: Research and Applications, 4th European Semantic Web Conference, ESWC 2007, Innsbruck, Austria, June 3–7, 2007, Proceedings*, volume 4519 of *Lecture Notes in Computer Science.* Springer, Berlin, 2007, pages 384–398.
26. Tobias Matzner and Pascal Hitzler. Any-world access to OWL from Prolog. In J. Hertzberg, M. Beetz, and R. Englert, editors, *KI 2007: Advances in Artificial Intelligence, 30th Annual German Conference on AI, KI 2007, Osnabrck, Germany, September 2007, Proceedings*, volume 4667 of *Lecture Notes in Computer Science.* Springer, Berlin, 2007, pages 84–98.
27. Jing Mei, Zuoquan Lin, and Harold Boley. *LC*: An integration of description logic and general rules. In Marchiori et al., editors, *Web Reasoning and Rule Systems, First International Conference, RR 2007, Innsbruck, Austria, June 7–8, 2007, Proceedings*, volume 4524 of *Lecture Notes in Computer Science.* Springer, Berlin, 2007, pages 163–177.
28. Boris Motik. *Reasoning in Description Logics using Resolution and Deductive Databases.* PhD thesis, Universität Karlsruhe (TH), Germany, 2006.
29. Boris Motik, Ian Horrocks, Riccardo Rosati, and Ulrike Sattler. Can OWL and logic programming live together happily ever after? In Isabel F. Cruz, Stefan Decker, Dean Allemang, Chris Preist, Daniel Schwabe, Peter Mika, Michael Uschold, and Lora Aroyo, editors, *International Semantic Web Conference*, volume 4273 of *Lecture Notes in Computer Science.* Springer, Berlin, 2006, pages 501–514.
30. Boris Motik and Riccardo Rosati. A faithful integration of description logics with logic programming. In *Proceedings of the Twentieth International Joint Conference on Artificial Intelligence (IJCAI-07).* AAAI, New York, 2007, pages 477–482.

31. Boris Motik, Ulrike Sattler, and Rudi Studer. Query answering for OWL-DL with rules. In *International Semantic Web Conference*, 2004, pages 549–563.
32. Boris Motik, Ulrike Sattler, and Rudi Studer. Query answering for OWL-DL with rules. *Journal of Web Semantics*, 3(1):41–60, 2005.
33. Boris Motik, Rob Shearer, and Ian Horrocks. Optimized Reasoning in Description Logics using Hypertableaux. In *Proc. of the 21st Conference on Automated Deduction (CADE-21)*, volume 4603 of *Lecture Notes in Artificial Intelligence*, Bremen, Germany, July 17–20. Springer, Berlin, 2007, pages 67–83.
34. Riccardo Rosati. Tight integration of description logics and disjunctive datalog. In *Proceedings KR2006*. AAAI, New York, 2006, pages 68–78.
35. Sebastian Rudolph, Markus Krötzsch, and Pascal Hitzler. All elephants are bigger than all mice. In *Proc. 21st Int. Workshop on Description Logics (DL-08), Dresden, Germany*, 2008.
36. Sebastian Rudolph, Markus Krötzsch, Pascal Hitzler, Michael Sintek, and Denny Vrandečić. Efficient OWL reasoning with logic programs – Evaluations. In *Proceedings of the First International Conference on Web Reasoning and Rule Systems, RR2007, Innsbruck, Austria, June 2007*, volume 4524 of *Lecture Notes in Computer Science*. Springer, Berlin, 2007, pages 370–373.
37. Roman Schindlauer. *Answer-Set Programming for the Semantic Web*. PhD thesis, Vienna University of Technology, Austria, 2006.
38. Julian Seidenberg and Alan L. Rector. Representing transitive propagation in OWL. In David W. Embley, Antoni Olivé, and Sudha Ram, editors, *ER*, volume 4215 of *Lecture Notes in Computer Science*. Springer, Berlin, 2006, pages 255–266.
39. York Sure, Stephan Bloehdorn, Peter Haase, Jens Hartmann, and Daniel Oberle. The SWRC ontology – Semantic Web for Research Communities. In Carlos Bento, Amilcar Cardoso, and Gael Dias, editors, *Proceedings of the 12th Portuguese Conference on Artificial Intelligence – Progress in Artificial Intelligence (EPIA 2005)*, volume 3803 of *LNCS*, Covilha, Portugal. Springer, Berlin, 2005, pages 218–231.
40. Raphael Volz. *Web Ontology Reasoning With Logic Databases*. PhD thesis, Universität Fridericiana zu Karlsruhe (TH), Germany, 2004.

Part II

Ontology Engineering

Ontology Engineering Methodology

York Sure[1], Steffen Staab[2], and Rudi Studer[3]

[1] SAP Research, CEC Karlsruhe, Karlsruhe, Germany, york.sure@sap.com
[2] Research Group ISWeb, University of Koblenz-Landau, Koblenz, Germany, staab@uni-koblenz.de
[3] Institute AIFB, University of Karlsruhe (TH) and FZI Research Center for Information Technology, Karlsruhe, Germany, studer@aifb.uni-karlsruhe.de

Summary. In this chapter we present a methodology for introducing and maintaining ontology based knowledge management applications into enterprises with a focus on Knowledge Processes and Knowledge Meta Processes. While the former process circles around the usage of ontologies, the latter process guides their initial set up. We illustrate our methodology by an example from a case study on skills management. The methodology serves as a scaffold for Part B "Ontology Engineering" of the handbook. It shows where more specific concerns of ontology engineering find their place and how they are related in the overall process.

1 Introduction

Ontologies constitute valuable assets that are slowly, but continuously gaining recognition and use throughout a set of disciplines – as becomes visible in Part C of this book. Ontologies frequently being a complex asset, their creation and management does neither come by coincidence nor does it come for free. Rather, the objectives pursued with their development as well as the development itself must be critically assessed by the organization or – rarely – the individual who is pushing for their creation and maintenance. The discipline that investigates the principles, methods and tools for initiating, developing and maintaining ontologies is "ontology engineering" which is the topic of this part of the handbook. "Ontology engineering methodology" as a part of ontology engineering deals with the process and methodological aspects of ontology engineering, i.e. with the issues of how to provide guidelines and advice to (potential) developers of ontologies.

It is the purpose of this chapter to introduce a rather generic ontology engineering methodology to the reader and to indicate where this methodology links to more specific topics discussed mostly, but obviously not completely, in the remainder of part B of this handbook. Such as software engineering methodologies cannot be described in isolation from actual software

S. Staab and R. Studer (eds.), *Handbook on Ontologies*, International Handbooks on Information Systems, DOI 10.1007/978-3-540-92673-3,

engineering activities, the purpose of ontology engineering methodologies can only be understood in the context of actual ontology engineering experiences.

The methodology presented here, has been derived from several case studies of building and using ontologies in the realm of knowledge management. Knowledge management deals with the thorough and systematic management of knowledge within an enterprise and between several cooperating enterprises. Knowledge management is a major issue for human resource management, enterprise organization and enterprise culture – nevertheless, information technology (IT) constitutes a crucial enabler for many aspects of knowledge management and ontologies frequently turn out to be valuable assets for knowledge management in order to target core knowledge management issues such as search, information integration, or mapping of knowledge assets. As a consequence, knowledge management is an inherently interdisciplinary subject and ontologies used for knowledge management play a central role, but at the same time they are by no means the single factor to determine success or failure of the overall system. Thus, we may derive our rationale that the objective of knowledge management constitutes a typical, yet comprehensive blueprint for issues that arise when developing complex ontologies. Therefore, we have chosen the knowledge management setting described below in order to report on a generic ontology engineering methodology.

IT-supported KM solutions are frequently built around some kind of organizational memory [1] that integrates informal, semi-formal and formal knowledge in order to facilitate its access, sharing and reuse by members of the organization(s) for solving their individual or collective tasks [7]. In such a context, knowledge has to be modelled, appropriately structured and interlinked for supporting its flexible integration and its personalized presentation to the consumer. Ontologies may provide such structuring and modeling of problems by providing a formal conceptualization of a particular domain that is shared by a group of people in an organization [14,22].

There exist various proposals for methodologies that support the systematic introduction of KM solutions into enterprises and with it the construction of ontologies. A classical approach for introducing knowledge management systems – including ontologies – is CommonKADS that puts emphasis on an early feasibility study as well as on constructing several models that capture different kinds of knowledge needed for realizing a KM solution [26].

Re-engineering earlier approaches, we found that methodologies must distinguish two processes in order to achieve a clear identification of issues [27]: whereas the first process addresses aspects of introducing a new ontology-based system into an organization as well as maintaining it (the so-called "Knowledge Meta Process"), the second process addresses the management of knowledge using the developed ontology (or ontologies), i.e. the so-called "Knowledge Process" (see Fig. 1). E.g. in the approach described in [25], one may recognize the intermingling of the two aspects from the different roles that, e.g. "knowledge identification" and "knowledge creation" play.

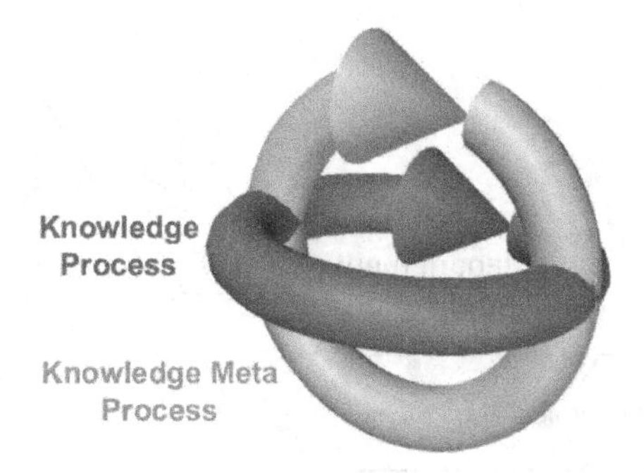

Fig. 1. Two orthogonal processes with feedback loops

The Knowledge Meta Process would certainly have its focus on knowledge identification and the Knowledge Process would rather stress knowledge creation.

The generic methodology presented here has been developed and applied in the EU project On-To-Knowledge[1] [6]. We now describe some general issues when implementing and launching ontology-based knowledge management applications. Then we focus on the knowledge meta process and the knowledge process and illustrate the instantiation of the knowledge meta process by an example from a skills management case study of the On-To-Knowledge project. During the description of the process we will point to more specific topics of ontology engineering dealt with in further chapters of this handbook.

2 Implementation and Launch of KM Applications

To implement and launch a KM application, one has to consider different processes (cf. Fig. 2). We have dealt with three major processes occurring in our case study, i.e. "Knowledge Meta Process", "Human Issues" and "Software Engineering". The processes are not completely separate but they do also overlap and interfere. As mentioned before, KM is an inherently interdisciplinary subject which should not be dominated by information technology (IT) alone, but which needs to take human and organizational issues into account. Hence, the targeted solution must trade off between problems to be solved by automated IT solutions and problems to be taken care of by human actors and through organizational processes. As a rule of thumb KM experts at a "Dagstuhl Seminar on Knowledge Management"[2] (cf. [23]) estimated that IT support cannot cover more than 10–30% of KM concerns.

Human issues (HI) and the related cultural environment of organizations heavily influence the acceptance of KM. It is often mentioned in discussions that the success of KM – and especially KM applications – strongly depends

[1] http://www.ontoknowledge.org
[2] http://dagstuhl-km-2000.aifb.uni-karlsruhe.de/

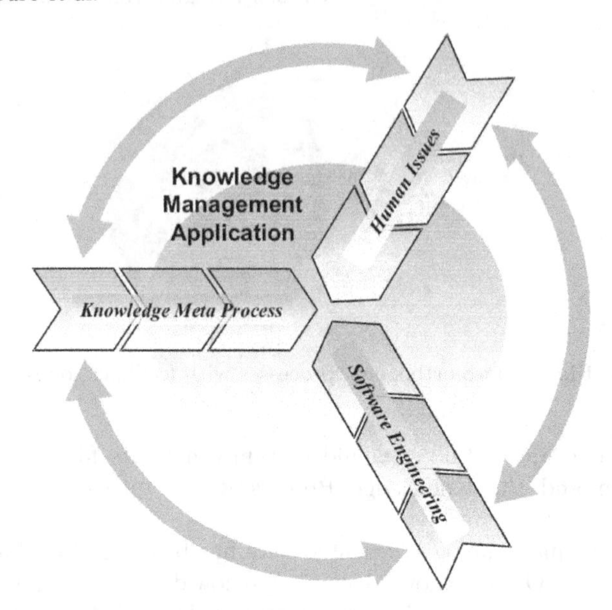

Fig. 2. Relevant processes for developing and deploying KM applications

on the acceptance by the involved people. As a consequence, "quick wins" are recommended for the initial phase of implementing any KM strategy. The aim is to quickly convince people that KM is useful for them and adds value to their daily work.

Software engineering (SE) for knowledge management applications has to accompany the other processes. The software requirements coming from the knowledge processes need to be reflected in the planning and management of the overall system design and implementation.

In the following sections we will now focus on the Knowledge Meta Process as the core process of ontology engineering and we will mention some cross-links to the other processes as well as to more specific ontology engineering issues.

3 Knowledge Meta Process

The Knowledge Meta Process (cf. Fig. 3) consists of five main steps. Each step has numerous sub-steps, requires a main decision to be taken at the end and results in a specific outcome. The main stream indicates steps (phases) that finally lead to an ontology-based KM application. The phases are "Feasibility Study", "Kickoff", "Refinement", "Evaluation" and "Application and Evolution". Below every box depicting a phase the most important sub-steps are listed, e.g. "Refinement" consists of the sub-steps "Refine semi-formal ontology description", "Formalize into target ontology" and "Create prototype", etc. Each document-flag above a phase indicates major outcomes of the step, e.g. "Kickoff" results in an "Ontology Requirements Specification Document

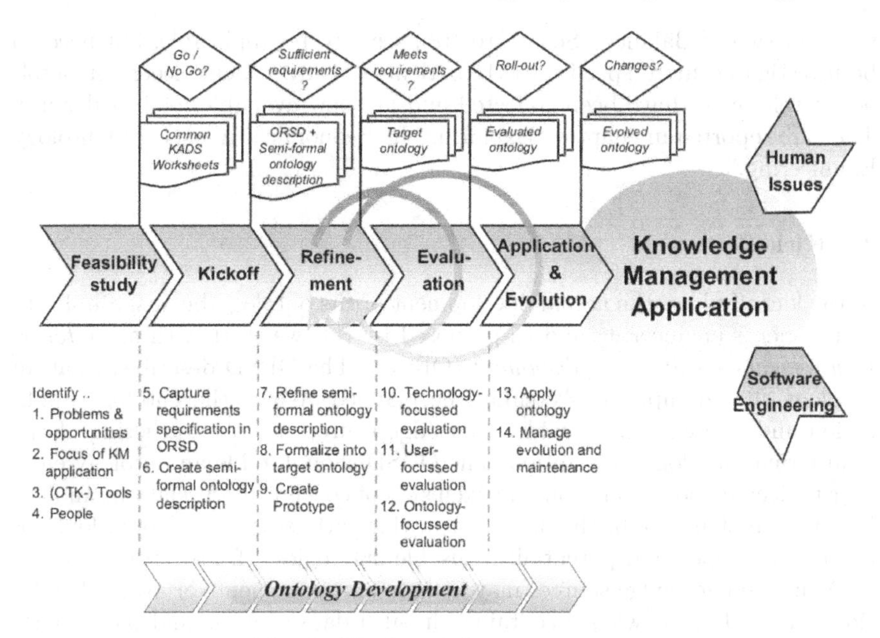

Fig. 3. The knowledge meta process

(ORSD)" and the "Semi-formal ontology description", etc. Each node above a flag represents the major decisions that have to be taken at the end to proceed to the next phase, e.g. whether in the Kickoff phase one has captured sufficient requirements. The major outcomes typically serve as decision support for the decisions to be taken. The phases "Refinement–Evaluation–Application and Evolution" typically need to be performed in iterative cycles. One might notice that the development of such an application is also driven by other processes, e.g. software engineering and human issues. We will only briefly mention some human issues in the example section.

3.1 Feasibility Study

Any knowledge management system may function properly only if it is seamlessly integrated in the organization in which it is operational. Many factors other than technology determine success or failure of such a system. To analyze these factors, we initially start with a *feasibility study* [26], e.g. to identify problem/opportunity areas and potential solutions. In general, a feasibility study serves as a decision support for economical, technical and project feasibility, determining the most promising focus area and target solution.

Considering ontology engineering specifically, there is a need to consider the return on investment of developing ontologies as an asset. So far, the accounting of ontology as value assets has not been undertaken to our knowledge. Methods of accounting other intangible assets, such as [8] which builds on the approach of Balanced Scorecard [18], seem to be applicable, but need to be investigated more specifically. Experiences of investment needs for ontology development have been collected and are now available for broader use. They are reported in chapter "Exploring the Economical Aspects of Ontology Engineering."

3.2 Kickoff

In the kickoff phase the actual development of the ontology begins. Similar to requirements engineering and as proposed by [11] we start with an *ontology requirements specification document* (ORSD). The ORSD describes what an ontology should support, sketching the planned area of the ontology application and listing, e.g. valuable knowledge sources for the gathering of the semi-formal ontology description. The ORSD should guide an ontology engineer to decide about inclusion and exclusion of concepts and relations and the hierarchical structure of the ontology. In this early stage one should look for already developed and potentially reusable ontologies (cf. [31] on reuse).

Valuable knowledge sources may include text documents or available relational data. The knowledge contained in such data sources, and particularly in text, may be unlocked by means of ontology learning methods (cf. chapter "Ontology and the Lexicon"). A specific techniques, which is sometimes used for ontology learning, is the analysis of concept properties allowing for the derivation of hierarchical relationships by means of formal concept analysis (cf. chapter "Formal Concept Analysis").

The *outcome* of this phase is (beside the ontology requirement specification document (ORSD)) a semi-formal description of the ontology, i.e. a graph of named nodes and (un-)named, (un-)directed edges, both of which may be linked with further descriptive text, e.g. in form of mind maps (cf. [4, 32]). If the requirements are sufficiently captured, one may proceed with the next phase. The *decision* is typically taken by ontology engineers in collaboration with domain experts. "Sufficiently" in this context means, that from the current perspective there is no need to proceed with capturing or analyzing knowledge. However, it might be the case that in later stages gaps are recognized. Therefore, the ontology development process is cyclic.

3.3 Refinement

During the kick-off and refinement phase one might distinguish in general two concurrent approaches for modeling, in particular for refining the semi-formal ontology description by considering relevant knowledge sources: top–down and bottom–up. In a *top–down*-approach for modeling the domain one starts by modeling concepts and relationships on a very generic level. Subsequently these items are refined. This approach is typically done manually and leads to a high-quality engineered ontology. Available top-level ontologies (cf. chapter

"Foundational Choices in DOLCE") may here be reused and serve as a starting point to develop new ontologies. In our example scenario we encountered a *middle-out* approach, i.e. to identify the most important concepts which will then be used to obtain the remainder of the hierarchy by generalization and specialization. However, with the support of an automatic document analysis (cf. chapter "Ontology and the Lexicon"), a typical *bottom–up*-approach may be applied. There, relevant concepts are extracted semi-automatically from available documents. Based on the assumption that most concepts and conceptual structures of the domain as well the company terminology are described in documents, applying knowledge acquisition from text for ontology design helps building ontologies automatically.

To *formalize* the initial semi-formal description of the ontology into the target ontology, ontology engineers firstly form a taxonomy out of the semi-formal description of the ontology and add relations other than the "is-a" relation which forms the taxonomical structure. The ontology engineer adds different types of relations as analyzed, e.g. in the competency questions to the taxonomic hierarchy. However, this step is cyclic in itself, meaning that the ontology engineer now may start to interview domain experts again and use the already formalized ontology as a base for discussions. It might be helpful to visualize the taxonomic hierarchy and give the domain experts the task to add attributes to concepts and to draw relations between concepts (e.g. we presented them the taxonomy in form of a mind map as mentioned in the previous section). The ontology engineer should extensively document the additions and remarks to make ontological commitments made during the design explicit. The application of design patterns for ontologies (cf. chapter "Ontology Design Patterns") may greatly improve the efficiency and effectiveness of the process as well as the quality of the ontology.

The *outcome* of this phase is the "target ontology", that needs to be evaluated in the next step. The major *decision* that needs to be taken to finalize this step is whether the target ontology fulfills the requirements captured in the previous kickoff phase. Typically an ontology engineer compares the initial requirements with the current status of the ontology. This decision will typically be based on the personal experience of ontology engineers. As a good rule of thumb we discovered that the first ontology should provide enough "flesh" to build a prototypical application. This application should be able to serve as a first prototype system for evaluation.

3.4 Evaluation

We distinguish between three different types of evaluation: (1) technology-focussed evaluation, (2) user-focussed evaluation and (3) ontology-focused evaluation.

Our evaluation framework for *technology-focussed evaluation* consists of two main aspects: (1) the evaluation of properties of ontologies generated by development tools, (2) the evaluation of the technology properties, i.e. tools and applications which includes the evaluation of the evaluation tool properties themselves. In an overview these aspects are structured as follows: (1) Ontology properties (e.g. language conformity (Syntax), consistency (Semantics)) and (2) technology properties (e.g. interoperability, turn around ability, scalability etc.).

The framework shown above concentrates on the technical aspects of ontologies and related ontologies. However, the aspect of *user-focussed evaluation* remains open. The most important point from our perspective is to evaluate whether users are satisfied by the KM application. More specific, whether an ontology based application is at least as good as already existing applications that solve similar tasks.

Beside the above mentioned process oriented and pragmatic evaluation methods, one also need to *formally evaluate ontologies*. One of the most prominent approaches here is the OntoClean approach (cf. chapter "An Overview of OntoClean"), which is based on philosophical notions. Another well-known approach (cf. chapter "Ontology Engineering Environments") takes into account the normalization of an ontology. Applying such approaches helps avoiding common modelling errors and leads to more correct ontologies.

The *outcome* of this phase is an evaluated ontology, ready for the roll-out into a productive system. However, based on our own experiences we expect in most cases several iterations of "Evaluation–Refinement–Evaluation" until the outcome supports the decision to roll-out the application. The major *decision* that needs to be taken for finalizing this phase is whether the evaluated ontology fulfills all evaluation criteria relevant for the envisaged application of the ontology.

3.5 Application and Evolution

The *application* of ontologies in productive systems, or, more specifically, the usage of ontology based systems, is being described in the following Sect. 4 that illustrates the knowledge process.

The *evolution* of ontologies is primarily an organizational process. There have to be rules to the update, insert and delete processes of ontologies (cf. [29]). We recommend, that ontology engineers gather changes to the ontology and initiate the switch-over to a new version of the ontology after thoroughly testing all possible effects to the application. Most important is therefore to clarify *who* is responsible for maintenance and *how* it is performed and in *which time intervals* is the ontology maintained. However, there also exist technical approaches for the consistent evolution of ontologies (cf. [16,17,30]).

A current topic for research and practice is the use of evolutionary knowledge management technologies that frequently build on Web2.0 technology and that decentralize the responsibility of knowledge management processes and meta processes to the individuals in the (virtual) organization with a corresponding need to decentralize ontology engineering (cf. [3, 28] on decentralized and evolutionary knowledge management and chapter "Ontology

Engineering and Evolution in a Distributed World Using DILIGENT" on decentralized, evolutionary ontology engineering).

The *outcome* of an evolution cycle is an evolved ontology, i.e. typically another version of it. The major *decision* to be taken is when to initiate another evolution cycle for the ontology.

4 Knowledge Process

Once a KM application is fully implemented in an organization, knowledge processes essentially circle around the following steps (cf. Fig. 4):

- *Knowledge creation* and/or *import* of documents and meta data, i.e. contents need to be created or converted such that they fit the conventions of the company, e.g. to the knowledge management infrastructure of the organization.
- then knowledge items have to be *captured* in order to elucidate importance or interlinkage, e.g. the linkage to conventionalized vocabulary of the company by the creation of relational metadata.
- *retrieval of* and *access to knowledge* satisfies the "simple" requests for knowledge by the knowledge worker;
- typically, however, the knowledge worker will not only recall knowledge items, but she will process it for further *use* in her context.

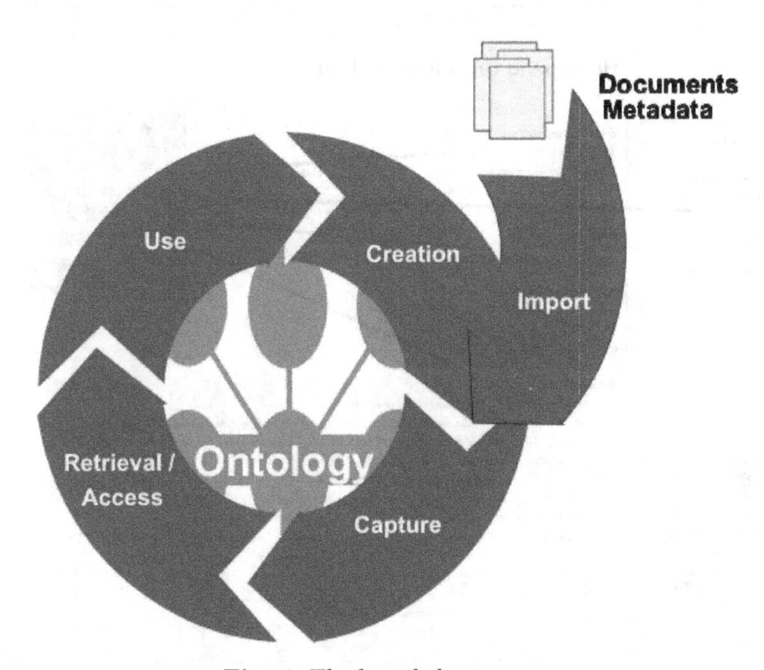

Fig. 4. The knowledge process

5 Example: Skills Management at Swiss Life

We now give an example of the Knowledge Meta Process instantiation of a skills management case study at Swiss Life (cf. [19]). Skills management makes skills of employees explicit. Within the case study existing skill databases and documents (like, e.g. personal homepages) are integrated and expanded. Two aspects are covered by the case study: first, explicit skills allow for an advanced expert search within the intranet. Second, one might explore his/her future career path by matching current skill profiles vs. job profiles. To ensure that all integrated knowledge sources are used in the same way, ontologies are used as a common mean of interchange to face two major challenges. Firstly, being an international company located in Switzerland, Swiss Life has internally four official languages, viz. German, English, French and Italian. Secondly, there exist several spellings of same concepts, e.g. "WinWord" vs. "MS Word". To tackle these problems, ontologies offer external representations for different languages and allow for representation of synonymity. Figure 5 shows a screenshot from the skills management application. The prototype enables any employee to integrate personal data from numerous distributed and heterogeneous sources into a single coherent personal homepage.

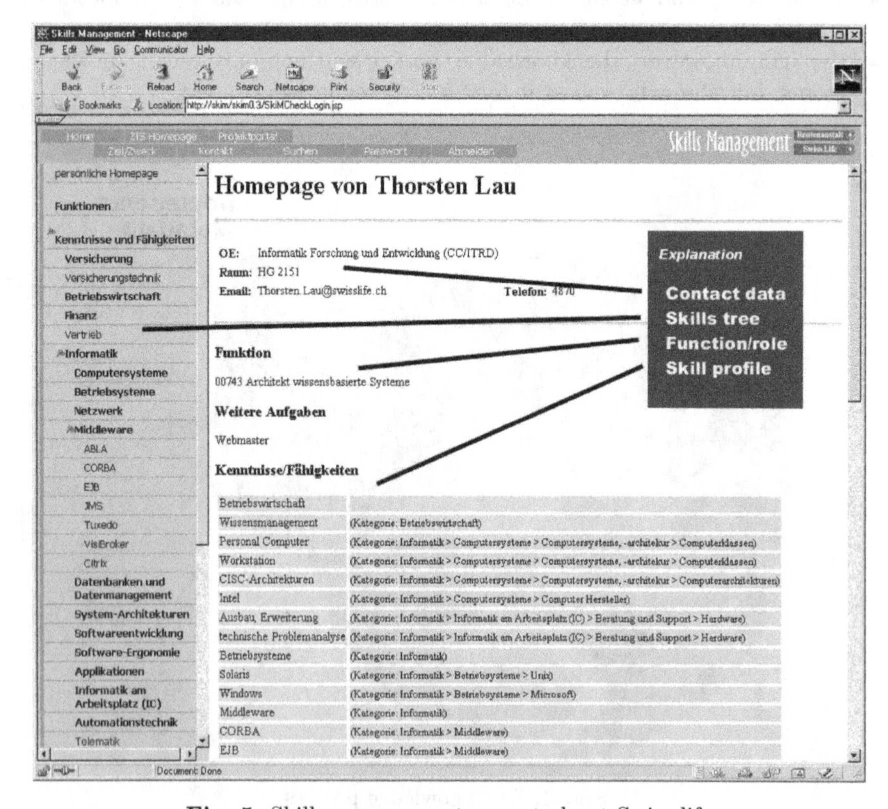

Fig. 5. Skills management case study at Swiss life

5.1 Feasibility Study

For identifying factors which can be central for the success or failure of the ontology development and usage we made a requirement analysis of the existing skills management environment and evaluated the needs for a new skills management system. We identified mainly the human resources department and the management level of all other departments as actors and stakeholders for the skills management. After finding the actors and stakeholders in the skills management area, we named the ontology experts for each department, which are preferably from the associated training group of each department.

5.2 Kickoff

The departments private insurance, human resources and IT constitute three different domains that were the starting point for an initial prototype. The task was to develop a skills ontology for the departments containing three trees, viz. for each department one. The three trees should be combined under one root with cross-links in between. The root node is the abstract concept "skills" (which means in German "Kenntnisse/Faehigkeiten") and is the starting point to navigate through the skills tree from the top.

During the *kickoff* phase two workshops with three domain experts[3] were held. The first one introduced the domain experts to the ideas of ontologies. Additional potential knowledge sources were identified by the domain experts, that were exhaustively used for the development of the ontologies, e.g. a book of the Swiss Association of Data Processing ("Schweizerischer Verband fuer Datenverarbeitung") describing professions in the computing area in a systematic way similar to an ontology. Obviously, this was an excellent basis to manually build the skills ontology for the IT domain. First experiments with extracting an ontology semi-automatically by using information extraction tools did not satisfy the needs for a clearly structured and easily understandable model of the skills. The domain experts and potential users felt very uncomfortable with the extracted structures and rather chose to build the ontology by themselves "manually". To develop the first versions of the ontologies, we used a mind mapping tool ("MindManager"). It is typically used for brainstorming sessions and provides simple facilities for modelling hierarchies very quickly. The early modelling stages for ontologies contain elements from such brainstorming sessions (e.g. the gathering of the semi-formal ontology description).

During this stage a lot of "concept islands" were developed, which were isolated sets of related terms. These islands are subdomains of the corresponding domain and are self-contained parts like "operating systems" as sub domain in the IT domain. After developing these concept islands it was necessary to

[3] Thanks to Urs Gisler, Valentin Schoeb and Patrick Shann from Swiss Life for their efforts during the ontology modelling.

combine them into a single tree. This was a more difficult part than assembling the islands, because the islands were interlaced and for some islands it was possible to add them to more than one other island, which implies awkward skills trees that contain inconsistencies after merging. For each department one skills tree was built in separate workshops. A problem that came up very early was the question where to draw the line between concepts and instances. E.g. is the programming language Java instantiated by "jdk1.3" or is "jdk1.3" so generic that it still belongs to the concept-hierarchy? Another problem was the size of the ontology. What is the best depth and width of each skills tree? Our solution was, that it depends on the domain and should be determined by the domain expert.

As *result* of the kick-off phase we obtained the semi-formal ontology descriptions for the three skills trees, which were ready to be formalized and integrated into a single skills ontology. At this stage the skills trees reached a maturity that the combination of them caused no major changes for the single skills trees.

5.3 Refinement

During the *refinement* phase we formalized and integrated the semi-formal ontology descriptions into a single coherent skills ontology. An important aspect during the formalization was (1) to give the skills proper names that uniquely identify each skill and (2) to decide on the hierarchical structure of the skills. We discussed two different approaches for the hierarchical ordering: we discovered that categorization of skills is typically not based on an `is-a`-taxonomy, but on a much weaker HASSUBTOPIC relationship that has implications for the inheritance of attached relations and attributes. However, for our first prototype this distinction made no difference due to missing cross-taxonomical relationships. But, according to [15], subsumption provided by `is-a` taxonomies is often misused and a later formal evaluation of the skills ontology according to the proposed OntoClean methodology possibly would have resulted in a change of the ontology.

In a second refinement cycle we added one more relation type, an "associative relation" between concepts. They express relations outside the hierarchic skills tree, e.g. a relation between "HTML" and "JSP", which occur not in the same tree, but correspond with each other, because they are based on the same content. "HTML" is in the tree "mark-up languages", while the tree "scripting languages" contains "JSP". This is based on the basic characteristics and the history of both concepts, which changed over time. But in reality they have a close relationship, which can be expressed with the associative relation.

The other task in this phase was to integrate the three skills ontologies into one skills ontology and eliminate inconsistencies in the domain ontology parts and between them. Because the domain ontologies were developed separately, the merger of them caused some overlaps, which had to be resolved. This happened for example in the computer science part of the skills trees, where the departments IT and private insurance have the same concepts like "Trofit" (which is a Swiss Life specific application). Both departments use this concept, but each uses a different view. The IT from the development and the private insurance from the users view. Additionally the personal skills of any employee are graded according to a generic scale of four levels: basic knowledge, practical experience, competency, and top specialist. The employees will grade their own skills themselves. As known from personal contacts to other companies (e.g. Credit Suisse, ABB and IBM), such an approach proved to produce highly reliable information.

As a *result* at the end of the refinement phase the "target skills ontology" consisted of about 700 concepts, which could be used by the employees to express their skill profile.

5.4 Application and Evolution

The *evaluation* of the prototype and the underlying ontology was unfortunately skipped due to internal restructuring at Swiss Life which led to a closing down of the whole case study.

Still, we considered the following aspects for the *evolution* of our skills management application: The competencies needed from employees are a moving target. Therefore the ontologies need to be constantly evaluated and maintained by experts from the human resource department. New skills might be suggested by the experts themselves, but mainly by employees. Suggestions include both, the new skill itself as well as the position in the skills tree where it should be placed. While employees are suggesting only new skills, the experts decide which skills should change in name and/or position in the skills tree and, additionally, decide which skill will be deleted. This was seen as necessary to keep the ontology consistent and to avoid that, e.g. similar if not the same concept appear even in the same branch. For each ontology (and domain) there should exist a designated ontology manager who decides if and how the suggested skill is integrated.

6 Related Work on Methodologies

A first overview on methodologies for ontology engineering can be found in [9]. Within OntoWeb[4] there have been joint efforts of members, who produced an extensive state-of-the-art overview of methodologies for ontology engineering (cf. [10,13]). There exist also deliverables on guidelines and best practices for industry (cf. [20,21]) with a focus on applications for E-Commerce, Information Retrieval, Portals and Web Communities.

---

[4] OntoWeb, a European thematic network, see http://www.ontoweb.org for further information.

CommonKADS [26] is not *per se* a methodology for ontology development. It covers aspects from corporate knowledge management, through knowledge analysis and engineering, to the design and implementation of knowledge-intensive information systems. CommonKADS has a focus on the initial phases for developing knowledge management applications, we therefore relied on CommonKADS for the early feasibility stage. E.g. a number of worksheets is proposed that guide through the process of finding potential users and scenarios for successful implementation of knowledge management.

The *Enterprise Ontology* [37] [38] proposed three main steps to engineer ontologies: (1) to identify the purpose, (2) to capture the concepts and relationships between these concepts, and the terms used to refer to these concepts and relationships, and (3) to codify the ontology. In fact, the principles behind this methodology influenced many approaches in the ontology community. These principles are also reflected and appropriately extended in the steps kickoff and refinement of our methodology.

TOVE [36] proposes a formalized method for building ontologies based on competency questions. We found the approach of using competency questions, that describe the questions that an ontology should be able to answer, very helpful and integrated it in our methodology.

METHONTOLOGY [11, 12] is a methodology for building ontologies either from scratch, reusing other ontologies as they are, or by a process of re-engineering them. The framework enables the construction of ontologies at the "knowledge level". The framework consists of: identification of the ontology development process where the main activities are identified (evaluation, configuration, management, conceptualization, integration implementation, *etc.*); a lifecycle based on evolving prototypes; and the methodology itself, which specifies the steps to be taken to perform each activity, the techniques used, the products to be output and how they are to be evaluated. METHONTOLOGY is partially supported by WebODE. Our combination of the On-To-Knowledge Methodology and OntoEdit (cf. [32, 33]) is quite similar to the combinations of METHONTOLOGY and WebODE (cf. [2]. In fact, they are the only duet that has reached a comparable level of integration of tool and methodology.

More recently, the DILIGENT methodology has been developed that addresses the decentralized engineering of ontologies [24]. The development of DILIGENT is driven by the fact that in a lot of application scenarios a geographically dispersed group of ontology engineers, domain experts, and ontology users that are often distributed across different organizations, has to develop and maintain a shared ontology for knowledge management. DILIGENT puts special emphasis on supporting the argumentation process that is needed in agreeing on updates of a shared ontology [35]. Obviously, these techniques would be a valuable support for the refinement and evolution phases of our methodology. A detailed description of DILIGENT is given in chapter "Ontology Engineering and Evolution in a Distributed World Using DILIGENT."

Currently, the NeOn methodology for engineering networked ontologies is under development as part of the NeOn [5] project [31]. This methodology supports among others the reuse of ontologies as well as of non-ontological resources as part of the engineering process. The NeOn methodology also provides detailed guidelines for executing its various activities. This includes the usage of ontology design patterns as described in chapter "Ontology Design Patterns." Thus, the NeOn methodology would provide additional methods for the kickoff and refinement phases of our methodology.

7 Conclusion

The described methodology was developed and applied in the On-To-Knowledge project and influenced work, e.g. in the SEKT and the NEON projects. One of the core contributions of the methodology that could not be shown here is the linkage of available tool support with case studies by showing when and how to use tools during the process of developing and running ontology based applications in the case studies (cf. [34]).

Lessons learned during setting up and employing the methodology in the On-To-Knowledge case studies include: (1) different processes drive KM projects, but "Human Issues" might dominate other ones (as already outlined by Davenport [5]), (2) guidelines for domain experts in industrial contexts have to be pragmatic, (3) collaborative ontology engineering requires physical presence *and* advanced tool support and (4) brainstorming is very helpful for early stages of ontology engineering, especially for domain experts not familiar with modelling (more details on be found, e.g. in [32, 33]).

In this chapter we have shown a process oriented methodology for introducing and maintaining ontology based knowledge management systems. Core to the methodology are Knowledge Processes and Knowledge Meta Processes. While Knowledge Meta Processes support the setting up of an ontology based application, Knowledge Processes support its usage. Still, there are many open issues to solve, e.g. how to handle a distributed process of emerging and aligned ontologies that is likely to be the scenario in the semantic web.

References

1. A. Abecker, A. Bernardi, K. Hinkelmann, O. Kuehn, and M. Sintek. Toward a technology for organizational memories. *IEEE Intelligent Systems*, 13(3):40–48, 1998.
2. J. C. Arpírez, O. Corcho, M. Fernández-López, and A. Gómez-Pérez. WebODE: A scalable workbench for ontological engineering. In *Proceedings of the First International Conference on Knowledge Capture (K-CAP) Oct. 21–23, 2001, Victoria, BC, Canada*, 2001.

[5] EU Integrated Project NeOn Lifecycle Support for Networked Ontologies, see www.neon-project.org for further information.

3. M. Bonifacio, T. Franz, and S. Staab. *A Four-Layer Model for Information Technology Support of Knowledge Management*, chapter 6. Advances in Management Information Systems. M.E. Sharpe, Armonk, NY, 2008.
4. T. Buzan. *Use your head.* BBC Books, 1974.
5. T. H. Davenport and L. Prusak. *Working Knowledge – How organisations manage what they know.* Havard Business School Press, Boston, MA, 1998.
6. J. Davies, D. Fensel, and F. van Harmelen, editors. *On-To-Knowledge: Semantic Web enabled Knowledge Management.* Wiley, New York, 2002.
7. R. Dieng, O. Corby, A. Giboin, and M. Ribiere. Methods and tools for corporate knowledge management. *International Journal of Human-Computer Studies*, 51(3):567–598, 1999.
8. A. M. Fairchild. Knowledge management metrics via a balanced scorecard methodology. In *HICSS. Proceedings of the 35th Annual Hawaii International Conference on System Sciences, 2002*, pages 3173–3180, 2002.
9. M. Fernández-López. Overview of methodologies for building ontologies. In *Proceedings of the IJCAI-99 Workshop on Ontologies and Problem-Solving Methods: Lessons Learned and Future Trends.* CEUR, Aachen, Germany, 1999.
10. M. Fernandéz-López, A. Gómez-Pérez, J. Euzenat, A. Gangemi, Y. Kalfoglou, D. M. Pisanelli, M. Schorlemmer, G. Steve, L. Stojanovic, G. Stumme, and Y. Sure. A survey on methodologies for developing, maintaining, integrating, evaluating and reengineering ontologies. OntoWeb deliverable 1.4, Universidad Politecnia de Madrid, 2002.
11. M. Fernández-López, A. Gómez-Pérez, J. P. Sierra, and A. P. Sierra. Building a chemical ontology using Methontology and the Ontology Design Environment. *Intelligent Systems*, 14(1), 1999.
12. A. Gómez-Pérez. A framework to verify knowledge sharing technology. *Expert Systems with Application*, 11(4):519–529, 1996.
13. A. Gómez-Pérez, M. Fernandéz-López, O. Corcho, T. T. Ahn, N. Aussenac-Gilles, S. Bernardos, V. Christophides, O. Corby, P. Crowther, Y. Ding, R. Engels, M. Esteban, F. Gandon, Y. Kalfoglou, G. Karvounarakis, M. Lama, A. López, A. Lozano, A. Magkanaraki, D. Manzano, E. Motta, N. Noy, D. Plexousakis, J. A. Ramos, and Y. Sure. Technical roadmap. OntoWeb deliverable 1.1.2, Universidad Politecnia de Madrid, 2002.
14. T. R. Gruber. Towards principles for the design of ontologies used for knowledge sharing. *International Journal of Human-Computer Studies*, 43(5/6):907–928, 1995.
15. N. Guarino and C. Welty. Evaluating ontological decisions with OntoClean. *Communications of the ACM*, 45(2):61–65, 2002.
16. Peter Haase, Frank van Harmelen, Zhisheng Huang, Heiner Stuckenschmidt, and York Sure. A framework for handling inconsistency in changing ontologies. In Y. Gil, E. Motta, V. R. Benjamins, and M. A. Musen, editors, *Proceedings of the Fourth International Semantic Web Conference (ISWC2005)*, volume 3729 of *LNCS*, pages 353–367. Springer, Berlin, 2005.
17. P. Haase, J. Vlker, and Y. Sure. Management of dynamic knowledge. *Journal of Knowledge Management*, 9(5):97–107, 2005.
18. R. S. Kaplan and D. P. Norton. The balanced scorecard – Measures that drive performance. *Harvard Business Review*, 71ff, January–February 1992.
19. T. Lau and Y. Sure. Introducing ontology-based skills management at a large insurance company. In *Proceedings of the Modellierung 2002*, pages 123–134, Tutzing, Germany, March 2002.

20. A. Léger, H. Akkermans, M. Brown, J.-M. Bouladoux, R. Dieng, Y. Ding, A. Gómez-Pérez, S. Handschuh, A. Hegarty, A. Persidis, R. Studer, Y. Sure, V. Tamma, and B. Trousse. Successful scenarios for ontology-based applications. OntoWeb deliverable 2.1, France Télécom R&D, 2002.
21. A. Léger, Y. Bouillon, M. Bryan, R. Dieng, Y. Ding, M. Fernandéz-López, A. Gómez-Pérez, P. Ecoublet, A. Persidis, and Y. Sure. Best practices and guidelines. OntoWeb deliverable 2.2, France Télécom R&D, 2002.
22. D. O'Leary. Using AI in knowledge management: Knowledge bases and ontologies. *IEEE Intelligent Systems*, 13(3):34–39, 1998.
23. D. O'Leary and R. Studer. Knowledge management: An interdisciplinary approach. *IEEE Intelligent Systems, Special Issue on Knowledge Management*, 16(1), 2001.
24. H. S. Pinto, S. Staab, and C. Tempich. Diligent: Towards a fine-grained methodology for distributed, loosely-controlled and evolving engineering of ontologies. In *ECAI*, pages 393–397, 2004.
25. G. Probst, K. Romhardt, and S. Raub. *Managing Knowledge.* Wiley, New York, 1999.
26. G. Schreiber, H. Akkermans, A. Anjewierden, R. de Hoog, N. Shadbolt, W. van de Velde, and B. Wielinga. *Knowledge Engineering and Management – The CommonKADS Methodology.* MIT, Cambridge, MA, 1999.
27. S. Staab, H.-P. Schnurr, R. Studer, and Y. Sure. Knowledge processes and ontologies. *IEEE Intelligent Systems, Special Issue on Knowledge Management*, 16(1):26–34, 2001.
28. S. Staab and H. Stuckenschmidt, editors. *Semantic Web and Peer-to-Peer.* Springer, Berlin, 2006.
29. L. Stojanovic, A. Maedche, B. Motik, and N. Stojanovic. User-driven ontology evolution management. In A. Gómez-Pérez and V. R. Benjamins, editors, *Proc. of EKAW-2002*, volume 2473 of *LNCS*, pages 285–300, Siguenza, Spain. Springer, Berlin, 2002.
30. L. Stojanovic. *Methods and Tools for Ontology Evolution.* Ph.D. thesis, Universität Karlsruhe (TH), Universität Karlsruhe (TH), Institut AIFB, D-76128 Karlsruhe, 2004.
31. M. C. Suárez-Figueroa, G. Aguado de Cea, C. Buil, K. Dellschaft, M. Fernández-López, A. García, A. Gómez-Pérez, G. Herrero, E. Montiel-Ponsoda, M. Sabou, B. Villazon-Terrazas, and Z. Yufei. Neon methodology for building contextualized ontology networks. Technical report, NeOn Deliverable D5.4.1, February 2008.
32. Y. Sure, M. Erdmann, J. Angele, S. Staab, R. Studer, and D. Wenke. OntoEdit: Collaborative ontology development for the semantic web. In I. Horrocks and J. A. Hendler, editors, *Proc. of International Semantic Web Conference 2002*, volume 2342 of *Lecture Notes in Computer Science (LNCS)*, pages 221–235, Sardinia, Italy, 2002. Springer, Berlin, 2002.
33. Y. Sure, S. Staab, and J. Angele. OntoEdit: Guiding ontology development by methodology and inferencing, volume 2519 of *Lecture Notes in Computer Science (LNCS)*, pages 1205–1222, University of California, Irvine, USA, 2002. Springer, Berlin, 2002.
34. Y. Sure and R. Studer. On-To-Knowledge Methodology – Final version. On-To-Knowledge deliverable 18, Institute AIFB, University of Karlsruhe, 2002.
35. C. Tempich, E. Simperl, M. Luczak, H. S. Pinto, and R. Studer. Argumentation-based ontology engineering. *IEEE Intelligent Systems*, 22(6):52–59, 2007.

36. M. Uschold and M. Grueninger. Ontologies: Principles, methods and applications. *Knowledge Sharing and Review*, 11(2), 1996.
37. M. Uschold and M. King. Towards a methodology for building ontologies. In *Workshop on Basic Ontological Issues in Knowledge Sharing, held in conjunction with IJCAI-95*, Montreal, Canada, 1995.
38. M. Uschold, M. King, S. Moralee, and Y. Zorgios. The enterprise ontology. *Knowledge Engineering Review*, 13(1):31–89, 1998.

Ontology Engineering and Evolution in a Distributed World Using DILIGENT

H. Sofia Pinto[1], C. Tempich[2], and Steffen Staab[3]

[1] Dep. de Engenharia Informática, Instituto Superior Técnico, Av. Rovisco Pais, 1049-001 Lisboa, Portugal, sofia.pinto@dei.ist.utl.pt
[2] Institute AIFB, University of Karlsruhe (TH), 76128 Karlsruhe, Germany, tempich@aifb.uni-karlsruhe.de
[3] ISWeb, University of Koblenz Landau, 56016 Koblenz, Germany, staab@uni-koblenz.de

Summary. Existing mature ontology engineering approaches are based on some basic assumptions that are often neglected in practice.

Ontologies often need to be built in a decentralized way, ontologies must be given to a community in a way such that individuals have partial autonomy over them, ontologies have a life cycle that involves an iteration back and forth between construction/modification and use and ontologies should support the participation of non-expert users in ontology engineering processes.

While recently there have been some initial proposals to consider these issues, they lack the appropriate rigor of mature approaches. i.e. these recent proposals lack the appropriate depth of methodological description, which makes the methodology usable, and they lack a proof of concept by concrete cases studies. In this paper, we describe the DILIGENT methodology that takes decentralization, partial autonomy, iteration and non-expert builders into account and we demonstrate its proof-of-concept in two real-world organizational case studies.

1 Introduction and Motivation

Ontologies are used to improve the quality of communication between computers, between humans and computers as well as between humans. Therefore an ontology should result from an agreement between its different stakeholders and this agreement must be reached in a comprehensive ontology engineering process. There are several mature methodologies that have been proposed to structure this process and thus to facilitate it (cf. chapter "Ontology Engineering Methodology" and [4, 17, 24]) and their success has been demonstrated in a number of applications. Nevertheless, these methodologies make some basic assumptions about the way the ontology engineering process takes place and

S. Staab and R. Studer (eds.), *Handbook on Ontologies*, International Handbooks on Information Systems, DOI 10.1007/978-3-540-92673-3,

about the way the resulting ontologies are used. In practice, we thus observe that these methodologies neglect some important issues:

1. *Decentralization*: Existing methodologies do not take into account that even a medium size group of stakeholders of an ontology is often quite *distributed* and does not necessarily meet often or easily. These methodologies approach ontology engineering in the same style that knowledge-based systems were approached in the past: while the user group of a resulting ontology may be large, its development is performed by a comparatively small group of (1) domain experts who *represent* the user community and (2) ontology engineers who *help structuring* that knowledge.
   In contrast, we have observed that ontology-based applications tend to be built and used in a more widely distributed fashion. By distributed we mean, not only geographically dispersed, but also involving a large number of interested parties from different organizations, with different areas of expertise and competence, different kinds of users with different requirements, etc. For instance, the Gene Ontology (GO), as reported in its web page,[1] is a result of a consortium with 99 members from 18 organizations distributed worldwide, and statistics show above 1,000 hits per week of the GO download web page, on average. Therefore, it almost seems a characteristic of ontologies, that they are more useful if the systems that they support are reaching out over several locations, several independent information systems and several, if not many, independent groups of users. However, applications that are heavily distributed, e.g. applications for virtual organizations[2] or ontology-based Peer-to-Peer applications[3] or Semantic Web applications, have people and organizations frequently leaving or joining a network. Therefore, ontology engineering processes targeting more traditional, centralized knowledge structures do not provide a representative picture of what the stakeholders of the ontology require. In such a scenario, the ontology development process needs to integrate a wider group of stakeholders, and take into account that stakeholders will hardly ever gather in one place – not even in a virtual space.
   Therefore, ontology engineering methodologies need to consider *decentralization* in depth and provide corresponding methodological support.
2. *Partial Autonomy*: We have had the experience that potential users of an ontology are typically forced to use an ontology as is, but that they are commonly not able to influence its development and have to forget about it if it does not fit their needs exactly. A typical situation that we have encountered was that people want to retain a part of the shared ontology and *modify it locally*, i.e. personalize it [13].

[1] `http://www.geneontology.org`
[2] `http://www.virtuelle-fabrik.com`
[3] `http://swap.semanticweb.org`

There have been very few approaches that have touched upon the issue of adaptation to individual purposes [10, 14]. Most of these approaches have targeted this question by considering the re-use of (parts of) ontologies for constructing a new and rather independent ontology, while in the setting of individual adaptation one rather needs to construct a living view onto an existing ontology that is augmented by individual, idiosyncratic extensions.
Thus, existing methodologies have not really dealt with users adapting ontologies for personal use.
3. *Iteration*: Existing ontology engineering methodologies mention the problem of evolving the ontology, but the actual cases that support the methodologies are typically cases where the ontology construction phase strictly precedes its usage phase.
In contrast, we often see the need for interleaving ontology construction and use [13]. Moreover, there is a lack of case studies that support hypotheses about how to iterate in the ontology *evolution* process.
Therefore, evolution needs to be addressed in real, and long run case studies.
4. *Non-expert builders*: Existing ontology engineering methodologies have been derived in a style useful for knowledge engineers. These methodologies propose check lists to guide the engineering process which have been shaped by the needs of knowledge engineers to cope with a complex process and to come up with an often intricate resulting system or ontology. In contrast, in the distributed, evolving cases we consider, the participation of a knowledge engineer is often restricted to a, possibly complex, core ontology. Beyond the core, typical application cases involve extensive participation and, comparatively simple, concept formation by *domain experts* and/or *users*. Support for their participation is mostly lacking in these methodologies.

These issues arise naturally for many ontologies, e.g. [15] or GO and one might claim for all ontologies in the Semantic Web! Recently a number of other approaches that touch these issues have been proposed [1, 6]. However, *none* goes very far from a methodological point of view, namely they do not provide elaborated methodological support, or were extensively used in concrete case studies with regard to these four issues, such as actions to take, their input and output, etc.

Therefore, to account for some of the differences between classical knowledge engineering and ontology engineering methodologies derived from there, we thus have started to develop DILIGENT, a methodology for:

1. DIstributed
2. Loosely-controlled, and
3. evolvInG Engineering of oNTologies that is able to
4. support non-expert ontology builders

While developing DILIGENT, we also had to consider two general methodological objectives:

First, we wanted to provide guidance to the knowledge engineer, the ontology engineer and the non-expert ontology builders that was as fine-grained as possible to make the sequence of tasks as concrete and re-producible by novices as possible.

Second, we needed to check DILIGENT by some concrete case studies to show that it can live up to its promises. Clearly, it is very difficult to near impossible to match any methodology, which constitutes an abstraction of many processes, onto an instantiated process in detail. Nevertheless without a reasonable substantiation of the proposed steps in concrete case studies a proposal like DILIGENT would remain vacuous.

We will therefore describe DILIGENT in detail and some experiences where it was shown how it maps onto comprehensive case studies. Nevertheless, it will not be possible to describe the finest grain size of DILIGENT. At the finest grain size of methodological support, we have proposed an argumentation framework, an argumentation ontology, technical support and several case studies to investigate only these aspects. Including all these investigations in depth as required by a sound scientific presentation would have doubled the size of this paper, hence we only refer to this work here [12,19,20,23] and sketch it briefly in Sect. 4.

In the following, we present our ontology engineering methodology, DILIGENT. In Sect. 2 we give an overview of how we have proceeded to design and validate DILIGENT. In Sect. 3 we describe DILIGENT elaborating the hierarchical task structure in detail. In Sect. 4, we briefly describe how we have applied DILIGENT in some comprehensive case studies, i.e. a distributed knowledge management scenario supported by an ontology-based peer-to-peer knowledge sharing platform and supported by wikis. Eventually, we compare with related work in Sect. 5 and conclude.

2 Developing the DILIGENT Ontology Engineering Methodology

In order to arrive at a sound Ontology Engineering (OE) methodology we have proceeded in five steps to develop DILIGENT.

Around 2000, ontology engineering efforts with a clear distributed, loosely-controlled and dynamic flavor were taking place. For instance SUMO[4] was being collaboratively developed by a group of worldwide distributed researchers, in a loosely-controlled and evolutionary fashion. No particular methodologies were being followed to control these new features, but these processes were clearly following different process models from the ones that were being tackled by the methodologies available at that time. These new efforts [13] provided the initial ideas to conceive our initial DILIGENT framework.

[4] http://suo.ieee.org/

Second, the first step in DILIGENT consists of the construction of a core ontology (cf. Sect. 3). In this step DILIGENT does not introduce any special or new requirements for the core ontology when compared to the ones dealt with by existing methodologies (cf. Sect. 1). Therefore, with regard to this step, we have decided not to develop a new methodology, but to borrow from existing work. We expect that any mature methodology can be used. In our case studies, we have exploited the OTK-methodology (chapter "Ontology Engineering Methodology").

Third, in order to validate the combined methodology we proceed in two fronts. On the one hand, we analyzed its potential for the past and ongoing development process of the biological taxonomy of living species. When we analyze its evolution since 1735 one can realize that it completely follows the 5-step DILIGENT process, as briefly described in Sect. 3. On the other hand, we conducted a lab experiment case study to specifically investigate whether some argumentation structures dominate the progress in the ontology engineering task and should therefore be accounted for in a fine-grained methodology. Our experiments [12] provide strong indication – though not full-fledged evidence – that a restriction of arguments can enhance the ontology engineering effort in a distributed environment. Moreover it also shows us that proper social management procedures and tool support helps to reach consensus in a smoother way (cf. [2]).

Fourth, we started a real-life, cross-organizational case study in the tourism industry. We reported about its initial state supporting means in [11]. In this case study, the process template was realized in a decentralized, autonomous and collaborative setting with high personalization requirements. The process was supported by a peer-to-peer system and tools were specifically developed to support non-ontology engineering experts. Two rounds following DILIGENT were monitored over a 3 month period.

Fifth, by the sum of these initial process templates,[5] cases and experiments, we arrived at the new and refined DILIGENT methodology that we present here. The focus of the refinement has been on decentralization, iteration and partial autonomy as well as on guiding users who are not ontology engineering experts. The methodology has been validated by the iterative case study presented in Sect. 4 and others reported in the literature [23, 25]. Thus, we have repeatedly switched between hypothesis formulation and validation.

3 The DILIGENT Methodology

In order to give the necessary context for the detailed process description as described in Sect. 3.2 we start by summarizing the overall DILIGENT process model.

[5] In our terminology, a methodology for an engineering artefact is a tested and validated process template abstracting over all possible successful engineering processes for engineering the artefact.

The *DILIGENT* process [11] supports its participants, in collaboratively building one shared ontology. In DILIGENT we assume that there are several participants, with different and complementary skills, which, in most of the cases, are geographically distributed, and which have genuine interest in collaboratively building or using one ontology. For instance, in a virtual organization, the different participants may be in a "coopetition" relationship: on the one hand they may be from different but similar organizations that compete for the same resources, but on the other, to compete against external threats, they should cooperate to improve their chances of success. In this case, it may be important, for instance to promote interoperability between their applications, that they all agree on a given ontology, the shared ontology, and use it as a common ground of understanding.

There are different kinds of participants in the DILIGENT process: (1) domain experts, that know about the domain that is targeted (2) ontology engineers, that know how to build ontologies (3) knowledge engineers, that know how to build knowledge or information systems based on ontologies, and (4) users, that use the ontology resulting from the process in their systems for their own uses. The participants directly involved in building the ontology, may or may not use the ontology. However, most ontology users will typically not build or modify the given ontology. DILIGENT supports trained ontology engineers as well as typical users of information systems likewise. The ontology engineers perform the defined activities with more accuracy and awareness of the process, while the non-ontology-engineering-expert users will tend to follow them implicitly guided by the provided tools. At some points of the process there is a subset of participants that plays a special role and has added responsibilities: the board. As in the other steps of the process, the composition of the board is not fixed, that is members can enter or leave, although it should have a more stable composition than that of the participants involved in the DILIGENT process. This board is responsible for the shared ontology: in the beginning it builds the initial version of the ontology, in the iterations that follow it is responsible for the evolution of the shared ontology.

3.1 General Process

The process comprises five main activities: (1) *build*, (2) *local adaptation*, (3) *analysis*, (4) *revision*, (5) *local update*, Fig. 1. The process starts by having *domain experts*, *users*, *knowledge engineers* and *ontology engineers building* an initial ontology. The team involved in building the initial ontology should be relatively small, in order to more easily find a small and consensual first version of the shared ontology. At this point, it is not required to arrive at an initial ontology that covers the complete domain.

Once the initial ontology is made available, users can start using it and *locally adapting* it for their own purposes. Typically, due to new business requirements or user and organization changes, their local ontologies evolve.

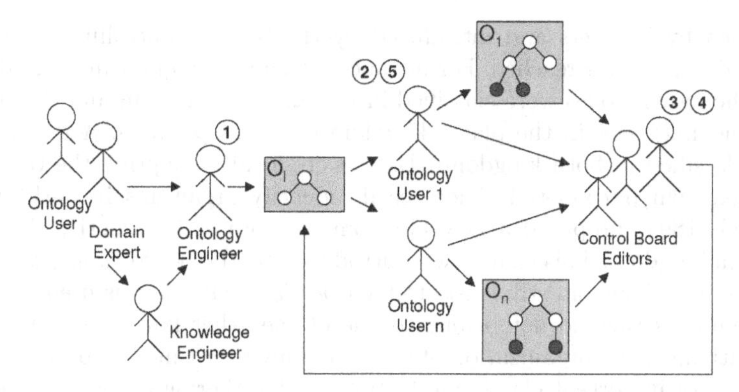

Fig. 1. Distributed, loosely controlled, evolving Ontology Engineering

In DILIGENT there are two kinds of ontologies: The *shared ontology* and *local ontologies.* The shared ontology is available to all users and cannot be changed directly except by the board. Users are free to change, in their local environments, a copy of the shared ontology. The ontology resulting from the changes of a user is the user local ontology.

A board of ontology stakeholders *analyzes* the local ontologies and the users' requests and tries to identify similarities in their ontologies. At this point it is not intended to merge all local ontologies. Instead, changes to local ontologies will be analyzed by the board in order to decide which changes introduced or requested by the users should be introduced in the shared ontology. Therefore, a crucial activity of the board is deciding which changes are going to be introduced in the next version. A balanced decision that takes into account the different needs of user's evolving requirements has to be found.

The board should regularly *revise* the shared ontology, so that the parts of the local ontologies overlapping the domain of the shared ontology do not diverge too far from it. Therefore, the board should have a well-balanced and representative participation of the different kinds of participants involved in the process, which includes *ontology engineers, domain experts* and *users.* Of course, these are roles that may overlap.

Once a new version of the shared ontology is released, users can *update* their own *local* ontologies to better use the knowledge represented in the new version. The last four stages of the process are performed in a cyclic manner: when a new common ontology is available a new round starts again.

There are evidences that this process template can be used in different areas and therefore understanding and better supporting it is important. For instance, the taxonomy of life on earth has been evolving since 1735 following a DILIGENT like 5-step process. It was initially proposed by Linnaeus (*build*) based on phenetics (observable features). Considering the "most general" level, initially, two kingdoms were proposed: animals and plants. As more and more detailed knowledge about them was discovered, new kingdoms were proposed by its users and introduced by the boards controlling them once some consensus was reached. For instance, when microorganisms were discovered the moving ones were classified in the animals kingdom and the colored (non-moving) ones in the plants kingdom (*local adaptation*). A few of them were classified in both kingdoms. Users were locally adapting the taxonomy for their own purposes. To more easily identify organisms in both classes, Haeckel (1894) proposed a new kingdom to more easily identify them, the Protista kingdom. This change was introduced by the board (*analysis and revision*). This kingdom still exists today (*locally update*) and is used to gather all organisms that do not belong to one of the other kingdoms. The major force driving the reorganization of the taxonomy over time has been the identification of important classifying features and gathering all beings sharing a given value for that feature into that class. The parallel between DILIGENT template process and the development of the taxonomy of life on earth is far more deep than described here. For other examples see [12].

3.2 DILIGENT Process Stages

In order to facilitate the application of DILIGENT ontology engineering processes and provide guidance to its participants in real settings, DILIGENT had to be more detailed. For this purpose, we have analyzed the different process stages in detail. For each stage we have identified (1) major roles, (2) input, (3) decisions, (4) actions, (5) available tools, and (6) output information. One should stress that this elaboration is rather a recipe or check list than an algorithm or integrated tool set. In different contexts it may have to be adapted or further refined to fit particular needs and settings. Tools may need to be integrated or customized to match requirements of the application context. In Fig. 2 we sketch our results, which are presented in the following. For the sake of brevity we refer the reader to [20] that includes an even more detailed process description.

Build

As mentioned before, DILIGENT focuses on distributed ontology development and ontology evolution, but borrows from established methodologies (chapter "Ontology Engineering Methodology" and [4]). This is particularly true at this stage. The goal is to quickly build an ontology that is going to be used in an application. At this stage one can follow different approaches and even approaches inspired from software engineering methodologies, such as rapid prototyping, extreme programming and open source guidelines. The motto is: get something small and useful and give it to the users, as early as possible. Therefore, there is no need for completeness, although usability and usefulness are crucial.

Roles: Usually, there are three roles: knowledge engineer, ontology engineer and domain expert. The domain expert provides both knowledge and ontology engineers with the required domain knowledge and knowledge sources.

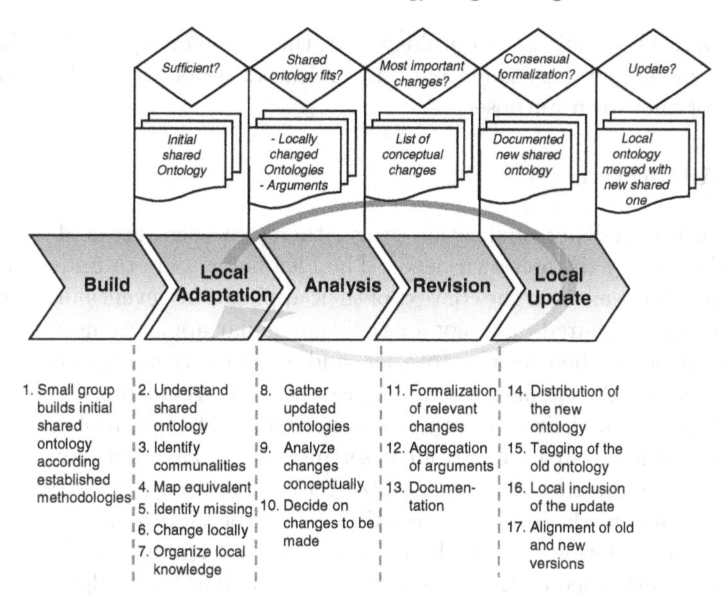

Fig. 2. Process stages (1–5), actions (1–17) and structures

The knowledge engineer creates a conceptual model of the domain from the knowledge extracted from these sources. The ontology engineer generates a machine readable ontology from the conceptual model. Quite often the knowledge engineer and ontology engineer are roles performed by the same person. Additionally to these classical roles we also propose the involvement of users. At this stage, usually the actors involved as users are also involved in the process in one of the other more classical roles. Most of those involved in the build stage are initial board members.

Input: Since this stage borrows from traditional OE the usual pre-development activities are performed. Given our analysis of existing methodologies [21] we recommend the adoption of the OTK methodology (chapter "Ontology Engineering Methodology", since it is the one providing more guidance and has a more detailed and complete set of activities. However, the use of other methodologies is not excluded.

Decisions: The usual decisions of a classical OE process need to be taken. In contrast to common OE methodologies we do not require completeness of the ontology at this stage. It is particularly important that the ontology is clear and easily understandable by possible users.

Actions: As in classical OE development, common core activities are conceptualization, formalization, and implementation.[6] Integral activities like knowledge acquisition, evaluation, reuse (comprising fusion and composition), and documentation are complemented in DILIGENT with a recommendation for *Argument provision.*

[6] Maintenance is supported by later stages of DILIGENT.

Output: The result is an ontology with the main concepts of the domain. Once an initial ontology is (1) *built* and released, users will start to adapt it locally for their own purposes.

Local Adaptation

This is a use and personalization stage, therefore users use and adapt the released ontology to their own needs. The idea is for users to understand the shared ontology, use it in the context of their applications, eventually find some problems in the shared ontology for their particular applications that require customization on their local ontologies, and accordingly modify these to best suit their needs. All changes should be justified with arguments. Their changes will only apply to their local copies and not to the shared ontology that was made available to all users. In ideal settings, users can also have access to other users' ontologies, when customizing the shared ontology (either under the same framework or from external sources) therefore reuse of ontologies may also be performed.[7] One should stress that all traditional OE activities are usually performed by the users at this stage, such as knowledge acquisition, conceptualization, formalization, evaluation, integration, etc. Once in a while a new shared ontology is made available to users.

Roles: The actors involved in the local adaptation step are users of the ontology. However, they usually do not have an OE background. They use the ontology, e.g. to retrieve documents which are related to certain topics modeled in the ontology or more structured data like the projects an employee was involved in.

Input: Besides the common shared ontology, in the local adaptation step the information available in the local information space is used. This can be existing databases, ontologies or folder structures and documents. Moreover, external knowledge sources or ontologies can also be reused as well as other user's ontologies.

Decisions: The users decide which changes they want to make to their local ontology, hence they must decide if and where new concepts are needed and which relations a concept should have. They should provide reasons for their changes.

Actions: To achieve the desired output the user performs different groups of actions namely: Analyze the shared ontology; Change and integrate the shared and local ontologies; and Use the shared ontology. The last two actions of the process step are performed in a cyclic manner until a new shared ontology is available and the entire process step starts again.

One important issue is the fact that this stage can either be performed immediately after a build or after a local update stages. In both cases, the shared ontology is available: in the first case, it is the only ontology users have had so far, in the second they have already their own local ontologies

[7] With naive users this usually does not occur often.

that were somewhat connected (or not) to the shared ontology. Users then start adapting the shared ontology to their own purposes. Although these two situations are not different from a conceptual point of view, from a practical point of view they are different since in the second case users usually are not going to simply discard their local ontologies and build them again so that they can be connected to the new version of the shared ontology. Therefore, it is important to assure that there can be a smooth transition.

We now describe in detail each one of the proposed actions:

The Analysis of shared ontology usually involves (2) *Understand shared ontology* and (3) *Identify similarities between own and shared conceptualization.* An ontology should represent a shared conceptualization of part of the world. At this point the analysis is mainly the identification of similarities and mismatches between the available shared ontology and either the conceptual model of the domain users have in their minds or the local ontologies they already developed in previous iterations of the process.

(2) Understand the shared ontology The user must learn where the different concepts are located in the ontology and how they are interrelated. The ontology can be very complex, thus understanding the ontology depends mainly on its visualization, and good naming conventions.

(3) Identify similarities between own and shared conceptualization Following the comprehension of the ontology, the user can realize the similarities and differences between the own and shared conceptualizations

Change and integration of shared and local ontologies usually involves (4) *Map equivalent conceptualizations of different actors* (5) *Identify mismatches in conceptualizations*, and (6) *Change conceptualization.*

(4) Map equivalent conceptualizations of different actors: After the identification of similarities they should be made explicit, otherwise the system will not be able to make use of these findings in later stages. This is particularly important when the user is identifying similarities between his local concepts and the new concepts in the shared ontology. Different implementations may add specialized adds-on. Mappings have the advantage, that they leave the original local structures unchanged. Of course users may also decide to change their local structures in favor of the common structure. In this case the changes must be traceable, so that the user can retain his old version, whenever he wants.

(5) Identify mismatches in conceptualizations: The techniques to identify similarities can also be applied in the subsequent step to support the user in identifying missing conceptualizations. Depending on the scenario, the user might have access to other users' ontologies and use their local adaptations as further input to identify missing concepts in his own conceptual model.

(6) Change conceptualization: After identifying missing or unwanted conceptualizations the user must be enabled to introduce them. This is a customization phase and of course, evaluation is also performed here. Users should assure that their changes are adequate both from a domain and a representation point of view. Since later on the board analyzes the changes performed

or requested by the users, users must provide reasons for each change and/or request for change, so that the board can understand them. To support the user in providing reasons, the argumentation framework focuses the user on the relevant arguments, [19].

Ontology use typically involves that users (7) *Organize local knowledge according to the new conceptualization.* At this point the local ontology should reflect the user's conceptualization. Now he can use the ontology in his application. In our case studies ontologies were used in information retrieval scenarios therefore, ontology use typically involved the organization of local knowledge according to the local conceptualization. Therefore, the user instantiated the ontology with the information available locally and hence contributed to the collective knowledge.

Output: One output of the process step is a locally changed ontology which better reflects the user's needs. Each change is supported by arguments explaining the reasons for a change. At this point changes are not propagated to the shared ontology. Moreover, users can send requests for changes directly to the control board, which should also be duly justified. Only in the *analysis step* the board gathers all ontology changes and requests and their corresponding arguments to be able to evolve the common shared ontology in a user driven *revision step.*

Analysis

In this stage, the board analyzes incoming requests and observations of changes.[8] The idea is for the board to identify which changes should be made to the ontology based on the changes made or requested by the users. The frequency of this analysis is determined based on the frequency and volume of changes to the local ontologies. The board analyzes and decides which changes would the users most benefit from and would most like to see implemented. At this stage the new requirements for the future version of the shared ontology are identified. At this stage, work is conducted at a conceptual level. This activity borrows from classical ontology reuse processes, but is simpler since local ontologies are available in the same environment and language.

Roles: In the analysis stage we can distinguish three roles played by board members: (1) The domain expert decides which changes to the common ontology are relevant from the domain point of view and which are relevant for smaller communities only. (2) Representatives from the users explain different requirements from the usability perspective. (3) The ontology engineers analyze the proposed changes from a knowledge representation point of view foreseeing whether the requested changes could later be formalized and implemented.[9]

[8] Ideally the board should have access to all users' ontologies. However, in some settings it may only have access to requests for changes.

[9] In the revision stage.

Input: The analysis stage takes as input the ontology changes requested and/or made by the participating actors. To be able to understand their changes and requests, users should have provided their reasons. Both manual and automated methods can be used in the previous stages, therefore besides of arguments by ontology stakeholders, one may here consider rationales generated by automated methods, e.g. ontology learning. The arguments underlying the proposed changes constitute important input for the board to achieve a well balanced decision about which changes to adopt.

Decisions: The board must decide which changes to introduce into the new shared ontology at the conceptual level. Metrics to support this decision are (1) the number of users who introduced a change in proportion to all users who made changes. (2) The number of queries including certain concepts. (3) The number of concepts adapted by the users from previous rounds.

Actions: To achieve the desired output the board takes different actions namely (8) *Gather locally updated ontologies and corresponding arguments*, (9) *Analyze the introduced changes* and (10) *Decide on changes to be made.*

We now describe in detail each one of the proposed actions:

(8) Gather locally updated ontologies and corresponding arguments: Depending on the deployed application the gathering of the locally updated ontologies can be more or less difficult. It is important that the board has access to the local changes from users and their corresponding arguments to be able to analyze them. It may also be interesting not only to analyze the current users' ontologies, but also its evolution. However, with an increasing number of participants this in-depth analysis may be unfeasible. Since usually analysis takes place at the conceptual level, reverse engineering is usually an important technique to get the conceptual model from the formalized model [4]. To support users providing their reasons, the argumentation framework focuses the users on the relevant arguments, [19].

(9) Analyze introduced changes: In this action the board tries to identify the parts of the shared ontology which should be modified. As the number of change requests may be large and also contradictory, first the board must identify the different areas in which changes took place. Within analysis the board should bear in mind that changes of concepts should be analyzed before changes of relations and these before changes of axioms. Good indicators for changes relevant to the users are (1) overlapping changes and (2) their frequency. Furthermore, the board should analyze (3) the queries made to the ontology. This should help to find out which parts of the ontology are more often used. Since actors instantiate the ontology locally, (4) the number of instances for the different proposed changes can also be used to determine the relevance of certain adaptations.

(10) Decide on changes to be made: Having analyzed the changes and having grouped them according to the different parts of the ontology they belong to, the board has to identify the most relevant changes, that is identify changes presumably relevant for a significant share of all actors. Based on the provided arguments the board must decide which changes should be

introduced. Depending on the quality of the arguments the board itself might argue about different changes. For instance, the board may decide to introduce a new concept that better abstracts several specific concepts introduced by users, and connect it to the several specific ones. Therefore, the final decisions entail some form of evaluation from a domain and a usage point of view.

Output: The outcome of this action is a reduced and structured list of changes that are to be implemented in the shared ontology that were agreed by the board. Arguments should be provided for each one of them. All changes which should not be introduced into the shared ontology are filtered. Arguments justifying the decisions to leave them out should also be provided. At this stage it is not required to decide on the final modeling of the shared ontology.

Revision

The revision and analysis stages are closely related. While in the previous stage the new requirements for the shared ontology are identified, in this stage they are formalized and implemented. In the end the new version of the shared ontology is distributed to its users.

Roles: The ontology engineers from the board judge the changes from an ontological perspective more exactly at a formalization level. Some changes may be relevant for the common ontology, but may not be correctly formulated by the users. The domain experts from the board should judge and decide wether new concepts/relations should be introduced into the common ontology even though they were not requested by the users

Input: The input for the revision phase is a list of changes at a conceptual level which should be included into the ontology and the arguments underlying them.

Decisions: The main decisions in the revision phase are formal ones. All intended changes identified during the analysis phase should be included into the common ontology. In the revision phase the ontology engineer decides how the requested changes should be formalized. Evaluation of the decisions is performed by comparing the changes on the conceptual level with the final formal decisions. The differences between the original formalization by the users and the final formalization in the shared ontology should be kept to a minimal basis.

Actions: To achieve the desired output the members of the board, mainly its ontology engineers, perform different actions namely (11) *Formalization of the decided changes*, (12) *Aggregation of corresponding arguments*, (13) *Documentation*, and (14) *Distribution of the new ontology to all actors.*

We now describe in detail each one of the proposed actions:

(11) Formalization of the decided changes: As in classical OE development, the requested changes must be formalized with respect to the expressivity of the ontology representation language. Before their actual implementation, the agreed changes should be analyzed from a knowledge representation point of

view. This evaluation is somehow similar to the one performed when reusing an ontology according to classical reuse methodologies. The goal is to determine how the changes identified in the previous step should be formalized. Once this is done, the actual changes are formalized and the quality of the resulting ontology is again assured through evaluation. All required activities are addressed by classical OE methodologies.

(12) Aggregation of arguments: As arguments play a major roll in the decision process we expect that the changes which are eventually included into the common ontology are supported by good arguments. One of the reasons for keeping track of the arguments is to enable users to better understand why certain decisions have been made. Therefore, the board should summarize and aggregate understandable, pedagogical and the most convincing arguments underlying each change. The user should be able to retrieve them.

(13) Documentation: With the help of the arguments, the introduced changes are already well documented. However, we assume that some arguments may only be understandable by the domain experts and not users. Hence, we expect that the changes should be documented to a certain level.

(14) Distribution of the ontology to all actors: Analogously to stage (1) the shared ontology must be distributed to the different participants. Depending on the overall system architecture different methods can be applied here. Moreover, the board should assure version and release management.

Output: The new version of the shared ontology together with its arguments and documentation is the result of this stage. This documentation is essential for users to understand the new shared ontology when a new cycle begins.

Local Update

In the local update stage the new shared ontology is released and put to use by its users. They decide which changes they will adopt. Part of this stage is similar to local adaptation: users must get familiar with the new version and identify which parts of their local ontologies they will discard in favor of the new shared ontology and which ones they will retain.

Roles: The local update phase involves only users. They perform different actions to include the new common ontology into their local system before they start a new round of local adaptation.

Input: The new formalized shared ontology is the input for this step. We also require as input the documentation and arguments justifying those changes. For a better understanding the user should be allowed to request a delta to the original version.

Decisions: The user must decide which changes he will introduce locally. This depends on the differences between the own and the new shared conceptualization. The user does not need to update his entire ontology. This stage interferes a lot with the next local adaptation stage. We do not exclude the possibility of conflicts and/or ambiguity between local and shared ontologies, which may entail reduced precision if the ontology is being used in IR applications.[10]

Actions: To achieve the desired output the user takes different actions namely *Analysis of the new shared ontology*; and *Integration of new shared version with current user's local one.*

After the local update, the iteration continues with local adaptation. During the next analysis step the board reviews which changes were actually accepted by the users.

We now describe in detail each one of the proposed actions:

Analysis of the new shared ontology: The goal is to understand the new shared ontology. The user scans for the changes introduced by the board that are relevant for his use, and controls whether his change proposals were implemented. He must further identify wether the benefits of updating to the new version outweight its effort. Issues to be analyzed include: concepts introduced by other users, consistency of new shared version with local version, maintenance of interoperability with other users.

Integration of new shared version with current user's local one: In this action the user reuses or not the new version of the shared ontology. If the new shared ontology is not of use the system should allow the user to retain the outdated version. In this case the user will have to perform (15) *Tagging of the outdated version.* In case the user finds the new shared version of use two further subactions can be performed: (16) *Inclusion of the updated version* and (17) *update local adaptations not included in the common ontology.*

(15) Tagging of the outdated version: To ensure user satisfaction, the system must enable the user to retain his old version of the ontology or parts of it. The user may later realize that the new updated version of the common ontology does not represent his needs anymore and thus want to leave the update cycle out and return to the old version. To reach a better acceptance this must be possible and is foreseen in the methodology. The user can always balance between the advantages of using a shared ontology or using his own conceptual model. Therefore, the old version should be stored for possible later reuse.

(16) Inclusion of the updated version: The system must support the user to easily integrate the new version into his local system. It must be guaranteed that all annotations made for the old version of the ontology are available in the new version. It may require restructuring and adaptation of instantiations to stay in line with the new model.

(17) Update of local adaptations which are not included in the common ontology: The update of the local ontology can lead to different kinds of conflict. Changes proposed by the user may indeed have found their way into the common ontology. Hence, the user should be enabled to use from now on

[10] Ideally one should be able to blacken out the ambiguous parts like in multilevel databases. This has not been transferred to OE yet.

the shared model instead of his own identical model. Furthermore, the board might have included a change based on arguments the user was bringing forward, but has drawn different conclusions. Here the user can decide wether he prefers the shared interpretation.

Other options may emerge in the course of further case studies.

Output: Ideally the output of the local update phase is an updated local ontology which includes all changes made to the shared ontology. However, since not all users may want to completely change to the new version, we do not require the users to adopt all changes proposed by the board. So, the output is not mandatory since the actors could change the new ontology back to the old one in the local adaptation stage.

4 Applying DILIGENT in Case Studies

In this section we describe briefly how we specifically investigated how a distributed, loosely controlled and evolving ontology engineering process following DILIGENTcould be implemented. For more detailed descriptions refer to the relevant bibliography referred in each subsection.

4.1 The IBIT Case Study

The first running case study took place under the SWAP project. In this project, the challenges were how the process template could be implemented in a multi-organizational setting with non-expert ontology engineering users, and which finer grained support could be provided to these users.

In the SWAP project, the IBIT case study was in the tourism domain of the Balearic Islands. The needs of the tourism industry there, which accounts for 80% of the islands' economy, are best described by the term "coopetition". On the one hand the different organizations *compete* for customers against each other. On the other hand, they must *cooperate* in order to provide high quality for regional issues like infrastructure, facilities, clean environment, or safety – that are critical for them to be able to compete against other tourism destinations. To collaborate on regional issues a number of organizations now collect and share information about *indicators* reflecting the impact of growing population and tourist fluxes in the islands, their environment, and their infrastructures. Moreover, these indicators can be used to make predictions and help planning. For instance, organizations that require *Quality & Hospitality Management* use the information to better plan, e.g. their marketing campaigns. As another example, the governmental agency IBIT,[11] the Balearic Government's co-ordination center of telematics, provides the local industry with information about *New Technologies* that can help the tourism industry to better perform their tasks.

[11] `http://www.ibit.org`

Due to the different working areas and goals of the collaborating organizations, it proved impossible to build a centralized knowledge management system or even a centralized ontology satisfying all user requirements. The users emphasized the need for local control over their ontologies. They asked explicitly for a system without a central server, where knowledge sharing was integrated into the normal work, but where different kinds of information, like files, emails, bookmarks and addresses could be shared with others. To this end the SWAP consortium – including us at University of Karlsruhe, IBIT, Free Univ. Amsterdam, Meta4, and Empolis – developed the SWAP generic P2P platform and built a concrete application on top that allowed the satisfaction of the information sharing needs just elaborated using local ontologies, which were linked to a shared ontology. A case study was set up. The main goals were the evaluation of the DILIGENT process and the developed peer-to-peer platform. The case study lasted for 3 months. Moreover, a set of tools were also specifically developed [18] to support the participants in the case study. However, most of the tools were being developed at the same time as the process was taking place. Therefore, the administrator had a major role in bridging the gap between our real users and the weaknesses of the tools, for instance by doing local adaptations for the users since the tools were not error-proof.

Regarding the methodology we had four hypothesis: (1) DILIGENT supports collaborative development of a shared ontology; (2) ontologies in use need to evolve; (3) non-ontology engineering experts can participate in ontology engineering processes, and (4) the organizational structure DILIGENT suggests fits the organizational setting found in the IBIT case study, a peer-to-peer setting.

The first round of our OE process started with the distribution of the three modules of the common ontology to all users. In both rounds, users – during the local adaptation stage – and the board – in the revision stage – could perform ontology change operations (concepts/relations/instances). Most frequently the concept hierarchy was changed.

The first month of the case study, corresponded to the first round of the DILIGENT process. One organization with seven peers participated. This organization can be classified as a rather loose one. In the first round we had seven users, six of which had no OE background. In general, the users added concepts to the shared ontology to represent the topics of their core working area. They did not share all their local information, but selected the documents which they thought would be interesting for the group. In the interviews they commented, that they would share more files at a later stage, when they would feel more confident with the system. In this organization most of the users were very active and did local adaptations to best serve their own needs. They also add access to other user's ontologies. Moreover, the board received by e-mail requests to modify the shared ontology. The first round of the process resulted in seven adapted ontologies.

In Analysis, the board consisted of two ontology engineers and two domain experts/users, the same that were involved in the build stage. The local adaptations from seven users were collected. Additionally the board had access to their folder structures. All changes introduced were motivated by the users' requests and changes. They all made sense and were not contradictory on the conceptual level. Then, the new shared ontology was distributed.

In Local update all users decided to use the new shared ontology as it covered more domain knowledge and they found their requests integrated to it. As a result of this stage the new shared ontology was commonly used and the users' folders were aligned with the new shared ontology.

In the second round the case study was extended to four organizations with 21 peers. The users participating in the first round had more experience and were still active. One of the new organizations was very hierarchical. None of the new 14 users had OE experience. The experienced users started with the result of the local update stage, while the new users received only the new shared ontology. All users shared the local information which they thought relevant for the group. The new users behaved in a similar way as the users in the first stage and did not share many folders, as they wanted to gain confidence in the system first. The experienced users, however, published more information, and adapted the local ontologies accordingly. The second local adaptation stage resulted in 14 adapted ontologies. The rest of the users did not make changes. Although, some did not change the shared ontology directly, they submitted change requests to their supervisor, thus they delegated the modeling task. The supervisor then communicated the requests to the board.

In Analysis, in this round the board consisted of one domain expert and two ontology engineers. Additionally two users were invited to answer questions to clarify the changes they introduced. The 21 local ontologies of the users were the input to the second round. This time the board had to perform reverse engineering on the formal local ontologies from users in order to get their conceptual models. As in the first round the board included all change requests from users. Again, as in the first round, only very few concepts in the common ontology were never used. All conceptual requests could be modelled in the ontology, providing the next version.

The case study ended after the distribution of the new shared ontology. We collected feedback from the users w.r.t. to their impressions on the new version. They emphasized that the new version represented their requirements at that time. The users commented that they appreciated being involved in the development process, although they recognized that they were not experienced in ontology engineering. They did not object to the modeling decisions of the board and understood the reasons for the differences between their change requests and the final modeling.

However, updating to the new version was still a problem, since some instances of the ontology might have to be newly annotated to the new concepts of the shared ontology. In our case, documents needed a new classification. This problem can be partly overcome with the help of technology [7].

For more detailed descriptions on this project refer to [20, 21].

4.2 The Judges Case Study

The Judges case study took place under the SEKT project. It aimed at providing an intelligent Frequently Asked Questions system, Iuriservice, that offers help to newly appointed judges in Spain. Although judges had a strong and thorough education and became experts in their domain, they still often seek the help of senior judges or tutors regarding procedural questions. The system focuses on such procedural knowledge, which is often neglected, as it is very hard to externalize. Examples for procedural questions are: How should I organize a round of recognition of suspects if there are no people available? Which are the actual functions and competences of the judge as compared to those of the secretaries?

In this regard, the design of legal ontologies requires not only to represent the legal, normative language of written documents (decisions, judgments, rulings, partitions, etc.) but also those chunks of professional knowledge from the daily practice at courts. One of the main features of this professional legal knowledge is that it is context-sensitive. In this sense, it implies: (1) the ability to discriminate among related but different situations; (2) the practical attitude or disposition to rule, judge or make a decision; (3) the ability to relate new and past experiences of cases; (4) the ability to share and discuss these experiences with the group of peers.

In this case, the argumentation framework developed under the DILIGENTmethodology, together with a wiki system proved an invaluable tool that promoted discussion and allowed finding good solutions for the problems newly appointed judges faced.

For more detailed descriptions on this project refer to [20, 22].

5 Related Work

In the past, there have been OE case studies involving dispersed teams, such as $(KA)^2$ ontology [1] or [13]. However, they usually involved tight control of the ontology, of its development process, and of a small team of ontology engineering experts that could cope with the lack of precise guidelines.

Established methodologies for ontology engineering summarized in [4, 17, 24], focus on the centralized development of static ontologies, i.e. they do not consider iteration between construction/modification and use. *METHONTOLOGY* [4] and the *OTK methodology* [17] are good examples for this approach. They offer guidance for building ontologies either from scratch, reusing other ontologies as they are, or re-engineering them. They divide OE processes into several stages which produce an evaluated ontology for a specific domain.

Holsapple et al. [5] focus their methodology on the collaborative aspects of ontology engineering but still aim at a static ontology. A knowledge engineer defines an initial ontology which is extended and modified based on the feedback from a panel of domain experts. *HCOME* is a methodology which integrates argumentation and ontology engineering in a distributed setting [6]. It supports the development of ontologies in a decentralized setting and allows for ontology evolution. It introduces three different spaces in which ontologies can be stored: In the *Personal Space* users can create and merge ontologies, control ontology versions, map terms and word senses to concepts and consult the top ontology. The evolving personal ontologies can be shared in the *Shared Space*. The *Shared Space* can be accessed by all participants. In the shared space users can discuss ontological decisions. After some discussion and agreement, the ontology is moved into the *Agreed space*. However, they have neither reported that their methodology had been applied in a case study nor do they provide any detailed description of the defined process stages.

There are a number of technical solutions to tackle problems of remote collaboration, e.g. ontology editing with mutual exclusion [3], inconsistency detection with a voting mechanism [9], collaborative ontology editing [8, 16] or evolution of ontologies by different means [7]. All these solutions address the issue of keeping an ontology consistent. Obviously, none supports (and do not intend to) the work process of the ontology engineers by way of a methodology.

6 Conclusion

Decentralization can take different forms. One can have more loose or more hierarchical organizations. We observed and supported both kinds of organizations. Therefore, the first finding is the fact that this process can be adapted both to hierarchical and to more loose organizations. DILIGENT processes cover both traditional OE processes and more Semantic Web-oriented OE processes, that is with strong decentralization and partial autonomy requirements.

The process helped non-OE-expert users to conceptualize, specialize and refine their domain. The agreement met with the formalized ontologies was high, as shown by people willing to change their folder structures to better use the improved domain conceptualization. In spite of the technical challenges, user feedback was very positive.

The DILIGENT process proved to be a natural way to have different people from different organizations collaborate and change the shared ontology. The set-up phase for DILIGENT was rather fast, and users could profit from their own proposals (local adaptations) immediately. The result was much closer to the user's own requirements. Moreover, other users profited from them in a longer term. Finally, the case studies clearly have shown the need for evolution. Users performed changes and adaptations.

The development of ontologies in centralized settings is well studied and there are established methodologies. However, current experiences from projects suggest that ontology engineering should be subject to continuous improvement rather than a one-time effort and that ontologies promise the most benefits in decentralized rather than centralized systems. To this end we have conceived the DILIGENT methodology. DILIGENT supports domain experts, users, knowledge engineers and ontology engineers in collaboratively building a shared ontology in a distributed setting. Moreover, the methodology guides the participants in a fine grained way through the ontology evolution process, allowing for personalization. We have demonstrated the applicability of our process model in a cross-organizational case study in the realm of tourism industry and another in the judicial domain. Real users were using the ontology to satisfy their information needs for an extended period of time.

References

1. V. R. Benjamins, D. Fensel, S. Decker, and A. Gómez-Pérez. $(KA)^2$: Building ontologies for the internet. *International Journal of Human-Computer Studies (IJHCS)*, 51(1):687–712, 1999.
2. K. Dellschaft, H. Engelbrecht, J. M. Barreto, S. Rutenbeck, and S. Staab. Cicero: Tracking design rationale in collaborative ontology engineering. In *ESWC*, volume 5021 of *Lecture Notes in Computer Science*, pages 782–786. Springer, Berlin, 2008.
3. A. Farquhar et al. The ontolingua server: A tool for collaborative ontology construction. Technical report KSL 96–26, Stanford, 1996.
4. A. Gómez-Pérez, M. Fernández-López, and O. Corcho. *Ontological Engineering*. Advanced Information and Knowlege Processing. Springer, Berlin, 2003.
5. C. W. Holsapple and K. D. Joshi. A collaborative approach to ontology design. *Communications of the ACM*, 45(2):42–47, 2002.
6. K. Kotis, G. A. Vouros, and Jerónimo Padilla Alonso. HCOME: Tool-supported methodology for collaboratively devising living ontologies. In *SWDB'04: Second International Workshop on Semantic Web and Databases 29–30 August 2004 Co-located with VLDB*. Springer, Berlin, 2004.
7. A. Maedche, B. Motik, and L. Stojanovic. Managing multiple and distributed ontologies on the semantic web. *The VLDB Journal*, 12(4):286–302, 2003.
8. N. Noy, A. Chugh, W. Liu, and M. A. Musen. A framework for ontology evolution in collaborative environments. In *International Semantic Web Conference*, volume 4273 of *Lecture Notes in Computer Science*, pages 544–558. Springer, Berlin, 2006.
9. A. Pease and J. Li. Agent-mediated knowledge engineering collaboration. In L. van Elst, V. Dignum, and A. Abecker, editors, *Agent-Mediated Knowledge Management International Symposium AMKM 2003 Stanford, CA, USA*, Lecture Notes in Artificial Intelligence (LNAI) 2926, pages 405–415. Springer, Berlin, 2003.
10. H. S. Pinto and J. P. Martins. A methodology for ontology integration. In *Proceedings of the First International Conference on Knowledge Capture (K-CAP2001)*, pages 131–138. ACM Press, New York, 2001.

11. H. S. Pinto, S. Staab, Y. Sure, and C. Tempich. OntoEdit empowering SWAP: A case study in supporting DIstributed, Loosely-controlled and evolvInG Engineering of oNTologies (DILIGENT). In C. Bussler, J. Davies, D. Fensel, and R. Studer, editors, *First European Semantic Web Symposium, ESWS 2004*, volume 3053 of *LNCS*, pages 16–30, Heraklion, Crete, Greece, May. Springer, Berlin, 2004.
12. H. S. Pinto, S. Staab, and C. Tempich. DILIGENT: Towards a fine-grained methodology for DIstributed, Loosely-controlled and evolvInG Engineering of oNTologies. In R. L. de Mántaras and L. Saitta, editors, *Proceedings of the 16th European Conference on Artificial Intelligence (ECAI 2004)*, pages 393–397, Valencia, Spain, August 2004. IOS Press, Amsterdam, 2004.
13. H. Sofia Pinto and J. P. Martins. Evolving Ontologies in Distributed and Dynamic Settings. In D. Fensel, F. Giunchiglia, D. L. McGuiness, and M.-A. Williams, editors, *KR2002 Proceedings*. Morgan Kaufmann, San Fransisco, CA, 2002.
14. T. Pirlein and R. Studer. An environment for reusing ontologies within a knowledge engineering approach. *International Journal of Human-Computer Studies*, 43(5):945–965, 1995.
15. M. I. Sucasas, C. Caracciolo, C. Baldassarre, and Y. Jaques. Revised specifications of user requirements for the Fisheries case study. NeOn deliverable 7.1.2, Food and Agriculture Organization of the United Nations (FAO), 2008.
16. Y. Sure, M. Erdmann, J. Angele, S. Staab, R. Studer, and D. Wenke. Ontoedit: Collaborative ontology development for the semantic web. In *International Semantic Web Conference*, volume 2342 of *Lecture Notes in Computer Science*, pages 221–235. Springer, Berlin, 2002.
17. Y. Sure, S. Staab, and R. Studer. On-to-knowledge methodology. In S. Staab and R. Studer, editors, *Handbook on Ontologies in Information Systems*. Springer, Berlin, 2004.
18. C. Tempich, H. S. Pinto, S. Staab, and Y. Sure. A case study in supporting DIstributed, Loosely-controlled and evolvInG Engineering of oNTologies (DILIGENT). In K. Tochtermann and H. Maurer, editors, *Proceedings of the 4th International Conference on Knowledge Management (I-KNOW'04)*, pages 225–232, Graz, Austria, June 30–July 02 2004. Journal of Universal Computer Science (JUCS).
19. C. Tempich, H. S. Pinto, Y. Sure, and S. Staab. An argumentation ontology for DIstributed, Loosely-controlled and evolvInG Engineering processes of oNTologies (DILIGENT). In C. Bussler, J. Davies, D. Fensel, and R. Studer, editors, *Second European Semantic Web Conference, ESWC 2005*, LNCS, Heraklion, Crete, Greece, May. Springer, Berlin, 2005.
20. C. Tempich. *Ontology Engineering and Routing in Distributed Knowledge Management Applications*. PhD thesis, Karlsruhe University, 2006.
21. C. Tempich, H. S. Pinto, and S. Staab. Ontology engineering revisited: An iterative case study. In *Proceedings of the 3rd European Semantic Web Conference*, 2006.
22. C. Tempich, H. S. Pinto, Y. Sure, D. Vrandecic, N. Casellas, and P. Casanovas. Evaluating DILIGENT Ontology Engineering in a Legal Case Study. In P. Casanovas, P. Noriega, D. Bourcier, and V. R. Benjamins, editors, *IVR 22nd World Congress – Law and Justice in a Global Society*. International Association for Philosophy of Law and Social Philosophy, 2005.

23. C. Tempich, E. Simperl, H. S. Pinto, M. Luczak, and R. Studer. Argumentation-based ontology engineering. *IEEE Intelligent Systems*, 22:52–29, 2007.
24. M. Uschold and M. King. Towards a methodology for building ontologies. In *Workshop on Basic Ontological Issues in Knowledge Sharing, held in conjunction with IJCAI-95*, Montreal, Canada, 1995.
25. D. Vrandečić, H. S. Pinto, Y. Sure, and C. Tempich. The diligent knowledge processes. *Journal of Knowledge Management*, 9(5):85–96, 2005.

Formal Concept Analysis

Gerd Stumme[1,2]

[1] Hertie Chair for Knowledge & Data Engineering, University of Kassel, Wilhelmshöher Allee 73, 34121 Kassel, Germany, http://www.kde.cs.uni-kassel.de

[2] Research Center L3S, Appelstr. 9a, 30167 Hannover, Germany, http://www.l3s.de

Summary. Formal concept analysis (FCA) is a mathematical theory about concepts and concept hierarchies. Based on lattice theory, it allows to derive concept hierarchies from datasets. In this survey, we recall the basic notions of FCA, including its relationship to folksonomies. The survey is concluded by a list of FCA based knowledge engineering solutions.

1 Introduction

Formal concept analysis (FCA) [71] is a mathematical theory for concepts and concept hierarchies that reflects an understanding of "concept" which is first mentioned explicitly in the Logic of Port Royal [2] in 1668 and has been established in the German standard 'DIN 2330 – Concepts and terms; general principles' [19]. FCA explicitly formalises extension and intension of a concept, their mutual relationships, and the fact that increasing intent implies decreasing extent and vice versa. Based on lattice theory, it allows to derive a concept hierarchy from a given dataset. FCA complements thus the usual ontology engineering approach, where the concept hierarchy is modeled manually.

FCA differs from other knowledge representation formalisms (like RDF (see chapter "Resource Description Framework"), description logics (see chapter "Description Logics"), OWL (see chapter "Web Ontology Language: OWL"), or conceptual graphs [53]). The standard DIN 2330 [19] helps us pointing out the difference. It distinguishes three levels: object level, concept level, and representation level (see Fig. 1). There is no immediate relationship between objects and names. This relationship is rather provided by concepts. On the concept level, the objects under discussion constitute the extension of the concept, while their shared properties constitute the intension of the

S. Staab and R. Studer (eds.), *Handbook on Ontologies,* International Handbooks on Information Systems, DOI 10.1007/978-3-540-92673-3,

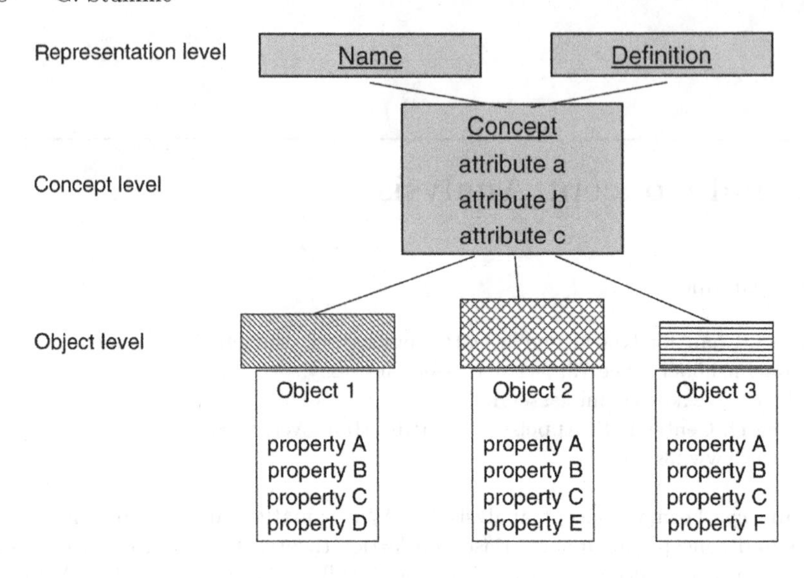

Fig. 1. Object level, concept level, and representation level according to DIN 2330

concept. On the representation level, a concept is specified by a definition and is referred to by a name.[1]

While formalisms like description logics, conceptual graphs, RDF or OWL focus on the representation level, the focus of FCA is on the concept level: in FCA, concepts consist of extension and intension, while concept names and definitions are not within the core notions of FCA. FCA, on the other hand, considers the extensional and the intensional part equally important, whereas the other formalisms are biased towards the intensional aspect of concepts. FCA is thus complementing other conceptual knowledge representations; and the combination of FCA with other representations has been the topic of many publications. For instance, several approaches combined FCA with description logics (e.g. [3,4,46,58]) and with conceptual graphs (e.g. [29,74,75]).

In the mid 1990s, a triadic version of FCA has been launched. Beside its extension and its intension, a tri-concept additionally contains a set of conditions under which extension and intension match. During the last dozen years, triadic FCA has been primarily of academic interest. With the rise of social bookmarking systems in the Web 2.0, however, they gained increased interest, as the data structures of social bookmarking systems – so-called *folksonomies* – match exactly the theory of triadic FCA.

[1] After a discussion of the three levels, DIN 2330 focusses on guidelines for good namings and definitions. It is thus a valuable resource for ontology engineers. An alternative source is the related international standard 'ISO 704: Terminology Work – Principles and Methods' [41].

1.1 Organisation of the Survey

In Sects. 2–5, the theory of FCA is presented, including basic notions like "formal context" and "concept lattice", visualisations with (nested) line diagrams, conceptual scaling, and the relationship to association rule mining. Section 6 presents an FCA based approach for knowledge acquisition. Section 7 discusses the relationship between triadic FCA and folksonomies in detail. Section 8 concludes the paper with a selection of FCA based knowledge engineering applications.

2 Formal Concept Analysis: A Theory About Concepts and Concept Hierarchies

This section presents the basic notions of FCA. Good starting points for a more in depth lecture are the textbooks [12,28,30], the proceedings of the Intl. Conferences on FCA[2] and the Intl. Conferences on Conceptual Structures[3] (both series are published as Springer Lecture Notes), as well as the collection of FCA publications at BibSonomy.[4]

To allow for a definition of concepts, FCA starts with a *(formal) context.*

Definition 1. *A* (formal) context *is a triple* $\mathbb{K} := (G, M, I)$*, where* G *is a set whose elements are called* objects*,* M *is a set whose elements are called* attributes*, and* I *is a binary relation between* G *and* M *(i.e.* $I \subseteq G \times M$*).* $(g, m) \in I$ *is read "the object* g has *the attribute* m*".*

Example 1. The left part of Fig. 2 shows a formal context developed for an educational movie about living beings and water ([66], see also [30]). The object set G comprises the eight living beings that were discussed in the movie, and the attribute set M lists features that distinguish the living beings. The binary relation I is given by the cross table and describes which living being has which of the attributes.

For a given context (G, M, I), we can define two *derivation operators*, both denoted by the $'$ symbol. They are used for defining formal concepts.

Definition 2. *For* $A \subseteq G$*, let* $A' := \{m \in M \mid \forall g \in A\colon (g, m) \in I\}$ *and, for* $B \subseteq M$*, let* $B' := \{g \in G \mid \forall m \in B\colon (g, m) \in I\}$*. A* (formal) concept *of a formal context* (G, M, I) *is a pair* (A, B) *with* $A \subseteq G$*,* $B \subseteq M$*,* $A' = B$ *and* $B' = A$*.*[5] *The sets* A *and* B *are called the* extent *and the* intent *of the formal concept* (A, B)*, respectively. The* subconcept–superconcept relation *is*

[2] http://www.informatik.uni-trier.de/~ley/db/conf/icfca/

[3] http://www.informatik.uni-trier.de/~ley/db/conf/iccs/

[4] http://www.bibsonomy.org/tag/fca

[5] This is equivalent to requiring that $A \times B \subseteq I$ such that neither A nor B can be enlarged without validating this condition.

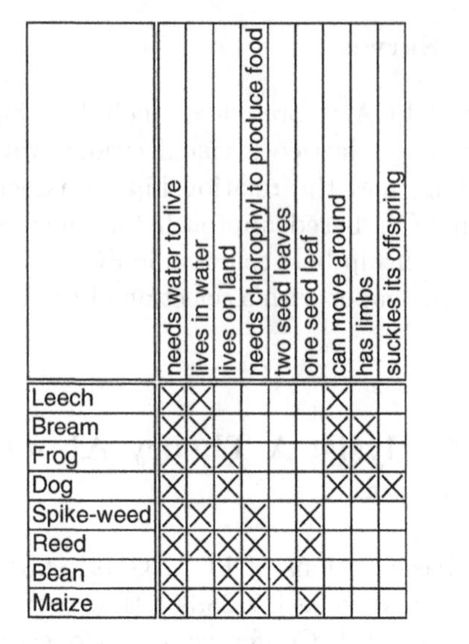

	needs water to live	lives in water	lives on land	needs chlorophyl to produce food	two seed leaves	one seed leaf	can move around	has limbs	suckles its offspring
Leech	×	×					×		
Bream	×	×					×	×	
Frog	×	×	×				×	×	
Dog	×		×				×	×	×
Spike-weed	×	×		×		×			
Reed	×	×	×	×		×			
Bean	×		×	×	×				
Maize	×		×	×		×			

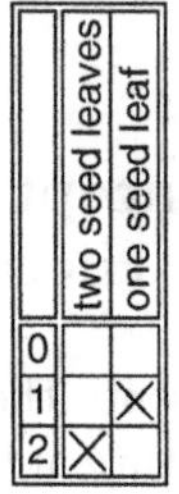

	two seed leaves	one seed leaf
0		
1		×
2	×	

Fig. 2. *Left*: a formal context about living beings. *Right*: a conceptual scale for the many-valued attribute "Number of seed leaves"

formalised by $(A_1, B_1) \leq (A_2, B_2) :\Longleftrightarrow A_1 \subseteq A_2 \quad (\Longleftrightarrow B_1 \supseteq B_2)$*. The set of all formal concepts of a context* $\mathbb{K}$ *together with this order relation is always a complete lattice, called the* concept lattice *of* $\mathbb{K}$ *and denoted by* $\underline{\mathfrak{B}}(\mathbb{K})$.

Theorem 1 ([30]). *The concept lattice* $\underline{\mathfrak{B}}(\mathbb{K})$ *of a context* $\mathbb{K} := (G, M, I)$ *is a complete lattice*[6] *in which infimum* ($\bigwedge$) *and supremum* ($\bigvee$)*of a set* $\{(A_t, B_t) | t \in T\}$ *of concepts (where* T *is any index set) are given by*

$$\bigwedge_{t\in T}(A_t, B_t) = \left(\bigcap_{t\in T} A_t, \left(\bigcup_{t\in T} B_t\right)''\right) \quad \text{and} \quad \bigvee_{t\in T}(A_t, B_t) = \left(\left(\bigcup_{t\in T} A_t\right)'', \bigcap_{t\in T} B_t\right).$$

A complete lattice $\mathbb{V}$ *is isomorphic to* $\underline{\mathfrak{B}}(\mathbb{K})$ *if and only if there are mappings* $\tilde{\gamma}\colon G \to \mathbb{V}$ *and* $\tilde{\mu}\colon M \to \mathbb{V}$ *such that* $\tilde{\gamma}(G)$ *is supremum-dense in* $\mathbb{V}$ *[i.e. each element in* $\mathbb{V}$ *is supremum of some subset of* $\tilde{\gamma}(G)$*] and* $\tilde{\mu}(M)$ *is infimum-dense in* $\mathbb{V}$*. In particular,* $\mathbb{V} \cong \underline{\mathfrak{B}}(\mathbb{K})$ *for* $\mathbb{K} = (\mathbb{V}, \mathbb{V}, \leq)$.

Concept lattices can be visualised as *line diagrams*. Line diagrams follow the conventions for the visualisation of hierarchical concept systems as established in the German standard DIN 2331 [18]. In a line diagram, each node represents a formal concept. A concept $\mathfrak{c}_1$ is a subconcept of a concept $\mathfrak{c}_2$ if and only if there is a path of descending edges from the node representing $\mathfrak{c}_2$ to the node representing $\mathfrak{c}_1$.

[6] That is, each subset of concepts has a unique greatest common subconcept (called its *infimum*) and a unique least common superconcept (called its *supremum*).

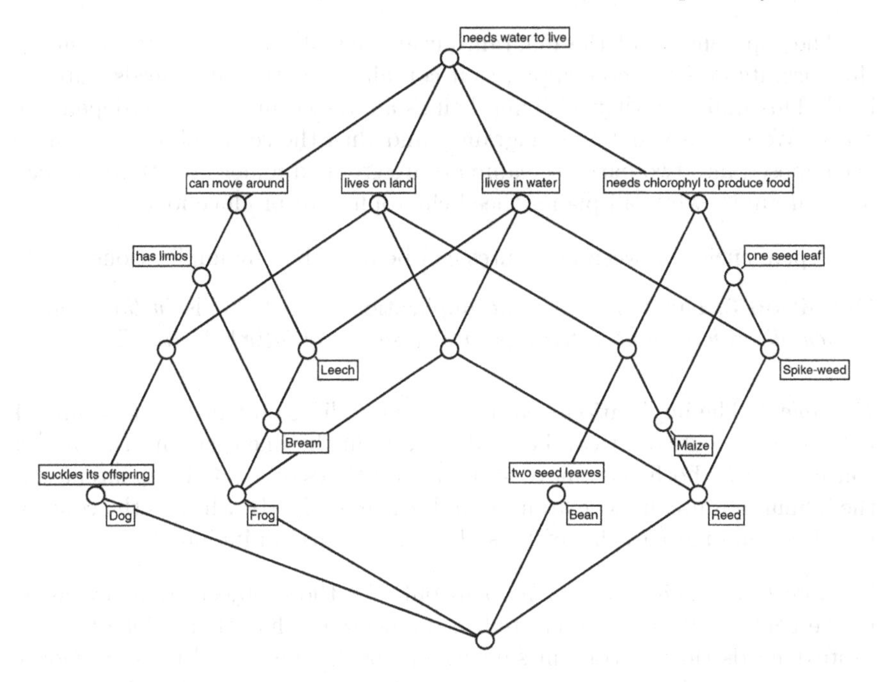

Fig. 3. The concept lattice of the context in Fig. 2

Example 2. Figure 3 shows the concept lattice of the context in Fig. 2 as a line diagram. We describe below how to read it.

The second part of Theorem 1 provides an efficient visualisation of concept lattices via line diagrams, as it states that a diagram is unambiguous even if each object name and each attribute name is displayed only once: in the line diagram, the name of an object $g \in G$ is attached to the node representing the object concept $\gamma(g) := (\{g\}'', \{g\}')$, and the name of an attribute $m \in M$ is attached to the node representing the attribute concept $\mu(m) := (\{m\}', \{m\}'')$. This means that the name of an object g is always attached to the node representing the smallest concept with g in its extent; dually, the name of an attribute m is always attached to the node representing the largest concept with m in its intent. We can read the context relation from the diagram because an object g has an attribute m if and only if the concept labeled by g is a subconcept of the one labeled by m. The extent of a concept consists of all objects whose labels are attached to subconcepts, and, dually, the intent consists of all attributes attached to superconcepts.

Example 3. For example, the concept in the very middle of Fig. 3 has {Frog, Reed} as extent, and {lives on land, lives in water, needs water to live} as intent. It is a direct subconcept of the two concepts ({lives on land}, {Dog, Frog, Maize, Reed, Bean}) and ({lives in water}, {Leech, Bream, Frog, Spike-weed, Reed}).

The top concept of the diagram always has all objects in its extent. In this case, its intent is non-empty, as it contains the attribute "needs water to live". This indicates that all living beings addressed in the movie depend on water. We see also that the diagram – and thus the set of objects – can be decomposed into two parts: the animals are grouped under the attribute "can move around", while all plants "need chlorophyll to produce food".

Dependencies between attributes can be described by implications.

Definition 3. *For* $X, Y \subseteq M$*, the* implication $X \to Y$ holds *in the context, if each object having all attributes in* X *also has all attributes in* Y*.*

Example 4. The implication {can move around, lives on land} $\to$ {has limbs} holds in this context. It can be read directly in the line diagram: the largest concept having both "can move around" and "lives on land" in its intent (i.e. the infimum of μ(can move around) and μ(lives on land), which is the concept that is second-most to the left) also has "has limbs" in its intent.

Note that these implications hold only for those objects that are listed in the context. If one is interested in implications that "hold globally", the context needs thus to contain sufficiently many "typical" objects. Section 6 discusses a knowledge acquisition process for interactively determining these typical objects.

3 Nested Line Diagrams

Nested line diagrams are used for visualising larger concept lattices. For their construction, the formal context is vertically split, and for each part a separate line diagram of its concept lattice is drawn. The line diagrams are then combined as shown in Fig. 4.

Example 5. Figure 4 shows a nested line diagram for the living being example. The outer diagram consists of four nodes, and results from the first three attributes of the left context in Fig. 2. It shows for instance that Dog, Bean, and Maize all live on land, while Frog and Reed live both on land and in water. The inner diagram results from the remaining six attributes of the left context in Fig. 2. It shows for instance that Maize, Reed and Spike-weed all have one seed leaf.

The combination of the two diagrams represents the direct product of both concept lattices. Its order relation can be read by replacing each of the four lines of the outer diagram by eight parallel lines linking corresponding nodes in the inner diagrams. The concept lattice given in Fig. 3 consists of the large nodes only. The concept mentioned above (the most central one in Fig. 3) is for instance represented in the lowest node of the outer diagram by the uppermost large circle of the inner diagram. The different shades of grey indicate

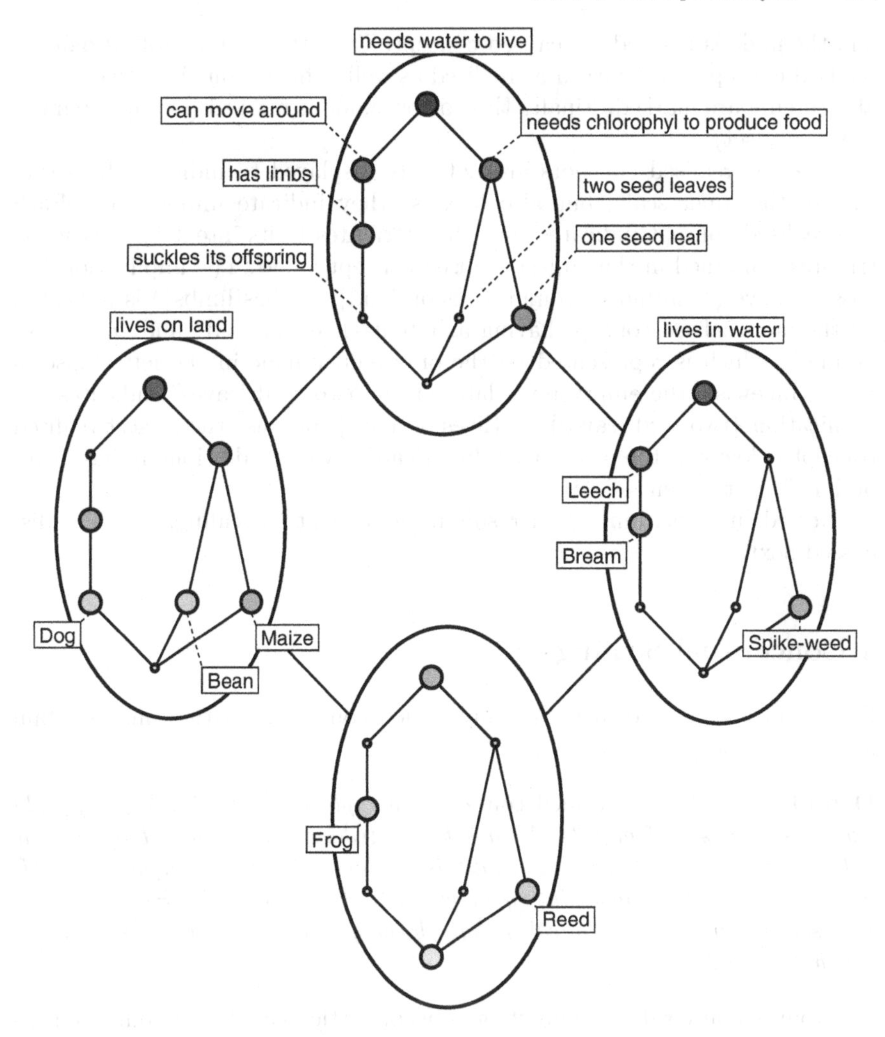

Fig. 4. A nested line diagram of the concept lattice in Fig. 3

the number of objects belonging to the respective concept extents. The 19 concepts represented by larger nodes are referred to as "realised concepts", as their intents correspond to concept intents of the non-nested line diagram in Fig. 3.

It can be shown [30], that the set of realised concepts is always a complete supremum–semilattice of the direct product, which is generated by the object concepts. This means that the supremum of any set of realised concepts is always a realised concept as well, and gives us a construction method for the nested line diagram: First, the direct product is drawn. Each object g is attached to the node representing the smallest concept with g in its extent, and the node is marked as realised concept. Then the suprema of all pairs of realised concepts are marked as realised as well. This is done iteratively until all suprema are marked. Finally the bottom node is marked (as supremum of the empty set).

The non-realised concepts are not only displayed to indicate the structure of the inner scale, but also because they indicate implications: Each non-realised concept indicates that the attributes in its intent imply the attributes contained in the largest realised concept below. The implication discussed above ({can move around, lives on land} → {has limbs}) is indicated by the non-realised concept having as intent "lives on land" and "can move around", which is represented by the only empty node in the left ellipse in Fig. 4. Likewise, the empty node labeled by "two seed leaves" indicates the implication {two seed leaves} → {lives on land}, because the largest realised concept below it is the one labeled by "Bean" – which additionally has "lives on land" in its intent.

Nested line diagrams often result from conceptual scaling, which is discussed next.

4 Conceptual Scaling

FCA is also able to deal with many-valued contexts, i.e. they may contain attribute-value pairs.

Definition 4. *A* many-valued context *is a tuple* $\mathbb{K} := (G, M, (W_m)_{m \in M}, I)$ *where G is a set of objects, M a set of attributes, W_m the set of possible values for the attribute $m \in M$, and I is a relation $I \subseteq G \times \{(m, w) \mid m \in M, w \in W_m\}$ with the constraint $(g, m, w_1) \in I, (g, m, w_2) \in I \implies w_1 = w_2$ imposed. $(g, m, w) \in I$ indicates that object $g \in G$ has value $w \in W_m$ for attribute $m \in M$.*

From a many-valued context, a concept lattice cannot be computed directly. One has to transform it first into a a one-valued context. This transformation is called *conceptual scaling* [26].

Definition 5. *A* conceptual scale *for a subset $B \subseteq M$ of attributes is a (one-valued) formal context $\mathbb{S}_B := (G_B, M_B, I_B)$ with $G_B \subseteq \times_{m \in B} W_m$.*

Let $\mathfrak{S}$ be the set of conceptual scales for the many-valued context $\mathbb{K} := (G, M, (W_m)_{m \in M}, I)$. For any subset $\mathcal{S} \subseteq \mathfrak{S}$ of scales, we can now translate the many-valued context into a one-valued one:

Definition 6. *The* derived context $\mathbb{K}_\mathcal{S}$ *is defined by* $\mathbb{K}_\mathcal{S} := (G, \bigcup_{\mathbb{S}_B \in \mathcal{S}} M_B, I_\mathcal{S})$ *with $(g, n) \in I_\mathcal{S}$ if there exists a scale $\mathbb{S}_B \in \mathcal{S}$ with $m \in M_B$ and $w \in W_m$ where $(g, m, w) \in I$ and $(g, n) \in I_B$.*

Example 6. The two attributes "one seed leaf" and "two seed leaves" in our running example have been derived from a many-valued attribute "Number of seed leaves", which has as set of possible values $W_{\text{Number of seed leaves}} := \{0, 1, 2\}$. Applying the conceptual scale $\mathbb{S}_{\{\text{Number of seed leaves}\}}$ that is displayed at the right of Fig. 2 results in the two columns in the context in the left of Fig. 2 that are labeled by "one seed leaf" and "two seed leaves".

The concept lattice of the derived context can canonically be visualised in a nested line diagram. For each scale, its line diagram is pre-computed. The nested line diagram for any combination of scales can then be combined online. This combination of conceptual scaling and nested line diagrams is implemented in the open source software ToscanaJ.[7]

Conceptual scaling covers typical scales as known from measurement theory: Let the set G of objects of the scale w. l. o. g. be $\{0, \ldots, n\}$. The *nominal scale* has then $\{= 0, \ldots, = n\}$ as set of attributes; and is applied to many-valued attributes with incomparable values (e.g. eye colors). The *ordinal scale* has $\{\leq 0, \ldots, \leq n\}$ as set of attributes; and is used for attributes where the values are ordered, like weights or costs. The *inter-ordinal scale* extends the ordinal scale. It has $\{\leq 0, \ldots, \leq n, > 0, \ldots, > n\}$ as set of attributes; and is applied when intervals are of interest. (The relations of all these scales are defined in the obvious way.)

Example 7. The scale for the number of seed leaves is a nominal scale, with the difference that the value 0 has been suppressed, as it was not considered meaningful for this application. The choice of the appropriate scale is the task of the knowledge engineer. Depending on his aim, he could have chosen an ordinal or inter-ordinal scale instead.

5 Iceberg Concept Lattices and Bases of Association Rules

The concept lattice of a formal context can be considered as a conceptual, hierarchical clustering of the set of objects, with the concept extents being the clusters and the intents being their descriptions [11, 40, 56]. In comparison to other conceptual clustering approaches, concept lattices have structural properties which can be stated explicitly: they do not depend on any parameters (whose semantics are often difficult to interpret), nor on the order in which the input is presented to the algorithm, nor on any particularities of the implementation. Another distinction to other hierarchical clustering results is that they allow for multiple inheritance (and not only for trees), so that all

[7] http://toscanaj.sourceforge.net/, see also [37] for its foundations and a description of its predecessor TOSCANA.

potentially interesting specialisation paths are contained in the resulting hierarchy. The trade-off is a high complexity, as the concept lattice can be – in the worst case – exponential in the size of the input.

To alleviate this complexity problem, the notion of *iceberg concept lattices* has been introduced in [64]. For a given threshold *minsupp* $\in [0,1]$, the iceberg concept lattice of a formal context $\mathbb{K}$ contains all concepts (A, B) of $\mathbb{K}$ whose support $supp(A,B) := \frac{|A|}{|G|}$ $(= \frac{|B'|}{|G|})$ is larger than *minsupp*. Iceberg concept lattices show thus only the top-most part of a concept lattice.

Example 8. In Fig. 5, an iceberg concept lattice of the MUSHROOMS data set from the *UCI KDD Archive*[8] is shown. The context consists of 8,416 mushrooms as objects, and of 80 attributes (which were derived from 22 many-valued attributes with nominal scales); and its complete concept lattice consists of 32,086 concepts. The minimum support threshold in Fig. 5 is set to 0.7, i.e. all concepts whose extents do not comprise at least 70% of all mushrooms are pruned. Instead of the objects names, the diagram displays the support of each concept. From the diagram, we can read three implications: {} → {veil type: partial}, {ring number: one, veil color: white} → {gill attachment: free}, and {ring number: one, gill spacing: close} → {veil color: white, gill attachment: free}. These implications hold for the whole dataset.

Iceberg concept lattices were developed as an answer of FCA to the association rule mining problem [1]. *Association rules* are statements of the type "67% of the customers buying cereals and sugar also buy milk (where 7% of all customers buy all three items)". The task of mining association rules is to determine all rules whose *confidences* (67% in the example) and *supports* (7% in the example) are above user-defined thresholds.

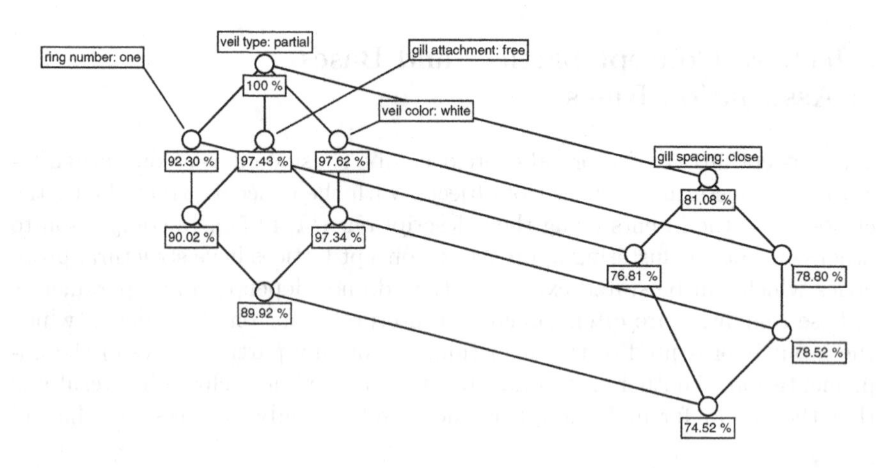

Fig. 5. Iceberg concept lattice for the MUSHROOMS data set for *minsupp* = 70%

[8] http://kdd.ics.uci.edu/

The expensive step of computing association rules is to compute all frequent itemsets, i.e. in terms of FCA, all subsets $B \subseteq M$ of the attribute set M of a context (G, M, I) with $supp(B) := \frac{|B'|}{|G|} \geq minsupp$. Lakhal et al. [42], Zaki [77], and Stumme [61] discovered independently that it is sufficient to compute only the frequent concept intents. In the data mining community, they were then called frequent *closed* itemsets, because the set of concept intents forms a closure system (which means that the intersection of any set of concept intents is again a concept intent). The corresponding closure operator is the function which maps each itemset $B \subseteq M$ to its *closure*, the itemset B''.

The support of any itemset $B \subseteq M$ equals the support of the smallest concept intent containing it (which is just B''). Hence, for computing all association rules, it is sufficient to consider only frequent concept intents instead of all frequent itemsets.

One can go even one step further and compute not all association rules, but only a *basis*, i.e. a non-redundant subset from which all other rules can be derived [43, 65, 76]. The smallest basis for exact rules (i.e. those with 100% confidence; called implications above) is the so-called *Duquenne–Guigues basis*, see [21, 24, 65] for details. The exact rules can also be read directly from the line diagram as described above. The *Luxenburger basis* ([65], based on work by M. Luxenburger [39]) is a basis for the approximate rules (those with confidence < 100%) which can be visualised in the concept lattice.

Example 9. The Luxenburger basis for the MUSHROOMS database with a minimum support of 70% and a minimum confidence of 95% is shown in Fig. 6. Every edge (read downwards) is one rule of the basis, with its confidence listed. Its support equals the support of the concept the arrow is pointing to, and can be read from Fig. 5.

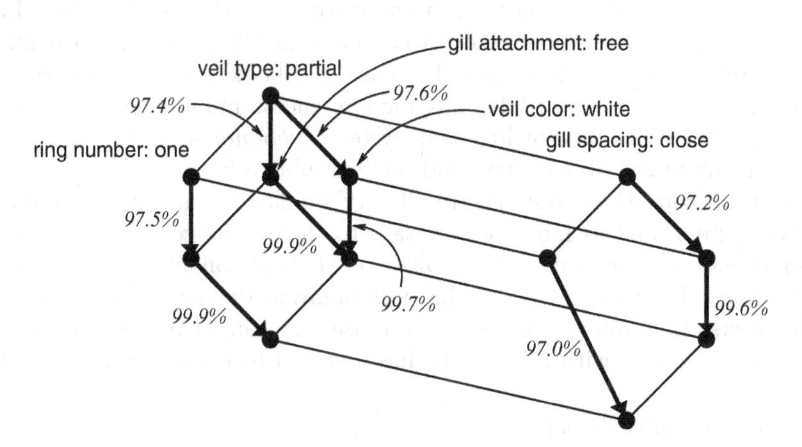

Fig. 6. Luxenburger basis for the Mushrooms data set

The first results on concept intents/closed itemsets launched the search for other condensed representations of the set of frequent itemsets, for instance key sets [5] = free sets [8] = minimal generators [43], non-derivable itemsets [10], disjunction free sets [9], and k-free sets [47]. These presentations have not yet been extended to bases of association rules (with the exception of minimal generators, see [43]).

6 Knowledge Acquisition with Formal Concept Analysis

Knowledge Acquisition aims at supporting the acquisition of knowledge from humans and its transformation into a formal model. The most prominent FCA technique is B. Ganter's *Attribute Exploration* [24] (see also [30]). It addresses the problem of a formal context where the object set is not completely known a priori, or too large to be completely listed. In an interactive, iterative approach, the Duquenne–Guigues basis is computed. Each implication is suggested to the user. She has then either to accept it (i.e. she excludes potential objects) or to provide a counter-example (i.e. she provides a (typical) object), until the basis – and thus the concept lattice – is completely determined.

Concept Exploration extends this approach to situations where both the object set and the attribute set of the context are not completely known a priori or too large [57,60]. An overview over interactive knowledge acquisition techniques based on FCA can be found in [59].

7 Folksonomies and Triadic Concept Analysis

Social resource sharing systems are Web 2.0 systems that allow users to upload their resources, and to label them with arbitrary words, so-called *tags*. Each system has a specific type of resources it supports. Flickr, for instance, allows the sharing of photos, del.icio.us the sharing of bookmarks, CiteULike[9] and Connotea[10] the sharing of bibliographic references, and 43Things[11] even the sharing of goals in private life. Our system *BibSonomy*[12] ([33], see Fig. 7) allows to share both bookmarks and bibliographic references.

In their core, these systems are all very similar. Users can add resources to the system, and assign arbitrary tags to them. The collection of a users assignments is his *personomy*, the collection of all personomies constitutes the *folksonomy*. The user can explore his personomy, as well as the personomies of other users, in all dimensions: for a given user one can see all resources he has uploaded, together with the tags he has assigned to them (see Fig. 7); when

[9] http://www.citeulike.org
[10] http://www.connotea.org
[11] http://www.43things.com
[12] http://www.bibsonomy.org

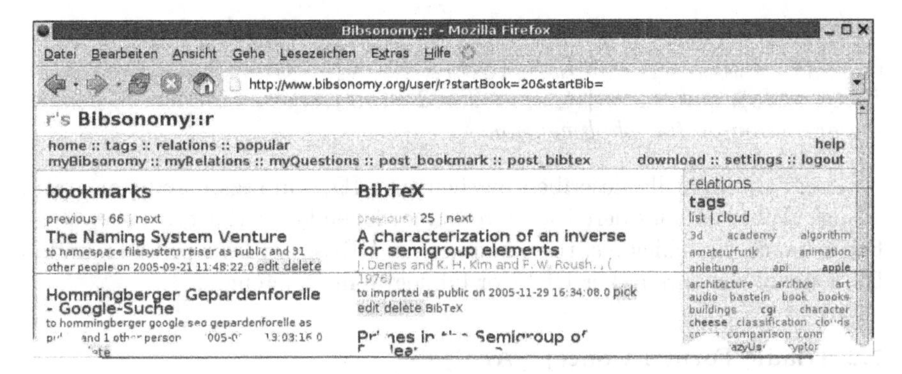

Fig. 7. BibSonomy displays bookmarks and bibliographic references simultaneously

clicking on a resource one sees which other users have uploaded this resource and how they tagged it; and when clicking on a tag one sees who assigned it to which resources.

With the emergence of social bookmarking systems, the interest in a triadic extension of FCA raised again. This extension was introduced by R. Wille and F. Lehmann in [38] in 1995, where a formal context was extended by a third dimension: the sets of objects and of attributes were complemented with a third set, containing so-called conditions. The notion of concepts and of concept lattices was lifted to the third dimension as well.

Structurally, triadic contexts are equal to folksonomies. Hence the whole theory of triadic FCA can be applied directly to folksonomies.

In the first part of this section, we present a formal definition of folksonomies, before giving an overview over triadic FCA, and connecting its notions to those of folksonomies.

7.1 Folksonomies

The word "folksonomy" is a blend of the words "taxonomy" and "folk", and stands for conceptual structures created by the people. Folksonomies are thus a bottom-up complement to more formalised Semantic Web technologies, as they rely on *emergent semantics* [54,55] which result from the converging use of the same vocabulary. A folksonomy describes the users, resources, and tags, and the user-based assignment of tags to resources.

Definition 7 ([34]). *A* folksonomy *is a tuple* $\mathbb{F} := (U, T, R, Y, \prec)$ *where:*

- U*,* T*, and* R *are finite sets, whose elements are called* users, tags *and* resources, *resp.*
- Y *is a ternary relation between them, i.e.* $Y \subseteq U \times T \times R$*, called tag assignments (TAS for short)*
- $\prec$ *is a user-specific subtag/supertag-relation, i.e.* $\prec \subseteq U \times T \times T$*, called* subtag/supertag relation

The personomy $\mathbb{P}_u$ *of a given user* $u \in U$ *is the restriction of* $\mathbb{F}$ *to* u*, i.e.* $\mathbb{P}_u := (T_u, R_u, I_u, \prec_u)$ *with* $I_u := \{(t,r) \in T \times R \mid (u,t,r) \in Y\}$, $T_u := \pi_1(I_u)$, $R_u := \pi_2(I_u)$, *and* $\prec_u := \{(t_1,t_2) \in T \times T \mid (u,t_1,t_2) \in \prec\}$, *where* π_i *denotes the projection on the ith dimension.*

Users are typically described by their user ID, and tags may be arbitrary strings. What is considered as a resource depends on the type of system. For instance, in del.icio.us, the resources are URLs, in Flickr pictures, and in BibSonomy they are either URLs or bibliographic references.

7.2 Triadic Formal Concept Analysis

Inspired by the pragmatic philosophy of Charles S. Peirce with its three universal categories [44], Rudolf Wille and Fritz Lehmann extended FCA in 1995 with a third category [38].

Definition 8 ([38]). *A* triadic formal context *is a quadruple* $\mathbb{K} := (G, M, B, Y)$ *where* G, M, *and* B *are sets, and* Y *is a ternary relation between* G, M, *and* B, *i.e.* $Y \subseteq G \times M \times B$. *The elements of* G, M, *and* B *are called* objects, attributes, *and* conditions, *resp, and* $(g, m, b) \in Y$ *is read "object* g *has attribute* m *under condition* b*".*

In terms of FCA, a folksonomy (without its is-a relation $\prec$ on the tag set) is thus just a triadic formal context $\mathbb{F} := (U, T, R, Y)$. In the remainder of this paper, we will thus use the two terms synonymously.

Triadic concepts are defined in a manner analogue to the dyadic case.

Definition 9 ([38]). *A* tri-concept *of* $\mathbb{K}$ *is a triple* (A_1, A_2, A_3) *with* $A_1 \subseteq G$, $A_2 \subseteq M$, $A_3 \subseteq B$, *and* $A_1 \times A_2 \times A_3 \subseteq Y$ *such that none of its three components can be enlarged without violating this condition.* A_1 *is called the* extent, A_2 *the* intent, *and* A_3 *the* mode *of the tri-concept* (A_1, A_2, A_3).

We define three quasi-orders $\lesssim_1$, $\lesssim_2$, *and* $\lesssim_3$ *on the set of all tri-concepts, one for each dimension* G, M, *and* B: $(A_1, A_2, A_3) \lesssim_i (B_1, B_2, B_3)$ *iff* $A_i \subseteq B_i$, *for* $i = 1, 2, 3$.

Lemma 1. *For two tri-concepts* $\mathfrak{a}$ *and* $\mathfrak{b}$, $\mathfrak{a} \lesssim_i \mathfrak{b}$ *and* $\mathfrak{a} \lesssim_j \mathfrak{b}$ *imply* $\mathfrak{b} \lesssim_k \mathfrak{a}$, *for* $\{i, j, k\} = \{1, 2, 3\}$.

Lehmann and Wille present in [38] an extension of the theory of ordered sets and (concept) lattices to the triadic case, and discuss structural properties. This approach initiated research on the theory of *concept trilattices*, e.g. [6, 7, 17, 25, 72] see BibSonomy[13] for more references. References [38] and [17] present several ways to project a triadic context to a dyadic one. Reference [62] presents a model for navigating a triadic context by visualising concept lattices of such projections.

[13] http://www.bibsonomy.org/tag/triadic+fca

7.3 Analysing Folksonomies with Triadic FCA

The notion of iceberg concept lattices/frequent closed itemsets has been lifted to the triadic case by Jäschke et al. [35]. For each of the three dimensions, one can provide a minimum support threshold. A tri-concept is then said to be frequent, if the cardinalities of its extent, intent, and mode are all above the respective thresholds. The authors also provide an efficient algorithm for mining iceberg tri-concept lattices, based on the Next Closure algorithm presented above.

Example 10. Figure 8 shows the iceberg tri-concept lattice for the publication part of the social bookmark and publication management system

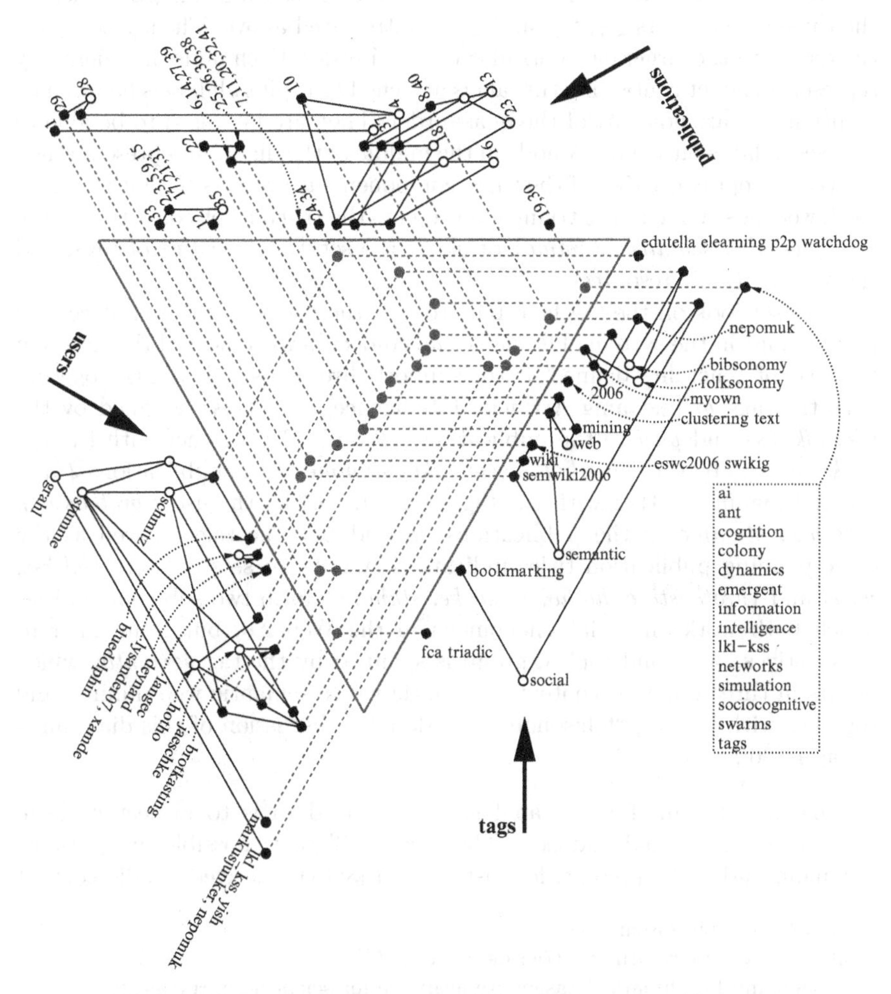

Fig. 8. All frequent tri-concepts of the BibSonomy publications

BibSonomy.[14] It includes all publications entered until November 23, 2006, excluding the content of the DBLP computer science bibliography[15] The resulting snapshot contains $|Y| = 44,944$ tag assignments built by $|U| = 262$ users, containing $|R| = 11,101$ publication references tagged with $|T| =$ 5,954 distinct tags.[16] The complete tri-concept lattice consists of 13,992 tri-concepts. Figure 8 shows all 21 tri-concepts which contain at least three users, two tags and two publication entries. The publications in the figure are substituted by numbers for space reasons. They can be found under http://www.bibsonomy.org/group/kde/trias_example?items=50, and are numbered in alpabetic order of the paper title.

The 21 nodes in the center of the triangle represent the 21 frequent tri-concepts. The sets of users, tags, and resources composing a tri-concept can be read off the three sides of the triangle. There, three Hasse diagrams display the three quasi-orders $\lesssim_1$, $\lesssim_2$, and $\lesssim_3$ as introduced above. The arrows guide the reader to the larger elements of each quasi-order. Each node in a hierarchy represents the set containing the labels attached to it plus all labels below. The empty nodes are not part of the quasi-order. They are just used to be able to place each label once only. A node in the middle of the diagram represents then the tri-concept consisting of the three components it projects to. For instance, the lower-most node in the triangle represents the tri-concept consisting of the set {*jaeschke, schmitz, stumme*} of users, the set {*fca, triadic*} of tags, and the set {*1, 37*} of resources.

A closer look on the tag hierarchy reveals the content of the most central publications in the system. The tag *social* co-occurs with most of the tags. On the level of generality defined by the support thresholds, this tag is (together with the tags *ai* [meaning Artificial Intelligence], ..., *tags*) assigned by the users *lkl_kss* and *yish* to the publications *19* and *30*, (together with the tag *bookmarking*) by the users *hotho, jaeschke, stumme* to the publications *4* and *28*, and (again together with the tag *bookmarking*) by the users *brotkasting, jaeschke, stumme* to the publications *28* and *29*. The tags as well as the corresponding publication titles indicate that the two sets of users {*lkl_kss, yish*} and {*brotkasting, hotho, jaeschke, stumme*} form two sub-communities which both work on social phenomena in the Web 2.0, but from different perspectives. A second topical group is spanned by the tag *semantic*, which occurs in three different contexts: semantic wikis, semantic web mining, and together with the tag "folksonomy". A detailed discussion of this diagram is given in [36].

One way to simplify the analysis of triadic data is to project it down to a two-dimensional dataset. There are different possible projections. Lehmann and Wille present, for instance, in [38] the derived dyadic context

[14] http://www.bibsonomy.org

[15] http://www.informatik.uni-trier.de/~ley/db/

[16] BibSonomy benchmark datasets are available for scientific purposes, see http://www.bibsonomy.org/faq.

$\mathbb{K}^{(1)} := (G, M \times B, Y^{(1)})$ with $(g,(m,b)) \in Y^{(1)} :\iff (g,m,b) \in Y$, and its two symmetric variations. The set B was also used to define two modal operators for generating projections [17]. A survey over the potential projections is given in [62], where also a model for navigating a triadic context by visualising concept lattices of such projections is given.

We discussed in [49], how to compute association rules from a triadic context based on such projections. A first step towards truly "triadic association rules" has been done by Ganter and Obiedkov [25].

The focus of this section was on folksonomy research that is related to Formal Concepts Analysis. A general survey over theory and applications of folksonomies is out of the scope of this article. A collection of folksonomy related publications can be found at http://www.bibsonomy.org/tag/folksonomies.

8 Ontology Engineering with Formal Concept Analysis

We conclude this survey on FCA with a short discussion of some FCA based ontology engineering solutions. A more detailed survey can be found in [15].

8.1 Browsing Ontologies

Ferré and Ridoux use FCA for building a virtual file system in which the files are considered as the objects of a formal context, and descriptors (in first order logic) are its attributes [22]. The extents of the formal concepts are then used as virtual folders. A similar idea is exploited in the Conceptual Email Manager[17] [16], which is an FCA based email management system that allows to classify emails simultaneously in more than one folder of a folder/concept hierarchy. The application of FCA results in the visualisation of a concept lattice which additionally contains all possible intersections (in terms of logic: conjunctions) of folders as new, virtual folders. The ImageSleuth system [20] follows this approach for browsing collections of images. Another system along this line is the Courseware Watchdog [69], which allows for browsing also the non-hierarchical relations of an ontology. A formal framework of this approach is presented by Tane [67,68], based on the notion of a multi-context [73], which is roughly equivalent to a set of RDF statements (without reification).

8.2 Non-atomic Attributes

The attributes of a formal context can be considered as unary predicates. In the standard context, they are atomic, but no one prevents us from defining them in a more complex knowledge representation. In this sense, first order logic and SQL were used for conceptual scaling [45], and the description logic $\mathcal{ALC}$ for defining one-valued attributes of formal contexts [46]. These

[17] See http://www.mail-sleuth.com/ for a commercial follow-up.

approaches were followed independently by Chaudron and Maille [13] and Ferre and Ridoux [23], using first order logic. The latter was then exploited for the virtual file system [22] described above.

8.3 Ontology Learning and Merging

The Attribute Exploration procedure described in Sect. 6 was used by Baader et al. and Rudolph for computing the hierarchy of all conjunctions of all concepts of a given description logic knowledge base [3,4,48] . Ganter used it for interactively refining the concept hierarchy of a given ontology [27].

The method FCA–MERGE [63] for merging ontologies follows a bottom-up approach for merging ontologies. For the source ontologies, it extracts instances from a given set of domain-specific text documents by applying natural language processing techniques. The resulting concept lattice provides a conceptual clustering of the concepts of the source ontologies. It is explored and interactively transformed to the merged ontology by the ontology engineer. Cimiano et al. follow a similar approach for ontology learning [14]. They use text mining methods to extract a formal context from a set of documents. Its concept lattice is then transformed automatically into an ontology.

8.4 Class Design in Software Engineering

The task of designing class hierarchies in object-oriented (OO) programming has many similarities with the construction of an ontology. There are several FCA applications in this area. Reference [32] use FCA for deriving class hierarchies from use cases. Reference [31] uses concept lattices for structuring OO class hierarchies. Reference [50] builds a conceptual hierarchy from source code, based on preprocessor commands. Reference [52] is the most complex application in this domain. It describes a semantics-preserving refactoring of a class hierarchy using FCA. Surveys of FCA support for software engineering are given in [51] and [70].

Remark. An extended list of references can be found at BibSonomy.[18]

References

1. R. Agrawal, T. Imielinski, and A. Swami. Mining association rules between sets of items in large databases. In *Proceedings of the 1993 ACM SIGMOD International Conference on Management of Data (SIGMOD'93)*, pages 207–216. ACM, New York, 1993.
2. A. Arnauld and P. Nicole. *La logique ou l'art de penser – Contenant, outre les règles communes, plusieurs observations nouvelles, propres à former le jugement.* Ch. Saveux, 1668.

[18] http://www.bibsonomy.org/user/stumme/OntologyHandbook+FCA

3. F. Baader. Computing a minimal representation of the subsumption lattice of all conjunctions of concepts defined in a terminology. In *Proceedings of the International Symposium on Knowledge Retrieval, Use, and Storage for Efficiency, KRUSE 95*, pages 168–178, Santa Cruz, USA, 1995.
4. F. Baader, B. Ganter, B. Sertkaya, and U. Sattler. Completing description logic knowledge bases using formal concept analysis. In M. M. Veloso, editor, *Proc. IJCAI 2007*, pages 230–235, 2007.
5. Y. Bastide, R. Taouil, N. Pasquier, G. Stumme, and L. Lakhal. Mining frequent patterns with counting inference. *SIGKDD Explorations, Special Issue on Scalable Algorithms*, 2(2):71–80, 2000.
6. K. Biedermann. How triadic diagrams represent conceptual structures. In D. Lukose, H. S. Delugach, M. Keeler, L. Searle, and J. F. Sowa, editors, *Conceptual Structures: Fulfilling Peirce's Dream*, number 1257 in LNAI, pages 304–317. Springer, Heidelberg, 1997.
7. K. Biedermann. Triadic Galois connections. In K. Denecke and O. Lders, editors, *General Algebra and Applications in Discrete Mathematics*, pages 23–33. Shaker, Aachen, 1997.
8. J.-F. Boulicaut, A. Bykowski, and C. Rigotti. Approximation of frequency queries by means of free-sets. In *PKDD '00: Proceedings of the 4th European Conference on Principles of Data Mining and Knowledge Discovery*, pages 75–85. Springer, London, 2000.
9. A. Bykowski and C. Rigotti. A condensed representation to find frequent patterns. In *PODS '01: Proceedings of the Twentieth ACM SIGMOD-SIGACT-SIGART Symposium on Principles of Database Systems*, pages 267–273. ACM Press, New York, 2001.
10. T. Calders and B. Goethals. Mining all non-derivable frequent itemsets. In *PKDD*, pages 74–85, 2002.
11. C. Carpineto and G. Romano. GALOIS : An order-theoretic approach to conceptual clustering. In *Proceedings of the 10th International Conference on Machine Learning (ICML'90)*, pages 33–40, July 1993.
12. C. Carpineto and G. Romano. *Concept Data Analysis*. Wiley, New York, 2004.
13. L. Chaudron and N. Maille. Generalized formal concept analysis. In B. Ganter and G. W. Mineau, editors, *ICCS*, volume 1867 of *Lecture Notes in Computer Science*, pages 357–370. Springer, Berlin, 2000.
14. P. Cimiano, A. Hotho, and S. Staab. Learning concept hierarchies from text corpora using formal concept analysis. *Journal on Artificial Intelligence Research*, 24:305–339, 2005.
15. P. Cimiano, A. Hotho, G. Stumme, and J. Tane. Conceptual knowledge processing with formal concept analysis and ontologies. In P. Eklund, editor, *Concept Lattices*, volume 2961 of *LNAI*, pages 189–207, Heidelberg, 2004. Second International Conference on Formal Concept Analysis, ICFCA 2004, Springer, Berlin, 2004.
16. R. J. Cole, P. W. Eklund, and G. Stumme. Document retrieval for email search and discovery using formal concept analysis. *Journal of Applied Artificial Intelligence (AAI)*, 17(3):257–280, 2003.
17. F. Dau and R. Wille. On the modal understanding of triadic contexts. In R. Decker and W. Gaul, editors, *Classification and Information Processing at the Turn of the Millennium*, Proc. Gesellschaft für Klassifikation, 2001.
18. Deutsches Institut für Normung. DIN 2331: Begriffssysteme und ihre Darstellung, 1980.

19. Deutsches Institut für Normung. DIN 2330: Begriffe und Benennungen - Allgemeine Grundsätze, 1993.
20. J. Ducrou, B. Vormbrock, and P. W. Eklund. FCA-based browsing and searching of a collection of images. In *Proc. of the 14th Int. Conference on Conceptual Structures*. Springer, Berlin, 2006.
21. V. Duquenne and J.-L. Guigues. Famille minimale d'implications informatives résultant d'un tableau de données binaires. *Mathématiques et Sciences Humaines*, 24(95):5–18, 1986.
22. S. Ferré and O. Ridoux. A file system based on concept analysis. In J. W. Lloyd, V. Dahl, U. Furbach, M. Kerber, K.-K. Lau, C. Palamidessi, L. M. Pereira, Y. Sagiv, and P. J. Stuckey, editors, *Computational Logic*, volume 1861 of *Lecture Notes in Computer Science*, pages 1033–1047. Springer, Berlin, 2000.
23. S. Ferré and O. Ridoux. A logical generalization of formal concept analysis. In G. Mineau and B. Ganter, editors, *Int. Conf. Conceptual Structures*, LNCS 1867, pages 371–384. Springer, Berlin, 2000.
24. B. Ganter. Algorithmen zur Formalen Begriffsanalyse . In B. Ganter, R. Wille, and K. E. Wolff, editors, *Beiträge zur Begriffsanalyse*, pages 241–254. BI-Wissenschaftsverlag, Mannheim, 1987.
25. B. Ganter and S. A. Obiedkov. Implications in triadic contexts. In *Conceptual Structures at Work: 12th International Conference on Conceptual Structures*, volume 3127 of *Lecture Notes in Computer Science*, pages 186–195. Springer, Berlin, 2004.
26. B. Ganter, J. Stahl, and R. Wille. Conceptual measurement and many-valued contexts. In W. Gaul and M. Schader, editors, *Classification as a Tool of Research*, pages 169–176. North-Holland, Amsterdam, 1986.
27. B. Ganter and G. Stumme. Creation and merging of ontology top-levels. In A. de Moor, W. Lex, and B. Ganter, editors, *Conceptual Structures for Knowledge Creation and Communication*, volume 2746 of *LNAI*, pages 131–145. Springer, Heidelberg, 2003.
28. B. Ganter, G. Stumme, and R. Wille, editors. *Formal Concept Analysis – Foundations and Applications*, volume 3626 of *LNAI*. Springer, Heidelberg, 2005.
29. B. Ganter and R. Wille. Contextual attribute logic. In W. M. Tepfenhart and W. R. Cyre, editors, *ICCS*, volume 1640 of *Lecture Notes in Computer Science*, pages 377–388. Springer, Berlin, 1999.
30. B. Ganter and R. Wille. *Formal Concept Analysis: Mathematical Foundations*. Springer, 1999. Translation of: *Formale Begriffs analyse: Mathematische Grundlagen*. Springer, Heidelberg, 1996.
31. R. Godin and P. Valtchev. *Formal Concept Analysis-Based Class Hierarchy Design in Object-Oriented Software Development*, volume 3626 of *LNAI*, pages 304–323. Springer, Berlin, 2005.
32. W. Hesse and T. A. Tilley. *Formal Concept Analysis used for Software Analysis and Modelling*, volume 3626 of *LNAI*, pages 288–303. Springer, Berlin, 2005.
33. A. Hotho, R. Jäschke, C. Schmitz, and G. Stumme. BibSonomy: A social bookmark and publication sharing system. In *Proc. of the ICCS 2006 Conceptual Structures Tool Interoperability Workshop*, 2006.
34. A. Hotho, R. Jäschke, C. Schmitz, and G. Stumme. Information retrieval in folksonomies: Search and ranking. In *Proceedings of the 3rd European Semantic Web Conference*, Lecture Notes in Computer Science. Springer, Berlin, 2006.

35. R. Jäschke, A. Hotho, C. Schmitz, B. Ganter, and G. Stumme. Trias – An algorithm for mining iceberg tri-lattices. In *Proceedings of the 6th IEEE International Conference on Data Mining (ICDM 06)*, pages 907–911. IEEE Computer Society, Hong Kong, December 2006.
36. R. Jäschke, A. Hotho, C. Schmitz, B. Ganter, and G. Stumme. Discovering shared conceptualizations in folksonomies. *Journal of Web Semantics*, 6(1):38–53, 2008.
37. W. Kollewe, M. Skorsky, F. Vogt, and R. Wille. TOSCANA – ein Werkzeug zur begrifflichen Analyse und Erkundung von Daten. In R. Wille and M. Zickwolff, editors, *Begriffliche Wissensverarbeitung-Grundfragen und Aufgaben*, pages 267–288. BI-Wissenschaftsverlag, Mannheim, 1994.
38. F. Lehmann and R. Wille. A triadic approach to formal concept analysis. In G. Ellis, R. Levinson, W. Rich, and J. F. Sowa, editors, *Conceptual Structures: Applications, Implementation and Theory*, volume 954 of *Lecture Notes in Artificial Intelligence*, pages 32–43. Springer, Berlin, 1995.
39. M. Luxenburger. Implications partielles dans un contexte. *Mathématiques, Informatique et Sciences Humaines*, 29(113):35–55, 1991.
40. G. Mineau and R. Godin. Automatic structuring of knowledge bases by conceptual clustering. *IEEE Transactions on Knowledge and Data Engineering*, 7(5):824–829, 1985.
41. I. O. of Standardization. ISO 704. Terminology Work – Principles and Methods, 2000.
42. N. Pasquier, Y. Bastide, R. Taouil, and L. Lakhal. Pruning closed itemset lattices for association rules. In *Actes des 14èmes journées Bases de Donnes Avancées (BDA'98)*, pages 177–196, Octobre 1998.
43. N. Pasquier, R. Taouil, Y. Bastide, G. Stumme, and L. Lakhal. Generating a condensed representation for association rules. *Journal of Intelligent Information Systems*, 24(1):29–60, 2005.
44. C. S. Peirce. *Collected Papers of Charles Sanders Peirce*. Harvard University Press, Cambridge, 1931–1935, 1958.
45. S. Prediger. Logical scaling in formal concept analysis. In D. Lukose, H. Delugach, M. Keeler, L. Searle, and J. F. Sowa, editors, *Conceptual Structures: Fulfilling Peirce's Dream*, number 1257 in Lecture Notes in Artificial Intelligence. Springer, Berlin, 1997.
46. S. Prediger and G. Stumme. Theory-driven logical scaling. In E. F. et al. editors, *Proc. 6th Intl. Workshop Knowledge Representation Meets Databases (KRDB'99)*, volume CEUR Workshop Proc. 21, 1999. Also in P. Lambrix et al. editors, *Proc. Intl. Workshop on Description Logics (DL'99)*. CEUR Workshop Proc. 22, 1999 http://ceur-ws.org/Vol-21.
47. F. Rioult. *Extraction de connaissances dans les bases de données comportant des valeurs manquantes ou un grand nombre d'attributs*. PhD thesis, Université de Caen Basse-Normandie, 2005.
48. S. Rudolph. *Relational Exploration – Combining Description Logics and Formal Concept Analysis for Knowledge Specification*. Universitätsverlag Karlsruhe, 2006. Dissertation.
49. C. Schmitz, A. Hotho, R. Jäschke, and G. Stumme. Mining association rules in folksonomies. In *Proceedings of the IFCS 2006 Conference*, Lecture Notes in Computer Science. Springer, July 2006.
50. G. Snelting. Reengineering of configurations based on mathematical concept analysis. *ACM Trans. Softw. Eng. Methodol.*, 5(2):146–189, 1996.

51. G. Snelting. *Concept Lattices in Software Analysis*, volume 3626 of *LNAI*, pages 272–287. Springer, Berlin, 2005.
52. G. Snelting and F. Tip. Understanding class hierarchies using concept analysis. *ACM Trans. Program. Lang. Syst.*, 22(3):540–582, 2000.
53. J. F. Sowa. *Conceptual Structures: Information Processing in Mind and Machine.* Addison-Wesley, Reading, MA, 1984.
54. S. Staab, S. Santini, F. Nack, L. Steels, and A. Maedche. Emergent semantics. *Intelligent Systems, IEEE [see also IEEE Expert]*, 17(1):78–86, 2002.
55. L. Steels. The origins of ontologies and communication conventions in multi-agent systems. *Autonomous Agents and Multi-agent Systems*, 1(2):169–194, October 1998.
56. S. Strahringer and R. Wille. Conceptual clustering via convex-ordinal structures. In O. Opitz, B. Lausen, and R. Klar, editors, *Information and Classification*, pages 85–98. Springer, Berlin, 1993.
57. G. Stumme. Knowledge acquisition by distributive concept exploration. In G. Ellis, R. Levinson, W. Rich, and J. F. Sowa, editors, *Conceptual Structures: Applications, Implementation and Theory*, number 954 in Lecture Notes in Artificial Intelligence. Springer, Berlin, 1995.
58. G. Stumme. The concept classification of a terminology extended by conjunction and disjunction. In N. Foo and R. Goebel, editors, *PRICAI'96: Topics in Artificial Intelligence. Proc. PRICAI'96*, volume 1114 of *LNAI*, pages 121–131. Springer, Heidelberg, 1996.
59. G. Stumme. Exploration tools in formal concept analysis. In E. Diday, Y. Lechevallier, and O. Opitz, editors, *Ordinal and Symbolic Data Analysis. Proc. OSDA'95. Studies in Classification, Data Analysis, and Knowledge Organization 8*, pages 31–44. Springer, Heidelberg, 1996.
60. G. Stumme. Concept exploration – A tool for creating and exploring conceptual hierarchies. In D. Lukose, H. Delugach, M. Keeler, L. Searle, and J. F. Sowa, editors, *Conceptual Structures: Fulfilling Peirce's Dream*, Lecture Notes in Artificial Intelligence, volume 1257, pages 318–331. Springer, Berlin, 1997.
61. G. Stumme. Conceptual knowledge discovery with frequent concept lattices. FB4-Preprint 2043, TU Darmstadt, 1999.
62. G. Stumme. A finite state model for on-line analytical processing in triadic contexts. In B. Ganter and R. Godin, editors, *Proc. 3rd Intl. Conf. on Formal Concept Analysis*, volume 3403 of *LNCS*, pages 315–328. Springer, Berlin, 2005.
63. G. Stumme and A. Maedche. FCA-Merge: Bottom-up merging of ontologies. In B. Nebel, editor, *Proc. 17th Intl. Conf. on Artificial Intelligence (IJCAI '01)*, pages 225–230, Seattle, WA, USA, 2001.
64. G. Stumme, R. Taouil, Y. Bastide, N. Pasqier, and L. Lakhal. Computing iceberg concept lattices with Titanic. *Journal on Data and Knowledge Engineering*, 42(2):189–222, 2002.
65. G. Stumme, R. Taouil, Y. Bastide, N. Pasquier, and L. Lakhal. Intelligent structuring and reducing of association rules with formal concept analysis. In F. Baader, G. Brewker, and T. Eiter, editors, *KI 2001: Advances in Artificial Intelligence*, volume 2174 of *LNAI*, pages 335–350. Springer, Heidelberg, 2001.
66. V. Takcs. Two applications of Galois graphs in pedagogical research. Manuscript of a lecture given at TH Darmstadt, 60 pages, February 1984.
67. J. Tane. Using a query-based multicontext for knowledge base browsing. In *Formal Concept Analysis, Third International Conf., ICFCA 2005-Supplementary Volume*, pages 62–78. IUT de Lens, Universite d Artois, FEB 2006.

68. J. Tane, P. Cimiano, and P. Hitzler. Query-based multicontexts for knowledge base browsing: An evaluation. In H. Schärfe, P. Hitzler, and P. Øhrstrøm, editors, *Proc. 14th Intl. Conf. on Conceptual Structures*, volume 4068 of *LNCS*, pages 413–426. Springer, Berlin, 2006.
69. J. Tane, C. Schmitz, and G. Stumme. Semantic resource management for the Web: An e-learning application. In *Proc. 13th International World Wide Web Conference (WWW 2004)*, pages 1–10, 2004.
70. T. A. Tilley, R. J. Cole, P. Becker, and P. W. Eklund. *A Survey of Formal Concept Analysis Support for Software Engineering Activities*, volume 3626 of *LNAI*, pages 250–271. Springer, Berlin, 2005.
71. R. Wille. Restructuring lattice theory: An approach based on hierarchies of concepts. In I. Rival, editor, *Ordered Sets*, pages 445–470. Reidel, Dordrecht, 1982.
72. R. Wille. The basic theorem of triadic concept analysis. *Order*, 12:149–158, 1995.
73. R. Wille. Conceptual structures of multicontexts. In P. W. Eklund, G. Ellis, and G. Mann, editors, *Conceptual Structures: Representation as Interlingua*, pages 23–39. Springer, Berlin, 1996.
74. R. Wille. Restructuring mathematical logic: An approach based on peirce's pragmatism. In A. Ursini and P. Agliano, editors, *Logic and Algebra*, pages 267–281. Marcel Dekker, New York, 1996.
75. R. Wille. Conceptual graphs and formal concept analysis. In D. Lukose, H. Delugach, M. Keeler, L. Searle, and J. F. Sowa, editors, *Conceptual Structures: Fulfilling Peirce's Dream*, volume 1257 of *Lecture Notes in Artificial Intelligence*, pages 290–303. Springer, Heidelberg, 1997.
76. M. J. Zaki. Generating non-redundant association rules. In *Proc. KDD 2000*, pages 34–43, 2000.
77. M. J. Zaki and C.-J. Hsiao. Charm: An efficient algorithm for closed association rule mining. technical report 99–10. Technical report, Computer Science Dept., Rensselaer Polytechnic, October 1999.

An Overview of OntoClean*

Nicola Guarino[1] and Christopher A. Welty[2]

[1] Laboratory for Applied Ontology (ISTC-CNR) Via alla Cascata 56/C, 38100 Trento, Italy, nicola.guarino@cnr.it
[2] IBM Watson Research Center 19 Skyline Dr., Hawthorne, NY 10532, USA, welty@us.ibm.com

Summary. OntoClean is a methodology for validating the ontological adequacy and logical consistency of taxonomic relationships. It is based on highly general ontological notions drawn from philosophy, like *essence*, *identity*, and *unity*, which are used to elicit and characterize the intended meaning of properties, classes, and relations making up an ontology. These aspects are represented by formal metaproperties, which impose several constraints on the taxonomic relationships between concepts. The analysis of these constraints helps in evaluating and validating the choices made. In this chapter we present an informal overview of the philosophical notions involved and their role in OntoClean, review some common ontological pitfalls, and walk through the example that has appeared in pieces in previous papers and has been the basis of numerous tutorials and talks.

1 Introduction

The OntoClean methodology was first introduced in a series of conference-length papers in 2000 [4–7, 12], and received much attention and use in subsequent years. The main contribution of OntoClean was the beginning of a formal foundation for ontological analysis. Alan Rector, a seasoned veteran at ontological analysis in the medical domain, said of OntoClean, "...what you have done is reduce the amount of time I spend arguing with doctors that the way I want to model the world is right..." [10]. A similar comment came from the CYC people attending our AAAI-2000 tutorial: "You showed why the heuristic choices we adopted were right." Most experienced domain modelers can see the correct way to, e.g., structure a taxonomy, but are typically unable to justify themselves to others. OntoClean has provided a logical basis for arguing against the most common modeling pitfalls, and arguing for what we have called "clean ontologies."

* This is a slightly edited version of the paper with the same title published in the previous edition of this volume.

S. Staab and R. Studer (eds.), *Handbook on Ontologies*, International Handbooks on Information Systems, DOI 10.1007/978-3-540-92673-3,

In this chapter we present an informal overview of the four basic notions *essence, identity, unity,* and *dependence*, and their role in OntoClean, review the basic ontology pitfalls, and walk through the example that has appeared in pieces in previous papers and has been the basis of numerous tutorials and talks beginning with AAAI-2000.

1.1 Background

The basic notions in OntoClean were not new, but existed in philosophy for some time. Indeed, the practice of modeling the world for information systems has many parallels in philosophy, whose scholars have been trying to describe the universe in a formal, logical way since the time of Aristotle. Philosophers have struggled with deep problems of existence, such as God, life and death, or whether a statue and the marble from which it is made are the same entity (see [11] for a classic text on the notions touched here). While these problems may seem irrelevant to the designer of an information system, we found that *the conceptual analysis and the techniques used to attack these problems* are not, and form the basis of our methodology.

1.2 Properties, Classes, and Subsumption

Many terms have been borrowed by computer science from mathematics and logic, but unfortunately this borrowing has resulted often in a skewed meaning. In particular, the terms *property* and *class* are used in computer science with often drastically different meanings from the original. The use of the term *property* in RDF is an example of such unfortunate deviation from the usual logical sense.

In this chapter, we shall consider properties as the *meanings* (or *intensions*) of expressions like *being an apple* or *being a table*, which correspond to unary predicates in first-order logic. Given a particular state of affairs (or *possible world*, if you prefer), we can associate to each property a *class* (its *extension*), which is the set of entities that exhibit that property in that particular situation. The members of this class will be called *instances* of the property. Classes are therefore sets of entities that share a property in common; they are the extensional counterpart of properties. In the following, we shall refer most of the time to properties rather than classes or predicates, to stress the fact that their ontological nature (characterized by means of *meta-properties*) does not depend on syntactic choices (as it would be for predicates), nor on specific states of affairs (as it would be for classes).

The independence of properties from states of affairs gives us the opportunity to make clear the meaning of the term *subsumption* we shall adopt in this paper. A property p subsumes q if and only if, *for every possible state of affairs*, all instances of q are also instances of p. On the syntactic side, this corresponds to what is usually held for description logics, P subsumes Q if and only if there is no model of $Q \wedge \neg P$.

2 The Basic Notions

2.1 Essence and Rigidity

A property of an entity is *essential* to that entity if it *must* be true of it in every possible situation, i.e., if it *necessarily holds* for that entity. For example, the property of *having a brain* is essential to human beings. Every human *must* have a brain in every possible situation.

A special form of essentiality is rigidity; a property is *rigid* if it is essential to all its possible instances; an instance of a rigid property cannot stop being an instance of that property in a different situation. For example, while having a brain may be essential to humans, it is not essential to, say, scarecrows in the *Wizard of Oz*. If we were modeling the world of the *Wizard of Oz*, the property of *having a brain* would not be rigid, though still essential to humans. On the other hand, the property *being a human* is typically rigid, every human is necessarily so.

The fact that we said "typically" in the previous statement requires an immediate clarification. The point of OntoClean is *not* to help people deciding about the ontological nature of a certain property; this choice depends on the way the domain at hand is *conceptualized* [3], and cannot be forced in advance. What OntoClean offers is, rather, a formal framework for expressing (some of) the ontological assumptions lying behind a certain conceptualization (its so-called *ontological commitment*). Rigidity is the first ingredient of this framework: expressing (by means of meta-properties) whether it holds or not for the relevant properties of our conceptualization helps clarifying the *ontological commitment* of such conceptualization.

When a property is *non-rigid*, it can acquire or lose (some of) their instances depending on the situation at hand. Within non-rigid properties, we distinguish between properties that are essential to *some* entities and not essential to others (*semi-rigid*), and properties that are not essential to *all* their instances (*anti-rigid*). For example, the property *being a student* is typically anti-rigid – *every* instance of student can cease to be such in a suitable situation, whereas the property *having a brain* in our *Wizard of Oz* world is semi-rigid, since there are instances that must have a brain as well as others that consider a brain just as a (useful) optional.

Rigidity and its variants are important meta-properties, every property in an ontology should be labeled as rigid, non-rigid, or anti-rigid. In addition to providing more information about what a property is intended to mean, these meta-properties impose constraints on the subsumption relation, which can be used to check the ontological consistency of taxonomic links. One of these constraints is that anti-rigid properties cannot subsume rigid properties. For example, the property *being a student* cannot subsume *being a human* if the former is anti-rigid and the latter is rigid. To see this, consider that, if p is an anti-rigid property, all its instances can cease to be such. This is certainly the case for *student*, since any student may cease being a student. However, no instance of *human* can cease to be a human, and if all humans would be necessarily students (the meaning of subsumption), then no person could cease to be a student, creating therefore an inconsistency.

2.2 Identity and Unity

Although very subtle and difficult to explain without experience, identity and unity are perhaps the most important notions we use in our methodology. These two things are often confused with each other; in general, *identity* refers to the problem of being able to recognize individual entities in the world as being the same (or different), and *unity* refers to being able to recognize all the parts that form an individual entity.

Identity *criteria* are the criteria we use to answer questions like, "is that my dog?" In point of fact, identity criteria are conditions used to *determine* equality (sufficient conditions) and that are *entailed by* equality (necessary conditions).

It is perhaps simplest to think of identity criteria over time (*diachronic* identity criteria), e.g., how do we recognize people we know as the *same* person even though they may have changed? It is also very informative, however, to think of identity criteria at a single point in time (*synchronic* identity criteria). This may, at first glance, seem bizarre. How can you ask, "are these *two* entities the same entity?" If they are the same then there is one entity, it does not even make sense to ask the question.

The answer is not that difficult. One of the most common decisions that must be made in ontological analysis concerns identifying circumstances in which one entity is actually two (or more). Consider the following example, drawn from actual experience: somebody proposed to introduce a property called *time duration* whose instances are things like *one hour* and *two hours*, and a property *time interval* referring to specific intervals of time, such as "1:00–2:00 next Tuesday" or "2:00–3:00 next Wednesday." The proposal was to make *time duration* subsume *time interval*, since all time intervals are time durations. Seems to make intuitive sense, but how can we evaluate this decision?

In this case, an analysis based on the notion of identity can be informative. According to the identity criteria for time durations, two durations of the same length are the same duration. In other words, all one-hour time durations are identical – they are the *same* duration and therefore there is only one "one hour" time duration. On the other hand, according to the identity criteria for time intervals, two intervals of the same duration occurring at the same time are the same, but two intervals occurring at different times, even if they are the same duration, are different. Therefore the two example intervals above would be different intervals. This creates a contradiction: if all instances of *time interval* are also instances of *time duration* (as implied by the subsumption relationship), how can they be two instances of one property and a single instance of another?

This is one of the most common confusions of natural language when used for describing the world. When we say "all time intervals are time durations" we really mean "all time intervals *have* a time duration" – the duration is a component of an interval, but it is not the interval itself. In this case we cannot model the relationship as subsumption, time intervals have durations (essentially) as *qualities*. More examples of such confusions are provided at the end of this article.

One of the distinctions proposed by OntoClean is between properties that *carry an identity criterion* and properties that do not. The former are labeled with an ad hoc meta-property, +**I**. Since criteria of identity are inherited along property subsumption hierarchies, a further distinction is made to mark those properties that *supply* (rather just *carrying*) some "own" identity criteria, which are not inherited from the subsuming properties. These properties are marked with the label +**O** (where **O** stands for "own").

Unfortunately, despite their relevance, recognizing identity criteria may be extremely hard. However, in many cases identity analysis can be limited to detecting the properties that are just *necessary* for keeping the identity of a given entity, i.e., what we have called the *essential properties*. Obviously, if two things do not have the same essential properties they are not identical. Take for instance the classical example of the statue and the clay: is the statue identical to the clay it is made of? Let us consider the essential properties: having (more or less) a certain shape is essential for the statue, but not essential for the clay. Therefore, they are different: we can say they have different identity criteria, even without knowing exactly what these criteria are. In practice, we can say that "sharing the essential property P," where P is essential for all the instances of a property Q different from P, is the weakest form of an identity criterion carried by Q. Such criterion can be used to make conclusions about non-identity, if not about identity.

A second notion that is extremely useful in ontological analysis is *Unity*. Unity refers to the problem of describing the parts and boundaries of objects, such that we know in general what is part of the object, what is not, and under what conditions the object is *whole*.

Unity can tell us a lot about the intended meaning of properties in an ontology. Certain properties pertain to wholes, that is, all their instances are wholes, others do not. For example, *being (an amount of) water* does not have wholes as instances, since each amount can be arbitrarily scattered or confused with other amounts. In other words, knowing it is an amount of water does not tell us anything about its parts, and recognizing it as a single entity. On the other hand, *being an ocean* is a property that picks up whole objects, as its instances, such as "the Atlantic Ocean" is recognizable as a single entity. Of course, one might observe that oceans have vague boundaries, but this is not an issue here: the important difference with respect to the previous example is that in this case we have a criterion to tell, at least, what is *not* part of the Atlantic Ocean, and still part of some other ocean. This is impossible for amounts of water.

In general, in addition to specifying whether or not properties have wholes as instances, it is also useful to analyze the specific conditions that must hold among the parts of a certain entity in order to consider it a whole. We call these conditions *unity criteria* (UC). They are usually expressed in terms of a suitable *unifying relation*, whose ontological nature determines different kinds of wholes. For example, we may distinguish *topological wholes* (a piece of coal), *morphological wholes* (a constellation), *functional wholes* (a hammer, a bikini). As these examples show, nothing prevents a whole from having parts that are themselves wholes (under different unifying relations). Indeed, a *plural whole* can be defined as a whole that is a mereological sum of wholes.

In OntoClean, we distinguish with suitable meta-properties the properties *all* whose instances *must* carry a *common* UC (such as *ocean*) from those that do not. Among the latter, we further distinguish properties all of whose instances must be wholes, although with different UCs, from properties all of whose instances are not necessarily wholes. An example of the former kind may be *legal agent*, if we include both people and companies (with different UCs) among its instances. *Amount of water* is usually an example of the latter kind, since none of its instances *must* be wholes (this is compatible with the view that a particular amount of water may become a whole for a short while, e.g., while forming an iceberg. We say that *ocean* carries unity (+**U**), *legal agent* carries no unity (−**U**), and *amount of water* carries anti-unity (~**U**).

The difference between unity and anti-unity leads us again to interesting problems with subsumption. It may make sense to say that "Ocean" is a subclass of "Water," since all oceans are water. However, if we claim that instances of the latter must not be wholes, and instances of the former always are, then we have a contradiction. Problems like this again stem from the ambiguity of natural language, oceans are not "kinds of" water, they are *composed* of water.

2.3 Constraints and Assumptions

A first observation descending immediately from our definitions regards some *subsumption constraints*. Given two properties, p and q, when q subsumes p the following constraints hold:

1. If q is anti-rigid, then p must be anti-rigid
2. If q carries an identity criterion, then p must carry the same criterion
3. If q carries a unity criterion, then p must carry the same criterion
4. If q has anti-unity, then p must also have anti-unity
5. If q is dependent on property c, then p is dependent on property c

Finally, we make the following assumptions regarding identity (adapted from Lowe [8]):

- *Sortal Individuation.* Every domain element must instantiate some property carrying an IC ($+I$). In this way we satisfy Quine's dicto "No entity without identity" [9].

- *Sortal Expandability.* If something is an instance of different properties (for instance related to different times), then it must be also instance of a more general property carrying a criterion for its identity.

Together, the two assumptions imply that every entity must instantiate a *unique* most general property carrying a criterion for its identity.

3 An Extended Example

In this section we provide a walk-through of the way the OntoClean analysis can be used. This example is based on those presented at various tutorials and invited talks.

We begin with a set of classes arranged in a taxonomy, as shown in Fig. 1. The taxonomy we have chosen makes intuitive sense *prima facie*, and in most cases the taxonomic pairs were taken from existing ontologies such as Wordnet[1], Pangloss[2], and the 1993 version of CYC.[3]

We have chosen, following our previous papers, to use a shorthand notation for indicating meta-property choices on classes. Rigidity is indicated by **R**, identity by **I**, unity by **U**, and dependence by **D**. Each letter is preceded

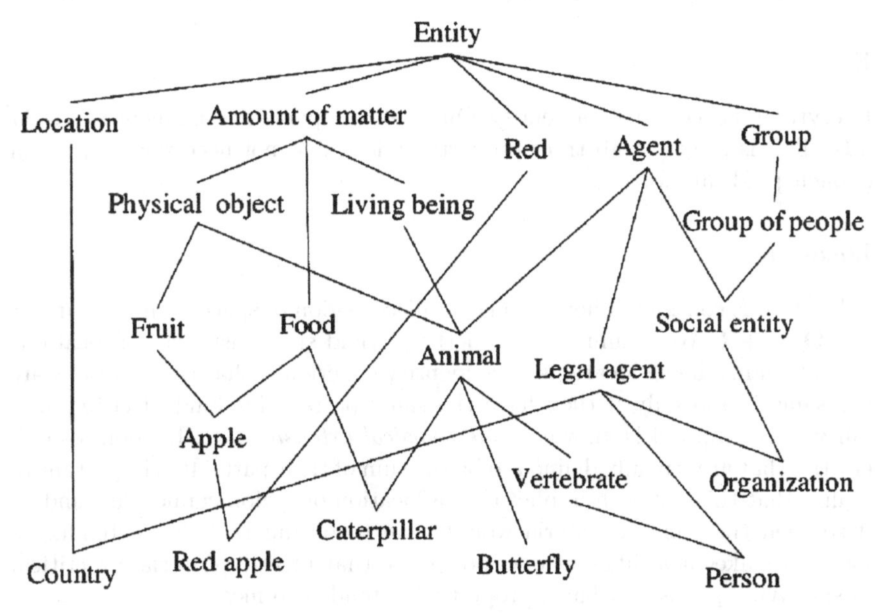

Fig. 1. An uncleaned taxonomy

[1] URL...

[2] URL...

[3] The current version of Cyc no longer contains these errors.

by +, − or ∼, to indicate the positive, negative, or *anti* meta-property, e.g., being rigid (+**R**), carrying an identity criterion (+**I**), carrying a common unity criterion (+**U**); not rigid (−**R**), not carrying an identity criterion (−**I**), not carrying a common unity criterion (−**U**); being anti-rigid (∼**R**) and having anti-unity (∼**U**). We also used (+**O**) to indicate when a property carries its *own* identity criterion, as opposed to inheriting one from a more general property.

3.1 Assigning Meta-Properties

The first step is to assign the meta-properties discussed above to each property in the taxonomy. When designing a new ontology, this step may occur first, before arranging the properties in a taxonomy. Note that the assignments discussed here are not meant to be definitive at all: rather, these represent *prima facie* decisions reflecting our intuitions about the meaning ascribed to the terms used. The point of this exercise is not so much to discuss the ontological nature of these properties, but rather *to explore and demonstrate the logical consequences of making these choices.* As we shall see, in some cases they will result contradictory with respect to the formal semantics of our meta-properties, although intuitive at a first sight. In our opinion, this proves the utility of a formal approach to ontology analysis and evaluation.

Entity

Everything is necessarily an entity. Our meta-properties assignment is −**I**−**U** +**R**. This is the most abstract property, indeed it is not necessary having an explicit predicate for it.

Location

A location is considered here as a generalized region of space. Our assignment is +**O**∼**U**+**R**. We assume the property as rigid since instances of locations cannot change being locations. The identity criterion is that two locations are the same if and only if they have the same parts. This kind of criterion is fairly common, and is known as *mereological extensionality.* It applies to all entities that are trivially defined to be the sum of their parts. It is important to realize that this criterion implies that a location or region cannot "expand" – if so then the identity criteria would have to be different. So, extending a location makes it a different one. So we see that identity criteria are critical in specifying precisely what a property is intended to mean.

Amount of Matter

We conceptualize an amount of matter as a clump of unstructured or scattered "stuff" such as a liter of water or a kilogram of clay. Amounts of matter should not be confused with *substances*, such as water or clay; an amount of matter is a particular amount of the substance. Therefore, amounts of matter are mereologically extensional, so we assign +**O** to this property. As discussed above, they are not necessarily wholes, so our assignment is ~**U**. Finally, every amount of matter is necessarily so, therefore the property is +**R**.

Red

What we have in mind here is the property of being a red *thing*, not the property of being a particular shade color. We see in this case that it is useful to ask ourselves *what* the instances of a certain property are. Do we have oranges and peppers in the extension of this property, or just their colors? Red entities share no common identity criteria, so our assignment is -**I**. A common confusion here regarding identity criteria concerns the fact that all instances of *red* are colored red, therefore we have a clear *membership criterion*. Membership criteria are not identity criteria, as the latter gives us information about how to distinguish entities from each other. Having a color red is common to all instances of this property, and thus is not informative at all for identity.

A red amount of matter would be an instance of this property, which is not a whole, as would a red ball, which is a whole. Therefore we must choose -**U**, indicating that there is no common unity criterion for all instances.

Finally, we choose -**R** since some instances of *Red* may be necessarily so, and most will not. This weak and unspecific combination of meta-properties indicates that this property is of minimal utility in an ontology, we call them *attributions* [12].

Agent

We intend here an entity that plays a causal part in some event. Just about anything can be an agent, a person, the wind, a bomb, etc. Thus there is no common identity nor unity criterion for all instances, and we choose -**I**-**U**. No instance of *agent* is necessarily an agent, thus the property is ~**R**. Clearly this assignment of meta-properties selects a particular meaning of *agent* among the many possible ones. See for example [2] for a discussion on the meaning of *causal agent* in WordNet.

Group

We see here a group as an *unstructured* finite collection of wholes. Instances of *group* are mereologically extensional as they are defined by their members, thus +**O**. Since, given a group, we have no way to isolate it from other groups, no group is *per se* a whole, thus ~**U**. In any case, like many general terms, *Group* is fairly ambiguous, and once again this choice of identity criteria and anti-unity exposes the choice we have made. Finally, it seems plausible to assume that every instance of group is necessarily so, thus +**R**.

Physical Object

We think here of physical objects as isolated material entities, i.e., something that can be "picked up and thrown" (at a suitable scale, since a planet would be considered an instance of a physical object as well...). Under this vision, what characterizes physical objects is that they are *topological wholes* – so we assign +**U** to the corresponding property.

For the sake of simplicity, we assume here that no two instances of this property can exist in the same spatial location at the same time. This is an identity criterion, so we assign +**O** to this property. Note that this is a *synchronic* identity criterion (see identity and unity, above) – we do not assume a common diachronic identity criterion for all physical objects.

Physical object is a rigid property, so we have +**R**. To see this, consider the alternative: there must be some instance of the property that can, possibly, *stop* being a physical object, yet still exist and retain its identity. By assigning rigidity to this property, we assert that there is no such instance, and that every instance of *Physical Object* ceases to exist if it ceases to be a physical object.

Living Being

Instances of *living being* must be wholes according to some common biological unity criterion. We do not need to specify it to assign +**U** to this property.

For identity, it is difficult to assume a single criterion that holds for all instances of living being. The way we, e.g., distinguish people may be different from the way we distinguish dogs. However, a plausible diachronic criterion could be *having the same DNA* (although only-necessary, since it does not help in the case of clones). Moreover, we can easily think of essential properties that characterize living beings (e.g., the need of taking nutrients from the environment), and this is enough for assigning them +**O**.

We assume *living being* to be a rigid property (+**R**), so if an entity ceases to be living then it ceases to exist. Notice that this is a precise choice that is totally dependent on our conceptualization: nothing would exclude considering life as a *contingent* (non-rigid) property; by considering it as rigid, we are indeed *constructing* a new kind of entity, justified by the fact that this property is very relevant for us.

Food

Nothing is necessarily food, and just about anything is possibly food. In a linguistic sense, "food" is a role an entity may play in an eating event. Considering that anything that is food can also possibly *not* be food, we assign ~**R** to this property. We also assume that any quantity of food is an amount of matter and inherits its extensional identity criterion, thus +**I** and ~**U**.

Animal

Like for *living being*, the identity criteria for *animal* may be difficult to characterize precisely, but we can devise numerous essential properties that apply only to them, or only-sufficient conditions that act as heuristics especially for diachronic identity criteria. Humans, in particular, are quite good at recognizing most individual animals, typically based on clues present in their material bodies. The undeniable fact is that we do recognize "the same" animal over time, so there must be some way that is accomplished. Therefore, we assign +**O**.

The property is clearly rigid (+**R**); moreover, being subsumed by *living being*, it clearly carries unity (+**U**).

Legal Agent

This is an agent that is recognized by law. It exists only because of a legal recognition. Legal agents are entities belonging to the so-called *social reality*, insofar their existence is the result of social interaction. All legal systems assign well-defined identity criteria to legal agents, based on, for example, an id number. Therefore, it seems plausible to assign +**O**. Concerning unity, if we include companies (as well as persons) among legal agents, then probably there is no unity criteria shared by all of them, so we assign −**U**. Finally, since nothing is necessarily a legal agent, we assign ~**R**. For instance, we may assume that a typical legal entity, such as a person, becomes such only after a certain age.

Group of People

A special kind of *Group* all of whose members are instances of *Person*. Identity and unity criteria are the same as *Group*, and thus we have +**I**~**U**. Finally, we consider *Group of People* to be rigid, since any entity which is a group of people must necessarily be such.

Social Entity

A group of people together for social reasons. Such as the "Bridge Club" (i.e., people who play cards together). We cannot imagine a common identity criteria for this property, however we assume it is rigid and carries unity. −**I**+**U**+**R**.

Organization

Instances of this property are intended to be things like companies, departments, governments, etc. They are made up of people with play specific roles according to some structure. Like people, organizations seem to carry their own identity criterion, and are wholes with a functional notion of unity, so we assign +**O**+**U**+**R**.

Fruit

We are thinking here of individual fruits, such as oranges or bananas. We assume they have their own essential properties, and can clearly be isolated from each other. Therefore, **+O+U+R** seems to be an obvious assignment.

Apple

This likely adds its on essential properties to those of fruits, so we assign it **+O+U+R**.

Red-Apple

Red apples do not have essential meta-properties in addition to apples. Moreover, no red apple is necessarily red, therefore we assign **+I+U~R**.

Vertebrate

This property is actually intended to be vertebrate-*animal*. This is a biological classification that adds new *membership* criteria to *Animal* (has-backbone), but apparently no new identity criteria: **+I+U+R**.

Person

Like *Living Entity* and *Animal*, the *Person* property is **+I+U**. It seems clear that specializing from *Vertebrate* to *Person* we add some further essential properties, thus we assume that *Person* has its own identity criteria, and we assign **+O**.

Butterfly and Caterpillar

Like *Animal*, *Butterfly* and *Caterpillar* have **+I+U**. However, every instance of *Caterpillar* can possibly become a non-caterpillar (namely a butterfly), and every instance of *Butterfly* can possibly be (indeed, must have been) a non-butterfly (namely a caterpillar), thus we assign **~R** to each.

Country

Intuitively, a country is a place recognized by convention as having a certain political status. Identity may be difficult to characterize precisely, but some essential properties seem to be clearly there, so **+O**. Countries are certainly wholes, so **+U**. Interestingly, it seems clear that some countries, like Prussia, still exist but are no longer countries, so we must assign **~R**.

3.2 Analyzing Rigid Properties

The Backbone Taxonomy

We now focus our analysis on what we have called the *backbone taxonomy*, that is, the rigid properties in the ontology, organized according to their subsumption relationships. These properties are the most important to analyze first, since they represent the invariant aspects of the domain. Our sortal expandability and individuation principles guarantee that no element of the domain is "lost" due to this restriction, since *every element must instantiate at least one of the backbone properties*, that supplies an identity criterion for it.

The backbone taxonomy based on the initial ontology is shown in Fig. 2.

After making the initial decisions regarding meta-properties and arranging the properties in a taxonomy, we are then in a position to verify whether any constraints imposed by the meta-properties are violated in the backbone. These violations have proven to be excellent indicators of misunderstandings and improperly constructed taxonomies. When a violation is encountered, we must reconsider the assigned meta-properties and/or the taxonomic link. and take some corrective action.

Living beings are not amounts of matter. The first problem we encounter is between Amount of Matter and Living Being. The problem is that a ~**U** property cannot subsume one with +**U**. While it certainly seems to make sense to say that all living beings are amounts of matter, based on the meaning

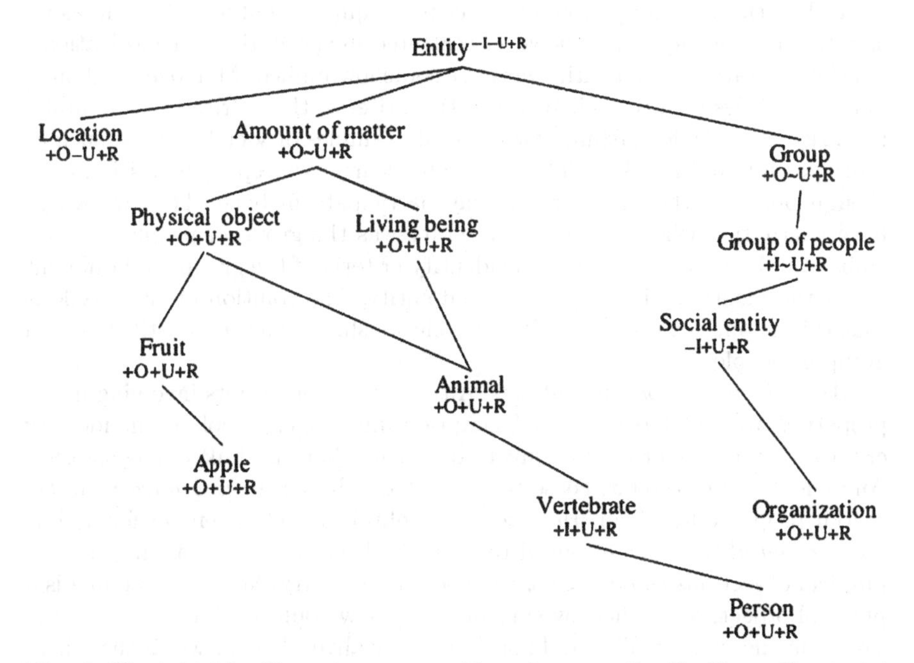

Fig. 2. The initial backbone taxonomy with meta-properties Backbone Constraint Violations

we have assigned there is an inconsistency: every amount of matter can be arbitrarily scattered, but this is certainly not the case for living beings. A further reason against this subsumption link is in the identity criteria: amounts of matter have an extensional identity, that is, they are different if any of their parts is substituted or annihilated – if you remove some clay from a lump of clay, it is a different amount. Living beings, on the other hand, can change parts and still remain the same – when you cut your fingernails off you do not become a different person.

This is one of the most common modeling problems we have. Living beings are *constituted* of amounts of matter, they are not themselves the matter. Natural language convention fails to capture this subtle distinction, but it is a violation of the intended meaning to claim that all living beings are mereologically extensional.

The solution here is to remove the subsumption link between these two properties, and represent the relationship as one of constitution.

Physical objects are not amounts of matter. Again, we see a violation since a $\sim$**U** property cannot subsume one with +**U**. This is yet another example of constitution being confused with subsumption. Physical objects are not themselves amounts of matter, they are constituted of matter. The solution is to make *Physical Object* subsumed directly by *Entity*.

Social entities are not groups of people. Another $\sim$**U**/+**U** violation, as well as a violation of identity criteria. Social entities are constituted of people, but, as with other examples here, they are not merely groups of people, they are more than that. A group of people does not require a unifying relation, as we assume these people can be however scattered in space, time, or motivations. On the contrary, a social entity *must* be somehow unified. Moreover, although both properties supply their own identity criteria, these criteria are mutally inconsistent. Take for instance two typical examples of social entities, such as a bridge club and a poker club. These are clearly two separate entities, even though precisely the same people may participate in both. Thus we would have a situation where, if the social entity was the group of people, the two clubs would be the same under the identity criteria of the group, and different under the identity criteria of the social entity. The solution of the puzzle is that this is, once again, a constitution relationship: a club is constituted by a group of people.

Animals are not physical objects. Although no constraints involving meta-properties are violated in this subsumption link, a closer look at the identity criteria of the two properties involved reveals that the link is inconsistent. Animals, by our account, cease to exist at death, since *being alive* is an essential property for them. However their physical bodies remain for a time after: *being alive* is not essential to them. Indeed, under our assumption *no* physical object has *being alive* as an essential property. Now, if an animal is a physical object, as implied by subsumption, how could it be that it is at the same time necessarily alive and not necessarily alive? The answer is that there must be two entities, related by a form of constitution, and the subsumption link should be removed.

In this example, it is not the meta-properties, but the methodology requiring to make identity criteria explicit in terms of essential properties that reveals the error.

3.3 Analyzing Non-rigid Properties

Let us now turn our attention to the *non-rigid* properties, which – so to speak – "flesh out" the backbone taxonomy. In [12] we have discussed a taxonomy of property kinds based on an analysis of their meta-properties, which distinguishes three main cases of non-rigid properties: *phased sortals, roles,* and *attributions.* All these cases appear in our example, and are discussed below.

Among other things, the differences among these property kinds are based on a meta-property not discussed here, based on the notion of *dependence.* A proper grasping of this notion (which is rather difficult to formalize) is not essential for an introductory understanding of the OntoClean methodology, so we shall rely on intuitive examples only.

Phased Sortals

The notion of a phased sortal was originally introduced by Wiggins [13]. A phased sortal is a property whose instances are allowed to change certain of their identity criteria during their existence, while remaining the same entity. The canonical example is a caterpillar. The intuition here is that when the caterpillar changes into a butterfly, something fundamental about the way it may be recognized and distinguished has changed, even though it is still the same entity. Phased sortals are recognized in our methodology by the fact that they are independent, anti-rigid, and supply identity criteria.

In the typical case, phased sortals come into clusters of at least two properties – an instance of a phased sortal (e.g., *Caterpillar*) should be able to "phase" into another one (e.g., *Butterfly*), and these clusters should have a common subsuming property providing an identity criterion for across phases, according to the sortal individuation principle.

Caterpillars and butterflies. Consider now our example. *Caterpillar* and *Butterfly* appear in out initial taxonomy, but there is no single property that subsumes *only* the phases of the same entity. Our formal analysis shows that there *must* be such property. After some thinking, we find what we need: it is the property *Lepidopteran*, which is **+O+U+R**. This is what supplies the identity criteria needed to recognize the same entity across phases.

Countries. The property *Country* does not, prima facie, appear to be a phased sortal, yet it meets our definition (**+O~R**). This is an example where reasoning on the meta-properties assignments and their consequences helps us pushing our ontological analysis further: what are we talking of, here? Is it a region that occasionally becomes a country, and in this case acquires some extra (yet temporary) identity criteria? What happens when something is not a country any more? Does it *cease to exist*, or does it just undergo the change of a property, like changing from being sunny and being shady? While answering to these questions, we realize we are facing a common problem in building ontologies, that of lumping together multiple meanings of a term into a single property. It seems there are two different interpretations of "country," one as a geographical region, and another as a geopolitical entity. It is the latter that ceases to exist when the property does not hold any more.

So there are two entities: the *Country* Prussia and the *Geographical Region* Prussia. These two entities are related to each other (e.g., countries occupy regions), but are not the same, and therefore we must break the current property into two.

We assign **+O+U+R** to Country, and **+I−U+R** to *Geographical Region.* The intuition is that countries have their own identity criteria, while geographical regions inherit the identity of locations. Countries have clearly a unity, while this is not the case for arbitrary geographical regions. Both properties are now rigid. Interestingly enough, we replaced an anti-rigid property with two rigid properties.

Roles

After analyzing phased sortals, we end up with the taxonomy shown in Fig. 3, and we are now ready to consider adding *roles* back into the taxonomy. Roles are properties that characterize the way something participates to a *contingent* event or state of affairs. It is because of such contingency that these properties are anti-rigid. Differently from phase sortals, roles do not supply identity criteria.

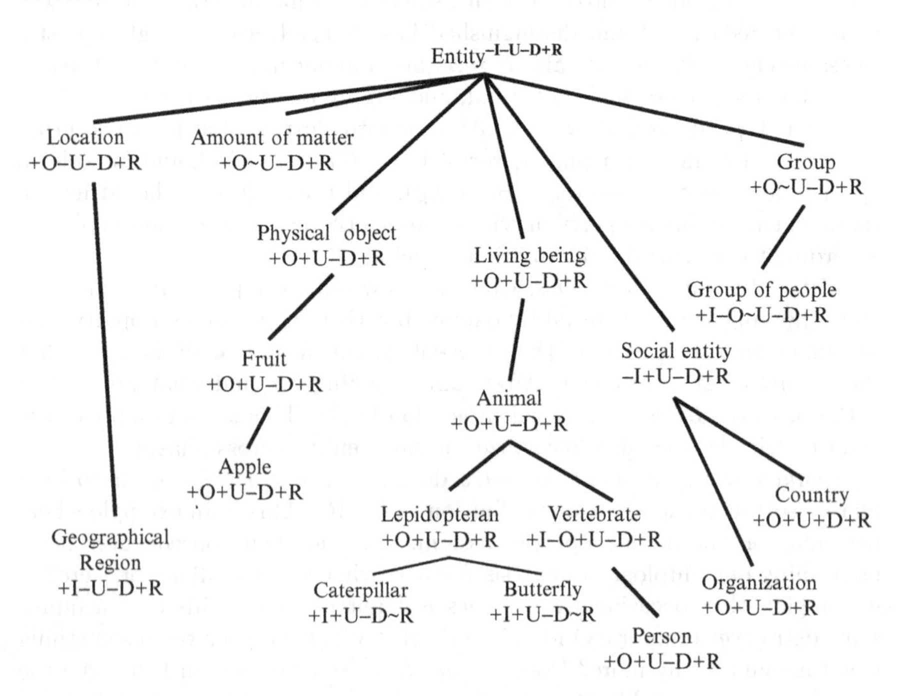

Fig. 3. The taxonomy after backbone and phased sortals

Agent. The analysis of roles often exposes subsumption violations concerning rigidity, in particular that a property with ~**R** cannot subsume a property with +**R**. Indeed, when we add the *Agent* property back to the backbone we see that it originally subsumed two classes, *Animal* and *Social Entity.* These subsumption links (shown to the right as dotted lines) should be removed, as they are incorrect.

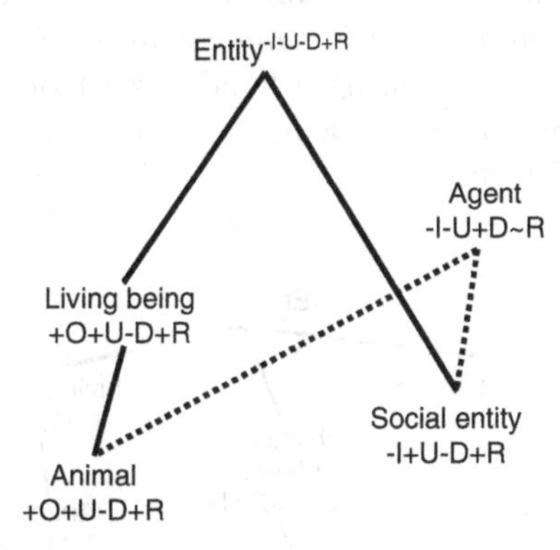

This is a different kind of problem in which subsumption is being used to represent a type restriction. The modeler intends to mean, not that all animals are agents, but that animals *can be* agents. This is a very common misuse of subsumption, often employed by object-oriented programmers. The correct way to represent this kind of relationship is with a covering, i.e., all agents are either animals or social entities. Clearly this is a different notion than subsumption. The solution is to remove the subsumption links and represent this information elsewhere.

Legal Agent. The next problem we encounter is when the role *Legal Agent* is added below *Agent*, with its subsuming links to *Person*, *Organization*, and *Country.* Again, as with the previous example, we have a contradiction, an anti-rigid property cannot subsume a rigid one, so these subsumption links (shown as dotted lines at right) must be removed.

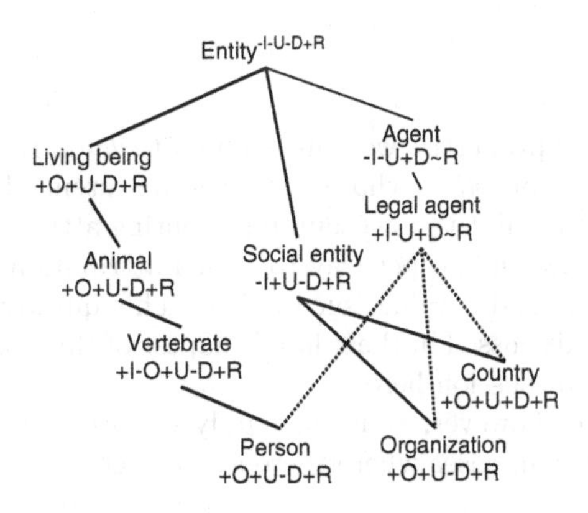

As with the *Agent* role, being forced to remove these links forces us to reconsider the meaning of the *Legal Agent* property. A legal agent is simply an entity recognized by law as an agent in some transaction or contract. Again, as with the *Agent* example, this is not a true subsumption link, but rather another type restriction. The links should be removed and replaced with a covering axiom.

Food. We chose to model the notion of food as a role, that is a property of things that may or can be food *in some situation.* So nothing is essentially food – even a stuffed turkey during a holiday feast or an enormous bowl of pasta with pesto sauce may avoid being eaten and end up not being food (it is *possible*, however unlikely).

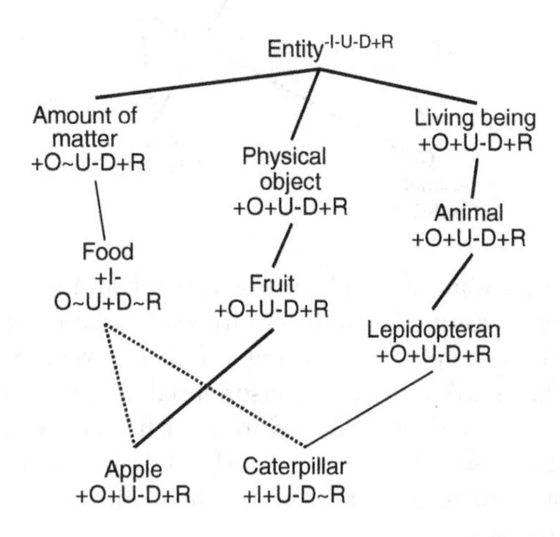

While our notion of what an apple means may seem to be violated by removing the subsumption link to *food*, the point is that we have chosen to represent the property in a particular way, as a role, and this link is inconsistent with that meaning and should be removed. In this case, the links are probably being used to represent *purpose* (see, e.g., [1]), not subsumption.

Attributions

The final category of properties we consider are *attributions*. We have one such property in our example, *Red*, whose instances are intended to be red things. We think that in general it is not useful representing attributions explicitly in a taxonomy, and that the proper way to model attributions is with a simple attribute, like color, and a value, such as red. This quickly brings us to the notion of *qualities*, discussed in the related chapter of this handbook on Dolce, and we avoid that discussion here.

Attributions do, however, come in handy on occasions. Their practical utility is often found in cases where there are a large number of entities that need to be partitioned according to the value of some attribute. We may have apples and pears, for example, and decide we need to partition them into red and green ones. Ontologically, however, the notion of red-thing does not have much significance, since there is nothing we can necessarily say of red-things, besides their color. This seems to us a very good reason for not consider attributions as part of the backbone. In other words, the backbone taxonomy helps in focusing on the more important classes for understanding the invariant aspects of domain structure, whereas attributions help in organizing the instances on an ad-hoc, temporary basis.

4 Conclusion

The final, cleaned, taxonomy is shown in Fig. 4. The heavier lines indicate subsumption relationships between members of the backbone taxonomy. Although it is not always the case, the cleaned taxonomy has far fewer "multiple inheritance" links than the original. The main reason for this is that subsumption is often used to represent things other than subsumption, that can be described in language using "is a." We may quite naturally say, for example, that an animal is a physical object, however we have shown in this

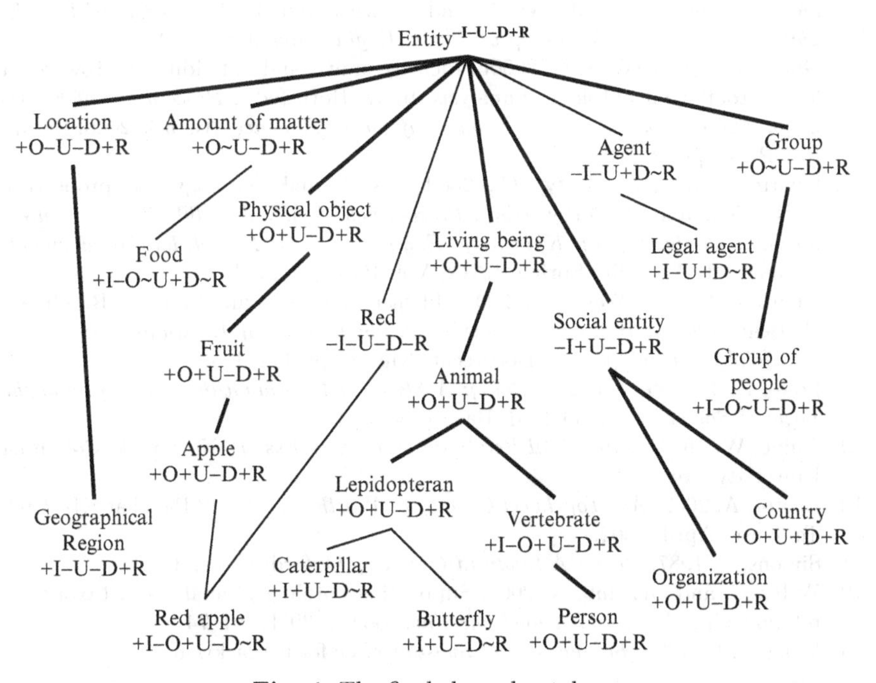

Fig. 4. The final cleaned ontology

chapter that this kind of linguistic use of "is a" is not logically consistent with the subsumption relationship. This results in many subsumption relationships being removed after analysis.

Acknowledgments

Many people have made useful comments on OntoClean, and have participated in its refinement. We would like to thank in particular Mariano Fernandez Lopez, Aldo Gangemi, Giancarlo Guizzardi, Claudio Masolo, Alessandro Oltramari.

References

1. Fan, J., Barker, K., Porter, B., and Clark, P. 2001. Representing roles and purpose. In *Proceedings of the 1st International Conference on Knowledge Capture (K-Cap'01)*. Vancouver: ACM.
2. Gangemi, A., Guarino, N., Masolo, C., and Oltramari, A. 2003. Restructuring Wordnet's Top-level. *AI Magazine,* 40: 235–244.
3. Guarino, N. 1998. Formal ontology in information systems. In N. Guarino (ed.), *Formal Ontology in Information Systems. Proceedings of FOIS'98, Trento, Italy, 6–8 June 1998.* Amsterdam: IOS Press, pp. 3–15.
4. Guarino, N. and Welty, C. 2000a. Towards a methodology for ontology based model engineering. In J. Bezivin and J. Ernst, (eds.), *Proceedings of IWME-2000: International Workshop on Model Engineering.* June, 2000.
5. Guarino, N. and Welty, C. 2000b. Identity, unity, and individuality: Towards a formal toolkit for ontological analysis. In W. Horn (ed.), *Proceedings of ECAI-2000: The European Conference on Artificial Intelligence,* August, 2000. Berlin: IOS Press, pp. 219–223.
6. Guarino, N. and Welty, C. 2000c. A formal ontology of properties. In R. Dieng and O. Corby (eds.), *Proceedings of EKAW-2000: The 12th International Conference on Knowledge Engineering and Knowledge Management,* October, 2000. Berlin: Spring, LNCS Vol. 1937, pp. 97–112.
7. Guarino, N. and Welty, C. 2002. Identity and subsumption. In R. Green, C. Bean, and S. H. Myaeng (eds.), *The Semantics of Relationships: An Interdisciplinary Perspective.* Dordrecht: Kluwer, pp 111–125.
8. Lowe, E. J. 1989. *Kinds of Being: A Study of Individuation, Identity, and the Logic of Sortal Terms.* Oxford: Basil Blackwell.
9. Quine, W. 1969. *Ontological Relativity and Other Essays.* New York: Columbia University Press.
10. Rector, A. 2002. *Are Top-Level Ontologies Worth the Effort?* Panel at KR-2002. Toulouse, April, 2002.
11. Simons, P. 1987. *Parts: A Study in Ontology.* Oxford: Clarendon.
12. Welty, C. and Guarino, N. 2001. Supporting ontological analysis of taxonomic relationships. *Data and Knowledge Engineering,* 39(1): 51–74.
13. Wiggins, D. 1980. *Sameness and Substance.* Oxford: Blackwell.

Ontology Design Patterns

Aldo Gangemi and Valentina Presutti

Institute for Cognitive Sciences and Technology (ISTC-CNR), Rome, Italy,
aldo.gangemi@cnr.it, valentina.presutti@istc.cnr.it

Summary. Computational ontologies in the context of information systems are artifacts that encode a description of some world, for some purpose. Under the assumption that there exist classes of problems that can be solved by applying common solutions (as it has been experienced in software engineering), we envision small, task-oriented ontologies with explicit documentation of design rationales. In this chapter, we describe components called Ontology Design Patterns (OP), and methods that support pattern-based ontology design.

We present a typology of OPs, and then focus on Content Ontology Design Patterns in terms of their background, definition, communication means, related work beyond ontology engineering, exemplification, creation, and usage principles. At the time of chapter's final version, recently performed experiments of pattern-based ontology design show remarkable quality improvement within some sample ontology design projects, specially in terms of compliance to tasks expressed as competency questions or scenarios.

1 Introduction

Computational ontologies in the context of information systems are artifacts that encode a description of some world (actual, possible, counterfactual, impossible, desired, etc.), for some purpose. They have a (primarily logical) structure, and must match both domain and task: they allow the description of entities whose attributes and relations are of concern because of their relevance in a domain for some purpose, e.g. query, search, integration, matching, explanation, etc.

Like any artifact, ontologies have a lifecycle: they are designed, implemented, evaluated, fixed, exploited, reused, etc. (cf. chapter "Ontology Engineering Methodology" for an in-depth examination of ontology engineering methodologies).

In this chapter, we focus on patterns for ontology design [14, 18].

Despite the original ontology engineering approach, when ontologies were seen as "portable" components [22], and its enormous impact on Semantic

S. Staab and R. Studer (eds.), *Handbook on Ontologies*, International Handbooks on Information Systems, DOI 10.1007/978-3-540-92673-3,

Web and interoperability, today one of the most challenging and neglected areas of ontology design is *reusability*. The possible reasons include at least: *size* and *complexity* of the major reusable ontologies, *opacity* of design rationales in most ontologies, *lack of criteria* in the way existing knowledge resources (e.g. thesauri, database schemata, lexica) can be reengineered, and *brittleness* of tools that should assist ontology designers.

Nowadays, an average user that is trying to build or reuse an ontology, or an existing knowledge resource, is typically left with just some limited assistance in using unfriendly logical structures, some large, hardly comprehensible ontologies, and a bunch of good practices that must be discovered from the literature. A typical usage scenario includes, e.g. a large set of web ontologies that are evaluated (usually in an implicit way) against the intended domain and tasks. The selected ontology (if any) is reused, and then an adaptation process is started in order to cope with the implicit requirements from an ontology project. This scenario is costly in many cases, and automatic selection mechanisms do not help with the adaptation process. Another typical scenario includes so-called "reference" or "core" ontologies that are supposed to be directly reused and specialized. Unfortunately, even if well designed, they are usually large and cover more knowledge than what a designer might need. In this case, it is hard to reuse only the "useful pieces" of the ontology, and consequently the cost of reuse is higher than developing a new ontology from scratch.

On the other hand, the success of very simple and small ontologies like FOAF [6] and SKOS [31] shows the potential of really portable, or "sustainable" ontologies. The lesson learnt supports the new approach to ontology design, which is sketched here.

Under the assumption that there exist classes of problems that can be solved by applying common solutions (as it has been experienced in software engineering), we propose to support reusability on the design side specifically. We envision small (or cleverly modularized) ontologies with explicit documentation of design rationales, and best reengineering practices. These components need specific functionalities in order to be implemented in repositories, registries, catalogues, open discussion and evaluation forums, and ultimately in new-generation ontology design tools. In this chapter, we describe small, motivated ontologies that can be used as *building blocks* in ontology design. A formal framework for (collaborative) ontology design that justifies the use of building blocks with explicit rationales is presented in [18].

We call the basic *building blocks* to be used in ontology design *Content Ontology Design Patterns* (CP) [14]. CPs are small ontologies that mediate between use cases (problem types) and design solutions. They are used as modelling components: ideally, an ontology results from a composition of CPs, with appropriate dependencies between them, plus the necessary design expansion based on specific needs.

Throughout experiences in ontology engineering projects[1] as well as in other ongoing international projects that have experimented with these ideas, typical conceptual patterns have emerged out of different domains, for different tasks, and while working with experts having heterogeneous backgrounds. For example, a simple CP called *participation* (including objects taking part in events) emerges in domain ontologies as different as enterprise models [23], legal norms [19], sofware management [34], biochemical pathways [16], and fishery techniques [17]. Other, more complex CPs have also emerged in the same disparate domains.

Moreover, since CPs are strictly related to small use cases, they are transparent with respect to the rationales applied to the design of a certain ontology. CPs are therefore an additional tool to achieve tasks such as ontology evaluation, matching, modularization, etc. For example, an ontology can be evaluated against the presence of certain patterns (which act as *unit tests* for ontologies, cf. [50] and chapter "Ontology Engineering Environments") that are typical of the tasks addressed by a designer. Furthermore, mapping and composition of CPs can facilitate ontology mapping and alignment/merging. Two ontologies drafted according to CPs can be mapped in an easier way: CP hierarchies will be more stable and well-maintained than local, partial, scattered ontologies. Finally, CPs can be also used in training and educational contexts for ontology engineers.

CPs are a very beneficial kind of patterns for ontology design, because they provide solutions to domain-oriented problems, and are directly reusable. On one hand, CPs are comparable to software engineering (SE) design patterns for what concerns the way they are documented and communicated. On the other hand, the intuition behind their usage is analogous to that of software engineering (object oriented) reusable libraries, e.g. Java libraries. A similar intuition is at the base of approaches to modularization of ontologies, e.g. [8], where the typical distinction between interface and implementation is used in order to distinguish between the module interface and the ontologies that a module encapsulates. CPs are compliant with this approach, and can be encapsulated in modules. However, this aspect is not key to the purpose of this chapter, and does not impact on their expected usage.

There are other types of ontology design patterns (OPs) that are beneficial for different purposes and targeted at different types of users. A typology of OPs will be also introduced in this chapter.

In principle, OPs do not depend on any specific representation language.[2] In this context, we focus mainly on CPs; in order to provide the readers with concrete examples and a closer view on their exploitation on the Semantic Web, we have decided to refer to OWL CPs (cf. chapter "Web Ontology

[1] For example, in the projects *FOS*: http://www.fao.org/agris/aos/, *WonderWeb*: http://wonderweb.semanticweb.org, *Metokis*: http://metokis.salzburgresearch.at, and *NeOn*: http://www.neon-project.org

[2] With the exception of Logical OPs.

Language: OWL" for details on OWL). In fact, CPs fit well with Semantic Web requirements for reuse and interoperability of ontologies and data, and as part of our work we have set up the ontologydesignpatterns.org web portal, which collects and makes them available on the Web [36].

Chapter's content is organized as follows: Sect. 1.1 gives some background notions; Sect. 2 introduces the types of OPs, defines them, and provides the reader with some examples; Sect. 3 presents a sample catalogue of CPs; Sect. 4 describes ways to create and work with CPs, and Sect. 5 presents an example of their application. Finally, Sect. 6 provides some conclusions and remarks.

1.1 Background

In the seventies, the architect and mathematician Christopher Alexander introduced the term "design pattern" for shared guidelines that help solve design problems. In [1] he argues that a good (architectural) design can be achieved by means of a set of rules that are "packaged" in the form of patterns, such as "courtyards which live", "windows place", or "entrance room". Design patterns are then assumed as archetypal solutions to design problems in a certain context.

Taking seriously the architectural metaphor, the notion has been eagerly endorsed by software engineering [12, 21, 29], and DBMS applications with so-called data model patterns [27]. In these areas, *pattern* is used as a general term for formatted guidelines in software reuse, and, more recently, has also appeared in requirements analysis, conceptual modelling, and ontology engineering [7, 11, 24, 39, 44, 48].[3] Traditionally, design patterns appear more like a collection of shortcuts and suggestions related to a class of context-bound problems and success stories. Software engineering patterns are largely used for documenting software [26], and there is software support for automatic code generation based on them (see, e.g the Eclipse functionality for generating factory methods,[4] and the *Whole platform.*[5]) Furthermore, there is recent work going towards a more formal encoding of design patterns (notably [3,24,30]), and even towards using ontology patterns to encode software engineering patterns [34].

Ontology engineering literature has tackled the notion of design pattern at least since [7, 39], while in the context of Semantic Web research and application, where OPs are now a hot topic, the notion has been introduced by [16, 38, 45, 48] and has been approached also by the W3C Semantic Web Best Practices and Deployment Group.[6] In particular, [16, 48] take a foundational approach that anticipates that presented in [14, 37] (which are closely related

[3] In software engineering, formal approaches to design patterns, based on dedicated ontologies, are being investigated, e.g. in so-called *semantic middleware* [34].

[4] Eclipse (http://www.eclipse.org/) is a programming environment used for developing Java projects.

[5] http://whole.sourceforge.net/

[6] See http://www.w3.org/2001/sw/BestPractices/

to this chapter). Some work [4] has also attempted a learning approach (by using case-based reasoning) to derive and rank patterns with respect to user requirements. The research has also addressed domain-oriented best practices and patterns, e.g. to express sequences in OWL [10], for content objects and multimedia [2] (cf. chapter "Ontologies for Cultural Heritage"), software components (cf. chapter "COMM: A Core Ontology for Multimedia Annotation"), business modelling and interaction [20], medical [43, 46] (cf. chapter "An Ontology for Software").

2 Types of Ontology Design Patterns

An ontology design pattern (OP) is a modelling solution to solve a recurrent ontology design problem. We have identified several types of OPs, and have grouped them into six families (cf. Fig. 1): *Structural* OPs, *Correspondence* OPs, *Content* OPs (CPs), *Reasoning* OPs, *Presentation* OPs, and *Lexico-Syntactic* OPs.

Although this chapter mainly focuses on CPs, in this section we give an overview of the OP families, with some examples. For more details, the reader can refer to [37].

Structural OPs

Structural OPs include Logical OPs and Architectural OPs. Logical OPs are compositions of logical constructs that solve a problem of expressivity, while Architectural OPs affect the overall shape of the ontology either internally or externally.

Logical OPs are only expressed in terms of a logical vocabulary, because their signature (the set of predicate names, e.g. the set of classes and properties in an OWL ontology) is empty (with minor exceptions, e.g. the default inclusion of `owl:Thing` in OWL). On one hand, Logical OPs are independent from a specific domain of interest (i.e. they are content-independent), on the other hand, they depend on the expressivity of the logical formalism that is used for representation. In other words, Logical OPs help to solve design problems where the primitives of the representation language do not directly support certain logical constructs. For example, if the representation language is OWL, and a designer needs to represent a relation between more than two

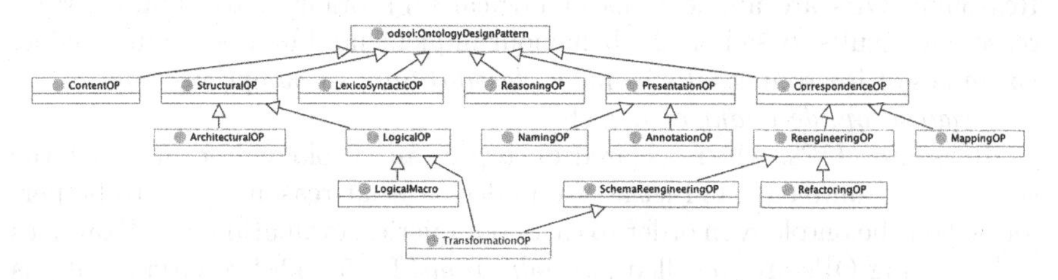

Fig. 1. Ontology design pattern types

elements, a Logical OP is needed in order to express an n-ary relation semantics by only using class and binary relation primitives. The root of *Logical OPs* can be found in [5], where so-called description logics were conceived as a way of representing knowledge in a structural manner by singling out the most relevant and tractable patterns from first-order logic (and beyond). The first proposal for a library of Semantic Web logical patterns is [45]. We can informally divide Logical OPs into two types:

Logical macros provide a shortcut to model a recurrent intuitive logical expression, e.g. the combination of *owl:allValuesFrom* restriction with *owl:someValuesFrom* restriction.

Transformation patterns translate a logical expression from a logical language into another, which approximates the semantics of the first, in order to find a trade-off between requirements and expressivity. For example, the so called *n-ary relation* pattern, documented in [33] with respect to OWL, is a transformation pattern from first-order logic to OWL DL. Other Logical OPs are documented in [33, 37, 47].

The application of Logical OPs has consequences on the results and efficiency of reasoning procedures. They can be used in order to document design choices and are particularly suitable for teaching good practices of ontology design as they provide designers with solutions to represent complex logical expressions.

Architectural OPs affect the overall shape of the ontology: their aim is to constrain "how the ontology should look like". They can be of two types: (i) *internal*, defined in terms of collections of Logical OPs that have to be exclusively employed when designing an ontology, e.g. an OWL species (cf. chapter "Web Ontology Language: OWL"), or the varieties of description logics (cf. chapter "Description Logics"); (ii) *external*, defined in terms of meta-level constructs, e.g. the *modular architecture* consists of an ontology network, where the involved ontologies play the role of modules (according to definitions given in [25]). The modules are connected by the *import* operation.

Architectural OPs emerged as design choices motivated by specific needs, e.g. computational complexity constraints. Such OPs are also useful as reference documentation for those initially approaching the design of an ontology.

Reasoning OPs

Reasoning OPs are applications of Logical OPs oriented to obtain certain reasoning results, based on the behaviour implemented in a reasoning engine. Examples of Reasoning OPs include: *classification*, *subsumption*, *inheritance*, *materialization*, *de-anonymizing*, etc.

Reasoning OPs, when declared on top of an ontology, inform about the state of that ontology, and let a system decide what reasoning has to be performed on the ontology in order to carry out queries, evaluation, etc. Examples of Reasoning OPs are so called *normalizations*. In [51, 52] five normalizations have been identified (cf. chapter "Ontology Engineering Environments").

Correspondence OPs

Correspondence OPs include Reengineering OPs and Mapping OPs.

Reengineering OPs provide designers with solutions to the problem of transforming a conceptual model, which can even be a non-ontological resource, into a new ontology. Mapping OPs are patterns for creating semantic associations between two existing ontologies.

Reengineering OPs are transformation rules applied in order to create a new ontology (*target* model) starting from elements of a *source* model. The target model is an ontology, while the source model can be either an ontology, or a non-ontological resource, e.g. a thesaurus concept, a data model pattern, a UML model, a linguistic structure, etc. Reengineering OPs are described in terms of metamodel transformation rules. We distinguish two types of Reengineering OPs.

Schema reengineering patterns are rules for transforming, e.g. a non-OWL DL metamodel into an OWL DL ontology. For example, consider the use of SKOS [31] for Knowledge Organization Systems (KOS) reengineering to a knowledge base (an OWL ABox), based-on the SKOS TBox. Transformation Logical OPs are a kind of schema reengineering patterns. In principle, all modelling problems can be represented as higher-order logical expressions, and if we have to represent them, e.g. in OWL DL, we implicitly apply a schema reengineering pattern in order to stay within the expressivity of OWL DL. However, we also (pragmatically) distinguish between transformation and schema reengineering patterns because of the different intention of the designer. In the first case, the designer wants to directly represent a modelling solution in a certain representation formalism, e.g. OWL DL,[7] while in the second case the designer wants to reengineer, e.g. an existing non-OWL DL model into an OWL DL ontology.

Refactoring patterns provide designers with rules for transforming, i.e. *refactoring*, e.g. an existing OWL DL source ontology into a new OWL DL target ontology. In this case, the transformation rule has the effect of changing the type of the ontology elements that are involved in the refactoring. For example, let us consider the case in which an ontology defines an object property for representing the relation of *preparing a coffee*, which holds between *agents* and *coffees*. Now, let us consider a change of requirements, so that a designer has to represent that the coffee is prepared by an agent at a certain time by using a certain tool. In order to address such a change in OWL DL, a designer has to apply an *n-ary relation* Logical OP, because *preparing a coffee* has now four arguments: *agent*, *coffee*, *time interval*, and *tool*. The *n-ary relation* Logical OP plus the description of how to apply it in order to replace an object property from an existing ontology is a Refactoring OP.

---

[7] In the pragmatics of an ontology designer, the fact that all modelling solutions are representable as higher-order logic expressions is hardly relevant, and such implicit reengineering has been never documented as actually happening.

Mapping ontology design patterns Mapping OPs refer to the possible semantic relations between mappable elements, as defined in [25]. There are three basic semantic relations that are used for mapping assertions: *equivalence*, *containment*, and *overlap*. They can be supplemented by their negative counterparts, i.e. *not equivalent*, *not contained*, and *not overlap* or *disjoint*, respectively. Mapping OPs provide designers with solutions to relate two ontologies without changing the logical types (e.g. owl:Class) of the ontology elements involved.

Presentation OPs

Presentation OPs deal with usability and readability of ontologies from a user perspective. They are meant as good practices that support the reuse of ontologies by facilitating their evaluation and selection. Examples are Naming OPs and Annotation OPs. The former are conventions about how to create names for namespace, files, and ontology elements in general (classes, properties, etc.). They are good practices that boost ontology readability and understanding by humans, by supporting homogeneity in naming procedures. Annotation OPs provide annotation properties or annotation property schemas that can be used in order to improve the understandability of ontologies and their elements.

An example of Naming OP relates to *namespace declared for ontologies*. It is recommended to use the base URI of the organization that publishes the ontology (e.g. `http://www.w3.org` for the W3C, `http://www.fao.org` for the FAO, `http://www.loa-cnr.it` for the Laboratory for Applied Ontologies (LOA) etc.) followed by a reference directory for the ontologies (e.g. `http://www.loa-cnr.it/ontologies/`). Additionally, it is also important to choose an approach for encoding versioning, either on the name, or on the reference directory.

Lexico-Syntactic OPs

Lexico-Syntactic OPs are linguistic structures or schemas that consist of certain types of words following a specific order, and that permit to generalize and extract some conclusions about the meaning they express. They are useful for associating simple Logical and Content OPs with natural language sentences, e.g. for didactic purposes.

Content Ontology Design Patterns (CPs)

CPs encode *conceptual*, rather than *logical* design patterns. In other words, while Logical OPs solve design problems independently of a particular conceptualization, CPs propose patterns for solving design problems for the domain classes and properties that populate an ontology, therefore addressing *content* problems [14]. CPs are instantiations of Logical OPs (or of compositions of Logical OPs), featuring a non-empty signature. Hence, they have an explicit non-logical vocabulary for a specific domain of interest (i.e. they are content-dependent). CPs provide solutions to domain modelling problems and affect only the specific region of the ontology dealing with such domain modelling problems. They are typically reused by applying specialization, extension, and composition to them. In principle, CPs do not depend on any specific language, however in order to reuse them as *building blocks*, they have to be implemented in some way. In the context of this chapter, we deal with CPs in a Semantic Web context. Hence, we use OWL as a reference formalism for representation.

3 Towards a Catalogue and Repository of CPs

In this section we focus on CPs. We define them, and explain the dependencies between CPs and use cases (Sect. 3.1). Section 3.2 lists the characteristics that differentiate CPs as special ontologies (such characteristics cross the boundaries between ontology engineering, cognitive science, and linguistics). Finally, we describe two CPs (Sect. 3.3).

The way to document OPs can be compared to the typical way followed for SE patterns. The mainstream approach for describing SE patterns is to use a template, although there is no standard format. A description of the most well-known SE pattern templates can be found at Martin Fowler's web site.[8] The templates used for describing SE patterns follow quite closely that suggested by Alexander [1]: given an *artifact type*, the pattern provides *examples* of it, its *context*, the *problem* addressed by the pattern, the involved "*forces*" (requirements and constraints), and a *solution*.

In order to describe CPs, we follow a similar approach: each CP is associated with a *catalogue entry* including the following set of information fields.

Name provides a name for the pattern; *Intent* describes the *Generic Use Case* addressed by the pattern; *Competency questions* contains examples of competency questions that the knowledge base associated with the CP needs to address; *Also Known as* provides other names (if any) with which the pattern is known; *Scenarios* provides examples of requirements, expressed in natural language, which can be modeled by using the pattern; *Diagram* depicts a UML class diagram representing the pattern; *Elements* describes the elements (classes and relations) included in the pattern, and their role within the pattern; *Consequences* provides a description of the benefits and/or possible trade-offs when using the pattern; *Known uses* gives examples of realistic ontologies where the pattern is used, *Extracted from/Reengineered from* provides the reference ontology/conceptual schema (if any), from which the pattern has been extracted/reused; *Related patterns* indicates other patterns (if any) that are either a *specialization*, *generalization*, *composition*, or *component* of the pattern being described. Furthermore, this field may indicate other

[8] http://www.martinfowler.com/articles/writingPatterns.html#CommonPatternForms

patterns that are typically used in conjunction with the described one. Important similarities and differences with other patterns can be also described here; *Building block* provides references to implementations of the pattern, a URI. In the case of CPs for Semantic Web ontologies, this field provides the URI of an OWL file (containing an implementation of the pattern).

Section 3.3 contains two examples of CPs that are described by means of a simplified version of the catalogue template. Such a catalogue can be found at the ontologydesignpatterns.org web portal [36], a dedicated wiki site through which a *lightweight* repository of CPs can be accessed. In fact, [36] allows users to download, propose, and discuss CPs. Furthermore, each CP includes a set of annotations[9] that can be exploited by Semantic Web applications. The reader can refer to chapter "Ontology Repositories" for more details on ontology repositories.

3.1 CPs and Competency Questions

CPs are reusable solutions to recurrent modelling problems. As known from a long time in conceptual modelling (cf. the difference between class and use case diagrams in UML) and knowledge engineering (cf. the distinction between domain and task ontologies in UPML [32]), these problems have two components: a domain and a use case (or task). A same domain can have many use cases (e.g. different scenarios in a clinical information context), and a same use case can be found in different domains (e.g. different domains with a same "competence finding" scenario).

Ontologies are usually considered models for a domain, but their use case is usually unknown. As reusable solutions, CPs must explicitly encode both a domain and a use case. Since use cases are extremely diversified, a catalogue of CPs requires the notion of a "Generic Use Case" (GUC), i.e. a generalization of use cases that can be provided as examples for an issue of domain modelling. A GUC is the expression of a recurrent scenario in different domain ontology projects.

Being generic at the use case level allows us to divide, or to refactor the design problems of a use case, by composing different GUCs. We can hierarchically organize GUCs from the most generic to the most specific ones, and from the "purest" (e.g. "which objects take part in a certain event?") to the most articulated and applied ones (e.g. "what protein is involved in the Jack/Stat biochemical pathway?").

The intuition underlying GUC hierarchies is based on a methodological observation: ontologies must be built out of domain tasks that can be captured by means of *competency questions* [23]. A competency question is a typical query that an expert might want to submit to a knowledge base of its target domain, for a certain task. In principle, an accurate domain ontology should specify *all and only* the conceptualizations required in order to answer all

[9] http://www.ontologydesignpatterns.org/schema/cpannotationschema.owl

the competency questions formulated by, or acquired from, experts. A GUC cannot do much as a guideline, unless we are able to find formal patterns that encode it. CPs are the solution to this issue. Based on the above assumptions, we define a CP as:

> CPs are distinguished ontologies. They address a specific set of competency questions, which represent the problem they provide a solution for. Furthermore, CPs show certain characteristics, i.e. they are: computational, small, autonomous, hierarchical, cognitively relevant, linguistically relevant, and best practices.

3.2 General Characteristics of CPs

CPs are components that represent, and possibly help solving a modelling problem arising across different use cases. E.g. the *agent-role* pattern provides a solution to represent agents that play some role. We have sketched their theoretical basis in Sect. 2, and explained their dependance on use cases (Sect. 3.1). Before providing a sample list of CPs against an example use case (Sect. 3.3), we now describe a more inclusive set of general, pragmatic features of CPs. These features, besides positioning CPs in a wider scientific context, give hints on how to discover or to extract CPs from existing knowledge resources.

Computational components. CPs are language-independent, and should be encoded in a higher-order representation language. Nevertheless, their (sample) representation in OWL is needed in order to (re)use them as building blocks over the Semantic Web.

Small, autonomous components. Regardless of the particular way a CP has been created, it is a *small, autonomous* ontology. Smallness (typically two to ten classes with relations defined between them) and autonomy of CPs facilitate ontology designers: composing CPs enable them to govern the complexity of the whole ontology, because of the explicit rationales and the amount of know-how provided by the users of a same CP library. Smallness also allows diagrammatical visualizations that are aesthetically acceptable and easily memorizable.

Hierarchical components. A CP can be an element in a partial order, where the ordering relation requires that at least one of the classes or properties in the pattern is specialized. A hierarchy of CPs can be built by specializing or generalizing some of the elements (either classes or relations).

Inference-enabling components. There are combinations of ontology elements that do not allow any useful inference, e.g. a taxonomy with two sibling classes, an object property alone, etc. A CP allows some form of inference, e.g. a taxonomy with two sibling disjoint classes, a property with explicit domain and range set, a property and a class with a universal restriction on that property, etc.

Cognitively relevant components. CP visualization must be intuitive and compact, and should catch relevant, "core" notions of a domain [14].

Linguistically relevant components. Many CPs nicely match linguistic patterns called *frames.* A frame can be described as a lexically founded OP. The richest repository of frames is FrameNet [3]. Frames can be used for validating CPs with respect to lexical coverage, for lexicalizing them, and can be reengineered as CPs.

Best practice components. A CP should be used to describe a "best practice" of modelling. Best practices are intended here as *local*, thus derived from experts, emerging from real applications. The quality of CPs is currently based on the personal experience and taste of the proposers, or on the provenance of the knowledge resource where the pattern comes from. However, evidence from reusability across different projects, large-scale applications, and open rating systems will provide a good base for CP evaluation.

3.3 Samples of CP Catalogue Entries

In this section we show two CPs taken from [36], Each CP is presented in a catalogue-like way, and with reference to the OWL language. For space reasons, we describe each CP with a simplified catalogue entry composed of: the *Name* (including possible alternative names), the *Intent* (i.e., the GUC), *Competency questions*, some *Examples* of its application, the *Diagram* describing its structure, the *Elements* and the role they play in the pattern, and some *General Remarks* that indicate general guidelines about how to use it, including relations to other CPs. The complete entry[10] also contains a field named *building block* that provides references to implementations of the pattern, i.e. repository of reusable components. In the case of CPs for Semantic Web ontologies, this field provides the URI of an OWL file (containing an implementation of the pattern). We have used TopBraid Composer[11] in order to produce the OWL encoding. With the same tool, we automatically generated a diagrammatical visualization based on a UML profile for OWL. UML classes (boxes) are used in order to depicts OWL classes. Two kinds of OWL classes can be visualized in a diagram: named classes (`owl:Class`, in white boxes) and anonymous classes (in grey boxes), e.g. `owl:Restriction` with `owl:someValuesFrom`. UML generalization (arrow with a large end) corresponds to `rdfs:subClassOf`, while UML association (arrow with a small end) corresponds to `owl:ObjectProperty`. Finally, UML class attributes (statements inside white boxes) are used in order to indicate either `rdfs:domain` and `rdfs:range`, or `owl:Restriction` with `owl:allValuesFrom`. When a class name is preceded by a prefix, e.g. `sit:`, it is interpreted as a class imported (e.g. by `owl:imports`) from another (typically more general) CP that is indexed by means of that prefix.

In the rest of this section we use the OWL terminology in order to describe the proposed design solutions, e.g. object property, datatype property, etc.

[10] See [36,37].

[11] http://www.topbraidcomposer.com/

The information realization CP

The *information realization* CP is extracted from the Dolce+DnS Ultra Lite ontology,[12] and represents the relations between information objects like poems, songs, formulas, etc., and their physical realizations like printed books, registered tracks, physical files, etc..

The *information realization CP* is associated with information according to the catalogue entry fields reported below:

Intent: to represent relations between information objects and their physical realizations.

Competency questions: which physical object realizes a certain information object? Which information object is realized by a certain physical object?

Diagram: Fig. 2 shows a UML diagram of the information realization CP.

Elements:

- `InformationObject`: a piece of information, such as a musical composition, a text, a word, a picture, independently from how it is concretely realized.
- `InformationRealization`: a concrete realization of an InformationObject, e.g. the written document containing the text of a law.
- `realizes`: a relation between an information realization and an information object, e.g. the paper copy of the Italian Constitution realizes the text of the Constitution.
- `isRealizedBy`: a relation between an information object and an information realization, e.g. the text of the Constitution is realized by the paper copy of the Italian Constitution.

General remarks: this CP[13] allows to distinguish between information encoded in an object and the possible physical representations of it. The Multimedia ontology (cf. chapter "Ontologies for Cultural Heritage") uses this CP.

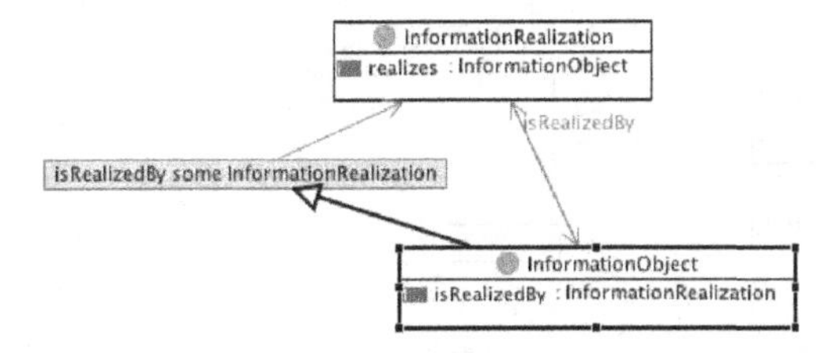

Fig. 2. The information realization CP UML graphical representation

[12] http://www.ontologydesignpatterns.org/ont/dul//DUL.owl
[13] http://www.ontologydesignpatterns.org/cp/owl/informationrealization.owl

The Time Indexed Person Role CP

The *time indexed person role* is a CP that represents time indexing for the relation between persons and roles they play, e.g. George W. Bush was the president of the United States in 2007. This CP is also extracted from the Dolce+DnS Ultra Lite ontology.

According to its associated catalogue entry, the main information associated with this CP are the following:

Intent: to represent time indexing for the relation between persons and roles they play.

Competency questions: who was playing a certain roles during a given time interval? When did a certain person play a specific role?

Diagram: see Fig. 3, the elements which compose the CP are described in the *Elements* field.

Elements:

- `Entity`: anything: real, possible, or imaginary, which some modeller wants to talk about for some purpose.
- `Person`: persons in commonsense intuition, i.e. either as physical agents (humans) or social persons.
- `Role`: a Concept that classifies a Person.
- `TimeInterval`: any region in a dimensional space that aims at representing time.
- `TimeIndexedPersonRole`: a situation that expresses time indexing for the relation between persons and roles they play.
- `hasRole`: a relation between a Role and an Entity, e.g. "John is considered a typical rude man"; your last concert constitutes the achievement of a lifetime; "20-year-old means she's mature enough".
- `isRoleOf`: a relation between a Role and an Entity, e.g. the Role "student" classifies a Person "John".

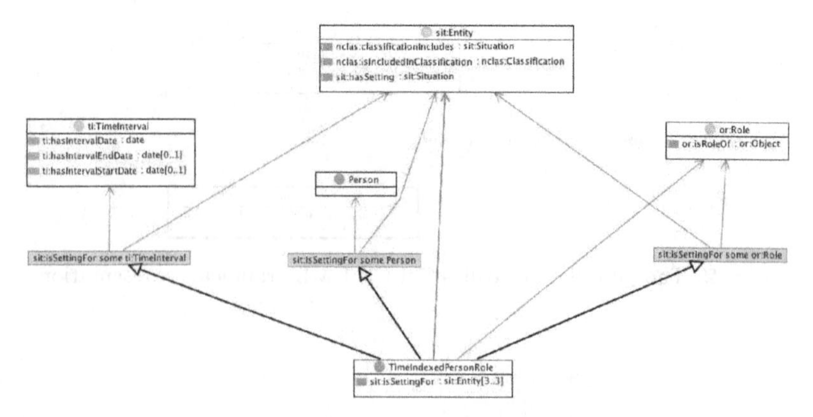

Fig. 3. The time indexed person role CP UML graphical representation

- `isSettingFor`: a relation between time indexed role situations and related entities, e.g. "I was the director between 2000 and 2005", i.e. the situation in which I was a director is the setting for a the role of director, me, and the time interval.
- `hasSetting`: the inverse relation of `isSettingFor`.

General remarks: this CP[14] allows to assign a time interval to roles played by people.

4 Creating and Working with CPs

This section discusses how CPs can be created, and provides guidelines on how they can be practically (re)used. Section 4.1 describes four approaches to create CPs, while Sect. 4.2 shows the main operations that are performed for reusing a CP, and describes the possible situations of CP selection and usage that can occur in practice.

4.1 Where do Content Ontology Design Patterns Come from?

CP creation and usage rely on a common set of operations.

Import: consists of including a CP in the ontology under development. This is the basic mechanism for reusing CPs (and ontologies in general). By importing a CP, the importing ontology ensures the set of inferences allowed by the CP in its corresponding knowledge base. Elements of an imported CP cannot be modified.

Specialization: can be referred to ontology elements or to CPs. Specialization between ontology elements of a CP consists of creating sub-classes of some CP's class and/or sub-properties of some CP's properties. A CP c_1 is a specialization of a CP c if c_1 imports c, and at least one ontology element from c_1 specializes an ontology element from c.

Generalization: A CP c_1 is a generalization of a CP c if c_1 imports c, and at least one ontology element from c_1 generalizes an element from c.

Composition: consists of associating classes (properties) of one CP with classes (properties) of other CPs, by means of some OWL axiom.

Expansion: consists of adding new classes, properties and axioms to the ontology to the aim of covering the requirements that are not addressed by the reused CPs.

CPs come from the experience of ontology engineers in modelling *foundational* (cf. chapter "Foundational Choices in DOLCE"), *upper-level*, *core* [15], or *domain* ontologies. Informally, the distinction between these kinds of ontologies relates to the degree by which an ontology covers the domain of interest,

[14] http://ontologydesignpatterns.org/cp/owl/timeindexedpersonrole.owl

cf. chapter "What Is an Ontology" for details. Assuming the above distinctions, there are four main ways of creating CPs, which can be summarized as follows:

Reengineering from other data models. A CP can be the result of a reengineering process applied to different conceptual modelling languages, primitives, and styles. Knowledge resources that can be reengineered to produce candidate CPs are database schemas, knowledge organization systems such as thesauri, and lexica. For more references, the reader can refer to [20] that describes a reengineering approach for creating CPs starting from UML diagrams [35], workflow patterns [49], and data model patterns [27].

Specialization/Composition of other CPs. A CP can be created by composing other CPs, or by specializing another CP, (both composition and specialization can be combined with expansion, see below).

Extraction from reference ontologies. A CP can be *extracted* from an existing ontology, which acts as the "source" ontology. In this case, the CP corresponds to a fragment of the source ontology, which constitutes its axiomatic background context. A CP is axiomatized according to the fragment it extracts. E.g. the *co-participation* CP depends on a set of axioms from the DOLCE ontology [9], which state that an event has at least one participant, that co-participation requires two participants in a same event, that participants must participate at least partly at the same time, etc. If a modeller specializes the *co-participation* CP for representing, e.g. an academic lecture or a football match, the reasoning services will operate with reference to the *co-participation* axioms, without the need for encoding them again. However, a CP is autonomous, and only the axioms that have been extracted from the reference ontology are actually used by an ontology that reuses a CP. Therefore, reasoning services do not need to also process the general axiomatic context from the reference ontology.

Creation by combining extraction, specialization, generalization, and expansion The definition of a CP can be the result of an extraction (see above), followed by specialization and/or generalization of some ontology elements, and expansion.[15]

4.2 How to Use Content Ontology Design Patterns

Supporting reuse and alleviating difficulties in ontology design activities are the main goals of setting up a catalogue of CPs. In order to be able to reuse CPs, two main functionalities must be ensured: *selection* and *application*.

Selection of CPs corresponds to finding the most appropriate CP for the actual domain modelling problem. Hence, selection includes search and evaluation of available CPs. This task can be performed by applying procedures for ontology selection [28, 41] and evaluation [13] (cf. chapter "Ontology Engineering Environments").

[15] See [37] for more details.

Informally, a GUC, i.e. the *intent* of a CP, must match an actual use case. Once a CP has been selected, it has to be applied to the domain ontology. Typically, application is performed by means of import, specialization, composition, or expansion (see Sect. 4.1). In realistic design projects, the operations are usually combined as it is shown by the example of Sect. 5.

Several situations of matching between GUCs and actual use cases can occur, each associated with a different approach to using CPs. The following summary assumes a *manual* (re)use of CPs. However, an initial library of CPs is already available [36], and tool support to their selection and usage can take into account the principles informally explained in the summary below as base requirements. *Precise or redundant matching.* The CP matches a GUC, which is either more complex or directly usable to describe the local use case: the CP has only to be *imported* in the domain ontology.

Broader matching. The CP matches a GUC that is more general than the local use case: the CP's catalogue entry may contain reference to less general CPs that specialize it. If none of them is appropriate, the CP has firstly to be *imported*, then it has to be *specialized* in order to cover the domain part to be represented.

Narrower matching. The CP matches a GUC that is more specific than the local use case: the CP's catalogue entry may contain references to more general CPs. If none of them is appropriate, a the CP has firstly to be *imported*, then it has to be *generalized* according to the local requirements.

Partial matching. The CP partly matches a GUC that does not cover all aspects of the local use case (it is simpler): the CP's catalogue entry may contain references to CPs it is a component of. If none of such compound CPs is appropriate, the local use case has to be partitioned into smaller pieces. One of these pieces will be covered by the selected CP. For the other pieces, other CPs have to be selected. All selected CPs have to be *imported* and *composed.*

In all the above situation, *expansion* is performed when needed.

5 Use Case Example in the Music Industry Domain

As an example of usage we design a small fragment of an ontology for the music industry domain. The ontology fragment has to address the following competency questions:

Which recordings of a certain song do exist in our archive?
Who did play a certain musician role in a given band during a certain period?

The first competency question requires to distinguish between a song and its recording, while the second competency question highlights the issue of assigning a given musician role, e.g. singer, guitar player, etc., to a person who is member of a certain band, at a given period of time. The intent of the *information realization* is related to the first competency question with a *broader matching.* The intent of the *time indexed person role* partially and broadly matches the second competency question. Hence, we select these two CPs as building blocks for our ontology.[16]

We proceed by importing and composing the two selected CPs in our ontology (the *information realization* CP is associated with the prefix *ir:*; the *time indexed person role* CP is associated with the prefix *tipr:*). Additionally, we might want to import the *time interval* CP[17] that allows us to assign a date to the time interval. In order to complete our ontology fragment we create: the class `Song` that specializes `ir:InformationObject`, the class `Recording` that specializes `ir:InformationRealization`, the class `MusicianRole` that specializes `tipr:Role`, the class `Band`, and the object property `memberOf` (and its inverse) with explicit domain, i.e. `tipr:Person`, and range, i.e. `Band`. A screenshot of the resulting ontology fragment is shown in Fig. 4.[18] On the left side of the picture ontology classes are shown, on the right side there are ontology properties, while at the bottom there are the imported CPs. Notice that CPs can be very useful when they address issues in a specific domain. For this reason, an ontology fragment like this one might be proposed as a CP[19] if it is associated with a successful application in an ontology design project.

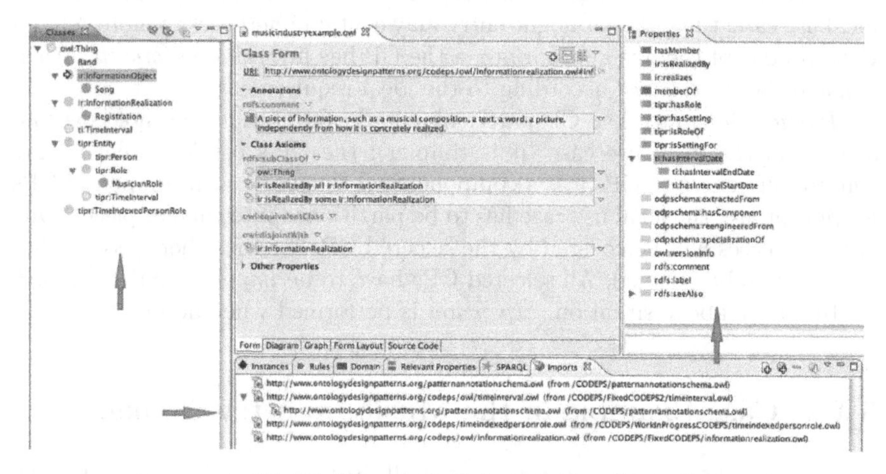

Fig. 4. The music industry example

---

[16] Notice that the second requirement would also require to represent membership relation between a person and a band. The *collection entity* CP available at http://www.ontologydesignpatterns.org/cp/owl/collectionentity.owl addresses membership. We do not include the description of this CP and its usage for the sake of brevity.

[17] Available at http://www.ontologydesignpatterns.org/cp/owl/timeinterval.owl

[18] The screenshot shows the TopBraid Composer interface, see http://www.topbraidcomposer.com

[19] See [36] area of proposed CP.

6 Conclusion and Remarks

Ontology design is a crucial research area for semantic technologies. Many bottlenecks in the wide adoption of semantic technologies depend on the difficulty of understanding ontologies and on the scarcity of tools supporting their lifecycle, from creation to adaptation, reuse, and management. The lessons learnt until now, either from the early adoption of Semantic Web solutions or from local, organizational applications, put a lot of emphasis on the need for simple, modular ontologies that are accessible and understandable by typical computer scientist and field experts, and on the dependability of these ontologies on existing knowledge resources.[20]

In this chapter, we have described a breed of components, called Ontology Design Patterns, and tools that will support ontology design at the level that is more natural to domain experts and laymen, i.e. the level at which small, expertise-aware components can be assembled as easy-to-apply, easy-to-customize building blocks.

The quality of these components is expected to be evaluated with respect to known good practices, as well as in the large testbed of organizational or web-scale open rating systems. In order to allow the maximum transparency and flexibility, OPs are supplied with a rich set of metadata for their explanation, rationale declaration, use case history, evaluation criteria, etc. In this chapter, we have sketched a typology of OPs, then focused on Content Ontology Design Patterns (which are most beneficial to ontology design) in terms of their background, definition, communication means, related work beyond ontology engineering, exemplification, creation, and usage principles.

There is still a lot of work to be carried out for populating repositories of patterns, discoverying or extracting them from existing ontologies, assisting users in their application, defining a robust semantics and algebra for them, etc. (cf. [42]). The larger context of ontology design research is still very young, and many ideas are just emerging, for example in (semi-)automatizing the creation and evaluation of ontologies, based only on informal documentation from users, a set of software components, and a repository of design patterns. In a larger report [37] and by setting up the ontologydesignpatterns.org web portal [36], we make some steps towards the open issues. Moreover, we have designed a set of experiments that are going to be performed in order to show OPs' effectiveness, e.g. in teaching ontology design, lower the cost of ontology projects, etc. However, some initial experiences with PhD students classes, and the employment of CPs in of our own recent projects have provided us with concrete proof of the benefits deriving from their application.

---

[20] An interesting review of evaluation, selection and reuse methods in ontology engineering is in [40].

References

1. Christopher Alexander. *The Timeless Way of Building.* Oxford Press, 1979.
2. Richard Arndt, Raphael Troncy, Steffen Staab, Lynda Hardman, and Miroslav Vacura. COMM: Designing a Well-Founded Multimedia Ontology for the Web. In *Proceedings of the 4th European Semantic Web Conference (ISCW'07)*, Busan Korea, November 2007. Springer.
3. Collin F. Baker, Charles J. Fillmore, and John B. Lowe. The Berkeley FrameNet project. In Christian Boitet and Pete Whitelock, editors, *Proceedings of the Thirty-Sixth Annual Meeting of the Association for Computational Linguistics and Seventeenth International Conference on Computational Linguistics*, pages 86–90, San Francisco, California, 1998. Morgan Kaufmann.
4. Eva Blomqvist. Fully automatic construction of enterprise ontologies using design patterns: Initial method and first experiences. In Robert Meersman, Zahir Tari, Mohand-Said Hacid, John Mylopoulos, Barbara Pernici, Özalp Babaoglu, Hans-Arno Jacobsen, Joseph P. Loyall, Michael Kifer, and Stefano Spaccapietra, editors, *OTM Conferences (2)*, volume 3761 of *Lecture Notes in Computer Science*, pages 1314–1329, 2005. Springer.
5. Ronald J. Brachman. A Structural Paradigm for Representing Knowledge. PhD thesis, Harvard University, USA, 1977.
6. Dan Brickley and Libby Miller. FOAF Vocabulary Specification. Working draft, 2005.
7. Peter Clark, John Thompson, and Bruce Porter. Knowledge patterns. In Anthony G. Cohn, Fausto Giunchiglia, and Bart Selman, editors, *KR2000: Principles of Knowledge Representation and Reasoning*, pages 591–600, San Francisco, 2000. Morgan Kaufmann.
8. Mathieu d'Aquin, Peter Haase, Sebastian Rudolph, Jerome Euzenat, Antoine Zimmermann, Martin Dzbor, Marta Iglesias, Yves Jacques, Caterina Caracciolo, Carlos Buil Aranda, and Jose Manuel Gomez. NeOn Formalisms for Modularization: Syntax, Semantics, Algebra. Deliverable D1.1.3, NeOn project, 2008.
9. DOLCE - Project Home Page. http://dolce.semanticweb.org.
10. Nicholas Drummond, Alan Rector, Robert Stevens, Georgina Moulton, Matthew Horridge, Hai Wang, and Julian Sedenberg. Putting owl in order: Patterns for sequences in owl. In *OWL Experiences and Directions (OWLEd 2006)*, Athens Georgia, 2006.
11. Didier Dubois, Christopher A. Welty, and Mary-Anne Williams, editors. *Principles of Knowledge Representation and Reasoning: Proceedings of the Ninth International Conference (KR2004), Whistler, Canada, June 2–5, 2004.* AAAI Press.
12. Erich Gamma, Richard Helm, Ralph E. Johnson, and John Vlissides. *Design Patterns. Elements of Reusable Object-Oriented Software*, March 1995. Addison-Wesley. ISBN-10: 0201633612 ISBN-13: 978-0201633610.
13. Aldo Gangemi, Carola Catenacci, Massimiliano Ciaramita, and Jos Lehmann. Modelling Ontology Evaluation and Validation. In *Proceedings of the Third European Semantic Web Conference*, 2006. Springer.
14. Aldo Gangemi. Ontology Design Patterns for Semantic Web Content. In M. Musen et al. editors, *Proceedings of the Fourth International Semantic Web Conference*, Galway, Ireland, 2005. Springer.

15. Aldo Gangemi and Stefano Borgo, editors. *Proceedings of the EKAW*04 Workshop on Core Ontologies in Ontology Engineering, Northamptonshire (UK)*, volume 118. CEUR-WS, October 2004.
16. Aldo Gangemi, Carola Catenacci, and Massimo Battaglia. Inflammation ontology design pattern: An exercise in building a core biomedical ontology with descriptions and situations. In Domenico Maria Pisanelli, editor, *Ontologies in Medicine*, 2004. IOS Press.
17. Aldo Gangemi, Frehiwot Fisseha, Johannes Keizer, Jos Lehmann, Anita Liang, Ian Pettman, Margherita Sini, and Marc Taconet. A Core Ontology of Fishery and its Use in the FOS Project. In Aldo Gangemi and Stefano Borgo, editors, *Proceedings of the EKAW*04 Workshop on Core Ontologies in Ontology Engineering*, volume 118. CEUR-WS, 2004.
18. Aldo Gangemi, Jos Lehmann, Valentina Presutti, Malvina Nissim, and Carola Catenacci. C-ODO: An OWL meta-model for collaborative ontology design. In Harith Alani, Natasha Noy, Gerd Stumme, Peter Mika, York Sure, and Denny Vrandecic, editors, *Workshop on Social and Collaborative Construction of Structured Knowledge (CKC 2007) at WWW 2007*, Banff, Canada, 2007.
19. Aldo Gangemi, Domenico Maria Pisanelli, and Geri Steve. An ontological framework to represent norm dynamics. In R. Winkels, editor, *Proceedings of the 2001 Jurix Conference, Workshop on Legal Ontologies*, Amsterdam, 2001.
20. Aldo Gangemi and Valentina Presutti. Ontology design for interaction in a reasonable enterprise. In Peter Rittgen, editor, *Handbook of Ontologies for Business Interaction*. IGI Global, Hershey, PA, November 2007.
21. J.-M. Le Goff and I. Willers. Design patterns for description-driven systems, 1999.
22. Thomas R. Gruber. A translation approach to portable ontology specifications. *Knowledge Acquisition*, 5(2):199–220, 1993.
23. Michael Gruninger and Mark Fox. The role of competency questions in enterprise engineering, 1994.
24. Giancarlo Guizzardi and Gerd Wagner. A unified foundational ontology and some applications of it in business modeling. In *CAiSE Workshops (3)*, pages 129–143, 2004.
25. Peter Haase, Saartje Brockmans, Raul Palma, Jerome Euzenat, and Mathieu d'Aquin. D1.1.2 updated version of the networked ontology model. Deliverable D1.1.2, Neon Project, 2007.
26. Neil B. Harrison, Paris Avgeriou, and Uwe Zdlin. Using patterns to capture architectural decisions. *Software*, 24(4):38–45, 2007.
27. David C. Hay. *Data Model Patterns*, 1996. Dorset House Publishing.
28. Tzung-Pei Hong, Wen-Chang Chang, and Jiann-Horng Lin. A two-phased ontology selection approach for semantic web. In Rajiv Khosla, Robert J. Howlett, and Lakhmi C. Jain, editors, *KES (4)*, volume 3684 of *Lecture Notes in Computer Science*, pages 403–409, 2005. Springer.
29. David Mapelsden, John Hosking, and John Grundy. Design pattern modelling and instantiation using dpml. In *CRPIT '02: Proceedings of the Fortieth International Conference on Tools Pacific*, pages 3–11, Darlinghurst, Australia, Australia, 2002. Australian Computer Society, Inc.
30. David Maplesden, John G. Hosking, and John C. Grundy. A visual language for design pattern modelling and instantiation. In *HCC*, pages 338–339, 2001. IEEE Computer Society.

31. Alistar Miles and Dan Brickley. SKOS Core Vocabulary Specification. Technical report, World Wide Web Consortium (W3C), November 2005. http://www.w3.org/TR/2005/WD-swbp-skos-core-spec-20051102/.
32. Enrico Motta and Wenjin Lu. A library of components for classification problem solving. ibrow project ist-1999-19005: An intelligent brokering service for knowledge-component reuse on the world-wide web. Technical report, KMI, 2000.
33. Natasha Noy and Alan Rector. Defining N-ary Relations on the Semantic Web: Use With Individuals. Technical report, W3C, 2005. http://www.w3.org/TR/swbp-n-aryRelations/ (2004).
34. Daniel Oberle, Peter Mika, Aldo Gangemi, and Marta Sabou. Foundations for service ontologies: Aligning OWL-S to DOLCE. In *Proceedings of the World Wide Web Conference (WWW2004)*, volume Semantic Web Track, 2004.
35. Object Management Group (OMG). Unified modeling language specification: Version 2, revised final adopted specification (ptc/04-10-02), 2004.
36. Ontology design patterns. http://www.ontologydesignpatterns.org.
37. Valentina Presutti, Aldo Gangemi, Stefano David, Guadalupe Aguado de Cea, Mari-Carmen Suarez Figueroa, Elena Montiel-Ponsoda, and María Poveda. Library of design patterns for collaborative development of networked ontologies. Deliverable D1.1.3, NeOn project, 2008.
38. Alan Rector and Jeremy Rogers. Patterns, properties and minimizing commitment: Reconstruction of the galen upper ontology in owl. In Aldo Gangemi and Stefano Borgo, editors, *Proceedings of the EKAW*04 Workshop on Core Ontologies in Ontology Engineering*. CEUR, 2004.
39. Jacqueline Rene Reich. Ontological design patterns: Metadata of molecular biological ontologies, information and knowledge. In Mohamed T. Ibrahim, Josef Küng, and Norman Revell, editors, *DEXA*, volume 1873 of *Lecture Notes in Computer Science*, pages 698–709, 2000. Springer.
40. Marta Sabou, Sofia Angeletou, Mathieu dAquin, Jesus Barrasa, Klaas Dellschaft, Aldo Gangemi, Jos Lehmann, Holger Lewen, Diana Maynard, Dunja Mladenic, Malvina Nissim, WimPeters, Valentina Presutti, and BorisVillazon. Methods for selection and integration of reusable components from formal or informal user specifications. Deliverable D2.2.1, NeOn project, 2007.
41. Marta Sabou, Vanessa Lopez, and Enrico Motta. Ontology selection for the real semantic web: How to cover the queen's birthday dinner? In Steffen Staab and Vojtech Svatek, editors, *EKAW*, volume 4248 of *Lecture Notes in Computer Science*, pages 96–111, 2006. Springer.
42. Jorge Santos and Steffen Staab. Fonte: Factorizing ontology engineering complexity. In *K-CAP '03: Proceedings of the 2nd International Conference on Knowledge Capture*, pages 146–153, New York, NY, 2003. ACM.
43. Stefan Schulz, Anand Kumar, and Thomas Bittner. Biomedical ontologies: What part-of is and isn't. *Journal of Biomedical Informatics*, 39(3):350–361, 2006.
44. Dmitri Soshnikov. Ontological design patterns for distributed frame hierarchy. In *Proceedings of the 5th International Workshop on Computer Science and Information Technologies*, Ufa, Russia, 2003.
45. S. Staab, M. Erdmann, and A. Maedche. Engineering ontologies using semantic patterns, 2001.
46. Robert Stevens, na Aranguren Mikel Ega Katy Wolstencroft, Ulrike Sattler, Nick Drummond, Matthew Horridge, and Alan Rector. Using owl to model

biological knowledge. *International Journal of Human–Computer Studies*, 65(7): 583–594, 2007.
47. Mari Carmen Suarez-Figueroa, Saartje Brockmans, Aldo Gangemi, Asuncion Gomez-Perez, Jos Lehmann, Holger Lewen, Valentina Presutti, and Marta Sabou. Neon modelling components. Deliverable D5.1.1, NeOn project, 2007.
48. Vojtech Svatek. Design patterns for semantic web ontologies: Motivation and discussion. In *Proceedings of the 7th Conference on Business Information Systems*, Poznan, 2004.
49. Wil M. P. Van Der Aalst, Arthur H. M. Ter Hofstede, Bartek Kiepuszewski, and Alistair P. Barros. Workflow Patterns. *Distributed and Parallel Databases*, 14:5–51, 2003.
50. Denny Vrandecic and Aldo Gangemi. Unit tests for ontologies. In Mustafa Jarrar, Claude Ostyn, Werner Ceusters, and Andreas Persidis, editors, *Proceedings of the 1st International Workshop on Ontology content and evaluation in Enterprise*, LNCS, Montpellier, France, OCT 2006. Springer.
51. Denny Vrandecic and York Sure. How to design better ontology metrics. In Wolfgang May and Michael Kifer, editors, *Proceedings of the 4th European Semantic Web Conference (ESWC'07)*, Innsbruck, Austria, June 2007, volume 4519, pages 311–325. Springer.
52. Denny Vrandecic, York Sure, Raul Palma, and Francisco Santana. Ontology repository and content evaluation. Deliverable D1.2.10v2, KnowledgeWeb project, 2007.

Ontology Learning

Philipp Cimiano[1], Alexander Mädche[2], Steffen Staab[3], and Johanna Völker[1]

[1] Institute AIFB, University of Karlsruhe, Karlsruhe, Germany, cimiano.voelker@aifb.uni-karlsruhe.de
[2] SAP AG, Walldorf, Germany, alexander.maedche@sap.com
[3] ISWEB Group, University of Koblenz-Landau, Koblenz, Germany, staab@uni-koblenz.de

Summary. Ontology learning techniques serve the purpose of supporting an ontology engineer in the task of creating and maintaining an ontology. In this chapter, we present a comprehensive and concise introduction to the field of ontology learning. We present a generic architecture for ontology learning systems and discuss its main components. In addition, we introduce the main problems and challenges addressed in the field and give an overview of the most important methods applied. We conclude with a brief discussion of advanced issues which pose interesting challenges to the state-of-the-art.

1 Introduction

Ontology engineering is slowly changing its status from an art to a science and in fact, during the last decade, several ontology engineering methodologies (see chapters "Ontology Engineering Methodology" and "Ontology Engineering and Evolution in a Distributed World Using DILIGENT") have been examined. But still, as pointed out in chapter "Exploring the Economical Aspects of Ontology Engineering", the task of engineering an ontology remains a resource-intensive and costly task. Therefore, techniques which support the task of ontology engineering are necessary to reduce the costs associated with the engineering and maintenance of ontologies. As data in various forms (textual, structured, visual, etc.) is massively available, many researchers have developed methods aiming at supporting the engineering of ontologies by data mining techniques, thus deriving meaningful relations which can support an ontology engineer in the task of modeling a domain. Such data-driven techniques supporting the task of engineering ontologies have become to be known as *ontology learning*. Ontology learning has indeed the potential to reduce the cost of creating and, most importantly, maintaining an ontology. This is the reason why a plethora of ontology learning frameworks have been developed in the last years and integrated with standard ontology engineering tools. Text-ToOnto [55], for example, was originally integrated into the KAON ontology

S. Staab and R. Studer (eds.), *Handbook on Ontologies*, International Handbooks on Information Systems, DOI 10.1007/978-3-540-92673-3,

engineering environment [27], OntoLT [11] was integrated with Protégé and Text2Onto [22] has been recently integrated with the NeOn Toolkit.[1]

There are three kinds of data to which ontology learning techniques can be applied: structured (such as databases), semi-structured (HTML or XML, for example) as well as unstructured (e.g., textual) documents. The methods applied are obviously dependent on the type of data used. While highly structured data as found in databases facilitates the application of pure machine learning techniques such as Inductive Logic Programming (ILP), semi-structured and unstructured data requires some preprocessing, which is typically performed by natural language processing methods.

Ontology Learning builds upon well-established techniques from a variety of disciplines, including natural language processing, machine learning, knowledge acquisition and ontology engineering. Because the fully automatic acquisition of knowledge by machines remains in the distant future, the overall process is considered to be semi-automatic with human intervention.

Organization

This chapter is organized as follows: Sect. 2 introduces a generic architecture for ontology learning and its relevant components. In Sect. 3 we introduce various complementary basic ontology learning algorithms that may serve as a basis for ontology learning. Section 4 describes ontology learning frameworks and tools which have been implemented in the past. In particular, we also discuss our own system, Text2Onto, the successor of the TextToOnto framework [55].

2 An Architecture and Process Model for Ontology Learning

The purpose of this section is to introduce a generic ontology learning architecture and its major components. The architecture is graphically depicted in Fig. 1. In general, the process of ontology learning does not differ substantially from a classical data mining process (e.g., [15]) with the phases of *business and data understanding*, *data preparation*, *modeling*, *evaluation* and *deployment*. The key components of an architecture for ontology learning are the following: an *ontology management*, a *coordination*, a *resource processing* and an *algorithm library component*. We describe these components in more detail in the following.

2.1 Ontology Management Component

The ontology engineer uses the ontology management component to manipulate ontologies. Ontology management tools typically facilitate the import,

[1] http://www.neon-toolkit.org

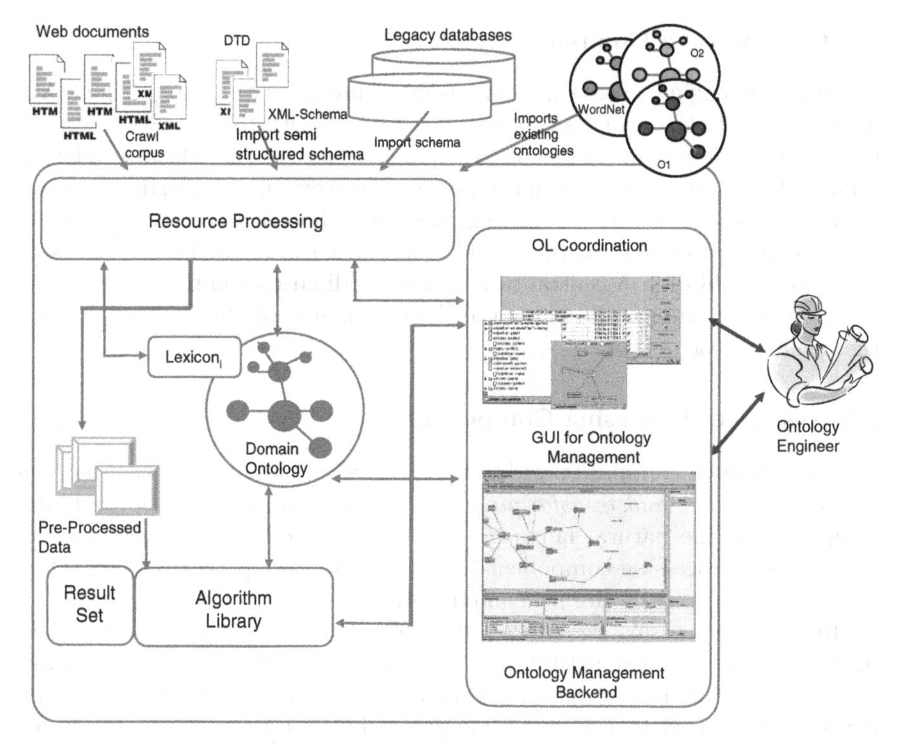

Fig. 1. Ontology learning conceptual architecture

browsing, modification, versioning as well as evolution of ontologies. However, the main purpose of the ontology management component in the context of ontology learning is to provide an interface between the ontology and the learning algorithms. When learning new concepts, relations or axioms, the learning algorithms should add them into the ontology model accessing the Application Programming Interface (API) of the ontology management component. Thus, the ontology management API should at least contain methods for creating new concepts, relations, axioms, individuals, etc. Most available APIs indeed fulfill this requirement. Further important functionalities for ontology learning are: *evolution*, *reasoning* and *evaluation*. Techniques for ontology evolution as presented in [52] or [35] are very important for ontology learning as it is an inherently dynamic process. As the underlying data changes, the learned ontology should change as well and this requires not only incremental ontology learning algorithms, but also some support for ontology evolution at the ontology management level. Reasoning and evaluation play a crucial role in guiding the ontology learning process. In case the ontology learning system faces several alternatives, it should definitely choose that alternative which preserves the consistency of the underlying ontology [36] or the one which maximizes certain quality criteria.

2.2 Coordination Component

The ontology engineer uses this component to interact with the ontology learning components for resource processing as well as with the algorithm library. Comprehensive user interfaces should support the user in selecting relevant input data that are exploited in the further discovery process. Using the coordination component, the ontology engineer also chooses among a set of available resource processing methods and among a set of algorithms available in the algorithm library. A central task of the coordination component is further to sequentially arrange and apply the algorithms selected by the user, passing the results to each other.

2.3 Resource Processing Component

This component contains a wide range of techniques for *discovering, importing, analyzing and transforming* relevant input data. An important subcomponent is the natural language processing system. The general task of the resource processing component is to generate a pre-processed data set as input for the algorithm library component.

Resource processing strategies differ depending on the type of input data made available. Semi-structured documents, like dictionaries, may be transformed into a predefined relational structure. HTML documents can be indexed and reduced to free text. For processing free text, the system must have access to language-specific natural language processing systems. Nowadays, off-the-shelf frameworks such as GATE [24] already provide most of the functionality needed by ontology learning systems. The needed NLP components could be the following ones:

- A *tokenizer* and a *sentence splitter* to detect sentence and word boundaries.
- A *morphological analyser*. For some languages a lemmatizer reducing words to their base form might suffice, whereas for languages with a richer morphology (e.g., German) a component for structuring a word into its components (lemma, prefix, affix, etc.) will be necessary. For most machine learning-based algorithms a simple stemming of the word might be sufficient (compare [60]).
- A *part-of-speech (POS) tagger* to annotate each word with its syntactic category in context, thus determining whether it is a noun, a verb, an adjective, etc. An example for a POS tagger is the TreeTagger [63].
- *Regular expression matching* allowing to define regular expressions and match these in the text. This functionality is provided for example by GATE's Java Annotation Pattern Engine (JAPE).
- A *chunker* in order to identify larger syntactic constituents in a sentence. Chunkers are also called *partial parsers*. An example of a publicly available chunker is Steven Abney's CASS [1].
- A *syntactic parser* determining the full syntactic structure of a sentence might be needed for some ontology learning algorithms (compare [20]).

Example 1. Given a sentence such as *The man drove a bike to Buxton.* we would for example yield the following tokenization, lemmatization as well as result of the POS tagging:

Tokens	The	man	drove	the	bike	to	Buxton.
Lemma	The	man	drive	the	bike	to	Buxton.
POS	DT	NN	VBD	DT	NN	IN	NP.

where we use the Penn Treebank[2] tagset in which DT stands for a determiner, NN for a singular noun, IN for a preposition, NP for a proper noun and VBD for a past tense verb. The parse tree for the above sentence looks as follows:

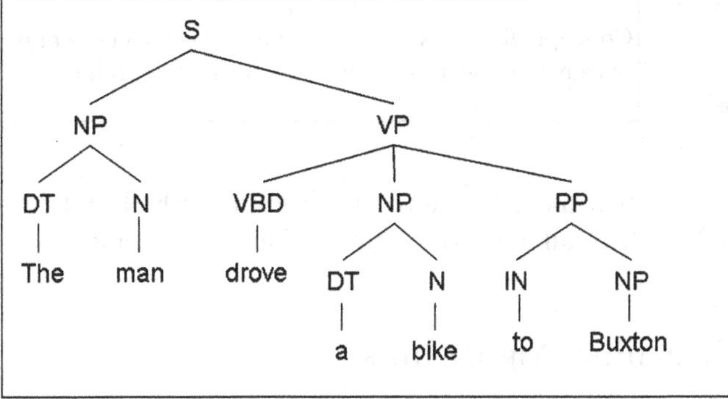

2.4 Algorithm Library Component

This component acts as the algorithmic backbone of the framework. A number of algorithms are provided for the extraction and maintenance of the ontology modeling primitives contained in the ontology model. Thus, the algorithm library contains the actual algorithms applied to learning. In particular, the algorithm library consists mainly of machine learning algorithms and versions of these customized for the purpose of ontology learning. In particular, machine learning algorithms typically contained in the library are depicted in Table 1.

Most of these machine learning algorithms can be obtained off-the-shelf in various versions from standard machine learning frameworks such as WEKA [76]. Additionally, the library should also contain a comprehensive number of implemented distance or similarity measures such as Jaccard, Dice, the cosine measure, the Kullback–Leibler divergence, etc. (compare [49]) to support semantic clustering. In addition, the algorithm library could also contain traditional measures for discovering collocations between words known from computational linguistics research (e.g., [43]). In order to be able to combine the extraction results of different learning algorithms, it is necessary to standardize the output in a common way. In general, a common result structure for all learning methods is needed. In the Text2Onto system [22],

[2] See http://www.cis.upenn.edu/~treebank

Table 1. Typical machine learning algorithms in the algorithm library

Algorithm	Generic use	Use in ontology learning
Association rule discovery (e.g., [2])	Discovery of "interesting" transactions in itemsets (e.g., customer data)	Discovery of interesting associations between words
(Hierarchical) Clustering	Discovery of groups in data (unsupervised)	Clustering of words
Classification (e.g., SVMs, Naive Bayes, kNN, etc.)	Prediction (supervised)	Classification of new concepts into an existing hierarchy
Inductive logic programming ([48])	Induction of rules from data (supervised)	Discovery of new concepts from extensional data
Conceptual clustering (e.g., FCA – see chapter "Formal Concept Analysis")	Concept discovery (extension and intension)	Learning concepts and concept hierarchies

for example, there is a blackboard-style result structure – the POM (Possible Ontologies Model) – where all algorithms can update their results.

3 Ontology Learning Algorithms

The various tasks relevant in ontology learning have been previously organized in a layer diagram showing the conceptual dependencies between different tasks. This *ontology learning layer cake* was introduced in [18] and is shown in Fig. 2. It clearly focuses on learning the TBox part of an ontology. With respect to information extraction techniques to populate the ABox of an ontology, the interested reader is referred to chapter "Information Extraction". The layers build upon each other in the sense that results of tasks at lower layers typically serve as input for the higher layers. For example, in order to extract relations between concepts, we should consider the underlying hierarchy to identify the right level of generalization for the domain and range of the relation. The two bottom layers of the layer cake correspond to the lexical level of ontology learning. The task in this part of the layer is to detect the relevant terminology as well as groups of synonymous terms, respectively. The extracted terms and synonym groups can then form the basis for the formation of concepts. Concepts differ from terms in that they are ontological entities and thus abstractions of human thought in the sense of Ganter and Wille [32]. According to our formalization, concepts are triples c:=$< i(c), [\![c]\!], Ref_c >$ consisting of an intensional description $i(c)$, an extension $[\![c]\!]$ and a reference function Ref_c representing how the concept is symbolically realized in a text corpus, an image, etc. (see [10]). At higher levels of the layer cake, we find the layers corresponding to the tasks of learning a concept hierarchy, relations, a relation hierarchy as well as deriving arbitrary rules and axioms. The top two

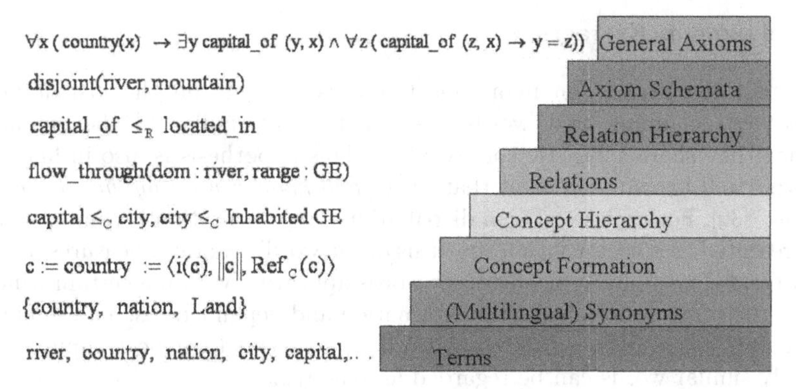

Fig. 2. Ontology learning layer cake from [18]

layers correspond certainly to the most challenging task as in principle there is no limit on the type and complexity of axioms and rules to be learned. In practice, however, as we commit to a specific knowledge representation language – the Web Ontology Language (OWL) for example (see chapter "Web Ontology Language: OWL") – the types of axioms that are allowed are more restricted. In what follows, we discuss the various tasks layer by layer and point the reader to relevant approaches in the literature of the field.

3.1 Term Extraction

The task at the lexical layers is to extract terms and arrange these into groups of synonymous words. A simple technique for extracting relevant terms that may indicate concepts is counting frequencies of terms in a given set of (linguistically preprocessed) documents, the corpus $\mathcal{D}$. In general this approach is based on the assumption that a frequent term in a set of domain-specific texts indicates the occurrence of a relevant concept. Research in information retrieval has shown that there are more effective methods of term weighting than simple counting of frequencies. Weighting measures well-known from information retrieval such as tf.idf (see [6]) might also be applied here.

Further, the computational linguistics community has proposed a wide range of more sophisticated techniques for term extraction. An interesting measure is the C-value/NC-value measure presented in [31] which does not only take into account the frequency of terms but also the fact that terms can be nested into each other. Further, the approach also takes into account contextual clues which are strong indicators of the *"termhood"* of some sequence of words. Overall, while the field of term extraction seems quite mature and a plethora of techniques have been suggested and examined, there is not yet a clear understanding of which measures work best for which purpose. Clearly, specific domains, such as genomics, medicine or E-commerce need corresponding adaptations of tools and methods with respect to their specific characteristics.

3.2 Synonym Extraction

In order to extract synonyms, most approaches rely on the *distributional hypothesis* claiming that words are semantically similar to the extent to which they share syntactic contexts [39]. This hypothesis is also in line with Firth's well known statement that *"you shall know a word by the company it keeps"* [30]. For each word w, a distributional representation is computed and represented as a vector $\mathbf{v_w}$ on the basis of the word's context. Features used to represent a word are typically other words appearing within a certain window from the target word, syntactic information and dependencies. The similarity in vector space between different word vectors can then be computed and highly similar words can be regarded as synonyms.

Example 2. Assuming that we parse a text corpus and identify, for each noun, the verbs for which it appears at the object position, we can construct a matrix as follows:

	Book_{obj}	Rent_{obj}	Drive_{obj}	Ride_{obj}	Join_{obj}
Hotel	x				
Apartment	x	x			
Car	x	x	x		
Bike	x	x	x	x	
Excursion	x				x
Trip	x				x

Each row represents the context of a word, while each column corresponds to one dimension of the context representation, in our case the different verbs that the nouns appear at the object position. Assuming the representation as binary vectors shown in the matrix above, we can for example calculate the similarity between the different terms using the Jaccard coefficient which compares the sets A and B of the non-negative dimensions of the vector representations two words a and b: $Jaccard := \frac{|A \cap B|}{|A \cup B|}$. The resulting similarities are thus:

	Hotel	Apartment	Car	Bike	Excursion	Trip
Hotel	1.0	0.5	0.33	0.25	0.5	0.5
Apartment		1.0	0.66	0.5	0.33	0.33
Car			1.0	0.75	0.25	0.25
Bike				1.0	0.2	0.2
Excursion					1.0	1.0
Trip						1.0

Important approaches along these lines include the work of Grefenstette [33] as well as Lin et al. [50]. Some researchers have also combined different similarity extractors using ensemble methods [25]. Other techniques for extracting synonyms include the application of statistical methods to the Web (cf. [7, 69]) or the calculation of semantic relatedness with respect to a taxonomy or semantic network such as WordNet (compare [61]) or more recently also the Wikipedia categories (see [67]). WordNet [29] is a lexical database organizing words in terms of synonym sets (synsets) and providing lexical relations between these synsets, i.e., hypernymy/hyponymy ("is a kind of") as well as meronymy/holonymy ("part of") relations. In addition, WordNet provides *glosses*, which are natural language definitions of the synsets. Turney [69] for example relies on the well-known Pointwise Mutual Information (PMI) measure to extract synonyms. The pointwise mutual information of two events x and y is defined as:

$$PMI(x, y) := log_2 \frac{P(x, y)}{P(x)\ P(y)}$$

where $P(x, y)$ is the probability for a joint occurrence of x and y and $P(x)$ is the probability for the event x. The PMI is thus in essence the (logarithmic) ratio of the joint probability and the probability under the assumption of independence. In fact, if $P(x, y) \leq P(x)P(y)$, we will have a negative (or zero) value for the PMI, while in case $P(x, y) > P(x)P(y)$, we will have a positive PMI value. The PMI can be calculated using *Google* and counting hits as follows:

$$PMI_{Web}(x, y) := log_2 \frac{Hits(x\ AND\ y)\ MaxPages}{Hits(y)\ Hits(y)}$$

where *MaxPages* is an approximation for the maximum number of English web pages. This measure can thus be used to calculate the statistical dependence of two words on the Web. If they are highly dependent, we can assume they are synonyms or at least highly semantically related. This approach to discover synonyms has been successfully applied to the TOEFL test (see [69]).

3.3 Concept Learning

In this section we focus on approaches inducing concepts by clearly defining the intension of the concept. We will distinguish the following three paradigms:

- Conceptual clustering
- Linguistic analysis
- Inductive methods

Conceptual Clustering

Conceptual clustering approaches such as Formal Concept Analysis ([32], chapter "Formal Concept Analysis") have been applied to form concepts and to order them hierarchically at the same time. Conceptual clustering approaches typically induce an intensional description for each concept in terms of the attributes that it shares with other concepts as well as those that distinguish it from other concepts.

Linguistic Analysis

Linguistic analysis techniques can be applied to derive an intensional description of a concept in the form of a natural language description. The approach of Velardi et al. [70] for example relies on WordNet to compositionally interpret a compound term and as a byproduct produce a description on the basis of the WordNet descriptions of the single terms constituting the compound [70]. The definition of the term *knowledge management practices*: *"a kind of practice, knowledge of how something is customarily done, relating to the knowledge of management, the process of capturing value, knowledge and understanding of corporate information, using IT systems, in order to maintain, re-use and de-ploy that knowledge."* is compositionally determined on the basis of the definitions of *knowledge management*[3] and *practice.*[4] For this purpose, disambiguation with respect to the different senses of a word with respect to its several meanings in a lexical database (such as WordNet) is required. Further, a set of rules is specified which drive the above compositional generation of definitions.

Finally, given a populated knowledge base, approaches based on inductive learning such as Inductive Logic Programming can be applied to derive rules describing a group of instances intentionally. Such an approach can for example be used to reorganize a taxonomy or to discover gaps in conceptual definitions (compare [51]).

3.4 Concept Hierarchy

Different methods have been applied to learn taxonomic relations from texts. In what follows we briefly discuss approaches based on matching *lexico-syntactic patterns*, *clustering*, *phrase analysis* as well as *classification.*

Lexico-Syntactic Patterns

In the 1980s, people working on extracting knowledge from machine readable dictionaries already realized that regularities in dictionary entries could be exploited to define patterns to automatically extract hyponym/hypernym and other lexical relations from dictionaries (compare [3, 4, 13]). This early work was continued later in the context of the ACQUILEX project (e.g., [23]).

In her seminal work, Hearst [40] proposed the application of so-called lexico-syntactic patterns to the task of automatically learning hyponym relations from corpora. In particular, Hearst defined a collection of patterns

[3] Knowledge management: the process of capturing value, knowledge and understanding of corporate information, using IT systems, in order to mantain, re-use and re-deploy that knowledge.

[4] Practice: knowledge of how something is customarily done.

indicating hyponymy relations. An example of a pattern used by Hearst is the following:

$$\text{such } NP_0 \text{ as } NP_1,...,NP_{n-1} \text{ (or|and) other } NP_n$$

where NP_i stands for a noun phrase. If such a pattern is matched in a text, according to Hearst we could derive that for all $0 < i \leq n$ *hyponym*(NP_i,NP_0).[5] For example, from the sentence *Such injuries as bruises, wounds and broken bones...*, we could derive the relations: *hyponym(bruise, injury)*, *hyponym(wound, injury)* and *hyponym(broken bone, injury)*.

The patterns used by Hearst are the following:

Hearst1: NP_{hyper} such as $\{NP_{hypo},\}^*$ {(and | or)} NP_{hypo}
Hearst2: such NP_{hyper} as $\{NP_{hypo},\}^*$ {(and | or)} NP_{hypo}
Hearst3: NP_{hypo} $\{,NP\}^*$ {,} or other NP_{hyper}
Hearst4: NP_{hypo} $\{,NP\}^*$ {,} and other NP_{hyper}
Hearst5: NP_{hyper} including $\{NP_{hypo},\}^*$ NP_{hypo} {(and | or)} NP_{hypo}
Hearst6: NP_{hyper} especially $\{NP_{hypo},\}^*$ {(and|or)} NP_{hypo}

Overall, lexico-syntactic patterns have been shown to yield a reasonable precision for extracting *is-a* as well as *part-of* relations (e.g., [14, 16, 59]).

Clustering

Clustering can be defined as the process of organizing objects into groups whose members are similar in some way based on a certain representation, typically in the form of vectors (see [46]). In general, there are three major styles of clustering:

1. *Agglomerative:* In the initialization phase, each term is defined to constitute a cluster of its own. In the growing phase, larger clusters are iteratively generated by merging the most similar/least dissimilar ones until some stopping criterion is reached. Examples of uses of agglomerative clustering techniques in the literature are [8, 20, 28].
2. *Divisive*: In the initialization phase, the set of all terms constitutes a cluster. In the refinement phase, smaller clusters are (iteratively) generated by splitting the largest cluster or the least homogeneous cluster into several subclusters. Examples for divisive clustering can be found in [20, 58].
   Both agglomerative and divisive clustering techniques are used to produce hierarchical descriptions of terms. Both rely on notions of (dis-)similarity, for which a range of measures exist (e.g., Jaccard, Kullback–Leibler divergence, L1-norm, cosine; cf. [49]).
3. *Conceptual:* Conceptual clustering builds a lattice of terms by investigating the exact overlap of descriptive attributes between two represented terms. In the worst case, the complexity of the resulting concept lattice is exponential in n. Thus, people either just compute a sublattice [68] or rely

[5] From a linguistic point of view, a term t_1 is a hyponym of a term t_2 if we can say *a* t_1 *is a kind of* t_2. Correspondingly, t_2 is then a hypernym of t_1.

on certain heuristics to explore and/or prune the lattice. Examples of applications of conceptual clustering techniques to ontology learning can be found in [20, 38].

Either way one may construct a hierarchy of term clusters for detailed inspection by the ontology engineer.

Example 3. Using hierarchical agglomerative clustering, we can build a cluster tree for the objects in Example 2. Let us assume we are using *single linkage* as measure of the similarity between clusters. First, we cluster *excursion* and *trip* as they have a similarity of 1.0. We then cluster *bike* and *car* as this is the next pair with the highest degree of similarity. We then build a cluster consisting of *bike, car* and *apartment.* Next, we either join the latter cluster with *hotel* or build a cluster between *hotel* and the already created cluster consisting of *excursion* and *trip.* Assuming that we traverse the similarity matrix from the upper left corner to the lower right one, we can add *hotel* to the cluster consisting of *bike, car* and *apartment.* At the top level we then join the clusters {*hotel, apartment, bike, car*} and {*excursion, trip*} producing a universal cluster containing all elements. The corresponding cluster tree would then look as follows:

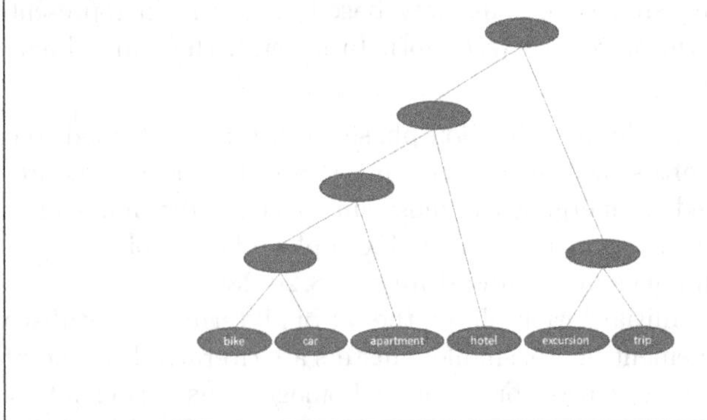

Phrase Analysis

Some approaches rely on the fact that the internal structure of noun phrases can be used to discover taxonomic relations (compare [11, 62], but also [21]). In essence, these methods build on the heuristic that additional modifiers (adjectival or nominal) added to the front of a noun typically define a subclass of the class denoted by the noun. That means, for example, that *focal epilepsy* is interpreted as a subclass of *epilepsy.* This is in essence the approach implemented in the OntoLT system (see below and [11]). Sanchez and Moreno [62] exploit this heuristic in a web setting to find terms which occur to the left of a term to be refined into subclasses. A measure inspired in the Pointwise Mutual Information (PMI) is used to assess the degree of correlation between the term in question and the modifier to the left.

Classification

When a substantial hierarchy is already given, e.g., by basic level categories from a general resource like WordNet [29], one may rather decide to refine the taxonomy by classifying new relevant terms into the given concept hierarchy. The distributional representation described above is then used to learn a classifier from a training corpus and the set of predefined concepts with their lexical entries. Afterwards, one may construct the distributional representations of relevant, unclassified terms and let the learned classifier propose a node to which to classify the new term. While many researchers have considered lexical databases such as WordNet to test such algorithms (e.g., [41] and [75], other researchers have indeed considered domain-specific ontologies see, e.g., the work of Pekar and Staab [57]). Pekar and Staab have in particular considered different algorithms to classify a new term into an existing concept hierarchy without testing all concepts, e.g., by exploiting the hierarchical structure using tree-ascending or tree-descending algorithms.

3.5 Relations

In order to discover arbitrary relations between words, different techniques from the machine learning and statistical natural language processing community have found application in ontology learning. In order to discover "anonymous" associations between words, one can look for a strong co-occurrence between words within a certain boundary, i.e., a window of words, a sentence or a paragraph. Mädche et al. [53] apply the well-known association discovery algorithm and represent co-occurrences of words within a sentence as transactions. This representation allows to calculate the support and confidence for binary transactions and thus to detect anonymous binary associations between words.

In the computational linguistics community, the task of discovering strong associations between words is typically called *collocation discovery*. In essence, the idea is to discover words which co-occur beyond chance in a statistically significant manner. Statistical significance is typically checked using some test such as the Student's t-test or the χ^2-test (compare [43,56]). Other researchers have aimed at learning labeled relations by relying on linguistic predicate-argument dependencies. Typically, verb structures are considered for this purpose (compare [17,19,64]). When learning relations, a crucial issue is to find the right level of abstraction with respect to the concept hierarchy for the domain and range of the relation in question. This issue can be addressed in different ways. While Mädche et al. [53] incorporate the concept hierarchy into the association discovery process, Ciaramita et al. [17] as well as Cimiano et al. [19] formulate this as a problem of generalizing along a hierarchy as long as the statistical significance does not diminish.

3.6 Axioms and Rules

Ontology learning approaches so far have focused on the acquisition of rather simple taxonomic hierarchies, properties as well as lexical and assertional knowledge. However, the success of OWL which allows for modeling far more expressive axiomatizations has led to some advances in the direction of learning complex ontologies and rules.

Völker et al. [71] propose an approach for generating formal class descriptions from natural language definitions extracted, e.g., from online glossaries and encyclopedias. The implementation of this approach is essentially based on a syntactic transformation of natural language definitions into OWL DL axioms in line with previous work on lexico-syntactic patterns (cf. Sect. 3.4) and lexical entailment.

One of the first methods for learning disjointness axioms relies on a statistical analysis of enumerations which has been implemented as part of the Text2Onto framework [37]. Völker et al. [73] developed a supervised learning approach based on an extended set of methods yielding both lexical and logical evidence for or against disjointness.

3.7 Pruning/Domain Adaptation

One relatively straightforward approach towards generating an appropriate domain ontology given a corpus is to prune an existing general ontology. Along these lines, Buitelaar et al. [12] propose a method by which WordNet synsets can be ranked by relevance with respect to the corpus in question on the basis of a frequency-based measure. The techniques used to prune WordNet, thus adapting it to a certain domain and corpus, can be also applied to prune a given ontology. Kietz et al. [47] for example present a method which additionally uses a general corpus as contrast. They only select a concept as relevant in case it appears a factor c times more relevant in the domain-specific than in the general corpus. Hereby, c is a user-specified constant and the relevance measure used is tf.idf.

4 Ontology Learning Systems

In the last years, many different tools and frameworks for ontology learning have emerged. Needless to say that it is out of the scope of this chapter to discuss them all. Instead, we will provide a rather subjective snapshot of the current tool landscape. Some well-known and frequently cited tools are for example: OntoLearn [70], OntoLT [11], Terminae [5] as well as TextToOnto [55] and its successor Text2Onto [22]. All these tools implement various and different methods, such that a detailed discussion and comparison is out of the scope of this chapter. OntoLearn for example integrates a word sense disambiguation component to derive intensional descriptions of complex domain-specific terms, which are assumed to denote concepts, on the basis of WordNet glosses (compare Sect. 3.3). In this sense, OntoLearn also induces intensionally defined domain concepts and ingeniously exploits the knowledge available in general resources for a specific domain. OntoLT , which is available as a plug-in to the Protégé ontology editor [34], allows for term extraction using various measures such as tf.idf and extraction of taxonomic relations relying on interpreting modifiers (nominal or adjectival) as introducing subclasses (compare Sect. 3.4).

TextToOnto [55] is a framework containing various tools for ontology learning. It includes standard term extraction using a number of different measures, the algorithm for mining relations based on association rules described in [53] (see Sect. 3.5) as well as hierarchical clustering algorithms based on Formal Concept Analysis (compare Sect. 3.4). Its successor, Text2Onto, besides implementing most of the algorithms available also in TextToOnto, abstracts from a specific knowledge representation language and stores the learned ontology primitives in the form of a meta-model called *Possible Ontologies Model* (POM), which can then be translated to any reasonably expressive knowledge representation language, in particular to OWL and RDFS. On the other hand, it implements a framework for data-driven and incremental learning in a sense that changes in the underlying corpus are propagated to the algorithms, thus leading to explicit changes to the POM. The advantage is that these changes can be easily traced back to the original corpus changes, which gives more control to the ontology engineer.

5 Advanced Issues

In this section, we briefly discuss some advanced and open issues in ontology learning that are still under research. This section should help newcomers to get a feeling for the open questions and allow for a quicker entry into the field.

5.1 Methodology

Certainly, besides providing tool support for ontology learning methods, it is crucial to define how ontology learning methods can be integrated into the process of engineering an ontology. Blueprints in this direction can be found in the work of Simperl et al. [65] as well as Aussenec-Gilles et al. [5]. Simperl et al. for example provide a methodology defining the necessary activities and roles for ontology engineering supported by ontology learning methods. In particular, they argue that without a clear methodology to be followed by an ontology engineering project, ontology learning techniques can not reasonably support ontology engineering activities. Aussenec-Gilles et al. [5] have also conducted research on methodological issues in the context of their Terminae method. In particular, they have argued that knowledge models, in particular ontologies, need to be anchored in language. Therefore, they emphasize the role of language in the process of ontology engineering. In general, though first attempts have been provided, there is still much further work to do in order to clarify the benefits and drawbacks of different methodologies for integrating ontology learning into available ontology engineering methodologies. As argued by Simperl et al., ontology learning tools need to improve on their usability and intuitiveness in order to be useful for the purpose of ontology engineering.

5.2 Evaluation

A crucial part of ontology learning is to evaluate how good the learned ontologies actually are. Such an evaluation can in turn guide and control the ontology learning process in the search towards an "optimal" ontology. However, the evaluation of ontology learning tools is a quite delicate issue as it is not clear what one could compare to. The critical issue in many cases is to define a *gold standard* which we can regard as ground truth an one can compare with (see [26]). However, it is well known that there is no ground truth for ontologies as different people will surely come up with very different ontologies when asked to model a certain domain (see, e.g., the experiments described in [54]).

Other approaches aim at approximating the appropriateness of some ontology by other means. Brewster et al. [9], for example, try to measure the "corpus fit" of the ontology by considering the frequency with which the terms in the ontology appear in the corpus. A completely different way to check the quality of an ontology is pursued by the AEON framework [72], that aims to automatize the application of the OntoClean methodology (see chapter "An Overview of OntoClean"), hence ensuring the formal consistency of an ontology. Finally, an integration of ontology learning and evaluation is proposed by Haase et al. [37]. They describe an approach to exploiting contextual information such as OntoClean meta-properties, or confidence and relevance values for resolving logical inconsistencies in learned ontologies, and to optimize the outcome of the ontology learning process.

5.3 Expressivity

Many people argue that the main benefits of using ontologies for knowledge modeling become most evident in reasoning-based applications. Inferring new knowledge and drawing conclusions beyond explicit assertions is an important aspect of "intelligent" applications. However, the power of reasoning largely depends on the expressivity of the underlying knowledge representation formalism and its instantiation by means of a concrete ontology.

The vast majority of today's lexical ontology learning focuses on the extraction of simple class descriptions and axioms, i.e., atomic concepts, subsumption and object properties, as well as ABox statements expressing concept or property instantiation. The expressivity of ontologies generated by lexical approaches, e.g., based on natural language processing techniques is mostly restricted to $\mathcal{ALC}$ (Attributive Language with Complements) or similar DL fragments such as $\mathcal{AL}$-*log*. These rather simple, often informal ontologies have proven to be useful for many applications, or as Jim Hendler has put it "A little semantics goes a long way" (see [42]). But semantic applications relying on reasoning over very complex domains such as bioinformatics or medicine require more precise and accurate knowledge representation.

Learning more expressive ontologies greatly facilitates the acquisition and evaluation of complex domain knowledge. But it also brings new challenges, e.g., with respect to logical inconsistencies that may arise as soon as any kind of negation or cardinality constraints are introduced into learned ontologies [45]. Methods for debugging, consistent ontology evolution, or inconsistency reasoning will be required to face these challenges.

Finally, a tighter integration of lexical and logical ontology learning approaches will be required in order to prevent problems resulting from different semantic interpretations, e.g., of lexical and ontological relations (see discussion in [71]). A first approach in this line is the RELExO framework by Völker and Rudolph [74], which combines a lexical approach to the acquisition of complex class descriptions with the FCA-based technique of relational exploration.

5.4 Combination of Evidence

As it is very unlikely that we will be able to derive high-quality ontologies from one single source of evidence and using one single approach, a few researchers have addressed the challenge of learning ontologies by considering multiple sources of evidence. Cimiano et al. [21] have for example presented a classification-based approach in which a classifier is trained with features derived from various approaches and data sources. The approach is shown to outperform any of the single algorithms considered. Snow et al. [66] have phrased the problem in probabilistic terms and considered the task of adding new concepts (synsets) to the WordNet taxonomy. These approaches have so far focused only on learning taxonomic relations with the notable exception of initial approaches to the automatic generation of disjointness axioms [73]. In general, there is a lot of further research needed in this direction.

5.5 Dynamics and Evolution

Most ontology learning approaches assume that there is one static corpus from which to learn. However, collecting such a corpus can sometimes be a non-trivial task. Currently, some researchers are attempting to frame ontology learning as the task of keeping an *equilibrium* between a (growing) corpus, a (growing) set of extracted ontological primitives and a (changing) set of extraction patterns. While it seems very hard to define such an equilibrium in such a way that certain actions are triggered towards restoring it, first attempts in this direction can be found in the work of Iria et al. [44]. Further, as the underlying corpus can and will evolve, it is an important question to explicitly track changes in the ontology model with respect to changes in the corpus, thus enhancing the transparency of the ontology learning process and allowing human inspection. From a performance point of view, an incremental approach to ontology learning has moreover the benefit that the whole corpus will not need to be processed each time it changes. A first approach in this direction has been implemented in the Text2Onto system (compare [22]).

6 Conclusion

Ontology learning is a challenging and exciting research field at the intersection of machine learning, data and text mining, natural language processing and knowledge representation. While fully automatic knowledge acquisition techniques are not yet feasible (and possibly will nor should ever be), ontology learning techniques have a high potential to support ontology engineering activities. In fact, according to our view, ontology engineering can not be considered without the automatic or semi-automatic support of ontology learning methods. Future work should and will surely aim at developing a new generation of intuitive ontology learning tools which are able to learn expressive ontologies, but at the same time hide their internal complexity from the user. These new generation of tools should feature intuitive user interfaces as well as smoothly integrate into existing methodologies for ontology engineering.

References

1. S. Abney. Partial parsing via finite-state cascades. In *Proceedings of the ESSLLI '96 Robust Parsing Workshop*, pages 8–15, 1996.
2. R. Agrawal and R. Srikant. Fast algorithms for mining association rules. In *Proceedings of the 20th International Conference on Very Large Databases (VLDB)*, 1994.
3. H. Alshawi. Processing dictionary definitions with phrasal pattern hierarchies. *Computational Linguistics*, 13(3–4):195–202, 1987. Special Issue of the Lexicon.
4. R. A. Amsler. A taxonomy for english nouns and verbs. In *Proceedings of the 19th Annual Meeting of the Association for Computational Linguistics (ACL)*, pages 133–138, 1981.
5. N. Aussenac-Gilles, S. Despres, and S. Szulman. The TERMINAE method and platform for ontology engineering from text. In P. Buitelaar and P. Cimiano, editors, *Bridging the Gap between Text and Knowledge: Selected Contributions to Ontology Learning and Population from Text*, volume 167 of *Frontiers in Artificial Intelligence*. IOS Press, 2007.

6. R. Baeza-Yates and B. Ribeiro-Neto. *Modern Information Retrieval.* Addison-Wesley, 1999.
7. M. Baroni and S. Bisi. Using cooccurrence statistics & the web to discover synonyms in a technical language. In *Proceedings of the 4th International Conference on Language Resources and Evaluation (LREC)*, pages 1725–1728, 2004.
8. G. Bisson, C. Nédellec, and L. Ca namero. Designing clustering methods for ontology building – The Mo'K workbench. In *Proceedings of the ECAI Ontology Learning Workshop*, pages 13–19, 2000.
9. C. Brewster, H. Alani, S. Dasmahapatra, and Y. Wilks. Data-driven ontology evaluation. In *Proceedings of the 4th International Conference on Language Resources and Evaluation*, Lisbon, 2004. European Language Resources Association.
10. P. Buitelaar, T. Declerck, A. Frank, S. Racioppa, M. Kiesel, M. Sintek, R. Engel, M. Romanelli, D. Sonntag, B. Loos, V. Micelli, R. Porzel, and P. Cimiano. Linginfo: Design and applications of a model for the integration of linguistic information in ontologies. In *Proceedings of the OntoLex06 Workshop at LREC*, 2006.
11. P. Buitelaar, D. Olejnik, and M. Sintek. A Protégé plug-in for ontology extraction from text based on linguistic analysis. In *Proceedings of the 1st European Semantic Web Symposium (ESWS)*, pages 31–44, 2004.
12. P. Buitelaar and B. Sacaleanu. Ranking and selecting synsets by domain relevance. In *Proceedings of the NAACL Workshop on WordNet and Other Lexical Resources: Applications, Extensions and Customizations*, 2001.
13. N. Calzolari. Detecting patterns in a lexical data base. In *Proceedings of the 22nd Annual Meeting of the Association for Computational Linguistics (ACL)*, pages 170–173, 1984.
14. S. Cederberg and D. Widdows. Using LSA and noun coordination information to improve the precision and recall of automatic hyponymy extraction. In *Conference on Natural Language Learning (CoNNL)*, pages 111–118, 2003.
15. P. Chapman, R. Kerber, J. Clinton, T. Khabaza, T. Reinartz, and R. Wirth. The CRISP-DM process model. Discussion Paper, March 1999.
16. E. Charniak and M. Berland. Finding parts in very large corpora. In *Proceedings of the 37th Annual Meeting of the Association for Computational Linguistics (ACL)*, pages 57–64, 1999.
17. M. Ciaramita, A. Gangemi, E. Ratsch, J. Šarić, and I. Rojas. Unsupervised learning of semantic relations between concepts of a molecular biology ontology. In *Proceedings of the 19th International Joint Conference on Artificial Intelligence (IJCAI)*, pages 659–664, 2005.
18. P. Cimiano. *Ontology Learning and Population from Text: Algorithms, Evaluation and Applications.* Springer, 2006.
19. P. Cimiano, M. Hartung, and E. Ratsch. Finding the appropriate generalization level for binary ontological relations extracted from the Genia corpus. In *Proceedings of the International Conference on Language Resources and Evaluation (LREC)*, 2006.
20. P. Cimiano, A. Hotho, and S. Staab. Learning concept hierarchies from text corpora using formal concept analysis. *Journal of Artificial Intelligence Research (JAIR)*, 24:305–339, 2005.
21. P. Cimiano, L. Schmidt-Thieme, A. Pivk, and S. Staab. Learning taxonomic relations from heterogeneous evidence. In P. Buitelaar, P. Cimiano, and B. Magnini,

editors, *Ontology Learning from Text: Methods, Applications and Evaluation*, number 123 in Frontiers in Artificial Intelligence and Applications, pages 59–73. IOS Press, 2005.
22. P. Cimiano and J. Völker. Text2onto – A framework for ontology learning and data-driven change discovery. In E. Métais, A. Montoyo, and R. Muñoz, editors, *Proceedings of the 10th International Conference on Applications of Natural Language to Information Systems (NLDB)*, volume 3513 of *Lecture Notes in Computer Science*, pages 227–238, 2005.
23. A. Copestake. An approach to building the hierarchical element of a lexical knowledge base from a machine readable dictionary. In *Proceedings of the 1st International Workshop on Inheritance in Natural Language Processing*, pages 19–29, 1990.
24. H. Cunningham, K. Humphreys, R.J. Gaizauskas, and Y. Wilks. GATE – A general architecture for text engineering. In *Proceedings of Applied Natural Language Processing (ANLP)*, pages 29–30, 1997.
25. J. Curran. Ensemble methods for automatic thesaurus construction. In *Proceedings of the Conference on Empirical Methods in Natural Language Processing (EMNLP)*, pages 222–229, 2002.
26. K. Dellschaft and S. Staab. On how to perform a gold standard based evaluation of ontology learning. In *Proceedings of the International Semantic Web Conference*, pages 228–241, 2006.
27. E. Bozsak et al. KAON – Towards a large scale Semantic Web. In *Proceedings of the Third International Conference on E-Commerce and Web Technologies (EC-Web)*. Springer Lecture Notes in Computer Science, 2002.
28. D. Faure and C. Nédellec. A corpus-based conceptual clustering method for verb frames and ontology. In P. Velardi, editor, *Proceedings of the LREC Workshop on Adapting lexical and corpus resources to sublanguages and applications*, pages 5–12, 1998.
29. C. Fellbaum. *WordNet, an electronic lexical database*. MIT Press, 1998.
30. J. Firth. *A synopsis of linguistic theory 1930–1955*. Studies in Linguistic Analysis, Philological Society, Oxford. Longman, 1957.
31. K. Frantzi and S. Ananiadou. The C-value/NC-value domain independent method for multi-word term extraction. *Journal of Natural Language Processing*, 6(3):145–179, 1999.
32. B. Ganter and R. Wille. *Formal Concept Analysis – Mathematical Foundations*. Springer, 1999.
33. G. Grefenstette. SEXTANT: Exploring unexplored contexts for semantic extraction from syntactic analysis. In *Meeting of the Association for Computational Linguistics*, pages 324–326, 1992.
34. W. Grosso, H. Eriksson, R. Fergerson, J. Gennari, S. Tu, and M. Musen. Knowledge modelling at the millenium: The design and evolution of Protégé. In *Proceedings of the 12th International Workshop on Knowledge Acquisition, Modeling and Management (KAW'99)*, 1999.
35. P. Haase and L. Stojanovic. Consistent evolution of owl ontologies. In A. Gomez-Perez and J. Euzenat, editors, *Proceedings of the Second European Semantic Web Conference*, volume 3532 of *LNCS*, pages 182–197, 2005.
36. P. Haase and J. Völker. Dealing with uncertainty and inconsistency. In P. C. G. da Costa, K. B. Laskey, K. J. Laskey, and M. Pool, editors, *Proceedings of the Workshop on Uncertainty Reasoning for the Semantic Web (URSW)*, pages 45–55, 2005.

37. P. Haase and J. Völker. Ontology learning and reasoning – Dealing with uncertainty and inconsistency. In P. C. G. da Costa, K. B. Laskey, K. J. Laskey, and M. Pool, editors, *Proceedings of the Workshop on Uncertainty Reasoning for the Semantic Web (URSW)*, pages 45–55, 2005.
38. H.-M. Haav. An application of inductive concept analysis to construction of domain-specific ontologies. In *Proceedings of the VLDB Pre-conference Workshop on Emerging Database Research in East Europe*, 2003.
39. Z. S. Harris. *Mathematical Structures of Language.* Wiley, 1968.
40. M. A. Hearst. Automatic acquisition of hyponyms from large text corpora. In *Proceedings of the 14th International Conference on Computational Linguistics (COLING)*, pages 539–545, 1992.
41. M. A. Hearst and H. Schütze. Customizing a lexicon to better suit a computational task. In *Proceedings of the ACL SIGLEX Workshop on Acquisition of Lexical Knowledge from Text*, 1993.
42. J. Hendler. On beyond ontology. Keynote Talk at the International Semantic Web Conference (ISWC), 2004.
43. G. Heyer, M Läuter, U. Quasthoff, T. Wittig, and C. Wolff. Learning relations using collocations. In *Proceedings of the IJCAI Workshop on Ontology Learning*, 2001.
44. J. Iria, C. Brewster, F. Ciravegna, and Y. Wilks. An incremental tri-partite approach to ontology learning. In *Proceedings of the Language Resources and Evaluation Conference (LREC-06), Genoa, Italy 22–28 May*, 2006.
45. P. Haase J. Völker and P. Hitzler. Learning expressive ontologies. In P. Buitelaar and P. Cimiano, editors, *Bridging the Gap between Text and Knowledge: Selected Contributions to Ontology Learning and Population from Text*, volume 167 of *Frontiers in Artificial Intelligence.* IOS Press, 2007.
46. L. Kaufman and P. Rousseeuw. *Finding Groups in Data: An Introduction to Cluster Analysis.* Wiley, 1990.
47. J.-U. Kietz, R. Volz, and A. Mädche. Extracting a domain-specific ontology from a corporate intranet. In *Proceedings of the 2nd Learning Language in Logic (LLL) Workshop*, 2000.
48. N. Lavrac and S. Dzeroski. *Inductive Logic Programming: Techniques and Applications.* Ellis Horwood, 1994.
49. L. Lee. Measures of distributional similarity. In *Proceedings of the 37th Annual Meeting of the Association for Computational Linguistics (ACL)*, pages 25–32, 1999.
50. D. Lin. Automatic retrieval and clustering of similar words. In *Proceedings of the 36th Annual Meeting of the Association for Computational Linguistics and 17th International Conference on Computational Linguistics (COLING-ACL)*, pages 768–774, 1998.
51. F. A. Lisi and F. Esposito. Two orthogonal biases for choosing the intensions of emerging concepts in ontology refinement. In G. Brewka, S. Coradeschi, A. Perini, and P. Traverso, editors, *Proceedings of the 17th European Conference on Artificial Intelligence (ECAI)*, pages 765–766. IOS Press, 2006.
52. A. Mädche, B. Motik, and L. Stojanovic. Managing multiple and distributed ontologies in the semantic web. *VLDB Journal*, 12(4):286–302, 2003.
53. A. Mädche and S. Staab. Discovering conceptual relations from text. In *Proceedings of the 14th European Conference on Artificial Intelligence (ECAI)*, pages 321–325, 2000.

54. A. Mädche and S. Staab. Measuring similarity between ontologies. In *Proceedings of the European Conference on Knowledge Acquisition and Management (EKAW)*, pages 251–263, 2002.
55. A. Mädche and R. Volz. The Text-To-Onto ontology extraction and maintenance system. In *Workshop on Integrating Data Mining and Knowledge Management, collocated with the 1st International Conference on Data Mining*, 2001.
56. C. Manning and H. Schütze. *Foundations of Statistical Language Processing.* MIT Press, 1999.
57. V. Pekar and S. Staab. Taxonomy learning: Factoring the structure of a taxonomy into a semantic classification decision. *Proceedings of the 19th Conference on Computational Linguistics (COLING)*, 2:786–792, 2002.
58. F. Pereira, N. Tishby, and L. Lee. Distributional clustering of english words. In *Proceedings of the 31st Annual Meeting of the Association for Computational Linguistics (ACL)*, pages 183–190, 1993.
59. M. Poesio, T. Ishikawa, S. Schulte im Walde, and R. Viera. Acquiring lexical knowledge for anaphora resolution. In *Proceedings of the 3rd Conference on Language Resources and Evaluation (LREC)*, 2002.
60. M. F. Porter. An algorithm for suffix stripping. *Program*, 14(3):130–137, 1980.
61. P. Resnik. Semantic similarity in a taxonomy: An information-based measure and its application to problems of ambiguity in natural language. *Journal of Artificial Intelligence Research (JAIR)*, 11:95–130, 1999.
62. D. Sanchez and A. Moreno. Web-scale taxonomy learning. In C. Biemann and G. Pass, editors, *Proceedings of the Workshop on Extending and Learning Lexical Ontologies using Machine Learning Methods*, 2005.
63. H. Schmid. Probabilistic part-of-speech tagging using decision trees. In *Proceedings of the International Conference on New Methods in Language Processing*, 1994.
64. A. Schutz and P. Buitelaar. RelExt: A tool for relation extraction from text in ontology extension. In *Proceedings of the International Semantic Web Conference (ISWC)*, pages 593–606, 2005.
65. E. Simperl, C. Tempich, and D. Vrandecic. A methodology for ontology learning. In P. Buitelaar and P. Cimiano, editors, *Bridging the Gap between Text and Knowledge: Selected Contributions to Ontology Learning and Population from Text*, volume 167 of *Frontiers in Artificial Intelligence*. IOS Press, 2007.
66. R. Snow, D. Jurafsky, and Y. Ng. Semantic taxonomy induction from heterogenous evidence. In *Proceedings of the 21st International Conference on Computational Linguistics and the 44th annual meeting of the ACL*, pages 801–808, 2006.
67. M. Strube and S. Paolo Ponzetto. Wikirelate! computing semantic relatedness using wikipedia. In *Proceedings of the National Conference on Artificial Intelligence (AAAI)*, pages 1419–1424, 2006.
68. G. Stumme, R. Taouil, Y. Bastide, N. Pasqier, and L. Lakhal. Computing iceberg concept lattices with titanic. *Journal of Knowledge and Data Engineering (KDE)*,, 42(2):189–222, 2002.
69. P. D. Turney. Mining the web for synonyms: PMI-IR versus LSA on TOEFL. In *Proceedings of the 12th European Conference on Machine Learning (ECML)*, pages 491–502, 2001.
70. P. Velardi, R. Navigli, A. Cuchiarelli, and F. Neri. Evaluation of OntoLearn, a methodology for automatic population of domain ontologies. In P. Buitelaar,

P. Cimiano, and B. Magnini, editors, *Ontology Learning from Text: Methods, Applications and Evaluation*, number 123 in Frontiers in Artificial Intelligence and Applications, pages 92–106. IOS Press, 2005.
71. J. Völker, P. Hitzler, and P. Cimiano. Acquisition of OWL DL axioms from lexical resources. In *Proceedings of the 4th European Semantic Web Conference (ESWC'07)*, pages 670–685, 2007.
72. J. Völker, D. Vrandecic, and Y. Sure. Automatic evaluation of ontologies (AEON). In Y. Gil, E. Motta, V. R. Benjamins, and M. A. Musen, editors, *Proceedings of the 4th International Semantic Web Conference (ISWC)*, volume 3729 of *LNCS*, pages 716–731. Springer, 2005.
73. J. Völker, D. Vrandecic, Y. Sure, and A. Hotho. Learning disjointness. In *Proceedings of the 4th European Semantic Web Conference (ESWC'07)*, pages 175–189, 2007.
74. J. Völker and S. Rudolph. Lexico-logical acquisition of OWL DL axioms – An integrated approach to ontology refinement. In R. Medina and S. Obiedkov, editors, *Proceedings of the 6th International Conference on Formal Concept Analysis (ICFCA'08)*, volume 4933 of *Lecture Notes in Artificial Intelligence*, pages 62–77. Springer, 2008.
75. D. Widdows. Unsupervised methods for developing taxonomies by combining syntactic and statistical information. In *Proceedings of the Human Language Technology Conference/North American Chapter of the Association for Computational Linguistics (HLT/NAACL)*, pages 276–283, 2003.
76. I. H. Witten and E. Frank. *Data Mining: Practical Machine Learning Tools and Techniques with Java Implementations.* Morgan Kaufmann, 1999.

[illegible] P. Cimiano, and B. Motik. [illegible] Ontology Learning from Text. [illegible] applications [illegible] In [illegible] Artificial Intelligence and Applications [illegible]
71. J. Völker, P. Haase, and P. Hitzler. [illegible] In Proceedings of the [illegible] on Semantic Web [illegible] LNCS [illegible]
[illegible] A. [illegible] Benjamins, and [illegible] In Proceedings of the [illegible] International [illegible] Conference [illegible] LNCS [illegible] Springer, 2006.
72. J. Völker, D. Vrandečić, Y. Sure, and A. Hotho. Learning disjointness. In Proceedings of the European Semantic Web Conference [illegible]
[illegible]
74. [illegible]
[illegible]
[illegible]
[illegible]
[illegible]
75. [illegible]
[illegible]
[illegible]
[illegible]
76. [illegible]
[illegible]

Ontology and the Lexicon

Graeme Hirst

Department of Computer Science, University of Toronto, Toronto, ON,
Canada M5S 3G4, gh@cs.toronto.edu

Summary. A lexicon is a linguistic object and hence is not the same thing as an ontology, which is non-linguistic. Nonetheless, word senses are in many ways similar to ontological concepts and the relationships found between word senses resemble the relationships found between concepts. Although the arbitrary and semi-arbitrary distinctions made by natural languages limit the degree to which these similarities can be exploited, a lexicon can nonetheless serve in the development of an ontology, especially in a technical domain.

1 Lexicons and Lexical Knowledge

1.1 Lexicons

A *lexicon* is a list of words in a language – a *vocabulary* – along with some knowledge of how each word is used. A lexicon may be general or domain-specific; we might have, for example, a lexicon of several thousand common words of English or German, or a lexicon of the technical terms of dentistry in some language. The words that are of interest are usually *open-class* or *content* words, such as nouns, verbs, and adjectives, rather than *closed-class* or *grammatical function* words, such as articles, pronouns, and prepositions, whose behaviour is more tightly bound to the grammar of the language. A lexicon may also include multi-word expressions such as fixed phrases (*by and large*), phrasal verbs (*tear apart*), and other common expressions (*merry Christmas!; teach ⟨someone⟩'s grandmother to suck eggs; Elvis has left the building*).

Each word or phrase in a lexicon is described in a *lexical entry*; exactly what is included in each entry depends on the purpose of the particular lexicon. The details that are given (to be discussed further in Sects. 2.1 and 3.2 below) may include any of its properties of spelling or sound, grammatical behaviour, meaning, or use, and the nature of its relationships with other words. A lexical entry is therefore a potentially large record specifying many aspects of the linguistic behaviour and meaning of a word.

S. Staab and R. Studer (eds.), *Handbook on Ontologies*, International Handbooks on Information Systems, DOI 10.1007/978-3-540-92673-3,

Hence a lexicon can be viewed as an index that maps from the written form of a word to information about that word. This is not a one-to-one correspondence, however. Words that occur in more than one syntactic category will usually have a separate entry for each category; for example, *flap* would have one entry as a noun and another as a verb. Separate entries are usually also appropriate for each of the senses of a *homonym* – a word that has more than one unrelated sense even within a single syntactic category; for example, the noun *pen* would have distinct entries for the senses writing instrument, animal enclosure, and swan. *Polysemy* – related or overlapping senses – is a more-complex situation; sometimes the senses may be discrete enough that we can treat them as distinct: for example, *window* as both opening in wall and glass pane in opening in wall (*fall through the window; break the window*). But this is not always so; the word *open*, for example, has many overlapping senses concerning unfolding, expanding, revealing, moving to an open position, making openings in, and so on, and separating them into discrete senses, as the writers of dictionary definitions try to do, is not possible (see also Sects. 2.3 and 3.1 below).

On the other hand, morphological variants of a word, such as plurals of nouns and inflected forms of verbs, will not normally warrant their own complete lexical entry. Rather, the entry for such forms need be little more than a pointer to that for the base form of the word. For example, the entries for *takes, taking, took,* and *taken* might just note that they are inflected forms of the base-form verb *take*, and point to that entry for other details; and conversely, the entry for *take* will point to the inflected forms. Similarly, *flaps* will be connected both to the noun *flap* as its plural and to the verb *flap* as its third-person singular. The sharing of information between entries is discussed further in Sect. 2.2 below.

A lexicon may be just a simple list of entries, or a more-complex structure may be imposed upon it. For example, a lexicon may be organized hierarchically, with default inheritance of linguistic properties (see Sect. 2.2 below). However, the structures that will be of primary interest in this chapter are semantic, rather than morphological or syntactic; they will be discussed in Sect. 3.2 below.

1.2 Computational Lexicons

An ordinary dictionary is an example of a lexicon. However, a dictionary is intended for use by humans, and its style and format are unsuitable for computational use in a text or natural language processing system without substantial revision. A particular problem is the dictionary's explications of the senses of each word in the form of definitions that are themselves written in natural language; computational applications that use word meanings usually require a more-formal representation of the knowledge. Nonetheless, a dictionary in a machine-readable format can serve as the basis for a computational lexicon, as in the ACQUILEX project [8] – and it can also serve as the basis for a semantic hierarchy (see Sect. 5.2 below). (An alternative or complementary source of lexical information is inference from the usage observed in text corpora; see, e.g. [7].)

Perhaps the best-known and most widely used computational lexicon of English is WordNet [25]. The primary emphasis of WordNet is on semantic relationships between words; it contains little syntactic and morphological data and no phonetic data. The basic lexical entry in WordNet is the *synset* (for "synonym set"), which groups together identical word senses. For example, the synonymous nouns *boarder, lodger,* and *roomer* are grouped together in a synset. WordNet includes an extensive network of relationships between synsets; this will be discussed in detail in Sect. 3.2. Following the success of WordNet for English, wordnets with a similar (but not necessarily identical) structure have been (or are being) developed for a large number of other languages (some as part of the EuroWordNet project [67]), including Basque, Dutch, French, Hindi, and Tamil (see `www.globalwordnet.org`).

Some other important general-purpose lexicons include CELEX [5], which is a set of large, detailed lexicons of Dutch, German, and English, and the PAROLE project (`www.ub.es/gilcub/SIMPLE/simple.html`) and its successor SIMPLE [40], which are large, rich lexicons for 12 European languages.

Two important sources for obtaining lexicons are these:

ELDA: The Evaluations and Language resources Distribution Agency (`www.elda.org`) distributes many European-language general-purpose and domain-specific lexicons, both monolingual and multilingual, including PAROLE and EuroWordNet.

LDC: The Linguistic Data Consortium (`ldc.upenn.edu`), although primarily a distributor of corpora, offers CELEX and several other lexicons.

In addition, English WordNet is available free of charge from the project's Web page (`wordnet.princeton.edu`).

2 Lexical Entries

2.1 What is in a Lexical Entry?

Any detail of the linguistic behaviour or use of a word may be included in its lexical entry: its phonetics (including pronunciations, syllabification, and stress pattern), written forms (including hyphenation points), morphology (including inflections and other affixation), syntactic and combinatory behaviour, constraints on its use, its relative frequency, and, of course, all aspects of its meaning. For our purposes in this chapter, the word's semantic properties, including relationships between the meanings of the word and those of other words, are the most important, and we will look at them in detail in Sect. 3.2 below.

Thus, as mentioned earlier, a lexical entry is potentially quite a large record. For example, the CELEX lexicons of English, Dutch, and German [5] are represented as databases whose records have 950 fields. And in an *explanatory combinatorial dictionary* (ECD) (e.g. [46, 47]), which attempts to explicate literally every aspect of the knowledge that a speaker needs to have in order to use a word correctly, lexical entries can run to many pages. For example, Steele's [61] ECD-style entry for eight senses of *hope* (noun and verb) is 28 book-sized pages long, much of which is devoted to the combinatory properties of the word – for example, the noun *hope* permits *flicker of* to denote a small amount (whereas *expectation*, in contrast, does not).

Many linguistic applications will require only a subset of the information that may be found in the lexical entries of large, broad-coverage lexicons. Because of their emphasis on detailed knowledge about the linguistic behaviour of words, these large, complex lexicons are sometimes referred to as *lexical knowledge bases*, or *LKBs*. Some researchers distinguish LKBs from lexicons by regarding LKBs as the larger and more-abstract source from which instances of lexicons for particular applications may be generated. In the present chapter, we will not need to make this distinction, and will just use the term *lexicon*.

2.2 Inheritance of Linguistic Properties

Generally speaking, the behaviour of words with respect to many non-semantic lexical properties in any given language tends to be regular: words that are phonetically, morphologically, or syntactically similar to one another usually exhibit similar phonetic, morphological, or syntactic behaviour. For example, in English most verbs form their past tense with either *-ed* or *-d*, and even most of those that do not do so fall into a few small categories of behaviour; and quite separately, verbs also cluster into a number of categories by their *alternation* behaviour (see Sect. 4.3 below).

It is therefore possible to categorize and subcategorize words by their behaviour – that is, build an ontology of lexical behaviour – and use these categories to construct a lexicon in which each word, by default, inherits the properties of the categories and subcategories of which it is a member. Of course, idiosyncratic properties (such as many of the combinatory properties listed in an ECD) will still have to be specified in each word's entry. Inheritance of properties facilitates both economy and consistency in a large lexicon. A hierarchical representation of lexical knowledge with property inheritance is really just a special case of this style or method of knowledge representation. Accordingly, the inheritance of properties in the lexicon and the design of formal languages for the representation of lexical knowledge have been areas of considerable study (e.g. [8, 28]; for an overview, see [17]; for the DATR language for lexical knowledge representation, see [22]).

It should be clear that a hierarchical representation of similarities in lexical behaviour is distinct from any such representation of the *meaning* of words; knowing that *boy* and *girl* both take *-s* to make their plural form whereas *child* does not tells us nothing about the relationship between the meanings of those words. Relationships between meanings, and the hierarchies or other structures that they might form, are a separate matter entirely; they will be discussed in Sect. 3.2.

2.3 Generating Elements of the Lexicon

Even with inheritance of properties, compiling a lexicon is a large task. But it can be eased by recognizing that because of the many regularities in the ways that natural languages generate derived words and senses, many of the entries in a lexicon can be automatically predicted.

For example, at the level of inflection and affixation, from the existence of the English word *read*, we can hypothesize that (among others) *reading, reader, unreadable,* and *antireadability* are also words in the lexicon, and in three out of these four cases we would be right. Viegas et al. [64] present a system of *lexical rules* that propose candidate words by inflection and affixation (an average of about 25 from each base form), automatically generating lexical entries for them; a lexicographer must winnow the proposals. In their Spanish lexicon, about 80% of the entries were created this way. But a lexicon can never anticipate nonce words, neologisms, or compounds that are easily created from combinations of existing words in languages such as German and Dutch; additional word-recognition procedures will always be needed.

At the level of word sense, there are also regularities in the polysemy of words. For example, the senses of the word *book* include both its sense as a physical object and its sense as information-content: *The book fell on the floor; The book was exciting.* (A problem for natural language processing, which need not concern us here, is that both senses may be used at once: *The exciting book fell on the floor.*) In fact, the same polysemy can be seen with any word denoting an information-containing object, and if a new one comes along, the polysemy applies automatically: *The DVD fell on the floor; The DVD was exciting.* There are many such regularities of polysemy; they have been codified in Pustejovsky's [54] theory of the *generative lexicon.* Thus it is possible to write rules that generate new lexical entries reflecting these regularities; if we add an entry for *DVD* to the lexicon as an information-containing object, then the other sense may be generated automatically [9]. (*A fortiori*, the theory of the generative lexicon says that a purely enumerative lexicon – one that is just a list of pre-written entries – can never be complete, because the generative rules always permit new and creative uses of words.)

3 Word Senses and the Relationships Between Them

Most of the issues in the relationship between lexicons and ontologies pertain to the nature of the word senses in the lexicon and to relationships between those senses – that is, to the semantic structure of the lexicon.

3.1 Word Senses

By definition, a *word sense*, or the "meaning" of a word, is a semantic object – a *concept* or *conceptual structure* of some kind, though exactly what kind is a matter of considerable debate, with a large literature on the topic. Among other possibilities, a word sense may be regarded as a purely *mental object*; or as a structure of some kind of *primitive units of meaning*; or as the *set of all the things in the world that the sense may denote*; or as a *prototype* that other objects resemble to a greater or lesser degree; or as an *intension* or *description* or *identification procedure* – possibly in terms of necessary and sufficient conditions – of all the things that the sense may denote.

Word senses tend to be fuzzy objects with indistinct boundaries, as we have seen already with the example of *open* in Sect. 1.1 above. Whether or not a person may be called *slim*, for example, is, to some degree, a subjective judgement of the user of the word. To a first approximation, a word sense seems to be something like a category of objects in the world; so the word *slim* might be taken to denote exactly the category of slim objects, with its fuzziness and its subjectivity coming from the fuzziness and subjectivity of the category in the world, given all the problems that are inherent in categorization (see also [38]). Indeed, some critics have suggested that word senses are *derived, created,* or *modulated* in each context of use, and cannot just be specified in a lexicon [37, 58].

Nonetheless, one position that could be taken is that a word sense *is* a category. This is particularly appealing in simple practical applications, where the deeper philosophical problems of meaning may be finessed or ignored. The problems are pushed to another level, that of the ontology; given some ontology, each word sense is represented simply as a pointer to some concept or category within the ontology. In some technical domains this may be entirely appropriate (see Sect. 5.1 below). But sometimes this move may in fact make matters worse: all the problems of categorization remain, and the additional requirement is placed on the ontology of mirroring some natural language or languages, which is by no means straightforward (see Sect. 4 below); nonetheless, an ontology may act as an interpretation of the word senses in a lexicon (see Sect. 5.4 below).

In addition to the *denotative* elements of meaning that refer to the world, word senses also have *connotation*, which may be used to express the user's attitude: a speaker who chooses the word *sozzled* instead of *drunk* is exhibiting informality, whereas one who chooses *inebriated* is being formal; a speaker who describes a person as *slim* or *slender* is implying that the person's relative narrowness is attractive to the speaker, whereas the choice of *skinny* for the same person would imply unattractiveness.

3.2 Lexical Relationships

Regardless of exactly how one conceives of word senses, because they pertain in some manner to categories in the world itself, *lexical relationships* between word senses mirror, perhaps imperfectly, certain relationships that hold between the categories themselves. The nature of lexical relationships and the degree to which they may be taken as ontological relationships are the topics of most of the rest of this chapter. In the space available, we can do no more than introduce the main ideas of lexical relationships; for detailed treatments, see [15, 23, 30].

The "classical" lexical relationships pertain to identity of meaning, inclusion of meaning, part–whole relationships, and opposite meanings. *Identity of meaning* is *synonymy*: Two or more words are synonyms (with respect to one sense of each) if one may substitute for another in a text without changing the meaning of the text. This test may be construed more or less strictly; words may be synonyms in one context but not another; often, putative synonyms will vary in connotation or linguistic style (as in the *drunk* and *slim* examples in Sect. 3.1 above), and this might or might not be considered significant. More usually, "synonyms" are actually merely near-synonyms (see Sect. 4.1 below).

The primary *inclusion* relations are *hyponymy* and its inverse *hypernymy* (also known as *hyperonymy*) [15, 16]. For example, *noise* is a hyponym of *sound* because any noise is also a sound; conversely, *sound* is a hypernym of *noise*. Sometimes names such as *is-a* and *a-kind-of* are used for hyponymy and *subsumption* for hypernymy; because these names are also used for ontological categories, we avoid using them here for lexical relationships. The inclusion relationship between verbs is sometimes known as *troponymy*, emphasizing the point that verb inclusion tends to be a matter of "manner"; *to murmur* is *to talk* in a certain manner [26]. Inclusion relationships are transitive, and thus form a *semantic hierarchy*, or multiple hierarchies, among word senses; words without hyponyms are leaves and words without hypernyms are roots. (The structures are more usually networks than trees, but we shall use the word *hierarchy* to emphasize the inheritance aspect of the structures.)

The *part–whole* relationships *meronymy* and *holonymy* may be glossed roughly as *has-part* and *part-of*, but we again avoid these ontologically biased terms. The notion of part–whole is overloaded; for example, the relationship between *wheel* and *bicycle* is not the same as that of *professor* and *faculty* or *tree* and *forest*; the first relationship is that of functional component, the second is group membership, and the third is element of a collection. For analysis of part–whole relationships, see [15, 36, 53].

Words that are opposites, generally speaking, share most elements of their meaning, except for being positioned at the two extremes of one particular dimension. Thus *hot* and *cold* are opposites – *antonyms*, in fact – but *telephone* and *Abelian group* are not, even though they have no properties in common (that is, they are "opposite" in every feature or dimension). Cruse [15] distinguishes several different lexical relations of oppositeness, including *antonymy* of gradable adjectives, *complementarity* of mutually exclusive alternatives (*alive–dead*), and directional opposites (*forwards–backwards*).

These "classical" lexical relationships are the ones that are included in the WordNet lexicon. Synonymy is represented, as mentioned earlier, by means of synsets: if two words have identical senses, they are members of the same synset. Synsets are then connected to one another by pointers representing inclusion, part–whole, and opposite relations, thereby creating hierarchies.

There are many other kinds of lexical relationships in addition to the "classical" ones. They include temporal relationships such as *happens-before* (*marry–divorce*) [12] and relationships that may be broadly thought of as deriving from *association* or *typicality* [49]; for example, the relationship between *dog* and *bark* is that the former is a frequent and typical agent of the latter. Other examples of this kind of relationship include typical instrumentality (*nail–hammer*), cause (*leak–drip*), and location (*doctor–hospital*).

Synonymy, inclusion, and associative relations are often the basis of the structure of a *thesaurus*. While general-purpose thesauri, such as *Roget's* [57], leave the relationships implicit, others, especially those used in the classification of technical documents, will make them explicit with labels such as *equivalent term, broader term, narrower term*, and *related term*.

4 Lexicons are not (Really) Ontologies

The obvious parallel between the hypernymy relation in a lexicon and the subsumption relation in an ontology suggests that lexicons are very similar to ontologies. It even suggests that perhaps a lexicon, together with the lexical relations defined on it, *is* an ontology (or is a kind of ontology in the ontology of ontologies). In this view, we identify word senses with ontological categories and lexical relations with ontological relations. The motivation for this identification is clear from the preceding discussion (Sect. 3.2).

Nonetheless, a lexicon, especially one that is not specific to a technical domain (see Sect. 5.1 below), is not a very good ontology. An ontology, after all, is a set of categories of objects or ideas in the world, along with certain relationships among them; it is not a linguistic object. A lexicon, on the other hand, depends, by definition, on a natural language and the word senses in it. These give, at best, an ersatz ontology, as the following sections will show.

4.1 Overlapping Word Senses and Near-Synonymy

It is usually assumed in an ontology that subcategories of a category are disjoint (cf. [66]). For example, if the category domesticated-mammal subsumes the categories dog and cat, among others, then dog $\cap$ cat is empty: nothing can be both a dog and a cat. This is not always so for the hyponymy relation in lexicons, however; rather, two words with a common hypernym will often overlap in sense – that is, they will be *near-synonyms.*

Consider, for example, the English words *error* and *mistake*, and some words that denote kinds of mistakes or errors: *blunder, slip, lapse, faux pas, bull, howler,* and *boner*. How can we arrange these in a hierarchy? First we need to know the precise meaning of each and what distinguishes one from another. Fortunately, lexicographers take on such tasks, and the data for this group of words is given in *Webster's New Dictionary of Synonyms* [29]; an excerpt appears in Fig. 1; it lists both denotative and connotative distinctions, but here we need consider only the former. At first, we can see some structure: *faux pas* is said to be a hyponym of *mistake*; *bull, howler,* and *boner* are apparently true synonyms – they map to the same word sense, which is a hyponym of *blunder*. However, careful consideration of the data shows that a strict hierarchy is not possible. Neither *error* nor *mistake* is the more-general term; rather, they overlap. Neither is a hypernym of the other, and both, really, are hypernyms of the more-specific terms. Similarly, *slip* and *lapse* overlap, differing only in small components of their meaning. And a *faux pas*, as a mistake in etiquette, is not really a type of mistake or error distinct from the others; a faux pas could also be a lapse, a blunder, or a howler.

Error implies a straying from a proper course and suggests guilt as may lie in failure to take proper advantage of a guide ...

Mistake implies misconception, misunderstanding, a wrong but not always blameworthy judgment, or inadvertence; it expresses less severe criticism than *error*.

Blunder is harsher than *mistake* or *error*; it commonly implies ignorance or stupidity, sometimes blameworthiness.

Slip carries a stronger implication of inadvertence or accident than *mistake*, and often, in addition, connotes triviality.

Lapse, though sometimes used interchangeably with *slip*, stresses forgetfulness, weakness, or inattention more than accident; thus, one says a *lapse* of memory or a *slip* of the pen, but not vice versa.

Faux pas is most frequently applied to a mistake in etiquette.

Bull, howler, and **boner** are rather informal terms applicable to blunders that typically have an amusing aspect.

Fig. 1. An entry (abridged) from *Webster's New Dictionary of Synonyms* [29]

This example is in no way unusual. On the contrary, this kind of cluster of near-synonyms is very common, as can be seen in *Webster's New Dictionary of Synonyms* and similar dictionaries in English and other languages. Moreover, the differences between the members of the near-synonym clusters for the same broad concepts are different in different languages. The members of the clusters of near-synonyms relating to errors and mistakes in English, French, German, and Japanese, for example, do not line up neatly with one another or translate directly [20]; one cannot use these word senses to build an ontology of errors.

These observations have led to the proposal [19, 20] that a fine-grained hierarchy is inappropriate as a model for the relationship between the senses of near-synonyms in a lexicon for any practical use in tasks such as machine translation and other applications involving fine-grained use of word senses. Rather, what is required is a very coarse-grained conceptual hierarchy that represents word meaning at only a very coarse-grained level, so that whole clusters of near-synonyms are mapped to a single node: their *core meaning*. Members of a cluster are then distinguished from one another by explicit differentiation of any of the *peripheral concepts* that are involved in the fine-grained aspects of their denotation (and connotation). In the example above, *blunder* might be distinguished on a dimension of severity, while *faux pas* would be distinguished by the domain in which the mistake is made.

4.2 Gaps in the Lexicon

A lexicon, by definition, will omit any reference to ontological categories that are not *lexicalized* in the language – categories that would require a (possibly long) multi-word description in order to be referred to in the language. That is, the words in a lexicon, even if they may be taken to represent categories, are merely a subset of the categories that would be present in an ontology covering the same domain. In fact, every language exhibits *lexical gaps* relative to other languages; that is, it simply lacks any word corresponding to a category that is lexicalized in some other language or languages. For example, Dutch has no words corresponding to the English words *container* or *coy*; Spanish has no word corresponding to the English verb *to stab* "to injure by puncturing with a sharp weapon"; English has no single word for the German *Gemütlichkeit* "combination of cosiness, cheerfulness, and social pleasantness" or for the French *bavure* "embarrassing bureaucratic error". On the face of it, this seems to argue for deriving a language-independent ontology from the union of the lexicons of many languages (as attempted by Emele et al. [21]); but this is not quite feasible.

Quite apart from lexical gaps in one language relative to another, there are many categories that are not lexicalized in *any* language. After all, it is clear that the number of categories in the world far exceeds the number of word senses in a language, and while different languages present different inventories of senses, as we have just argued, it nonetheless remains true that, by and large, all will cover more or less the same "conceptual territory", namely the concepts most salient or important to daily life, and these will be much the same across different languages, especially different languages of similar cultures. As the world changes, new concepts will arise and may be lexicalized, either as a new sense for an existing word (such as *browser* "software tool for viewing the World Wide Web"), as a compositional fixed phrase (*road rage*), or as a completely new word or phrase (*demutualization* "conversion of a mutual life insurance company to a company with shareholders", *proteomics, DVD*). That large areas remain unlexicalized is clear from the popularity of games and pastimes such as Sniglets ("words that do not appear in the dictionary but should") [32] and Wanted Words [24], which derive part of their humour from the identification of established concepts that had not previously been articulated and yet are immediately recognized as such when they are pointed out.

But even where natural languages "cover the same territory", each different language will often present a different and mutually incompatible set of word senses, as each language lexicalizes somewhat different categorizations or perspectives of the world. It is rare for words that are *translation equivalents* to be completely identical in sense; more usually, they are merely cross-lingual near-synonyms (see Sect. 4.1 above).

An area of special ontological interest in which the vocabularies of natural languages tend to be particularly sparse is the upper ontology (see chapter "Foundational Choices in DOLCE"). Obviously, all natural languages need to be able to talk about the upper levels of the ontology. Hence, one might have thought that at this level we would find natural languages to be in essential agreement about how the world is categorized, simply because the distinctions seem to be so fundamental and so basic to our biologically based, and therefore presumably universal, cognitive processes and perception of the world. But natural languages instead prefer to concentrate the richest and most commonly used parts of their vocabulary in roughly the middle of the hierarchy, an area that has come to be known as the *basic-level categories*; categories in this area maximize both informativeness and distinctiveness [50]. A standard example: in the context *Be careful not to trip over the X*, in most situations one is more likely to choose the word *dog* for *X* than *entity, living thing, animal, mammal*, or *Beddlington terrier*, even though the alternatives are ontologically equally correct. Certainly, all languages have words similar to the English *thing, substance,* and *process*; but these words tend to be vague terms and, even here, vary conceptually from one language to another. That this is so is clear from the difficulty of devising a clear, agreed-on top-level ontology, a project that has exercised many people for many years. That is, we have found that we cannot build a satisfactory top-level ontology merely by looking at the relevant vocabulary of one or even several natural languages; see, for example, the extensive criticisms by Gangemi et al. [27] of the top level of WordNet as an ontology. From this, we can conclude that the upper levels of the lexical hierarchy are a poor ontology.

4.3 Linguistic Categorizations That are not Ontological

And yet, even though natural languages omit many distinctions that we would plausibly want in an ontology, they also make semantic distinctions – that is, distinctions that are seemingly based on the real-world properties of objects – that we probably would not want to include in an ontology. An example of this is semantic categorizations that are required for "correct" word choice within the language and yet are seemingly arbitrary or unmotivated from a strictly ontological point of view. For example, Chinese requires that a noun be preceded by an appropriate *classifier* in contexts involving numbers and certain quantifiers:

> In the Chinese expression *liang tiao yu* ('two fish'), the classifier *tiao*, which has a semantic indication for "long and rope-like" objects, must be present between the number (*two*) and the head noun (*fish*). Since *tiao* also occurs with other nouns in a quantifying structure, we can assume that these nouns belong to one class by sharing similar semantic features denoted by the classifier *tiao*: *she* 'snake', *tui* 'leg', *kuzi* 'pair of pants', *he* 'river', *bandeng* 'bench'. (Zhang [70], pp. 43–44, glosses simplified)

There are about 900 such classifiers in Chinese; they are based on characteristics such as shape, aggregation, and value [70]. But while characteristics such as "long and rope-like" are semantic, it is unlikely that fish and pants, for example, will be closely related in a practical ontology. Many other languages of the world, including Japanese and Korean, also have a noun classification system; Aikhenvald [1] describes in detail the kinds of semantic features that various languages use in their classifications.

Often, such linguistic categorizations are not even a reliable reflection of the world. For example, many languages distinguish in their syntax between objects that are discrete and those that are not: *countable* and *mass nouns*. This is also an important distinction for many ontologies; but one should not look in the lexicon to find the ontological data, for in practice, the actual linguistic categorization is rather arbitrary and not a very accurate or consistent reflection of discreteness and non-discreteness in the world. For example, in English, *spaghetti* is a mass noun, but *noodle* is countable; the English word *furniture* is a mass noun, but the French *meuble* and (in some uses) the German *Möbel* are countable. Similarly, in Chinese, the classifier *tiao* mentioned above is not a reliable indicator of a long and rope-like shape: because it applies to pants it also applies, by extension, to any piece of clothing one puts one's legs through, such as *youyongku* "swimming trunks" [70].

A particularly important area in which languages make semantic distinctions that are nonetheless ontologically arbitrary is in the behaviour of verbs in their *diathesis alternations* – that is, alternations in the optionality and syntactic realization of the verb's arguments, sometimes with accompanying changes in meaning [41]. Consider, for example, the English verb *to spray*:

(1) Nadia sprayed water on the plants.
(2) Nadia sprayed the plants with water.
(3) Water sprayed on the plants.
(4) *The plants sprayed with water.

(The "*" on (4) denotes syntactic ill-formedness.) These examples (from [41]) show that *spray* permits the *locative alternation* (examples 1 and 2), with either the medium or the target of the spraying (*water* or *the plants*) being realized as the syntactic object of the verb, and the second case (example 2) carrying the additional implication that the entire surface of the target was affected; moreover, the agent of spraying (*Nadia*) is optional (the *causative alternation*) in the first case (example 3) but not the second (example 4).

In view of the many different possible syntactic arrangements of the arguments of a verb, and the many different possible combinations of requirement, prohibition, and optionality for each argument in each position, a large number of different kinds of alternations are possible. However, if we classify verbs by the syntactic alternations that they may and may not undergo, as Levin [41] has for many verbs of English, we see a semantic coherence to the classes. For example, many verbs that denote the indirect application of a liquid to a surface behave in the same manner as *spray*, including *shower, splash,* and *sprinkle.* Nonetheless, the semantic regularities in alternation behaviour often seem ontologically unmotivated, and even arbitrary. For example, verbs of destruction that include in their meaning the resulting physical state of the affected entity (*smash, crush, shatter*) fall into a completely different behaviour class from verbs that just report the fact of the destruction (*destroy, demolish, wreck*) (Levin [41], p. 239).

Even what is perhaps the most basic and seemingly ontological distinction made by languages, the distinction between nouns, verbs, and other syntactic categories, is not as ontologically well-founded as it might seem. From the viewpoint of *object-dominant* languages [62] such as English (and the majority of other languages), we are used to the idea that nouns denote physical and abstract objects and events (*elephant, Abelian group, running, lunch*) and verbs denote actions, processes, and states (*run, disembark, glow*). But even within European languages, we find that occasionally what is construed as an action or state in one language is not in another; a commonly cited example is the English verb *like* translating to an adverb, a quality of an action, in German: *Nadia likes to sing: Nadia singt gern.* But there are *action-dominant* languages in which even physical objects are referred to with verbs:

> For example, in a situation in which English might say *There's a rope lying on the ground*, Atsugewi [a language of Northern California] might use the single polysynthetic verb form *ẃoswalak·a* ... [This can] be glossed as 'a-flexible-linear-object-is-located on-the-ground because-of-gravity-acting-on-it'. But to suggest its nounless flavor, the Atsugewi form can perhaps be fancifully rendered in English as: "it gravitically-linearizes-aground". In this example, then, Atsugewi refers to two physical entities, a ropelike object and the ground underfoot, without any nouns. (Talmy [62], p. 46)

4.4 Language, Cognition, and the World

All the discussion above on the distinction between lexicon and ontology is really nothing more than a few examples of issues and problems that arise in discussions of the relationship between language, cognition, and our view of the world. This is, of course, a Big Question on which there is an enormous literature, and we cannot possibly do more than just allude to it here in order to put the preceding discussion into perspective. Issues include the degree of mutual causal influence between one's view of the world, one's culture, one's thought, one's language, and the structure of cognitive processes. The *Sapir–Whorf hypothesis* or *principle of linguistic relativity*, in its strongest form, states that language determines thought:

> We dissect nature along lines laid down by our native languages. The categories and types that we isolate from the world of phenomena we do not find there because they stare every observer in the face; on the contrary, the world is presented in a kaleidoscopic flux of impressions which has to be organized by our minds – and this means largely by the linguistic systems in our minds. We cut nature up, organize it into concepts, and ascribe significances as we do, largely because we are parties to an agreement to organize it in this way – an agreement that holds throughout our speech community and is codified in the patterns of our language. The agreement is, of course, an implicit and unstated one, *but its terms are absolutely obligatory*; we cannot talk at all except by subscribing to the organization and classification of data which the agreement decrees. (Whorf [69])

> No two languages are ever sufficiently similar to be considered as representing the same social reality. The worlds in which different societies live are distinct worlds, not merely the same world with different labels attached. (Sapir [59])

These quotations imply a pessimistic outlook for the enterprise of practical, language-independent ontology (or even of translation between two languages, which as a distinct position is often associated with Quine [55]); but conversely, they imply a bright future for ontologies that are strongly based on a language, although such ontologies would have to be limited to use within that language community. But taken literally, linguistic relativity is certainly not tenable; clearly, we can have thoughts for which we have no words. The position is more usually advocated in a weaker form, in which language strongly influences worldview but does not wholly determine it. Even this is not broadly accepted; a recent critic, for example, is Pinker [52], who states bluntly, "There is no scientific evidence that languages dramatically shape their speakers' ways of thinking" (p. 58). Nonetheless, we need to watch out for the un-dramatic shaping.

From a practical standpoint in ontology creation, however, while an overly language-dependent or lexicon-dependent ontology might be avoided for all the reasons discussed above, there is still much in the nature of natural languages that can help the creation of ontologies: it might be a good strategy to adopt or adapt the worldview of a language into one's ontology, or to merge the views of two different languages. For example, languages offer a rich analysis in their views of the structure of events and of space that can serve as the basis for ontologies; see, for example, the work of Talmy [63], in analyzing and cataloguing these different kinds of views. (For an overview of the more-general matter of learning ontologies from natural language text, see chapter "Ontology and the Lexicon".) And, conversely, languages are crucial for human comprehension of ontologies:

> In fact, an ontology without natural language labels attached to classes or properties is almost useless, because without this kind of grounding it is very difficult, if not impossible, for humans to map an ontology to their own conceptualization, i.e. the ontology lacks human-interpretability. (Völker et al. [65])

5 Lexically Based Ontologies and Ontologically Based Lexicons

Despite all the discussion in the previous section, it is possible that a lexicon with a semantic hierarchy might serve as the basis for a useful ontology, and an ontology may serve as a grounding for a lexicon. This may be so in particular in technical domains, in which vocabulary and ontology are more closely tied than in more-general domains. But it may also be the case for more-general vocabularies when language dependence and relative ontological simplicity are not problematic or are even desirable – for example if the ontology is to be used primarily in general-purpose, domain-independent text-processing applications in the language in question and hence inferences from the semantic properties of words have special prominence over domain-dependent or application-dependent inferences. In particular, Dahlgren [18] has argued for the need to base an ontology for intelligent text processing on the linguistic distinctions and the word senses of the language in question.

5.1 Technical Domains

In highly technical domains, it is usual for the correspondence between the vocabulary and the ontology of the domain to be closer than in the case of everyday words and concepts. This is because it is in the nature of technical or scientific work to try to identify and organize the concepts of the domain clearly and precisely and to name them unambiguously (and preferably with minimal synonymy). In some fields of study, there is a recognized authority that maintains and publishes a categorization and its associated nomenclature. For example, in psychiatry, the *Diagnostic and Statistical Manual* of the American Psychiatric Association [3] has this role. In botanical systematics, so vital is unambiguous communication and so enormous is the pool of researchers that a complex system of rules [45] guides the naming of genera, species, and other taxa and the revision of names in the light of new knowledge.

Obviously, the construction of explicit, definitive ontologies, or even explicit, definitive vocabularies, does not occur in all technical domains. Nor is there always general consensus in technical domains on the nature of the concepts of the domain or uniformity in the use of its nomenclature. On the contrary, technical terms may exhibit the same vagueness, polysemy, and near-synonymy that we see exhibited in the general vocabulary. For example, in the domain of ontologies in information systems, the terms *ontology*, *concept*, and *category* are all quite imprecise, as may be seen throughout this volume; nonetheless, they are technical terms: the latter two are used in a more-precise way than the same words are in everyday speech.

However, in technical domains where explicit vocabularies exist (including glossaries, lexicons, and dictionaries of technical terms, and so on, whether backed by an authority or not), an ontology exists at least implicitly, as we will see in Sect. 5.2 below. And where an explicit ontology exists, an explicit vocabulary certainly does; indeed, it is often said that the construction of any domain-specific ontology implies the parallel construction of a vocabulary for it; e.g. Gruber ([31], p. 909): "Pragmatically, a common ontology defines the vocabulary with which queries and assertions are exchanged among agents".

An example of a technical ontology with a parallel vocabulary is the Unified Medical Language System (UMLS) (e.g. [42]; `www.nlm.nih.gov/research/umls`; see also chapter "An Ontology for Software"). The concepts in the Metathesaurus component of the UMLS, along with their additional interpretation in the Semantic Net component, constitute an ontology. Each concept is annotated with a set of terms (in English and other languages) that can be used to denote it; this creates a parallel vocabulary. Additional linguistic information about many of the terms in the vocabulary is given in the separate Specialist Lexicon component.

5.2 Developing a Lexically Based Ontology

It has long been observed that a dictionary implicitly contains an ontology, or at least a semantic hierarchy, in the genus terms in its definitions. For example, if *automobile* is defined as *a self-propelled passenger vehicle that usually has four wheels and an internal-combustion engine*, then it is implied that *automobile* is a hyponym of *vehicle* and even that automobile IS-A vehicle; semantic or ontological part–whole relations are also implied.

Experiments on automatically extracting an ontology or semantic hierarchy from a machine-readable dictionary were first carried out in the late 1970s. Amsler [4], for example, derived a "tangled hierarchy" from *The Merriam-Webster Pocket Dictionary* [48]; Chodorow et al. [13] extracted hierarchies from *Webster's Seventh New Collegiate Dictionary* [68]. The task requires parsing the definitions and disambiguating the terms used [11]; for example *vehicle* has many senses, including *a play, role, or piece of music used to display the special talents of one performer or company*, but this is not the sense that is used in the definition of *automobile*. In the analysis of the definition, it is also necessary to recognize the semantically significant patterns that are used, and to not be misled by so-called "empty heads": apparent genus terms that in fact are not, such as *member* in the definition of *hand* as *a member of a ship's crew* [2,44]. Perhaps the largest project of this type was MindNet [56].

Often, the literature on these projects equivocates on whether the resulting hierarchies or networks should be thought of as purely linguistic objects – after all, they are built from words and word senses – or whether they have an ontological status outside language. If the source dictionary is that of a technical domain, the claim for ontological status is stronger. The claim is also strengthened if new, non-lexically derived nodes are added to the structure. For example, in The Wordtree, a complex, strictly binary ontology of transitive actions by Burger [10], the nodes of the tree were based on the vocabulary of English (for example, to sweettalk is to flatter and coax), but names were manually coined for nodes where English fell short (to goodbadman is to reverse and spiritualize; to gorilla is to strongarm and deprive). A different approach was taken in creating the lexically based ontology Omega [51], which was built not from a dictionary but by merging the WordNet lexicon (see Sect. 1.2 above) with Mikrokosmos [43], a less lexically oriented ontology. Following Cooper [14] (in contrast to the remarks in Sect. 3.1 above), Omega distinguishes between word senses and ontological concepts, taking the former to be much more fine-grained than the latter. Hovy [34] describes a linguistically based methodology for deriving a suitable inventory of concepts from an initial set of word senses from a lexicon.

5.3 Finding Covert Categories

One way that a hierarchy derived from a machine-readable dictionary might become more ontological is by the addition of categories that are unlexicalized in the language upon which it is based. Sometimes, these categories are implicitly reified by the presence of other words in the vocabulary, and, following Cruse [15], they are therefore often referred to as *covert categories*. For example, there is no single English word for things that can be worn on the body (including clothes, jewellery, spectacles, shoes, and headwear), but the category nonetheless exists "covertly" as the set of things that can substitute for X in the sentence *Nadia was wearing (an) X*. It is thus reified through the existence of the word *wear* as the category of things that can meaningfully serve as the object of this verb.

Barrière and Popowich [6] showed that these covert categories (or some of them, at least) can be identified and added as supplementary categories to a lexically derived semantic hierarchy (such as those described in Sect. 5.2 above). Their method relies on the definitions in a children's dictionary, in which the language of the definitions is simple and, unlike a regular dictionary, often emphasizes the purpose or use of the definiendum over its genus and differentia; for example, *a boat carries people and things on the water.* The central idea of Barrière and Popowich's method is to find frequently recurring patterns in the definitions that could signal the reification of a covert category. The first step is to interpret the definitions into a conceptual-graph representation [60]. Then, a graph-matching algorithm looks in the conceptual-graph representations for subgraph patterns whose frequency exceeds an experimentally determined threshold. For example, one frequent subgraph is

[X]←(agent)←[carry]→(object)→(person),

which could be glossed as "things that carry people". This pattern occurs in the definitions of many words, including *boat, train, camel,* and *donkey.* It thus represents a covert category that can be named and added to a semantic hierarchy as a new hypernym (or subsumer, now) of the nodes that were derived from these words, in addition to any other hypernym that they already had. The name for the covert category may be derived from the subgraph, such as carry-object-person-agent for the example above. The hierarchy thus becomes more than just lexical relations, although less than a complete ontology; nonetheless, the new nodes could be helpful in text processing. The accuracy of the method is limited by the degree to which polysemy can be resolved; for example, in the category of *things that people play*, it finds, among others, *music, baseball,* and *outside*, representing different senses of *play.* Thus the output of the method must be regarded only as suggestions that require validation by a human.

Although Barrière and Popowich present their method as being for general-purpose, domain-independent hierarchies and they rely on a particular and very simple kind of dictionary, their method might also be useful in technical domains to help ensure completeness of an ontology derived from a lexicon by searching for unlexicalized concepts.

5.4 Ontologies for Lexicons

As mentioned in Sect. 3.1, most theories of what a word sense is relate it in some way to the world. Thus, an ontology, as a non-linguistic object that more-directly represents the world, may provide an interpretation or grounding of word senses. A simple, albeit limited, way to do this is to map between word senses and elements of or structures in the ontology. Of course, this will work only to the extent that the ontology can capture the full essence of the meanings. We noted in Sect. 5.1 above that the UMLS grounds its Metathesaurus this way.

In machine translation and other multilingual applications, a mapping like this could act as an interlingua, enabling the words in one language to be interpreted in another. However, greater independence from any particular language is required; at the very least, the ontology should not favour, say,

Japanese over English if it is to be used in translation between those two languages. In the 12-language SIMPLE lexicon [40], a hand-crafted upper ontology of semantic types serves as an anchor for lexical entries in all the languages [39]. The semantic types are organized into four *qualia roles*, following the tenets of generative lexicon theory (see Sect. 2.3 above).

Hovy and Nirenburg [35] have argued that complete language-independence is not possible in an ontologically based interlingua for machine translation, but some degree of *language-neutrality* with respect to the relevant languages can nonetheless be achieved; and as the number of languages involved is increased, language-independence can be asymptotically approached. Hovy and Nirenburg present a procedure for merging a set of language-dependent ontologies, one at a time, to create an ontology that is neutral with respect to each. Near-synonyms across languages (Sect. 4.1 above) are just one challenge for this approach. (See also Hovy [33] and chapter "Ontology Mapping".)

6 Conclusion

In this chapter, we have discussed the relationship between lexicons, which are linguistic objects, and ontologies, which are not. The relationship is muddied by the difficult and vexed relationship between language, thought, and the world: insofar as word-meanings are objects in the world, they may participate in ontologies for non-linguistic purposes, but they are inherently limited by their linguistic heritage; but non-linguistic ontologies may be equally limited when adapted to applications such as text and language processing.

Acknowledgements

The preparation of this chapter was supported by a grant from the Natural Sciences and Engineering Research Council of Canada. I am grateful to Eduard Hovy, Jane Morris, Nadia Talent, Helena Hong Gao, and Ulrich Germann for helpful discussions and examples. I am also grateful to the anonymous reviewers of this chapter, who made many excellent suggestions for improvement; I regret that space limitations precluded me from adopting all of them.

References

1. Aikhenvald, Alexandra Y. (2000): *Classifiers: A Typology of Noun Categorization Devices*. Oxford University Press.
2. Alshawi, Hiyan (1987): Processing dictionary definitions with phrasal pattern hierarchies. *Computational Linguistics*, 13(3/4), 195–202. `www.aclweb.org/anthology/J87-3001`

3. American Psychiatric Association (2000): *Diagnostic and Statistical Manual of Mental Disorders: DSM-IV-TR* (4th edition, text revision). American Psychiatric Association, Washington, DC.
4. Amsler, Robert A. (1981): A taxonomy for English nouns and verbs. *Proceedings of the 19th Annual Meeting of the Association for Computational Linguistics*, Stanford, 133–138. `www.aclweb.org/anthology/P81-1030`
5. Baayen, Harald R., Piepenbrock, Richard, and Gulikers, Leon (1995): *The CELEX Lexical Database.* CD-ROM, Linguistic Data Consortium, University of Pennsylvania, Philadelphia, PA. `www.ldc.upenn.edu/Catalog/CatalogEntry.jsp?catalogId=LDC96L14`
6. Barrière, Caroline and Popowich, Fred (2000): Expanding the type hierarchy with nonlexical concepts. In: Hamilton, Howard J. (ed.), *Advances in Artificial Intelligence* (Proceedings of the 13th Biennial Conference of the Canadian Society for Computational Studies of Intelligence, Montreal, May 2000), Lecture Notes in Artificial Intelligence, volume 1822, pages 53–68. Springer, Berlin. `link.springer.de/link/service/series/0558/tocs/t1822.htm`
7. Boguraev, Branimir and Pustejovsky, James (eds.) (1996): *Corpus Processing for Lexical Acquisition.* MIT, Cambridge, MA.
8. Briscoe, Ted, de Paiva, Valeria, and Copestake, Ann (eds.) (1993): *Inheritance, Defaults, and the Lexicon.* Cambridge University Press.
9. Buitelaar, Paul (1998): CoreLex: An ontology of systematic polysemous classes. In: Guarino, Nicola (ed.), *Formal Ontology in Information Systems*, pages 221–235. IOS, Amsterdam.
10. Burger, Henry G. (1984): *The Wordtree.* The Wordtree, Merriam, KS.
11. Byrd, Roy J., Calzolari, Nicoletta, Chodorow, Martin S., Klavans, Judith L., Neff, Mary S., and Rizk, Omneya A. (1987): Tools and methods for computational lexicography. *Computational Linguistics*, 13(3/4), 219–240. `www.aclweb.org/anthology/J87-3003`
12. Chklovski, Timothy and Pantel, Patrick (2004): VerbOcean: Mining the Web for Fine-Grained Semantic Verb Relations. *Proceedings of the Conference on Empirical Methods in Natural Language Processing (EMNLP-04)*, Barcelona, Spain, 33–40. `acl.ldc.upenn.edu/acl2004/emnlp`
13. Chodorow, Martin S., Byrd, Roy J., and Heidorn, George E. (1985): Extracting semantic hierarchies from a large on-line dictionary. *Proceedings of the 23rd Annual Meeting of the Association for Computational Linguistics*, Chicago, 299–304. `www.aclweb.org/anthology/P85-1034`
14. Cooper, Martin C. (2005): A mathematical model of historical semantics and the grouping of word meanings into concepts. *Computational Linguistics*, 31(2), 227–248.
15. Cruse, D. Alan (1986): *Lexical Semantics.* Cambridge University Press.
16. Cruse, D. Alan (2002): Hyponymy and its varieties. In: Green, Bean, and Myaeng (eds.), *The Semantics of Relationships*, pages 3–21. Kluwer, Dordrecht.
17. Daelemans, Walter, De Smedt, Koenraad, and Gazdar, Gerald (1992): Inheritance in natural language processing. *Computational Linguistics*, 18(2), 205–218. `www.aclweb.org/anthology/J92-2004`
18. Dahlgren, Kathleen (1995): A linguistic ontology. *International Journal of Human–Computer Studies*, 43(5/6), 809–818.

19. Edmonds, Philip and Hirst, Graeme (2000): Reconciling fine-grained lexical knowledge and coarse-grained ontologies in the representation of near-synonyms. *Proceedings of the Workshop on Semantic Approximation, Granularity, and Vagueness*, Breckenridge, Colorado. www.cs.toronto.edu/compling/Publications
20. Edmonds, Philip and Hirst, Graeme (2002): Near-synonymy and lexical choice. *Computational Linguistics*, 28(2), 105–144. www.aclweb.org/anthology/J02-2001
21. Emele, Martin, Heid, Ulrich, Momma, Stefan, and Zajac, Rémi (1992): Interactions between linguistic constraints: Procedural vs. declarative approaches. *Machine Translation*, 7(1/2), 61–98. www.springerlink.com/content/100310/
22. Evans, Roger and Gazdar, Gerald (1996): DATR: A language for lexical knowledge representation. *Computational Linguistics*, 22(2), 167–216. www.aclweb.org/anthology/J96-2002
23. Evens, Martha Walton (ed.) (1988): *Relational Models of the Lexicon.* Cambridge University Press.
24. Farrow, Jane (2000): *Wanted Words: From Amalgamots to Undercarments.* Stoddart, Toronto.
25. Fellbaum, Christiane (1998): *WordNet: An Electronic Lexical Database.* MIT, Cambridge, MA.
26. Fellbaum, Christiane (2002): On the semantics of troponymy. In: Green, R., Bean, C., Myaeng, S. (eds.), *The Semantics of Relationships*, pages 23–34. Kluwer, Dordrecht.
27. Gangemi, Aldo, Guarino, Nicola, and Oltramari, Alessandro (2001): Conceptual analysis of lexical taxonomies: The case of WordNet top-level. In: Welty, Chris and Smith, Barry (eds.), *Formal Ontology in Information Systems: Collected Papers from the Second International Conference*, pages 285–296. ACM, New York.
28. Gazdar, Gerald and Daelemans, Walter (1992): Special issues on Inheritance. *Computational Linguistics*, 18(2) and 18(3). acl.ldc.upenn.edu/J/J92/
29. Gove, Philip B. (ed.) (1973): *Webster's New Dictionary of Synonyms.* G. & C. Merriam Company, Springfield, MA.
30. Green, Rebecca, Bean, Carol A., and Myaeng, Sung Hyon (eds.) (2002): *The Semantics of Relationships: An Interdisciplinary Perspective.* Kluwer Academic, Dordrecht.
31. Gruber, Thomas R. (1993): Toward principles for the design of ontologies used for knowledge sharing. *International Journal of Human–Computer Studies*, 43(5/6), 907–928.
32. Hall, Rich (1984): *Sniglets (Snig'lit): Any Word That Doesn't Appear in the Dictionary, but Should.* Collier Books.
33. Hovy, Eduard (1998): Combining and standardizing large-scale, practical ontologies for machine translation and other uses. *Proceedings of the 1st International Conference on Language Resources and Evaluation (LREC)*, Granada, Spain. www.isi.edu/natural-language/people/hovy/publications.html
34. Hovy, Eduard (2005): Methodologies for the reliable construction of ontological knowledge. In: Dau, Frithjof, Mugnier, Marie-Laure, and Stumme, Gerd (eds.), *Conceptual Structures: Common Semantics for Sharing Knowledge*, pages 91–106. Springer, Berlin. www.isi.edu/natural-language/people/hovy/publications.html

35. Hovy, Eduard and Nirenburg, Sergei (1992): Approximating an interlingua in a principled way. *Proceedings of the DARPA Speech and Natural Language Workshop*, Hawthorne, NY. www.isi.edu/natural-language/people/hovy/publications.html
36. Iris, Madelyn Anne, Litowitz, Bonnie E., and Evens, Martha (1988): Problems of the part–whole relation. In: Evens, M. (Ed.), *Relational Models of the Lexicon: Representing Knowledge in Semantic Networks*, pages 261–288. Cambridge University Press.
37. Kilgarriff, Adam (1997): I don't believe in word senses. *Computers and the Humanities*, 31(2), 91–113. www.springerlink.com/content/100251
38. Lakoff, George (1987): *Women, Fire, and Dangerous things: What Categories Reveal About the Mind*. The University of Chicago Press, Chicago.
39. Lenci, Alessandro (2001): Building an ontology for the lexicon: Semantic types and word meaning. In: Jensen, Per Anker and Skadhauge, Peter (eds.), *Ontology-Based Interpretation of Noun Phrases: Proceedings of the First International OntoQuery Workshop*, University of Southern Denmark, pages 103–120. www.ontoquery.dk/publications/
40. Lenci, Alessandro et al. (2000). SIMPLE: A general framework for the development of multilingual lexicons. *International Journal of Lexicography*, 13(4), 249–263.
41. Levin, Beth (1993): *English Verb Classes and Alternations: A preliminary investigation*. The University of Chicago Press.
42. Lindberg, Donald A. B., Humphreys, Betsy L., and McCray, Alexa T. (1993): The Unified Medical Language System. *Methods of Information in Medicine*, 32(4), 281–289.
43. Mahesh, Kavi, and Nirenburg, Sergei (1995): A situated ontology for practical NLP. *Proceedings of the Workshop on Basic Ontological Issues in Knowledge Sharing, International Joint Conference on Artificial Intelligence (IJCAI-95)*, Montreal.
44. Markowitz, Judith, Ahlswede, Thomas, and Evens, Martha (1986): Semantically significant patterns in dictionary definitions. *Proceedings of the 24th Annual Meeting of the Association for Computational Linguistics*, New York, pages 112–119. www.aclweb.org/anthology/P86-1018
45. McNeill, J., Barrie, F. R., Burdet, H. M., Demoulin, V., Hawksworth, D. L., Marhold, K., Nicolson, D. H., Prado, J., Silva, P. C., Skog, J. E., Wiersema, J., and Turland, N. J. (eds.) (2006): *International Code of Botanical Nomenclature (Vienna Code)*. A.R.G. Gantner, Ruggell, Liechtenstein.
46. Mel'čuk, Igor (1984): *Dictionnaire explicatif et combinatoire du français contemporain*. Les Presses de l'Université de Montréal.
47. Mel'čuk, Igor and Zholkovsky, Alexander (1988): The explanatory combinatorial dictionary. In: Evens (ed.), *Relational Models of the Lexicon: Representing Knowledge in Semantic Networks*, pages 41–74. Cambridge University Press.
48. *The Merriam-Webster Pocket Dictionary*. G.&C. Merriam Company, Springfield, MA.
49. Morris, Jane and Hirst, Graeme (2004): Non-classical lexical semantic relations. *Proceedings, Workshop on Computational Lexical Semantics*, Human Language Technology Conference of the North American Chapter of the Association for Computational Linguistics, Boston. Reprinted in: Hanks, Patrick (ed.), *Lexicology: Critical Concepts in Linguistics*, Routledge, 2007. www.aclweb.org/anthology/W04-2607

50. Murphy, Gregory L. and Lassaline, Mary E. (1997): Hierarchical structure in concepts and the basic level of categorization. In: Lamberts, Koen and Shanks, David (eds.), *Knowledge, Concepts, and Categories*, pages 93–131. MIT, Cambridge, MA.
51. Philpot, Andrew, Hovy, Eduard, and Pantel, Patrick (2005): The Omega ontology. *Proceedings, IJCNLP workshop on Ontologies and Lexical Resources (OntoLex-05)*. Jeju Island, South Korea. www.aclweb.org/anthology/I05-7009
52. Pinker, Steven (1994): *The Language Instinct.* William Morrow and Company, New York.
53. Pribbenow, Simone (2002): Meronymic relationships: From classical merology to complex part–whole relationships. In: Green, R., Bean, C. and Myaeng, S. (eds.), *The Semantics of Relationships*, pages 35–50. Kluwer, Dordrecht.
54. Pustejovsky, James (1995): *The Generative Lexicon.* MIT, Cambridge, MA.
55. Quine, Willard Van Orman (1960): *Word and Object.* MIT, Cambridge, MA.
56. Richardson, Stephen D., Dolan, William B., and Vanderwende, Lucy (1998): MindNet: Acquiring and structuring semantic information from text. *Proceedings, 36th Annual Meeting of the Association for Computational Linguistics and the 17th International Conference on Computational Linguistics (COLING-98)*, Montreal, pages 1098–1104. www.aclweb.org/anthology/P98-2180
57. Roget, Peter Mark. *Roget's Thesaurus.* Many editions and variant titles.
58. Ruhl, Charles (1989): *On Monosemy: A Study in Linguistic Semantics.* State University of New York Press, Albany, NY.
59. Sapir, Edward (1929/1964): The status of linguistics as a science. *Language*, 5, 207–214. Reprinted in: Mandelbaum, David G. (ed.), *Culture, Language, and Personality: Selected Essays of Edward Sapir*, University of California Press, Berkeley.
60. Sowa, John F. (2000): *Knowledge Representation: Logical, Philosophical, and Computational Foundations.* Brooks/Cole, Pacific Grove, CA.
61. Steele, James (1990): The vocable *hope*: A family of lexical entries for an explanatory combinatorial dictionary of English. In: Steele, James (ed.), *Meaning–Text Theory: Linguistics, Lexicography, and Implications*, pages 131–158. University of Ottawa Press.
62. Talmy, Leonard (ed.) (2000a): The Relation of Grammar to Cognition. In: *Toward a Cognitive Semantics*, pages I-21–96. MIT, Cambridge, MA.
63. Talmy, Leonard (2000b): *Toward a Cognitive Semantics*, two volumes. MIT, Cambridge, MA.
64. Viegas, Evelyn, Onyshkevych, Boyan, Raskin, Victor, and Nirenburg, Sergei (1996): From *submit* to *submitted* via *submission*: On lexical rules in large-scale lexicon acquisition. *Proceedings, 34th Annual Meeting of the Association for Computational Linguistics,* Santa Cruz, pages 32–39. www.aclweb.org/anthology/P96-1005
65. Völker, Johanna, Hitzler, Pascal, and Cimiano, Philipp (2007): Acquisition of OWL DL axioms from lexical resources. In: Franconi, Enrico, Kifer, Michael, and May, Wolfgang (eds.), *The Semantic Web: Research and Applications (Proceedings of the 4th European Semantic Web Conference, ESWC 2007)*, Lecture Notes in Computer Science, volume 4519, pages 670–685. Springer, Berlin. www.springerlink.com/content/5716324h77684v9g/fulltext.pdf

66. Völker, Johanna, Vrandečić, Denny, Sure, York, Hotho, Andreas (2007): Learning disjointness. In: Franconi, Enrico, Kifer, Michael, and May, Wolfgang (eds.), *The Semantic Web: Research and Applications (Proceedings of the 4th European Semantic Web Conference, ESWC 2007)*, Lecture Notes in Computer Science, volume 4519, pages 175–189. Springer, Berlin. `www.springerlink.com/content/n711wk1t3348482/fulltext.pdf`
67. Vossen, Piek (ed.) (1998): Special issue on EuroWordNet. *Computers and the Humanities*, 32(2–3), 73–251. Reprinted as a separate volume: *EuroWordNet: A multilingual database with lexical semantic networks.* Kluwer Academic, Dordrecht. `www.springerlink.com/content/100251/`
68. *Webster's Seventh New Collegiate Dictionary* (1963): G.&C. Merriam Company, Springfield, MA.
69. Whorf, Benjamin Lee (1940/1972): Science and linguistics. *Technology Review*, 42(6), 227–231, 247–248. Reprinted in: Carroll, John B. (ed.), *Language, Thought and Reality: Selected writings of Benjamin Lee Whorf*, MIT, Cambridge, MA.
70. Zhang, Hong (2007): Numeral classifiers in Mandarin Chinese. *Journal of East Asian Linguistics*, 16(1), 43–59. `www.springerlink.com/content/102930/`

Ontology Evaluation

Denny Vrandečić

Institut AIFB, Universität Karlsruhe (TH), Karlsruhe, Germany,
denny@aifb.uni-karlsruhe.de

Summary. The evaluation of ontologies is still an emerging field. A set of preliminary ideas and frameworks have been suggested in the literature. This chapter collects ontology quality criteria and lays out a common framework for aspects of ontology evaluation. It will present in depth descriptions of these ontology aspects and how to evaluate them. The techniques and ideas collected and presented here will help to uncover errors in ontologies. This chapter concentrates on the automatic, domain- and task-independent evaluation of an ontology.

1 Introduction

Today's software systems are growing in size and complexity. They often consist of a big number of software components, developed by heterogeneous groups at different times, with varying skills and goals. These components are still supposed to exchange and share data, and to cooperate for the advantage of the users and their communities. Such software systems are increasingly connected to each other, communicating on behalf of their users between different platforms and for changing purposes. The biggest such system is the Semantic Web [6], an extension of the current web, that is used by aforementioned components in order to cooperate on a world-wide scale.

Ontologies are used in order to specify in a standard way the knowledge that is exchanged and shared between the different systems, and within the systems by the various components. Ontologies are engineering artifacts that define the formal semantics of the terms used, and the relations between these terms. They provide an "explicit specification of a conceptualization" [22]. Ontologies ensure that the meaning of the data, that is exchanged between and within systems, is consistent and shared – both by computers (expressed by formal models) and humans (as given by their conceptualization). Ontologies make sure that all participants "speak a common language".

Like any engineering artifact, an ontology needs to be thoroughly evaluated. But the evaluation of ontologies poses a number of unique challenges:

S. Staab and R. Studer (eds.), *Handbook on Ontologies*, International Handbooks on Information Systems, DOI 10.1007/978-3-540-92673-3,

due to the declarative nature of ontologies developers cannot just compile and run them like most other software artifacts. They are data that has to be shared between different components and used for potentially different tasks. Within the context of the Semantic Web, often ontologies may be used in ways unexpected by the original creators of the ontology. Ontologies are expected to enable the simple and serendipitous reuse and integration of heterogeneous data sources. Such goals are hard to test in advance.

This chapter discusses the evaluation of web ontologies, i.e. ontologies specified in one of the standard web ontology languages (for now RDF(S) [31] and the different flavours of OWL [48]) and published on the web, so that they can be used and extended in ways not expected by the creators of the ontology, outside of a central control mechanism. Some of the discussion in this chapter will also apply to other ontology languages, and also for ontologies within a better controlled environment than the web. Many problems discussed in earlier work on ontology evaluation do not apply in the context of web ontologies: since the properties of the ontology language with regards to monotonicity, expressivity, and other features are known, they need not to be evaluated for each ontology anymore. This chapter will focus on domain- and task-independent automatic evaluations (which does not mean that the ontology has to be domain-independent or generic, but rather the evaluation approach itself). We will discuss other types of evaluations briefly in Sect. 10.

Web ontologies as defined by the RDF or OWL standards do not include only terminological knowledge – the terms used to describe data, and the formal relations between these terms – but may also include the knowledge bases themselves, i.e. terms describing individuals and ground facts asserting the state of affairs between these individuals. In many cases such knowledge bases are not regarded as being proper ontologies [38], but for the remainder of this chapter we follow the OWL standard and regard ontologies as artifacts encompassing both the terminological as well as the assertional knowledge.

The next section will discuss a number of ontology quality criteria as provided by literature. This will offer a frame of reference for the evaluation methods described in the rest of the chapter. As we will see though, the quality criteria will not be trivially mappable to evaluation methods. Section 3 defines several aspects of an ontology, so that the following sections can work on these aspects: vocabulary (Sect. 4), syntax (Sect. 5), structure (Sect. 6), semantics (Sect. 7), representation (Sect. 8), and finally the context in which the ontology is used (Sect. 9). The chapter closes with a discussion of further approaches and an outlook at future work in this area.

2 Criteria

Ontology evaluation can target a number of several different criteria. In this section we will list criteria from literature, namely five papers describing principles for good ontologies [17,19,22,23,38]. A good ontology will not perform equally well with regards to all these criteria – some of the criteria are even contradicting, like *minimal ontological commitment* and *completeness.* The first task of the evaluator is therefore to choose the criteria relevant for the given evaluation and then to choose the proper evaluation methods to assess how well the ontology meets these criteria.

Gómez-Pérez [19] introduces the two terms ontology verification and validation for describing ontology evaluation: *ontology verification* deals with building the ontology correctly, that is, ensuring that its definitions implement correctly the requirements. *Ontology validation* refers to whether the meaning of the definitions really models the real world for which the ontology was created. Or, to put it slightly different: ontology verification answers if the ontology was built in the right way, whereas ontology validation answers if the right ontology was built. The majority of this chapter will deal with ontology verification.

A complementing overview article dealing with ontology validation is provided by Obrst et al. [38]. It provides a concise overview of many evaluation approaches that are not discussed here. These approaches include the alignment with upper level ontologies for evaluation purposes, human assessment, natural language evaluation techniques, using reality as a benchmark, and ontology accreditation, certification, and maturity models. Ontology validation is an important part of assessing the quality of an ontology, and usually the only way to assure the correctness of the knowledge encoded in the ontology. But most validation approaches require the close cooperation of domain and ontology engineering experts. Validation often can not be performed automatically. Since this chapter focuses on automatic evaluation approaches, we will not repeat the approaches discussed by Obrst et al. [38].

Besides the two basic terms of verification and validation, the following quality criteria are discussed in the literature. In order to arrive at a concise description we take the liberty to collapse similar criteria.

- *Accuracy* [38]: Do the axioms *comply to the expertise* of one or more users [17]? Does the ontology capture and represent correctly aspects of the real world [38]?
- *Adaptability* [38]: Does the ontology anticipate its uses? Does it offer a conceptual foundation for a range of anticipated tasks? Can the ontology be extended and specialized monotonically, i.e. without the need to remove axioms? How does the ontology react to small changes in the axioms [19]? Does the ontology comply to procedures for extension, integration, and adaptation [17]? (also named *expandability* and *sensitiveness* by [19], *extendibility* by [22], and *flexibility* by [17])
- *Clarity* [22]: Does the ontology communicate effectively the intended meaning of the defined terms? Are the definitions objective and independent of context? Does the ontology use definitions or partial descriptions? Are the definitions documented? Is the ontology understandable? (also named *cognitive ergonomics*, *transparency* [17], and *intelligibility* [38])

- *Completeness* [19]/*competency* [23]: Is the domain of interest appropriately covered? Are competency questions defined? Can the ontology answer them? Does the ontology include all relevant concepts and their lexical representations? (also called *richness* and *granularity* [38])
- *Computational efficiency* [17, 38]: How easy and successful can reasoners process the ontology? How fast can the usual reasoning services (satisfiability, instance classification, querying, etc.) be applied to the ontology?
- *Conciseness* [19]: Does the ontology include irrelevant axioms with regards to the domain to be covered (i.e. a book ontology including axioms about African lions)? Does it include redundant axioms? Does it impose a *minimal ontological commitment* [22], i.e. does it specifying the weakest theory possible and define only essential terms? How weak are the assumptions regarding the ontology's underlying philosophical theory about reality [38]?
- *Consistency* [19]/*coherence* [22]: Do the axioms lead to contradictions (logical consistency)? Are the formal and informal description of the ontology consistent, i.e. does the documentation match the specification? Does the translation from the knowledge level to the encoding show a *minimal encoding bias*? Are any representation choices made purely for the convenience of notation or implementation [22]? (covers also *meta-level integrity*, i.e. following ordering principles [17] like OntoClean (see chapter "An Overview of OntoClean")
- *Organizational fitness* [17]/*commercial accessibility*: Is the ontology easily deployed within the organization? Do ontology-based tools within the organization put constraints upon the ontology? Was the proper process for creating the ontology used? Was it certified, if required? Does it meet legal requirements? Is it easy to access? Does it align to other ontologies already in use? Is it well shared among potential stakeholders?

These quality criteria define a good ontology. But just like the answer to the question *how good is the ontology?* is usually not a simple one, the same holds true for these quality criteria: they are all desiderata, goals to guide the creation and evaluation of the ontology. None of them can be directly measured.

Concrete evaluation methods are required in order to assess specific features of an ontology. The relationship between criteria and methods is complex: criteria provide justifications for the methods, whereas the result of a method will provide an indicator for how well one or more criteria are met. Most methods provide indicators for more than one criteria, therefore criteria are a bad choice to structure evaluation methods. In the following sections we describes evaluation methods. In order to give some structure for the description of the methods, we first introduce different aspects of an ontology. These aspects provide guidance for the rest of the chapter.

3 Aspects

An ontology is a complex, multi-layered information resource. In this section we will identify different aspects that are amenable to the automatic, domain- and task-independent evaluation of an ontology. Based on the evaluations of the different ontology aspects, evaluators can then integrate the different evaluation results in order to achieve an aggregated, qualitative ontology evaluation. For each aspect we will show evaluation approaches within the following sections.

Each aspect of an ontology that can be evaluated must represent a degree of freedom (if there is no degree of freedom, there can be no evaluation since it is the only choice). So each aspect describes some choices that have been made during the design of the ontology.

- *Vocabulary.* The vocabulary of an ontology is the set of all names in that ontology, be it URI references or literals, i.e. a value with a datatype or a language identifier. This aspect deals with the different choices with regards to the used URIs or literals.
- *Syntax.* Web ontologies can be described in a number of different surface syntaxes like RDF/XML [4], N-Triples [20], OWL Abstract Syntax [41], the Manchester Syntax [27], or many else. Often the syntactic description within a certain syntax can differ widely. This aspect is about the different serializations in the various syntaxes.
- *Structure.* A web ontology describes an RDF graph. The structure of an ontology is this graph. The structure can vary highly even describing semantically the same ontology. These variances are evaluated when regarding this aspect.
- *Semantics.* A consistent ontology describes a non-empty, usually infinite set of possible models. The semantics of an ontology are the common characteristics of all these models. This aspect is about the semantics features of the ontology.
- *Representation.* This aspect captures the relation between the structure and the semantics. Representational aspects are usually evaluated by comparing metrics calculated on the simple RDF graph with features of the possible models as specified by the ontology.
- *Context.* This aspect is about the features of the ontology when compared with other artifacts in its environment, which may be, e.g. an application using the ontology, a data source that the ontology describes, a different representation of the data within the ontology, or formalized requirements towards the ontology in form of competency questions.

Note that in this chapter we assume that logical consistency or coherence of the ontology is given, i.e. that any inconsistencies or incoherences have been previously resolved using other methods. There is a wide field of work discussing these logical properties, and also well-developed and active research in debugging inconsistency and incoherence. Ontologies are inconsistent if they do not allow any model to fulfill the axioms of the ontology. Incoherent ontologies have classes with an empty intension [24]. Regarding the evaluation aspects, note that the vocabulary, syntax, and structure of the ontology can be evaluated even when dealing with an inconsistent ontology. This also holds true for some parts of the context. But semantic aspects – and thus also representational and some contextual aspects – can not be evaluated if the ontology does not have a model.

4 Vocabulary

The *vocabulary* of a web ontology is the set of all names used in it. Names can be either URIs or literals. The set of all URIs of an ontology is called the *signature* of the ontology (and is thus the subset of the vocabulary without the literals). *Literals* are either typed (i.e. they consist of a tuple with a literal value and a URI identifying the datatype of the value) or untyped. Untyped literals may have a language tag (based on [42]).

Most of the nodes in the ontology graph are named with URIs [5]. Unlike URLs, URIs are not limited to identifying things that have network locations, or use other computer access mechanisms. They can be used to identify anything, from a person over an abstract idea to a simple information resource on the web. Most URIs in web ontologies are using a protocol that can be resolved by the machine in order to fetch further information about the given URI, if set up appropriately [46] – which is one of the major strengths of the Semantic Web. Most commonly this is achieved by using the HTTP protocol [16].

A further issue concerns the naming of the URIs: even though the URI standard [5] states that URIs should be treated opaque and no meaning should be read into them besides their usage with their appropriate protocols, it is obvious that a URI like *http://aifb.de/person/Rudi_Studer* will invoke a certain denotation in the human reader: the user will assume that this is the URI of Rudi Studer, and it would be quite surprising if it were the URI of the movie Casablanca.

URIs should also, if possible, reuse a common URI for a specific resource instead of introducing a new one. In order to enable easier sharing, exchange, and aggregation of information on the Semantic Web, the reuse of commonly used URIs instead of inventing new ones will be helpful. At the time of writing most domains do not have yet a lexicon of URIs, but it is expected that projects like *Swoogle* [13], *Sindice* [39], or *Semantic Wikipedia* [33] will change that soon.

In web ontologies, the type of a name may often be inferred, e.g. a triple like `ex:i rdf:type ex:A` will let us infer that `ex:i` is an `owl:Individual` and `ex:A` an `owl:Class`. This automatism leads to the problem that it is impossible for a reasoner to discern if, e.g. `ex:Adress` is a new entity or merely a typo of `ex:Address`. This can be checked by *declaring* names, so that a tool can check if all used names are properly declared. This further brings the additional benefit of a more efficient parsing of ontologies [36].

Labelling and commenting the URIs should follow a style guide: all URIs should have labels, the languages that need to be supported are defined [42], it should be clear if classes are labelled with a plural or singular noun, if properties are labelled with nouns or verbs (which can be tested using a resource like WordNet [15]), and under what circumstances comments should be used. It is obvious that this will be usually broken in an environment where ontologies are assembled on the fly from smaller ontologies or ontology parts [1]. It is rather improbable that a web wide style guide will ever become ubiquitous. But then again, certain style guides may be created, and an ontology could specify which style guide it follows by means of an adequate ontology property.

Besides URIs, an ontology often contains *data values* like numbers, strings or dates. They are given in the form of *typed literals*, i.e. a value and a URI identifying the datatype of the value. Usually, one of a number of datatypes given in the XML Schema Definition [14] standard are chosen. An ontology should be checked if all used datatypes in the ontology can be dealt with by the tools being used, and if all literals are syntactically correct.

5 Syntax

Web ontologies are serialized in a big (and growing) number of different surface syntaxes. They may be separated into different groups: the ones that describe a graph (which in return describes the ontology), and the ones that describe the ontology directly (a graph can still be calculated from the ontology based on the transformation described in [41]). Examples of the former group are RDF/XML [3] or NTriples [20], examples of the latter are the Manchester Syntax [27], OWL Abstract Syntax [41], or the OWL XML Presentation Syntax [25].

All these different syntaxes should be transformable automatically from and into each other. Note that many of the syntaxes allow for *comments* which are not part of the semantics of the ontology. For example, an XML comment like `<!-- Created with Protege -->` will usually be lost when transforming syntaxes. Often such comments can be expressed as statements with an explicit semantics, like an ontology property describing the used tool. This way the content of the comment would be available to tools using the ontology.

Common features that can be evaluated over most of the syntaxes in a fairly uniform way are the proper *indentation* in the file and the *order* of the triples (for graph-based syntaxes) or axioms (for ontology-based syntaxes). Triples forming complex axioms should be grouped together, as well as groups of axioms forming an ontology pattern (see chapter "Ontology Design Patterns"). Often axioms should precede facts, and facts about the same individual should be grouped together.

A speciality of the RDF/XML syntax is that it can be used to let ontologies, particularly simple knowledge bases, resemble traditional XML documents and even be accompanied by a schema in DTD [8], XSD [14], or RelaxNG [11]. A prime example of such a serialization is given by RSS 1.0 documents [49]. RSS 1.0 files are proper RDF files, but they still have a DTD describing the document structure. This requires a specially written serialization module, but has the unique advantage that incoming files can be checked not only with regards to their syntax but also with regards to their *data-completeness* (see Sect. 7). A schema can describe required fields for an entity, and thus files that are considered valid against the schema will guarantee to have certain properties filled out. This is a very elegant way to check for data completeness with an existing and well tested tool infrastructure.

Most serialization formats include mechanisms to abbreviate URI references. They are often based on known XML based approaches, like *entities* [8] or *XML namespaces* [7]. When defining abbreviations in an ontology, it should be taken care to bind well known abbreviations with the appropriate URIs, e.g. `foaf` should always be defined as `http://xmlns.com/foaf/0.1/` [10]. A service like *Swoogle* [13] can be used to search for common usages of namespaces.

6 Structure

The most widely explored measures used on ontologies are structural measures. Graph measures are applied to the complete or partial RDF graph describing the ontology. An example of an extensively investigated subgraph would be the one consisting only of edges with the name `rdfs:subClassOf` and the nodes connected by these edges (i.e. the explicit class hierarchy). This subgraph can be checked to see if the explicit class hierarchy is a tree, a set of trees, or if it has circularities, etc. If it is indeed a tree, the depth and breadth of the tree can be measured. Current literature defines more than forty different metrics ([17, 18, 35, 50]) that measure the structure of the ontology. Structural measures have a number of advantages:

- They can be calculated effectively from the ontology graph. Graph metrics libraries are readily available and reusable for this task.
- They yield simple numbers. This makes tracking the evolution of the ontology easy, because even in case the meaning of the result itself is not well understood, its change often is.
- Their results can be checked automatically against constraints, e.g. constraining the maximal number of outgoing edges of the type *rdfs:subClassOf* from a single node to 5 can be checked on each commit of the ontology to a version control system. Upon violation, an appropriate message can be created.
- They can be simply visualized and reported.

Due to these advantages, quite some ontology toolkits provide ready access to a number of these metrics (see chapter "Exploring the Economical Aspects of Ontology Engineering"). Also a number of ontology repositories provide annotations of ontologies with such metrics (see chapter "Ontology Repositories").

Structural metrics are often not well-defined. That is, based on their definitions in literature, it is hard to implement them unambiguously. Also there is often confusion with regards to their meaning: for example, measure (M29) in [17] is called the *Class/relation ratio*, suggesting that it returns the ratio between classes and relations. But applying the definition yields the ratio between the number of nodes representing classes and the number of nodes representing relations, which will be a different number since a number of nodes, and thus names, can all denote the same class or relation. We will return to this difference in the following section.

Besides measures counting structural features of the ontology, the structure can also be investigated with regards to certain patterns. The best known example is to regard cycles within the taxonomic structure of the ontology as an error [19]. But also more subtle patterns (or anti-patterns) and heuristics can be used to discover such errors: Disjointness axioms between classes that are very distant in the taxonomic structure [34], as well as certain usages of the universal quantifier [54]. Chapter "Ontology Design Patterns" deals with patterns in more detail.

Even though the complexity of the OWL DL language is known, this does not give much information on particular ontologies. The expressivity of the used language fragment merely defines an upper bound on the complexity that applies to the reasoning tasks. A simple list of the constructs used within the ontology allows to further refine the used DL fragment, and thus a possibly lower complexity bound. For example, OWL DL is known to correspond to the description logic $\mathcal{SHOIN}(\mathbf{D})$ [28], and thus reasoning tasks like satisfiability checks are known to be NExpTime-Complete [47]. But most OWL ontologies do not use the more expressive constructs [12,59]. Ontology editors like SWOOP [30] show the language fragment that is actually being used in a given ontology. Furthermore, even if more expressive constructs are used, queries to the ontology can often be answered much more efficiently than the theoretical upper bound suggests. Experiments indicate that a priori estimates of *resoning speed* can be pursued based purely on structural features [58].

7 Semantics

Most current metrics do not take the semantics of the ontology being described by the RDF graph into account, but consider only the structural properties described in the previous section. But the structure of an ontology is often of less interest than its actual semantics, especially when merging ontologies.

In order to measure the properties of the models described by the structure, *normalization* can be used [56]. Normalization transforms the structure of an ontology to make certain features of its semantics explicit within the structure while retaining the semantics. Then the previously introduced structural measures can be used to measure these explicit semantic features.

Five normalization steps are defined:

1. Name all classes, so no anonymous complex class descriptions are left
2. Name anonymous individuals
3. Materialize the subsumption hierarchy and normalize names (so that classes, properties and individuals that have more than one name use only one of them, see the example in Fig. 1)
4. Instantiate the deepest possible class or property for each individual and property instance (there may be more than one such class or property)
5. Normalize property instances (i.e. materialize symmetric and inverse properties and clean the transitivity graph)

Normalization offers the advantage that metrics are much easier defined on the normalized ontology since some properties are guaranteed: the ontology graph will have no cycles, the number of normal names and classes will correspond, and problems of mapping (see chapter "Ontology Mapping") and redundancy are dealt with. For example, most metrics defining the depth of an ontology (like metric (M3) in [17]) will yield ∞ when encountering a cycle in the subsumption graph, whereas on the normalized ontology the very same metrics will usually yield the result that they intuitively describe.

Another aspect of semantic metrics is their *stability* with regards to the open world assumption of OWL DL ontologies [56]. The question is, how does the metric fare when further axioms are added to the ontology? For example, a taxonomy may have a certain depth, but new axioms could be added that declare the equivalence of all leaves of the taxonomy with its root, thus leading to a depth of 1. This often will not even raise an inconsistency, but is still an indicator for a weak ontology.

When the original ontology was built, the engineer knew the minimal depth of the ontology. But the stable metric measuring the minimal depth remained 1 because no axioms prevented the collapse of the taxonomy. By adding further axioms to strengthen the taxonomy (for example complete partitions) the minimal depth will raise, indicating a more robust ontology with regards to future changes. Stable metrics are indicators for stable ontologies.

Stability with regards to the knowledge base can also be used by closing certain classes. In some cases we know that a knowledge base offers complete coverage: for example, we may publish a complete list of all members of a certain work group, or a complete list of all countries. In this case we can use enumerations to close off the class and declare its completeness. But note that such a closure often has undesirable computational side effects in many reasoners.

Completeness is one of the ontology quality criteria. There are different types of completeness. Here we will consider language completeness. Two more types – data completeness and domain completeness will be discussed in Sects. 9 and 10.

Language completeness is defined on a certain ontology with regards to a specific ontology language (or subset of the language). Given a specific signature (i.e. set of names), language completeness measures the ratio between the knowledge that can be expressed and the knowledge that is actually given. For example, if we have an ontology with the signature *Adam*, *Eve*, *Apple*, *knows*, *eats*, *Person*, we can ask which of the individuals are persons, and which of the individuals know each other. Thus assuming a simple assertional language like RDF, language completeness with regards to that language (or short: assertional completeness) is achieved by knowing about all possible ground facts that can be described by the ontology (i.e. for each fact $\{C(i)|\forall C \in O, \forall i \in O\} \cup \{R(i,j)|\forall R \in O, \forall i \in O, \forall j \in O\}$ we can say if it is true or not). An expressive ontology language allows numerous more questions to be asked besides this ground facts: is the domain of *knows* a Person? Is it its range? Is *eats* a subproperty of *knows*? In order to have a language complete ontology with regards to the more expressive language, the ontology must offer defined answers for all questions that can be asked with the given language. Relational exploration is a method to explore language fragments of higher expressivity, and to calculate the smallest set of questions that have to be answered in order to achieve a complete ontology [44].

8 Representation

Representational aspects of the ontology deal with the relation between the semantics and the structure, i.e. how are the semantics structurally represented? This will often uncover mistakes and omissions within the relation between the *formal specification* and the *shared conceptualization* – or at least the models which are supposedly isomorphic to the conceptualizations.

In order to evaluate features of the representation, we compare the results of the structural measures to the results of the semantic measures. Using the normalization described in Sect. 7, we can even often use the same (or a very similar) metric, applied before and after normalization, and compare the respective results. Any deviations between the results of the two measurements indicate elements of the ontology that require further investigation. For example, consider the ontology given in Fig. 1. The number of classes before normalization is 5, and after normalization 3. This difference shows that several classes collapse into one, which may be an error or done by intention. In case this is an error, it needs to be corrected. If this is done intentionally, the rationale for this design decision should be documented in the ontology.

By contrasting the two ontology structures in Fig. 1 we see that the right one is a more faithful representation of the semantics of the ontology. Both

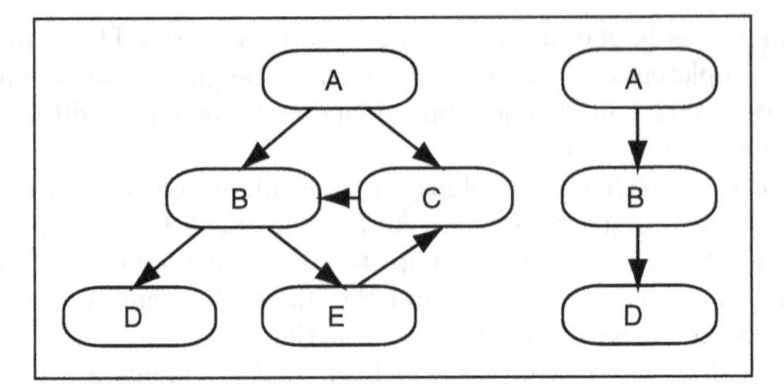

Fig. 1. An simple taxonomy before (*left*) and after (*right*) normalization. The *arrow* denotes subsumption

structures have the same semantics, i.e. allow the same sets of models. The right one is more concise, and for most cases more suitable than the left one. Evaluation methods dealing with representational aspect can uncover such differences and indicate problematic parts of an ontology [56].

9 Context

There are a number of approaches in literature describing the creation and definition of artifacts accompanying the ontology. An evaluating tool can load both the additional artifact and the ontology and then perform further evaluations. The additional artifact thus provides a formalized context.

One of the earliest approaches toward ontology evaluation was the introduction of *competency questions*, i.e. questions that the ontology should be able to answer [23]. In order to enable the automatic evaluation with regards to competency questions, they need to be formalized in a query language that can be used with an appropriate tool. The query language has to be expressive enough to encode the competency questions appropriately. If it is not, the relevance of the competency questions needs to be assessed: if the ontology-based tool cannot ask the question, why should the correct answer be important? Often the specific competency question can be dropped. The additional artifact, in this case, would be the set of formalized competency questions and the required correct answers.

Unit tests for ontologies [55] test if certain axioms can or can not be derived from the ontology. This is especially useful in the case of evolving or dynamic ontologies: we can test certain assumptions with regards to the ontology O. This is done by introducing two accompanying ontologies T^+ and T^- so that $O \models A_i^+ \forall A_i^+ \in T^+$ and $O \not\models A_i^- \forall A_i^- \in T^-$, i.e. every axiom in T^+ can be inferred from the tested ontology O, and on the other hand no axiom in T^- can be inferred from O. The main purpose of unit tests for ontologies is similar to their purpose in software engineering: whenever an error is encountered with an axiom which is falsely inferred or respectively incorrectly not inferred), the ontology maintainer may add this piece of knowledge to the appropriate test ontology. Whenever the ontology is changed, the changed ontology can be automatically checked against the test ontology, containing the formalized knowledge of previously encountered errors. There exists an extension for Protégé that allows for some of the unit tests described here [26].

Certain language constructs, or their combination, may increase reasoning time considerably. In order to avoid this, ontologies are often kept simple. But instead of abstaining from the use of more complex constructs, an ontology could also be modularized with regards to its complexity: a highly axiomatized ontology may come in different versions, one just defining the used vocabulary, and maybe their explicitly stated taxonomic relations, and a second ontology adding much more knowledge, like disjointness or domain and range declarations. If the simple ontology is properly build, it can lead to an ontology which often yields the same results to queries as the complete ontology (depending, naturally, on the actual queries). The additional axioms of the highly axiomatized ontology can be used in order to check the consistency of the ontology and the knowledge base with regards to the higher axiomatized version, but for querying them the simple ontology may suffice.

This idea can be extended to include ontology extensions to the original ontology that are formalized in more expressive languages than the current web standards. For example, [37] introduces the notion of a CBox (constraint box), that includes axioms that check the ontology with regards to additional constraints. This way, for example, it is possible not only to declare domain and ranges like in OWL, which means that they get inferred in case they are not already given in a property instantiation, but also to check if property instantiations follow domain and range *constraints*. Domain and ranges as they are defined in RDFS and OWL are often a confusing element of ontology engineering, especially for programmers who are used to declaring types for parameters in order to check incoming parameters for type correctness.

Another approach is to describe what kind of information an ontology is supposed to have, using *autoepistemic constructs* like the **K**- and **A**-operators [21]. In this way we can, for example, define that every person in our knowledge base needs to have a known name and an address, and put this into a description. Note that this is different than just using the existential construct: whereas an axiom as $Human \sqsubseteq \exists parent.Human$ tells us that every human has to have a human parent, using the **K** operator we could require that every humans' parent has to be explicitly given in the knowledge base or else an inconsistency should be raised inconsistency. Autoepistemic operators allow for a more semantic way to test the completeness of knowledge bases than the syntactic XML based schema definitions described in Sect. 5. This way the ontology can be evaluated with regards to its *data completeness*. Data completeness is defined with regards to a tool that uses the data. The tool needs to explicate which properties it expects when being confronted with an individual belonging to a certain class. The tool can define its data completeness conditions within an autoepistemically extended ontology.

A further example for the use of more expressive ontology languages complementing an existing ontology is the application of *higher order consistency rules*. The best known example of such an approach is the OntoClean methodology (see chapter "An Overview of OntoClean"). OntoClean defines constraints with regards to the subsumptions of classes tagged with certain metaproperties. One approach to using second order semantics on a given ontology is to reify the ontology in OWL DL [57], add additional constraints using the reification language, and then check for consistency (as done using the AEON system [53]).

10 Other Aspects

As stated in the introduction to this chapter, we focused on the domain- and task-independent evaluation of the ontology. This ignored other relevant evaluation methods that will be briefly discussed and referenced in this section.

An ontology is usually defined as a social artifact It specifies the shared agreement of a group of stakeholders. Thus one quality criteria of the ontology will be its *sharedness*, the agreement of the stakeholders with the conceptualization specified in the ontology. This is relevant with regards to vocabulary (is it complete with regards to the domain? are terms missing?) and semantics (do they represent the relations between the terms correctly?). Aspects like syntax, structure, and representation are usually of a lesser concern to the stakeholders. Modern ontology engineering methodologies like DILIGENT [51] and HCOME [32] take into account the fundamental role of discourse between different stakeholders (see also chapter "Ontology Engineering and Evolution in a Distributed World Using DILIGENT" on distributed ontology engineering).

A related question is how to actually ground the terms in an ontology [29], i.e. how will the terms in the ontology be understood by the users of the ontology (either directly, or via a certain tool). Since there is no way to encode the meaning of a term (besides the very weak understanding of meaning as the model-theoretic formal semantics) we need to make sure that a term like *foaf:Person* is well grounded, usually through the documentation and shared understanding. The meaning can be completely computer-based: since computers are able to recognize and handle XML files, for example, the concept of *XML files* can be completely grounded in the computer. But for concepts relating to the world outside of the computer this remains a challenge.

The bigger a group that commits to an ontology (and the shared conceptualization it purports), the harder it is to reach a consensus – but also the bigger the potential benefit. Thus the status of the ontology with regards to relevant standardization bodies in the given domain is a major criteria when evaluating an ontology. Ontologies may be *standardized or certified* by a number of bodies like W3C, Oasis, IETF, and other organizations that may standardize ontologies in their area of expertise – or just a group of peers that declare an ontology a standard (cf. the history of RSS [49] or FOAF [10]). In certain cases, *adoption* by relevant peers (especially business partners) may be even more important than standardization. Tools like Swoogle [13] allow to check how often an ontology is instantiated, and thus to measure the adoption of the ontology on the web.

There are several approaches for *task-dependent evaluations* [9,43,45], and *ontology ranking and selection* with regards to certain tasks [2, 40]. Sabou et al. [45] create custom-tailored ontologies on the fly from the formulation of a task [1] and evaluate them afterwards. Several problems are encountered, ranging from broken links to incompatible axioms due to different contexts and points of views. This strengthens the hypothesis that ontology evaluation will become increasingly important in the context of networked ontologies.

Domain completeness is given when an ontology covers the complete domain of interest. This can be only measured automatically if the complete domain is accessible automatically and can be compared to the ontology. A way to assess the completeness of an ontology with respect to a certain text corpus is to use ontology learning techniques to construct an ontology from the corpus [52]. The learned ontology is then compared to the evaluated ontology to see if the learned ontology covers knowledge not expressed in the evaluated ontology. Ontology learning is discussed in chapter "Ontology and the Lexicon".

Often ontologies are tightly interwoven with an application, so that the ontology cannot be simply exchanged. It may drive parts of the user interface, the internal data management, and parts of it may be hard-coded into the application. On the other hand, the user never accesses an ontology directly, but always through some application. Often the application needs to be evaluated with the ontology, regarding the ontology as merely another component of the used tool. Such a situation has the advantage that well-known software evaluation methods can be applied, since the system can be regarded as an integrated system where the fact that an ontology is used is of less importance.

11 Conclusions

In this chapter we discussed a number of evaluation methods for the different aspects of an ontology. It should be clear that none of these methods, neither alone nor in combination, can guarantee a good ontology. But they can at least recognize problematic parts of an ontology, i.e. they can tell you when an ontology is *not* good.

As we have seen, on the one hand, a number of quality criteria have been suggested in literature. On the other hand, a number of evaluation methods have been designed, implemented, and experimentally verified. Their actual relation is only badly understood and superficially investigated, if at all.

Whereas Gangemi et al. [17] discuss a number of quality criteria and how some of the metrics may serve as indicators for certain criteria, either positive or negative, they do not report on any experiments investigating the correlations between the methods and the criteria. Recent experiments actually point to some indeed counterintuitive relations, like a higher *tangledness* actually increasing efficiency when using the ontology in a browsing task [60]. The lack of such experimental evaluations matching methods and criteria will hinder meaningful ontology evaluations. Figuring out the effect of certain features on the quality of the ontology remains an open challenge for now.

It is obvious that domain- and task-independent evaluation techniques, as discussed here, provide some common and minimum quality level, but can only go a certain way. In order to properly evaluate an ontology, the evaluator always needs to come up with methods appropriate for the domain and task at hand, and decide on the relative importance of the evaluation criteria. But the minimum quality level discussed here will at least provide the ontology engineer with the confidence that they eliminated many errors and can publish the ontology.

It seems that a proper ontology evaluation – being able to tell that an ontology will be good for a certain task and in a certain environment – will remain a task for a human level intelligence. But the more problems ontology evaluation techniques can eliminate before-hand, the more will humans be able to concentrate on the tasks they are best at: comparing the formal specification in the computer with their human conceptualization, and thus enabling computers and humans to speak a common language.

Acknowledgements

The ideas presented in this chapter have been collected and explored in discussions with many others. I want to especially thank York Sure, Aldo Gangemi, Philipp Cimiano, Johanna Völker, Enrico Motta, Elena Paslaru Bontas Simperl, Marta Sabou, Markus Krötzsch, Anupriya Ankolekar, Boris Motik, Stephan Grimm, Hans-Jörg Happel, Malvina Nissim, Jos Lehmann, Valentina Presutti, Joey Lam, Christoph Tempich, Peter Haase, Dan Connolly, Rinke Hoekstra, Sebastian Rudolph, Chris Welty, Aleks Jakulin, and the participants of the EON 2006 workshop in Edinburgh, Scotland, and the EON 2007 workshop in Busan, South Korea.

References

1. Harith Alani. Position paper: Ontology construction from online ontologies. In Les Carr, David De Roure, Arun Iyengar, Carole A. Goble, and Michael Dahlin, editors, *Proceedings of the 15th International Conference on World Wide Web, WWW 2006, Edinburgh, Scotland, UK, May 23–26, 2006*, pages 491–495. ACM, New York, 2006.

2. Harith Alani and Christopher Brewster. Metrics for ranking ontologies. In Denny Vrandečić, Mari del Carmen Suárez-Figueroa, Aldo Gangemi, and York Sure, editors. *Proceedings of the 4th International Workshop on Evaluation of Ontologies for the Web (EON2006) at the 15th International World Wide Web Conference (WWW 2006)*, Edinburgh, Scotland, May 2006, pages 24–30.
3. Sean Bechhofer, Frank van Harmelen, Jim Hendler, Ian Horrocks, Deborah L. McGuinness, Peter F. Patel-Schneider, and Lynn Andrea Stein. OWL Web Ontology Language Abstract Reference, 2004. W3C Rec. 10 February 2004.
4. Dave Beckett. RDF/XML syntax specification (revised). W3C Recommendation, February 2004.
5. Tim Berners-Lee. Universal Resource Identifiers in WWW: A Unifying Syntax for the Expression of Names and Addresses of Objects on the Network as used in the World-Wide Web. Technical Report 1630, Internet Engineering Task Force, June 1994.
6. Tim Berners-Lee, James Hendler, and Ora Lassila. The Semantic Web. *Scientific American*, 284(5):34–43, 2001.
7. Tim Bray, Dave Hollander, Andrew Layman, and Richard Tobin. Namespaces in XML 1.0 (second edition), 2006. W3C Rec. 16 August 2006.
8. Tim Bray, Jean Paoli, C. Michael Sperberg-McQueen, Eve Maler, and Francois Yergeau. Extensible markup language (XML) 1.0 (fourth edition), 2006. W3C Rec. 16 August 2006.
9. Christopher Brewster, Harith Alani, Srinandan Dasmahapatra, and Yorick Wilks. Data-driven ontology evaluation. In *Proceedings of the Language Resources and Evaluation Conference (LREC 2004)*, pages 164–168, Lisbon, Portugal, 2004. European Language Resources Association.
10. Dan Brickley and Libby Miller. The Friend Of A Friend (FOAF) vocabulary specification, July 2005.
11. James Clark and Makoto Murata. RELAX NG Specification, December 2001. OASIS committee specification.
12. Mathieau d'Aquin, Claudio Baldassarre, Larian Gridinoc, Sofia Angeletou, Marta Sabou, and Enrico Motta. Characterizing knowledge on the semantic web with watson. In Denny Vrandečić, Raúl García-Castro, Asunción Gómez-Pérez, York Sure, and Zhisheng Huang, editors, *Proceedings of the Workshop on Evaluation of Ontologies and Ontology-Based Tools, 5th International EON Workshop (EON2007) at ISWC/ASWC'07*, pages 1–10, Busan, Korea, November 2007.
13. Li Ding, Tim Finin, Anupam Joshi, Rong Pan, R. Scott Cost, Yun Peng, Pavan Reddivari, Vishal Doshi, and Joel Sachs. Swoogle: A search and metadata engine for the semantic web. In *CIKM'04: Proceedings of the Thirteenth ACM International Conference on Information and Knowledge Management*, pages 652–659. ACM, New York, NY, 2004.
14. David C. Fallside and Priscilla Walmsley. XML schema part 0: Primer second edition, 2004. W3C Rec. 28 October 2004.
15. Christine Fellbaum. *WordNet: An Electronic Lexical Database (Language, Speech, and Communication)*. MIT, New York, 1998.
16. Roy Fielding, James Gettys, Jeffrey Mogul, Henrik Frystyk, Larry Masinter, Paul Leach, and Tim Berners-Lee. Hypertext Transfer Protocol – HTTP/1.1. RFC 2616, June 1999.
17. Aldo Gangemi, Carola Catenacci, Massimiliano Ciaramita, and Jens Lehmann. Ontology evaluation and validation: An integrated formal model

for the quality diagnostic task. Technical report, Laboratory of Applied Ontologies – CNR, Rome, Italy, 2005. `http://www.loa-cnr.it/Files/OntoEval4OntoDev_Final.pdf`.

18. Aldo Gangemi, Carola Catenaccia, Massimiliano Ciaramita, and Jos Lehmann. Qood grid: A metaontology-based framework for ontology evaluation and selection. In Denny Vrandečić, Mari del Carmen Suárez-Figueroa, Aldo Gangemi, and York Sure, editors. *Proceedings of the 4th International Workshop on Evaluation of Ontologies for the Web (EON2006) at the 15th International World Wide Web Conference (WWW 2006)*, Edinburgh, Scotland, May 2006, pages 8–15.
19. Asunción Gómez-Pérez. Ontology evaluation. In Steffen Staab and Rudi Studer, editors, *Handbook on Ontologies in Information Systems, First Edition*, International Handbooks on Information Systems, chapter 13, pages 251–274. Springer, Berlin, 2004.
20. Jan Grant and Dave Beckett. RDF test cases. W3C Recommendation, February 2004.
21. Stephan Grimm and Boris Motik. Closed world reasoning in the semantic web through epistemic operators. In Bernardo Cuenca Grau, Ian Horrocks, Bijan Parsia, and Peter Patel-Schneider, editors, *Second International Workshop on OWL: Experiences and Directions (OWLED 2006)*, Galway, Ireland, 2005.
22. Thomas R. Gruber. Towards principles for the design of ontologies used for knowledge sharing. *International Journal of Human-Computer Studies*, 43(5/6):907–928, 1995.
23. Michael Grüninger and Mark S. Fox. Methodology for the design and evaluation of ontologies. In *IJCAI95 Workshop on Basic Ontological Issues in Knowledge Sharing*, Montreal, 1995.
24. Peter Haase and Guilin Qi. An analysis of approaches to resolving inconsistencies in DL-based ontologies. In *Proceedings of International Workshop on Ontology Dynamics (IWOD'07)*, pages 97–109, June 2007.
25. Masahiro Hori, Jérôme Euzenat, and Peter F. Patel-Schneider. OWL Web Ontology Language XML presentation syntax, 2003. W3C Note 11 June 2003.
26. Matthew Horridge. The Protégé OWL unit test framework, 2005. Website at `http://www.co-ode.org/downloads/owlunittest/`.
27. Matthew Horridge, Nick Drummond, John Goodwin, Alan Rector, Robert Stevens, and Hai Wang. The manchester owl syntax. In *OWLED2006 Second Workshop on OWL Experiences and Directions*, Athens, GA, USA, 2006.
28. Ian Horrocks and Peter F. Patel-Schneider. Reducing OWL Entailment to Description Logic Satisfiability. *Journal of Web Semantics*, 1(4):7–26, 2004.
29. Aleks Jakulin and Dunja Mladenić. Ontology grounding. In *Proceedings of 8th International Multi-conference Information Society IS-2005*, pages 170–173, 2005.
30. Aditya Kalyanpur, Bijan Parsia, Evren Sirin, Bernardo Cuenca Grau, and James Hendler. Swoop: A web ontology editing browser. *Journal of Web Semantics: Science, Services and Agents on the World Wide Web*, 4(2):144–153, June 2006.
31. Graham Klyne and Jeremy Carroll. Resource Description Framework (RDF): Concepts and abstract syntax. W3C Recommendation 10 February 2004.
32. Konstantinos Kotis, George A. Vouros, and Jerónimo P. Alonso. HCOME: A tool-supported methodology for engineering living ontologies. *Semantic Web and Databases*, pages 155–166, 2005.

33. Markus Krötzsch, Denny Vrandečić, Max Völkel, Heiko Haller, and Rudi Studer. Semantic wikipedia. *Journal of Web Semantics*, 5:251–261, 2007.
34. Joey Lam. *Methods for Resolving Inconsistencies in Ontologies.* PhD thesis, University of Aberdeen, 2007.
35. Adolfo Lozano-Tello and Asunción Gómez-Pérez. OntoMetric: A method to choose the appropriate ontology. *Journal of Database Management Special Issue on Ontological analysis, Evaluation, and Engineering of Business Systems Analysis Methods*, 15(2), 2004.
36. Boris Motik and Ian Horrocks. Problems with OWL syntax. In *OWLED2006 Second Workshop on OWL Experiences and Directions*, Athens, GA, 2006.
37. Boris Motik, Ian Horrocks, and Ulrike Sattler. Adding Integrity Constraints to OWL. In Christine Golbreich, Aditya Kalyanpur, and Bijan Parsia, editors, *Third International Workshop on OWL: Experiences and Directions 2007 (OWLED 2007)*, Innsbruck, Austria, June 6–7, 2007.
38. Leo Obrst, Werner Ceusters, Inderjeet Mani, Steve Ray, and Barry Smith. The evaluation of ontologies. In Christopher J.O. Baker and Kei-Hoi Cheung, editors, *Revolutionizing Knowledge Discovery in the Life Sciences*, chapter 7, pages 139–158. Springer, Berlin, 2007.
39. Eyal Oren, Renaud Delbru, Michele Catasta, Richard Cyganiak, Holger Stenzhorn, and Giovannia Tummarello. Sindice.com: A document-oriented lookup index for open linked data. *International Journal of Metadata, Semantics and Ontologies*, 3(1), 2008.
40. Chintan Patel, Kaustubh Supekar, Yugyung Lee, and E. K. Park. OntoKhoj: A semantic web portal for ontology searching, ranking and classification. In *Proceedings of Fifth ACM International Workshop on Web Information and Data Management*, pages 58–61, New York, NY, 2003.
41. Peter Patel-Schneider, Patrick Hayes, and Ian Horrocks. OWL Web Ontology Language Abstract Syntax and Semantics, 2004. W3C Rec. 10 February 2004.
42. Addison Phillips and Mark Davis. Tags for Identifying Languages. Technical Report 4646, Internet Engineering Task Force, September 2006.
43. Robert Porzel and Rainer Malaka. A task-based approach for ontology evaluation. In Paul Buitelaar, Siegrfried Handschuh, and Bernardo Magnini, editors, *Proceedings of ECAI 2004 Workshop on Ontology Learning and Population*, Valencia, Spain, August 2004.
44. Sebastian Rudolph, Johanna Völker, and Pascal Hitzler. Supporting lexical ontology learning by relational exploration. In Uta Priss, Simon Polovina, and Richard Hill, editors, *Conceptual Structures: Knowledge Architectures for Smart Applications, Proc. ICCS 2007*, volume 4604 of *LNAI*, pages 488–491. Springer, Sheffield, UK, 2007.
45. Marta Sabou, Jorge Gracia, Sofia Angeletou, Mathieu d'Aquin, and Enrico Motta. Evaluating the semantic web: A task-based approach. In *Proceedings of the 6th International Semantic Web Conference (ISWC'07)*, pages 423–437, November 2007.
46. Leo Sauermann and Richard Cyganiak. Cool URIs for the semantic web. Interest group note, W3C, March 2008.
47. Andrea Schaerf. Reasoning with individuals in concept languages. *Data and Knowlegde Engineering*, 13(2):141–176, 1994.

48. Michael K. Smith, Chris Welty, and Deborah McGuinness. OWL Web Ontology Language Guide, 2004. W3C Rec. 10 February 2004.
49. Aaron Swartz. RDF site summary (RSS) 1.0, December 2000. Official specification.
50. Samir Tartir, I. Budak Arpinar, Michael Moore, Amit P. Sheth, and Boanerges Aleman-Meza. OntoQA: Metric-based ontology quality analysis. In *Proceedings of IEEE Workshop on Knowledge Acquisition from Distributed, Autonomous, Semantically Heterogeneous Data and Knowledge Sources*, pages 45–53, 2005.
51. Christoph Tempich, Helena Sofia Pinto, York Sure, and Steffen Staab. An argumentation ontology for DIstributed, Loosely-controlled and evolvInG Engineering processes of oNTologies (DILIGENT). In Asunción Gómez-Pérez and Jérôme Euzenat, editors, *Proceedings of the Second European Semantic Web Conference*, volume 3532, pages 241–256, Heraklion, Crete, Greece. Springer, Berlin, 2005.
52. Paola Velardi, Roberto Navigli, Alessandro Cucchiarelli, and Francesca Neri. Evaluation of OntoLearn, a methodology for automatic population of domain ontologies. In Paul Buitelaar, Philipp Cimiano, and Bernardo Magnini, editors, *Ontology Learning from Text: Methods, Applications and Evaluation.* IOS Press, Amsterdam, 2005.
53. Johanna Völker, Denny Vrandečić, and York Sure. Automatic evaluation of ontologies (AEON). In Yolanda Gil, Enrico Motta, V. Richard Benjamins, and Mark A. Musen, editors, *Proceedings of the Fourth International Semantic Web Conference (ISWC'05)*, volume 3729 of *LNCS*. Springer, Berlin, 2005.
54. Denny Vrandečić. Explicit knowledge engineering patterns with macros. In Chris Welty and Aldo Gangemi, editors, *Proceedings of the Ontology Patterns for the Semantic Web Workshop at the ISWC 2005*, Galway, Ireland, November 2005.
55. Denny Vrandečić and Aldo Gangemi. Unit tests for ontologies. In Mustafa Jarrar, Claude Ostyn, Werner Ceusters, and Andreas Persidis, editors, *Proceedings of the 1st International Workshop on Ontology Content and Evaluation in Enterprise*, LNCS, Montpellier, France, October. Springer, Berlin, 2006.
56. Denny Vrandečić and York Sure. How to design better ontology metrics. In Wolfgang May and Michael Kifer, editors, *Proceedings of the 4th European Semantic Web Conference (ESWC'07)*, Innsbruck, Austria. Springer, Berlin, 2007.
57. Denny Vrandečić, Johanna Völker, Peter Haase, Duc Thanh Tran, and Philipp Cimiano. A metamodel for annotations of ontology elements in OWL DL. In York Sure, Saartje Brockmans, and Jürgen Jung, editors, *Proceedings of the Second Workshop on Ontologies and Meta-Modeling*, Karlsruhe, Germany, October 2006. GI Gesellschaft für Informatik.
58. Taowei David Wang and Bijan Parsia. Ontology performance profiling and model examination: First steps. In Karl Aberer, Key-Sun Choi, Natasha Fridman Noy, Dean Allemang, Kyung-Il Lee, Lyndon J. B. Nixon, Jennifer Golbeck, Peter Mika, Diana Maynard, Riichiro Mizoguchi, Guus Schreiber, and Philippe Cudré-Mauroux, editors, *Proceedings of the 6th International Semantic Web Conference/2nd Asian Semantic Web Conference (ISWC/ASWC'07)*, volume 4825 of *Lecture Notes in Computer Science*, pages 595–608, Busan, Korea, November. Springer, Berlin, 2007.
59. Taowei David Wang, Bijan Parsia, and James A. Hendler. A survey of the web ontology landscape. In Isabel Cruz, Stefan Decker, Dean Allemang Chris Preist, Daniel Schwabe, Peter Mika, Michael Uschold, and Lora Aroyo, editors,

Proceedings of the Fifth International Semantic Web Conference (ISWC'06), volume 4273 of *Lecture Notes in Computer Science*, pages 682–694, Athens, Georgia, November 2006. Springer.

60. Jonathan Yu, James A. Thom, and Audrey Tam. Ontology evaluation using wikipedia categories for browsing. In *Proceedings of the ACM Sixteenth Conference on Information and Knowledge Management (CIKM)*, Lisboa, Portugal, November 2007. ACM.

Ontology Engineering Environments

Riichiro Mizoguchi and Kouji Kozaki

The Institute of Scientific and Industrial Research, Osaka University, Osaka, Japan, miz@ei.sanken.osaka-u.ac.jp, kozaki@ei.sanken.osaka-u.ac.jp

Summary. In this chapter we discuss trends of ontology engineering environments and their characteristics through comparison between some tools. After a summarization of the recent trends of them, the authors enumerate factors which characterize those environments. Then we take up OntoEdit, Hozo, WebODE, SWOOP and Protégé, and compare them according to the factors.

1 Introduction

In order to discuss ontology engineering environments, we first need to clarify what we mean by ontology engineering. Ontology engineering is a successor of knowledge engineering which has been considered as a key technology for building knowledge-intensive systems. Although knowledge engineering has contributed to eliciting expertise, organizing it into a computational structure, and building knowledge bases, AI researchers have noticed the necessity of a more robust and theoretically sound engineering which enables knowledge sharing/reuse and formulation of the problem solving process itself. Knowledge engineering technology has thus developed into "ontology engineering" where "ontology" is the key concept to investigate.

There is another story concerning the importance of ontology engineering. It is the Semantic Web. The Semantic Web strongly requires semantic interoperability among metadata which are made using semantic tags defined in different ontologies. The issue here is to build good ontologies to come up with meaningful sets of tags which are made interoperable by ontology alignment.

Although the importance of ontology is well-understood, it is also known that building a good ontology is a hard task. This is why there have been developed some methodologies for ontology development [Chapter 6, 9] and have been built a number of ontology representation and editing tools.

This chapter discusses factors of an ontology engineering environment thorough comparison of some tools. The purpose is not to rank them but to discuss characteristics of them intended to give a guideline for users to

S. Staab and R. Studer (eds.), *Handbook on Ontologies*, International Handbooks on Information Systems, DOI 10.1007/978-3-540-92673-3,

choose an appropriate tool for their purpose. While over a hundred tools are developed to date, because of the space limitation, this chapter takes up OntoEdit [22, 23], Hozo [12, 13], WebODE [1], SWOOP [10, 11] and Protégé [17] which cover a wide range of ontology development process rather than being single-purpose tools which are covered elsewhere. After discussing the recent trends of ontology engineering tools, the authors compare some of them.

2 Trends of Ontology Engineering Environment

In the 1990s, several ontology engineering environments, such as Ontolingua Server, WebOnto, Ontosaurus, have been developed as the advancement of ontology engineering. Reference [3] surveys features of six ontology development tools at that time and found all tools did not have common ontology representation language and they were implemented based on their own ontological theories and representation models.

In the 2000s, OIL, DAML and DAML+OIL, which are the predecessors of OWL [Chapter 4], were published, and ontology engineering tools for those languages were developed. The representatives of them are OilEd, OntoEdit, Protégé and so on. After RDF(S) [Chapter 3], and OWL were published, these tools supported them as well as many other tools did. In Ontology Tools Survey[1] on XML.com, 52 tools were listed at November 06, 2002, and 93 tools were listed at September 14, 2004. At the present, the authors could find about 150 ontology development tools on the web[2] (Table 1). This shows a rapid increase of ontology engineering environments. According to the observation of these tools, the authors summarize the trends of ontology engineering environments as follows:

Domain-specific environments: In several domains, such as the Semantic Web, bioinformatics, medical science, agent technology, and software development, ontology development tools specialized to each domain are developed. For instance, OBO-Edit and DAG-Edit are ontology editors for Gene Ontology (GO) in bioinformatics, CliniClue is an ontology browsing tool for SNOMED CT in the medical domain, and Zeus is an agent development tool kit including an ontology editing tool.

Integrated environment for ontology development and use: Several tools are developed as an integrated environment which supports all processes for ontology construction to use them for development of ontology-based applications. Such environments provide users with an ontology editor, an ontology management tool, API for ontologies and so on. For instance, IODT (IBM Integrated Ontology Development Toolkit) developed by IBM includes an Eclipse-based ontology-engineering environment and OWL

[1] http://www.xml.com/pub/a/2004/07/14/onto.html

[2] Some of them are listed in the web sites such as ESW Wiki SemanticWebTools (http://esw.w3.org/topic/SemanticWebTools), Ontology Tool Survey and so on.

Table 1. List of ontology engineering environments (portion)

@@name of tools	@@web site
Domain specific environments	
OBO-Edit	http://geneontology.sourceforge.net/
DAG-Edit	http://amigo.geneontology.org/dev/
CliniClue	http://www.cliniclue.com/
Zeus	http://labs.bt.com/projects/agents/zeus/
ArgoUML	http://argouml.tigris.org/
COE	http://cmap.ihmc.us/coe/
CoGui	http://www.lirmm.fr/cogui/
Cypher	http://www.monrai.com/products/cypher
Integrated environment for ontology development and use	
KAON2	http://kaon2.semanticweb.org/
IODT	http://www.alphaworks.ibm.com/tech/semanticstk
WSMO Studio	http://www.wsmostudio.org/
Supporting system for ontology development based on various techniques	
OntoBilder(OntoX, etc)	http://iew3.technion.ac.il/OntoBuilder/
OntoGen	http://ontogen.ijs.si/
DODDLE-OWL	http://doddle-owl.sourceforge.net/
commercial tools	
OntoStudio	http://www.ontoprise.de/
IODE	http://www.ontologyworks.com/
TopBraid	http://www.topbraidcomposer.com/

The detailed list is available at http://www.hozo.jp/OntoTools/

ontology storage with an inference system based on RDBMS. WSMO Studio is Eclipse-based integrated environments to edit the Semantic Web service for WSMO[3] (Web Service Modeling Ontology). It can be used with other tools for WSMO such as a reasoner, a validator, API for web services and so on.

Supporting system for ontology development based on various techniques: Many researchers propose methods to support ontology development based on various techniques such as Natural Language Processing, Machine Learning [Chapter 11], and Search Engine. For instance, OntoBilder supports ontology development by extracting terms form web pages, OntoGen is a semi-automatic ontology construction system based on machine learning and text mining algorithms, and DODDLE-OWL supports construction of domain ontology by extracting valuable information from existing lexical databases or ontologies such as Word-Net[4]. GINO (a guided input natural language ontology editor) uses controlled natural language to edit and query ontologies.

[3] http://www.wsmo.org/

[4] http://wordnet.princeton.edu/

Increase of commercial tools: Recently, commercial tools for ontology development are increasing. About 30 tools among 150 which the authors found are commercial software. Most of them support large scale construction for development of enterprise system. OntoStudio powered by Ontoprise is a successor of OntoEdit which was developed in the early days of the Semantic Web research. It supports RDF(S), OWL as ontology language and F-Logic for the rule processing. And it can connect to databases, file-systems, applications and web-serves thorough many connecters. Ontology Works provides integrated environments for ontology construction and uses such as modeling tools, databases server and information integration software. Their central technology is the Integrated Ontology Development Environment (IODE^{TM}). It supports construction and management of high-fidelity domain ontologies. TopBraid Composer^{TM} is eclipse-based platform for developing web ontologies and the Semantic Web applications. It supports the Semantic Web standards and other components for applications such as Geography and Location Mapping, Ontology-Driven Forms, UML-like Class Diagrams and so on.

3 Factors of an Ontology Engineering Environment

A comprehensive evaluation of ontology engineering tools is found in [3, 6] in which the major focus is put on static characteristics of tools. The evaluation in this chapter is done focusing on dynamic aspects of the tools. We consider that a lifecycle of ontology engineering process consists of ontology development phase, ontology use phase and ontology refinement and evaluation phase. We concentrate on characteristics of the ontology engineering process supported by the five environments. Let us enumerate factors by which an environment should be characterized for each phase.

Ontology development phase The first key task of ontology engineering is ontology construction. It includes constructions of class hierarchies, describing definitions of classes, defining relations between classes, and so on. Ontology engineering environments should support the process with the following characteristics.

Development methodology Though an ontology development requires a sophisticated development methodology, a methodology itself is not sufficient. Developers need an integrated environment which helps them build an ontology in every phase of the building process. In other words, a computer system should navigate developers in the ontology building process according to a methodology.

Collaborative development Building an ontology is often done with collaboration of multiple developers who need help in orchestration of the collaborative activities.

Compliance with an ontological theory (Theory-awareness) An ontology is not just a set of concepts but at least a "well-organized" set of concepts. An environment is expected to guide users to a well-organized ontology which largely depends on the environment's discipline of what an ontology should be rather than an ad hoc classification of concepts or a frame representation. This is why an environment needs to be compliant with a sophisticated theory of ontology.

Ontology use phase Ontology use is the other key task of ontology engineering. Users need also effective support in how to share ontology with others, how to use/reuse an ontology and how to build an instance model based on an ontology.

Compliance with WWW standard There are many languages standardized by W3C: XML, RDF(S), DAML+OIL and OWL, etc. The environment is required to be compliant with these.

Ontology/Model(instance) server Ontologies and instance models should be available through internet.

Ontology evaluation and refinement phase To construct a well-organized ontology, evaluation and refinement of the developed ontology are repeated many times. An environment should support the process.

Evaluation methodology Many theories and methods for ontology evaluation are discussed [Chapter 13]. Tools should support them.

Inference service An inference engine is used to check the consistency of ontologies/instances.

Refinement mechanism It is important to manage version of ontologies and its change histories for maintenance of the consistency of ontologies. Debugging mechanisms and suggestion for modification are also useful for refinement of ontologies.

Software level issues

Usability GUI as well as functionality is essential to the usability of the environment.

Architecture of the environment An environment should be designed in an advanced and sophisticated architecture to make it usable.

Extensibility It is good if users easily extend the environment.

4 OntoEdit

OntoEdit [22, 23], professional version, is an ontology engineering environment to support the development and maintenance of ontologies. Ontology development process in OntoEdit is based on their own methodology, On-To-Knowledge [Chapter 6] which is originally based on Common KADS methodology and consists of major three steps such as requirement specification, refinement and evaluation processes. The requirement specification consists of description of the domain and the goal of the ontology, design guidelines, available knowledge sources, potential users and use cases, and applications

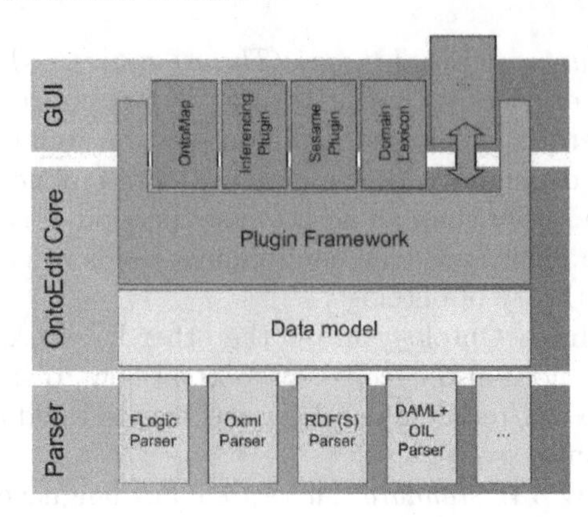

Fig. 1. Architecture of OntoEdit

supported by the ontology. The output of this phase is refined into a formal description in the next phase. Refinement is done usually collaboratively. In the evaluation phase, competency questions are used to evaluate if the ontology built can answer these questions.

Figure 1 shows the architecture of OntoEdit consisting of three layers: GUI, OntoEdit Core and Parser. It employs the plug-in architecture to make it easily extensible and customizable by the users. It is compliant with XML family standards in import and export the ontology. At the present, the technologies of OntoEdit are inherited to OntoStudio as a commercial tool. It has new features such as connectors to many kinds of resources, integrated rule management, mapping view between different ontologies and so on.

4.1 Ontology Development Phase

Requirement Specification Phase

Two tools, OntoKick and Mind2Onto, are prepared for supporting this phase of ontology capture. OntoKick is designed for computer engineers who are familiar with software development process and tries to build relevant structures for building informal ontology description by obtaining competency questions proposed in [8] which the resulting ontology and ontology-based applications have to answer. Examples of competency questions made by OntoKick include "which research groups exist at the institute?", "which teaching courses are offered by the insti-tute?", etc. Mind2Onto is a graphical tool for capturing informal relations between concepts. It is easy to use because it has a good visual interface and allows loose identification of relations between concepts. However, it is necessary to convert the map into a more formal organization to generate an ontology.

4.2 Ontology Evaluation and Refinement Phase

Refinement Phase [23]

This phase is for developers to use the editor to refine the ontological structure and the definition of concepts and relations. Like most of other tools, OntoEdit employs the client/server architecture where ontologies are managed in a server and multiple clients access and modify it. A sophisticated transaction control is introduced to enable concurrent development of an ontology in a collaborative manner. Because OntoEdit allows multiple users to edit the same class in an ontology at the same time, it needs a powerful lock mechanism of each class and devises Strict two Phase Locking protocol: S2PL to support arbitrary nested transactions.

Evaluation Phase

The key process in this phase is use of competency questions obtained in the first phase to see if the designed ontology satisfies the requirements. To do this, OntoEdit provides users with a function to form a set of instances and axioms used as a test set for evaluating the ontology against the competency questions. It also provides users with debugging tools for ease of identify and correct incorrect part of the ontology. It maintains the dependency between competency questions and concepts derived from them to facilitate the debugging process. This allows users to trace back to the origins of each concept. Another unique feature of this phase is that collaborative evaluation is also supported by introducing the name space so that the inference engine can process each of test sets given by multiple users. Further, it enables local evaluation corresponding to respective test sets followed by global evaluation using the combined test. Like WebODE, OntoEdit supports OntoClean [Chapter 9] methodology to build a better *is-a* hierarchy.

Inference

OntoEdit employs Ontobroker [2] and F-Logic[Chapter 2] as its inference engine. It is used to process axioms in the refinement and evaluation phases. Especially, it plays an important role in the evaluation phase because it processes competency questions to the ontology to prove that it satisfies them. It exploits the strength of F-logic in that it can express arbitrary powerful rules which quantify over the set of classes which Description logics cannot.

5 Hozo

Hozo[5] is an ontology engineering environment based on fundamental ontological theories [12,13]. It is composed of "Ontology Editor", "Onto-Studio," "Ontology Server" and "Ontology Manager." One of the most remarkable features of Hozo is that it can deal with Role based on a sophisticated ontological theory of Role [16].

When an ontology and its instance model seriously reflects the real world, users have to be careful not to confuse the Role such as teacher, mother, fuel, etc. with other basic concepts (natural type) such as human, water, oil, etc. Let us take an example: <*teacher is-a human*>. Assume John is a teacher of a school. Given the usual semantics of *is-a*, since John is an instance of teacher then he is also an instance of human at the same time. When he quits being a teacher, he cannot be an instance of teacher so that you need to delete the instance-of link between John and teacher. However, you have to restore an instance-of link between John and human, otherwise John dies. This problem would be difficult for a model with no idea of roles to represent changes in the roles played by John (e.g., teacher, husband, patient) according to contexts or aspects.

In Hozo, three different classes are introduced to deal with the concept of role appropriately.

Role-concept A concept representing a role dependent on a context (e.g., teacher role)

Basic concept A concept which does not need other concepts for being defined (e.g., human)

Role holder An entity of a basic concept which is holding the role (e.g., teacher)

A basic concept is used as the class constraint which indicates potential players who can play the role (role concepts). Then an instance that satisfies the class constraint plays the role and becomes a role holder. Hozo supports to define such a role concept as well as a basic concept.

5.1 Ontology Development Phase

Like other editors, Ontology Editor in Hozo provides users with a graphical interface through which they can browse and modify ontologies by simple mouse operations. How to deal with "role concept" and "relation" on the basis of fundamental consideration is discussed in [12]. This interface consists of the following four parts (Fig. 2):

1. *Navigation pane* provides several functionalities for browsing the ontology such as displaying the ontology in a hierarchical structure according to

[5] http://www.hozo.jp/

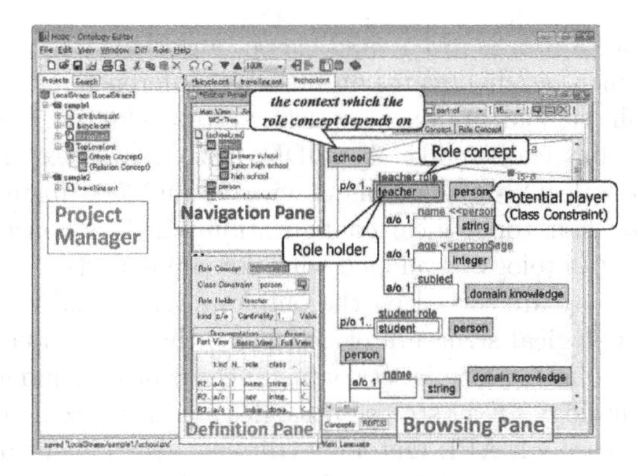

Fig. 2. GUI of ontology editor

only *is-a* relation between concepts, showing thumbnail of the ontology and searching for concepts.

2. *Browsing pane* displays the concept graphically, and the user can select concepts which he/she wants to edit.
3. *Definition pane* allows users to define and modify the selected concept in the browsing pane or in the *is-a* hierarchy browser.
4. *Project Manager* supports distributed development of ontologies.

Collaborative Development

Collaborative development of an ontology is supported in Hozo [21]. At the primitive level, the ontology server stores ontologies under version management and access control by lock/unlock mechanism. It allows users to sharing ontologies and to avoid conflict of modification by different users. Furthermore, Hozo allows users to divide an ontology into several components and manages the dependency between them to enable the concurrent development of the whole ontology. In the concurrent development, one of key issues is the maintenance of consistency among inter-dependent component ontologies. Hozo provides users with a module to maintain consistencies of the dependencies among ontologies. When a component ontology is updated, the system checks the change by comparing the modified ontology and its old version. Hozo shows users a list of changes with possible countermeasures for coping with each of the changes. These countermeasures are devised through our investigation on conceptual dependencies of ontologies and the change type of imported concepts.

5.2 Ontology Use Phase

Functionality and GUI of Hozo's instance editor is the same as the one for ontology. The consistency of all the instances with the ontology is automatically guaranteed, since a user is given valid classes and their slot value restrictions by the editor when he/she creates an instance. Inference mechanism of Hozo is not very sophisticated. Axioms are defined for each class but it works as a semantic constraint checker like WebODE. Hozo has an experience in modeling of a real-scale Oil-refinery plant with about 2000 instances including even pipes and their topological configuration which is consistent with the Oil-refinery plant ontology developed with domain experts [15]. The model as well as the ontology are served by the ontology server and can answer questions on the topological structure of the plant, the name of each device, etc. Any ontology can have multiple sets of instances which are independent of one another. The ontology server stores ontologies and instance models and serves them to clients through API. Ontology editor is also a client of the ontology server. The internal representation of Hozo is XML-based frame language and it generates RDF(S) and OWL code to export the ontology and instance.

6 WebODE

WebODE[6] [1] is a scalable and integrated workbench for ontology engineering and is considered as a Web evolution of ODE(Ontology Development Environment [4]). It supports building an ontology at the knowledge level, and translates it into different ontology languages. WebODE is designed on the basis of a general architecture shown in Fig. 3 and to cover most of the

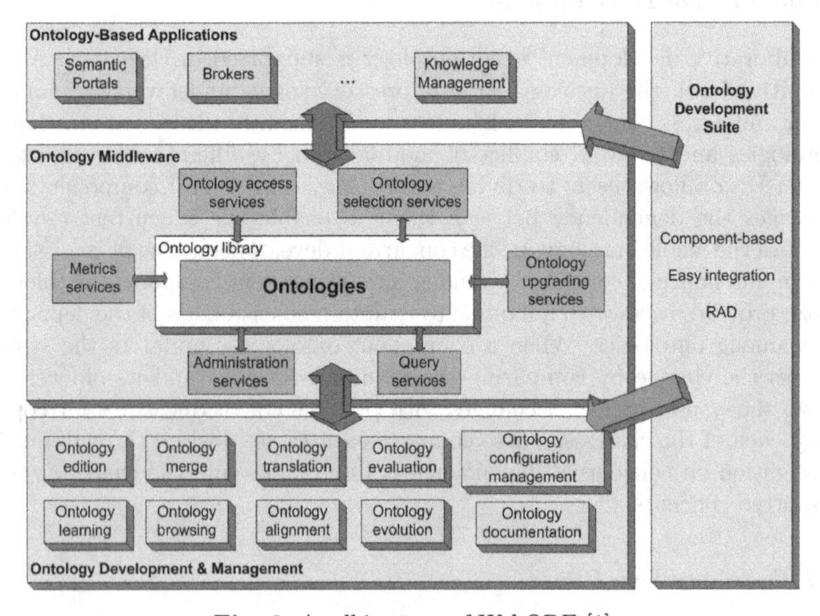

Fig. 3. Arcdhitecture of WebODE [1]

[6] http://webode.dia.fi.upm.es/WebODEWeb/index.html

processes appearing in the ontology lifecycle. WebODE is based on a client-server architecture which provides high extensibility and usability by allowing the addition of new services and the use of existing services. Ontology is stored in an SQL database to attain high performance in the case of a large ontology. It has export and import services from and into XML, and its translation services into and from various ontology specification languages such as OWL, RDF(S), OIL, DAML+OIL, UML, Prolog, X-CARIN, Jess and F-Logic. Like OntoEdit, WebODE's ontology editor allows the collaborative edition of ontologies. One of the most characteristic features of WebODE is that it is based on an ontology development methodology named METHONTOLOGY [4].

6.1 Ontology Development Phase

WebODE has ontology editing service, WAB: WebODE Axiom Builder service, inference engine service, interoperability service and ontology documentation service in this phase. The ontology editor provides users with form based and graphical user interfaces, WAB provides an easy graphical interface for defining axioms. It enables users to define an axiom by using templates given by the tool with simple mouse operations. Axioms are translated into Prolog. The inference engine is based on Prolog and OKBC protocol to make it implementation independent. Interoperability services provided by WebODE are of variety. It includes ontology access API, ontology export/import in XML-family languages, translation of classes into Java beans to enable Jess system to read them and OKBC compliance.

ODEClean [5]

Like OntoEdit, WebODE supports OntoClean methodology to build a more convincing *is-a* hierarchy. Ontology for OntoClean is composed of top level universal ontology developed by Guarino [Chapter 9], a set of meta-properties and OntoClean axioms which are translated into Prolog to be interpreted by WebODE inference engine. It is given to the ODEClean which works on the basis of it.

Collaborative Development

The collaborative editing of an ontology is supported by a mechanism that allows users to establish the type of access to the ontologies developed through the notion of groups of users. Synchronization mechanism is also introduced to enable several users to safely edit the same ontology. Ontologies are automatically documented in different formats such as HTML tables with Methontology's intermediate representations, concept taxonomies and XML.

6.2 Ontology Use Phase

To support the use process of ontology, WebODE has several functionalities. Like Hozo, it allows users to have multiple sets of instances for an ontology by introducing instance sets depending on different scenarios, and conceptual views from the same conceptual model, which allows creating and storing different parts of the ontology, highlighting and/or customizing the visualization of the ontology for each user. WebPicker is a set of wrappers to enable users to bring classification of products in the e-Commerce world into WebODE ontology. ODEMerge is a module for merging ontologies with the help of correspondence information given by the user. Methontology and ODE have been used for building many ontologies including chemical ontology [4].

WebODE is also used for developing some semantic web frameworks such as ODE SWS [7] and ODESeW. The ODE SWS is a framework for designing semantic web services at the knowledge level. It supports development of web services based on problem solving method ontology. The ODE SeW is a semantic web application framework to develop and manage web sites as a knowledge portal. In these ways, many applications have been developed using WebODE as workbench for ontology engineering.

7 SWOOP

SWOOP[7] [10,11] is an ontology browser and editor designed wholly for OWL, while many other tools (e.g., Hozo, WebODE and Protégé) support OWL as an extended feature. The architecture is based on the Model-View-Controller paradigm. SwoopModel component stores OWL ontologies loaded by a reasoner and other information related to them. They are visualized by renderers in multiple views. Controller is based on the plug-in architecture. Although the development of SWOOP is done by Mindswap project but has been terminated on August in 2006, its source code is available at URL.

7.1 Ontology Development Phase

A key feature of its design rationale is to realize user interface like the standard web browser. It consists of an address bar, history buttons (back, next), a navigation sidebar, bookmarks and so on (Fig. 4). In this GUI, URIs play a central role for understanding and constructing OWL ontologies. The users can load an OWL ontology by entering its URL in the address bar. If the ontology is importing other ontologies by owl:import property, SWOOP also loads the imported ontologies automatically. The loaded multiple ontologies are listed on the top of the navigation sidebar and their class/property hierarchies are shown at the bottom. The contents of selected ontology/entity (class,

[7] http://www.mindswap.org/2004/SWOOP/

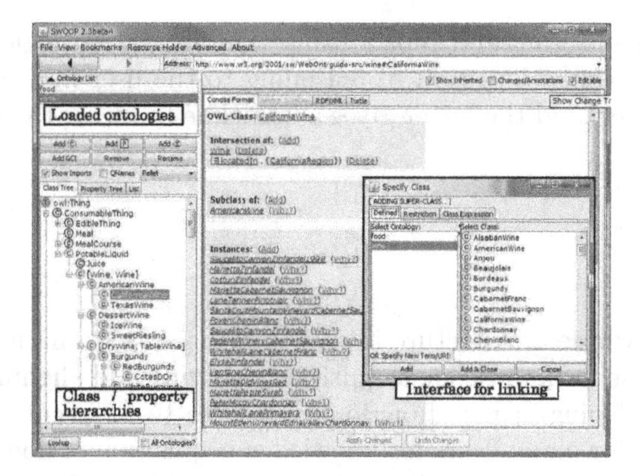

Fig. 4. Graphical User Interface of SWOOP

property and instance) are displayed on the center pane in the webpage-like format. In the pane, relationships between entities are represented by hyperlinks. It enables users to navigate the OWL ontology just like another web page. For linking entities in different ontologies, SWOOP provides a single common interface. It is displayed by clicking "Add" hyperlink in the center pane and shows the list of ontologies along with entities defined in them. Users can edit the ontology by selecting the entity to link. Through the editing process, external ontologies are modified as a local version and maintained separately. SWOOP also supports various presentation syntaxes for OWL such as RDF/XML, OWL Abstract Syntax and Turtle. Users can browse and edit[8] ontologies in these syntaxes.

Collaborative Development

For collaborative ontology development, SWOOP supports collaborative annotation and version control. The collaborative annotation is based on the standard W3C Annotea protocols. Users can share annotations about change of ontologies and discussions through a public Annotea server. The version control supports undo/redo with logging of changes and save of checkpoints. While the change logs can be used to track back the changes, the checkpoints are used as a snapshot of ontology at particular time.

7.2 Ontology Evaluation and Refinement Phase

SWOOP contains two reasoners: RDFS-like and Pellet. The former is a lightweight reasoner for RDFS, and the latter is a powerful reasoner for OWL-DL. Pellet is based on the tableaux algorithms and can be used to check inconsistencies of definition in ontologies. SWOOP provides functions

[8] Inline editing in RDF/XML and Turtle is supported by SWOOP ver.2.3.

for ontology debugging and repair using the description logic reasoner [10]. The former explains the result of reasoning to the user in a meaningful and readable manner, and the latter gives a guideline to repair the inconsistencies of ontologies.

8 Protégé

Protégé[9] [17] whose architecture is shown in Fig. 5 is strong in the use phase of ontology: Use for knowledge acquisition, merging and alignment of existing ontologies, and plug-in new functional modules to augment its usability. It has been used for many years for knowledge acquisition of domain knowledge and for domain ontology building in recent years. Its main features include:

1. Extensible knowledge model to enable users to redefine the representational primitives
2. A customizable output file format to adapt any formal language
3. A customizable user interface
4. Powerful plug-in architecture to enable integration with other applications

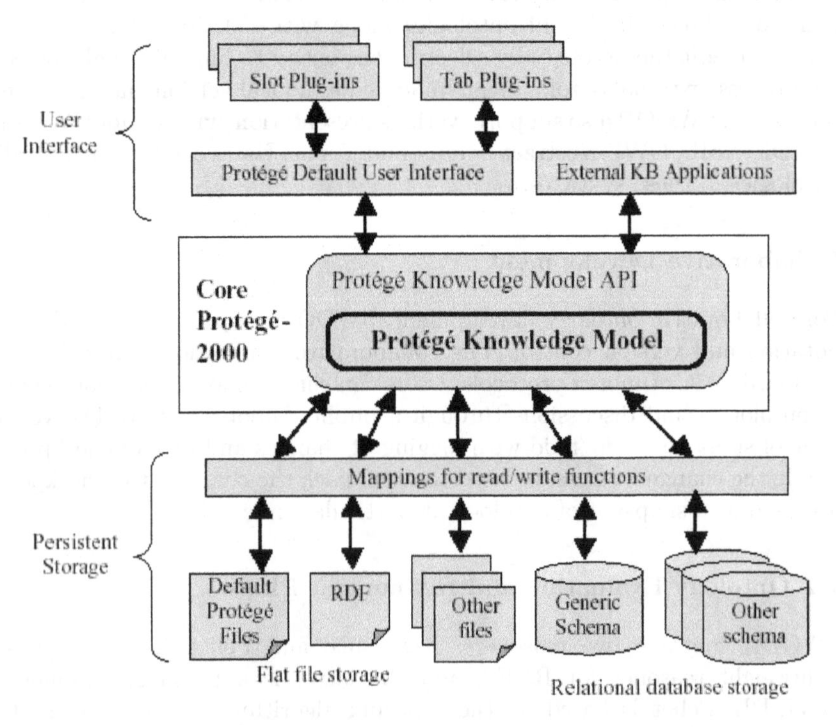

Fig. 5. Architecture of Protégé [17]

[9] http://protege.stanford.edu/

These features make Protégé a meta-tool for domain model building, since a user can easily adapt it to his/her own instance acquisition tool together with the customized interface. It is highly extensible thanks to its very sophisticated plug-in architecture and a Java-based API for development of knowledge based applications. Unlike the other three, Protégé assumes local installation rather than use through internet using client/server architecture. Its knowledge model is based on frame similar to other environments. Especially, the fact that Protégé generates its output in many ontology languages and its powerful customizability make it easy for users to change it to an editor of a specific language. For instance, the definition of a class of RDFS is defined as a subclass of standard class of Protégé. This "meta-tuning" can be easily done thanks to Protégé's declarative definition of all the meta-classes which play a role of a template of a class.

8.1 Ontology Development Phase

The system provides two main ways for ontology development such as Protégé-Frames and Protégé-OWL. The former supports frame-based knowledge model which is compatible to OKBC. And the latter is extensions of them using the Protégé OWL plug-in [14] for supporting OWL. It also supports to edit SWRL rules. It support owl:imports mechanism by ontology repository manager which manages URLs of imported ontologies.

Protégé has a semi-automatic tool for ontology merging and alignment named PROMPT [18]. It performs some tasks automatically and guides the user in per-forming other tasks. It also detects possible inconsistencies in the ontology, which result from the user's actions, and suggests ways to remedy them. For ontology evolution in collaborative environments [19], Protégé provides two functionalities: Change-management which stores a list of change with annotations and shows history of the change to the user, and Client-Server mode which support synchronous ontology editing by multiple users.

9 Comparison and Discussion

The five environments are compared according to the factors presented above. Table 2 summarizes the comparison.

9.1 Ontology Development Phase

Development Methodology

Philosophy of supporting ontology development is partly based on viewing an ontology as a software product. Common features of OntoEdit and WebODE

Table 2. Comparison of the five environments

	OntoEdit	Hozo	WebODE	SWOOP	Protégé
			Ontology development phase		
Methodological support	On-To-Knowledge	No	METHON TOLOGY	No	No
Collaboration support	Partly	Yes	Partly	Yes	Yes
Ontological theory	Ontoclean	Role theory	Ontoclean	Implicit	Implicit
			Ontology use phase		
Standards compliance	RDF(S), F-Logic	RDF(S), OWL (export only)	RDF(S),OWL F-Logic	RDF(S), OWL	RDF(S), OWL, SWRL
Ontology/ model server	High	High	High	Middle (Web server)	Middle
			Ontology evaluation and refinement phase		
Evaluation methodology	OntoClean	No	OntoClean	Debugging by reasoner	No
Inference service	OntoBroker	Constraint checking	Prolog, Jess	RDFS-like, Pellet	FaCT, Jess, F-Logic...
Refinement support	Debugging tool	Change checking	ODEClean	Debugging tool	PROMPT
			Software level issue		
Friendly GUI	GUI based editing	Graphically editing	GUI based editing	In line editting	GUI based editing
Architecture	Client /server	Client/ server	Client/ server	Standalone	Standa-lone
Extensibility	Plug-in	API	API/Plug-in	Plug-in	API/Plug-in

include management of the well-known steps in software development process, that is, requirement specification, conceptual design, implementation and evaluation. OntoEdit is based on the On-To-Knowledge methodology, and WebODE is based on the METHONTOLOGY. Others have no such a methodology.

Collaboration

Collaboration occurs in two different ways: (1) Construction a single ontology by different developers and (2) Construction several modularized ontologies in parallel by different developers. In the case of (1), because multiple persons might modify the same class at the same time, transaction control is one of the main issues in supporting collaboration. OntoEdit, WebODE, SWOOP and Protégé take this approach. While OntoEdit and WebODE only have some mechanisms for access management to ontologies, SWOOP and Protégé can manage histories of change with annotation for supporting collaborative construction.

On the other hand, Hozo mainly takes (2). Its main issue is to take care of the dependencies between the modularized ontologies because each developer constructs some modules under his responsibility. When building a large ontology, (2) is very useful because it allows users the concurrent development of an ontology like usual software development. To make the latter approach feasible, however, the system does need to provide developers with relevant information of changes done in other ontologies developed by others which might influence on the ontology they are developing. Hozo is designed to cope with all the possible situations developers encounter by analyzing possible patterns of influences propagated to each modularized ontology in a different module according to the type of the change. Although both approaches look different, they are complementary. The former can be incorporated in the latter. In fact, Hozo supports the former as well so that users can share a component ontology.

Theory-Awareness

Ontology building is not easy. This is partly because a good guideline is not available which people badly need when articulating the target world and organizing a taxonomic hierarchy of concepts. An ontology engineering environment has to be helpful also in this respect. WebODE and OntoEdit support Guarino's Ontoclean method. Guarino and his group have been investigating basic theories for ontology for several years and have come up with a sophisticated methodology which identifies inappropriate organization of *is-a* hierarchy of concepts. Developers who develop an ontology based on their intuition tend to misuse of *is-a* relation and to use it in more situations than are valid, which Guarino called "*is-a* overloading." Ontoclean is based on the idea of meta-property which contributes to proper categorization of concepts at the meta-level and hence to appropriate organization of *is-a* hierarchy.

OntoEdit and WebODE way of ontology cleaning can be said that post-processing way. On the contrary, Hozo tries to incorporate the fruits of ontological theories during the development process. One of the major causes of producing an inappropriate *is-a* hierarchy from Guarino's theory is lack of the concept of Role such as teacher, mother, food, etc. which has different characteristics from so-called basic concepts like human, tree, fish, etc. Ontology editor in Hozo incorporates a way of representing the concept of Role.

9.2 Ontology Use Phase

Standards Compliance

All the five have support standards ontology languages such as RDF(S) and OWL. Hozo only can export its ontology and model in RDF(S) and OWL. Protégé also supports Semantic Web Rule Language (SWRL).

Ontology/Model(Instance) Server

Hozo and WebODE has an ontology/model server which allows agents to access the ontologies and instance models through internet. OntoEdit and Protégé have an ontology server. SWOOP does not have a specific ontology server but can download ontologies in general web servers.

9.3 Ontology Evaluation and Refinement Phase

Evaluation Methodology

OntoEdit and WebODE support OntoClean methodology to build a better *is-a* hierarchy. SWOOP provides functions for ontology debugging and repair using the tableaux based reasoning. It explains the result of reasoning with a guideline to repair the inconsistencies of ontologies. Hozo and Protégé have no evaluation methodology.

Inference Service

All the five have inference mechanisms. Hozo supports only inference for constraint checking of own language.

Refinement Support

OntoEdit and WebODE have a debugging tool based on OntoClean. Hozo has a function to Check changes and suggest countermeasures for modification by comparison of ontologies. SWOOP provides a debugging tool based on reasoning and change management and a version control mechanism with logging of changes. Protégé also supports change monument and a semi-automatic ontology alignment tool named PROMPT.

9.4 Software Level Issue

Friendly GUI

All the five have sophisticated GUI such as visualization of class hierarchies and editing tool for constraints (axioms) of classes. It makes users free from coding using a complicated language. In Hozo, visualization of an ontology is default, and users can browse and edit it graphically. SWOOP supports in-line/GUI based editing functions and visualization of ontologies as just like a web page. Others provides mainly GUI based editing functions with some graphical visualization tool of ontologies. For instance, Protégé supports several ontology visualization pulig-ins such as OWL-Viz and Jambalaya.

Architecture and Extensibility

While WebODE and Hozo employ standardized API to the main ontology base, OntoEdit and SWOOP supports a plug-in architecture. Protégé provides both of API and plug-in. Both enable a module can easily added or deleted to make the environment extensible. WebODE, OntoEdit and Hozo are web-based, while Protégé is basically not. But Protégé has another mode to support server-client architecture.

10 Other Environments

In this section, other environments are summarized. We discuss four tools which have characteristic functionalities to aid user's ontology development.

10.1 OntoGen

OntoGen[10] is a system for semi-automatic ontology construction developed under SEKT project. The system has functionalities for keywords extraction form text data and suggestion using text mining and machine learning techniques. It helps users construct overview of ontologies from text documents. Some features of this tool are as follows:

Keywords extraction: In the system, two keyword extraction methods, the concept's centroid model and SVM linear model, are implemented. The extracted keywords by both methods are shown with related information about the number of related documents, average inner-cluster similarity measure, and so on.

Concept suggestion: The system suggests sub-concepts of the selected concept to the user based on two different approaches: an unsupervised approach and a supervised one. In the unsupervised approach, OntoGen supports four clustering methods: k-means, LSI, PH k-means, and a categorization according to the labels in the input data. The user can select one of the methods, and supervises the parameters for the method. Then the system suggests sub-concepts using the selected method with the parameters. The supervised approach is based on SVM active learning method. In the approach, the user enters a query the active learning system then the system asks if a particular document (instance) belongs to the selected concept and the user answers yes or no. After repetition of this learning process the system outputs most important keywords with information about positively classified into the concept.

Document management: The system manages documents related to concepts. When a new concept is added to ontology, it automatically assigns documents to it according to the similarity between documents. The system

[10] http://ontogen.ijs.si/

also has a functionality to detect if documents related to a concept belongs to its super concept. The user can know inconsistency between ontologies through the result.

10.2 CampTools Ontology Editor

CampTools ontology editor (COE) [9] is a tool for collaborative ontology development and reuse based on CampTools. CmapTools is software to construct, navigate and navigate a Concept map which is a knowledge representation model to display knowledge as a two-dimensional network of labeled nodes and links. COE supports editing, storing, and sharing Concept maps and ontologies, and the users can search for concepts and properties in them. The system also provides a cluster-based vicinity concepts view. In the view, the system shows concepts which relevant to selected concept based on multi-viewpoint clustering analysis (MVP-CA) software developed by Pragati, Inc..

10.3 OntoBilder

OntoBuilder [20] is a tool for extraction and matching of ontologies from web sources. The system extracts HTML form elements of web pages and relationships among them. A set of terms (vocabulary) associated with the extracted elements are regarded as an ontology in the system. The main feature of OntoBuilder is functionalities for ontology matching. It supports several matching algorithm such as term matching, value matching, precedence matching and so on. And the user can add another matching algorithm as plug-in.

10.4 KAON

KAON[11]: Karlsruhe Ontology and Semantic Web framework is a sophisticated plug-in framework with API and provides services for ontology and metadata management for E-Services. Its main focus is put on the enterprise application in the semantic web age. At the present, its new version, called KAON2, is available. It supports OWL-DL, SWRL and F-logic, and provides an API for ontology, an inference engine for answering SPARQL queries, a DIG interface and so on.

11 Concluding Remarks

A lot of ontology engineering environments have been developed. Although some are powerful as a software tool, but many are passive in the sense that few guidance or suggestion is made by the environment. Theory-awareness should be enriched further to make the environment more sophisticated. Especially,

[11] http://kaon.semanticweb.org/, http://kaon2.semanticweb.org/

more effective guidelines for appropriate class and relationship identification are needed. Collaboration support becomes more and more important as ontology building requirements increases. Ontology alignment is also crucial for reusing the existing ontologies and for facilitating their interoperability. Combination of the strong functions of each environment of the five would realize a novel and better environment, which suggests that we are heading right directions to go.

References

1. Corcho O, Fernandez-Lopez M, Gómez-Pérez A, Vicente O (2002) WebODE: An Integrated Workbench for Ontology Representation, Reasoning and Exchange, Proc. of EKAW2002, Springer LNAI 2473: 138–153.
2. Decker S, Erdmann M, Fensel D, Studer R (1999) Ontobroker: Ontology Based Access to Distributed and Semi-structured Information: 351–369.
3. Duineveld A, Weiden M, Kenepa B, Benjamis R (1999) WonderTools? A Comparative Study of Ontological Engineering Tools. Proc. of KAW99.
4. Fernandez-Lopez M, Gómez-Pérez A, Pazos Sierra J (1999) Building a Chemical Ontology Using Methontology and the Ontology Design Environment, IEEE Intelligent Systems, 14(1): 37–46.
5. Fernandez-Lopez M, Gómez-Pérez A, (2002) The Integration of OntoClearn in WebODE, Proc. of the 1st Workshop on Evaluation of Ontology-based Tools (EON2002): 38–52.
6. Gómez-Pérez A, Angele J, Fernandez-Lopez, et al. (2002) A Survey on Ontology Tools. OntoWeb Deliverable 1.3, Universidad Politecnia de Madrid.
7. Gómez-Pérez A, González-Cabero R, Manuel Lama (2004) ODE SWS: A Framework for Designing and Composing Semantic Web Services, IEEE Intelligent Systems 19(4): 24–31.
8. Gruninger R, Fox M (1994) The Design and Evaluation of Ontologies for Enterprise Engineering, Proc. of Comparison of implemented ontology: 105–128.
9. Hayes P, Eskridge T C, Mehrotra M, et al. (2005) COE: Tools for Collaborative Ontology Development and Reuse, Proc. of K-CAP2005.
10. Kalyanpur A, Parsia B, Sirin E, Cuenca-Grau B, Hendler J (2005) Swoop: A 'Web' Ontology Editing Browser, Journal of Web Semantics 4(2).
11. Kalyanpur A, Parsia B, Sirin E, Hendler J (2005) Debugging Unsatisfiable Classes in OWL Ontologies, Journal of Web Semantics 3(4).
12. Kozaki K, et al. (2000) Development of an Environment for Building Ontologies Which is Based on a Fundamental Consideration of "Relationship" and "Role", Proc. of the Sixth Pacific Knowledge Acquisition Workshop (PKAW2000): 205–221.
13. Kozaki K, Kitamura Y, Ikeda M, Mizoguchi R (2002) Hozo: An Environment for Building/Using Ontologies Based on a Fundamental Consideration of "Role" and "Relationship", Proc. of EKAW2002: 213–218.
14. Knublauch H, Fergerson R W, et al. (2004) The Protégé OWL Plugin: An Open Development Environment for Semantic Web Applications, Proc. of ISWC 2004: 229–243.
15. Mizoguchi R, Kozaki K, Sano T, Kitamura Y (2000) Construction and Deployment of a Plant Ontology, Proc. of EKAW2000: 113–128.

16. Mizoguchi R, et al. (2007) A Model of Roles in Ontology Development Tool: Hozo, Journal of Ontological Analysis and Conceptual Modeling, 2(2): 159–179.
17. Musen M A, Fergerson R W, Grosso W E, Crubezy M, Eriksson H, et al. (2003) The Evolution of Protégé: An Environment for Knowledge-Based Systems Development, International Journal of Human–Computer Studies, 58(1): 89–123.
18. Noy N F, Musen M A (2000) PROMPT: Algorithm and Tool for Automated Ontology Merging and Alignment, Proc. of the Seventeenth National Conference on Artificial Intelligence (AAAI-2000): 450–455.
19. Noy N, Chugh A, Liu W, Musen M (2006) A Framework for Ontology Evolution in Collaborative Environments, Proc. of ISWC2006, LNCS 4273: 544–558.
20. Roitman H, Gal A (2006) OntoBuilder: Fully Automatic Extraction and Consolidation of Ontologies from Web Sources Using Sequence Semantics. Proc. of EDBT Workshops 2006: 573–576.
21. Sunagawa E, Kozaki K et al. (2003) An Environment for Distributed Ontology Development Based on Dependency Management, Proc. of ISWC2003: 453–468.
22. Sure Y, Staab S, Angele J (2002) OntoEdit: Guiding Ontology Development by Methodology and Inferencing, Proc. of the Confederated International Conferences CoopIS, DOA and ODBASE: 1205–1222.
23. Sure Y, Staab S, Erdmann M, et al. (2002) OntoEdit: Collaborative Ontology Development for the Semantic web, Proc. of ISWC2002: 221–235.

Exploring the Economical Aspects of Ontology Engineering

Elena Simperl[1] and Christoph Tempich[2]

[1] STI Innsbruck, University of Innsbruck, Austria, elena.simperl@sti2.at
[2] Institute AIFB, University of Karlsruhe, 76128 Karlsruhe, Germany, tempich@aifb.uni-karlsruhe.de

Summary. A core requirement for the usage of ontologies within enterprizes is the availability of proved and tested techniques which guarantee an efficient engineering of high-quality ontologies, be that by reuse, manual building or automatic knowledge acquisition. Besides feasible technological support this includes in equal measure integrating ontology engineering within the more general framework of enterprize information architectures, and taking into account the economics of ontology engineering projects, in particular issues of cost effectiveness and profitability. This chapter addresses these two aspects. We discuss the role of ontology engineering in the context of enterprize architectures, arguing for the importance of cost-related measures as decision support in planning and controlling. Then we analyze approaches for reliably assessing the costs of building ontologies, and the usage of cost-related information to quantifiably support decisions arising during the life cycle of an ontology and to optimize the operation of associated processes. We account for the similarities and differences between software and ontology engineering in order to establish the appropriateness of applying methods with a long-standing tradition in this adjacent engineering field to ontologies. Building upon the results of this analysis we introduce ONTOCOM as the first cost model for ontologies and discuss different methods to improve its accuracy for a wide range of ontology engineering projects at public and corporate level.

1 Introduction

The dissemination of ontologies and ontology-based applications within enterprizes requires methods and tools which are able to deal with both the technical and the economic challenges of ontology engineering . In order for ontologies to be efficiently built and deployed at large scale one needs mature technologies supporting the entire ontology, as well as proved and tested means to plan and control the overall engineering process as part of more general IT-related processes within an enterprize. A wide range of ontology engineering methodologies have emerged in the Semantic Web community. Apart from minor differences in the level of detail adopted for the description of the

S. Staab and R. Studer (eds.), *Handbook on Ontologies*, International Handbooks on Information Systems, DOI 10.1007/978-3-540-92673-3,

process model they define ontology engineering as an iterative process, which shows major similarities to models emerged in the neighbored research field of software engineering. However existing methodologies do not consider the economic factors commonly related to every real-world engineering project, in particular the estimation of development and maintenance costs using predefined cost models, and the impact of such cost information on the operation of an engineering process.

With ONTOCOM we present the first existing approach in this newly emerging field of ontology engineering [17]. Estimating costs for ontology engineering is similar to estimating costs for software (or product) engineering as it requires the consideration of economic aspects for generic products and the processes they result of. Therefore, our approach largely benefits from the experiences made in estimating costs for software engineering. Through expert interviews we identified the most relevant cost drivers for a wide class of ontology engineering projects. In a large user study we acquired relevant data from existing ontology engineering projects and calibrated a parametric cost prediction equation with promising results. Further on, we analyzed the appropriateness of applying alternative approaches such as estimation by analogy and the Delphi method in the same context. Combing the three we were able to identify dimensions for further research and development towards a methodology for the creation of any kind of cost estimation model for ontologies, independently of the ontology life cycle, the cost estimation method, or the organizational setting it might be employed.

Cost estimation methods have a long-standing tradition in more mature engineering disciplines such as software engineering or industrial production [7,9,11,18,20,23]. Although the importance of cost issues is well-acknowledged in the community, as to the best knowledge of the authors, no cost estimation model for ontology engineering has been published so far. Analogue models for the development of knowledge-based systems (e.g., [8]) implicitly assume the availability of the underlying conceptual structures. Reference [15] provides a qualitative analysis of the costs and benefits of ontology usage in application systems, but does not offer any model to estimate the efforts. Reference [5] presents empirical results for quantifying ontology reuse. Reference [14] adjusts the cost drivers defined in a cost estimation model for Web applications w.r.t. the usage of ontologies. The cost drivers, however, are not adapted to the requirements of ontology engineering and no evaluation is provided. We present an evaluated cost estimation model, introducing cost drivers with a proved relevance for ontology engineering, which can be applied in early stages of arbitrary ontology development processes and further customized to specific needs of these processes to improve its prediction quality.

The outline of this chapter is as follows. We start by reviewing the economical aspects of ontology engineering in the context of corporate IT, motivating the need for cost estimation models in this field in Sect. 2. Provided reliable means to estimate costs during particular stages of the life cycle of an ontology, Sect. 3 introduces a series of use cases showing how cost information could be utilized to optimize the operation of an ontology-related project. The remaining sections are dedicated to the design and evaluation of a cost model for ontologies. We first analyze potential methods for cost prediction in Sect. 4. Section 5 introduces the ONTOCOM model based on the previously identified most promising methods. Details about its application in concrete ontology engineering projects are provided in Sect. 6. Section 7 summarizes the lessons learned from our research and the planned future work.

2 Economical Aspects of Information Technology

In this section we situate ontology engineering in the IT landscape of an enterprize. We discuss how the development and deployment of ontologies influences the enterprize information architecture and analyze the most important economical aspects related to this setting. As a result we argue for the necessity of reliable instruments for cost prediction for ontology engineering, which is an essential part of the data architecture of an enterprize.

2.1 Enterprize Information Architecture

An enterprize information architecture includes the products, procedures, organizational structures and IT systems of an enterprize. The design of enterprize architectures and their continuously adaptation to new environmental requirements are realized with the help of so-called *architecture development frameworks* such as the Zachmann framework [26] or the TOGAF framework.[1] They provide a comprehensive approach including methodologies, tools, best practices and standardized guidelines to develop a broad range of different IT architectures in an enterprize.

According to the latter an enterprize information architecture is typically modeled iteratively at four levels: Business, Application, Data, and Technology. The design process starts at the business level with the definition of a business strategy, followed by the specification of requirements which lead to an overall architecture vision and to a business, product and process architecture. The following steps are more technically oriented. IT specialists design the application architecture, the actual technology architecture implementing the application, and the *information system/data architecture*, which refers to the data models, or in our case ontologies, used in the application (cf. Fig. 1).

So far the ontology engineering research community has focused mainly on methods and tools for building and managing ontologies without reflecting on the implications arising from enterprize architecture development processes.[2]

[1] `http://www.opengroup.org/togaf/`

[2] An exception from this statement is the IDEF integrated definition method. The IDEF approach defines a function modeling method (IDEF0), information

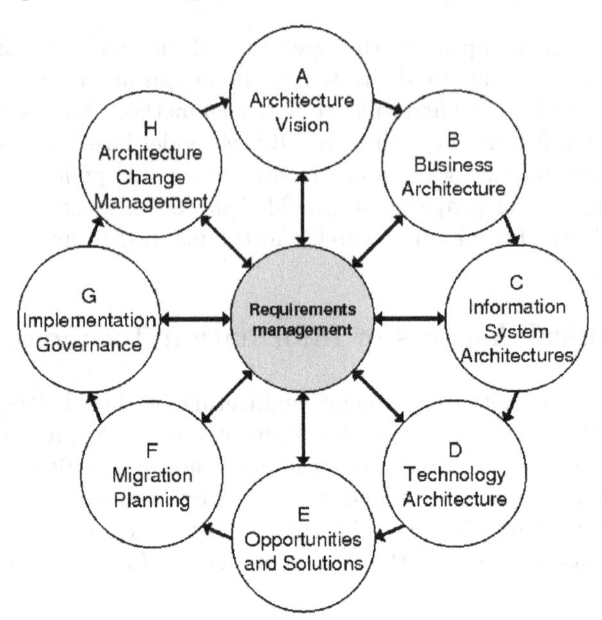

Fig. 1. Enterprize information architecture according to the TOGAF framework. Ontology engineering is related to (C) Information System Architectures

These implications are twofold: on the one hand the development of an enterprize information architecture results, among other things, in a series of domain and functional requirements regarding the scope and the utilization of the ontology; on the other hand this imposes non-functional boundaries such as the maximal development effort to be invested in the realization of the ontology. Consequently the requirements in the first category need to be matched to the estimated costs related to the development and the maintenance of the ontology.

2.2 Governance

A second essential aspect in the context of an enterprize IT infrastructure is the operation and maintenance of the underlying IT systems, tasks typically accomplished according to a governance framework evaluated and approved by dedicated institutes such as COBIT.[3] Governance frameworks cover the organization, control, steering and diffusion of corporate IT system development:

modeling method (IDEF1), data modeling method (IDEF1X), process description capture method (IDEF3), object oriented design method (IDEF4), and finally the ontology description capture method (IDEF5), which can be mapped to the aforementioned architectural models foreseen by enterprize architecture development frameworks. More information about IDEF is available at `http://www.idef.com/`

[3] `http://www.itgi.org/`

Organization Organizational aspects describe the different roles relevant for the development of an IT system in an organization, their responsibilities and decision procedures.

Steering Steering includes the definition of processes and activities in which the participants act in order to achieve the overall goals. The handling of intellectual property rights is specified as well in this step. This is an important aspect for ontology engineering as a systematic use of ontologies is an important pre-requisite for achieving application interoperability.

Control Control covers the definition of indicators in order to monitor the processes defined in the previous step and be able to detect unintended consequences of a system development or operation. From an ontology engineering perspective this step includes metrics for the characterization of ontologies and the associated development and maintenance processes.

Diffusion Finally, one of the most important aspects in large organizations is the definition of appropriate roll-out mechanisms in order to guarantee that the whole organization is able to follow the proposed processes.

In our work we focus on cost effectiveness, as one of the most important indicators at controlling level in an governance framework. Obviously the costs associated to the development of an IT system, of which ontology engineering is an important part, should be lower than the benefit expected to be obtained through its deployment. The estimation of the efforts related to engineering an ontology is crucial for the planning of data architecture change projects. Taking into account the operation of IT systems, it is worthwhile to pay attention to the reusability of an ontology. A reusable ontology is likely to be more user-friendly, thus reducing the training effort required in the roll-out phase of the associated IT system, while typically involving additional effort in the development. For monitoring purposes it is furthermore interesting to compare the estimated and the actual effort values in order to judge the ability of the organization to develop ontologies.

Provided these various usage scenarios, we argue for the necessity of extending existing IT-specific cost estimation approaches towards ontologies. In the next section we provide a second motivating scenario for this requirement, explaining how knowledge about the efforts implied by the development, maintenance and deployment of an ontology influences its life cycle.

3 Usage of Cost Information During the Life Cycle of an Ontology

The previous section was concerned with the relevance of cost information in the IT organization from a management perspective. This section focuses on the implementation side of ontologies, and thus on the relevance of cost information from a development and maintenance perspective. Figure 2 gives an overview how cost information is typically used within the life cycle of a

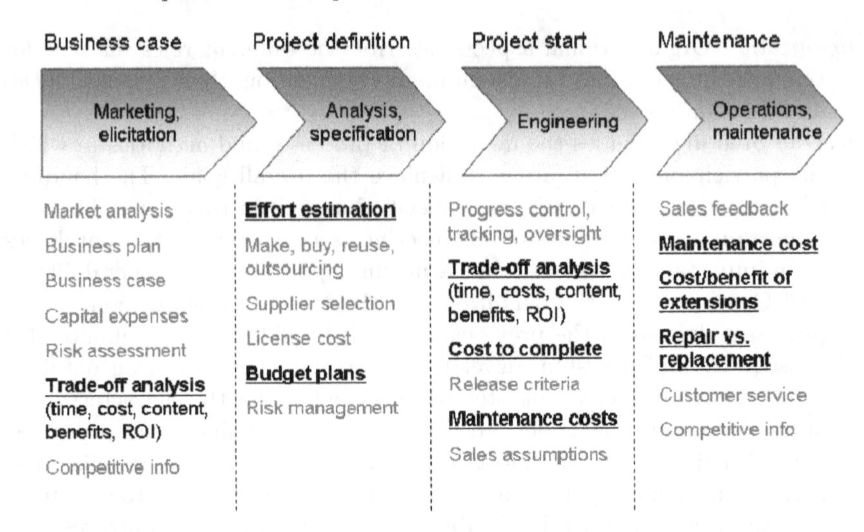

Fig. 2. Cost indicators during the life cycle of a product cf. [6]

product. Most product development projects start with the elaboration of a business case, of which the trade-off analysis is a major part. The trade-off analysis compares the expected benefit with the expected development costs. Turning back to IT systems, in order to make valid decisions based on the trade-off analysis, an accurate estimate of the expected costs and benefits of ontologies within the context of a particular application scenario is thus required in such early phases of a project.

The costs may be estimated with methods proposed in this chapter. The quantification of the benefits are covered by recent proposals to value IT development projects[4] which could be transferred to ontology development.

The effort related to ontology engineering is relevant in the initial project *planning* phase. In this phase the project manager assigns available resources to the planned tasks, taking into account the estimates of the effort associated to them, which are crucial for an on-time delivery of the project outcomes. These estimates are updated during the remaining project phases, as at the beginning of the project information about available skills and other influencing factors might not be available or reliable. Re-estimations during the project allow for adjustments in the project plan and provide a basis for the calculation of the total resources necessary to complete the project. In terms of ontology engineering the availability of cost information helps to make decisions related to the expected quality, size and granularity of the ontology to be developed. This is particularly of interest for ontology engineering methodologies following an iterative or rapid prototyping approach.

Effort estimates can be used in equal measure for *controlling* and *benchmarking* purposes. As such estimates are derived from previous project

[4] `http://www.isaca.org/Content/ContentGroups/Val_IT1/Val_IT.htm`

experiences, the comparison between planned and actual effort values is a benchmark against previous projects, either external or internal. If the variation exceeds certain thresholds the project manager can take early countermeasures or, at least, can thoroughly examine the project and detect the underlying reasons for the variation.

Towards the end of the ontology life cycle effort estimations for ontology *maintenance* can support repair versus replacement decisions. This of course requires knowledge about the total cost of ownership of an ontology and about the cost statements with respect to particular ontology management activities.

To summarize cost estimation models for ontologies are necessary in order to align the discipline of ontology engineering with common IT practice within enterprizes, as well as for shaping the life cycle of ontologies in an economically based fashion. The remaining sections describe how such cost models can be designed, evaluated and used in an organization.

4 Design of an Ontology Cost Estimation Model

In order to design a cost estimation model for ontologies we can resort to established approaches in the field of software measurement, which describe the steps required for this purpose and the way they can be effectively performed within a company (cf. Fig. 3).

Extract In this step the engineering team identifies the cost factors and decides upon the method(s) used to generate the estimates.

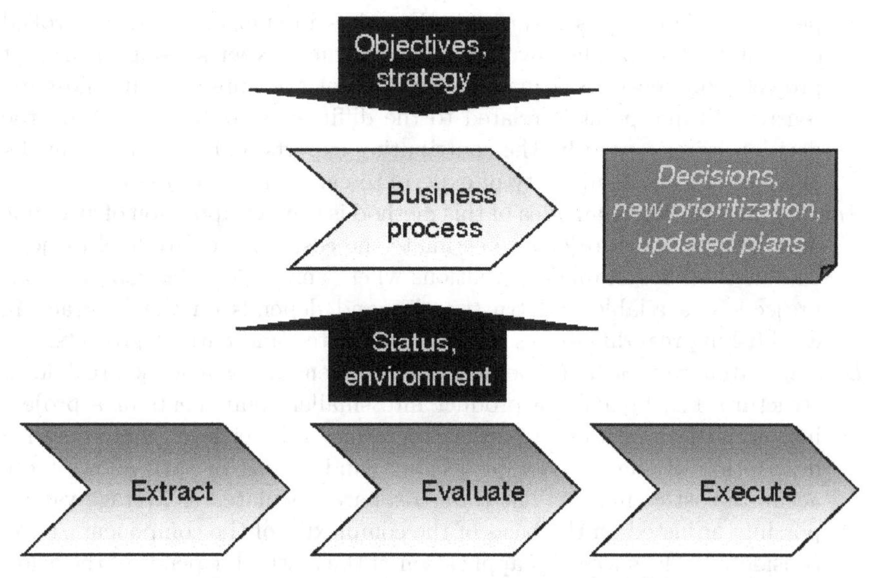

Fig. 3. General measurement process cf. [6]

Evaluate The model previously defined is evaluated and adapted according to a specific procedure (see below). For expert-based methods, the evaluation includes evaluation sessions among the participants. In case of mathematical prediction equations the evaluation uses data collected for this purpose from previous projects. In the latter case it is essential that the evaluation relies on a sufficient amount of historical data *from internal projects* in order to customize the model to the particularities of a given enterprize or department.

Execute Once a feasible quality of the predictions has been achieved the model is used at various stages of the life cycle of a product (in our case an ontology) and in relation to the more general enterprize architecture development process. In this context it is important that the employees understand the necessity of this additional workload and that they are trained to correctly use the model.

We now turn to a description of the first step. Information about the validation of a particular model (in our case based on a parametric approach) are available in [17].

4.1 Generic Methods

Estimating costs for engineering processes can be performed according to several methods, often used in conjunction in order to avoid individual limitations [1, 22].

Expert judgment/Delphi method The Delphi method is based on a structured process for collecting and distilling knowledge from a group of human experts by means of a series of questionnaires interspersed with controlled opinion feedback. The involvement of human experts using their past project experiences is a major advantage of the approach. Its most extensive critique point is related to the difficulties to explicitly state the decision criteria used by the contributing experts and to its inherent dependency of the availability of experts to carry on the process.

Analogy method The main idea of this method is the extrapolation of available data from similar projects to estimate the costs of the proposed project. The method is suitable in situations where empirical data from previous projects is available and trustworthy, and depends on the accuracy in establishing real differences between completed and current projects.

Decomposition method This method involves generating a work breakdown structure, i.e., breaking a product into smaller components or a project into activities and tasks in order to produce a lower-level, more detailed description of the product/project at hand, which in turn allows more accurate cost estimates. The total costs are calculated as average values, possibly adjusted on the basis of the complexity of the components/tasks considered. The successful application of the method depends of the availability of necessary information related to the work breakdown structure.

Parametric/algorithmic method This method involves the usage of mathematical equations based on research and historical data from previous projects. The method analyzes main cost drivers of a specific class of projects and their dependencies and uses statistical techniques to refine and customize the corresponding formulas. As in the case of the analogy method the generation of a proved and tested cost model using the parametric method is directly related to the availability of reliable and relevant data to be used in calibrating the initial core model.

Orthogonally to the aforementioned methods we mention two high-level approaches to cost estimation (cf. Table 1).

Bottom-up estimation This approach involves identifying and estimating costs of individual project components separately and subsequently summing up the outcomes to produce an estimation for the overall project. As such the bottom-up approach is at the core of the decomposition method introduced above.

Top-down estimation By contrast the top-down method relies on overall project parameters. For this purpose, the project is partitioned into lower-level components and life cycle phases beginning at the highest level. The approach produces are total project estimates, in which individual process tasks or product components are responsible for a proportion of the total costs.

The decomposition method is based on a bottom-up approach. Estimation by expert judgment, analogy or parametric equations can be carried out in a top-down or a bottom-up fashion, also depending of the stage of the project in

Table 1. Methods and approaches to cost estimation

	Bottom-up estimation	Top-down estimation
Expert judgement method	Experts estimate the costs of low-level components or activities	Experts estimate the total costs of a product or a project
Analogy method	Costs are calculated using analogies between low-level components or activities	Costs are estimated using a global similarity function for products or projects
Decomposition method	Costs are calculated as an average sum of the costs of lower-level units, whose development effort are known in advance	Not applicable
Parametric method	Costs are calculated using a statistic model which predicts the costs of lower-level units on the basis of historical data about the costs of developing such units	Costs are calculated using a statistic model which is calibrated using historical data about, and predicts the current value of the total development costs

which the estimates need to calculated. Top-down estimation is more applicable to early cost estimates when only global properties are known, but it can be less accurate due to the less focus on lower-level parameters and technical challenges usually predictable later in the process life cycle, at most. The bottom-up approach produces results of higher-quality, provided a realistic work breakdown structure and means to estimate the costs of the lower-level units the product/project has been decomposed into.

4.2 Methods Feasible for Ontology Engineering

In the following we examine the advantages and disadvantages of each of the aforementioned approaches given the product- and process-related characteristics of ontology engineering and the current state of the art in the field:

Expert judgment/Delphi method The expert judgement method seems to be appropriate for our goals since large amount of expert knowledge with respect to ontologies is already available in the Semantic Web community, while the costs of the related engineering efforts are not. Experts' opinion on this topic can be used to compliment the results of other estimation methods.

Analogy method The analogy method requires knowledge about the features of an ontology, or of an ontology development process, which are relevant for cost estimation purposes. Further on it assumes that an accurate comparison function for ontologies is defined, and that we are aware of cost information from previous projects. While several similarity measures for ontologies have already been proposed in the Semantic Web community, no case studies on ontology costs are currently available. There is a need to perform an in-depth analysis of the cost factors relevant for ontology engineering projects, as a basis for the definition of such an analogy function and its customization in accordance to previous experiences.

Decomposition method This method implies the availability of cost information with respect to single low-level engineering tasks, such as costs involved in the conceptualization of single concepts or in the instantiation of the ontology. Due to the lack of available information the decomposition method can not be applied yet to ontology engineering.

Parametric/algorithmic method Apart from the lack of costs-related information which should be used to calibrate cost estimation formula for ontologies, the analysis of the main cost drivers affecting the ontology engineering process can be performed on the basis of existing case studies on ontology building, representing an important step toward the elaboration of a predictable cost estimation strategy for ontology engineering processes. The resulting parametric cost model has to be constantly refined and customized when cost information becomes available. Nevertheless the definition of a fixed spectrum of cost factors is important for a controlled collection of existing real-world project data, a task which is

Table 2. Cost estimation methods and approaches potentially applicable to ontology engineering

	Bottom-up estimation	Top-down estimation
Expert judgment method	Currently not feasible	Feasible
Analogy method	Currently not feasible	Feasible
Decomposition method	Currently not feasible	Not applicable
Parametric method	Currently not feasible	Feasible

fundamental for the subsequent model calibration. This would also be useful for the design and customization of alternative prediction strategies, such as the aforementioned analogy approach.

Given the fact that cost estimation has been marginally explored in the Semantic Web community so far, and that little is known about the underlying cost factors , a bottom-up approach to the previously introduced methods is currently not practicable, though it would produce more accurate results. In turn, expert judgment, analogy and parametric cost estimates could be obtained in a top-down fashion, if the corresponding methods are clearly defined and customized in the context of ontology engineering. A summary of the results of this feasibility analysis is depicted in Table 2. Due to the incompleteness of the information related to cost issues, a combination of the three methods is likely to overcome certain limitations of single ones.

Section 5 introduces the ontology cost model ONTOCOM and discuss ways to improve its prediction quality. ONTOCOM follows a top-down approach to cost estimation, by identifying the cost drivers associated to the most important phases of the ontology life cycle and calculating a global effort estimate on the basis of different prediction methods. The current version of ONTOCOM investigates the usage of the parametric, the analogy and the Delphi methods to ontology engineering.

5 ONTOCOM: A Cost Model for Ontology Engineering

In this section we introduce the generic ONTOCOM cost estimation model. The model is generic in that it assumes a sequential ontology life cycle, according to which an ontology is conceptualized, implemented and evaluated, after an initial analysis of the requirements it should fulfill (see below). By contrast ONTOCOM does not consider alternative engineering strategies such as rapid prototyping or agile methods, which are based on different life cycles.[5] This limitation has been issued in previous work of ours, which describes how the generic model should be customized to suit such scenarios [16,21].

[5] Reference [10] for a discussion on the relation between this process model and the IEEE standards [12].

Table 3. Design of the ONTOCOM cost model (parametric, analogy and Delphi methods)

	Parametric method	Analogy method	Delphi method
Extract	Define work breakdown structure		
	Define cost drivers and ratings		Provide individual estimations
	–	Define similarities	
Evaluate	Collect data		Agree on estimates
	Perform statistical calibration	Calibrate weights	
Execute	Specify ratings of the cost drivers which correspond to the application at hand		Calculate final result
	Insert values to the prediction equation		
	Calculate estimate using equation		

The cost estimation model is realized as follows. First a *top-down* work breakdown structure for ontology engineering processes is defined in order to reduce the complexity of project budgetary planning and controlling operations down to more manageable units [1]. Then we can derive the global costs using various methods applicable for this top-down approach (cf. Table 3). Currently we are looking into three methods: the *parametric*, the *analogy* and the *Delphi* method, respectively.

For the parametric method the result of these steps is a statistical prediction model (i.e., a parameterized mathematical formula). Its parameters are given start values in pre-defined intervals, and are subsequently calibrated on the basis of previous project data. This empirical information complemented by expert opinions is used to evaluate and revise the predictions of the initial a priori *model*, thus creating a validated a posteriori *model*. The analogy method works similarly. It is based on a similarity equation, which is a mathematical formula aggregating similarity functions on the basic cost dimensions in a weighed fashion. The weights need to be specified according to empirical calibration and/or expert judgement, just as in the case of the parametric method. The Delphi method can be applied independently of any prediction formula or analogy function (see below).

The parametric equation has been carefully evaluated using statistical calibration and Bayes analysis as described in [17], whilst the analogy one is in the process of being customized. The Delphi method has been applied several times to derive specific initial inputs for the previous two methods.

5.1 The Work Breakdown Structure

The top-level partitioning of a generic ontology engineering process can be realized by taking into account available process-driven methodologies in this field [10,25] According to them ontology building consists of the following core steps (cf. Fig. 4):

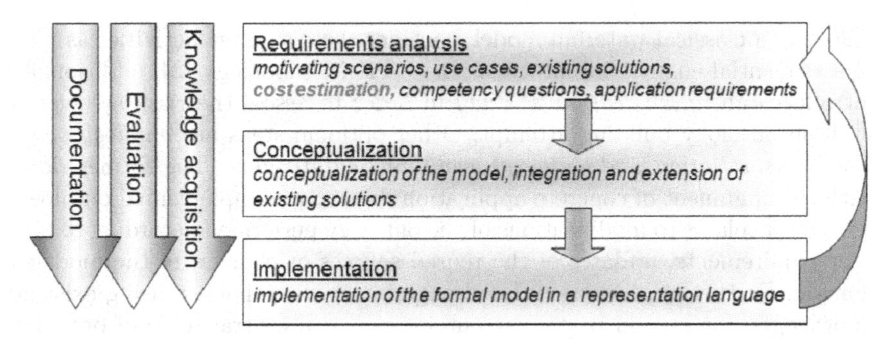

Fig. 4. Typical ontology engineering process

Requirements analysis The engineering team consisting of domain experts and ontology engineers performs a deep analysis of the project setting w.r.t. a set of pre-defined requirements. This step might also include *knowledge acquisition* activities in terms of the re-usage of existing ontological sources or by extracting domain information from text corpora, databases etc. If such techniques are being used to aid the engineering process, the resulting ontologies are to be subsequently customized to the application setting in the conceptualization/implementation phases. The result of this step is an ontology requirements specification document [24]. In particular this contains a set of competency questions describing the domain to be modeled by the prospected ontology, as well as information about its use cases, the expected size, the information sources used, the process participants and the engineering methodology.

Conceptualization The application domain is modeled in terms of ontological primitives, e.g. concepts, relations, axioms.

Implementation The conceptual model is implemented in a (formal) representation language, whose expressivity is appropriate for the richness of the conceptualization. If required reused ontologies and those generated from other information sources are translated to the target representation language and integrated to the final context.

Evaluation The ontology is evaluated against the set of competency questions. The evaluation may be performed automatically, if the competency questions are represented formally, or semi-automatically, using specific heuristics or human judgement. The result of the evaluation is reflected in a set of modifications/refinements at the requirements, conceptualization or implementation level.

Depending on the ontology life cycle underlying the process-driven methodology, the aforementioned four steps are to be seen as a sequential workflow or as parallel activities. Methontology [10], which applies prototypical engineering principles, considers *knowledge acquisition*, *evaluation* and *documentation* as being complementary *support activities* performed in parallel to the main development process. Other methodologies, usually following a classical waterfall model, consider these support activities as part of a sequential engineering process. The OTK-Methodology [24] additionally introduces an initial *feasibility study* in order to assess the risks associated with an ontology building attempt. Other optional steps are *ontology population/instantiation* and *ontology evolution/maintenance*. The former deals with the alignment of concrete application data to the implemented ontology. The latter relates to modifications of the ontology performed according to new user requirements, updates of the reused sources or changes in the modeled domain. Further on, likewise related engineering disciplines, reusing existing knowledge sources – in particular ontologies – is a central topic of ontology development. In terms of the process model introduced above, *ontology reuse* is considered a *knowledge acquisition* task.

We now introduce the cost drivers associated to this work breakdown structure.

5.2 The ONTOCOM Cost Drivers

The ONTOCOM cost drivers, which are proved to have a direct impact on the total development efforts, can be roughly divided into three categories [16,17]:

Product-related cost drivers account for the impact of the characteristics of the product to be engineered (i.e., the ontology) on the overall costs. The following cost drivers were identified for the task of ontology building:

- Domain Analysis Complexity (DCPLX) to account for those features of the application setting which influence the complexity of the engineering outcomes
- Conceptualization Complexity (CCPLX) to account for the impact of a complex conceptual model on the overall costs
- Implementation Complexity (ICPLX) to take into consideration the additional efforts arisen from the usage of a specific implementation language
- Instantiation Complexity (DATA) to capture the effects that the instance data requirements have on the overall process
- Required Reusability (REUSE) to capture the additional effort associated with the development of a reusable ontology item Evaluation Complexity (OE) to account for the additional efforts eventually invested in generating test cases and evaluating test results, and
- Documentation Needs (DOCU) to state for the additional costs caused by high documentation requirements

Personnel-related cost drivers emphasize the role of team experience, ability and continuity w.r.t. the effort invested in the engineering process:

- Ontologist/Domain Expert Capability (OCAP/DECAP) to account for the perceived ability and efficiency of the single actors involved in the process (ontologist and domain expert) as well as their teamwork capabilities

Table 4. The Conceptualization complexity cost driver *CCPLX*

Rating Level	*Description*
Very Low	Concept list
Low	Taxonomy, high nr. of patterns, no constraints
Nominal	Properties, general patterns available, some constraints
High	Axioms, few modeling patterns, considerable nr. of constraints
Very High	Instances, no patterns, considerable nr. of constraints

- Ontologist/Domain Expert Experience (OEXP/DEEXP) to measure the level of experience of the engineering team w.r.t. performing ontology engineering activities
- Language/Tool Experience (LEXP/TEXP) to measure the level experience of the project team w.r.t. the representation language and the ontology management tools
- Personnel Continuity (PCON) to mirror the frequency of the personnel changes in the team

Project-related cost drivers relate to overall characteristics of an ontology engineering process and their impact on the total costs:

- Support tools for Ontology Engineering (TOOL) to measure the effects of using ontology management tools in the engineering process, and
- Multisite Development (SITE) to mirror the usage of the communication support tools in a location-distributed team

The ONTOCOM cost drivers have been defined after extensively surveying recent ontology engineering literature and conducting expert interviews, and from empirical findings of numerous case studies in the field [16]. For each cost driver we specified in detail the decision criteria which are relevant for the model user in order for him to determine the concrete rating of the driver in a particular situation. For example for the cost driver CCPLX – accounting for costs produced by a particularly complex conceptualization – we pre-defined the meaning of the rating levels as depicted in Table 4. The decision criteria associated with a cost driver are typically more complex than in the previous example and might be sub-divided into further sub-categories, whose impact is aggregated to a final rating/value of the corresponding cost driver by means of normalized weights [16].

When using the model the project manager needs to specifies the current rating level for each cost driver according to the setting to which the estimation applies.

5.3 The Parametric Method

The parametric method integrates the efforts associated with each component of this work breakdown structure to a mathematical formula as described below:

$$PM = A * Size^{\alpha} * \prod CD_i \tag{1}$$

According to the parametric method the total development efforts are associated with cost drivers specific for the ontology engineering process and its main activities. Experiences in related engineering areas [1, 13, 18] let us assume that the most significant factor is the *size of the ontology* (in kilo entities) involved in the corresponding process or process phase. In Equation 1 the parameter $Size$ corresponds to the size of the ontology, i.e., the number of primitives which are expected to result from the conceptualization phase (including fragments built by reuse or other knowledge acquisition methods). The possibility of a non-linear behavior of the model w.r.t. the size of the ontology is covered by parameter α. The constant A represents a baseline multiplicative calibration constant in person months, i.e., costs which occur "if everything is normal." The *cost drivers* CD_i have a rating level (from Very Low to Very High) that expresses their impact on the development effort. For the purpose of a quantitative analysis each rating level of each cost driver is associated to a weight (*effort multiplier* EM_i). The *productivity range* PR_i of a cost driver (i.e., the ratio between the highest and the lowest effort multiplier of a cost driver $PR_i = \frac{max(EM_i)}{min(EM_i)}$) is an indicator for the relative importance of a cost driver for the effort estimation [1].

In order to determine the effort multipliers associated with the rating levels and to select non-redundant cost drivers we followed a three-stage approach: first experts estimated the a priori effort multipliers based on their experience as regarding ontology engineering. Second we applied linear regression to real world project data to obtain a second estimation of the effort multipliers.[6] Third we combined the expert estimations and the results of the linear regression in a statistically sound way using Bayes analysis [2]. More details on the calibration results are available in [17].

5.4 The Analogy Method

The analogy method has several advantages when compared to the parametric one, the most important being probably that its usage in a new measurement environment does not require additional calibration efforts, which potentially lead to varying accuracy levels for particular cost drivers. These advantages come, however, at the cost of significant computational power required to calculate similarities, therefore both methods can be seen as candidate techniques to be applied in conjunction [3].

The analogy method defines similarities for each of the cost drivers associated to the work breakdown structure and cumulates the results linearly in a weighed equation:

$$SIM = min \sum_{i,j=1,n} w_i * sim_i(CD_{i,current}, CD_{i,j}) \tag{2}$$

[6] Linear regression is a mathematical method to calculate the parameters of a linear equation so that the squared differences between the predictions from the linear equation and the observations are minimal [19].

In Equation 2 CD_i are ratings of the cost drivers elaborated above, including the size of the ontology to be built. sim_i is the similarity defined for ratings of the cost driver CD_i. The parameter w_i is the weight for this cost driver, all weights summing up to 1. Typically one uses the Euclidian distance as similarity function. The equation identifies the previous project with the closest values of the cost drivers as compared to the current project, and uses this overall similarity value to compute the estimate. j is an index of the size n of the project data set used for the comparisons.

5.5 The Delphi Method

The Delphi or expert judgement method for cost estimation [1] is suitable for ontology engineering projects in its generic form. Every Delphi process involves a moderator and a decision team of three to seven members, which meet two times in order to provide a consensual solution to a particular problem statement. In our context the experts are provided information about the current ontology engineering project and are asked to deliver an estimate of the development efforts according to their experience.

During the first brainstorming meeting the estimation team agrees upon the work breakdown structure, then the individual members provide estimates for the activities covered by this decomposition. In the second meeting the team aims at achieving a consensus on the final estimation by reviewing and revising the inputs of the members. This is an iterative process led by the moderator according to pre-defined rules. Once an agreement on the activity-based estimates has been achieved, the results are collected and compiled into a global figure, which can be used in the project.

As aforementioned such consensus-driven estimations can be used in combination with other methods and for particular cost drivers or activities in order to adjust the effects of data entries which might be unavailable, unreliable or skewed. For example, we used expert estimations of the productivity range of the ONTOCOM cost drivers for the calibration of the parametric equation [17]. A second important use case for such procedures is the estimation of the size of the prospected ontology, which is a core parameter of statistical methods. In terms of the analogy method, expert opinion is crucial for defining the similarity functions for each cost driver, for assigning a priori value to the weights and for evaluating the overall similarity equation.

6 Using ONTOCOM

Starting from a typical ontology building scenario, in which a domain ontology is created from scratch by the engineering team, we simulate the cost estimation process according to the parametric method underlying ONTOCOM. Given the top-down nature of our approach this estimation can be realized in the early phases of a project. In accordance to the process model introduced above the prediction of the arising costs can be performed during the feasibility study or, more reliably, during the requirements analysis. Many of the input parameters required to exercise the cost estimation are expected to be accurately approximated during this phase: the expected size of the ontology, the engineering team, the tools to be used, the implementation language etc.

The first step of the cost estimation is the specification of the size of the ontology to be built, expressed in thousands of ontological primitives (concepts, relations, axioms and instances): if we consider an ontology with 1,000 concepts, 200 relations (including is-a) and 100 axioms, the size parameter of the estimation formula will be calculated as follows:

$$Size = \frac{1{,}000 + 200 + 100}{1{,}000} = 1.3 \tag{3}$$

The next step is the specification of the cost driver ratings corresponding to the information available at this point (i.e., without reuse and maintenance factors, since the ontology is built manually from scratch). Depending on their impact on the overall development effort, if a particular activity increases the nominal efforts, then it should be rated with values such as High and Very high. Otherwise, if it causes a decrease of the nominal costs, then it would be rated with values such as Low and Very low. Cost drivers which are not relevant for a particular scenario, or are perceived to have a nominal impact on the overall estimate, should be rated with the nominal value 1, which does not influence the result of the prediction equation.

Assuming that the ratings of the cost drivers are those depicted in Table 5 these ratings are replaced by numerical values. The value of the DCPLX cost driver was computed as an equally weighted, averaged sum of a high-valued rating for the domain complexity, a nominal rating for the requirements complexity and a high effort multiplier for the information sources complexity (for details of other rating values see [17]). According to the formula 1 ($\alpha = 1$) the development effort of 11.44 PM would be calculated as follows:

$$PM = 2.92 * 1.3^1 * (1.26 * 1^{10} * 1.15 * 1.11 * 0.93 * 1.11 * 0.89 * 1.2 * 1.7) \tag{4}$$

The constant A has been set to 2.92 after the calibration of the model, while the economies of scale are so far not taken into consideration.

In order to use ONTOCOM in a particular setting (enterprize, business domain, types of ontologies, to name only a few criteria) the generic model should be customized according to the following steps:

- Refine and adapt the work breakdown structure in the light of the applied life cycle and process model followed when engineering the ontology
- Define the statistical prediction model (i.e., a parameterized mathematical formula)
- Calibrate the a priori model based on previous project data to create a valid (more accurate) a posteriori model
- Use the calibrated model to predict development costs

Table 5. Values of the cost drivers

Cost driver	*Effort*	*Value*	*Cost driver*	*Effort*	*Value*
Product factors			*Personnel factors*		
DCPLX	High	1.26	OCAP	High	1.11
CCPLX	Nominal	1	DCAP	Low	0.93
ICPLX	Low	1.15	OEXP	High	1.11
DATA	High	1	DEEXP	Very Low	0.89
REUSE	Nominal	1	LEXP	Nominal	1
DOCU	Low	1	TEXP	Nominal	1
OE	Nominal	1	PCON	Very High	1
Project factors					
TOOL	Very Low	1	SITE	Nominal	1

An example how the generic ONTOCOM model can be applied to a different ontology engineering methodology is described in [21]. Details about a similar enterprize based however on a particular type of ontologies can be found in [4].

7 Conclusions

Technologies related to the development, deployment and maintenance of ontologies have reached a maturity level that they become relevant for businesses. At this stage ontology engineering can no longer be accounted for in a stand alone manner, but should be integrated into the overall architecture and organization of an enterprize. We have shown how ontology engineering fits into existing architecture development frameworks: ontology engineering is an integral part of the information system architecture and influences the technology architecture of an enterprize. Companies complement their overall architecture with a governance framework setting the rules to organize, steer, control and diffuse its deployment. A major concern of IT governance is to timely identify changes in the architecture which are potentially of benefit and to control the realization of the expected benefits. In this context the availability of cost information related to the engineering of ontologies becomes important both at the beginning of an ontology engineering process and during its operation.

In this chapter we have focused on the estimation of costs related to ontology engineering for planning purposes. We have discussed different methods to derive cost information from the environmental setting an existing knowledge and selected three for a more detailed presentation. Following a top-down approach all methods start with a definition of the work breakdown structure. The Delphi method is based on consensual expert estimates aligned to this work breakdown structure, which are aggregated by the project manager towards a final effort prediction. The parametric and analogy method define cost drivers and rating levels as a basis for the mathematical equations customized according to historical project data.

The results from our case studies point in several directions. On the one hand incorporating cost-related aspects into ontology engineering practice is likely to facilitate the interaction between the ontology engineering community and business people. Cost information allow non-engineers to integrate ontology engineering into their management frameworks and makes the creation of ontologies more transparent from a business perspective. On the other hand the estimations are far from being precise yet. First results imply that the creation of glossary-like structures is well understood and the related effort predictable. By contrast the effort related to the creation of ontologies with a high axiomatization is hardly predictable and the exact correlations remain an open issue for future research.

Hence, we see in number of new research directions for the economics of ontology engineering. From a management perspective open issues remain in the areas of controlling and the applicability of the cost models for non-experts. Tool support and additional training materials are needed to ease non-experts the interaction with these models and to guarantee their correct usage. From a technical perspective, in the near future we intend to continue the data collection procedure in order to improve the quality of the generic model and its customizations. Much work needs to be done by many people, thus we see ONTOCOM as a seed for an urgently needed field of research, the cost estimation for ontologies. Any significant improvement in this field will substantially facilitate the uptake of semantic technologies for industrial projects. A second direction of research is related to the refinement of alternative methods for the estimation of critical input parameters such as the size of the prospected ontology. The analogy method seem to be a promising approach for this purpose.

Acknowledgments

This work has been partially supported by the European Network of Excellence "KnowledgeWeb-Realizing the Semantic Web" (FP6-507482), as part of the KnowledgeWeb researcher exchange program *T-REX*, and by the European project "Sekt-Semantically-Enabled Knowledge Technologies"(EU IP IST-2003-506826). Further information about ONTOCOM can be found under: `http://ontocom.ag-nbi.de`.

References

1. B. W. Boehm. *Software Engineering Economics*. Prentice-Hall, 1981.
2. G. Box and G. Tiao. *Bayesian Inference in Statistical Analysis*. Addison Wesley, 1973.

3. L. C. Briand, K. El Emam, D. Surmann, I. Wieczorek, and K. D. Maxwell. An assessment and comparison of common software cost estimation modeling techniques. In *ICSE '99: Proceedings of the 21st International Conference on Software Engineering*, pages 313–322, Los Alamitos, CA, USA, 1999. IEEE Computer Society.
4. T. Buerger, C. Ammendola, and E. Simperl. Evaluation of the economics of multimedia ontologies (salero deliverable d3.1.4). Technical report, STI Innsbruck, 2008.
5. P. R. Cohen, V. K. Chaudhri, A. Pease, and R. Schrag. Does prior knowledge facilitate the development of knowledge-based systems? In *AAAI/IAAI*, pages 221–226, 1999.
6. C. Ebert, R. Dumke, M. Bundschuh, and A. Schmietendorf. *Best Practices in Software Measurement.* Springer, 2005.
7. B. W. Boehm et al. *Software Cost Estimation with COCOMO II (with CD-ROM).* Prentice-Hall, 2000.
8. A. Felfernig. Effort estimation for knowledge-based configuration systems. In *Proc. of the 16th Int. Conf. of Software Engineering and Knowledge Engineering SEKE04*, 2004.
9. L. Fischman, K. McRitchie, and D. D. Galorath. Inside SEER-SEM. *The Journal of Defense Software Engineering*, 2005.
10. A. Gomez-Perez, M. Fernandez-Lopez, and O. Corcho. *Ontological Engineering – With examples form the areas of Knowledge Management, e-Commerce and the Semantic Web.* Springer, 2004.
11. W. S. Humphrey. Using a defined and measured personal software process. *IEEE Software*, 13(3):77–88, 1996.
12. IEEE Computer Society. IEEE Standard for Developing Software Life Cycle Processes. IEEE Std 1074-1995, 1996.
13. C. F. Kemerer. An Empirical Validation of Software Cost Estimation Models. *Communications of the ACM*, 30(5), 1987.
14. M. Korotkiy. On the effect of ontologies on web application development effort. In *Proc. of the Knowledge Engineering and Software Engineering Workshop*, 2005.
15. T. Menzies. Cost benefits of ontologies. *Intelligence*, 10(3):26–32, 1999.
16. E. P. Bontas and C. Tempich. How Much Does It Cost? Applying ONTOCOM to DILIGENT. Technical Report TR-B-05-20, Free University of Berlin, October 2005.
17. E. Simperl, C. Tempich, and Y. Sure. Ontocom: A cost estimation model for ontology engineering. In *Proceedings of the 5th International Semantic Web ISWC2006*, pages 625–639, Springer, 2006.
18. L. H. Putnam and W. Myers. *Measures for Excellence: Reliable Software on Time, Within Budget.* Yourdon, 1991.
19. G.A.F. Seber. *Linear Regression Analysis.* Wiley, 1977.
20. M. Shepperd, C. Schofield, and B. Kitchenham. Effort estimation using analogy. In *Proceedings of the 18th International Conference on Software Engineering ICSE1996*, pages 170–178, 1996.
21. E. Simperl, C. Tempich, and M. Mochol. Cost Estimation for Ontology Development: Applying the ONTOCOM Model. In *Technologies for Business Information Systems*, pages 327–340. Springer, 2007.
22. A. Stellman and J. Green. *Applied Software Project Management.* O'Reilly Media, 2005.

23. R. D. Stewart, R. M. Wyskida, and J. D. Johannes. *Cost Estimator's Reference Manual.* Wiley, 1995.
24. Y. Sure, S. Staab, and R. Studer. Methodology for development and employment of ontology based knowledge management applications. *SIGMOD Record*, 31(4), 2002.
25. Y. Sure, C. Tempich, and D. Vrandecic. Ontology engineering methodologies. In *Semantic Web Technologies: Trends and Research in Ontology-Based Systems.* Wiley, 2006.
26. J. A. Zachman. A framework for information systems architecture. *IBM Systems Journal*, 26(3), 1987.

Part III

Ontologies

Foundational Choices in DOLCE

Stefano Borgo and Claudio Masolo

Laboratory for Applied Ontology (ISTC-CNR), Trento, Italy, borgo@loa-cnr.it, masolo@loa-cnr.it

Summary. Foundational ontologies are ontologies that have a large scope, can be highly reusable in different modeling scenarios, are philosophically and conceptually well founded, and are semantically transparent.

After the analysis and comparison of alternative theories on general notions like 'having a property', 'being in time' and 'change through time', this paper shows how specific elements of these theories can be coherently integrated into a foundational ontology. The ontology is here proposed as an improvement of the core elements of the ontology DOLCE and is thus called DOLCE-CORE.

1 Introduction

Chapter "What is an *Ontology*?" analyses what ontologies are and their peculiarities with respect to other methods and technologies that exist in conceptual modeling and knowledge representation. *Foundational ontologies* are ontologies that: (1) have a large scope, (2) can be highly reusable in different modeling scenarios, (3) are philosophically and conceptually well founded, and (4) are semantically transparent and (therefore) richly axiomatized.

Foundational ontologies focus on very general and basic concepts (like the concepts of object, event, quality, role) and relations (like constitution, participation, dependence, parthood), that are not specific to particular domains but can be suitably refined to match application requirements. These notions have been largely investigated by philosophers and, even though foundational ontologies assume a modeling and engineering perspective (far from the absolutist view of most philosophical theories), one relies on philosophical considerations for the construction, comparison, organization, and assessment of the ontologies themselves.

To achieve semantic transparency, a careful choice of the primitives and a precise characterization of their meaning are needed. This goal requires a formal language with clear semantics and adequate expressive power (in this chapter we will use first-order logic). Unfortunately, application concerns

S. Staab and R. Studer (eds.), *Handbook on Ontologies*, International Handbooks on Information Systems, DOI 10.1007/978-3-540-92673-3,

lead to work with languages that are suitable for run-time reasoning and one often has to give up on expressivity and semantic clearness. For these reasons, foundational ontologies are used in applications only in approximated forms via partial translations into different application-oriented languages. Thus, the relevance of foundational ontologies does not rely in their direct impact on applications but in their ability to providing *conceptual handles* with which to carry out a coherent and structured analysis of the domains of interest.

The paper is organized as follows. Section 2 analyzes and compares alternative well founded theories on central notions like 'having a property', 'being in time' and 'change through time'. Then, in Sect. 3, we study how specific elements of these theories can be integrated into a foundational ontology that we call DOLCE-CORE and constitutes a first step in the revision of DOLCE[1] [17]. Other foundational ontologies are not discussed in this paper for lack of space.[2]

2 Foundational Distinctions

The literature on ontological choices is primarily of philosophical character. Several tenable positions for each issue have been individuated and some have been described to a rich level of detail. Unfortunately, there is no homogeneity in the depth of the analysis: some topics, like the theories of parthood and space, have been well studied others, e.g., the theories of dependence and unity still lack a stable landscape [26]. Perhaps more worryingly, there is no comprehensive list of ontological issues relevant to foundational ontologies.

2.1 Theories of Properties

The nature of properties, the explanation of what it means that an individual has a property, and, more specifically, of how different individuals can have the *same* property, have been widely discussed and investigated ([1,15,20] are good surveys). Intuitively, the term *individual* (or, alternatively *particular*) refers to entities that cannot have instances, that is, entities that cannot be predicated of others like Aristotle, the Tour Eiffel, the Mars planet. On the contrary, the term *property* denotes entities that can have instances, that is, entities that qualify other entities, e.g., Red (the color), Person (the kind), Fiat Panda (the car model). Traditionally, the notion of property has been formally represented in two ways. In the first, it is associated with the set-theoretical notion of *class*[3] and, in the latter, with the logical notion of *predicate*.

[1] `http://www.loa-cnr.it/DOLCE.html`

[2] See, for instance, BFO: `http://www.ifomis.org/bfo`; GFO: `http://www.onto-med.de/ontologies/gfo.html`; OPENCYC: `http://www.opencyc.org`; SUMO: `http://www.ontologyportal.org/`

[3] As usual in this area, we use the terms 'class' and 'set' interchangeably.

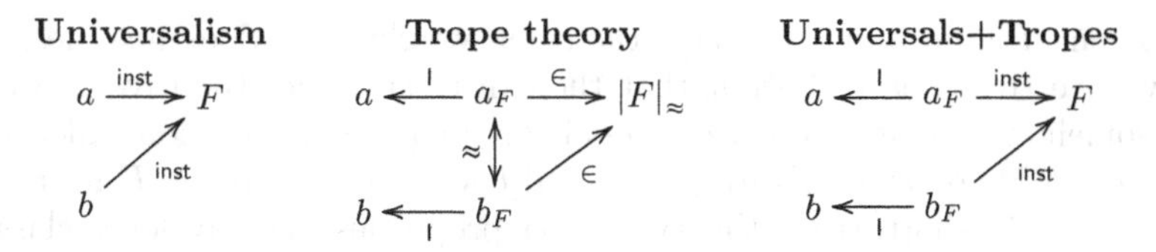

Fig. 1. Philosophical positions on properties

Our goal in this section is to briefly introduce a few alternative positions that are of particular interest in modeling. Consider the expression "the individuals a and b share the property F" (as exemplified by, say, "my car and my pen are both red").

Figure 1 illustrates three different ways to represent this expression.[4]

Universalism claims that both entities a and b *instantiate* (inst) the *universal* F which, in short, means that F is a repeatable and independent entity (a universal) that is *wholly present in* both a and b. Intuitively, one could rephrase this view by saying "may car and my pen both instantiate redness". The instance-of relation, inst, is different from the set-theoretical membership relation, $\in$, (exemplified by expression "my car and my pen both belong to the class of red things") for two reasons: (1) the latter is extensional (two different classes must have at least a different member) while the first might not (nothing prevents different universals to have exactly the same instances); and (2) classes are closed under union and intersection while nothing suggests that the union or intersection of two universals must be a universal itself. Universals are, so to speak, *sparse* and *minimal* since they cannot be generated by syntactic manipulations. They correspond to truly ontological distinctions that are present *in the world.*

The second diagram in Fig. 1 depicts the *trope theory* (see [5] for a good survey). This theory is based on the notion of *individual property* or *trope.* A trope *inheres in* (I) one single individual and it represents the distinct way an individual has a property ("my car is red" means that there is a specific individual property, a trope, of my car and this trope is classified red). If a and b are different individuals, then the way a is F (has property F) is necessarily different from the way b is F because a and b rely on different tropes. In Fig. 1, a_F is the F-trope of a and b_F the F-trope of b. This means that (1) the inherence relation between a trope and its bearer satisfies the *non migration principle*, i.e., tropes inhere in a unique bearer (a1), and that (2) tropes are existentially dependent on their bearers, i.e., tropes cannot exist without a bearer (a2). If we read $\mathsf{I}(x, y)$ as "x inheres in y" and $\text{TROPE}(t)$ as "t is a trope", then

a1 $\mathsf{I}(t, x) \wedge \mathsf{I}(t, y) \rightarrow x = y$
a2 $\text{TROPE}(t) \rightarrow \exists x(\mathsf{I}(t, x))$

[4] There are other positions like, e.g., the *bundle theory* [23].

Then, if John and Paul have the same weight, this does not mean that they have the same trope but that their distinct tropes (relative to weight) are somehow similar. Trope sameness is an a equivalent relation called *indistinguishability* or *resemblance* ($\approx$): a and b share the property F if and only if $a_F \approx b_F$. In short, trope theory reduces properties to equivalence classes of resembling tropes.

Note that trope theory and universalism are not antithetical. One can rely on tropes and the inherence relation while substituting the classes of resembling tropes by universals and membership (between tropes and classes) by instantiation (between tropes and universals). That is, the universalist view can be adopted to classify the tropes instead of the entities as in the third diagram in Fig. 1.

Basic Properties, Quality Kinds, and Spaces

People compare entities along a variety of aspects such as color, weight, smell, etc. For each aspect, similarities are established depending on the tools people dispose of, or on the specific analysis they are interested in. This knowledge disparity is often dismissed by philosophers as an epistemological or empirical issue: the entities, they say, have a completely determined shade of color even though in practice it is not accessible to the observer. This attitude somehow prevents the assessment of a philosophical analysis of this issue, of course, but the available philosophical notions still provide a good starting point for building a philosophically based and yet application oriented framework.

In [10, 11] an important *determinate-determinable* relation (dD) between properties has been suggested by combining subsumption and partitioning: $\mathsf{dD}(F, G)$ means that entities that have the property F also have the (more general) property G and entities that have the property G have at least one of the (more specific) properties that are the *determinates* of G, among which there is F. For example, "being crimson" and "being scarlet" are both determinates of "being red" and the latter is a determinate of "being colored". The dD relation induces a partial-order over properties. According to this ordering, properties about the same aspect of objects are organized in a tree the leaves of which are formed by the most specific properties, hereafter called *basic properties*. Then, any entity that has a property is claimed to have also a basic property in the corresponding tree. It is this basic property that makes the entity ontologically indistinguishable (with respect to the given aspect) from the other entities with the same basic property: two entities enjoying property "being $1m$ long" cannot be differentiated on the basis of their lengths. Vice versa, entities that have different basic properties are surely different. Sharing *non* basic properties indicates some form of similarity but has no direct import on the distinguishable/indistinguishable status of the entities.

In trope theory, sharing a basic property corresponds to having two *exactly resembling* tropes: two '$1m$ long' entities have *exactly similar* (yet distinct) length-tropes. If they resemble each other inexactly, it is said that their length-tropes resemble each other only *up to a degree*. One can add structure in the class of tropes by saying that $1m$ and $2m$ length-tropes have a higher degree of resemblance than the $1m$ and $30m$ length-tropes or analogously, that a scarlet-trope and a crimson-trope resemble each other better than a scarlet-trope and a turquoise-trope. In this view, non basic properties are built as classes of *inexactly* resembling tropes. Exploiting the degrees of resemblance, all the tropes can be collected in few large classes. However, if we put together a $1m$-trope and a 'red'-trope or a $1kg$-trope, we contradict the initial intuition that the comparison between entities has to be done for 'homogeneous' properties, i.e., properties on the same aspect of entities: the comparison between the length of an object with the color of another object is not really plausible.

General properties that identify specific aspects of entities (like "being colored", "being shaped", etc.) cannot be discharged: without these we cannot even conceive the functional laws of physics [2]. Ingvar Johansson [10] characterizes these general properties, hereafter called *quality kinds*, in terms of *maximal incompatibility and maximal comparability* of their determinates: (1) each entity that has a quality kind F must have just one basic property that is a determinate of F, and (2) all the basic properties that are determinate of F are qualitatively comparable. Summing up, each *quality kind* is a (non basic) property that corresponds to one aspect/dimension of comparison for entities, the property is partitioned into more specific properties that give different levels of distinctions for that aspect, the lowest level is that of the basic properties.

Properties in the same quality kind can be organized in taxonomies or in more sophisticated ways: from ordering (weight, length) to complex topological or geometrical relations (color splinter). Following [6] we call *spaces* these complex structures of properties. Sometimes properties can be combined together to model multi-dimensional or multi-aspectual properties like density, speed or force. The color property can also be seen as a multi-dimensional property since one can distinguish hue, saturation, and brightness as different quality kinds. These cases indicate that property spaces can combine to very specialized structures.

Often spaces are motivated by applications or epistemological considerations, it is quite natural to associate each quality kind to several spaces, each organizing properties (and thus objects) according to different principles, instruments of investigation, applications concerns, etc. These spaces rely on relative notions of resemblance that are discussed, adopted, and abandoned by (communities of) intentional agents. This view of spaces as generated (and eventually destroyed) structures leads to model spaces as temporal entities. Alternative spaces can differ on several aspects: their structure, the level of detail the adopted measuring tool can reach, or the point of view that motivate them. This variety of spaces can be partially ordered according to the level of detail they are capable of distinguishing, a notion often called *granularity*.

Concepts and Roles

The framework just introduced addresses two concerns: (1) representing intensional properties that are created (and eventually destroyed) by agents and (2) classifying qualities according to different points of view and granularities.

The first point is important independently of the need to organize properties in spaces. Take properties like 'being a student', 'being money', 'being a catalyst', etc. that we will call *concepts*. These have a clear conceptual and intensional nature – they are defined in terms of relationships with external entities, e.g., 'a person enrolled in a university', and do not depend on their instances – but do not present any special internal structure. The rich framework given by quality kinds and spaces is largely pointless for these concepts. A mechanism more tailored to these properties is needed.

Roles are a subclass of concepts. The nature and the representation of roles have been long discussed in a variety of fields: knowledge representation, conceptual modeling, multi-agent systems, linguistics, sociology, philosophy, and cognitive semantics (see [14,18,29]). These properties are intensional and *anti-rigid* (see Chapter "An Overview of OntoClean") in the sense that an entity may play a role for a limited time (and perhaps resume it in different periods) without changing its identity. Often in conceptual modeling roles are seen as classes but this approach has severe problems [29].

2.2 Being in Time

The entities that are mostly studied in applied ontology are entities that exist in time. Temporal existence is often modeled via a predicate like $\mathsf{PRE}(x, t)$, whose informal reading is 'x is present at time t'.

Since PRE is defined on *times*, these must be in the domain of quantification. However, this does not necessarily lead to strong ontological commitments on times: times could be constructed from events [12], 'being present at a time' can be reduced to 'being simultaneous with' other entities [27]. Of course, one can take the Newtonian view in which time is an independent container. In this case, PRE is a sort of localization relation in that container. In both cases, one can take times to be punctual or extended and even adopt different structures on them (discrete vs. continuous, linear vs. branching, etc.). Furthermore, there are different ways of being in time: existing in time vs. occurring in time (a distinction related to the contraposition between objects and events, see Sect. 2.4) or being wholly present vs. being partially present (relying on the contraposition between endurants and perdurants, see below).

We give for granted that some entities are present at different times, i.e., they are persisting through time. The explanation of this apparently obvious fact may be quite intricate. *Stage theory* [8] claims that all existing entities, called stages, are temporally instantaneous. In this perspective, 'persisting entity' is meaningless since no entity can exists at different times. Commonsense persistence is modeled by stage theory only at the *conceptual* level:

persisting entities are reconstructed as collections of stages and special rules, called *unity criteria* (see Chapter "An Overview of OntoClean"), are isolated to flag meaningful collections.

Two main philosophical positions accept the ontological existence of persisting entities: *endurantism* and *perdurantism*. Endurantists claim that one and the same entity is *wholly present* at different times (enduring) and read the formula $\mathsf{PRE}(a, t_1) \wedge \mathsf{PRE}(a, t_2)$ as "a is wholly present at both the times t_1 and t_2". 'Being *wholly* present' is often contrasted with 'being *partially* present', i.e., the rationale of perdurantism. Perdurantists claim that the persistence through time is analogous to the extension in space: an entity has different parts at different times (perduring). The previous formula is then read by perdurantists as claiming "a has a part at t_1 and a (different) part at t_2". Therefore, in addition to a, perdurantists commit to the existence of the parts of a that exist only at t_1 and at t_2, respectively.

Despite the disagreement between perdurantism and stage theory on the nature of persisting entities, both the theories associate each persisting entity with a sequel of other entities. Indeed, the following property holds in these systems (it may fail for endurantists):

a3 $\mathsf{PRE}(a, t_1) \wedge \mathsf{PRE}(a, t_2) \wedge t_1 \neq t_2 \rightarrow$
$$\exists b_1, b_2(\mathsf{PRE}(b_1, t_1) \wedge \forall t(\mathsf{PRE}(b_1, t) \rightarrow t = t_1) \wedge \mathsf{PRE}(b_2, t_2) \wedge \forall t(\mathsf{PRE}(b_2, t) \rightarrow t = t_2))$$

Provided one does not give up on expressive power, it is formally an advantage to have a core theory compatible with different philosophical positions since one can use the very same framework and specialize it, when needed, with the additional constraints of one or the other theory. In this perspective, without (a3) the formula $\mathsf{PRE}(a, t_1) \wedge \mathsf{PRE}(a, t_2)$ can be interpreted freely by endurantism, perdurantism, and stage theory.

2.3 Property Change

Persisting objects change through time by changing their properties: a may be red at time t_1 and green at t_2.[5] It should be clear by now that there are alternative views on properties and on persistence through time. However interesting these topics are, none is as debated as the issue of property change itself. Aiming at a wide-ranging presentation, we formally model properties in first order logic (FOL) via formulas of form $F(a, t)$ without committing to any ontological constraint beside those (fairly weak) of FOL itself. According to [21], $F(a, t)$ can be read in a very general way: "a exists at t and it has the property F when t is (was, will be) present". We will see alternative readings

[5] We limit this presentation to properties. The arguments, *mutatis mutandis*, hold for relations as well.

of $F(a,t)$ in terms of more committed theories. For the time being, let us begin with a minimal condition: since a at t has property F, a needs to exist at t.[6]

a4 $F(x,t) \rightarrow \mathsf{PRE}(x,t)$

Formula $F(a,t_1) \wedge G(a,t_2)$ formalizes the change of a property.

Following Sect. 2.1, universalists have three ways to model property change: (1) adding a temporal parameter to inst making it a ternary relation on entities, universals, and times as in (a5); (2) applying temporal modal operators to the binary inst, see (a6); (3) committing to temporal slices $x@t$ (the maximal part of x during t) as seen in perdurantism, see (a7). (Here we use the same letter for both the relational property and its nominalization: cfr. the occurrence of F on the left and on the right of $\leftrightarrow$, resp.ly, in (a5).)

a5 $F(x,t_1) \wedge G(x,t_2) \leftrightarrow \mathsf{inst}(x,F,t_1) \wedge \mathsf{inst}(x,G,t_2)$
a6 $F(x,t_1) \wedge G(x,t_2) \leftrightarrow \Box_{t_1} \mathsf{inst}(x,F) \wedge \Box_{t_2} \mathsf{inst}(x,G)$
a7 $F(x,t_1) \wedge G(x,t_2) \leftrightarrow \mathsf{inst}(x@t_1,F) \wedge \mathsf{inst}(x@t_2,G)$

A trope theorist explains change as trope substitution, (a8).[7] If one accepts both universals and tropes, trope substitution can be formulated as in (a9).

a8 $F(x,t_1) \wedge G(x,t_2) \leftrightarrow \exists f,g(\mathsf{I}(f,x) \wedge \mathsf{I}(g,x) \wedge f \in F \wedge g \in G \wedge \mathsf{PRE}(f,t_1) \wedge \mathsf{PRE}(g,t_2))$
a9 $F(x,t_1) \wedge G(x,t_2) \leftrightarrow \exists f,g(\mathsf{I}(f,x) \wedge \mathsf{I}(g,x) \wedge \mathsf{inst}(f,F) \wedge \mathsf{inst}(g,G) \wedge \mathsf{PRE}(f,t_1) \wedge \mathsf{PRE}(g,t_2))$

If both tropes and universals are considered, a notion of "tropes changing over time" becomes available, we call these *individual qualities*. An individual quality, like a trope, inheres in a unique bearer but, differently from tropes, it can change over time. In this case we can explain change according to the following schemata that are similar to (a5) and (a7), respectively[8]:

a10 $F(x,t_1) \wedge G(x,t_2) \leftrightarrow \exists q(\mathsf{I}(q,x) \wedge \mathsf{inst}(q,F,t_1) \wedge \mathsf{inst}(q,G,t_2))$
a11 $F(x,t_1) \wedge G(x,t_2) \leftrightarrow \exists q(\mathsf{I}(q,x) \wedge \mathsf{inst}(q@t_1,F) \wedge \mathsf{inst}(q@t_2,G))$

In these approaches the color, weight, shape, etc. of an object are each modeled by a different individual quality, and the changing through time of these qualities explain changes in the bearers: intuitively, it is the individual color of an object a that changes from, say, fuchsia to green and the individual weight (a different individual quality of a) that changes from some weight to

[6] Recall the notion of property given in Sect. 2.1. One should refrain from considering boolean combinations of predicates, like 'not being present', as possible values for F.
[7] We use the set-theoretical $\in$ predicate to indicate that here F stands for the class of tropes that satisfy F.
[8] We could do as in (a6) as well but we do not investigate this option here.

another. While (a10) is compatible with both an endurantist and a perdurantist reading about persistence of individual qualities, we see that (a11) is ontologically more demanding since it refers to temporal slices of individual qualities. On the other hand, (a11) has the advantage of being compatible with (a9) if we accept mereological sums of tropes. At the same time, (a9) is to be preferred to (a8) because in (a9) inst can be taken to be intensional.

Of course, one should have some advantage for introducing yet another type of entities like individual qualities. After all, why aren't (a5) and (a7) enough? The usefulness of individual qualities relies on the fact that they are associated to one quality kind only and the latter usually has different spaces associated. A change in the same individual quality is described differently by the different points of views encoded by the spaces. For example, a change in color can be described according to both a RGB and a CYMK color-space. Having a unique individual color-quality related to all the relevant spaces allows for expressing that it is the same aspect of the object (the color) that changes. In [16] alternative positions that avoid individual qualities are analyzed and it is shown that, if expressivity is to be maintained, these systems are technically and conceptually more complicated. These aspects may seem minor to a neophyte and yet they are crucial in setting a foundational ontology as we will see in Sect. 3.

Mereological Change: Endurantism vs. Perdurantism

The difference between the endurantist and perdurantist theories of persistence (Sect. 2.2) can be addressed in terms of the parthood relation. Classical endurantists think that "statements about what parts the object has must be made relative to some time or other" ([8], p. 26), which makes *temporary* parthood a primitive relation to endurantists. On the contrary, perdurantists can derive temporary parthood from the relations of parthood *simpliciter* and 'being present at a time' via schema (a7) (which is applicable since perdurantists accept temporal slices). In [24,25] Sider provides a direct comparison of these two positions by starting from a temporary parthood relation shaped to be acceptable to both endurantists and perdurantists (even though they would interpret it differently). Sider's formulation of perdurantism is given by the usual axioms for temporary parthood (see Sect. 3.2) plus the existence of temporal slices to characterize the notion of 'being partially present'. On the other hand, the notion of 'being wholly present' (that plays a central role in endurantism) remains somehow obscure and difficult to characterize notwithstanding some attempts have been made [4,9,19]. Both endurantists and perdurantists accept the usual axioms for temporary parthood, yet endurantists cannot accept that each entity has a temporal slice at each time at which the entity exists.[9] As noted in [25], either we assume that endurantism

[9] Endurantists do not refuse the existence of temporal parts and temporal slices in general. They do not accept that all the persistent entities necessarily have temporal slices at each time of their existence.

needs nothing more than the general axioms discussed before (therefore it is a theory less constrained than perdurantism), or we need to accept that the endurantist view still lacks a clear and formal characterization. After all, the intuitive notion of 'being wholly present at each instant' is trivially satisfied by temporary parthood even in the perdurantist axiomatization since all the parts of x at t are present at t.

From these observations, perdurantists may indifferently adopt temporary parthood or parthood simpliciter as the primitive relation, while endurantists must rely on temporary parthood. In the perspective of foundational ontology, this is an important result, exploited in Sect. 3.2, since it shows that one can construct a fairly general ontology that is compatible with both endurantism and perdurantism.

Parthood and Spatio-Temporal Inclusion

Perdurantists often see parthood as spatio-temporal inclusion and thus rely on extensional mereology (axioms (A1)-(A4) and definition (D2) of Sect. 3.2). This view pushes them to reject the existence of spatio-temporally coincident entities: if x and y have the same spatio-temporal extension then both $\mathsf{P}(x, y)$ and $\mathsf{P}(y, x)$ hold and consequently, due to antisymmetry of parthood, they are identical. This position is, however, more restrictive than the original proposal of Lesniewski [13]. Lesniewski proposed mereology as an alternative to set theory that avoids the cognitively obscure distinction between *urelements* and *sets* (not to mention the puzzling notion of empty set). The goal was to ensure that the entity $a+b$, obtained combining a and b, is nothing more than a *and* b (and not an abstract element like the set with members a and b). Indeed, in mereology the sum and the addenda have the same ontological status.

In its general perspective, extensional mereology is a purely formal theory and it applies to all kinds of entities (the spatio-temporal entities are just one case).[10] Parthood, when applied to spatio-temporal entities, is strictly related to spatio-temporal inclusion. Nonetheless, these relations must not be confused: philosophers and engineers like to apply parthood and mereological change even to entities like, e.g., mathematical theories, word meanings, beliefs and societies, i.e., entities that are said to be in time but not in space. On the other hand, it is unquestioned that two spatio-temporally extended entities, that are one part of the other, are also spatio-temporally included. The vice versa does not necessarily hold: some authors accept that some crete constituted a given statue and yet reject that crete is part of the statue [22].

2.4 Events and Objects

We can all distinguish *what changes* from the *changing* event itself. A lively and long discussion on the ontological status of events and on what distinguishes

[10] Analogously for temporary parthood even though, of course, this relation requires a notion of 'existence in time'.

them from objects has taken place especially in the philosophy of language [3]. Recently, philosophers have been discussing proposals to reduce events to other basic notions, while researchers from the cognitive, the common-sense, and the modeling perspectives are engaged in exploiting the strength and relevance of the category of events and its relationship with that of objects. There are formal and applicative advantages if events are part of the domain (quantifying over actions, predicating on causality, overcoming reductionist views.)

Several authors collapse the object vs. event distinction to the endurant vs. perdurant one by identifying objects with endurants and events with perdurants. The unification is endorsed by the observations that the 'life of John' is only *partially* present at each time at which it exists (it has distinct temporal parts at each time at which it exists) and 'John' is *wholly* present whenever it exists (the existence of temporal parts is not required). However, if this match were correct, classical perdurantism would not be able to embrace the object vs. event distinction. The reason is easily stated: as shown in Sect. 2.3, all the entities in a perdurantist view have temporal parts when they exist but distinct entities cannot have exactly the same spatio-temporal location. Thus, since 'John' and 'the life of John' have exactly the same spatio-temporal location, perdurantists must identity them. Furthermore, it is not really an option to insist that 'John' is part of the 'life of John' or viceversa. These observations pushed some philosophers to reject as naive the previous identification and to look for a separate (and perhaps more general) foundation of the distinction between objects and events.

Hacker [7] puts emphasis on the fact that events are *primarily* in (directly related to) time while (material) objects are *primarily* in (directly related to) space. This division is based on a series of observations among which:

- The properties (and qualities) that apply to material objects are different from those that apply to events. Typically, material objects have weight, size, shape, texture, etc. and are related by spatial relationships like congruence. Events, on the other hand, can be sudden, brief or prolonged, fast or slow, etc. and can occur before, after, simultaneously to other events.
- Space plays a role in the identification of material objects and in their unity criteria, time in that of events. Material objects that are simultaneously located at different places are different and events that have different temporal locations are different [30].

Of course, even though events are primarily in time and objects primarily in other dimensions, there are strong interrelationships between them. Several authors [7, 27] claim that events are not possible without objects and vice versa. Since technically there seems to be no real advantage in committing to a reductionist view (either choosing that events are the truly basic entities or, alternatively, attributing to objects this role), the most general option is to consider both events and objects as forming two primary and related categories: events need participants (objects) and objects need lives (events).

By means of the relationship between objects and events (aka *participation*), it is possible to say that an object a exists at a certain time t "if and because" its life exists at t [28], i.e., it is the life of a that is the truth-maker for the proposition 'a exists at t'. On the other hand, events are related to space only indirectly via the material objects participating in them.

3 DOLCE-CORE: The New Basis for DOLCE

DOLCE [17] is a foundational ontology developed with the vision that a unique universal ontology for knowledge representation cannot exist. The idea behind DOLCE is that an ontology should be philosophically consistent and transparent (i.e., embrace a clear ontological perspective) and promote its correct application (e.g., by describing explicitly the basic assumptions on which it relies). Furthermore, DOLCE puts much emphasis on interoperability, in particular with other ontological systems, and exploits the "no hidden choice" principle: if a philosophical or applicative position is compatible with the *explicit* commitments of an ontology, then this ontology can indeed be extended to formalize that position. DOLCE goes even further in this view by allowing coexistence of alternative ontological views via parametrization and other formal techniques.

The aim of DOLCE is to capture the intuitive and cognitive bias underlying common-sense while recognizing standard considerations and examples of linguistic nature. These claims are sustained by the accompanying documentation that carefully describes the foundational choices and motivates both the structure and the formalization of DOLCE. Generally speaking, DOLCE does not commit to a strong referentialist metaphysics (it does not make claims on the intrinsic nature of the world) nor to a scientific enterprise. Rather, it looks at reality from the mesoscopic and conceptual level aiming at a formal description of a specific conceptualization of the world. Technically, DOLCE is the result of a careful selection of constraints so to guarantee expressiveness, precision, and simplicity of use.

In the following, we resume our discussion in the previous sections to present the ontological choices made by DOLCE. The discussion is limited to a fragment of the whole ontology (the core formed by the most general categories) and, in some cases, it departs from the published version [17]. For this reason, we dub the ontology in these pages the 'core of DOLCE' or DOLCE-CORE, which forms the basis for the next version of the ontology. Due to lack of space, we will explain only major consequences of these changes.

3.1 Basic Categories

DOLCE-CORE is an ontology limited to entities that exist in time, called *temporal particulars*. While in DOLCE regions and spaces are abstract entities (i.e., entities that are outside time and space), DOLCE-CORE adopts a contextual perspective by introducing them as temporal entities that are created, adopted, abandoned, etc. Following [18], *concepts* (not considered in the original DOLCE) are treated similarly. These assumptions are somehow debatable but have the advantage of providing a general and comprehensive perspective on ontology which is well suited for applications. Abstract regions (and abstract entities in general) can of course exist in the full ontology. They are simply not discussed in the DOLCE-CORE fragment.

DOLCE-CORE partitions *temporal-particulars* (PT) (hereafter *particulars*) into six basic categories: *objects* (O), *events* (E), *individual qualities* (Q), *regions* (R), *concepts* (C), and *arbitrary sums* (AS). All these categories are rigid: an entity cannot change from one category to another over time. Following the observations in Sect. 2.4, the DOLCE's categories ED (endurant) and PD (perdurant) are, respectively, renamed O (object) and E (event). Individual qualities are themselves partitioned into *quality kinds* (Q_i). Each quality kind Q_i is associated to one or more *spaces* (S_{ij}): each individual quality in Q_i has location in (i.e., is associated to a region in each of) the associated spaces S_{ij}. Since we impose that the spaces are disjoint, regions are themselves partitioned into the spaces S_{ij}. For the sake of simplicity, we here consider a unique space T for (regions of) time.[11]

3.2 Parthood and Temporary Parthood

DOLCE-CORE carefully distinguishes spatio-temporal inclusion and parthood by adopting the axioms (A1)-(A4) of extensional mereology, see below. These axioms apply to all entities in the domain. The basic categories, with the exception of AS, are homogeneous: the parts and the sums of entities belonging to one category are still in the same category (see (A5) and (A6)). AS collects those *mixed* entities that are obtained as sum of elements in different basic categories. However, note that the ontology does not enforce any mereological sum of entities to exist. In particular, AS may very well be an empty category. It is left to the user to enforce this constraint (perhaps limited to specific kinds of entities) when needed.

In the following $\mathsf{P}(x,y)$ stands for 'x is part of y', $\mathsf{O}(x,y)$ for 'x overlaps with y', and $\mathsf{SUM}(z,x,y)$ for 'z is the mereological sum of x and y'.

D1 $\mathsf{O}(x,y) \triangleq \exists z(\mathsf{P}(z,x) \wedge \mathsf{P}(z,y))$ *(Overlap)*
D2 $\mathsf{SUM}(z,x,y) \triangleq \forall w(\mathsf{O}(w,z) \leftrightarrow (\mathsf{O}(w,x) \vee \mathsf{O}(w,y)))$ *(Binary Sum)*

A1 $\mathsf{P}(x,x)$ *(reflexivity)*
A2 $\mathsf{P}(x,y) \wedge \mathsf{P}(y,z) \rightarrow \mathsf{P}(x,z)$ *(transitivity)*
A3 $\mathsf{P}(x,y) \wedge \mathsf{P}(y,x) \rightarrow x = y$ *(antisymmetry)*
A4 $\neg\mathsf{P}(x,y) \rightarrow \exists z(\mathsf{P}(z,x) \wedge \neg\mathsf{O}(z,y))$ *(extensionality)*
A5 If ϕ is $\mathrm{O}, \mathrm{E}, \mathrm{Q}_i, \mathrm{S}_{jk}$, or C: $\phi(y) \wedge \mathsf{P}(x,y) \rightarrow \phi(x)$ *(dissectivity)*
A6 If ϕ is $\mathrm{O}, \mathrm{E}, \mathrm{Q}_i, \mathrm{S}_{jk}$, AS, or C: $\phi(x) \wedge \phi(y) \wedge \mathsf{SUM}(z,x,y) \rightarrow \phi(z)$ *(additivity)*

[11] All these statements are easily stated in logic. Here we omit their formal characterization.

As anticipated in Sect. 2.2 we introduce the primitive predicate 'being present at' (PRE) to identify at which times entities exist. No commitment to a specific notion of time is taken in DOLCE-CORE. Nonetheless, in Sect. 3.4 we will analyze different readings of this predicate depending on the category of entities it applies to. PRE is defined on times (A7) and it is dissective and additive over time ((A8) and (A9)).

A7 $\mathsf{PRE}(x,t) \rightarrow \mathrm{T}(t)$

A8 $\mathsf{PRE}(x,t) \wedge \mathsf{P}(t',t) \rightarrow \mathsf{PRE}(x,t')$ (*dissectivity*)

A9 $\mathsf{PRE}(x,t') \wedge \mathsf{PRE}(x,t'') \wedge \mathsf{SUM}(t,t',t'') \rightarrow \mathsf{PRE}(x,t)$ (*additivity*)

As stated in Sect. 3.1, all the entities considered in DOLCE-CORE exist in time:

A10 $\mathrm{PT}(x) \rightarrow \exists t(\mathsf{PRE}(x,t))$

To include entities not in time, one should add to DOLCE-CORE a more general category that includes both temporal and abstract particulars. In this general ontology, DOLCE-CORE provides the formalization of the subclass of temporal particulars.

DOLCE-CORE adopts a temporary extensional mereology, also denoted by P, which is based on axioms (A12)-(A15), i.e., those of extensional mereology adapted to the extra temporal parameter. Further mereological aspects are enforced via the notion of *time regular relation* (see below). Expression $\mathsf{P}(x,y,t)$ stands for 'x is part of y at time t', analogously for $\mathsf{O}(x,y,t)$.

D3 $\mathsf{O}(x,y,t) \triangleq \exists z(\mathsf{P}(z,x,t) \wedge \mathsf{P}(z,y,t))$ (*Temporary Overlap*)

A11 $\mathsf{P}(x,y,t) \rightarrow \mathsf{PRE}(x,t) \wedge \mathsf{PRE}(y,t)$ (*parthood implies being present*)

A12 $\mathsf{PRE}(x,t) \rightarrow \mathsf{P}(x,x,t)$ (*temporary reflexivity*)

A13 $\mathsf{P}(x,y,t) \wedge \mathsf{P}(y,z,t) \rightarrow \mathsf{P}(x,z,t)$ (*temporary transitivity*)

A14 $\mathsf{PRE}(x,t) \wedge \mathsf{PRE}(y,t) \wedge \neg\mathsf{P}(x,y,t) \rightarrow \exists z(\mathsf{P}(z,x,t) \wedge \neg\mathsf{O}(z,y,t))$ (*temporary extensionality*)

A15 If ϕ is $\mathrm{O}, \mathrm{E}, \mathrm{Q}_i, \mathrm{S}_{jk}$ or C: $\phi(y) \wedge \mathsf{P}(x,y,t) \rightarrow \phi(x)$ (*temporary dissectivity*)

Axiom (A3) implies that entities indistinguishable with respect to parthood are identical. Temporary coincidence (D4) provides a weaker form of identification: two entities x and y that are temporary coincident at time t, formally $\mathsf{CC}(x,y,t)$, are indistinguishable relatively to time t (they can still differ in general).[12] If $\mathsf{CC}(x,y,t)$ then all the properties of x at t are also properties of y at t and vice versa.[13] Yet, no constraint follows on properties of x and y at a time different from t.

[12] Perdurantists read $\mathsf{CC}(x,y,t)$ as the identity of the temporal slices $x@t$ and $y@t$.

[13] This claim has to be taken with a *grain of salt* since one should not consider properties that constrain x before or after t itself, e.g., 'being red an year after t' (provided this actually counts as a property).

Axiom (A16) states that in DOLCE-CORE parthood simpliciter can be defined on the basis of temporary parthood, i.e., temporary parthood is more informative. The opposite is true only committing to the existence of temporal parts that is not enforced here. This means that the axioms for temporary parthood are compatible with both the endurantist and perdurantist views of persistence through time. Note that axioms (A10) and (A16) make possible to define parthood simpliciter in terms of temporary parthood. Yet, we use two distinct primitives to avoid hidden commitments: in an extension of DOLCE-CORE that includes abstract entities, both the primitives are necessary (and the two axioms maintain their validity).

D4 $\mathsf{CC}(x,y,t) \triangleq \mathsf{P}(x,y,t) \wedge \mathsf{P}(y,x,t)$ (*Temp. Coincidence*)
D5 $\mathsf{CP}(x,y) \triangleq \exists t(\mathsf{PRE}(x,t)) \wedge \forall t(\mathsf{PRE}(x,t) \rightarrow \mathsf{P}(x,y,t))$ (*Const. Part*)

A16 $\exists t(\mathsf{PRE}(x,t)) \rightarrow (\mathsf{CP}(x,y) \leftrightarrow \mathsf{P}(x,y))$

Temporary parthood presents three main novelties with respect to the corresponding relationship of DOLCE: (1) it is defined on all the particulars that are in time; (2) the existence of sums is not guaranteed; (3) (A16) is new (in DOLCE it was given as a possible extension).

DOLCE-CORE makes use of a few relations that satisfy the following structural axioms:

$R(x,y,t) \wedge \mathsf{P}(t',t) \rightarrow R(x,y,t')$ (*dissectivity*)
$R(x,y,t') \wedge R(x,y,t'') \wedge \mathsf{SUM}(t,t',t'') \rightarrow R(x,y,t)$ (*additivity*)
$R(x,y,t) \wedge \mathsf{CC}(x',x,t) \wedge \mathsf{CC}(y',y,t) \rightarrow R(x',y',t)$ (*substitutivity*)

We can rephrase these constraints as follows: if the relation holds at a time, it holds at any sub-time; if the relation holds at two times, then it holds also at the time spanning the two (provided it exists); if the relation holds for two entities at t, then it holds for entities temporally coincident with them at t.

These constraints are important in setting the DOLCE-CORE framework and relations satisfying them are dubbed *time regular*. In particular, we enforce the temporal parthood of DOLCE-CORE to be *time regular*.

3.3 Properties

DOLCE-CORE offers three different options to represent properties and temporary properties. The first option is standard and consists in the introduction of an *extensional predicate*. With this choice one cannot represent whether the property is related to contextual or social constructions nor its intensional aspects. In addition, to model change through time one needs to add a temporal parameter as in expression $F(a,t)$, i.e., 'a has the property F at t'. This last solution allows to represent dynamics in the properties but, as anticipated, is not suited for roles [29]. For these reasons, predicates are adequate to model the *basic elements* of the user's conceptualization of the world as well as the categories and the primitive relations of DOLCE-CORE. The formalization of properties as extensional predicates is straightforward and requires no special formalism.

The second option consists in reifying properties, that is, in associating them to entities in the category of concepts, C. In order to deal with concepts and to relate concepts to an entity according to the properties the latter has, a (possibly intensional) 'instance-of' relation, called *classification* (CF), is introduced in the ontology. $\mathsf{CF}(x, y, t)$ stands for 'x classifies y as it is at time t' and is characterized in DOLCE-CORE as a *time regular* relation that satisfies also

A17 $\mathsf{CF}(x, y, t) \rightarrow \mathrm{C}(x)$
A18 $\mathsf{CF}(x, y, t) \rightarrow \mathsf{PRE}(y, t)$

The idea is to use concepts to represent properties for which the intensional, contextual, or dynamic aspects are important (as in the case of roles [18]): 'being a student', 'being a catalyst', 'being money'.[14] Since concepts are temporal entities, they can be created, destroyed, etc. Note, however, that they are mereologically constant i.e. they do not change through time with respect to parthood:

A19 $\mathrm{C}(x) \wedge \mathsf{PRE}(x, t) \wedge \mathsf{PRE}(x, t') \rightarrow \forall y(\mathsf{P}(y, x, t) \leftrightarrow \mathsf{P}(y, x, t'))$

The third option relies on the notions of *individual quality*, *quality kind* and (quality-)*space* introduced in Sect. 2.1. Each individual quality, say "the color of my car" or "the weight of John", and its host are in a special relationship called *inherence* (I). Formally, expression $\mathsf{I}(x, y)$, stands for "the individual quality x inheres in the entity y".[15] This relationship binds a specific bearer (A21) and each quality existentially depends on the entity that bears it (A22); in the previous examples the bearers are my car and John, respectively. Finally, axiom (A23) states that qualities exist during the whole life of their bearers.[16]

We anticipated that individual qualities are grouped into quality kinds, say Q_i is the color-quality kind, Q_j the weight-quality kind, etc. These constraints are simple and we do not report them explicitly except for axiom (A24) according to which an entity can have at most one individual quality for each specific quality kind. Axioms (A25) and (A26) say that if two particulars coincide at t then they need to have qualities of the same kind and

[14] Differently from [18], here we do not rely on logical definitions for concepts. The intensional aspect is (partially) characterized by explicitly stating when concepts are different.

[15] In the original version of DOLCE this relation is called *quality* and written **qt**.

[16] For those familiar with trope theory [5], qualities can be seen as sums of tropes. Indeed, one can interpret trope substitution as a change of quality location. The position adopted in DOLCE-CORE is compatible with trope theory without committing to the view that change corresponds to trope substitution.

these qualities also coincide at t. In other terms, entities coincident at t must have qualities that are indistinguishable at t. Axiom (A27) says that the sum of qualities of the same kind that inhere in two objects inheres in the sum of the objects (provided these sums exist).

A20 $\mathsf{I}(x,y) \rightarrow \mathrm{Q}(x)$
A21 $\mathsf{I}(x,y) \wedge \mathsf{I}(x,y') \rightarrow y = y'$
A22 $\mathrm{Q}(x) \rightarrow \exists y(\mathsf{I}(x,y))$
A23 $\mathsf{I}(x,y) \rightarrow \forall t(\mathsf{PRE}(x,t) \leftrightarrow \mathsf{PRE}(y,t))$
A24 $\mathsf{I}(x,y) \wedge \mathsf{I}(x',y) \wedge \mathrm{Q}_i(x) \wedge \mathrm{Q}_i(x') \rightarrow x = x'$
A25 $\mathsf{CC}(x,y,t) \rightarrow (\exists z(\mathsf{I}(z,x) \wedge \mathrm{Q}_i(z)) \leftrightarrow \exists z'(\mathsf{I}(z',y) \wedge \mathrm{Q}_i(z')))$
A26 $\mathsf{CC}(x,y,t) \wedge \mathsf{I}(z,x) \wedge \mathsf{I}(z',y) \wedge \mathrm{Q}_i(z) \wedge \mathrm{Q}_i(z') \rightarrow \mathsf{CC}(z,z',t)$
A27 $\mathsf{I}(x,y) \wedge \mathsf{I}(v,w) \wedge \mathrm{Q}_i(x) \wedge \mathrm{Q}_i(v) \wedge \mathsf{Sum}(z,x,v) \wedge \mathsf{Sum}(s,y,w) \rightarrow \mathsf{I}(z,s)$

The *location* relation (L) provides the link between qualities and spaces. First, we require regions (and in particular spaces) not to change over the time they exist (A28). Expression $\mathsf{L}(x,y,t)$ is used to state "at time t, region x is the location of the individual quality y" as enforced (at least in part) by axioms (A30) and (A31).[17] Each individual quality in Q_i must be located at least in one of the associated spaces S_{ij} (axioms (A34) and (A35)). The location in a single space is unique (A36) and a quality that has a location in a space needs to have some location in that space during its whole life (A37). (A38) says that two qualities coincident at t are also indistinguishable with respect to their locations. Together with (A25) and (A26), this axiom formalizes the substitutivity of temporary properties represented by qualities: two entities that coincide at t are indistinguishable at t with respect to their qualities.

Axioms (A32) and (A33) characterize the fact that the location of an individual quality at t is the mereological sum of all the locations the quality has *during* t, i.e., at all the sub-times of t. Note that if a is the region corresponding to a property value of $1kg$ and b corresponds to a property value of $2kg$, then the sum of a and b is the region including just the two mentioned and is distinguished from the region corresponding to the property value of $3kg$. The sum of locations must not be confused with the 'sum' of property values since, in general, the latter strictly depends on the space structure while the first does not.

A28 $\mathrm{R}(x) \wedge \mathsf{PRE}(x,t) \wedge \mathsf{PRE}(x,t') \rightarrow \forall y(\mathsf{P}(y,x,t) \leftrightarrow \mathsf{P}(y,x,t'))$
A29 $\mathrm{S}_{ij}(x) \wedge \mathrm{S}_{ij}(y) \wedge \mathsf{PRE}(x,t) \rightarrow \mathsf{PRE}(y,t)$
A30 $\mathsf{L}(x,y,t) \rightarrow \mathrm{R}(x) \wedge \mathrm{Q}(y)$
A31 $\mathsf{L}(x,y,t) \rightarrow \mathsf{PRE}(y,t)$

[17] In DOLCE this relation is called *quale* and written ql. In DOLCE there is also a distinction between the *immediate* quale (a non temporary relation) and the *temporary* quale. DOLCE-CORE uses one temporary relation only since the temporal qualities of an event e at t correspond to the temporal qualities of the maximal part of e that spans t.

A32 $\mathsf{L}(x,y,t) \wedge \mathsf{P}(t',t) \wedge \mathsf{L}(x',y,t') \wedge \mathrm{S}_{ij}(x) \wedge \mathrm{S}_{ij}(x') \rightarrow \forall t''(\mathsf{PRE}(x,t'') \rightarrow \mathsf{P}(x',x,t''))$

A33 $\mathsf{L}(x',y,t') \wedge \mathsf{L}(x'',y,t'') \wedge \mathsf{SUM}(t,t',t'') \wedge \mathsf{SUM}(x,x',x'') \wedge \mathrm{S}_{ij}(x') \wedge \mathrm{S}_{ij}(x'') \rightarrow \mathsf{L}(x,y,t)$

A34 $\mathsf{L}(x,y,t) \wedge \mathrm{Q}_i(y) \rightarrow \bigvee_j \mathrm{S}_{ij}(x)$

A35 $\mathrm{Q}(y) \wedge \mathsf{PRE}(y,t) \rightarrow \exists x(\mathsf{L}(x,y,t))$

A36 $\mathsf{L}(x,y,t) \wedge \mathsf{L}(x',y,t) \wedge \mathrm{S}_{jk}(x) \wedge \mathrm{S}_{jk}(x') \rightarrow x = x'$

A37 $\mathsf{L}(x,y,t) \wedge \mathsf{PRE}(y,t') \wedge \mathrm{S}_{jk}(x) \rightarrow \exists x'(\mathsf{L}(x',y,t') \wedge \mathrm{S}_{jk}(x'))$

A38 $\mathsf{L}(x,y,t) \wedge \mathsf{CC}(x',x,t) \wedge \mathsf{CC}(y',y,t) \rightarrow \mathsf{L}(x',y',t)$ (L-*substitutivity*)

3.4 Objects and Events

DOLCE-CORE characterizes the distinction between objects and events following the discussion in Sect. 2.4. In this approach events are primarily in time while objects are primarily in space (in the case of physical objects) or in other dimensions. Since by (A10) qualities, concepts, and regions are in time as well, their participation to events (like their creation or destruction) is plausible. One can investigate this position further and note that Q, C and R can be considered as specializations (subcategories) of O. However, to ensure generality, we made the assumption that qualities, concepts, and regions form categories disjoint from the category of objects.

The DOLCE-CORE unified framework relies on the *participation* relation (PC) to relate the temporal qualities of events and the atemporal qualities of objects. Participation is taken to be a *time regular* relation defined between objects and events: $\mathsf{PC}(x,y,t)$ stands for "the object x participates in the event y at t". Axioms (A40) and (A41) capture the mutual existential dependence between events and objects. Axioms (A42) and (A43) make explicit the fact that participation relies on unity criteria neither for objects nor for events [26]. This simply means that the participation relation is not bound by these unity criteria: an object does not participate to an event as a whole (its parts participate to it as well) and an event does not individuate its participants by the virtue of some special unity property (any larger event has those participants also). Participation, of course, can be used to define more specific relations that take into account unity criteria. Since these criteria often depend on the purposes for which one wants to use the ontology, they are not discussed here. Axiom (A44) makes explicit that a quality kind directly related to events cannot be also directly related to objects and vice versa. Note that the exact list of quality kinds that apply to objects and events are not fixed, they depend on the modeling interests of the user.

A39 $\mathsf{PC}(x,y,t) \rightarrow \mathrm{O}(x) \wedge \mathrm{E}(y)$

A40 $\mathrm{E}(x) \wedge \mathsf{PRE}(x,t) \rightarrow \exists y(\mathsf{PC}(y,x,t))$

A41 $\mathrm{O}(x) \wedge \mathsf{PRE}(x,t) \rightarrow \exists y(\mathsf{PC}(x,y,t))$

A42 $\mathsf{PC}(x,y,t) \wedge \mathsf{P}(y,y',t) \wedge \mathrm{E}(y') \rightarrow \mathsf{PC}(x,y',t)$

A43 $\mathsf{PC}(x,y,t) \wedge \mathsf{P}(x',x,t) \rightarrow \mathsf{PC}(x',y,t)$

A44 $\mathsf{I}(x,y) \wedge \mathrm{Q}_i(x) \wedge \mathrm{E}(y) \wedge \mathsf{I}(z,v) \wedge \mathrm{Q}_j(z) \wedge \mathrm{O}(v) \rightarrow \neg \mathrm{Q}_j(x) \wedge \neg \mathrm{Q}_i(z)$

Regarding the property of 'being primarily in time', we introduce the quality kind 'being time-located'.[18] Let us use TQ for the quality kind for time and recall that T, introduced in Sect. 3.1, is the unique space associated to TQ. DOLCE-CORE (as well as DOLCE) distinguishes *direct* qualities, i.e., properties that can be predicated of x because it has a corresponding individual quality, from *indirect* qualities, i.e., properties of x that are inherited from the relations x has with other entities. For instance, events have a direct temporal location, while objects are located in time just because they participate to events [28]. Analogously, physical objects have a direct spatial location, while events are indirectly located in space through the spatial location of their participants.

(A45) makes explicit the temporal nature of the parameter t in the location relation. (A46) guarantees that the events have a time-quality. These axioms, together with (A10) and the axioms on inheritance and location guarantee that, for events, 'being in time' reduces to having a time-quality located in T. In addition, together with (A41) and (A44) they show that objects are in time because of their participation in events.

A45 $\mathsf{L}(x, y, t) \wedge \text{TQ}(y) \rightarrow x = t$
A46 $\text{E}(x) \rightarrow \exists y(\text{TQ}(y) \wedge \mathsf{I}(y, x))$

Note that if we define the spatial location of events via the location of their participants, and the life of an object as the minimal event in which it (maximally) participates, we obtain that an object spatio-temporally coincides with its life. The distinction between participation and temporary parthood ensures that these two entities, although spatio-temporally coincident, are not identified.

4 Conclusions

In writing this introductory paper, we had three major goals: (1) to distinguish foundational studies from the rest of the ontology research, (2) to introduce topics and methodology typical of foundational ontology and (3) to show a concrete example of how these theoretical arguments can be used to build a foundational ontology. Unfortunately, in literature there is no good reference that presents this research area at length and any attempt to introduce these topics in the limited space of a paper are deemed to be unsatisfactory on several aspects. At least, we hope that the paper gives the average reader the opportunity to appreciate the goals of this area of research as well as the subtle interactions between philosophy, logic and representational issues. Finally, we are glad of the opportunity to present the DOLCE-CORE system of Sect. 3 which is the first step, after the release of the DOLCE ontology in 2002, toward a new version of this ontological system.

[18] Analogously, the ontology comprises the quality kind 'being space-located' which is not presented here.

References

1. D. M. Armstrong. *Universals: An Opinionated Introduction.* Westview Press, 1989.
2. D. M. Armstrong. *A World of States of Affairs.* Cambridge Studies in Philosophy. Cambridge University Press, Cambridge, 1997.
3. R. Casati and A. C. Varzi, editors. *Events.* Dartmund, Aldershot, 1996.
4. Thomas M. Crisp and Donald P. Smith. Wholly present defined. *Philosophy and Phenomenological Research*, 71:318–344, 2005.
5. Chris Daly. Tropes. In D.H. Mellor and A. Oliver, editors, *Properties*, pages 140–159. Oxford University Press, Oxford, 1997.
6. Peter Gärdenfors. *Conceptual Spaces: the Geometry of Thought.* MIT Press, Cambridge, Massachussetts, 2000.
7. P. M. S. Hacker. Events and objects in space and time. *Mind*, 91:1–19, 1982.
8. K. Hawley. *How Thing Persist.* Clarendon Press, Oxford, UK, 2001.
9. Christopher Hughes. More fuss about formulation: Sider (and me) on three- and four-dimensionalism. *Dialectica*, 59(4):463–480, 2005.
10. Ingvar Johansson. Determinables as universals. *The Monist*, 83(1):101–121, 2000.
11. W. E. Johnson. *Logic*, volume 1. Cambridge University Press, Cambridge, 1921.
12. H. Kamp. Events, istants and temporal reference. In R. Baüerle, U. Egli, and A von Stechow, editors, *Semantics from Different Points of View*, pages 376–417. Springer, Berlin, 1979.
13. S. Lesniewski. *Collected Works.* Kluwer, Dordrecht, 1991.
14. F. Loebe. Abstract vs. social roles - towards a general theoretical account of roles. *Applied Ontology*, 2(2):127–258, 2007.
15. Michael J. Loux, editor. *Universals and Particulars: Readings in Ontology.* University of Notre Dame Press, London, 1976.
16. C. Masolo and S. Borgo. Qualities in formal ontology. In P. Hitzler, C. Lutz, and G. Stumme, editors, *Foundational Aspects of Ontologies (FOnt 2005) Workshop at KI 2005*, pages 2–16, Koblenz, Germany, 2005.
17. Claudio Masolo, Stefano Borgo, Aldo Gangemi, Nicola Guarino, and Alessandro Oltramari. Wonderweb deliverable d18. Technical report, CNR, 2003.
18. Claudio Masolo, Laure Vieu, Emanuele Bottazzi, Carola Catenacci, Roberta Ferrario, Aldo Gangemi, and Nicola Guarino. Social roles and their descriptions. In D. Dubois, C. Welty, and M.-A. Williams, editors, *Ninth International Conference on the Principles of Knowledge Representation and Reasoning*, Whistler Canada, 2004.
19. Neil McKinnon. The endurance/perdurance distinction. *Australasian Journal of Philosophy*, 80(3):288–306, 2002.
20. D. H. Mellor and A. Oliver, editors. *Properties.* Oxford University Press, 1997.
21. Trenton Merricks. Endurance and indiscernibility. *Journal of Philosophy*, 91(4):165–184, 1994.
22. M. Rea, editor. *Material Constitution.* Rowman and Littlefield Publishers, 1996.
23. B. Russel. *Human Knowledge. Its Scope and Limits.* Allen and Unwin, London, 1948. parte IV, cap. 8; trad. it., La conoscenza umana, Longanesi, Milano 1963, parte IV, cap. 8, pp. 298-313.
24. T. Sider. Four-dimensionalism. *The Philosophical Review*, 106:197–231, 1997.
25. T. Sider. *Four-Dimensionalism. An Ontology of Persistence and Time.* Clarendon Press, Oxford, 2001.

26. P. Simons. *Parts: a Study in Ontology.* Clarendon Press, Oxford, 1987.
27. P. Simons. On being spread out in time: temporal parts and the problem of change. In Wolfang Spohn et al., editor, *Existence and Explanation*, pages 131–147. Kluwer Academic Publishers, 1991.
28. Peter Simons. How to exist at a time when you have no temporal parts. *The Monist*, 83(3):419–436, 2000.
29. Friedrich Steimann. On the representation of roles in object-oriented and conceptual modelling. *Data and Knowledge Engineering*, 35:83–106, 2000.
30. Eddy M. Zemach. Four ontologies. *Journal of Philosophy*, 67(8):231–247, 1970.

An Ontology for Software

Daniel Oberle[1], Stephan Grimm[2], and Steffen Staab[3]

[1] SAP Research, CEC Karlsruhe, 76131 Karlsruhe, Germany, d.oberle@sap.com
[2] Research Center for Information Technology, FZI, 76131 Karlsruhe, Germany, grimm@fzi.de
[3] University of Koblenz-Landau, ISWeb, 56016 Koblenz, Germany, staab@uni-koblenz.de

1 Introduction

The domain of software is a primary candidate for being formalized in an ontology. On the one hand, the domain is sufficiently complex with different paradigms (e.g., object orientation) and different aspects (e.g., security, legal information, interface descriptions, etc.). On the other hand, the domain is sufficiently stable, i.e., new paradigms and aspects occur rather seldom. Capturing this stable core in a reference ontology for software can be fruitful in order to prevent modeling from scratch. For example, the approaches described in the Chapter "Ontologies and Software Engineering" introduce individual formalizations of at least one paradigm or aspect although they share basic principles.

In this chapter, we present such a reference ontology for software, called the *Core Software Ontology*, which formalizes common concepts in the software engineering realm, such as data, software with its different shades of meaning, classes, methods, etc. As we cannot possibly formalize a complete and comprehensive view of software, the Core Software Ontology is designed for extensibility in different directions. In order to demonstrate the extensibility, the chapter presents three examples of how to extend the core ontology with the notions of libraries, policies, and software components.

The reference nature of such an ontology makes it important to clarify the intended meanings of its concepts and associations. Otherwise, users often have a hard time untangling the intended meanings. The prevailing type of ontologies, namely ones which are lightweight and quite often reduced to simple taxonomies, are not eligible for this purpose because they exhibit the following shortcomings (as identified in [10]):

Conceptual Ambiguity: We will consider an ontology to be conceptually ambiguous if it is difficult for users to understand the intended meaning

S. Staab and R. Studer (eds.), *Handbook on Ontologies*, International Handbooks on Information Systems, DOI 10.1007/978-3-540-92673-3,

of concepts, the associations between the concepts, and the relationships between concepts and modeled entities.

Poor Axiomatization: Even when an ontology is easy to understand by many or most of its users, it may have only a poor axiomatization. Such a poor axiomatization will lead to an unsatisfying restriction of possible logical models (cf. Chapter "What is an *Ontology*?").

Loose Design: An ontology is afflicted with loose design, if it contains modeling artifacts. Modeling artifacts are concepts and associations which do not bear ontological meaning.

Narrow Scope: An ontology exhibits narrow scope when it is unclear how a distinction could be made between the objects and events within an information system (regarding data and the manipulation of data) and the real-world objects and events external to such a system. As an example consider the distinction between a user account and its corresponding natural person(s).

In order to remedy the shortcomings, we build the *Core Software Ontology* on a foundational ontology (cf. Chapter "Foundational Choices in DOLCE") and apply content ontology design patterns (cf. Chapter "Ontology Design Patterns"). We demonstrate that the formalization of the software domain can greatly benefit from the use of the DOLCE foundational ontology and the content ontology design patterns extracted from Descriptions & Situations, the Ontology of Plans, as well as the Ontology of Information Objects.

The chapter is structured as follows: we start by presenting the origin and motivation of our ontology in Sect. 2 in order to understand which aspects have been taken into account and why. Subsequently, we sketch the formalization of the Core Software Ontology in Sect. 3 and some of its extensions in Sect. 4. For the complete formalization we refer the reader to [11,12]. Section 5 shows examples of how the four shortcomings are improved as a proof of concept. We give an overview of related work in Sect. 6 and conclude in Sect. 7.

2 Background

An initial ontology for software certainly cannot cover every single paradigm and aspect related to software. As an example, we limit ourselves to the concept of object orientation. In order to understand which aspects have been taken into account and why, we present here the origin and motivation of our ontology, viz., the work presented in [11]. The motivation for building the ontology in this work is the missing conceptual coherence of application server and Web service descriptors. We motivate that a careful and rigorous modeling of the computational domain is necessary to automate – or at least facilitate – some development and management tasks related to software components and Web services. Several use cases are identified that give us indications of what concepts a suitable ontology must contain.

The use cases relevant for developing and managing software components in application servers are: *libraries and their dependencies*, *conflicting licenses of libraries*, *capability descriptions*, *component classification and discovery*, *semantics of parameters*, *support in error handling*, *reasoning with transactional settings* and *reasoning with security settings.*

The use cases relevant for developing and managing Web services are: *analyzing message contexts*, *selecting service functionality*, *detecting loops in the interorganisational workflow*, *incompatible inputs and outputs*, *relating communication parameters*, *monitoring of changes*, *aggregating service information* and *quality of service.*

Altogether, the use cases let us derive a set of modeling requirements for deciding which aspects our ontology should formalize. The modeling requirements are: (i) *libraries*, *licenses*, *component profiles*, *component taxonomies*, *API descriptions*, *semantic API descriptions*, *access rights* and *workflow information* of software components and (ii) *service profiles*, *service taxonomies*, *policies*, *workflow information*, *API descriptions*, as well as *semantic API descriptions* of Web services.

We do not claim that the modeling requirements are exhaustive. However, they allow us to constrain the initial modeling horizon. As demonstrated in the following, the ontology is designed in an extensible way such that further modeling requirements can be met easily.

3 Formalization of the Software Domain

Our contribution starts in this section with the *Core Software Ontology (CSO)* which introduces fundamental concepts of the software domain such as software itself, data, classes, or methods. The purpose of the ontology is to provide a reference by specifying the intended meanings of software terms as precisely as possible, and to prevent the shortcomings mentioned in the introduction.

The contribution continues in Sect. 4 where we extend the Core Software Ontology in different directions, e.g., in the direction of software components, resulting in a Core Ontology of Software Components. All of the ontologies have been presented in detail in [11, 12] and are available at `http://cos.ontoware.org`.

Figure 1 shows that we reuse the foundational ontology DOLCE [9] as a modeling basis. DOLCE and its extensions Descriptions & Situations (DnS) [5], the Ontology of Plans (OoP) [4], and the Ontology of Information Objects (OIO) [4] provide us with content ontology design patterns which we apply for formalizing the software domain. For extensive running examples please refer to [11,12].

3.1 Software vs. Data

We start our discussion of the Core Software Ontology with a detailed discussion of software and data. In order to clarify both concepts, which are heavily

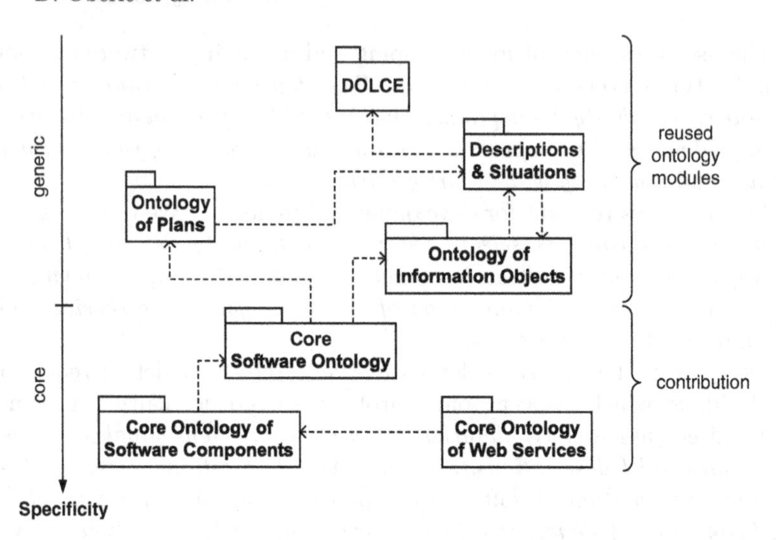

Fig. 1. Overview of the ontologies as UML package diagram. Packages represent ontologies; dotted lines represent dependencies between ontologies. An ontology O_1 depends on O_2 if it specializes concepts of O_2, has associations with domains and ranges to O_2 or reuses its axioms

inflicted by polysemy, it is necessary to identify and formalize the entities of the computational domain. The computational domain has a reality of its own, consisting of data manipulated by programs that implement algorithms. The programs that manipulate the data are usually referred to as software. Upon close inspection, it seems that the term software is overloaded and refers to at least three different concepts:

1. The encoding of an algorithm specification in some kind of representation. Encoding can be either in mind, on paper, or any other form. The Quicksort algorithm can be represented as Java or pseudo code, for instance. This is SoftwareAsCode (which we abbreviate to Software) and is a kind of OIO:InformationObject.[1]
2. The realization of the code in a concrete hardware. These realizations are the DOLCE:PhysicalEndurants that are stored on hard disc or residing in memory. Henceforth, we call them ComputationalObjects (a special kind of OIO:InformationRealization). This could be the appearance of the

[1] Throughout the chapter, concepts and associations are written in sans serif and are labelled in a namespace-like manner. Namespace-prefixes indicate the ontology where concepts and associations are defined. If no namespace is given, concepts and associations are assumed to be defined in the ontology currently discussed. With respect to the formulae given in the following, the reader might refer to Chapters "Description Logics, Ontologies in F-Logic, Resource Description Framework (RDF), Web Ontology Language: OWL, Ontologies and Rules" for the logic background.

Quicksort algorithm in main memory that can be interpreted and executed by the CPU. Hence, the difference between 1 and 2 is that 2 is physically present in some hardware.

3. The running system, which is the result of an execution of a ComputationalObject. This is the form of software which manifests itself in a sequence of activities in the computational domain, e.g., the increment of a variable, the comparison of data, the storage of data on the hard disc, etc. This form of software is a DOLCE:Perdurant which we call ComputationalActivity.

ComputationalObjects (item 2) are a specialization of OIO:InformationRealization (any entity that realizes an OIO:InformationObject) as introduced in the Ontology of Information Objects. ComputationalActivities (item 3) are a specialization of OoP:Activity as introduced in the Ontology of Plans. ComputationalObjects and ComputationalActivities are the entities that live in the computational domain.

ComputationalObjects are characterized by the fact that they are necessarily dependent on Hardware which is a DOLCE:PhysicalObject. A suitable dependence association is axiomatized in DOLCE and is called specificallyConstantlyDependsOn. A ComputationalObject is considered here as a spatiotemporally bounded entity, therefore it exists for the time a memory cell is realizing a certain Software, for instance. Copies of ComputationalObjects in the same or another Hardware are different, although related by some kind of "copy" association. For example, in the case of mobile agents, where people refer to a mobile agent as a piece of software that can move from machine to machine executing the "same" process, it is useful to make agents distinct because the "same" agent can perform differently from machine to machine. The similarity has to be caught via a specialized association, such as copy (which we do not define here) rather than via logical identity.

The execution of a ComputationalObject leads to ComputationalActivities. ComputationalActivities require at least one ComputationalObject as a participant. The definitions below formalize the described properties.

(D1) $\mathsf{ComputationalObject}(x) =_{def} \mathsf{OIO{:}InformationRealization}(x) \land$
$\forall y(\mathsf{DOLCE{:}participantIn}(x, y) \rightarrow \mathsf{ComputationalActivity}(y)) \land$
$\exists d(\mathsf{DOLCE{:}specificallyConstantlyDependsOn}(x, d) \land \mathsf{Hardware}(d))$

(D2) $\mathsf{ComputationalActivity}(x) =_{def} \mathsf{OoP{:}Activity}(x) \land$
$\forall y(\mathsf{DOLCE{:}participantIn}(y, x) \rightarrow \mathsf{ComputationalObject}(y)) \land$
$\exists c(\mathsf{DOLCE{:}participantIn}(c, x) \land \mathsf{ComputationalObject}(c))$

(D3) $\mathsf{DOLCE{:}specificallyConstantlyDependsOn}(x, y) =_{def}$
$(\exists t(\mathsf{DOLCE{:}presentAt}(x, t)) \land \forall t(\mathsf{DOLCE{:}presentAt}(x, t) \rightarrow$
$\mathsf{DOLCE{:}presentAt}(y, t)))$

(D4) $\mathsf{DOLCE{:}presentAt}(x, t) =_{def} \exists t'(\mathsf{DOLCE{:}ql}_T(t', x) \land \mathsf{DOLCE{:}part}(t, t'))$

Regarding item 1, we characterize Software as an OIO:InformationObject. Accordingly, we specialize the design pattern represented by the Ontology

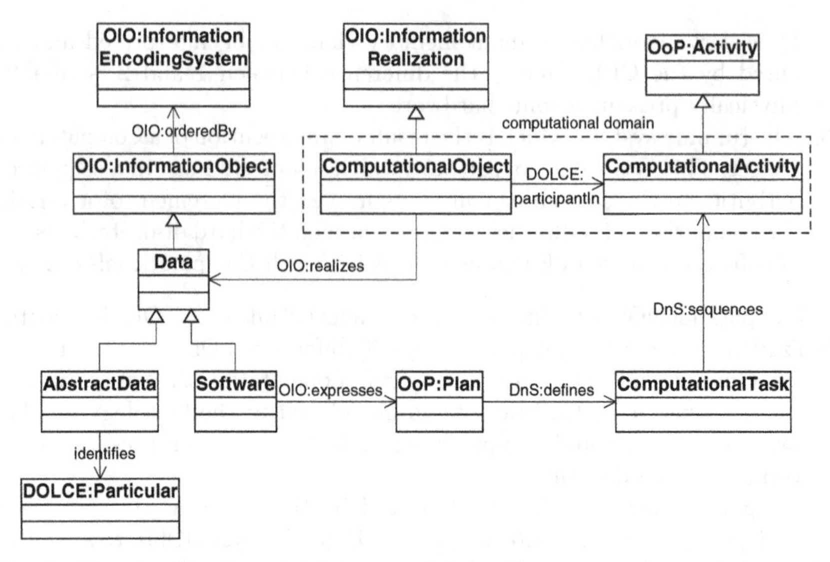

Fig. 2. The classification of software and data. Concepts and associations taken from DOLCE, Descriptions & Situations (DnS), the Ontology of Plans (OoP), the Ontology of Information Objects (OIO) are labelled with a namespace

of Information Objects [4]. First, we constrain the OIO:realizedBy association to ComputationalObjects. Second, we say that Software OIO:expresses an OoP:Plan (cf. Fig. 2 for an overview). The OoP:Plan consists of an arbitrary number of ComputationalTasks, which DnS:sequence ComputationalActivities (cf. Definition (D6) below). As explained in the Ontology of Plans [4], Tasks are the descriptive counterparts of OoP:Activities which are actually carried out. Definition (D5) below captures this intuition of software.

(D5) $\mathsf{Software}(x) =_{def} \mathsf{OIO{:}InformationObject}(x) \wedge \forall y(\mathsf{OIO{:}realizedBy}(x,y) \rightarrow \mathsf{ComputationalObject}(y)) \wedge \exists p,t(\mathsf{OoP{:}Plan}(p) \wedge \mathsf{OIO{:}expresses}(x,p) \wedge \mathsf{ComputationalTask}(t) \wedge \mathsf{DnS{:}defines}(p,t))$

(D6) $\mathsf{ComputationalTask}(x) =_{def} \mathsf{OoP{:}Task}(x) \wedge \forall y(\mathsf{DnS{:}sequences}(x,y) \rightarrow \mathsf{ComputationalActivity}(y))$

We consider the data that is manipulated by the programs as ComputationalObjects as well. This reflects the fact that the appearances in the main memory or on the hard disc can be interpreted as instructions for the CPU (i.e., as software) or can be treated as data from the viewpoint of another program. For example, the operating system manipulates application software (loading and unloading it into memory, etc.) much like application software manipulates application data.

Hence, Data can also be considered as a special kind of OIO:InformationObject. The difference to Software is that Data does not OIO:express an

OoP:Plan.[2] Furthermore, we introduce AbstractData as a special kind of Data that identifies something different from itself. An example for AbstractData might be a user account in a Unix operating system which has a physical counterpart in the real world. Thus, we say that AbstractData identifies a DOLCE:Particular (a natural person, a company, a physical object) [4]. The identifies association is a specialization of OIO:about. Definitions (D7), (D8), and (D9) capture these intuitions.

(D7) $\mathsf{Data}(x) =_{def} \mathsf{OIO{:}InformationObject}(x) \wedge \forall y(\mathsf{OIO{:}realizedBy}(x, y) \rightarrow \mathsf{ComputationalObject}(y))$

(D8) $\mathsf{AbstractData}(x) =_{def} \mathsf{Data}(x) \wedge \exists y(\mathsf{DOLCE{:}Particular}(y) \wedge \mathsf{identifies}(x, y))$

(D9) $\mathsf{identifies}(x, y) =_{def} \mathsf{OIO{:}about}(x, y) \wedge \mathsf{AbstractData}(x) \wedge \mathsf{DOLCE{:}Particular}(y) \wedge x \neq y$

The theorem (T1) below is an entailment of our axiomatization. (T1) states that Software must also be considered as Data. As discussed before, this is intuitively clear because an algorithm can be considered as Data from the viewpoint of a compiler, for example. Comparing (D5) and (D7), we find that Software additionally OIO:expresses an OoP:Plan with at least one ComputationalTask. Thus, Software is strictly more specific than Data.

(T1) $\mathsf{Software}(x) \rightarrow \mathsf{Data}(x)$

3.2 Interfaces, Classes, and Methods

Building on the fundamental notions of software and data introduced in the previous section, we now formalize the most important concepts of object orientation. We begin with a Class in Definition (D10) as a special kind of Software that encapsulates an arbitrary number of Data and an arbitrary number of Methods. Vice versa, a Method is defined as being a part of a Class, having input and output parameters and throwing exceptions.[3] The associations between Methods and their parameters and exceptions are established via methodRequires, methodYields, and methodThrows (cf. (D11), (A1), (A2), and (A3)). Exceptions are special kinds of Classes as defined in (D12). dataType relates Data with specific kinds of DOLCE:Regions in the case of simple datatypes, such as strings or integers, or with other Data in the case of complex datatypes, e.g., other classes (cf. Axiom (A4)).

(D10) $\mathsf{Class}(x) =_{def} \mathsf{Software}(x) \wedge \forall y(\mathsf{DOLCE{:}properPart}(y, x) \rightarrow (\mathsf{Data}(y) \vee \mathsf{Method}(y)))$

(D11) $\mathsf{Method}(x) =_{def} \mathsf{Software}(x) \wedge \forall y(\mathsf{DOLCE{:}properPart}(x, y) \rightarrow \mathsf{Class}(y))$

[2] The reader may note, that we occasionally use concept and association names (written in sans serif and preceded by a namespace to clarify their origin) as subjects, objects, and predicates of the sentences in the text.

[3] The OoP:Plan of the Class contains all Plans of its Methods as alternatives.

(D12) $\mathsf{Exception}(x) =_{def} \mathsf{Class}(x) \land \forall y(\mathsf{methodThrows}(y, x) \rightarrow \mathsf{Method}(y))$
(D13) $\mathsf{DOLCE{:}properPart}(x, y) =_{def} \mathsf{DOLCE{:}part}(x, y) \land \neg\mathsf{DOLCE{:}part}(y, x)$

(A1) $\mathsf{methodRequires}(x, y) \rightarrow \mathsf{Method}(x) \land \mathsf{Data}(y)$
(A2) $\mathsf{methodYields}(x, y) \rightarrow \mathsf{Method}(x) \land \mathsf{Data}(y)$
(A3) $\mathsf{methodThrows}(x, y) \rightarrow \mathsf{methodYields}(x, y) \land \mathsf{Exception}(y)$
(A4) $\mathsf{dataType}(x, y) \rightarrow \mathsf{Data}(x) \land (\mathsf{Region}(y) \lor \mathsf{Data}(y))$

We here introduce the notion of an Interface in order to group methods and parameters independently of the Classes they belong to (cf. (D14) and (A5) below). The Interface extends the notion of Java interfaces because it allows to grasp additional information as explained above. In our ontology, the Interface has to be classified as Data as it cannot be executed, i.e., it does not OIO:express an OoP:Plan. Different Classes may implement the same Interface as stated in (A6). In doing so, we are able to model that different Classes provide names for Methods with comparable functionality (e.g., `getPrice()` vs. `getCost()`).

(D14) $\mathsf{Interface}(x) =_{def} \mathsf{Data}(x) \land \forall m(\mathsf{inferfaceRequires}(x, m) \rightarrow$
$(\exists p(\mathsf{OIO{:}expresses}(m, p) \land \mathsf{OoP{:}Plan}(p)) \land \forall d(\mathsf{methodRequires}(m, d) \rightarrow$
$\exists e(\mathsf{DOLCE{:}Particular}(e) \land \mathsf{OIO{:}about}(d, e)))))$

(A5) $\mathsf{interfaceRequires}(x, y) \rightarrow$
$\mathsf{DOLCE{:}properPart}(y, x) \land \mathsf{Interface}(x) \land \mathsf{Method}(y)$
(A6) $\mathsf{implements}(x, y) \rightarrow \mathsf{Class}(x) \land \mathsf{Interface}(y) \land$
$\forall m_1 \exists m_2(\mathsf{interfaceRequires}(y, m_1) \rightarrow \mathsf{DOLCE{:}properPart}(x, m_2))$

3.3 Workflow Information

Workflow information, such as method invocations, also belong to the fundamental notions of software. In order to model such information, we use and specialize the ontology design pattern of the Ontology of Plans which in turn builds on Descriptions & Situations. We do so because the design pattern allows abstracting from concrete, i.e., actually executed, workflows. That means, we use ComputationalTasks, which are OoP:Tasks, to represent invocations, the addition of two integers, etc. rather than the actual executions of such tasks (which would be ComputationalActivities). ComputationalTasks are grouped and linked via the OoP:successor and OoP:predecessor associations in an OoP:Plan (a DnS:SituationDescription).[4]

The workflow information we need to model is twofold. First, we have to model invocations between software. Second, we also need to model the inputs and outputs of tasks because the Ontology of Plans does not provide such capabilities.

[4] The OoP:predecessor and OoP:successor associations hold between OoP:Tasks, and are different from OoP:precondition and OoP:postcondition associations, which hold between OoP:Plans and DnS:SituationDescriptions.

Invocations Between Software

We start with two associations, viz., executes and accesses, to formalize invocations between Software. Below, (D15) introduces executes as "shortcut" between Software, such as Class or Method, and a ComputationalTask. For example, the `doGet()` method of a servlet executes an invocation task.

(D16) introduces accesses as "shortcut" between the ComputationalTask and the Software or Data that is being called or modified by the task. The sequence of executes and accesses can be further abbreviated by invokes which is declared as being transitive (cf. (D17) and (A7)). Axioms (A8) and (A9) are introduced for convenience. Regarding (A8), we say that also a Class executes a ComputationalTask when one of its Methods executes this task. Regarding (A9), we state that invokes also holds when we have succeeding tasks.

(D15) $\mathsf{executes}(x, y) =_{def} \mathsf{Software}(x) \wedge \mathsf{ComputationalTask}(y) \wedge$
$\exists co, ca, p(\mathsf{ComputationalObject}(co) \wedge \mathsf{ComputationalActivity}(ca) \wedge$
$\mathsf{OoP{:}Plan}(p) \wedge \mathsf{OIO{:}realizedBy}(x, co) \wedge \mathsf{OIO{:}expresses}(x, p) \wedge$
$\mathsf{DnS{:}defines}(p, y) \wedge \mathsf{DnS{:}sequences}(y, ca) \wedge \mathsf{DOLCE{:}participantIn}(co, ca))$

(D16) $\mathsf{accesses}(x, y) =_{def}$
$\mathsf{ComputationalTask}(x) \wedge \mathsf{Data}(y) \wedge \exists ca, co(\mathsf{DnS{:}sequences}(x, ca) \wedge$
$\mathsf{ComputationalActivity}(ca) \wedge \mathsf{DOLCE{:}participantIn}(co, ca) \wedge$
$\mathsf{ComputationalObject}(co) \wedge \mathsf{OIO{:}realizes}(co, y))$

(D17) $\mathsf{invokes}(x, y) =_{def} \exists z(\mathsf{executes}(x, z) \wedge \mathsf{accesses}(z, y))$

(A7) $\mathsf{invokes}(x, z) \leftarrow \mathsf{invokes}(x, y) \wedge \mathsf{invokes}(y, z)$

(A8) $\mathsf{executes}(x, y) \leftarrow$
$(\mathsf{executes}(z, y) \wedge \mathsf{Method}(z) \wedge \mathsf{DOLCE{:}properPart}(z, x) \wedge \mathsf{Class}(x))$

(A9) $\mathsf{invokes}(x, z) \leftarrow \mathsf{executes}(x, y) \wedge \mathsf{OoP{:}successor}(y, t) \wedge \mathsf{accesses}(t, z)$

Inputs and Outputs

Besides invocations, we also need to model the Inputs and Outputs of tasks. The Ontology of Plans does not provide such capabilities. Inputs and Outputs are required when we want to represent the information of a WS-BPEL workflow, for instance. Inputs and Outputs are DnS:Roles which are both DnS:playedBy Data and DnS:definedBy an OoP:Plan (cf. (D18), (D19) and (A12)). The relationships between Inputs (Outputs) and ComputationalTasks are modeled by inputFor / outputFor, as specified in (A10), and (A11).[5] The difference between Inputs and Outputs is that the former must be present before the latter (cf. (A13)).

(D18) $\mathsf{Input}(x) =_{def} \mathsf{DnS{:}Role}(x) \wedge \forall y(\mathsf{DnS{:}playedBy}(x, y) \rightarrow \mathsf{Data}(y))$

(D19) $\mathsf{Output}(x) =_{def} \mathsf{DnS{:}Role}(x) \wedge \forall y(\mathsf{DnS{:}playedBy}(x, y) \rightarrow \mathsf{Data}(y))$

[5] Both are specializations of DnS:modalTarget, viz., the generic association holding between DnS:Roles and DnS:Courses.

(A10) $\mathsf{inputFor}(x, y) \rightarrow$
$\mathsf{DnS{:}modalTarget}(x, y) \wedge \mathsf{Input}(x) \wedge \mathsf{ComputationalTask}(y)$

(A11) $\mathsf{outputFor}(x, y) \rightarrow$
$\mathsf{DnS{:}modalTarget}(x, y) \wedge \mathsf{Output}(x) \wedge \mathsf{ComputationalTask}(y)$

(A12) $\mathsf{Input}(x) \vee \mathsf{Output}(x) \rightarrow \exists p(\mathsf{OoP{:}Plan}(p) \wedge \mathsf{DnS{:}defines}(p, x))$

(A13) $\mathsf{ComputationalTask}(ct) \rightarrow \forall d_1, d_2(\forall i, o(\mathsf{inputFor}(i, ct) \wedge$
$\mathsf{DnS{:}playedBy}(i, d_1) \wedge \mathsf{outputFor}(o, ct) \wedge \mathsf{DnS{:}playedBy}(o, d_2)) \rightarrow$
$\exists t_1, t_2(\mathsf{presentAt}(d_1, t_1) \wedge \mathsf{presentAt}(d_2, t_2) \wedge t_1 < t_2))$

4 Extensions to the Core Software Ontology

In this section, we continue our contribution of formalizing the software domain by extending the Core Software Ontology in three different directions. First, we start with the minor extension of libraries and licenses which is put in the Core Software Ontology itself. Second, our focus are access rights and policies which were originally put in the Core Software Ontology as well. However, [7] continued to extend in this direction and emancipated their formalization in a Core Policy Ontology. Third, an extra ontology module, i.e., the *Core Ontology of Software Components (COSC)*, is devoted to the paradigm of software componentry. The reader may note, that other extensions are possible, e.g., the Core Ontology of Web Services as presented in [12].

4.1 Libraries and Licenses

We introduce the concepts of SoftwareLibrary and License in (D20) and (D21) below. Both occur in many programming languages and are a common means in software engineering. A SoftwareLibrary consists of a number of CSO:Classes and is classified as CSO:Data because it cannot be executed as a whole. The concept License is a special kind of LegalContract as introduced in the Core Legal Ontology [3].

(D20) $\mathsf{SoftwareLibrary}(x) =_{def} \mathsf{CSO{:}Data}(x) \wedge \forall c(\mathsf{DOLCE{:}properPart}(x, c) \rightarrow$
$\mathsf{CSO{:}Class}(c))$

(D21) $\mathsf{License}(x) =_{def}$
$\mathsf{LegalContract}(x) \wedge \exists y(\mathsf{CSO{:}Software}(y) \wedge \mathsf{DnS{:}involves}(x, y))$

Very often there are functional dependencies between libraries that are revealed only during run time (e.g., by `ClassNotFoundExceptions` in Java). For example, a library `lib1.jar` might depend on `lib2.jar` which in turn depends on `lib3.jar` and so forth. It is a very tedious task to keep track of such dependencies and, additionally, to check whether there are conflicts between libraries in this dependency graph. In order to reason with such information, we introduce further associations and axioms, such as the transitive libraryDependsOn in (A14) and (A15) and the symmetric libraryConflictsWith in (A16) and (A17) below, while in (A18) we formalize indirect conflicts.

The existence of incompatible licenses further complicates the situation. Even if libraries in the dependency graph do not conflict, they might have incompatible licenses. In order to reason with such information, we further introduce the association releasedUnder between SoftwareLibraries and Licenses in (A19), as well as the symmetric licenseIncompatibleWith in (A20) and (A21).

(A14) libraryDependsOn$(x, y) \rightarrow$
DOLCE:specificallyConstantlyDependsOn$(x, y) \wedge$ SoftwareLibrary$(x) \wedge$ SoftwareLibrary(y)
(A15) libraryDependsOn$(x, z) \leftarrow$
libraryDependsOn$(x, y) \wedge$ libraryDependsOn(y, z)
(A16) libraryConflictsWith$(x, y) \rightarrow$ SoftwareLibrary$(x) \wedge$ SoftwareLibrary(y)
(A17) libraryConflictsWith$(x, y) \leftrightarrow$ libraryConflictsWith(y, x)
(A18) libraryConflictsWith$(x, z) \leftarrow$
libraryDependsOn$(x, y) \wedge$ libraryConflictsWith(y, z)
(A19) releasedUnder$(x, y) \rightarrow$
OIO:expresses$(x, y) \wedge$ SoftwareLibrary$(x) \wedge$ License(y)
(A20) licenseIncompatibleWith$(x, y) \rightarrow$ License$(x) \wedge$ License(y)
(A21) licenseIncompatibleWith$(x, y) \leftrightarrow$ licenseIncompatibleWith(y, x)

4.2 Access Rights and Policies

In general, access rights are required to state that access is granted for a specific user on a specific resource. Policies can be regarded as a generalization of access rights. They define high-level guidelines that constrain the behavior of an information system.

We use and specialize Descriptions & Situations for modeling access rights and policies. The design pattern represented by Descriptions & Situations provides us with the basic primitives of context modeling, such as the notion of roles, which allows us to talk about subjects and objects of a policy on the abstract level, i.e., independent of the entities that play such roles. As described in [5], Descriptions & Situations therefore distinguishes between *descriptive* entities and *ground* entities.

In a first step, it is necessary to introduce further *ground* entities which are required later on. (D22) below specifies a User as a special kind of AbstractData which identifies a DnS:Agent. The intuition behind User is a user account in an operating system. Hence, Users identify DnS:Agents which are either DOLCE:AgentivePhysicalObjects or DOLCE:AgentiveSocialObjects. In most cases, a natural person is associated with such an account. We aggregate Users to a UserGroup by exploiting DnS:Collection in (D23).

(D22) User$(x) =_{def}$ AbstractData$(x) \wedge \forall y($identifies$(x, y) \rightarrow$ DnS:Agent$(y))$
(D23) UserGroup$(x) =_{def}$ DnS:Collection$(x) \wedge \forall y($DnS:member$(x, y) \rightarrow$
User$(y))$

In a second step, we specialize the *descriptive* entities of Descriptions & Situations, viz., DnS:Roles, DnS:Courses, DnS:Parameters, and DnS:Situation-Descriptions as follows. First, we introduce two DnS:Roles to represent the subject and the object of a policy in (D24) and (D25). PolicySubjects are DnS:AgentiveRoles and can be DnS:playedBy Users or UserGroups. PolicyObjects are DnS:NonAgentiveRoles and can be DnS:playedBy Data. Second, we need to represent the predicate of a policy by a special kind of DnS:Course. (D6) already introduces ComputationalTask which meets this requirement. We further aggregate such tasks to TaskCollections in (D26). The intuition behind TaskCollections are the security "roles" in operating systems or database systems. This means that a TaskCollection groups ComputationalTasks, such as read, write, or execute. Third, we introduce Constraints as special kinds of DnS:Parameter. The ComputationalTask or TaskCollections can be constrained in some way, e.g., a Web service policy might state that an invocation is only possible with Kerberos or X509 authentication (cf. (D27)). Finally, we construct a PolicyDescription, viz., a special kind of DnS:SituationDescription, from the aforementioned concepts.[6] Axiom (A22) requires each PolicyDescription to have a PolicySubject, ComputationalTask, and a PolicyObject. Figure 3 provides an overview.

(D24) $\mathsf{PolicySubject}(x) =_{def} \mathsf{DnS{:}AgentiveRole}(x) \wedge \forall y(\mathsf{DnS{:}playedBy}(x,y) \rightarrow (\mathsf{User}(y) \vee \mathsf{UserGroup}(y))) \wedge \forall z(\mathsf{DnS{:}attitudeTowards}(x,z) \rightarrow (\mathsf{ComputationalTask}(z) \vee \mathsf{TaskCollection}(z)))$

(D25) $\mathsf{PolicyObject}(x) =_{def} \mathsf{DnS{:}NonAgentiveRole}(x) \wedge \forall y(\mathsf{DnS{:}playedBy}(x,y) \rightarrow \mathsf{Data}(y)) \wedge \forall z(\mathsf{DnS{:}attitudeTowards}(x,z) \rightarrow (\mathsf{ComputationalTask}(z) \vee \mathsf{TaskCollection}(z)))$

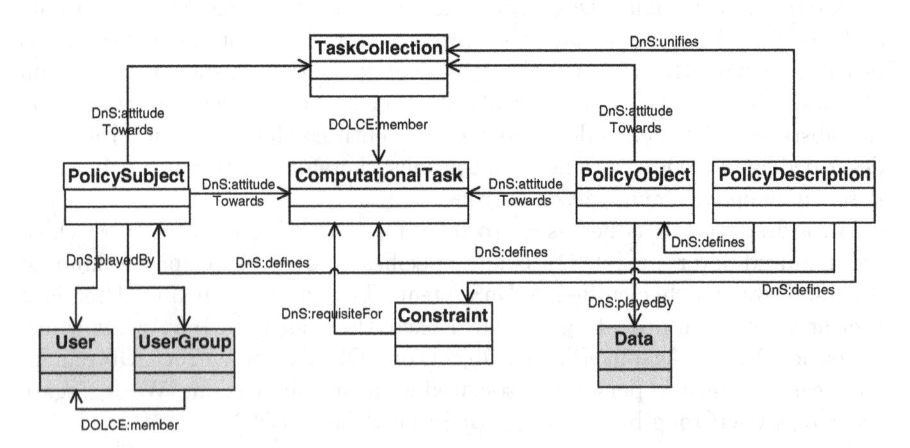

Fig. 3. The Policy Description as UML class diagram. Grey classes represent *ground* entities, white classes the *descriptive* entities of Descriptions & Situations or specializations

[6] Note that DnS:unifies is the generic association between DnS:SituationDescriptions and DnS:Collections.

(D26) $\mathsf{TaskCollection}(x) =_{def} \mathsf{DnS{:}Collection}(x) \wedge \forall y(\mathsf{DnS{:}member}(x,y) \rightarrow \mathsf{ComputationalTask}(y))$

(D27) $\mathsf{Constraint}(x) =_{def} \mathsf{DnS{:}Parameter}(x) \wedge \forall y(\mathsf{DnS{:}requisiteFor}(x,y) \rightarrow (\mathsf{ComputationalTask}(y) \vee \mathsf{TaskCollection}(y))) \wedge \forall z(\mathsf{DnS{:}defines}(z,x) \rightarrow \mathsf{PolicyDescription}(z))$

(D28) $\mathsf{PolicyDescription}(x) =_{def} \mathsf{DnS{:}SituationDescription}(x) \wedge \forall y(\mathsf{DnS{:}unifies}(x,y) \rightarrow \mathsf{TaskCollection}(y)) \wedge \forall z(\mathsf{DnS{:}defines}(x,z) \rightarrow \mathsf{Constraint}(z) \vee \mathsf{ComputationalTask}(z) \vee \mathsf{PolicySubject}(z) \vee \mathsf{PolicyObject}(z))$

(A22) $\mathsf{PolicyDescription}(x) \rightarrow \exists s,t,o(\mathsf{DnS{:}defines}(x,s) \wedge \mathsf{PolicySubject}(s) \wedge \mathsf{DnS{:}defines}(x,t) \wedge \mathsf{ComputationalTask}(t) \wedge \mathsf{DnS{:}defines}(x,o) \wedge \mathsf{PolicyObject}(o))$

It is worthwhile to spend some words on the DnS:attitudeTowards association between DnS:Roles and DnS:Courses. The DnS:attitudeTowards association is a special kind of DnS:modalTarget and can be considered the *descriptive* counterpart of the DOLCE:participantIn association. It is used to state attitudes, attention, or even subjection that an object can have with respect to an action or process. In our case, DnS:attitudeTowards is used to state the relationship between PolicySubjects, as well as PolicyObjects, and the ComputationalTask or TaskCollection. Descriptions & Situations provides us with three initial specializations of DnS:attitudeTowards, viz., DnS:rightTowards, DnS:empoweredTo, and DnS:obligedTo. We further refine DnS:rightTowards in (A23) below.

(A23) $\mathsf{computationalRightTowards}(x,y) \rightarrow \mathsf{DnS{:}rightTowards}(x,y) \wedge \mathsf{PolicySubject}(x) \wedge (\mathsf{ComputationalTask}(y) \vee \mathsf{TaskCollection}(y))$

(A24) $\mathsf{computationalRightTowards}(x,z) \leftarrow \mathsf{computationalRightTowards}(x,y) \wedge \mathsf{TaskCollection}(y) \wedge \mathsf{DnS{:}member}(y,z) \wedge \mathsf{ComputationalTask}(z)$

(A25) $(\mathsf{DnS{:}playedBy}(x,z) \wedge \mathsf{PolicySubject}(x) \wedge \mathsf{UserGroup}(z)) \rightarrow \exists y(\mathsf{DnS{:}member}(z,y) \wedge \mathsf{User}(y) \wedge \mathsf{DnS{:}playedBy}(x,y))$

(A24) and (A25) infer the closure of all resulting rights considering UserGroups and TaskCollections. A PolicySubject is granted rights on all tasks which are members of the TaskCollection. Similarly, a User is granted all access rights which are granted for his UserGroup.

4.3 Core Ontology of Software Components

Software componentry is a loosely defined term for a software technology proposing that software should be developed by glueing prefabricated components together as in the field of electronics or mechanics. Software componentry also proposes encapsulating software functionality for multiple use in a context-independent way, composable with other components, and as a unit of independent deployment and versioning.

Software components often take the form of object-oriented classes conforming to a framework specification. However, software components differ from classes. The basic idea in object-oriented programming is that software should be written according to a mental model of the actual or imagined objects it represents. Software componentry, by contrast, makes no such assumptions.

The framework specifications prescribe (1) interfaces that must be implemented by components and (2) protocols that define how components interact with each other. Examples of framework specifications are Enterprise JavaBeans (EJB) and the Component Object Model (COM) from Microsoft.

The definitions below formalize this intuition of software component as closely as possible. Assuming the object-oriented paradigm, (D31) below states that a SoftwareComponent is a special kind of CSO:Class that conforms to a FrameworkSpecification. According to the definition above, a FrameworkSpecification is (1) a DOLCE:Collection of CSO:Interfaces and (2) a special kind of OoP:Plan which specifies the interaction of components (cf. (D29)). Conformance means that at least one CSO:Interface prescribed by the FrameworkSpecification has to be implemented by the SoftwareComponent (cf. (D30)).

(D29) $\mathsf{FrameworkSpecification}(x) =_{def}$
$\mathsf{OoP{:}Plan}(x) \wedge \exists y(\mathsf{DOLCE{:}Collection}(y) \wedge \mathsf{DnS{:}unifies}(x, y) \wedge$
$\forall z(\mathsf{DOLCE{:}member}(y, z) \rightarrow \mathsf{CSO{:}Interface}(z)))$

(D30) $\mathsf{conforms}(x, y) =_{def} \mathsf{CSO{:}Class}(x) \wedge \mathsf{FrameworkSpecification}(y) \wedge$
$\exists i, c(\mathsf{CSO{:}Interface}(i) \wedge \mathsf{DOLCE{:}member}(c, i) \wedge \mathsf{DOLCE{:}Collection}(c) \wedge$
$\mathsf{DnS{:}unifies}(y, c) \rightarrow \mathsf{CSO{:}implements}(x, i))$

(D31) $\mathsf{SoftwareComponent}(x) =_{def}$
$\mathsf{CSO{:}Class}(x) \wedge \exists y(\mathsf{conforms}(x, y) \wedge \mathsf{FrameworkSpecification}(y))$

The Core Ontology of Software Components also introduces component profiles that group relevant information of a software component such as its interfaces, policy descriptions, or plans. We expect that such an aggregation makes browsing and querying for developers more convenient. The component profile is envisioned to act as the central information source for a specific software component rather than having bits and pieces all over the place.

(D32) and (A26) define a Profile as follows: First, it aggregates CSO:PolicyDescriptions, an OoP:Plan, the required SoftwareLibraries, the implemented Interfaces and additional Characteristics of a specific Software entity. Second, the link to the described Software is specified via the describes association. (D33) specializes this definition to ComponentProfile.

Often, we need to express certain capabilities or features of components, such as the version, transactional or security settings. For this purpose, we introduce Characteristics on a Profile in (D34). It is expected that ComponentProfiles are specialized and put into a taxonomy. For example, we might define a DatabaseConnectorProfile as a ComponentProfile that provides for specific Characteristics describing whether the underlying database supports transactions or SQL-99. A taxonomic structure further accommodates the developer in browsing and querying for ComponentProfiles in his system.

Finally, (A27) specifies the profiles association as a "catch-all" for DnS:defines, DnS:unifies, OIO:about, as well as OIO:expressedBy. This is done for convenience in order to relieve the developer from modeling details, who will certainly have to deal with such information.

(D32) $\text{Profile}(x) =_{def} \text{OIO:InformationObject}(x) \wedge \forall y(\text{profiles}(x, y) \rightarrow (\text{CSO:PolicyDescription}(y) \vee \text{SoftwareLibrary}(y) \vee \text{CSO:Interface}(y) \vee \text{OoP:Plan}(y) \vee \text{Characteristic}(y))) \wedge \forall z(\text{describes}(x, z) \rightarrow \text{Software}(z))$

(D33) $\text{ComponentProfile}(x) =_{def} \text{Profile}(x) \wedge \forall y(\text{describes}(x, y) \rightarrow \text{SoftwareComponent}(y))$

(D34) $\text{Characteristic}(x) =_{def} \text{DnS:Parameter}(x) \wedge \forall y(\text{DnS:defines}(y, x) \rightarrow \text{Profile}(y)) \wedge \forall z(\text{DnS:valuedBy}(x, z) \wedge \text{DOLCE:AbstractRegion}(z))$

(A26) $\text{describes}(x, y) \rightarrow \text{OIO:about}(x, y) \wedge \text{Profile}(x) \wedge \text{CSO:Software}(y)$

(A27) $\text{profiles}(x, y) \rightarrow \text{DnS:defines}(x, y) \vee \text{DnS:unifies}(x, y) \vee \text{OIO:about}(x, y) \vee \text{OIO:expressedBy}(x, y)$

5 Proof of Concept

In this section, we give some examples of how the Core Software Ontology and its extensions circumvent the four shortcomings, viz., conceptual ambiguity, poor axiomatization, loose design, and narrow scope, mentioned in the introduction. We argue that the use of the DOLCE foundational ontology as well as the use of content ontology design patterns help us here.

Conceptual Disambiguation

As mentioned in the introduction, lightweight ontologies typically suffer from conceptual ambiguity. A prominent example is the notion of OWL-S:Service in [8] which is defined twice and differently in the specification. In turn, both definitions stand in conflict with the axiomatization of the concept in the ontology. In [16], we have found a similar dilemma regarding the plethora of meanings and definitions of terms, such as component, software component, or software module. Typically, lightweight ontologies fail to convey their intended meanings of such terms and leave the interpretation to the ontology user.

In contrast to such commonly built ontologies we have captured the intended meanings of concepts and associations as precisely as possible. For this purpose, it proved to be rather helpful to capture the three different flavors of the term software via the information object content ontology design pattern, for instance.

While our definitions of the terms "software" and "software component" may not be the only ones, the fact that they are highly axiomatized allows comparing them to alternative definitions and allows fostering discussions on alternative conceptualizations.

Increased Axiomatization

Ontologies are often reduced to a simple taxonomy with domain and range restrictions on associations. This does not suffice to clarify the intended meaning of terms which is of central importance when building ontologies for reuse and reference purpose. As an example, consider control constructs, such as, fork or join, to specify workflow information. Many ontologies, such as OWL-S [8], omit a concise formalization of their intended meaning.

In our ontology we have made use of the Ontology of Plans which provides extensive axiomatization of OoP:Tasks and subconcepts thereof. OoP:Tasks are directly comparable to control constructs, but provide a heavyweight axiomatization. An example is *SynchroTask* (an instance of OoP:ControlTask) which matches the concept of a "join." A *SynchroTask* joins a set of tasks after a branching and waits for the execution of all tasks (except the optional ones) that are direct successors to a *ConcurrencyTask* or *AnyOrderTask*. Below we give the axiomatization of the *SynchroTask* as introduced in [4].

$\mathsf{ControlTask}(\mathit{SynchroTask}) \rightarrow \exists t_1, t_2, t_3 (t_1 = \mathit{ConcurrencyTask} \vee t_1 = \mathit{AnyOrderTask}) \wedge \mathsf{successor}(t_1, x) \wedge (\mathsf{ComplexTask}(t_2) \vee \mathsf{ActionTask}(t_2)) \wedge (\mathsf{ComplexTask}(t_3) \vee \mathsf{ActionTask}(t_3)) \wedge \mathsf{directSuccessor}(t_2, \mathit{SynchroTask}) \wedge \mathsf{directSuccessor}(t_3, \mathit{SynchroTask})$

Another example is the link between the control constructs and the process steps. Very often, the intended meaning of such links remains unclear. Is it a parthood association? And if yes, is it temporary, transitive, etc.? Our ontology is very specific with respect to such notions because it builds on the Ontology of Plans. The latter exploits the DOLCE:temporaryComponent association which has a firm foundation as a special kind of the more basic DOLCE:component mereological association and DOLCE:partlyCompresent temporally indexing association. Both are characterized by formal restrictions on their application to other basic concepts.

Improved Design

In our ontology, we propose to use contextualization as a design pattern. Contextualization allows us to move from software descriptions to the representation of different, possibly conflicting views with various granularity. The Descriptions & Situations ontology provides us with a corresponding content ontology design pattern with the basic primitives of context modeling such as the notion of roles, for instance. Roles allow us to talk about inputs and outputs on the abstract level, i.e., independent of the objects that play such roles.

This pattern applies clearly defined semantics and scoping provided by Descriptions & Situations where we want to express that the output of a process is the input to another process. In our ontology, inputs and outputs can be modeled as DnS:Roles which serve as variables. Thus, CSO:Data can play multiple roles within the same or different descriptions. It is natural to express that the given CSO:Data is output with respect to one process, but input to another (cf. Fig. 4).

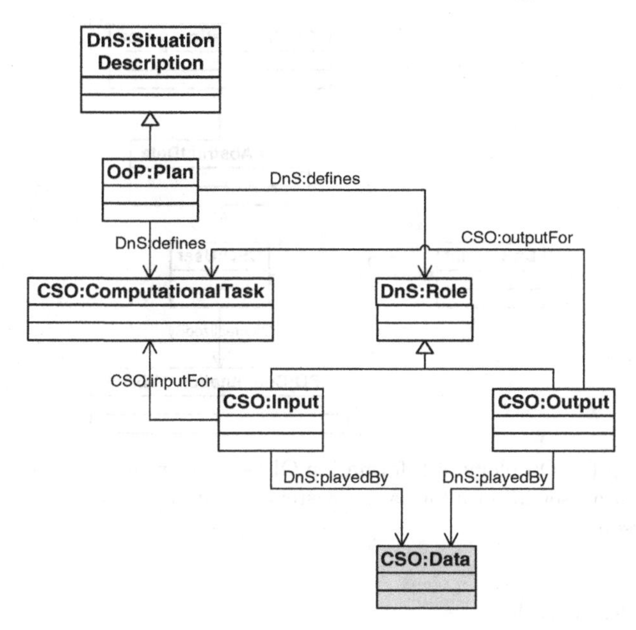

Fig. 4. Data can play both the role of an Input and an Output at the same time. Inputs and Outputs can be linked to ComputationalTasks in a Plan. White classes represent descriptive entities, grey classes represent ground entities

Wider Scope

Software resides on the boundary of the world inside an information system and the external world. Web services, in particular, may carry out operations to support a real-world service. Functionality, which is an essential property of a service, then arises from the entire process that comprises computational as well as real-world activities.

The distinction between information objects, events, and physical objects is not explicitly made in most ontologies. In our ontology, this separation naturally follows from the use of DOLCE and the Ontology of Information Objects, where the distinction is an important part of the characterization of concepts. In particular, it becomes possible to be more precise about the kinds of relationships that can occur among objects or between objects and events.

For example, we can distinguish a physical object (such as a natural person) from an information object (such as user account in an information system) and represent the link between the two. Figure 5 shows which capabilities our ontology provides to do so. It is worthwhile to make such differences explicit, e.g., when we want to infer the total of access rights granted for a natural person who might have several user accounts in and across information systems.

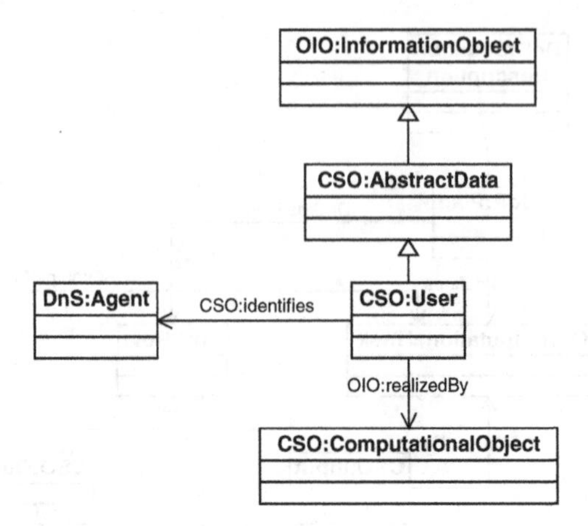

Fig. 5. Using the Ontology of Information Objects allows us to model the relationship between a user in an information system and its corresponding agent (e.g., a natural person)

6 Related Work

As already outlined in the introduction, the Chapter "Ontologies and Software Engineering" surveys a wealth of approaches that formalize at least one paradigm or aspect of software in an ontology. Although the respective ontologies share basic principles, they (1) rely on individual formalizations, and (2) are typically lightweight and not of a reference nature. Such approaches could benefit from our Core Software Ontology capturing the stable core and preventing modeling from scratch. The same proposition is valid for approaches like [13,14] that embed ontology modeling into the transformation processes of model-driven engineering in order to conjoin ontology modeling and software engineering.

Furthermore, there has been some work that overlaps with the ideas presented here. For example, the COHSE Java ontology[7] offers a formal schema for expressing a Java software project by an ontology. The open source project Introspector[8] is a back-end to the popular GNU compiler collection gcc,[9] which generates an RDF-defined ontology out of gcc compiled source code, and thus works with all languages supported by gcc, for example, C, C++, Java, Fortran, and others. [17] offers a more profound and sound ontology-based foundation, analyzing the constructs available when programming. All these works provide support for using ontologies in the area of software development, but on a much finer grained level than the work presented here. Thus, such ontologies could be used complementary to ours.

[7] `http://cohse.semanticweb.org/software.html`
[8] `http://introspector.sourceforge.net`
[9] `http://gcc.gnu.org`

An example of a higher level software component ontology in use is provided by [1]. The work focusses on the social and project-level management of Open Source software projects. Such aspects have not been considered by the Core Software Ontology yet.

Finally, there are some ontologies which focus on specific aspects, whereas our ontology tries to relate the different aspects in a larger focus. Examples are the Core Plan Representation (CPR) [15] and the Process Specification Language (PSL) [6] which are comparable to the Ontology of Plans. UPML, the Unified Problem-solving Method Development Language [2], has been developed to describe and implement intelligent broker architectures and components to facilitate semi-automatic reuse and adaptation.

7 Conclusion

The chapter has shown how to proceed in building a Core Software Ontology and extending it in different directions. The result is grounded in a foundational ontology and avoids the typical shortcomings of lightweight ontologies. Related, seminal approaches only weakly formalize the meaning of their terms and leave their disambiguation to the intuition of the reader, a situation that we here improve upon considerably.

The reader may note that what is presented here are the *reference ontologies* in this domain. For actual work, these reference ontologies need to be reduced to knowledge representation schemes that are more amenable to operation.

So far, the Core Ontology of Software Components has been applied in [11] for developing and managing software components in an application server. The Core Ontology of Web Services has been used by the EU project FUSION (cf. `http://www.fusionweb.org`) for information integration. The Core Policy Ontology has been emancipated, extended and used for service marketplaces by [7]. We expect many more fruitful applications and extensions of our ontologies in the future.

References

1. Anupriya Ankolekar, James Herbsleb, and Katia Sycara. Addressing Challenges to Open Source Collaboration With the Semantic Web. In Joseph Feller, Brian Fitzgerald, Scott Hissam, and Karim Lakhani, editors, *Proceedings of Taking Stock of the Bazaar: The 3rd Workshop on Open Source Software Engineering, the 25th International Conference on Software Engineering (ICSE)*, Washington, D.C., 2003. IEEE Computer Society.
2. Dieter Fensel, Richard Benjamins, Enrico Motta, and Bob J. Wielinga. UPML: A Framework for Knowledge System Reuse. In Thomas Dean, editor, *Proceedings of the 16th Int. Joint Conference on Artificial Intelligence, IJCAI 99, Stockholm, Sweden, 1999. 2 Volumes, 1450 pages*, pages 16–23. Morgan Kaufmann, 1999.

3. Aldo Gangemi, Maria-Teresa Sagri, and Daniela Tiscornia. A Constructive Framework for Legal Ontologies. Internal project report, EU 6FP METOKIS Project, Deliverable, 2004. `http://metokis.salzburgresearch.at`.
4. Aldo Gangemi, Stefano Borgo, Carola Catenacci, and Jos Lehmann. Task taxonomies for knowledge content. Metokis Deliverable D07, Sep 2004.
5. Aldo Gangemi and Peter Mika. Understanding the Semantic Web through Descriptions and Situations. In *DOA/CoopIS/ODBASE 2003 Confederated International Conferences DOA, CoopIS and ODBASE, Proceedings*, LNCS. Springer, 2003.
6. Michael Grüninger and Christopher Menzel. The Process Specification Language (PSL) Theory and Applications. *AI Magazine*, 24(3):63–74, 2003.
7. Steffen Lamparter, Anupriya Ankolekar, Daniel Oberle, Rudi Studer, and Christof Weinhardt. A Policy Framework for Trading Configurable Goods and Services in Open Electronic Markets. In *Proceedings of the 8th Int. Conference on Electronic Commerce (ICEC'06)*, pages 162–173, AUG 2006.
8. David Martin, Mark Burstein, Jerry Hobbs, Ora Lassila, Drew McDermott, Sheila McIlraith, Srini Narayanan, Massimo Paolucci, Bijan Parsia, Terry Payne, Evren Sirin, Naveen Srinivasan, and Katia Sycara. OWL-S: Semantic Markup for Web Services. `http://www.daml.org/services/owl-s/1.1/`, Nov 2004.
9. Claudio Masolo, Stefano Borgo, Aldo Gangemi, Nicola Guarino, and Alessandro Oltramari. Ontology Library (final). WonderWeb Deliverable D18, Dec 2003. `http://wonderweb.semanticweb.org`.
10. Peter Mika, Daniel Oberle, Aldo Gangemi, and Marta Sabou. Foundations for Service Ontologies: Aligning OWL-S to DOLCE. In *The 13th International World Wide Web Conference Proceedings*, pages 563–572. ACM, May 2004.
11. Daniel Oberle. *Semantic Management of Middleware*, volume I of *The Semantic Web and Beyond*. Springer, New York, Jan 2006.
12. Daniel Oberle, Steffen Lamparter, Stephan Grimm, Denny Vrandecic, Steffen Staab, and Aldo Gangemi. Towards Ontologies for Formalizing Modularization and Communication in Large Software Systems. *Journal of Applied Ontology*, 1(2):163–202, 2006.
13. F. Silva Parreiras, S. Staab, and A. Winter. On marrying ontological and metamodeling technical spaces. In *ESEC/ACM FSE-2007 — Proceedings of the 6th joint meeting of the European software engineering conference and the 14th ACM SIGSOFT symposium on Foundations of software engineering*, pages 439–448. ACM, September 2007.
14. F. Silva Parreiras, S. Staab, and A. Winter. Improving design patterns by description logics: An use case with abstract factory and strategy. In T. Kühne and F. Steimann, editors, *Proc. of Modellierung 2008*, LNI. GI e.V., March 2008.
15. Adam Pease. Core Plan Representation. Object Model Focus Group, Nov 1998.
16. Marta Sabou, Daniel Oberle, and Debbie Richards. Enhancing Application Servers with Semantics. In *1st Australian Workshop on Engineering Service-Oriented Systems (AWESOS 2004) Melbourne, Australia*, pages 7–15, 2004.
17. Christopher Welty. *An Integrated Representation for Software Development and Discovery*. PhD thesis, Rensselaer Polytechnic Institute Computer Science Department, 1995.

COMM: A Core Ontology for Multimedia Annotation

Richard Arndt[1], Raphaël Troncy[2], Steffen Staab[1], and Lynda Hardman*[2]

[1] ISWeb, University of Koblenz-Landau, Germany, rarndt@uni-koblenz.de, staab@uni-koblenz.de
[2] CWI, Amsterdam, The Netherlands, Raphael.Troncy@cwi.nl, Lynda.Hardman@cwi.nl

Summary. In order to retrieve and reuse non-textual media, media annotations must explain how a media object is composed of its parts and what the parts represent. Annotations need to link to background knowledge found in existing knowledge sources and to the creation and use of the media object. The representation and understanding of such facets of the media semantics is only possible through a formal language and a corresponding ontology. In this chapter, we analyze the requirements underlying the semantic representation of media objects, explain why the requirements are not fulfilled by most semantic multimedia ontologies and present COMM,[1] a core ontology for multimedia, that has been built re-engineering the current de-facto standard for multimedia annotation, i.e. MPEG-7, and using DOLCE as its underlying foundational ontology to support conceptual clarity and soundness as well as extensibility towards new annotation requirements.

1 Introduction

Multimedia objects are ubiquitous, whether found via web search (e.g. Google[2] or Yahoo![3] images), or via dedicated sites (e.g. Flickr[4] or YouTube[5]) or in the repositories of private users or commercial organizations (film archives, broadcasters, photo agencies, etc.). The media objects are produced and consumed by professionals and amateurs alike. Unlike textual assets, whose content can be searched for using text strings, media search is dependent on processes that have either cumbersome requirements for feature comparison (e.g. color

* Lynda Hardman is also affiliated with the Technical University of Eindhoven.
[1] This chapter is a revised and extended version of [1].
[2] `http://images.google.com/`
[3] `http://images.search.yahoo.com/`
[4] `http://www.flickr.com/`
[5] `http://www.youtube.com/`

S. Staab and R. Studer (eds.), *Handbook on Ontologies*, International Handbooks on Information Systems, DOI 10.1007/978-3-540-92673-3,

or texture) or rely on associated, more easily processable descriptions, selecting aspects of an image or video and expressing them as text, or as concepts from a predefined vocabulary. Individual annotation and tagging applications have not yet achieved a degree of interoperability that enables effective sharing of semantic metadata and that links the metadata to semantic data and ontologies found on the Semantic Web.

MPEG-7 [12, 13] is an international standard that specifies how to connect descriptions to parts of a media asset. The standard includes descriptors representing low-level media-specific features that can often be automatically extracted from media types. Unfortunately, MPEG-7 is not fully suitable for describing multimedia content, because (1) it is not open to standards that represent knowledge and make use of existing controlled vocabularies for describing the subject matter and (2) its XML Schema based nature has led to design decisions that leave the annotations conceptually ambiguous and therefore prevent direct machine processing of semantic content descriptions.

In order to avoid such problems, we advocate the use of Semantic Web languages and a core ontology for multimedia annotations, which is built based on rich ontological foundations provided by an ontology such as DOLCE (cf. Chapter "Foundational Choices in DOLCE") and sound ontology engineering principles. The result presented in this chapter is COMM, a core ontology for multimedia.

In the next section, we illustrate the main problems when using MPEG-7 for describing multimedia resources on the web. In Sect. 3, we review existing multimedia ontologies and show why previous proposals are inadequate for semantic multimedia annotation. Subsequently, we define the requirements that a multimedia ontology should meet (Sect. 4) before we present COMM, an MPEG-7 based ontology, and discuss our design decisions based on our requirements (Sect. 5). In Sect. 6, we demonstrate the use of the ontology with the scenario from Sect. 2 and then conclude.

2 Annotating Multimedia Assets on the Web

Let us imagine that Nathalie, a student in history, wants to create a multimedia presentation of the major international conferences and summits held in the last 60 years. Her starting point is the famous "Big Three" picture, taken at the Yalta (Crimea) Conference, showing the heads of government of the United States, the United Kingdom, and the Soviet Union during World War II. Nathalie uses an MPEG-7 compliant authoring tool for detecting and labeling relevant multimedia objects automatically. On the web, she finds three different face recognition web services which provide very good results for detecting Winston Churchill, Franklin D. Roosevelt and Josef Stalin, respectively. Having these tools, she would like to run the face recognition web services on images and import the extraction results into the authoring tool

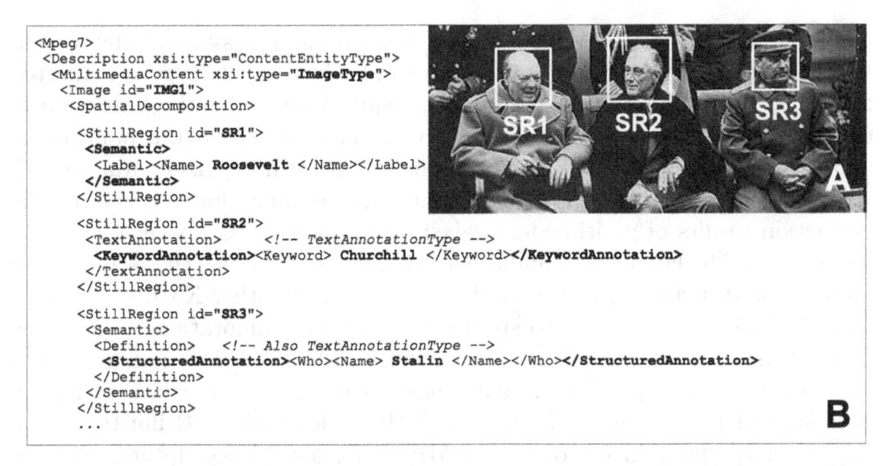
<Mpeg7>
<Description xsi:type="ContentEntityType">
<MultimediaContent xsi:type="ImageType">
<Image id="IMG1">
<SpatialDecomposition>
<StillRegion id="SR1">
<Semantic>
<Label><Name> Roosevelt </Name></Label>
</Semantic>
</StillRegion>
<StillRegion id="SR2">
<TextAnnotation> <!-- TextAnnotationType -->
<KeywordAnnotation><Keyword> Churchill </Keyword></KeywordAnnotation>
</TextAnnotation>
</StillRegion>
<StillRegion id="SR3">
<Semantic>
<Definition> <!-- Also TextAnnotationType -->
<StructuredAnnotation><Who><Name> Stalin </Name></Who></StructuredAnnotation>
</Definition>
</Semantic>
</StillRegion>
...
SR1
SR2
SR3
A
B

Fig. 1. MPEG-7 annotation example (Image adapted from Wikipedia), `http://en.wikipedia.org/wiki/Yalta_Conference`

in order to automatically generate links from the detected face regions to detailed textual information about Churchill, Roosevelt and Stalin (image in Fig. 1-A).

Nathalie would then like to describe a recent video from a G8 summit, such as the retrospective *A history of G8 violence* made by Reuters.[6] She uses again an MPEG-7 compliant segmentation tool for detecting the seven main sequences of this 2'26 minutes report: the various anti-capitalist protests during the Seattle (1999), Melbourne (2000), Prague (2000), Gothenburg (2001), Genoa (2001), St Petersburg (2006), Heiligendamm (2007) World Economic Forums, EU and G8 Summits. Finally, Nathalie plans to deliver her multimedia presentation in an Open Document Format (ODF) document embedding the image and video previously annotated. This scenario, however, causes several problems with existing solutions.

Fragment identification. Particular regions of the image need to be localized (anchor value in [6]). However, the current web architecture does not provide a means for uniquely identifying sub-parts of multimedia assets, in the same way that the fragment identifier in the URI can refer to part of an HTML or XML document. Indeed, for almost any other media type, the semantics of the fragment identifier has not been defined or is not commonly accepted. Providing an agreed upon way to localize sub-parts of multimedia objects (e.g. sub-regions of images, temporal sequences of videos or tracking moving objects in space and in time) is fundamental[7] [5]. For images, one can use either MPEG-7 or SVG snippet code to define the bounding box coordinates of specific regions. For temporal locations, one can use MPEG-7 code or

[6] `http://www.reuters.com/news/video/summitVideo?videoId=56114`

[7] See also the forthcoming W3C Media Fragments Working Group `http://www.w3.org/2008/01/media-fragments-wg.html`

the TemporalURI RFC.[8] MPEG-21 specifies a normative syntax to be used in URIs for addressing parts of any resource but whose media type is restricted to MPEG [11]. The MPEG-7 approach requires an indirection: an annotation is *about* a fragment of an XML document that *refers* to a multimedia document, whereas the MPEG-21 approach does not have this limitation [21].

Semantic annotation. MPEG-7 is a natural candidate for representing the extraction results of multimedia analysis software such as a face recognition web service. The language, standardized in 2001, specifies a rich vocabulary of multimedia descriptors, which can be represented in either XML or a binary format. While it is possible to specify very detailed annotations using these descriptors, it is not possible to guarantee that MPEG-7 metadata generated by different agents will be mutually understood due to the lack of formal semantics of this language [7, 18]. The XML code of Fig. 1-B illustrates the inherent interoperability problems of MPEG-7 : several descriptors, semantically equivalent and representing the same information while using different syntax can coexist [19]. As Nathalie used three different face recognition web services, the extraction results of the regions `SR1`, `SR2` and `SR3` differ from each other even though they are all syntactically correct. While the first service uses the MPEG-7`SemanticType` for assigning the `<Label>` *Roosevelt* to still region `SR1`, the second one makes use of a `<KeywordAnnotation>` for attaching the keyword *Churchill* to still region `SR2`. Finally the third service uses a `<StructuredAnnotation>` (which can be used within the `SemanticType`) in order to label still region `SR3` with *Stalin*. Consequently, alternative ways for annotating the still regions render almost impossible the retrieval of the face recognition results within the authoring tool since the corresponding XPath query has to deal with these syntactic variations. As a result, the authoring tool will not link occurrences of Churchill in the images with, for example, his biography as it does not expect semantic labels of still regions as part of the `<KeywordAnnotation>` element.

Web interoperability. Nathalie would like to link the multimedia presentation to historical information about the key figures of the Yalta Conference or the various G8 summits that is already available on the web. She has also found semantic metadata about the relationships between these figures that could improve the automatic generation of the multimedia presentation. However, she realizes that MPEG-7 cannot be combined with these concepts defined in domain-specific ontologies because of its closing to the web. As this example demonstrates, although MPEG-7 provides ways of associating semantics with (parts of) non-textual media assets, it is incompatible with (semantic) web technologies and has no formal description of the semantics encapsulated implicitly in the standard.

Embedding into compound documents. Finally, Nathalie needs to compile the semantic annotations of the images, videos and textual stories into a semantically annotated compound document. However, the current state of

[8] `http://www.annodex.net/TR/URI_fragments.html`

the art does not provide a framework which allows the semantic annotation of compound documents. MPEG-7 solves only partially the problem as it is restricted to the description of audiovisual compound documents. Bearing the growing number of multimedia office documents in mind, this limitation is a serious drawback.

3 Related Work

In the field of semantic image understanding, using a multimedia ontology infrastructure is regarded to be the first step for closing the, so-called, semantic gap between low-level signal processing results and explicit semantic descriptions of the concepts depicted in images. Furthermore, multimedia ontologies have the potential to increase the interoperability of applications producing and consuming multimedia annotations. The application of multimedia reasoning techniques on top of semantic multimedia annotations is also a research topic which is currently investigated [15]. A number of drawbacks of MPEG-7 have been reported [14, 17]. As a solution, multimedia ontologies based on MPEG-7 have been proposed.

Hunter [7] provided the first attempt to model parts of MPEG-7 in RDFS, later integrated with the ABC model. Tsinaraki et al. [22] start from the core of this ontology and extend it to cover the full Multimedia Description Scheme (MDS) part of MPEG-7, in an OWL DL ontology. A complementary approach was explored by Isaac and Troncy [10], who proposed a core audio-visual ontology inspired by several terminologies such as MPEG-7, TV Anytime and ProgramGuideML. Garcia and Celma [4] produced the first complete MPEG-7 ontology, automatically generated using a generic mapping from XSD to OWL. Finally, Simou et al. [2] proposed an OWL DL Visual Descriptor Ontology[9] (VDO) based on the Visual part of MPEG-7 and used for image and video analysis.

These ontologies have been recently compared with COMM according to three criteria: (1) the way the multimedia ontology is linked with domain semantics, (2) the MPEG-7 coverage of the multimedia ontology, and (3) the scalability and modeling rationale of the conceptualization [20]. Unlike COMM, all the other ontologies perform a one to one translation of MPEG-7 types into OWL concepts and properties. This translation does not, however, guarantee that the intended semantics of MPEG-7 is fully captured and formalized. On the contrary, the syntactic interoperability and conceptual ambiguity problems illustrated in Sect. 2 remain. Although COMM is based on a foundational ontology, the annotations proved to be no more verbose than those in MPEG-7.

Finally, general models for annotations of non-multimedia content have been proposed by librarians. The Functional Requirements for Bibliographic

[9] `http://image.ece.ntua.gr/~gstoil/VDO`

Records (FRBR)[10] model specifies the conventions for bibliographic description of traditional books. The CIDOC Conceptual Reference Model (CRM)[11] defines the formal structure for describing the concepts and relationships used in cultural heritage documentation (cf. Chapter "Using the PSL Ontology"). Hunter has described how an MPEG-7 ontology could specialize CIDOC-CRM for describing multimedia objects in museums [8]. Interoperability with such models is an issue, but interestingly, the design rationale used in these models are often comparable and complementary to foundational ontologies approach.

4 Requirements for Designing a Multimedia Ontology

Requirements for designing a multimedia ontology have been gathered and reported in the literature, e.g. in [9]. Here, we compile these and use our scenario to present a list of requirements for a web-compliant multimedia ontology.

MPEG-7 compliance. MPEG-7 is an existing international standard, used both in the signal processing and the broadcasting communities. It contains a wealth of accumulated experience that needs to be included in a web-based ontology. In addition, existing annotations in MPEG-7 should be easily expressible in our ontology.

Semantic interoperability. Annotations are only re-usable when the captured semantics can be shared among multiple systems and applications. Obtaining similar results from reasoning processes about terms in different environments can only be guaranteed if the semantics is sufficiently explicitly described. A multimedia ontology has to ensure that the intended meaning of the captured semantics can be shared among different systems.

Syntactic interoperability. Systems are only able to share the semantics of annotations if there is a means of conveying this in some agreed-upon syntax. Given that the (semantic) web is an important repository of both media assets and annotations, a semantic description of the multimedia ontology should be expressible in a web language (e.g. OWL, RDF/XML or RDFa).

Separation of concerns. Clear separation of subject matter (i.e. knowledge about depicted entities, such as the person Winston Churchill) from knowledge that is related to the administrative management or the structure and the features of multimedia documents (e.g. Churchill's face is to the left of Roosevelt's face) is required. Reusability of multimedia annotations can only be achieved if the connection between both ontologies is clearly specified by the multimedia ontology.

Modularity. A complete multimedia ontology can be, as demonstrated by MPEG-7, very large. The design of a multimedia ontology should thus be

[10] `http://www.ifla.org/VII/s13/frbr/index.htm`

[11] `http://cidoc.ics.forth.gr/`

made modular, to minimize the execution overhead when used for multimedia annotation. Modularity is also a good engineering principle.

Extensibility. While we intend to construct a comprehensive multimedia ontology, as ontology development methodologies demonstrate, this can never be complete. New concepts will always need to be added to the ontology. This requires a design that can always be extended, without changing the underlying model and assumptions and without affecting legacy annotations.

5 Adding Formal Semantics to MPEG-7

MPEG-7 specifies the connection between semantic annotations and parts of media assets. We take it as a base of knowledge that needs to be expressible in our ontology. Therefore, we re-engineer MPEG-7 according to the intended semantics of the written standard. We satisfy our semantic interoperability not by aligning our ontology to the XML Schema definition of MPEG-7, but by providing a formal semantics for MPEG-7. We use a methodology based on a foundational, or top level, ontology as a basis for designing COMM (cf. Chapter "Ontology Engineering Methodology"). This provides a domain independent vocabulary that explicitly includes formal definitions of foundational categories, such as processes or physical objects, and eases the linkage of domain-specific ontologies because of the shared definitions of top level concepts. We briefly introduce our chosen foundational ontology in Sect. 5.1, and then present our multimedia ontology, COMM, in Sects. 5.2 and 5.3. Finally, we discuss why our ontology satisfies all our stated requirements in Sect. 5.4.

COMM is available at `http://multimedia.semanticweb.org/COMM/`.

5.1 DOLCE as Modeling Basis

Using the review in [16], we select the Descriptive Ontology for Linguistic and Cognitive Engineering (DOLCE) (cf. Chapter "Foundational Choices in DOLCE") as a modeling basis. Our choice is influenced by two of the main design patterns: *Descriptions & Situations* (D&S) and *Ontology of Information Objects* (OIO) [3]. The former can be used to formalize contextual knowledge, while the latter, based on D&S, implements a semiotics model of communication theory. We consider that the annotation process is a *situation* (i.e. a reified context) that needs to be described.

5.2 Multimedia Patterns

The patterns for D&S and OIO need to be extended for representing MPEG-7 concepts since they are not sufficiently specialized to the domain of multimedia annotation. This section introduces these extended multimedia design patterns, while Sect. 5.3 details two central concepts underlying these patterns:

digital data and algorithms (cf. Chapter "Ontology Design Patterns"). In order to define design patterns, one has to identify repetitive structures and describe them at an abstract level. The two most important functionalities provided by MPEG-7 are: the *decomposition* of media assets and the (semantic) *annotation* of their parts, which we include in our multimedia ontology.

Decomposition. MPEG-7 provides descriptors for spatial, temporal, spatio-temporal and media source decompositions of multimedia content into segments. A segment is the most general abstract concept in MPEG-7 and can refer to a region of an image, a piece of text, a temporal scene of a video or even to a moving object tracked during a period of time.

Annotation. MPEG-7 defines a very large collection of descriptors that can be used to annotate a segment. These descriptors can be low-level visual features, audio features or more abstract concepts. They allow the annotation of the content of multimedia documents or the media asset itself.

In the following, we first introduce the notion of multimedia data and then present the patterns that formalize the decomposition of multimedia content into segments, or allow the annotation of these segments. The decomposition pattern handles the structure of a multimedia document, while the media annotation pattern, the content annotation pattern and the semantic annotation pattern are useful for annotating the media, the features and the semantic content of the multimedia document respectively.

Multimedia Data

This encapsulates the MPEG-7 notion of multimedia content and is a subconcept of digital-data[12] (introduced in more detail in Sect. 5.3). multimedia-data is an abstract concept that has to be further specialized for concrete multimedia content types (e.g. image-data corresponds to the pixel matrix of an image). According to the OIO pattern, multimedia-data is realized by some physical media (e.g. an image). This concept is needed for annotating the physical realization of multimedia content.

Decomposition Pattern

Following the D&S pattern, we consider that a decomposition of a multimedia-data entity is a situation[13] (a segment-decomposition) that satisfies a description, such as a segmentation-algorithm or a method (e.g. a user drawing a bounding box around a depicted face), which has been applied to perform the decomposition, see Fig. 2-B. Of particular importance are the roles that are defined by a segmentation-algorithm or a method. output-segment-roles express that some multimedia-data entities are segments of a multimedia-data entity that plays the role of an input segment (input-segment-role). These data

[12] Sans serif font indicates ontology concepts.

[13] Cf. Chapter "Foundational Choices in DOLCE".

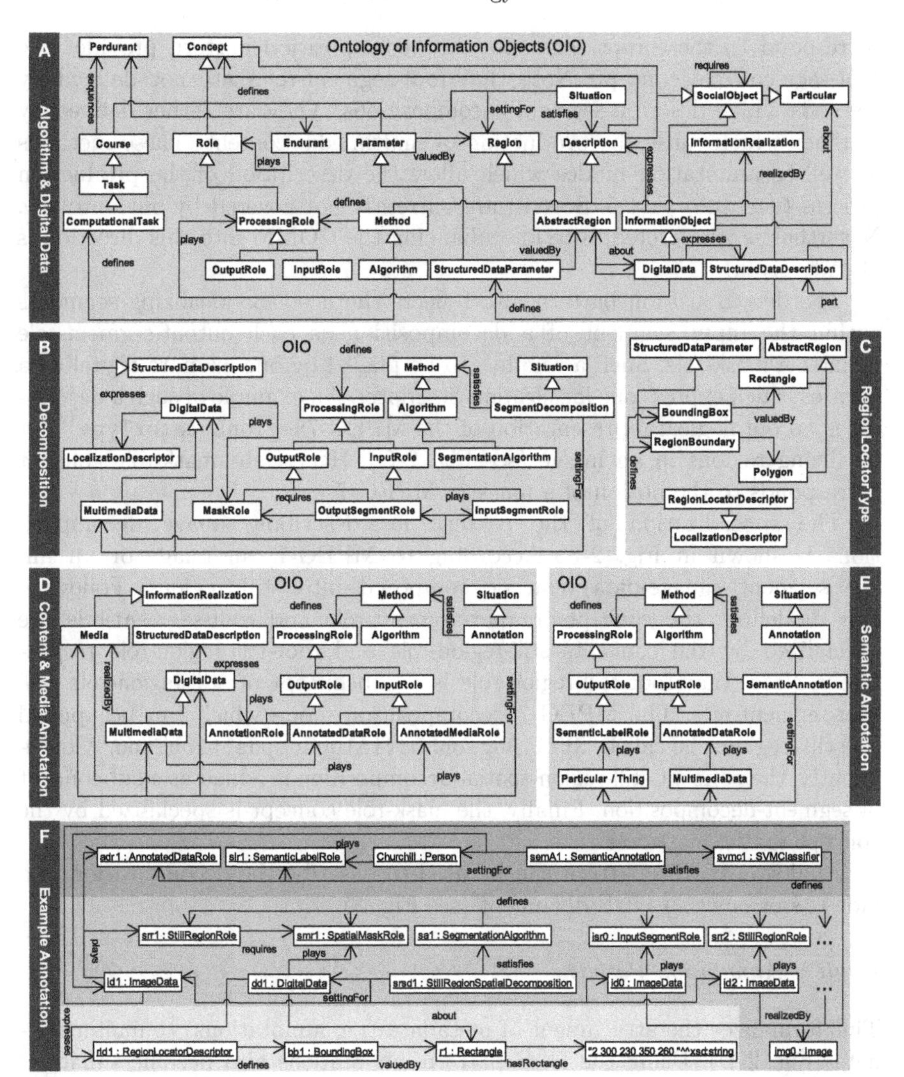

Fig. 2. COMM: Design patterns in UML notation: Basic design patterns (A), multimedia patterns (B, D, E) and modeling examples (C, F)

entities have as setting a segment-decomposition situation that satisfies the roles of the applied segmentation-algorithm or method. output-segment-roles as well as segment-decompositions are then specialized according to the segment and decomposition hierarchies of MPEG-7 ([12], part 5, Sect. 11). In terms of MPEG-7, unsegmented (complete) multimedia content also corresponds to a segment. Consequently, annotations of complete multimedia content start with a root segment. In order to designate multimedia-data instances that correspond to these root segments the decomposition pattern provides the root-segment-role concept. Note that root-segment-roles are not defined by methods which describe segment-decompositions. They are rather defined by methods which cause the production of multimedia content. These methods as well as annotation modes which allow the description of the production process (e.g. [12], part 5, Sect. 9) are currently not covered by our ontology. Nevertheless, the prerequisite for enhancing the COMM into this direction is already given.

The decomposition pattern also reflects the need for localizing segments within the input segment of a decomposition as each output-segment-role requires a mask-role. Such a role has to be played by one or more digital-data entities which express one localization-descriptor. An example of such a descriptor is an ontological representation of the MPEG-7`RegionLocatorType`[14] for localizing regions in an image (see Fig. 2-C). Hence, the mask-role concept corresponds to the notion of a mask in MPEG-7.

The specialization of the pattern for describing image decompositions is shown in Fig. 2-F. According to MPEG-7, an image or an image segment (image-data) can be composed into still regions. Following this modeling, the concepts output-segment-role and root-segment-role are specialized by the concepts still-region-role and root-still-region-role respectively. Note, that root-still-region-role is a subconcept of still-region-role *and* root-segment-role. The MPEG-7 decomposition mode which can be applied to still regions is called `StillRegionSpatialDecompositionType`. Consequently, the concept still-region-spatial-decomposition is added as a subconcept of segment-decomposition. Finally, the mask-role concept is specialized by the concept spatial-mask-role.

Analogously, the pattern can be used to describe the decomposition of a video asset or of an ODF document (see Fig. 3).

Content Annotation Pattern.

This formalizes the attachment of metadata (i.e. annotations) to multimedia-data (Fig. 2-D). Using the D&S pattern, annotations also become situations that represent the state of affairs of all related digital-data (metadata and annotated multimedia-data). digital-data entities represent the attached metadata by playing an annotation-role. These roles are defined by methods or algorithms. The former are used to express manual (or semi-automatic) annotation while the latter serve as an explanation for the attachment of automatically computed features, such as the dominant colors of a still region. It is mandatory that the multimedia-data entity being annotated plays an annotated-data-role.

The actual metadata that is carried by a digital-data entity depends on the structured-data-description that is expressed by it. These descriptions are formalized using the digital data pattern (see Sect. 5.3). Applying

[14] Italic type writer font indicates MPEG-7 language descriptors.

the content annotation pattern for formalizing a specific annotation, e.g. a dominant-color-annotation which corresponds to the connection of a MPEG-7DominantColorType with a segment, requires only the specialization of the concept annotation, e.g. dominant-color-annotation. This concept is defined by being a setting for a digital-data entity that expresses one dominant-color-descriptor (a subconcept of structured-data-description which corresponds to the DominantColorType).

Media Annotation Pattern

This forms the basis for describing the physical instances of multimedia content (Fig. 2-D). It differs from the content annotation pattern in only one respect: it is the media that is being annotated and therefore plays an annotated-media-role.

One can thus represent that some visual content (e.g. the picture of a digital camera) is realized by a JPEG image with a size of 462848 bytes, using the MPEG-7MediaFormatType. Using the media annotation pattern, the metadata is attached by connecting a digital-data entity with the image. The digital-data plays an annotation-role while the image plays an annotated-media-role. An ontological representation of the MediaFormatType, namely an instance of the structured-data-description subconcept media-format-descriptor, is expressed by the digital-data entity. The tuple formed with the scalar "462848" and the string "JPEG" is the value of the two instances of the concepts file-size and file-format respectively. Both concepts are subconcepts of structured-data-parameter.

Semantic Annotation Pattern

Even though MPEG-7 provides some general concepts (see [12], part 5, Sect. 12) that can be used to describe the perceivable content of a multimedia segment, independent development of domain-specific ontologies is more appropriate for describing possible interpretations of multimedia – it is useful to create an ontology specific to multimedia, it is not useful to try to model the real world within this. An ontology-based multimedia annotation framework should rely on domain-specific ontologies for the representation of the real world entities that might be depicted in multimedia content. Consequently, this pattern specializes the content annotation pattern to allow the connection of multimedia descriptions with domain descriptions provided by independent world ontologies (Fig. 2-E).

An OWL Thing or a DOLCE particular (belonging to a domain-specific ontology) that is depicted by some multimedia content is not directly connected to it but rather through the way the annotation is obtained. Actually, a manual annotation method or its subconcept algorithm, such as a classification algorithm, has to be applied to determine this connection. It is embodied through a semantic-annotation that satisfies the applied method. This description specifies that the annotated multimedia-data has to play an annotated-data-role and the depicted Thing/particular has to play a semantic-label-role. The pattern also allows the integration of features which might be evaluated in the context of a classification algorithm. In that case, digital-data entities that represent these features would play an input-role.

5.3 Basic Patterns

Specializing the D&S and OIO patterns for defining multimedia design patterns is enabled through the definition of basic design patterns, which formalize the notion of digital data and algorithm.

Digital Data Pattern

Within the domain of multimedia annotation, the notion of digital data is central – both the multimedia content being annotated and the annotations themselves are expressed as digital data. We consider digital-data entities of arbitrary size to be information-objects, which are used for communication between machines. The OIO design pattern states that descriptions are expressed by information-objects, which have to be about facts (represented by particulars). These facts are settings for situations that have to satisfy the descriptions that are expressed by information-objects. This chain of constraints allows the modeling of complex data structures to store digital information. Our approach is as follows (see Fig. 2-A): digital-data entities express descriptions, namely structured-data-descriptions, which define meaningful labels for the information contained by digital-data. This information is represented by numerical entities such as scalars, matrices, strings, rectangles or polygons. In DOLCE terms, these entities are abstract-regions. In the context of a description, these regions are described by parameters. structured-data-descriptions thus define structured-data-parameters, for which abstract-regions carried by digital-data entities assign values.

The digital data pattern can be used to formalize complex MPEG-7 low-level descriptors. Figure 2-C shows the application of this pattern by formalizing the MPEG-7`RegionLocatorType`, which mainly consists of two elements: a `Box` and a `Polygon`. The concept region-locator-descriptor corresponds to the `RegionLocatorType`. The element `Box` is represented by the structured-data-parameter subconcept BoundingBox while the element `Polygon` is represented by the region-boundary concept.

The MPEG-7 code example given in Fig. 1 highlights that the formalization of data structures, so far, is not sufficient – complex MPEG-7 types can include nested types that again have to be represented by structured-data-descriptions. In our example, the MPEG-7`SemanticType` contains the element `Definition` which is of complex type `TextAnnotationType`. The digital data pattern covers such cases by allowing a digital-data instance dd1 to be about a digital-data instance dd2 which expresses a structured-data-description that corresponds to a nested type (see Fig. 2-A). In this case the structured-data-description of instance dd2 would be a part of the one expressed by dd1.

Algorithm Pattern

The production of multimedia annotation can involve the execution of algorithms or the application of computer assisted methods which are used to produce or manipulate digital-data. The recognition of a face in an image region is an example of the former, while manual annotation of the characters is an example of the latter.

We consider algorithms to be methods that are applied to solve a computational problem (see Fig. 2-A). The associated (DOLCE) situations represent the work that is being done by algorithms. Such a situation encompasses digital-data[15] involved in the computation, regions that represent the values of parameters of an algorithm, and perdurants[16] that act as computational-tasks (i.e. the processing steps of an algorithm). An algorithm defines roles which are played by digital-data. These roles encode the meaning of data. In order to solve a problem, an algorithm has to process input data and return some output data. Thus, every algorithm defines at least one input-role and one output-role which both have to be played by digital-data.

5.4 Comparison with Requirements

We discuss now whether the requirements stated in Sect. 4 are satisfied with our proposed modeling of the multimedia ontology.

The ontology is **MPEG-7 compliant** since the patterns have been designed with the aim of translating the standard into DOLCE. It covers the most important part of MPEG-7 that is commonly used for describing the structure and the content of multimedia documents. Our current investigation shows that parts of MPEG-7 that have not yet been considered (e.g. navigation & access) can be formalized analogously to the other descriptors through the definition of further patterns. The technical realization of the basic MPEG-7 data types (e.g. matrices and vectors) is not within the scope of the multimedia ontology. They are represented as ontological concepts, because the about relationship which connects digital-data with numerical entities is only defined between concepts. Thus, the definition of OWL data type properties is required to connect instances of data type concepts (subconcepts of the DOLCE abstract-region) with the actual numeric information (e.g. xsd:string). Currently, simple string representation formats are used for

[15] digital-data entities are DOLCE endurants, i.e. entities which exist in time and space.

[16] Events, processes or phenomena are examples of perdurants. endurants participate in perdurants.

serializing data type concepts (e.g. rectangle) that are currently not covered by W3C standards. Future work includes the integration of the extended data types of OWL 1.1.

Syntactic and semantic interoperability of our multimedia ontology is achieved by an OWL DL formalization.[17] Similar to DOLCE, we provide a rich axiomatization of each pattern using first order logic. Our ontology can be linked to any web-based domain-specific ontology through the semantic annotation pattern.

A clear separation of concerns is ensured through the use of the multimedia patterns: the decomposition pattern for handling the structure and the annotation pattern for dealing with the metadata.

These patterns form the core of the modular architecture of the multimedia ontology. We follow the various MPEG-7 parts and organize the multimedia ontology into modules which cover (1) the descriptors related to a specific media type (e.g. visual, audio or text) and (2) the descriptors that are generic to a particular media (e.g. media descriptors). We also design a separate module for data types in order to abstract from their technical realization.

Through the use of multimedia design patterns, our ontology is also **extensible**, allowing the inclusion of further media types and descriptors (e.g. new low-level features) using the same patterns. As our patterns are grounded in the D&S pattern, it is straightforward to include further contextual knowledge (e.g. about provenance) by adding roles or parameters. Such extensions will not change the patterns, so that legacy annotations will remain valid.

6 Expressing the Scenario in COMM

The interoperability problem with which Nathalie was faced in Sect. 2 can be solved by employing the COMM ontology for representing the metadata of all relevant multimedia objects and the presentation itself throughout the whole creation workflow. The student is shielded from details of the multimedia ontology by embedding it in authoring tools and feature analysis web services.

The application of the Winston Churchill face recognizer results in an annotation RDF graph that is depicted in the upper part of Fig. 3 (visualized by an UML object diagram.[18]) The decomposition of Fig. 1-A, whose content is represented by id0, into one still region (the bounding box of Churchill's face) is represented by the lighter middle part of the UML diagram. The segment is represented by the image-data instance id1 which plays the still-region-role srr1. It is located by the digital-data instance dd1 which expresses the region-locator-descriptor rld1 (lower part of the diagram). Using the semantic

[17] Examples of the axiomatization are available on the COMM website.

[18] The scheme used in Fig. 3 is instance:Concept, the usual UML notation.

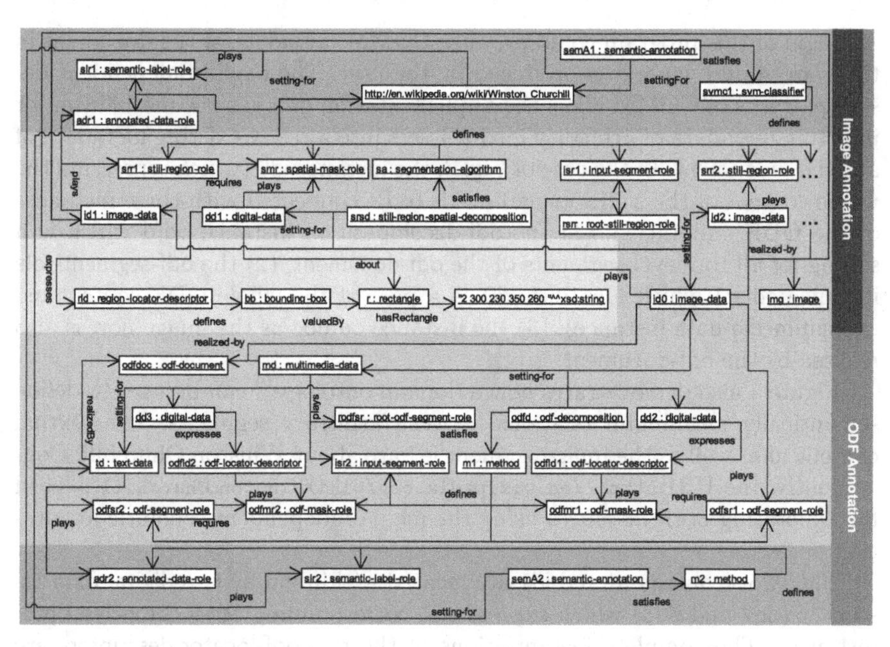

Fig. 3. Annotation of one segment of the Yalta picture and its embedding into an ODF document which contains a text segment that is also about Winston Churchill

annotation pattern, the face recognizer can annotate the still region by connecting it with the URI `http://en.wikipedia.org/wiki/Winston_Churchill`. An instance of an arbitrary domain ontology concept could also have been used for identifying the resource.

Running the two remaining face recognizers for Roosevelt and Stalin will extend the decomposition further by two still regions, i.e. the image-data instances id2 and id3 as well as the corresponding still-region-roles, spatial-mask-roles and digital-data instances expressing two more region-locator-descriptors (indicated at the right border of Fig. 3). The domain ontologies which provide the instances Roosevelt and Stalin for annotating id2 and id3 with the semantic annotation pattern do not have to be identical to the one that contains Churchill. If several domain ontologies are used, Nathalie can use the OWL sameAs and equivalentClass constructs to align the three face recognition results to the domain ontology that is best suited for enhancing the automatic generation of the multimedia presentation.

Decomposition of ODF documents is formalized analogously to image segmentation (see Fig. 2-F). Therefore, embedding the image annotation into an ODF document annotation is straightforward. The lower part of Fig. 3 shows the decomposition of a compound ODF document into textual and image content. This decomposition description could result from copying an image from the desktop and pasting it into an ODF editor such as OpenOffice.

A plugin of this program could produce COMM metadata of the document in the background while it is produced by the user. The media independent design patterns of COMM allow the implementation of a generic mechanism for inserting metadata of arbitrary media assets into already existing metadata of an ODF document. In the case of Fig. 3, the instance id0 (which represents the whole content of the Yalta image) needs to be connected with three instances of the ODF annotation: (1) the odf-decomposition instance odfd which is a setting-for all top level segments of the odf-document, (2) the odf-segment-role instance odfsr1 which identifies id0 as a part of the whole ODF content md (a multimedia-data instance), (3) the instance odfdoc as the image now is also realized-by the odf-document.

Figure 3 also demonstrates how a domain ontology[19] can be used to define semantically meaningful relations between arbitrary segments. The textual content td as well as the image segment id1 are about Winston Churchill. Consequently, the URI `http://en.wikipedia.org/wiki/Winston_Churchill` is used for annotating both instances using the media independent semantic annotation pattern.

The two segments td and id1 are located within md by two digital-data instances (dd2 and dd3) which express two corresponding odf-locator-descriptor instances. The complete instantiations of the two odf-locator-descriptors are not shown in Fig. 3. The modeling of the region-locator-descriptor, which is completely instantiated in Fig. 3, is shown in Fig. 2-C. The technical details of the odf-locator-descriptor are not presented. However, it is possible to locate segments in ODF documents by storing an XPath which points to the beginning and the end of an ODF segment. Thus, the modeling of the odf-locator-descriptor can be carried out analogously to the region-locator-descriptor.

In order to ease the creation of multimedia annotations with our ontology, we have developed a Java API[20] which provides an MPEG-7 class interface for the construction of meta-data at runtime. Annotations which are generated in memory can be exported to Java based RDF triple stores such as Sesame. For that purpose, the API translates the objects of the MPEG-7 classes into instances of the COMM concepts. The API also facilitates the implementation of multimedia retrieval tools as it is capable of loading RDF annotation graphs (e.g. the complete annotation of an image including the annotation of arbitrary regions) from a store and converting them back to the MPEG-7 class interface. Using this API, the face recognition web service will automatically create the annotation which is depicted in the upper part of Fig. 3 by executing the following code:

```
Image img0 = new Image();
StillRegion isr0 = new StillRegion();
```

[19] In this example, the domain ontology corresponds to a collection of wikipedia URI's.

[20] The Java API is available at `http://multimedia.semanticweb.org/COMM/api/`.

```
img0.setImage(isr0);
StillRegionSpatialDecomposition srsd1
   = new StillRegionSpatialDecomposition();
isr0.addSpatialDecomposition(srsd1);
srsd1.setDescription(new SegmentationAlgorithm());
StillRegion srr1 = new StillRegion();
srsd1.addStillRegion(srr1);
SpatialMask smr1 = new SpatialMask();
srr1.setSpatialMask(smr1);
RegionLocatorDescriptor rld1 = new RegionLocatorDescriptor();
smr1.addSubRegion(rld1);
rld1.setBox(new Rectangle(300, 230, 50, 30));
Semantic s1 = new Semantic();
s1.addLabel("http://en.wikipedia.org/wiki/Winston_Churchill");
s1.setDescription(new SVMClassifier());
srr1.addSemantic(s1);
```

7 Conclusion and Future Work

We have presented COMM, an MPEG-7 based multimedia ontology, well-founded and composed of multimedia patterns. It satisfies the requirements, as they are described by the multimedia community itself, for a multimedia ontology framework. The ontology is completely formalized in OWL DL and a stable version is available with its API at: `http://multimedia.semanticweb.org/COMM/`. It has been used in projects such as K-Space and X-Media.

The ontology already covers the main parts of the standard, and we are confident that the remaining parts can be covered by following our method for extracting more design patterns. Our modeling approach confirms that the ontology offers even more possibilities for multimedia annotation than MPEG-7 since it is interoperable with existing web ontologies. The explicit representation of algorithms in the multimedia patterns describes the multimedia analysis steps, something that is not possible in MPEG-7. The need for providing this kind of annotation is demonstrated in the algorithm use case of the W3C Multimedia Semantics Incubator Group.[21] The intensive use of the D&S reification mechanism causes that RDF annotation graphs, which are generated according to our ontology, are quite large compared to the ones of more straightforwardly designed multimedia ontologies. This presents a challenge for current RDF and OWL stores, but we think it is a challenge worth deep consideration as it is utterly necessary to overcome the isolation of current multimedia annotations and to achieve full interoperability for (nearly) arbitrary multimedia tools and applications.

[21] `http://www.w3.org/2005/Incubator/mmsem/XGR-interoperability/`

Acknowledgments

The research leading to this paper was partially supported by the European Commission under contract FP6-027026, Knowledge Space of semantic inference for automatic annotation and retrieval of multimedia content – K-Space, and under contract FP6-26978, X-Media, large scale knowledge sharing and reuse across media.

References

1. Richard Arndt, Raphaël Troncy, Steffen Staab, Lynda Hardman, and Miroslav Vacura. COMM: Designing a Well-Founded Multimedia Ontology for the Web. In *6th Int. Semantic Web Conference*, 2007.
2. S. Blöhdorn, K. Petridis, C. Saathoff, N. Simou, V. Tzouvaras, Y. Avrithis, S. Handschuh, Y. Kompatsiaris, S. Staab, and M. Strintzis. Semantic Annotation of Images and Videos for Multimedia Analysis. In *2nd European Semantic Web Conference*, 2005.
3. Aldo Gangemi, Stefano Borgo, Carola Catenacci, and Jos Lehmann. Task Taxonomies for Knowledge Content. Technical report, Metokis Deliverable 7, 2004.
4. Roberto Garcia and Oscar Celma. Semantic Integration and Retrieval of Multimedia Metadata. In *5th International Workshop on Knowledge Markup and Semantic Annotation*, 2005.
5. Joost Geurts, Jacco van Ossenbruggen, and Lynda Hardman. Requirements for practical multimedia annotation. In *Workshop on Multimedia and the Semantic Web*, 2005.
6. F. Halasz and M. Schwartz. The Dexter Hypertext Reference Model. *Communications of the ACM*, 37(2):30–39, 1994.
7. Jane Hunter. Adding Multimedia to the Semantic Web - Building an MPEG-7 Ontology. In *1st International Semantic Web Working Symposium*, pages 261–281, 2001.
8. Jane Hunter. Combining the CIDOC/CRM and MPEG-7 to Describe Multimedia in Museums. In *6th Museums and the Web Conference*, 2002. `http://www.archimuse.com/mw2002/papers/hunter/hunter.html`.
9. Jane Hunter and Liz Armstrong. A Comparison of Schemas for Video Metadata Representation. In *8th International World Wide Web Conference*, pages 1431–1451, 1999.
10. Antoine Isaac and Raphaël Troncy. Designing and Using an Audio-Visual Description Core Ontology. In *Workshop on Core Ontologies in Ontology Engineering*, 2004.
11. MPEG-21. Part 17: Fragment Identification of MPEG Resources. Standard No. ISO/IEC 21000-17, 2006.
12. MPEG-7. Multimedia Content Description Interface. Standard No. ISO/IEC 15938, 2001.
13. Frank Nack and Adam T. Lindsay. Everything you wanted to know about MPEG-7 (Parts I & II). *IEEE Multimedia*, 6(3-4), 1999.

14. Frank Nack, Jacco van Ossenbruggen, and Lynda Hardman. That Obscure Object of Desire: Multimedia Metadata on the Web (Part II). *IEEE Multimedia*, 12(1), 2005.
15. Bernd Neumann and Ralf Möller. *Cognitive Vision Systems*, chapter On Scene Interpretation with Description Logics, pages 247–275. Springer, 2006.
16. Daniel Oberle, S. Lamparter, S. Grimm, D. Vrandecic, Steffen Staab, and Aldo Gangemi. Towards Ontologies for Formalizing Modularization and Communication in Large Software Systems. *Journal of Applied Ontology*, 1(2):163–202, 2006.
17. Jacco van Ossenbruggen, Frank Nack, and Lynda Hardman. That Obscure Object of Desire: Multimedia Metadata on the Web (Part I). *IEEE Multimedia*, 11(4), 2004.
18. Raphaël Troncy. Integrating Structure and Semantics into Audio-visual Documents. In *2nd International Semantic Web Conference*, pages 566–581, 2003.
19. Raphaël Troncy, Werner Bailer, Michael Hausenblas, Philip Hofmair, and Rudolf Schlatte. Enabling Multimedia Metadata Interoperability by Defining Formal Semantics of MPEG-7 Profiles. In *1st International Conference on Semantics And digital Media Technology*, pages 41–55, 2006.
20. Raphaël Troncy, Óscar Celma, Suzanne Little, Roberto García, and Chrisa Tsinaraki. MPEG-7 based Multimedia Ontologies: Interoperability Support or Interoperability Issue? In *1st International Workshop on Multimedia Annotation and Retrieval enabled by Shared Ontologies*, 2007.
21. Raphaël Troncy, Lynda Hardman, Jacco van Ossenbruggen, and Michael Hausenblas. Identifying Spatial and Temporal Media Fragments on the Web. In *W3C Video on the Web Workshop*, 2007. `http://www.w3.org/2007/08/video/positions/Troncy.pdf`.
22. Chrisa Tsinaraki, Panagiotis Polydoros, Nektarios Moumoutzis, and Stavros Christodoulakis. Integration of OWL ontologies in MPEG-7 and TV-Anytime compliant Semantic Indexing. In *16th International Conference on Advanced Information Systemes Engineering*, 2004.

Using the PSL Ontology

Michael Grüninger

Department of Mechanical and Industrial Engineering, University of Toronto, Toronto, ON, Canada, gruninger@mie.utoronto.ca

1 Introduction

Representing activities and the constraints on their occurrences is an integral aspect of commonsense reasoning, particularly in manufacturing, enterprise modelling, and autonomous agents or robots. In addition to the traditional concerns of knowledge representation and reasoning, the need to integrate software applications in these areas has become increasingly important. However, interoperability is hindered because the applications use different terminology and representations of the domain. These problems arise most acutely for systems that must manage the heterogeneity inherent in various domains and integrate models of different domains into coherent frameworks. For example, such integration occurs in business process reengineering, where enterprise models integrate processes, organizations, goals and customers. Even when applications use the same terminology, they often associate different semantics with the terms. This clash over the meaning of the terms prevents the seamless exchange of information among the applications. translators between every pair of applications that must cooperate. What is needed is some way of explicitly specifying the terminology of the applications in an unambiguous fashion.

The Process Specification Language (PSL) ([6, 9]) has been designed to facilitate correct and complete exchange of process information among manufacturing systems.[1] Included in these applications are scheduling, process modeling, process planning, production planning, simulation, project management, workflow, and business process reengineering. This chapter will give an overview of the PSL Ontology, including its formal characterization as a set of theories in first-order logic and the range of concepts that are axiomatized in these theories.

[1] PSL has been published as the International Standard ISO 18629 by the International Organisation of Standardisation.

S. Staab and R. Studer (eds.), *Handbook on Ontologies*, International Handbooks on Information Systems, DOI 10.1007/978-3-540-92673-3,

2 How are Ontologies Used?

Applications of ontologies focus on their role as sharable and reusable representations of knowledge (Chapters "What is an *Ontology*?", "Ontology-Based Recommender Systems". Semantic heterogeneity is particularly acute problem for tasks that require correct and meaningful communication and integration among software systems, since different systems may ascribe disparate meanings to the same terms or use distinct terms to convey the same meaning. Ontologies support semantic integration through a shared understanding of the intended semantics of the terminology used by the software systems.

The reusability of an ontology is determined relative to the genericity of its axiomatization. In one sense, the axioms of the ontology can be instantiated within different domains; this leads to the notion of domain theories that capture the knowledge for particular problems. In another sense, the axioms of the ontology capture those properties of the world that are valid across multiple domains; new ontologies can then be constructed as more domain-specific extensions of the generic ontologies.

2.1 Specifying Domain Theories

Within the context of a process ontology , domain theories take the form of descriptions of processes as repeatable patterns of behaviour. The various forms of process representations are ubiquitous in industry: there is a plethora of business and engineering software applications – workflow, scheduling, discrete event simulation, process planning, business process modeling, and others – that are designed explicitly for the construction of process models of various sorts [6]. In addition, there are many concrete domains for process representations, including manufacturing, web services, and business processes.

A process ontology provides the underlying semantics for the process terminology that is common to the many disparate domains and software applications. This allows us to evaluate the consistency of process descriptions. In this way, ontologies can be used to support automated reasoning (such as theorem proving and constraint satisfaction) with the axioms of the ontology and domain theories alone.

Ontologies also provide guidance in the specification of domain theories. For example, each class of activities in the PSL Ontology is associated with a specific class of sentences that are the correct process descriptions for that class. The primary focus of this chapter will be a survey of the various classes of activities in the ontology together with examples of the corresponding process descriptions.

2.2 Semantic Integration

A semantics-preserving exchange of information between two software applications requires mappings between logically equivalent concepts in the ontology of each application. The challenge of semantic integration is therefore equivalent to the problem of generating such mappings (Chapter "Why Is Ontology Mapping Difficult?"), determining that they are correct, and providing a vehicle for executing the mappings, thus translating terms from one ontology into another.

The Twenty Questions approach ([3]) describes a technique for the semi-automatic generation of semantic mappings from application ontologies to the PSL Ontology, which can then be used to automatically derive direct mappings between application ontologies.

The work in [7] describes an example of using PSL as a common ontology to facilitate manufacturing process information exchange between two different software applications, ProCAP – a process modelling tool based upon the IDEF3 method of systems modelling and ILOG – a C++ library for constraint-based scheduling. In a typical scenario, a user of ProCAP describes the types or processes that are necessary to produce some product, specifies the order in which these processes must occur, and describes what types of resources are necessary for the creation of the product. Semantic mappings between the PSL Ontology and the terminology used in IDEF3 and ILOG process descriptions form the basis for translators between the software applications.

2.3 Building New Ontologies

An ontology with a consistent and complete axiomatization of its intended semantics can be used as a semantic foundation for either building a new ontology or for augmenting an ontology that has an incomplete axiomatization (Chapter "Foundational Choices in DOLCE"). For example, the process model of the semantic web services ontology OWL-S (Chapter "Semantic Web Services", [5]) contains a taxonomy of control constructs for specifying composite web services; however, the intended semantics of these constructs is expressed in natural language, since it cannot be axiomatized in OWL. The work in [2] provides a first-order axiomatization of these constructs using the PSL Ontology.

The Semantic Web Services Ontology (SWSO) ([11]) is an extension of the PSL Ontology with Web service-specific concepts which enables reasoning about the semantics underlying Web services and along with their interactions with each other and with the "real world". Because SWSO is an extension of the PSL Ontology, it also provides a first-order axiomatization of the intended semantics of the process model of OWL-S. This supports reasoning with the axioms of the ontology alone, rather than use extra-logical algorithms to guarantee that queries are entailed by the web service specifications.

3 Basic Ontological Distinctions

The PSL Ontology consists of a set of first-order logic theories within which there is a distinction between core theories and definitional extensions.[2] Core theories introduce new primitive concepts, while all terms introduced in a definitional extension that are conservatively defined using the terminology of the core theories.

All core theories within the ontology are consistent extensions of PSL-Core (T_{psl_core}), although not all extensions need be mutually consistent. Table 1 is a summary of the key terms in the lexicon of the eight core theories which will be used in this chapter.

3.1 Activity and Activity Occurrence

In general, business and engineering processes are described at the type level – a process specification characterizes a certain general pattern that might admit of many instantiations which might differ considerably from one another. For example, the specification of the manufacturing process for making a car will describe different sequences of tasks for building the components of the car and may even describe alternative ways of producing subassemblies. A robust foundation for process modelling, therefore, should be able to characterize both the general process pattern described by a specification as well as the class of possible instantiations of that pattern. Moreover, such a foundation must be able to clearly represent the constraints that a process specification places on something's counting as an instantiations of the process, that is, the constraints on process execution.

Within the PSL Ontology, an *activity* is a repeatable pattern of behaviour, while an *activity occurrence* corresponds to a concrete instantiation of this pattern. The relationship between activities and activity occurrences is represented by the $occurrence\_of(o, a)$ relation. Activities may have multiple occurrences, or there may exist activities which never occur. Any activity occurrence corresponds to a unique activity.

In contrast to many object-oriented approaches, activity occurrences are not considered to be instances of activities, since activities are not classes within the PSL Ontology. One can of course specify classes of activities in a process description. For example the term $pickup(x, y)$ can denote the class of activities for picking up some object x with manipulator y, and the term $move(x, y, z)$ can denote the class of activities for moving object x from location y to location z. Ground terms such as $pickup(Block1, LeftHand)$ and

[2] The complete set of axioms for the PSL Ontology can be found at `http://www.mel.nist.gov/psl/psl-ontology/`. Core theories are indicated by a .th suffix and definitional extensions are indicated by a .def suffix.

All axioms and definitions in the PSL Ontology are written in CLIF (Common Logic Interchange Format).

Table 1. Lexicon for core theories in the PSL Ontology

T_{psl_core}	$activity(a)$	a is an activity
	$activity\_occurrence(o)$	o is an activity occurrence
	$timepoint(t)$	t is a timepoint
	$object(x)$	x is an object
	$occurrence\_of(o, a)$	o is an occurrence of a
	$beginof(o)$	the beginning timepoint of o
	$endof(o)$	the ending timepoint of o
	$before(t_1, t_2)$	timepoint t_1 precedes timepoint t_2 on the timeline
$T_{subactivity}$	$subactivity(a_1, a_2)$	a_1 is a subactivity of a_2
	$primitive(a)$	a is a minimal element of the *subactivity* ordering
T_{atomic}	$atomic(a)$	a is either primitive or a concurrent activity
	$conc(a_1, a_2)$	the activity that is the concurrent composition of a_1 and a_2
$T_{occtree}$	$legal(s)$	s is an element of a legal occurrence tree
	$earlier(s_1, s_2)$	s_1 precedes s_2 in an occurrence tree
T_{disc_state}	$holds(f, s)$	the fluent f is true immediately after the activity occurrence s
	$prior(f, s)$	the fluent f is true immediately before the activity occurrence s
$T_{complex}$	$min\_precedes(s_1, s_2, a)$	the atomic subactivity occurrence s_1 precedes the atomic subactivity occurrence s_2 in an activity tree for a
	$root(s, a)$	the atomic subactivity occurrence s is the root of an activity tree for a
	$next\_subocc(s_1, s_2, a)$	the atomic subactivity occurrence s_1 is by the atomic subactivity occurrence s_2 in an activity tree for a
T_{actocc}	$subactivity\_occurrence(o_1, o_2)$	o_1 is a subactivity occurrence of o_2
	$root\_occ(o)$	the initial atomic subactivity occurrence of o
	$leaf\_occ(s, o)$	s is the final atomic subactivity occurrence of o
$T_{duration}$	$timeduration(d)$	d is a timeduration
	$duration(t_1, t_2)$	the timeduration whose value is the "distance" from timepoint t_1 to timepoint t_2

$move(Shipment1, Seattle, Chicago)$ are instances of these classes of activities, and each instance can have different occurrences. Furthermore, there may be classes of activity occurrences that do not correspond to activities, e.g. that class of activity occurrences that finish by Friday.

3.2 Time

Underlying the intuition that activity occurrences are the instantiations of activities is the notion that each activity occurrence is associated with unique timepoints that mark the begin and end of the occurrence. The PSL Ontology introduces two functions $beginof$ and $endof$ for this purpose.

The set of timepoints is linearly ordered, forwards into the future, and backwards into the past. Within the PSL Ontology, the extension of the $before$ relation captures this linear ordering. There are also different ontological commitments about time that are not made within the PSL Ontology, such as the denseness of the timeline; any such commitments must be axiomatized within a theory that extends the PSL Ontology.

There are some approaches (e.g. [4]) that do not distinguish between timepoints and activity occurrences, so that activity occurrences form a subclass of timepoints. However, activity occurrences have preconditions and effects, whereas timepoints do not. Other approaches hold that timepoints are primitives but activity occurrences are not; for example, approaches such as [10] claim that one can derive timepoints as "ticks" of a clock activity; however, such an approach ties the temporal ontology too closely to the process ontology.

The core theory $T_{duration}$ for duration adds a metric to the timeline by mapping every pair of timepoints to a new sort called *timeduration* that satisfies the axioms of algebraic fields. Of course, the duration of an activity occurrence is of most interest, and is equal to the duration between the endof and beginof timepoints of the activity occurrence.

3.3 Objects

Many debates have erupted within philosophy over the distinction between objects that are *continuants* (that is, they exist whole and entire at different times) and objects that are *occurrents* (that is, they have different parts existing at different times).[3] Although the PSL Ontology tries to avoid making any commitments that would preclude one position or another in this debate, activity occurrences can be considered to be occurrents, while continuants are represented by *objects*. The ternary relation $participates\_in(x, o, t)$ is used to tie the two approaches together by specifying that object x participates in activity occurrence o at timepoint t.

[3] This terminology is used in [1]. The treatment of objects as continuants is also known as endurantism or 3D-ontology, while the treatment of objects as occurrents is also known as perdurantism or 4D-ontology.

3.4 Composition

A ubiquitous feature of process formalisms is the ability to compose simpler activities to form new complex activities (or conversely, to decompose any complex activity into a set of subactivities). The PSL Ontology incorporates this idea while making several distinctions between different kinds of composition that arise from the relationship between composition of activities and composition of activity occurrences.

Subactivities

The PSL Ontology uses the *subactivity* relation to capture the basic intuitions for the composition of activities. The core theory $T_{subactivity}$ axiomatizes this relation as a discrete partial ordering (such as Fig. 1), in which primitive activities are the minimal elements.

$T_{subactivity}$ alone does not specify any relationship between the occurrence of an activity and occurrences of its subactivities. For example, we can compose the primitive activities *press* and *punch* in Fig. 1 to make the complex activity $surfacing$ and we can also compose them to make a different complex activity *shaping*. However, this specification of subactivities alone does not allow us to say that $surfacing$ is a deterministic activity, or that *shaping* is a nondeterministic activity. The core theory $T_{complex}$ is therefore introduced to characterize the relationship between the occurrence of a complex activity and occurrences of its subactivities.

Concurrency

Concurrency involves more than the fact that two activities occur at the same time, since concurrent activities may have different preconditions and effects than any of the activities if they occur alone. In particular, the activities may have interfering preconditions, so that even if two activities can possibly occur separately, they cannot occur concurrently (e.g. the oven cannot be used to bake a cake and a turkey at the same time) or the effects of two activities may clobber each other, so that the effects of the concurrent activity are different

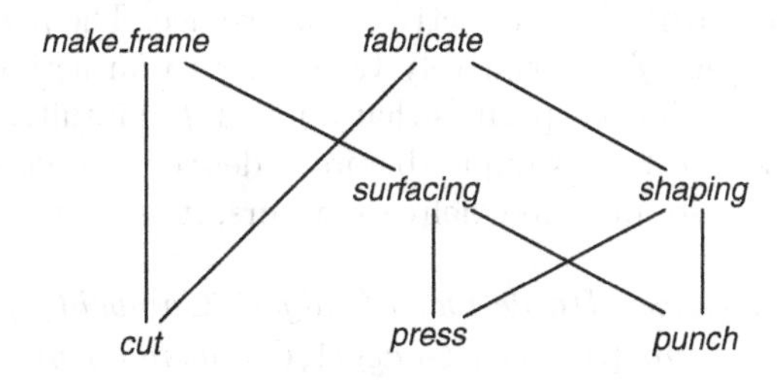

Fig. 1. Example of subactivities and their composition into different complex activities

than the effects of the two activities if they occur separately ([8]); for example, the effect of lifting only the right side or only the left side of a table has the effect that the table is touching the floor. Lifting both the right and left sides concurrently has the effect of lifting the entire table.

This observation leads to a notion of *atomic* activity which corresponds to some set of primitive activities. Concurrency is represented by the occurrence of concurrent activities rather than concurrent activity occurrences. The core theory T_{atomic} axiomatizes the *conc* function that specifies the aggregation of sets of primitive activities into concurrent activities.

Subactivity Occurrences

The core theory T_{actocc} axiomatizes the *subactivity_occurrence* relation, which is the composition relation over activity occurrences corresponding to the composition relation over activities. Occurrences of atomic activities are the minimal elements in this composition ordering – they do not have any nontrivial subactivity occurrences.

Following the intuition that activity occurrences are occurrents rather than continuants, one can consider the subactivity occurrence to be a temporal part of the complex activity occurrence. The axioms of T_{actocc} guarantee that any subactivity occurrence is "during" an occurrence of the complex activity.

3.5 State and Change

Many applications of process ontologies are used to represent dynamic behaviour in the world so that software systems may make predictions about the future and explanations about the past. In particular, these predictions and explanations are often concerned with the state of the world and how that state changes. The PSL core theory T_{disc_state} is intended to capture the basic intuitions about state and its relationship to activities.

Properties in the domain that can change are called *fluents* . Similar to the representation of activities, fluents can also be denoted by terms within the language. For example, $in\_stock(Gadget1, Cambridge)$ denotes the fluent that represents the property that the object $Gadget1$ is available in stock at the Cambridge warehouse.

Intuitively, a change in state is captured by the set of fluents that are either achieved or falsified by an activity occurrence. The $prior(f, o)$ relation specifies that a fluent f is intuitively true prior to an activity occurrence o and the $holds(f, o)$ relation specifies that a fluent f is intuitively true after an activity occurrence o. For example, before a delivery, $Gadget1$ is not in the Cambridge warehouse, but after delivery occurs, it is in stock:

$$\begin{aligned} &occurrence\_of(o, delivery(Gadget1, Cambridge)) \supset \\ &\neg prior(in\_stock(Gadget1, Cambridge), o) \\ &\wedge holds(in\_stock(Gadget1, Cambridge), o) \end{aligned}$$

A fluent is changed by the occurrence of activities, and a fluent can only be changed by the occurrence of activities. Thus, if some fluent holds after an activity occurrence, but after an activity occurrence later along the branch it is false, then an activity must occur at some point between that changes the fluent. This also leads to the requirement that the fluent holding after an activity occurrence will be the same fluent holding prior to any immediately succeeding occurrence, since there cannot be an activity occurring between the two by definition.

State does not change during the occurrence of an atomic activity. Consequently, the PSL Ontology cannot represent phenomena in which some feature of the world is changing as some continuous function of time (hence the name "Discrete State" for the extension). If state changes during an activity occurrence, then it must be an occurrence of a complex activity.

4 Process Descriptions for Atomic Activities

Within the taxonomy of the PSL Ontology, activities are classified according to the kinds of constraints that their occurrences satisfy. A process description for an activity in some class imposes constraints on activity occurrences corresponding to the definition of the class. Classes of atomic activities are defined with respect to constraints that arise from the following two questions:

- Under what conditions does an atomic activity occur?
- How do occurrences of atomic activities change fluents?

A detailed exposition of these constraints requires a closer look at the model theory of the core theory $T_{occtree}$, in particular, the notion of occurrence trees.

4.1 Occurrence Trees

An occurrence tree is a partially ordered set of atomic activity occurrences, such that for a given set of activities, all discrete sequences of their occurrences are branches of the tree. It is important to note that an occurrence tree contains all occurrences of *all* atomic activities; it is not simply the set of occurrences of a particular (possibly complex) activity. Because the tree is discrete, each activity occurrence in the tree has a unique successor occurrence of each activity.

Although occurrence trees characterize all sequences of activity occurrences, not all of these sequences will intuitively be physically possible within the domain. This leads to the notion of the legal occurrence tree, which is the subtree of the occurrence tree that consists only of *possible* sequences of activity occurrences; The $legal(o)$ relation specifies that the atomic activity occurrence o is an element of the legal occurrence tree.

4.2 Constraints on Legal Occurrence

The process descriptions for atomic activities constrain the legal occurrence tree. The general form of such a process description is:

$$(\forall o)\ occurrence\_of(o, a) \wedge legal(o) \supset \Phi(o) \tag{1}$$

where $\Phi(o)$ is a formula that specifies the constraint on the legal activity occurrence. Within the PSL Ontology, different classes of atomic activities correspond to different classes of formulae that are used to instantiate $\Phi(o)$ in the general process description. In particular, we consider cases in which the preconditions are based on state, time, or the occurrence of other activities.

State-Based Preconditions

The most prevalent kind of precondition are markovian preconditions, in which the possibility of occurrence depends only on the state that holds prior to an activity occurrence, e.g.

Mixing is not performed unless the moulding machine is clean.

In this case, the cleanliness of the machine is the state, and the occurrence of the mixing activity depends on whether or not this state holds:

$$(\forall o, x)\ occurrence\_of(o, mixing(x)) \wedge legal(o) \supset prior(clean(x), o) \tag{2}$$

Note that for this particular class of activities, the consequent of the sentence is a formula that contains only *prior* literals.

Time-Based Preconditions

In more general scenarios, there may be temporal preconditions that depend only on the time at which the activity is to occur, such as

The pre-heating operation can only be performed on Tuesday or Thursday.

which is axiomatized as

$$(\forall o, x)\ occurrence\_of(o, preheat(x)) \wedge legal(o) \supset$$
$$(beginof(o) = Tuesday) \vee (beginof(o) = Thursday) \tag{3}$$

The consequent of this process description is a formula that contains only *beginof* literals.

Occurrence Constraints

The possibility of an activity occurrence may depend on the occurrence of other activities. Consider the example:

If we do not fold the metal after fabrication, we need to reheat it which is axiomatized as

$$(\forall o_1, x)\ occurrence\_of(o_1, reheat(x)) \wedge legal(o_1) \supset$$
$$\neg(\exists o_2)\ occurrence\_of(o_2, fold(x)) \wedge earlier(o_2, o_1) \wedge legal(o_2) \quad (4)$$

In this case, an occurrence of the reheating activity will depend on the condition that there is no earlier legal occurrence of the folding activity.

Time-Based Occurrence Constraints

Preconditions may also take the form of periodic occurrences, e.g.
Drill bits are replaced every 10 min.

$$(\forall o_1, x_1)\ occurrence\_of(o_1, replace(x_1)) \wedge legal(o_1) \supset$$
$$(\exists o_2, x_2)\ occurrence\_of(o_2, replace(x_2)) \wedge earlier(o_2, o_1)$$
$$\wedge legal(o_2) \wedge (duration(beginof(o_2), beginof(o_1)) = 10) \quad (5)$$

In this example, occurrences of the replacement activity depend not only on the occurrence of an earlier replacement activity but also on the time at which that activity occurred.

4.3 Effects

Effects characterize the ways in which activity occurrences change the state of the world. Such effects may be context-free, so that all occurrences of the activity change the same states, or they may be constrained by other conditions. The general form of such a process description is:

$$(\forall o)\ occurrence\_of(o, a) \wedge \Phi(o) \supset holds(f, o) \quad (6)$$

where $\Phi(o)$ is a formula that specifies the constraint on the effects of the activity occurrence.

State-Based Effects

The most common constraint is state-based effects that depend on some context:
If the object is fragile, then it will break when dropped; if the object is elastic, then it will bounce when dropped.

$$(\forall o, x)\ occurrence\_of(o, drop(x)) \wedge prior(fragile(x), o)$$
$$\supset holds(broken(x), o) \quad (7)$$

Time-Based Effects

Although process descriptions for the effects of atomic activities are most often specifying state-based effects, other kinds of constraints also arise in practice, such as time-based effects:

If the rental car is returned after the due date, then the cost includes a late fee

which is axiomatized by

$$(\forall o, x)\ occurrence\_of(o, rental(x)) \wedge before(DueDate, endof(o)) \\ \supset holds(late\_fee(x), o) \quad (8)$$

The effects of the activity occurrence depend only on timepoints – the time at which the activity occurrence ends and the timepoint that is the due date of the rental.

Occurrence-Based Effects

In some cases, the effects depend not only on when the activity occurs, but also on the timepoints at which other activity occurrences begin or end. For example,

If we remove the coffee pot before the brewing activity completes, then the burner will be wet

is axiomatized by

$$(\forall o_1, o_2, x, y)\ occurrence\_of(o_1, brew(x, y)) \wedge occurrence\_of(o_2, remove(x, y)) \\ \wedge before(beginof(o_2), beginof(o_1)) \supset holds(wet(y), o_1) \quad (9)$$

and in this case, the formula in the process description contains multiple variables denoting different activity occurrences, as well as $before$ literals.

Duration-Based Effects

For some classes of atomic activities, the effects are dependent on the duration of the activity occurrences. For example,

The time on the clock display will change after holding the button for 3 s

is axiomatized by

$$(\forall o, x)\ occurrence\_of(o, press(x)) \wedge duration(endof(o), beginof(o)) = 3 \\ \supset holds(display(x), o) \quad (10)$$

The effects do not depend on the time at which the activity occurs, so that the formula does not contain any $before$ literals.

5 Process Descriptions for Complex Activities

Classes of complex activities are defined with respect to the following two questions:

- What is the relationship between the occurrence of the complex activity and occurrences of its subactivities?
- Under what conditions does a complex activity occur?

An activity may have subactivities that do not occur; the only constraint is that any subactivity occurrence must correspond to a subtree of the activity tree that characterizes the occurrence of the activity.

5.1 Activity Trees

The basic structure that characterizes occurrences of complex activities is the *activity tree*, which is a subtree of the legal occurrence tree that consists of all possible sequences of atomic subactivity occurrences beginning from a root subactivity occurrence. Each branch of an activity tree corresponds to a possible sequence of occurrences of subactivities of the complex activity.

In a sense, an activity tree is a microcosm of the occurrence tree, in which we consider all of the ways in which the world unfolds *in the context of an occurrence of the complex activity.* For example, consider the occurrence tree in Fig. 2, and suppose that an occurrence of the complex activity $make\_frame$

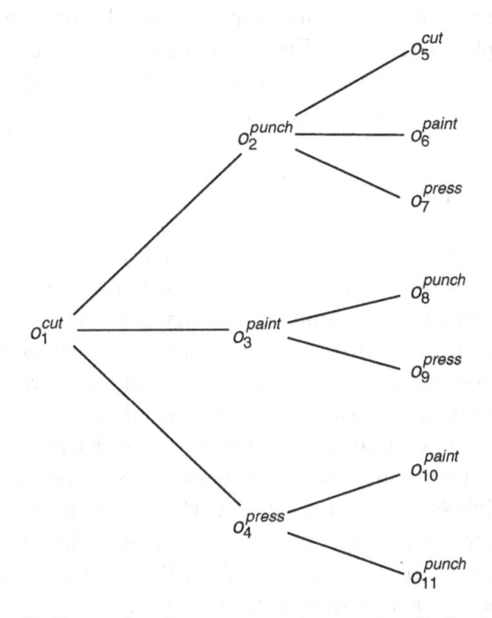

Fig. 2. Example of occurrence tree and activity trees

consists of an occurrence of *cut* followed by occurrences of *punch* and *press*. The subtree consisting of

$$\{o_1^{cut}, o_2^{punch}, o_7^{press}, o_4^{press}, o_{11}^{punch}\}$$

is a possible activity tree for *make_frame*.

The models of any process description for a complex activity consists of a set of activity trees within an occurrence tree. Each branch of an activity tree is a sequence of atomic subactivity occurrences that satisfies the process description.

Three relations in particular are used in process descriptions for complex activities. The $root(o, a)$ relation specifies that the atomic subactivity occurrence o is the root of the activity tree. The *min_precedes* relation is the ordering relation over the atomic subactivity occurrences in the activity tree. In Fig. 2, the activity tree for *make_frame* satisfies the process description

$$(\forall o)\ occurrence\_of(o, make\_frame) \supset (\exists o_1, o_2, o_3)\ occurrence\_of(o_1, cut)$$
$$\wedge occurrence\_of(o_2, punch) \wedge occurrence\_of(o_3, press)$$
$$\wedge root(o_1, make\_frame)$$
$$\wedge min\_precedes(o_1, o_2, make\_frame) \wedge min\_precedes(o_1, o_3, make\_frame)$$

The axioms of T_{actocc} guarantees that there is a one-to-one correspondence between branches of activity trees and complex activity occurrences. The axioms for *subactivity_occurrence* relation guarantee that the branches of the activity trees for a subactivity are contained in the branches of the activity tree for the complex activity. In Fig. 2, the branch $\{o_1^{cut}, o_2^{punch}, o_7^{press}\}$ of the activity tree corresponds to an occurrence $o_{12}^{make\_frame}$ of *make_frame*, and each element of the branch is a subactivity occurrence of $o_{12}^{make\_frame}$.

5.2 Branch Structure

Different subactivities may occur on different branches of the activity tree – different occurrences of an activity may have different subactivity occurrences or different orderings on the same subactivity occurrences.

In this sense, branches of the activity tree characterize the nondeterminism that arises from different ordering constraints or iteration. For example, the *surfacing* activity is intuitively nondeterministic; the activity trees for *surfacing* contain two branches, one branch consisting of an occurrence of *polish* and one branch consisting of an occurrence of *paint*.

Complex activities can be classified with respect to symmetries of its activity trees. Concretely, these are axiomatized by relationships between the different branches of an activity tree. We will now take a closer look at the process descriptions for activities in these classes.

Permuted Activities

For permuted activities, each branch of the activity tree is a different permutation of the same set of subactivity occurrences. For example, the informal process description

Making the frame consists of cutting, punching, and pressing. can be formally written as

$$(\forall o, x)\ occurrence\_of(o, make\_frame(x))$$
$$\supset (\exists o_1, o_2, o_3)\ occurrence\_of(o_1, cut(x))$$
$$\wedge occurrence\_of(o_2, punch(x)) \wedge occurrence\_of(o_3, press(x)) \quad (11)$$

If we consider the activity trees that satisfy this sentence (Fig. 3), we can see that each branch contains an occurrence of each subactivity.

Activities may also be nondeterministic; for example, there could be alternative process plans to produce the same product depending on the customer, such as the constraint

Fabrication consists of cutting the metal together with either pressing or punching. which is formally written as

$$(\forall o, x)\ occurrence\_of(o, fabricate(x))$$
$$\supset (\exists o_1, o_2)\ subactivity\_occurrence(o_1, o) \wedge subactivity\_occurrence(o_2, o)$$
$$\wedge occurrence\_of(o_1, cut(x))$$
$$\wedge (occurrence\_of(o_2, press(x)) \vee occurrence\_of(o_2, punch(x))) \quad (12)$$

The activity tree in Fig. 4 that satisfies this sentence has branches that contain occurrences of different subactivities.

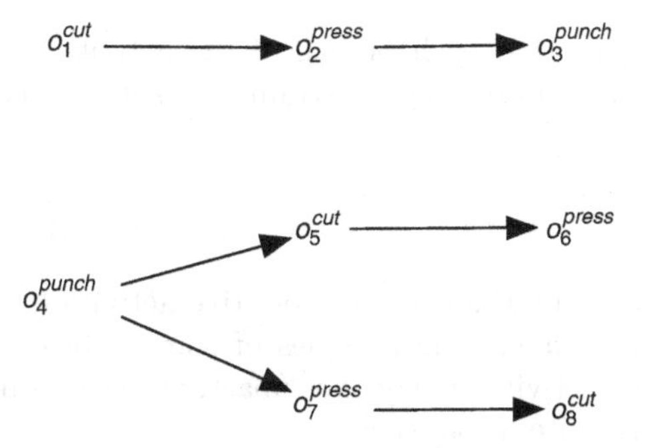

Fig. 3. Activity trees for permuted activities

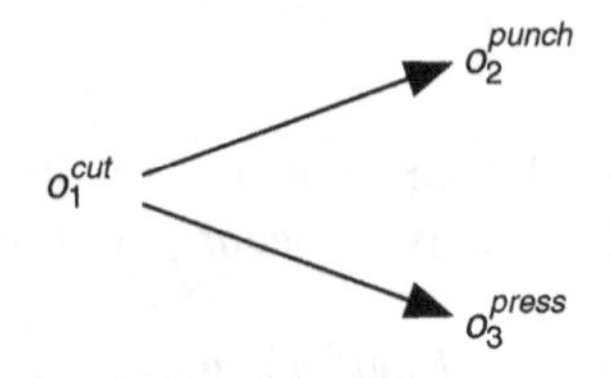

Fig. 4. Activity tree for a nondeterministic activity

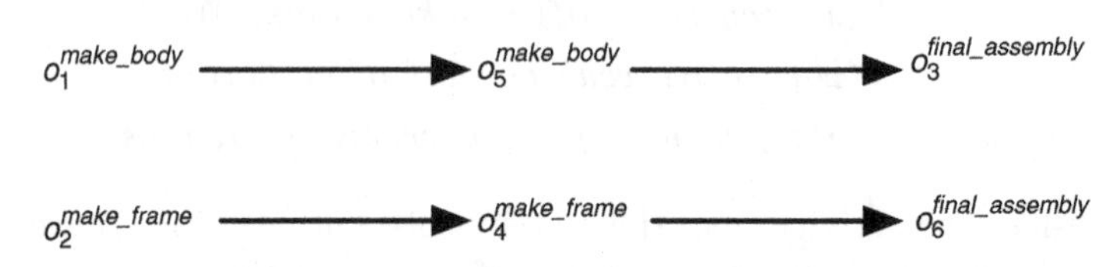

Fig. 5. Activity trees for permuted activities

Ordering Constraints

One of the most common intuitions about processes is the notion of process flow, or the specification of some ordering over the subactivities of an activity, such as

Making the car chassis involves making the body and making the frame in parallel, followed by final assembly.

which is axiomatized by the process description

$$\begin{aligned}(\forall o, o_1, o_2, o_3, x, y)\ occurrence\_of(o, make\_chassis(x, y))\\ \wedge occurrence\_of(o_1, make\_body(y)) \wedge occurrence\_of(o_2, make\_frame(x))\\ \wedge occurrence\_of(o_3, final\_assembly(x, y))\\ \supset min\_precedes(o_1, o_3, make\_chassis(x, y))\\ \wedge min\_precedes(o_2, o_3, make\_chassis(x, y))\end{aligned} \tag{13}$$

In Fig. 5, we can see that each branch of the activity tree for this activity satisfies the same set of ordering constraints on subactivity occurrences.

Iteration

Iteration is captured by the class of repetitive activities, in which the activity tree can be decomposed into copies of some subtree (which intuitively corresponds to the activity tree of the subactivity that is being iterated).

Nondeterministic iteration, such as

Occurrences of painting consist of multiple occurrences of coating is axiomatized by a process description of the form

$$(\forall o_1)\ occurrence\_of(o_1, painting) \supset$$
$$((\forall o_2, s_1)\ occurrence\_of(o_2, coating) \wedge subactivity\_occurrence(o_2, o_1)$$
$$\wedge leaf\_occ(s_1, o_2) \supset (leaf\_occ(s_1, o_1) \vee (\exists o_3, s_2)\ occurrence\_of(o_3, coating)$$
$$\wedge (s_2 = root\_occ(o_3)) \wedge next\_subocc(s_1, s_2, painting)) \qquad (14)$$

This process description says that for every occurrence of *coating* in an activity tree for *painting*, either there exists a next occurrence of *coating* or the leaf subactivity occurrence of the occurrence of *coating* is also the leaf occurrence of the occurrence of *painting*.

Complex activities in which the number of iterations depends on achieving some state (analogous to **while** loops) is a property of a set of activity trees, as we shall see in the next section.

5.3 Spectrum and Variation

A complex activity will in general have multiple activity trees within an occurrence tree, and not all activity trees for an activity need be isomorphic to each other. This property leads to the notion of the spectrum of an activity, which is the set of equivalence classes of isomorphic activity trees. While the former classes of activities compared branches within the same activity tree, we can also define classes with respect to the spectrum of the activity.

The notion of variation within the PSL Ontology characterizes the conditions under which activity trees for a complex activity are isomorphic to each other. Different activity trees for the same activity can have different subactivity occurrences, or the activity trees may differ on the ordering of the subactivity occurrences.

For conditional activities, the fluents that hold prior to the activity occurrence determine which subactivities occur, as in the constraint

Within the painting activity, if the surface of the product is rough, then sand the product:

which is written as

$$(\forall s, o_1, x)\ occurrence\_of(o_1, paint(x)) \wedge root\_occ(o_1) = s \wedge (prior(rough(x), s)$$
$$\supset (\exists o_2)\ occurrence\_of(o_2, sand(x)) \wedge subactivity\_occurrence(o_2, o_1)$$
$$\wedge (root\_occ(o_2) = s) \qquad (15)$$

Alternatively, the ordering over subactivity occurrences of an activity may depend on state, as in the constraint

If the machine is not ready, then perform the painting before final assembly

which can be written as

$$(\forall o, o_1, o_2, x, y)\ occurrence\_of(o, assembly(x, y))$$
$$\wedge occurrence\_of(o_1, paint(x)) \wedge occurrence\_of(o_2, final(x))$$
$$\wedge \neg prior(ready(y), root\_occ(o))$$
$$\supset min\_precedes(root\_occ(o_1), root\_occ(o_2), assembly(x)) \quad (16)$$

Notice how this is distinct from conditional activities, since both painting and final assembly will occur; the different activity trees in this case arise from the ordering of the occurrences of these activities.

5.4 Distribution

The preceding two sections have presented some of the classes in the ontology that are defined with respect to the relationship between occurrences of complex activities and occurrences of their atomic subactivities. We now turn to the classes of complex activities that arise from constraints under which complex activities themselves occur.

There may be branches of a subtree of the occurrence tree that are isomorphic to branches of an activity tree, yet they do not correspond to occurrences of the activity. For example, in Fig. 2, $\{o_1^{cut}, o_8^{punch}\}$ need not be an activity tree for $make\_frame$, even though it is isomorphic to a branch of an activity tree.

The general form for process descriptions related to distribution is:

$$(\forall s)\ \Phi(s) \supset (\exists o)\ occurrence\_of(o, a) \wedge s = root\_occ(o) \quad (17)$$

For triggered activities such as

Deliver the product when we have received three orders.

State determines when an activity must occur, so that the process description is written as

$$(\forall s, x)\ prior(order\_quantity(x, 3), s) \supset$$
$$(\exists o)\ occurrence\_of(o, deliver(x)) \wedge s = root\_occ(o) \quad (18)$$

For launched activities such as

Deliver the product at 1,000.

Time determines when an activity must occur, leading to the process description

$$(\forall s)\ (beginof(s) = 1000) \supset$$
$$(\exists o, x)\ occurrence\_of(o, deliver(x)) \wedge s = root\_occ(o) \quad (19)$$

In either case, models of the process description specify the distribution of activity trees within the occurrence tree.

5.5 Embedding Constraints

The PSL Ontology does not force the existence of complex activities; there may be subtrees of the occurrence tree that contain occurrences of subactivities, yet not be activity trees. We can exploit this property to represent the existence of activity attempts, intended effects, and temporal constraints; subtrees that do not satisfy the desired constraints will simply not correspond to activity trees for the activity.

External Activity Occurrences

For a given complex activity, there may be external activities (that is, activities that are not subactivities) whose occurrence either interfere with the complex activity or which are necessary for the activity to occur. Examples of such necessary activities include either activities performed by external agents (such as a courier delivery or pickup) or it may be an activity such as setup. In the constraint *To produce the chassis, first drill the series of 1 cm holes, followed by drilling the series of 2 cm holes*, the activity that changes the drill bit fixture is not a subactivity of the process plan for producing the chassis, but is a setup activity that must occur between drilling the two sets of holes.

Interruptability

Closely related to external activity occurrences is the notion of interruptability and activity attempts. With an interruptable activity, an external activity may occur without interfering with the original activity. For example, interruptable activities may be preempted or suspended:

The assembly of computers for one customer can be halted to work on a rush order for another customer

$$\begin{aligned}&(\forall s_1, x_1, x_2)\, root(s_1, assemble(x_1)) \wedge occurrence\_of(s_3, assemble(x_2))\\&\qquad \wedge legal(s_3) \wedge earlier(s_1, s_3)\\&\supset (\exists s_2)\, leaf(s_2, assemble(x_1)) \wedge min\_precedes(s_1, s_2, assemble(x_1)) \quad (20)\end{aligned}$$

while noninterruptable activities may not:
Pouring of metal from the furnace cannot be stopped once initiated.

$$\begin{aligned}&(\forall s_1, s_2)\, root(s_1, pour\_metal) \wedge leaf(s_2, pour\_metal)\\&\qquad \wedge min\_precedes(s_1, s_2, pour\_metal)\\&\supset \neg(\exists s_3)\, occurrence\_of(s_3, stop) \wedge earlier(s_1, s_3) \wedge earlier(s_3, s_2) \quad (21)\end{aligned}$$

In this latter example, if for some reason the metal pouring does stop, then we would intuitively consider this to be an activity attempt, rather than an occurrence of the activity.

Intended Effects

There are many circumstances in which we want to make a distinction between the intended effects of an activity and the actual effects of the activity. For example, the manufacturing process plan for making some product in a steel company is defined with respect to the properties specified by customer and quality requirements (such as grade, surface properties, width, and thickness), but due to external nondeterministic factors, not every occurrence of the process will provide products that satisfy these requirements. Quality problems arise from this divergence of actual effects from intended effects.

For example, informal process descriptions such as *Bake the soup until it is opaque* or *Heat the solution until reaches 50 C* can be formalized by sentences of the form

$$(\forall s)\ leaf(s, a) \supset holds(f, s) \tag{22}$$

In both of these examples, it is possible to terminate the activity occurrence before the intended state is achieved, but in the context of the intended effects, the activity occurrence will terminate only when the state is achieved.

Temporal Constraints

With temporal constraints, subactivities are not allowed to occur at arbitrary times during occurrences of the activity. Examples of such constraints include schedules, which specify the possible times at which the subactivities may occur:

The part will arrive 10 days after placing the order request

$$(\forall o, s_1, s_2)\ min\_precedes(s_1, s_2, a) \wedge occurrence\_of(s_1, a_1) \wedge occurrence\_of(s_2, a_2)$$

$$\supset duration(endof(s_2), endof(s_1)) = 10 \tag{23}$$

In this example, the possible occurrences of the activity are restricted to those whose subactivities satisfy the temporal constraints.

6 Summary

Within the increasingly complex environments of enterprise integration, electronic commerce, and the Semantic Web, where process models are maintained in different software applications, standards for the exchange of this information must address not only the syntax but also the semantics of process concepts. PSL draws upon well-known mathematical tools and techniques to provide a robust semantic foundation for the representation of process information. This foundation includes first-order theories for concepts together with complete characterizations of the soundness and completeness of these theories. In this chapter, we have seen how the PSL Ontology can be used to specify process descriptions for a broad range of problems and provide the semantic foundations for new ontologies.

References

1. Grenon, P. and Smith, B. (2004) SNAP and SPAN: Towards dynamic spatial ontology. *Spatial Cognition and Computation*, 4(1):69-104, 2004.
2. Gruninger, M. (2003) Applications of PSL to Semantic Web Services, *Workshop on Semantic Web and Databases.* Very Large Databases Conference, Berlin.
3. Gruninger, M. and Kopena, J. (2004) Semantic Integration through Invariants, *AI Magazine*, 26:11-20, 2004.
4. Hayes, P. (1996) *A Catalog of Temporal Theories.* Artificial Intelligence Technical Report UIUC-BI-AI-96-01, University of Illinois at Urbana-Champaign.
5. McIlraith, S., Son, T.C. and Zeng, H. (2001) Semantic Web Services, *IEEE Intelligent Systems*, Special Issue on the Semantic Web. 16:46–53, March/April, 2001.
6. Menzel, C. and Gruninger, M. (2001) A formal foundation for process modeling, *Second International Conference on Formal Ontologies in Information Systems*, Welty and Smith (eds), 256-269.
7. Ciocoiu, M., Gruninger M., and Nau, D. (2001) Ontologies for integrating engineering applications, *Journal of Computing and Information Science in Engineering*, 1:45-60.
8. Pinto, J. and Reiter, R. (1993) Temporal reasoning in logic programming: A case for the situation calculus. *Proceedings of the 10th International Conference on Logic Programming*, Budapest, Hungary, June 1993.
9. Schlenoff, C., Gruninger, M., Ciocoiu, M., (1999) The Essence of the Process Specification Language, *Transactions of the Society for Computer Simulation* vol.16 no.4 (December 1999) pages 204-216.
10. Sowa, J. (2000) *Knowledge Representation: Logical, Philosophical, and Computational Foundations.* Brooks/Cole Publishing.
11. Semantic Web Services Framework (SWSF) Overview W3C Member Submission 9 September 2005.

Ontologies for Formal Representation of Biological Systems

Nigam Shah and Mark Musen

Stanford Medical Informatics, Stanford, CA, USA 94305, nigam@stanford.edu, musen@stanford.edu

1 Introduction

This chapter provides an overview of how the use of ontologies may enhance biomedical research by providing a basis for a formalized, and shareable descriptions, of *models* of biological systems.

A wide variety of artifacts are labeled as "ontologies" in the Biomedical domain, leading to much debate and confusion. The most widely used ontological artifact are controlled vocabularies (CVs). A CV provides a list of terms whose meanings are specifically defined. Terms from a CV are usually used for indexing records in a database. The Gene Ontology (GO) is the most widely used CV in databases serving biomedical researchers. The GO provides term for declaring the molecular function (MF), biological process (BP) and cellular component (CC) of gene products. The statements comprising these MF, BP and CC declaration are called annotations [51], which are predominantly used to interpret results from high throughput gene expression experiments [27,53]. Arguably, CVs provide the most value for effort in terms of facilitating database search and interoperability.

The second most prevalent kind of artifact is an information model (or data model). An information model provides an organizing structure to information pertaining to a domain of interest, such as microarray[1] data, and describes how different parts of the information at hand, such as the experimental condition and sample description, relate to each other. In biomedical research, Microarray Gene Expression Object Model (MAGE-OM) is an example of

[1] An automated technique for simultaneously analyzing thousands of different DNA sequences or proteins affixed to a thumbnail-sized "chip" of glass or silicon. DNA microarrays can be used to monitor changes in the expression levels of genes in response to changes in environmental conditions or in healthy vs. diseased cells. Protein arrays can be used to study protein expression, protein–protein interactions, and interactions between proteins and other molecules. From – www.niaaa.nih.gov/publications/arh26-3/165-171.htm

S. Staab and R. Studer (eds.), *Handbook on Ontologies*, International Handbooks on Information Systems, DOI 10.1007/978-3-540-92673-3,

a widely known information model. MAGE-OM, along with the controlled terms that are used to populate the information model is referred to as the Microarray Gene Expression Data (MGED) Ontology. The MGED Ontology is used to describe the minimum information about a microarray experiment that is essential to make sense of the numbers comprising the microarray data.

The third kind of artifact is an ontology in its true sense, which is increasingly being used for knowledge representation in Biomedicine. In this interpretation, an ontology is a specification of entities (or concepts) and relationships among them in a domain of discourse; along with declarations of the properties of each relationship, and, in some cases, a set of explicit axioms defined for those relations and entities. In biomedical research, several ontologies are striving towards this goal. The foremost is the Foundational Model of Anatomy (FMA), which is a computer-based knowledge source for anatomy and represents classes and relationships necessary for the symbolic modeling of the structure of the human body in a form that is understandable to humans and is also navigable, parseable, and interpretable by machine-based systems [44]. The biomedical research community is perhaps the farthest along in recognizing the need and starting an organized effort for the creation of ontologies that serve as formal knowledge representations [47].

1.1 Uses of Ontologies in Biomedical Research

With the advent of high-throughput technologies,[2] biomedical research is undergoing a revolution in terms of the amount and types of data available to the scientist. On the one hand, there is an abundance of individual data types such as gene and protein sequences, gene expression data, protein structures, protein interactions and annotations. On the other hand, there is a shortage of tools and methods that can handle this deluge of information and allow a scientist to draw meaningful inferences.

Currently, a significant amount of time and energy is spent in merely locating and retrieving information rather than thinking about what that information means. For example, a researcher trying to understand how the proteins participating in the cell cycle interact with each other, has to read several reviews to determine the list of proteins S/he should track, search databases such as Uniprot to retrieve annotations for the relevant proteins, follow the citations evidencing the annotations to determine the experiment/s that were performed on each protein and in some cases retrieve the actual data sets from special databases. All this information comes in different formats and from different sources. It is extremely difficult to manually search through the various sources and integrate this diverse information about biological systems to formulate hypotheses (or "Models") spanning a large number genes and proteins [28].

[2] High-throughput technologies are large-scale, usually automated, methods to purify, identify, and characterize DNA, RNA, proteins and other molecules. They allow rapid analysis of very large numbers of samples.

Until recently, the predominant use of ontologies in biomedicine has been to facilitate interoperability among databases by indexing them with standard terms to address the problem of locating and retrieving information. Even if the problem of locating information were solved, it is still difficult to formulate formal hypotheses and models comprising a large number genes and proteins [28]. The difficulty arises primarily because there is no shared formalism – akin to engineering drawings – in which to express such hypotheses or models and the interpretation they convey. Lack of such a formalism also makes it difficult to determine whether the hypotheses are consistent internally or with data, to refine inconsistent hypotheses and to verify the implications of complicated hypotheses in 'what if' thought experiments [11, 24]. This situation needs to be rectified and tools need to be developed that utilize formal methods to assist in querying and interpreting the information at hand [11, 15, 52].

Besides using ontologies for enhancing interoperability among databases and enabling data exchange, researchers have also used ontologies to create knowledge bases that store large amounts of knowledge in a structured manner [22, 23]. For example, EcoCyc is a comprehensive source of structured knowledge on metabolic pathways in E. Coli. When used to create knowledge bases, an ontology enables the declaration and storage of a theory – an experimentally testable explanation of the interactions in a biological system [54]. If the ontologies are well-designed, then the resulting knowledge bases can be used to retrieve relevant facts, to organize and interpret disparate knowledge, to infer non-obvious relationships, and to evaluate hypotheses posited by scientists [4, 31, 41].

The emerging trend in the use of ontologies in biomedical research is that, at the outset, ontology terms are used to name things, gradually proceeding toward naming connections between things – first to create information models and then progressing towards the creation of a formal representation[3] [15, 52] which allows the creation of formal (both qualitative and quantitative) models[4] of biological systems.

In this chapter we focus on the latter use. Chapter "Ontology-Based Recommender Systems" discusses the current applications of bio-ontologies that are focused around the theme of database interoperability and data integration. Bodenreider and Stevens [5] have recently reviewed in detail the current progress in biomedical ontologies and we do not review it again in this chapter. In the next sections, we discuss how the use of ontologies for formal representation of biological systems can aid in biomedical research, we then outline the hurdles facing the realization of such use and in the end discuss the possible role of the Semantic Web in advancing this particular use.

[3] For this current discussion, a formal representation means a computer-interpretable standardized form that can be the basis for creating unambiguous descriptions of biological systems 2.1.

[4] We use "models" to mean a schematic description of a system or phenomenon that accounts for its known or inferred properties and can be used for further study of its characteristics.

2 Constructing Hypotheses and Models of Biological Systems

The discovery process in biomedical research is cyclical; Scientists examine existing data to formulate models that explain the data, design experiments to test the hypotheses and develop new hypotheses that incorporate the data generated during experimentation. Currently, in order to advance this cycle, the experimentalist must perform several tasks: (1) gather information of many different types about the biological entities that participate in a biological process (2) formulate hypotheses (or models) about the relationships among these entities, (3) examine the different data to evaluate the degree to which his/her hypothesis is supported and (4) refine the hypotheses to achieve the best possible match with the data. In todays data-rich environment, this is a very difficult, time-consuming and tedious task. For example, even to evaluate a simple hypothesis such as "protein A is a transcriptional activator of genes X, Y and Z", the experimentalist must examine the literature for evidence showing that protein A is a transcription factor or exhibits protein sequence homology with known transcriptional factors. S/he must look for evidence indicating DNA binding activity for protein A and if found, examine the promoters of X, Y and Z for presence of binding sequences for protein A. Moreover, each of the preceding steps incorporate a set of implicit assumptions such as sequence homology implying similarity of function.

Finally, the refined hypotheses are subjected to experimental testing. Hypotheses that survive these tests – validated hypotheses – are published in scientific publications and represent the growing knowledge about biological entities, processes and relationships among them. Validated hypotheses are eventually synthesized into systems of relationships called "models" that account for the known behavior of the system and provide the grounds for further experimentation. Biologists' models are generally presented as diagrams showing the type, direction and strength of relationships among biological entities such as genes and proteins. Figure[5] 1 shows a simplified a model of regulation of the mitotic cell cycle in humans.[6]

Usually the goal of constructing a model of a biological system is to predict the outcome (either qualitative of quantitative) from the system at some point in the future. For the moment, for most biological systems, scientists must describe the workings of biological systems in a qualitative manner because there is not enough known to formulate quantitative modes [11]. Even for qualitative models, we believe that such predictive models, though essential, lie in the future because much of current research uses prior knowledge for

[5] Source public domain, non copy righted image.

[6] The cell cycle is a complicated biological process and comprises of the progression of events that occur in a cell during successive cell replication. The process can be described at varying level of details ranging from a high level qualitative description to a detailed system of differential equations. However, for most biological processes the representation is primarily in terms of qualitative interactions.

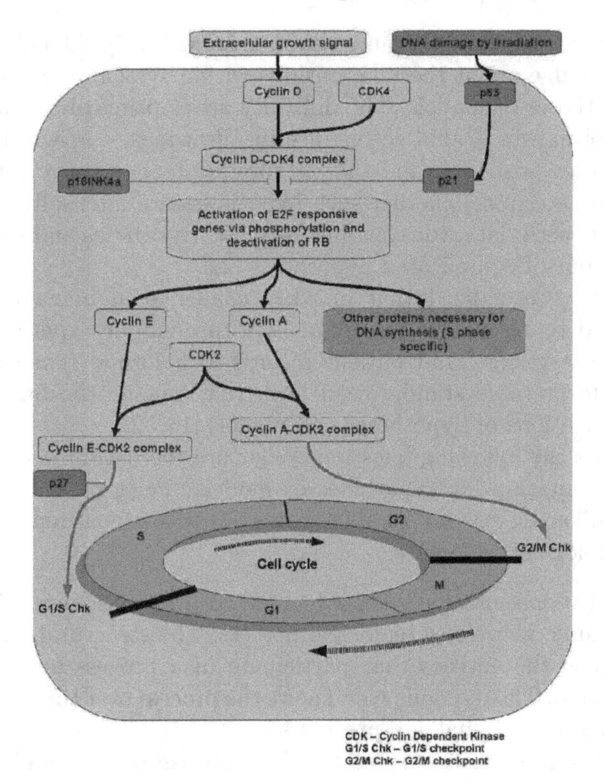

Fig. 1. A diagram showing a simple model of regulation of the mitotic cell cycle in humans. The filled rectangles in the cell (*gray box*) denote proteins and protein complexes that participate in the process. The green oval shows the phases of the cell cycle and the green arrows from the proteins shows the phase at which they function. The black arrows depict activating relationships and the red ones show inhibitory relationships. Note how temporal, logical and structural information is mashed together in one representation

interpreting data sets rather than applying prior knowledge as a set of axioms that will elicit new knowledge [56, 59]. In this situation, the most profitable manner to use models is to construct a model or a set of models and then test them for consistency with the available information and knowledge, revise models to minimize the inconsistencies and then pick the most consistent model as a basis for designing further experiments [39, 41, 52].

2.1 Creating a Formal Representation for Hypotheses and Models

If we accept the notion of an hypothesis (or a model) as the basis for an organizing framework for the data and information sources we wish to integrate and interpret, we immediately encounter several problems. As we have discussed, for a large number of participating entities such genes, proteins, clinical observations and laboratory data, it is extremely difficult to integrate current knowledge about the relationships within system under study to formulate hypotheses or models. The difficulty arises primarily because there is no shared formalism – akin to engineering drawings – in which to express such hypotheses or models and the interpretation they convey. Therefore, it is difficult to determine whether such hypotheses are internally consistent or are consistent with data, to refine inconsistent hypotheses and to understand the implications of complicated hypotheses [24].

It is widely recognized that one key challenge in managing this data overload is to represent the results of high-throughput experiments as well as clinical observations and patient records in a formal representation – a computer-interpretable standardized form that can be the basis for unambiguous descriptions of hypotheses and models [15, 52].

This raises the following question: What are the desirable properties of a formal representation for hypotheses or models? Peleg et al. [38], have suggested the following set of desirable properties in a formal representation for models of biological processes:

1. A formal representation should be able to present structural, functional and dynamic views of a biological process. The structural and functional views show the entities that participate in a process and relationships among them. The dynamic view shows the process over time, shows branch points and conditional sub-processes.
2. A formal representation should include an associated ontology that unambiguously identifies the entities and relationships in a process.
3. It should be able to represent biological processes at various scales and should allow hierarchical representation of sub processes to manage complexity.
4. The representation should be able to incorporate new data as they become available and should be extensible to allow new categories of information as they come in to existence.
5. The representation should have a corresponding conceptual (mathematical) framework that allows verification of system properties using simulation and/or logical inference mechanisms.
6. The representation should have an intuitive visual layout.

If we can devise a formal representation for hypotheses and the data at hand as well as the mechanisms to check the consistency of hypotheses with that data and prior knowledge, we can significantly streamline the task of interpreting diverse data. Moreover, if we develop such a formal representation, we can develop tools that can operate upon current data sets, information and existing knowledge to integrate them in an environment that supports the formulation and testing of hypotheses [4, 9, 14, 41, 54].

Developing such a formal representation is not a trivial task. Moreover it is unreasonable to expect one representation that will satisfy the needs of all users. However, the need for creating such representation for specific domain (or sub domain) of study has been proposed multiple times in Biomedicine [2,3,9,11,12,15,37]. In the next section we discuss the key issues in creating a formal representation of a domain (see Fig. 2 for a quick overview) and then in Sect. 3 we discuss the pivotal role ontologies can play in the process.

2.2 Challenges for Developing a Formal Representation

Knowledge Representation

The first challenge is the systematic representation of the various kinds of biological entities that participate in any given disease process and the many qualitatively different kinds of relationships among them. This requires the development of an ontology for unambiguously representing biological entities and interactions among them. Specifically, an ontology allows us to represent domain-specific entities along with their definitions, a set of relationships among them, properties of each relationship, and, in some cases, a set of explicit axioms defined for those relations and entities. We require different ontologies to represent biological processes at different levels of granularity because biological processes and the relevant data can be considered at varying levels of detail, ranging from molecular mechanisms to general processes such as cell division and from raw data matrixes to qualitative relationships [6].

Ontologies have gained a lot of popularity in molecular biology over the last several years. The earliest ontologies describe properties of 'objects' such as genes, gene products and small molecules. The later ontologies describe the 'processes' that gene, proteins and small molecules participate in.

Currently, there are several ontologies that allow representation of processes in a biological system by specifying relationships between biological entities for tasks ranging from modeling biological systems to extracting information from literature. At the simplest level, gene ontology BP annotations describe the processes a particular gene product might contribute to, which can be viewed as a minimal model of the biological process that does not contain any declaration of the specific relationships among its participants. At the other end is the Systems biology markup language, SBML, which can represent quantitative models of biochemical processes and pathways [20]. However for most systems such detailed information is not yet available.

There are multiple ontologies to represent biological processes, models and hypotheses at varying degrees of granularity between the two extremes of a GO BP annotation and a SBML representation. For example, EcoCyc's ontology, which is used to represent information about metabolic pathways for E. coli [22] provides an ontology of biological entity types and processes. An ontology developed by Rzhetsky et al. [49] allows representation signal transduction pathways at a granularity level that is optimal for programs that extract information about such pathways automatically from published literature [49,50].

In Sect. 3 we will discuss how ontologies enable the creation of a formal representation by enabling the knowledge representation task.

Conceptual Representation

The second problem is to represent the biological system conceptually. The conceptual framework for a biological system enables a user to reason about a biological system and perform thought experiments. Thought experiments serve two functions: prediction and verification. Prediction allows scientists to make future claims based on models of a system. Verification provides guidance and feedback about the accuracy of the models through comparison, manipulation, and evaluation against available data and knowledge. Prediction and verification enable scientists to ask 'what if' questions about a system, form explanations, as well as make and test predictions. Currently, one of the major limiting factor is the conceptual representation of the "mathematics" of biological systems [11].

A conceptual framework for representing biological systems must accommodate the modularity and temporal behavior of biological systems, as well as handle their non-linearity and redundancy. The conceptual frameworks used to represent biological models vary from ordinary differential equations to Boolean [1,30] and Bayesian [13,17,36] networks as well as Petri Nets [37,43] to qualitative process calculi [7], special logics [57] and rule systems [21].

The inability to represent disparate kinds of information, at different levels of detail, about biological systems in a common conceptual framework is a major limitation in the creation of formal representations of biological systems, and current efforts usually focus on a limited categories of information [52].

A promising approach is to represent the biological processes in a system as a sequence of 'events' that link particular 'states' of the system [40,42]. It has been shown before that complex processes, particularly those that exhibit non-linear behavior, are readily described by event-driven dynamics [19] because event dynamics allows description of the process in terms of observed effects of the non-linear behavior rather then requiring that the non-linear behavior itself be represented as a mathematical function. Moreover, event-based approaches can represent simple processes, such as protein phosphorylation, to complex ones, such as the cell cycle, allowing a wide range of resolution.

An event-based framework offers several other advantages as well: (1) It can explicitly represent states which allows for representing information such as commitment to a developmental pathway [38]; (2) It allows hierarchical representation of properties and hence avoid a rapid increase in the number of states that need to be represented [18]; (3) It can represent temporal constraints on when events occur; (4) It can readily accept new categories of information and represent information at different resolution levels.

It is unlikely that any conceptual framework will be adequate to represent all biological systems and it is much more productive to represent the data, information and knowledge at hand in an explicit ontology and then map the various entity-types and relationships in the ontology to a particular conceptual frameworks needed under specific situations. For example, Rubin et al have represented anatomic knowledge about the heart and the circulatory system in an ontology and then mapped the ontology to differential equation models for blood pressure as well as to a reasoning service to predict the effect of penetrating injuries to the heart [45, 46].

Knowledge and Data Acquisition

The final challenge is the gathering, storage, and encoding of existing information. Even if all of the above challenges are met, getting access to the information and converting (or encoding) it into the relevant ontology in an automated manner is a major challenge because the information resides in separate repositories, each with custom storage formats and diverse access methods. Moreover, most databases do not store information in an explicit ontology and groups that design ontologies capable of representing models of biological systems [10,20,38,41,50] do not store all relevant information structured in those ontologies. Hence, efforts aimed at building a unified formal representation need to convert existing information into their ontology. For most formal representations this conversion (or encoding) of existing information and knowledge remains an unsolved problem and is the most common bottleneck preventing the use of the representation.

All the challenges described above are strongly inter-related as shown in the Fig. 2 and need to solved in tandem for a particular domain of interest.

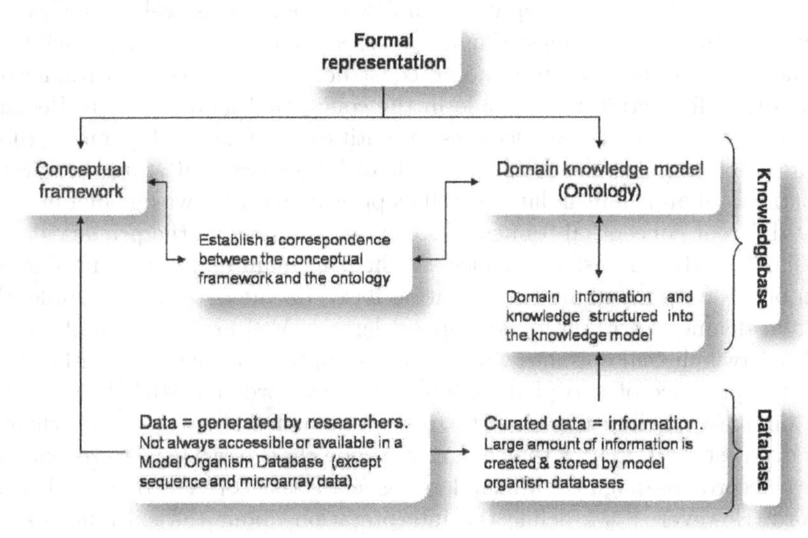

Fig. 2. Components of a Formal Representation: A formal representation is a computer-interpretable standardized form that can be the basis for unambiguous descriptions of hypotheses and models in a domain of discourse. Knowledge representation comprises the methods and processes of systematically representing the various biological entities and the different kinds of relationships between them. The conceptual framework for a biological system enables a user to reason about a biological system and perform *thought experiments*. The adoption and success of a formal representation depends critically on the ability to gather and encode existing information and knowledge

3 Ontologies Enable the Creation of a Formal Representation

As discussed till now, in order to create a unified formal representation, we need a conceptual framework that can represent models of biological systems. The conceptual framework should represent the temporal dynamics of the process and should not require a complete model rewrite on minor changes. The conceptual framework should provide systematic methods to evaluate, update, extend and revise models represented in that framework.

It is obvious that no conceptual framework will be adequate under all circumstances. We are better off representing the data, information and knowledge at hand in an explicit ontology and then mapping the various entity-types and relationships in the ontology to a particular conceptual framework needed under specific situations [45, 46].

The challenge then is to bridge the conceptual framework and the ontology to create the formal representation. For example, the relationship 'protein A activates protein B' is different from 'protein A activates gene X' though it may be described by the same words. When representing a biological process in a conceptual framework, it is essential to distinguish between the two meanings. An ontology can distinguish between the two relationships by providing different terms to represent the two meanings as well as by clearly specifying the two meanings. Having an associated ontology where each term in the ontology has a corresponding construct in the conceptual framework allows this distinction to be made in the conceptual model as well. Because an ontology unambiguously declares the entities and relationship among those entities, it can guide the design of the knowledgebases that store the various experimental and clinical data as well as prior domain knowledge in a manner that different conceptual frameworks can be overlayed on the primary data.

Particularly, the use ontologies will help in maintaining a strict distinction between data and an interpretation based on the data. For example, the current diagnosis criteria for Multiple Sclerosis (MS) are based on observing at least two clinical episodes with certain symptoms at least 3 months apart and the presence of two plaques (on the spinal cord) on MRI. If we design our database of clinical records to store the diagnostic code for MS, then, if the diagnostic criteria for that diagnostic code change, we have to re-examine every record, re-diagnose MS and re-associate the correct codes with each record. However, if we define the interpretation about the existence of MS separately from the data structure used to store observational data, then we can change our criteria for MS and still reliably identify all cases of MS. Such separation of the definition of a biomedical concept from the decision (such as recommending a treatment) and computation (such as searching for correlations with environmental factors) has been demonstrated to increase both maintainability and efficiency of computer reasoning [25, 32].

We believe that a particular conceptual framework along with the associated ontology is the optimal way to create a formal representation fit for a specific situation. For example, differential equations along with the Systems Biology Markup Language (SBML) create an appropriate formal representation for biochemical signaling pathways. Ontologies will play a central role in enabling such modularity and maintaining a separation between data, information and knowledge and the relevant conceptual (mathematical) framework.

The formal representation resulting from such separation will be easily extensible to incorporate new data types as they become available as well as to incorporate novel conceptual frameworks as required. The hypotheses, models and underlying data can become compatible with each other in the context of the relevant conceptual framework, making it possible to bring together the implications of many kinds of data and information in a unified manner [41, 54]. We will gain the ability to test complex interpretations as well as the ability to use data from unrelated research projects [35].

3.1 Unresolved Issues

Although, the use of ontologies in the creation of formal representations has a very strong case in its favor, there are several challenging issues that need to be addressed, which we discuss below:

Abstraction Levels

Biomedical researchers study biological systems at various scales, ranging from electron microscopy images to patient populations. No ontology can span all these and multiple ontologies already exist for different abstraction levels. We need to create a mechanism by which ontologies at different abstraction levels can be effectively mapped to each other.

Unambiguous Relationships

Within a particular abstraction level of representing a biological system, relationships need to be explicitly defined so that their interpretation is not subjective. The relationship ontology (RO) [55] is a step in this direction for defining relations for the molecular level. Although the RO provides explicit logical definitions, no computational implementation of the RO exists that actually allows a user to verify the correct use of the relations. General, well established mechanisms, such as Ontoclean (discussed in 9), to verify the clarity of relationships in ontologies exist. However, their use in the biomedical domain is minimal.

Consistency Across Abstraction Levels

Biomedical researchers cross multiple abstraction levels when describing biological systems. It is essential that relationships between entities at a particular abstraction levels can be consistently interpreted when we move to a different abstraction level (For example how does the mechanism of action of a drug at the protein level affect the efficacy of the drug in treating a patient).

Bindings with Conceptual Frameworks

We have suggested a separation between the ontology used to structure knowledge about a biological system and the conceptual framework used to model the system mathematically. However, currently the process of establishing a correspondence between constructs in an ontology and constructs in a conceptual framework is quite ad hoc. Usually the ontology is designed with one conceptual framework in mind (e.g. SBML [20] and differential equations or the Biological Process ontology and Petri nets [38]). The general problem of easily mapping a formal representation to conceptual models at different scales is still unsolved and promises to be an exciting research direction.

4 Role of the Semantic Web

The Semantic Web is an evolving extension of the World Wide Web in which web content can be expressed in a form that can be understood, interpreted and used by software agents (besides humans), thus permitting software agents to find, share and integrate information more easily.[7]

Given the heterogeneity of biological data both in form and location, the Semantic Web is of considerable interest to the life sciences community; particularly because key issues such as the need for consistent data and knowledge representation can be addressed using the Resource Description Framework (RDF) and Web Ontology Language (OWL) [16]. A variety of technologies have been built on this foundation of RDF and OWL that, together, support identifying, representing, and reasoning across a wide range of biomedical data [48, 58].

The expectation from the Semantic Web in life sciences is that relationships that exist implicitly in the minds of scientists will be explicitly declared (using OWL ontologies) and then used to aggregate genomic, proteomic, cellular, physiological, and chemical data. Semantic definitions will specify which objects are related to others and how. Such linking will enable semantic tools [34] that can pull together diverse information, render it in a manner defined by the user and possibly reason over the collated information to derive novel insights [8, 29, 33, 48].

However, not everyone is convinced that the Semantic Web will have such a revolutionizing effect on life sciences. There are implicit assumptions in the expected role of the Semantic Web, mainly that: (1) a simple syntax and the semantic of description logics will be sufficient (2) translation of existing information into the simple syntax as well as inferences on the simple semantics

[7] http://www.w3.org/2001/sw/SW-FAQ

will work right [26]. If these two assumptions are not met, the promise of the Semantic Web might not be realized in the field of biomedical research.

Currently there is immense excitement about the Semantic Web and its possible contribution to advancing biomedical research; it remains to be seen wether it bears out in practice.

5 Summary

In this chapter we have discussed how the use of ontologies for knowledge representation can aid in current biomedical research. We have argued that formally representing biological systems is necessary for advancing current biomedical research and that it is increasingly recognized that biologists need to use computational tools for performing thought experiments. We have described how biomedical ontologies can play a pivotal role in enabling that transition. We have outlined the hurdles facing the use of ontologies in creating formal representations that enable thought experiments. We have discussed the possible role of the Semantic Web in advancing this particular use of ontologies.

References

1. T. Akutsu, S. Miyano, and S. Kuhara. Algorithms for identifying boolean networks and related biological networks based on matrix multiplication and fingerprint function. *J Comput Biol*, 7(3-4):331–43, 2000. 1066-5277 Journal Article.
2. R. Altman, M. Buda, X. Chai, M. Carillo, R. Chen, and N. Abernethy. Riboweb: an ontology-based system for collaborative molecular biology. *Intelligent Systems, IEEE [see also IEEE Expert]*, 14(5):68–76, 1999. TY - JOUR.
3. G. An. Concepts for developing a collaborative in silico model of the acute inflammatory response using agent-based modeling. *J Crit Care*, 21(1):105–10; discussion 110–1, Mar 2006.
4. C. Baral, K. Chancellor, N. Tran, N. Tran, A. Joy, and M. Berens. A knowledge based approach for representing and reasoning about signaling networks. *Bioinformatics*, 20(suppl_1):15–22, 2004.
5. O. Bodenreider and R. Stevens. Bio-ontologies: current trends and future directions. *Brief Bioinform*, 7(3):256–274, Sep 2006.
6. A. Brazma. On the importance of standardisation in life sciences. *Bioinformatics*, 17(2):113–4, 2001. 21138228 1367-4803 Editorial.
7. L. Cardelli. Bioware languages. In A. Herbert and K. S. Jones, editors, *Computer Systems: Theory, Technology, and Applications*, pages 59–65. Springer, New York, 2005.
8. K. Cheung, P. Qi, D. Tuck, and M. Krauthammer. A semantic web approach to biological pathway data reasoning and integration. *Journal of Web Semantics*, 4:3, 2006.

9. T. Clark and J. Kinoshita. Alzforum and swan: the present and future of scientific web communities. *Brief Bioinform*, 8(3):163–171, May 2007.
10. E. Demir, O. Babur, U. Dogrusoz, A. Gursoy, A. Ayaz, G. Gulesir, G. Nisanci, and R. Cetin-Atalay. An ontology for collaborative construction and analysis of cellular pathways. *Bioinformatics*, 20(3):349–356, 2004.
11. N. Fedoroff, S. A. Racunas, and J. Shrager. Making biological computing smarter. *The Scientist*, 19(11):20–21, 2005.
12. C. Friedman, T. Borlawsky, L. Shagina, H. R. Xing, and Y. A. Lussier. Bio-ontology and text: bridging the modeling gap. *Bioinformatics*, 22(19): 2421–2429, Oct 2006.
13. N. Friedman, M. Linial, I. Nachman, and D. Pe'er. Using bayesian networks to analyze expression data. *J Comput Biol*, 7(3-4):601–20, 2000. 1066-5277 Journal Article.
14. Y. Gao, J. Kinoshita, E. Wu, E. Miller, R. Lee, A. Seaborne, S. Cayzer, and T. Clark. Swan: A distributed knowledge infrastructure for alzheimer disease research. *Journal of Web Semantics*, inpress, 2006.
15. D. K. Gifford. Blazing pathways through genetic mountains. *Science*, 293(5537):2049–51, 2001. 0036-8075 Journal Article.
16. B. M. Good and M. D. Wilkinson. The life sciences semantic web is full of creeps! *Brief Bioinform*, 7(3):275–286, Sep 2006.
17. A. J. Hartemink, D. K. Gifford, T. S. Jaakkola, and R. A. Young. Using graphical models and genomic expression data to statistically validate models of genetic regulatory networks. *Pac Symp Biocomput*, pages 422–33, 2001. Journal Article Validation Studies.
18. M. Heiner. On exploiting the analysis power of petri nets for the validation of discrete event systems. In *IMACS Symposium on Mathematical Modelling*, pages 171–176, Wien, 1997.
19. Y. C. Ho. Special issue on discrete event dynamical systems: Editorial. *Proc IEEE*, 77(1):24–38, 1989.
20. M. Hucka, A. Finney, H. M. Sauro, H. Bolouri, J. C. Doyle, H. Kitano, A. P. Arkin, B. J. Bornstein, D. Bray, A. Cornish-Bowden, A. A. Cuellar, S. Dronov, E. D. Gilles, M. Ginkel, V. Gor, I. Goryanin, W. J. Hedley, T. C. Hodgman, J. H. Hofmeyr, P. J. Hunter, N. S. Juty, J. L. Kasberger, A. Kremling, U. Kummer, N. Le Novere, L. M. Loew, D. Lucio, P. Mendes, E. Minch, E. D. Mjolsness, Y. Nakayama, M. R. Nelson, P. F. Nielsen, T. Sakurada, J. C. Schaff, B. E. Shapiro, T. S. Shimizu, H. D. Spence, J. Stelling, K. Takahashi, M. Tomita, J. Wagner, and J. Wang. The systems biology markup language (sbml): a medium for representation and exchange of biochemical network models. *Bioinformatics*, 19(4):524–31, 2003. 1367-4803 Evaluation Studies Journal Article.
21. T. R. Hvidsten, A. Laegreid, and J. Komorowski. Learning rule-based models of biological process from gene expression time profiles using gene ontology. *Bioinformatics*, 19(9):1116–23, 2003. Evaluation Studies Journal Article Validation Studies.
22. P. Karp. An ontology for biological function based on molecular interactions. *Bioinformatics*, 16(3):269–85–, 2000.
23. P. Karp, C. Ouzounis, C. Moore-Kochlacs, L. Goldovsky, P. Kaipa, D. Ahren, S. Tsoka, N. Darzentas, V. Kunin, and N. Lopez-Bigas. Expansion of the biocyc collection of pathway/genome databases to 160 genomes. *Nucleic Acids Res*, 33(19):6083–9–, 2005.

24. P. D. Karp. Pathway databases: a case study in computational symbolic theories. *Science*, 293(5537):2040–4, 2001. 0036-8075 Journal Article.
25. V. Kashyap, A. Morales, and T. Hongsermeier. On implementing clinical decision support: achieving scalability and maintainability by combining business rules and ontologies. *AMIA Annu Symp Proc*, pages 414–418, 2006.
26. T. Kazic. Putting semantics into the semantic web: how well can it capture biology? *Pac Symp Biocomput*, pages 140–151, 2006.
27. P. Khatri and S. Draghici. Ontological analysis of gene expression data: current tools, limitations, and open problems. *Bioinformatics*, 21(18):3587–95, 2005. 1367-4803 Journal Article.
28. A. Kuchinsky, K. Graham, D. Moh, M. Creech, K. Babaria, and A. Adler. Biological storytelling: a software tool for biological information organization based upon narrative structure. In *Advanced Visual Interfaces*, pages –, Trento, Italy, 2002.
29. H. Y. K. Lam, L. Marenco, T. Clark, Y. Gao, J. Kinoshita, G. Shepherd, P. Miller, E. Wu, G. T. Wong, N. Liu, C. Crasto, T. Morse, S. Stephens, and K.-H. Cheung. Alzpharm: integration of neurodegeneration data using rdf. *BMC Bioinformatics*, 8 Suppl 3:S4, 2007.
30. S. Liang, S. Fuhrman, and R. Somogyi. Reveal, a general reverse engineering algorithm for inference of genetic network architectures. *Pac. Symp. Biocomput.*, pages 18–29, 1998. in file.
31. J. P. Massar, M. Travers, J. Elhai, and J. Shrager. Biolingua: a programmable knowledge environment for biologists. *Bioinformatics*, 21(2):199–207, Jan 2005.
32. M. A. Musen. Scalable software architectures for decision support. *Methods Inf Med*, 38(4-5):229–238, Dec 1999.
33. E. Neumann. A life science semantic web: are we there yet? *Sci STKE*, 2005(283):pe22, 2005. 1525-8882 (Electronic) Journal Article Review.
34. E. Neumann. Biodash: A semantic web dashboard for drug development. In R. Altman, editor, *Pacific Symposium in Biocomputing*, volume 11, pages 176–187, Hawai, 2006.
35. M. J. O'Connor, D. L. Buckeridge, M. Choy, M. Crubezy, Z. Pincus, and M. A. Musen. Biostorm: a system for automated surveillance of diverse data sources. *AMIA Annu Symp Proc*, page 1071, 2003.
36. D. Pe'er, A. Regev, G. Elidan, and N. Friedman. Inferring subnetworks from perturbed expression profiles. *Bioinformatics*, 17(Suppl):S215–S224, 2001. 1367-4803 Journal article.
37. M. Peleg, S. Tu, A. Manindroo, and R. B. Altman. Modeling and analyzing biomedical processes using workflow/petri net models and tools. *Medinfo*, 2004:74–8, 2004. 1569-6332 Journal Article.
38. M. Peleg, I. Yeh, and R. B. Altman. Modelling biological processes using workflow and petri net models. *Bioinformatics*, 18(6):825–37, 2002. 22069932 1367-4803 Journal Article.
39. S. Racunas, C. Griffin, and N. Shah. A finite model theory for biological hypotheses. In *Computational Systems Bioinformatics Conference, 2004*, pages 585–589. IEEE, 2004. TY - CONF.
40. S. Racunas, N. Shah, and N. Fedoroff. A contradiction-based framework for testing gene regulation hypotheses. In *Computational Systems Bioinformatics Conference, 2003*, pages 634–638. IEEE, 2003. TY - CONF.

41. S. A. Racunas, N. H. Shah, I. Albert, and N. V. Fedoroff. Hybrow: a prototype system for computer-aided hypothesis evaluation. *Bioinformatics*, 20(suppl_1):257–264, 2004.
42. V. N. Reddy. Modeling biological pathways: A discrete event systems approach. Master's thesis, University of Maryland, College Park, 1994.
43. V. N. Reddy, M. L. Mavrovouniotis, and M. N. Liebman. Petri net representations in metabolic pathways. *Proc Int Conf Intell Syst Mol Biol*, 1:328–36, 1993. 96038982 Journal Article.
44. C. Rosse and J. L. V. Mejino. A reference ontology for biomedical informatics: the foundational model of anatomy. *J Biomed Inform*, 36(6):478–500, Dec 2003.
45. D. L. Rubin, Y. Bashir, D. Grossman, P. Dev, and M. A. Musen. Using an ontology of human anatomy to inform reasoning with geometric models. *Stud Health Technol Inform*, 111:429–435, 2005.
46. D. L. Rubin, D. Grossman, M. Neal, D. L. Cook, J. B. Bassingthwaighte, and M. A. Musen. Ontology-based representation of simulation models of physiology. *AMIA Annu Symp Proc*, pages 664–668, 2006.
47. D. L. Rubin, S. E. Lewis, C. J. Mungall, S. Misra, M. Westerfield, M. Ashburner, I. Sim, C. G. Chute, H. Solbrig, M.-A. Storey, B. Smith, J. Day-Richter, N. F. Noy, and M. A. Musen. National center for biomedical ontology: advancing biomedicine through structured organization of scientific knowledge. *OMICS*, 10(2):185–198, 2006.
48. A. Ruttenberg, T. Clark, W. Bug, M. Samwald, O. Bodenreider, H. Chen, D. Doherty, K. Forsberg, Y. Gao, V. Kashyap, J. Kinoshita, J. Luciano, M. S. Marshall, C. Ogbuji, J. Rees, S. Stephens, G. T. Wong, E. Wu, D. Zaccagnini, T. Hongsermeier, E. Neumann, I. Herman, and K.-H. Cheung. Advancing translational research with the semantic web. *BMC Bioinformatics*, 8 Suppl 3:S2, 2007.
49. A. Rzhetsky, I. Iossifov, T. Koike, M. Krauthammer, P. Kra, M. Morris, H. Yu, P. A. Duboue, W. Weng, W. J. Wilbur, V. Hatzivassiloglou, and C. Friedman. Geneways: a system for extracting, analyzing, visualizing, and integrating molecular pathway data. *J Biomed Inform*, 37(1):43–53, 2004. 1532-0464 Journal Article.
50. A. Rzhetsky, T. Koike, S. Kalachikov, S. Gomez, M. Krauthammer, S. Kaplan, P. Kra, J. Russo, and C. Friedman. A knowledge model for analysis and simulation of regulatory networks. *Bioinformatics*, 16(12):1120–8–, 2000.
51. N. Shah and M. M.A. Which annotation did you mean? Technical Report SMI-2007-1247, Stanford Medical Informatics, May 2007.
52. N. H. Shah. *Formal Methods for Genomic Data Integration.* PhD thesis, The Pennsylvania State University, University Park, 2005.
53. N. H. Shah and N. V. Fedoroff. Clench: a program for calculating cluster enrichment using the gene ontology. *Bioinformatics*, 20(7):1196–7, 2004. 1367-4803 Journal Article.
54. J. Shrager, R. Waldinger, M. Stickel, and J. P. Massar. Deductive biocomputing. *PLoS ONE*, 2:e339, 2007.
55. B. Smith, W. Ceusters, B. Klagges, J. Khler, A. Kumar, J. Lomax, C. Mungall, F. Neuhaus, A. L. Rector, and C. Rosse. Relations in biomedical ontologies. *Genome Biol*, 6(5):R46, 2005.
56. R. Stevens, C. A. Goble, and S. Bechhofer. Ontology-based knowledge representation for bioinformatics. *Brief Bioinform*, 1(4):398–414, 2000. 21357582 1467-5463 Journal Article Review Review, Tutorial.

57. C. Talcott, S. Eker, M. Knapp, P. Lincoln, and K. Laderoute. Pathway logic modeling of protein functional domains in signal transduction. *Pac Symp Biocomput*, pages 568–80, 2004. Journal Article.
58. X. Wang, R. Gorlitsky, and J. S. Almeida. From xml to rdf: how semantic web technologies will change the design of 'omic' standards. *Nat Biotechnol*, 23(9):1099–1103, Sep 2005.
59. L. Yue and W. C. Reisdorf. Pathway and ontology analysis: emerging approaches connecting transcriptome data and clinical endpoints. *Curr Mol Med*, 5(1):11–21, Feb 2005.

Ontologies for Cultural Heritage

Martin Doerr

ICS-FORTH, Greece, martin@ics.forth.gr

1 Introduction

In the cultural heritage domain information systems are increasingly deployed, digital representations of physical objects are produced in immense numbers and there is a strong political pressure on memory institutions to make their holdings accessible to the public in digital form. The sector splits into a set of disciplines with highly specialized fields. Due to the resulting diversity, one can hardly speak about a "domain" in the sense of "domain ontologies" [33]. On the other side, study and research of the past is highly interdisciplinary. Characteristically, archaeology employs a series of "auxiliary" disciplines, such as *archaeometry, archaeomedicine, archaeobotany, archaeometallurgy, archaeoastronomy*, etc., but also historical sources and social theories.

Interoperability between various highly specialized systems, integrated information access and information integration increasingly becomes a demand to support research, professional heritage administration, preservation, public curiosity and education. Therefore the sector is characterized by a complex schema integration problem of associating complementary information from various dedicated systems, which can be efficiently addressed by formal ontologies [14, 32, 33].

There is a proliferation of specialized terminology, but terminology is less used as a means of agreement between experts than as an intellectual tool for hypothesis building based on discriminating phenomena. Automated classification is a long established discipline of archaeology, but few terminological systems are widely accepted. The sector is, however, *more focused* on establishing knowledge about facts and context in the past than about classes of things and the laws of their behavior. Respectively, the concatenation of related facts by *co-reference* [56] to particulars, such as things, people, places, periods is a major open issue. *Knowledge Organisation Systems* (KOS, [62]) describing people and places are employed to a certain degree, and pose similar technical problems as ontologies, but the required scale is very large. In this chapter, we describe how ontologies are and could be employed to improve information management in the cultural heritage sector.

S. Staab and R. Studer (eds.), *Handbook on Ontologies*, International Handbooks on Information Systems, DOI 10.1007/978-3-540-92673-3,

2 The Cultural Heritage Domain

Layman may think of cultural heritage primarily as fine arts collections and regard the description and indexing of these objects as relatively straightforward and reasoning more as a matter of scholarly reflection about their ideal values than a matter of logic. In reality, cultural heritage is more than as a domain. It comprises a broad spectrum of functions about the study and preservation of physical evidence of the past of all sorts of human activities [19].

2.1 What is Cultural Heritage?

In a narrower sense, we may regard the cultural heritage as the things preserved by the *memory institutions*, i.e. museums, sites and monuments records ("SMR"), archives and libraries. Their international umbrella organizations are: the International Council of Museums (ICOM,[1]) the International Federation of Library Associations (IFLA,[2]) and the International Council of Archives (ICA.[3]) They maintain their specific documentation policies and standards.

Following ICOM, "A museum is a non-profit making, permanent institution in the service of society and of its development, and open to the public, which acquires, conserves, researches, communicates and exhibits, for purposes of study, education and enjoyment, material evidence of people and their environment" [60] and "Museums" hold primary evidence for establishing and furthering knowledge" [61]. SMRs are typically departments of a Ministry of Culture, pursuing similar goals as museums, but for immobile sites. Archives keep very large amounts of original material – mostly written and images – in their historical order, such as administrational records, letters from VIPs, photographic collections and others.

To a certain degree, libraries may also preserve cultural heritage when they keep unique books, however their focus is on mediating access to non-unique information sources. In contrast, most cultural heritage objects are a rather mute evidence of past events that acquire relevance from understanding the context of their origin and history. The object may appear less as an information source in its own right than as an "illustration" of the past. This distinction is important to understand the difference between library and cultural heritage information, and the immense complexity of the latter.

One can appreciate the diversity of cultural heritage from the following list of major kinds of collections:

- History of arts and modern arts (graphics, painting, photography, sculpture, architecture, manuscripts, religious objects),

[1] http://www.icom.org
[2] http://www.ifla.org
[3] http://www.ica.org

- Historical heirloom (treaties, letters, manuscripts, drawings, photos, films, personal objects, weapons),
- Archaeology (sherds, sculptures, tools, weapons, household items, human remains),
- Design (furniture, tableware, cars, etc.),
- Science and technology (machinery, tools, weapons, vehicles, famous experiments, discoveries),
- Ethnology (costumes, tools, weapons, household items, religious objects, etc.)
- Immobile sites (architecture, sculpture, rock art, caves)
- To a certain degree, natural history collections, such as paleontology, biodiversity, mineralogy are also evidence of human activities (i.e. research) and hence culture.

Handling information about all those kinds of things implies the use of very rich terminology, multilingual and often specific to particular communities or even to particular scholars. Agreement on common terminology is difficult and equivalent terms in other languages are often missing. It is an obvious challenge for employment of formal ontologies that poses not only technical problems, but also intellectual challenges in the approximation of intuitive or traditional concepts by logical definitions, such as the possible narrower and wider meanings of the same term, objective declaration of discriminating features or fuzzy transitions of instances from one class to another.

2.2 Functions of Cultural Heritage Information

One can distinguish kinds of cultural heritage information systems by their major functions. Those are:

- Collection management (acquisition, registration, "deaccession", inventory, loans, exhibitions, insurance, rights, protection zones) [29, 30]
- Conservation (diagnosis of deterioration, preventive measures, interventions, treatments and chemical agents)[78]
- Research (investigation, description, interpretation)
- Presentation (portals, teaching, publication)

In each of these four areas quite distinct and highly specialized information systems exist, created and maintained by different players. On the other side, information in all those systems overlaps and should be mutually accessible in order to do the job. One of the major challenges of cultural heritage information management is the interoperability of those system and integration of information across function and discipline.

Collection management systems are offered by several commercial vendors. They are mostly built on Relational or hierarchical database systems. Many customized systems are built on demand by IT experts. They support the technical management and administration of collections or sites and monuments. A comprehensive, internationally accepted definition of their functions can be found in [29]. *Curators* provide basic descriptions of the objects that serve their identification and handling, but do also research and justify their relevance, i.e. why the object is kept in a museum. Archival collections typically consist of millions of leaves. It is unusual to describe each item. Rather, curators document the historical context under which coherent sets of documents were created or brought together as finding aids for researchers, so-called collection-level metadata. Only recently, more fine-grained documentation is occasionally considered [23].

Conservation information may be part of the collection management or separate from it. It deals with the scientific, material analysis of the objects, preventive measures and interventions. Loan management and historical research may need those data. Art and monument conservation is an underestimated sector of financial importance. *Art conservators* are scientists who need, similar to doctors, to accumulate and exchange immense knowledge about diagnosis methods, treatments and side effects [78]. There are a few dedicated websites and systems for information exchange between experts [3,58] and learning [20], but there is still a wide market for such systems. Since they deal with *categorical* (general) knowledge, such systems should better be ontology-based.

Research information systems are highly specialized and mostly built on demand for specific projects. There are reference systems that list consolidated, uniform descriptions of all known items of a certain kind, such as Roman Inscriptions [15] or the Union List of Artist Names [12]. There are many systems[4] that integrate information from thousands of sources for statistical or other kinds of analysis. Particularly important became GIS-based reasoning systems, such as for archaeological site prediction, and systems for running automatic classification (see for instance, [24,38]). Unfortunately, idiosyncratic design and insufficient management of source references frequently make the reuse of the integrated information impossible after the project ends.[5] More effective means of data transformation and migration are still to be developed. Ontologies could play an important role in that.

Presentation systems give access to cultural heritage information to the general public or a community of subscribers, in particular teachers and academics (see Chapter "Ontology-Based Recommender Systems"). We estimate that more than 95% of museum objects are not in any exhibition, and archives are mostly closed to the public. Therefore there is a strong political pressure to make at least object descriptions from the collection management systems publicly accessible. Museum portals (see Chapter "Ontology-Based Recommender Systems") may present parts of collections. The scale is immense:

---

[4] For instance, those published by the conference series Computer Applications & Quantitative Methods in Archaeology `http://caa.leidenuniv.nl/proceedings/`

[5] Round Table discussion at the 8th EAA ANNUAL MEETING, 24–29 September 2002, Thessaloniki–Hellas, `http://www.symvoli.com.gr/eaa8/mple.htm#P5`

larger museums hold millions of objects. The Smithsonian Institutions hold over 100 million objects. Other presentation systems may take the form of an electronic exhibition, complementary information to a physical exhibition, or the form of an electronic publication that elaborates a particular subject matter. Ontologies play a major role to provide structured access points and to structure the subject matter itself in these systems.

Recent efforts deal with the capturing and preservation of *performing arts* and *oral tradition* [11, 34, 49]. Since there is no object to be described, traditional models of documentation are not appropriate, and new models are discussed.

3 The Schema Integration Problem

Most of the professional systems referred to above are based on fairly complex database schemata. For instance, CIDOC proposed until 1995 a standard Relational Schema for museums with more than 400 tables. As described above, cultural heritage information is distributed in many different systems which complement each other. One source may relate Roman names to Roman inscriptions, another Roman inscriptions to stones, another stones to place of finding, and another places to coordinates [25]. But still most efforts to integrate heritage information concentrate on finding minimal common description elements for objects as *finding aids* rather than documentation. This is motivated by practice from the library communities, in particular the Dublin Core Consortium.[6]

3.1 Metadata and Application Profiles

Since libraries and Digital Libraries hold objects that contain data, they use to call the descriptions of their objects "metadata", i.e. data about data. This term has also been adopted by museums for their object descriptions, even though their objects are not data. There is a plethora of attempts to structure metadata as flat lists of properties, which may be aggregated in so-called "application profiles" [9, 37], and the mapping and data transformation between different metadata formats may be called a "metadata crosswalk" [62]. The labels of metadata properties, such as "creator", "date", etc., remind concepts. Therefore several authors recently regard metadata schemata as "vocabularies" or a kind of ontologies. We regard this as a confusion of information models with ontologies, as elaborated by [71, 72]. It is to remind that formal ontologies were introduced to computer science to describe common conceptualization behind multiple schemata [32, 33], and not to become a synonym of schemata. Further, the reduction of complex object histories to a flat set of

[6] `http://dublincore.org/`

properties can only be achieved by semantic overloading of these properties, which conflicts with the definition of an ontology, as shown in [18, 43].

Nevertheless, numerous digitization projects of cultural objects create digital libraries with Dublin Core metadata elements as minimal standard. Also wide-spread is the use of MPEG7 ([8, 40] and Chapter "Ontologies for Cultural Heritage"), the metadata standard for multimedia objects, for obvious reasons, which is a far richer representation of the structure, history and subject of the object. There is serious research on automatic matching of metadata elements in order to support schema mapping and merging which is based on comparison of metadata elements with ontologies. The idea is to detect similarities between schema elements and the underlying concepts by the similarity of naming and properties. The underlying concepts are found in a formal ontology, such as WordNet [26].

3.2 ISO21127

Information integration based on finding aids for the objects actually fails to integrate the information about the wider historical contexts these objects illustrate and from which they get their relevance. If a serious integration of the relevant contents of cultural heritage information is intended, richer models must be employed. For instance, the Research Libraries Group in California successfully integrated in their Cultural Materials Initiative data from about a thousand cultural institutions encoded in about a hundred different schemata into a far richer schema, virtually without loss of information. This schema was derived from the CIDOC CRM ontology, now ISO21127, which is currently the most elaborated ontology for the integration of cultural heritage information.

The CIDOC CRM is a formal ontology [16] intended to facilitate the integration, mediation and interchange of heterogeneous cultural heritage information. It was developed by interdisciplinary teams of experts, coming from fields such as computer science, archaeology, museum documentation, history of arts, natural history, library science, physics and philosophy, under the aegis of the International Committee for Documentation (CIDOC) of the International Council of Museums (ICOM). Its development started *bottom up*, by reengineering and integrating the semantic contents of more and more database schemata and documentation structures from all kinds of museum disciplines, archives and recently libraries.

The development team applied strict principles to admit only concepts that serve the functionality of global information integration, and other, more philosophical restrictions about the kind of discourse to be supported (for more details see [19]). The application of these principles was successful in two ways. On the one side, the model became very compact without compromising *adequacy* [71]. The very first schema analyzed in 1996, the CIDOC Relational Data Model with more than 400 tables (described by [66]), could be reduced to a model of about 50 classes and 60 properties, with far wider applicability than the original schema. On the other side, the more schemata were analyzed, the fewer changes were needed in the model (see version history.[7]) The present model contains 80 classes and 132 properties, representing the semantics of may be hundreds of schemata. As a result of the successful reformulation of the original Relational model, CIDOC started the standardization process in collaboration with ISO in 2000. The model was accepted as ISO21127:2006 in September 2006.

Deliberately, the CIDOC CRM ontology is presented in a textual form to demonstrate independence from particular knowledge representation formats. There exists however a formal definition in TELOS [59]. The CRM distinguishes individual classes from properties (binary relationships). Properties are directed and bidirectional, with distinct labels for each direction. It employs strict multiple inheritance (without exceptions) for both classes and properties. It foresees multiple instantiation, i.e. one particular item can accidentally be instance of more than one class. Domain and range of properties are associated with the quantifiers zero, one or many. There exist valid equivalents in KIF, RDFS and OWL, to the degree the respective constructs are supported. Four ideas are central to the CRM (see Fig. 1):

1. The possible ambiguity of the relationship between entities and the identifiers ("Appellations") that are used to refer to the entities are a part of the historical reality to be described by the ontology rather than a problem to be resolved in advance. Therefore, the CRM distinguishes the nodes representing a real item from the nodes representing only the names of an item.
2. "Types" and classification systems are not only a means to structure information about reality from an external point of view, but also part of the historical reality in their nature as *human inventions*. As such they

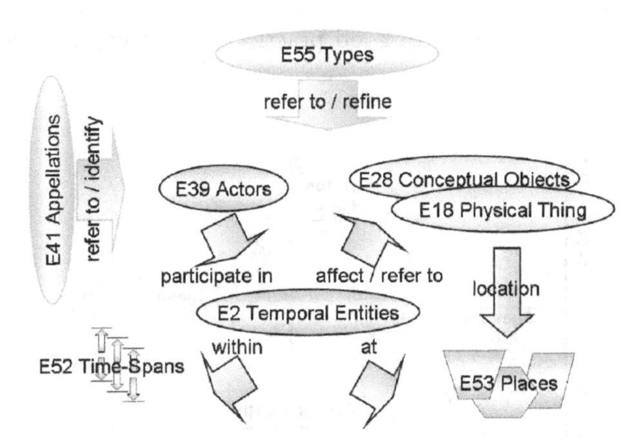

Fig. 1. Top-level entities of the CIDOC CRM

[7] `http://cidoc.ics.forth.gr`

fall under "Conceptual Objects", inheriting properties of creation, use, etc. Similarly, all documentation is seen as part of the reality, and may be described together with the documented content itself. This reification problem is not appropriately dealt with in current ontology languages. The CRM is forced to use some workarounds we do not analyze here further. Terminology, i.e. classes that are not contributing as *domain* or *range* to the relationships expressed in data structures, are not part of the core ontology itself but regarded as instances of "Type" for practical reasons.

3. The normal human way to analyze the past is to split up the evolution of matters into discrete events in space and time. Thus the documented past can be formulated as series of events involving "Persistent Items" (also called endurants, see [19]) like Physical Things and Persons. The involvement can be of different nature, but it implies at least the presence of the respective items. The linking of items, places and time through events creates a notion of "world-lines" of things meeting in space and time (see Fig. 2). Events, seen as processes of arbitrary scale, are generalized as "Periods" and further as "Temporal Entities" (also called perdurants [19]). Only the latter two classes are directly connected to space and time in the ontology. The Temporal Entities have fuzzy spatiotemporal boundaries which can be approximated by outer and inner bounds.
4. Immaterial objects ("Conceptual Objects") are items that can be created but can reside on more than one physical carrier at the same time, including human brains. Immaterial items can be present in events through the respective physical information carriers (see Fig. 3). Immaterial items cannot be destroyed, but they disappear when the last carrier is lost.

As a standard, the use of CRM concepts is not prescriptive, but provides a controlled language to describe common high-level semantics that allow for information integration at the schema level. It is intended to serve

1. As an intellectual guide to good practice of conceptual modeling in the sector.

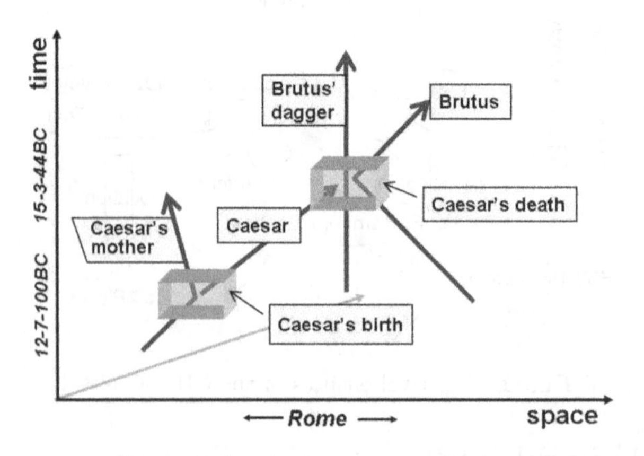

Fig. 2. Historical events as meetings

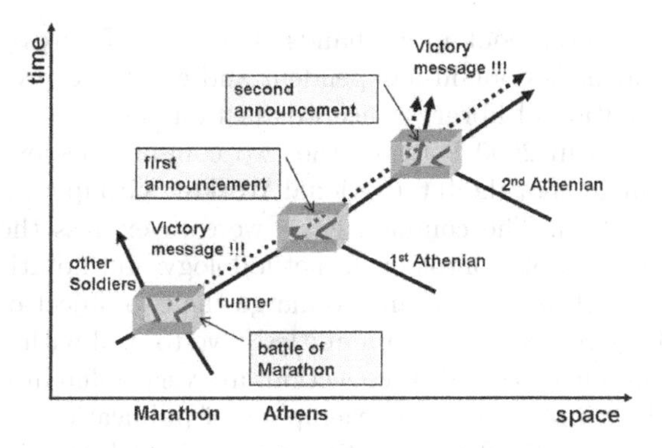

Fig. 3. Information exchange as meetings

2. As global model for information integration in a "Local as View" (LAV, [13]) or data warehouse manner.
3. As an intermediate model for data migration.

The coverage of the CRM for cultural heritage data has been validated by mappings from numerous data structures of the sector to the CRM. Even the common library format MARC ('Machine Readable Cataloguing') can be adequately mapped to it [49]. Such a mapping can be seen as an interpretation of the data structure elements in terms of the ontology. If the ontology is implemented as a schema (such as in RDFS), the mapping can also be seen as a specification for Local as View (LAV) schema integration. The examples of mappings from Dublin Core or EAD to the CRM [18, 43] show how well-defined common semantics can be associated with typical metadata formats. In particular they allow for describing explicitly the cases of semantic overloading (such as the use of DC.date for various events). Even MPEG7 has been aligned with the CRM [40]. The CRM is increasingly used in real integrated information environments for cultural heritage systems. A list of references can be found on [1, 2]. Due to the characteristic focus of the empirical base of the CRM, i.e. data structures used for collection descriptions, it is relatively poor in describing family relations, rights, and intellectual processes. The latter has been recently complemented by the FRBRoo model [7, 48].

3.3 FRBRoo and Performing Arts

The FRBR model ('Functional Requirements for Bibliographic Records') was designed as an entity-relationship model by a study group appointed by the International Federation of Library Associations and Institutions (IFLA) during the period 1991–1997 [68]. It was published in 1998. Its innovation is to cluster publications and other items around the notion of a common conceptual origin – the 'Work', in order to support information retrieval and to initiate a new bibliographic practice. It distinguishes four levels of abstraction from conception to the book in my hands: The Work, Expression, Manifestation, Item. Its focus is domain-independent and can be regarded as the most advanced formulation of library conceptualization [48].

Initial contacts in 2000 between the two communities eventually led to the formation in 2003 of the International Working Group on FRBR/CIDOC CRM Harmonisation. The common goals were to express the IFLA FRBR model with the concepts, ontological methodology and notation conventions provided by the CIDOC CRM, and to merge the two object-oriented models thus obtained. Although both communities have to deal with collections pertaining to cultural heritage, those collections are very different in nature: Most of library holdings are non-unique exemplars of publications, i.e. products of industrial processes. FRBR focuses therefore on the "abstract" characteristics that all copies of a single publication should typically display in order to be recognised as a copy of that publication. The history of individual copies and of the immaterial content is not regarded as particularly relevant in library catalogues and therefore widely ignored by FRBR. Of course, libraries do also hold unique items, such as manuscripts; but there are no internationally agreed standards how to deal with such materials, and FRBR mentions them but does not account for them in a very detailed way.

Museums, on the other hand, are mainly concerned with unique items – the uniqueness of which is counterpoised by a focus on the cultural circumstances under which they were produced and through which they are interrelated. CIDOC CRM highlights therefore the physical description of singular items, the context in which they were produced, and the multiple ways in which they can be related to other singular items, categories of items, or even just ideological systems or cultural trends. Of course, museums may also have to deal with exemplars of industrially produced series of artefacts, but CIDOC CRM covers that notion just with the multi-purpose E55 Type class. Museum objects may be referred to in literature kept in libraries. Museum objects may illustrate subjects described in literature. Literature and objects may be created by the same persons, in common events.

The Working Group has submitted the final draft of FRBRoo, i.e. the object-oriented version of FRBR, harmonized with CIDOC CRM, for public review by IFLA in February 2008. This formal ontology is intended to capture and represent the underlying semantics of bibliographic information and to facilitate the integration, mediation and interchange of bibliographic and museum information.

The major innovation of FRBRoo is a realistic, explicit model of the intellectual creation process (see Fig. 1), which should still be developed further in the future for the benefit of librarians and scholars from the various museum disciplines. FRBRoo makes a fundamental distinction between internal representations of our mind (Work), sets of signs or symbols human can interpret (Expression), and physical information carriers.

The idea is that products of our mind, as long as they stay in one person's mind only, are relatively volatile and not evident. In an event of first externalization, the "Expression Creation", concepts of a Work are made manifest by creating an Expression on a first physical carrier. This may be just another person's memory, as in the case of oral tradition, a paper manuscript or a computer disc. In its current draft version, FRBRoo includes a model of performing arts, connecting the interpretation of theatre plays with the recording and documentation of performances. It distinguishes and relates the three intellectual contributions (works) of the creation of the play, of the interpretation and the recording with the associated symbolic forms and physical carriers. This part of the model has been developed and tested in first examples in collaboration with the European funded project CASPAR on Digital Preservation. Even though there is a rising interest in documenting and preserving non-material culture, there are few other models about performing arts [11]. Jane Hunter has done interesting research on representing indigenous knowledge and its oral traditions [41].

3.4 Other Core Ontologies

Independent from the CRM, the European funded project IndeCs, a consortium of multimedia experts, developed around 1997 a core model to trace the provenance of contributions and associated intellectual property rights in multimedia products and implemented a respective information system. This model was taken up by the ABC ontology. The latter is an outcome of the Harmony Project, which was funded cooperatively by the Distributed Systems Technology Cooperative Research Centre (DSTC) (Australia), the Joint Information Systems Committee (JISC) (UK), and the National Science Foundation Digital Libraries Initiative (NSF DLI2) (US). The original goal and continuing motivation of the ABC work arose from the need to integrate information from multiple genres of multimedia information within digital libraries. The researchers working on the Harmony Project have each been involved in a number of metadata initiatives including Dublin Core and MPEG-7.

Complete details of the ABC ontology are described in [46]. It is far smaller than the CRM, just 13 classes and 14 properties. As the CIDOC CRM, ABC describes temporality in a first-class manner. Modeling change over time is critical to the description of digital content due to its inherent fluidity and the linkages of provenance to integrity or trust [45]. ABC includes both the notions of "events" and "situations", which respectively model transitions (i.e. verbs) and existential properties. The inspiration for these concepts lies in process models such as Petri Nets [63] and temporal extensions for first-order logic such as Situational Calculus [55]. Due to these temporal concepts, ABC is able to definitively model time periods during which certain properties of an object are static. It is also able to model events or transitions marking property modification, for example during the change of a version of a digital object. Finally, ABC builds on the concepts developed in the FRBR model [68]. These concepts – *works, expressions, manifestations,* and *items* – give ABC the ability to link entities that have common intellectual property origins. Work in the library community has proven the utility of these concepts [47].

Similarities between ABC and CRM aims and solutions were so striking, that both teams collaborated between 2001 and 2003 on a harmonization project, in which both ontologies adopted concepts from each other and re-arranged properties and IsA hierarchies, until a merged representation was possible [21]. The CRM did not adopt the concept of a *situation*: In the end, the representation of object history in ABC as a chain of states (*situation*) and state transitions (*events*) turned out to be redundant, making knowledge revision complex, and causing problems to integrate interconnected histories of multiple objects and agents. ABC has been mainly used in research. Another interesting core ontology is DOLCE [54]. It is product of careful reengineering of the core concepts of WordNet, a linguistic resource derived from dictionaries, enriched with theory-based foundational relationships such as participation, part-whole, constitution, etc. It is rigorously formulated in logic, making it rather difficult for domain experts to comprehend and use it. In contrast to the CRM, space and time are regarded as dependent properties of things, and not as things existing in a potentially empty space-time – the only, but deep incompatibility between both ontologies. Otherwise, many concepts exist in both ontologies. Some concepts in DOLCE are characteristic for other kinds of discourse than that found in data structures for heritage documentation. Interesting enough, museums are not much interested to analyze iconographic representations by discrete schema elements. With the aim of digital archive interoperability in mind, D'Andrea et al. [17] took the CIDOC CRM common reference model for cultural heritage and enriched it with the DOLCE D&S foundational ontology to better describe and subsequently analyze iconographic representations, from, in this particular work, scenes and reliefs from the meroitic time in Egypt.

3.5 Characteristics of Ontologies for Cultural Heritage

Ontologies that deal with semantics equivalent to those of data structures, as the ones presented above, contain few classes and are rich in relationships [19,51], in contrast to terminological ontologies for classifying individual items. Data structures can be seen as equivalents of propositions about a domain ("possible states of affairs", [33]). Therefore their semantics reveal characteristic parts of the discourse of a domain or sector. So what characterizes the discourse in cultural heritage as reflected in data structures and ontologies?

Cultural heritage can be seen as the material evidence of human activities of social relevance in the past. Therefore

- Information is mesoscopic, i.e. at a human scale, neither astronomic nor microscopic, except for microscopic analysis of traces and materials. Information is discrete. Processes are reported or become evident as discrete events involving discrete things, in contrast to geological or meteorological phenomena.

- Information is event centric. Things, people and ideas connect and relate via events.
- Its description is retrospective, in contrast to information to plan the future, such as for manufacturing.

Information is naturally incomplete at some scale. It can be complemented but not be completed. Its description serves a kind of detective work to reconstruct *possible pasts.* The distinction between evidence and conclusion is vital. Therefore information cannot be normalized and integrated on the basis of the assumed past, such as on absolute dates, geographic coordinates, cause-and-effect, states-and-state-transitions. The documentation of the process of observation is necessary to interpret correctly the observed evidence. Even the fact that some scholar classifies an object with a certain term is documented as a historical, intellectual process (this holds equally for biodiversity). Information is about material facts [35]. Observed individual facts are the basis to induce categorical behavior, such as "all Pharaohs were mummified".

The above characteristics hold equally well for other descriptive sciences, such as geography, biodiversity, paleontology, clinical observation and epidemic studies, but also for the documentation of experiments and observations in natural sciences. Whereas the latter formulate their conclusions about their observations in categorical theories ("F=m*a", or "any non-supported material object in the atmosphere of Earth will fall"), scholars interpreting cultural heritage would generally hesitate to formulate their categorical conclusions or hypotheses in a formal representation (see also [31]). Rather, interpretation is normally presented as text rendering implicitly a wealth of associated belief values. Therefore the presented ontologies are surprisingly domain and discipline independent. It is the retrospective discourse that determines their characteristic form. Ontologies describing the formal structure of iconographic subjects can be seen as an exception to this (see [17] and Sect. 4).

Also surprising is the fact that scholars hesitate to formulate in objective terms causes and causation [50]. Whereas in the domain of jurisdiction characteristic ontologies are being elaborated that detail contributions of individuals in activities, modern scholars prefer a more distant stance of multiple views and possible truths [39]. Noteworthy are the promising attempts of [27,67] to formally structure archaeological argumentation, which could lead to an ontology or better epistemology of cultural heritage argumentation, even though vehement arguments against this approach are not missing [39].

Particularly in ethnology and archaeology (as in biodiversity), some information is documented in a partially categorical form, such as: "The boomerang is a hunting weapon of the Australian Aborigines". I.e. a particular community is associated with characteristic kinds of things and kinds of activities. The described object is seen as example of the category and an illustration of the activity. There is neither currently a formal ontology nor a suitable ontology language which would give a realistic account of the relationship between such partially categorical statements and the individual facts as perceived by the domain expert, and there is no dedicated "metaontology" which could be instantiated with such partially categorical statements.

The CIDOC CRM makes a practical distinction between core classes and classes appearing as terminology motivated by the fact that they appear typically as data in data structures, in order to make fine distinctions between the kinds of the referred items. Even though knowledge representation does not distinguish between the two, it is an empirical fact that the sector uses to organize terminology differently, in vocabularies and thesauri, which may more and more be developed into formal ontologies in the proper sense. Consequently, the RDF schema SKOS [5], a W3C First Public Working Draft, suggests the encoding of terms from vocabularies and thesauri as particulars, and not as RDF classes. We follow this distinction here to structure this chapter. Cultural heritage terminology pertains mostly to classes detailing kinds of material things, which is quite similar and or even overlapping with product classification [69].

Other terminologies of the sector characteristically pertain to:

- Materials, conservation agents
- Information objects
- Processes, deterioration, activities
- Social roles
- Literary and iconographical subjects

In the following section we describe the role of terminology and the most important ontologies in the sector.

4 Terminology in Cultural Heritage

In many collaborations and discussions with museum curators and archaeologists we encountered a negative position towards the use of controlled vocabularies or even formal ontologies. Experts tend not to agree with the terminology used by colleagues [31]. This is in strong contrast to the library sector, which cannot exist without standardized terminology. We attribute this to the fact that in the cultural heritage sector terminology is less used as a means of agreement between experts than as an intellectual tool for hypothesis building based on discriminating phenomena. Consequently, automated classification is a long established discipline of archaeology, but few terminological systems are widely accepted. They are built ad-hoc for specific research questions.

The renowned archaeologist Franco Niccolucci posed the question, if archaeologists are "fuzzy" [38]. He discussed the notion of a neolithic tool, a scraper. From a given set of similar stone tools, several archaeologists did each classify a different subset as *scrapers*. The background of this disagreement is that the concept is used to deduce hypothetic function in the past from observed morphology. It represents already a debatable hypothesis. Many archaeologists develop their own *typologies*. There is a continuous demand for specialized reasoning systems (for instance, [24]).

4.1 Information Access by Terminology

The diversity and number of small ontologies, in the order of a hundred to a thousand terms each, puts interesting challenges to ontology matching and alignment. Only automated tools have a chance to exploit this expert terminology for retrieval and reasoning across local systems.

The task of librarians is not hypothesis building, but providing access to information. Quite naturally, they have a long tradition to agree on common terminology as access points. It is not easy for cultural heritage experts to appreciate the need for shared search terms (see for instance [31]), and there is still enough conviction work to be done. In contrast to [31], we assume that cultural heritage terminology could be separated into an upper, stable level suitable for search, and a lower volatile level supporting hypothesis building. This is motivated from our experience building ISO21127 and various information systems. The largest and stable thesaurus in the sector, the Art & Architecture Thesaurus (AAT, [64]), with more than 30,000 concepts, comes actually from a library background (see below).

A problem with classification of material objects are the different aspects (facets), under which the classification may be done. Dominant aspects are the function of the object, its shape or appearance, elements or principles of construction [22]. These three aspects are partially related. For instance, a typical hammer may have a classical shape and construction, but a motorized hammer may only share function, but not the other aspects. Other aspects are forms specific to historical periods or nations. The effect may confuse thesaurus and ontology editors when building IsA hierarchies, and may mislead users when they apply classification terms.

The so-called facet analysis tries to resolve this problem (e.g. [53, 74]) by systematic separation of the concepts for each facet. "Faceted classification", which goes back to the Indian Librarian Ranganathan (1965) [65], employs the systematic combination of classification terms for each relevant facet. For example, the AAT has removed the term "mills" because it can be constructed from "grinding & factories". The method greatly reduces complexity and depth of term hierarchies, and improves the quality of the ontology. On the other side, faceted classification can be seen as a precursor of employing Description Logic (DL) – simply the roles between the combined terms are implicit. It is assumed that the user has an intuitive understanding of the meaning, and that it is unambiguous. The use of DL in cultural heritage is still in the beginning.

It is standard for museum portals and other cultural information systems that provide information about material objects to offer faceted access by type of object, person, place, date. MuseumFinland [42] employs a faceted Finnish ontology, may be the most advanced system in terms of formal representation of terminological concepts. The UK national project FACET [75] with the UK Science Museum's collections database on Thesaurus-Based Query Expansion employs a combination of novel techniques with a faceted theasurus (the Getty Art and Architecture Thesaurus). Aroyo et al. [28] employ the VRA metadata scheme and encode terminology from the Getty Art and Architecture Thesaurus, Union List of Artist Names and ICONCLASS in SKOS [5].

All terminological systems contain very general terms as root elements of their hierarchies. These may vary considerably and cause unnecessary inconsistency between the ontologies, because the purpose of these ontologies is not to solve the philosophical questions these general terms are associated with. For instance, the AAT subsumes under *visual works* material and immaterial things, such as *paintings* and *electronic images*. In the CIDOC CRM, material and immaterial things are disjoint concepts, because reasoning differs considerably for the two. In integrated information systems depending on rich data structures, this incompatibility can interfere with schema integration. The use of a shared core ontologies to enable interoperability between different domain ontologies has been proposed a decade ago by [33]. The ongoing British STAR project [6] is now investigating cross search over different archaeological datasets and grey literature with the CIDOC CRM core ontology as an integrating framework for the datasets and domain thesauri.

4.2 Major Terminological Systems

The AAT is the most widespread ontology in cultural heritage. It has the form of a thesaurus, compatible with ISO2788. Its topic is art and architecture, but covers a wide range of archaeological and ethnological materials as well as any kinds of object that may be subject of art in some way, such as weapons. It was originally developed by merging culture-relevant subject keywords from several large library systems. It is built for faceted classification. Its major facets are: Activities, Agents, Materials, Objects, Physical Attributes, Styles & Periods, Associated Concepts.

The broader term and narrower term relationship are used in the sense of IsA. Its originally monohierarchical ("tree") generalization structure has been extended to polyhierarchical (directed acyclic graph). The AAT introduces so-called guide terms, (node labels in ISO2788) to group terms under minor facets, such as function or form, but there is no rigorous logic applied to this organization principle. The AAT has been translated into Spanish and Dutch.

English Heritage (EH) maintains also a very large thesaurus of terms for mobile and immobile objects for the United Kingdom, as well as the French MERIMEE thesaurus.

The multilingual thesaurus attached to the European HEREIN project [4] intends to offer a terminological standard for national policies dealing with architectural and archaeological heritage, integrating concepts from the above resources. Beyond just correlating concepts from different languages, the project decided to create for each language a new generalization-specialization hierarchy and to harmonize concepts manually. However, they did not preserve the concepts as found in other sources or link to them. We regard this as problematic, as an opportunity for interoperability seems to have been thrown away unnecessarily.

Remarkable is the successful use of SHIC [70], a classification system of human activities, for the description of museum objects by several British museums. Rather than characterizing the object, only the function or utility of an object for a human activity is regarded. This focus on one uniform aspect ("facet") avoids the ambiguity in the application of other terminological systems. In 1950, the Netherlands Institute for Art History (Rijksbureau voor Kunsthistorische Documentatie or RKD) began its collaboration with Henri Van der Waal on the development of ICONCLASS,[8] with the publication of mounted and annotated photographs of Dutch works of art, the so-called DIAL (Decimal Index of the Art of the Low Countries). From 1950 until 1982, 28 sets of 500 cards were produced, making for a total of 14,000 items. In the RKD images database, which can be consulted via the RKD website, a large number of Dutch works of art is made accessible with the help of ICONCLASS notations. In September 2006, the RKD acquired the rights for the ICONCLASS software from the Royal Academy of Arts and Sciences (KNAW) in Amsterdam. The ICONCLASS System is the only more widespread system for iconographic classification. It is a kind of faceted classification system with a hierarchy of concepts defined by decimal codes. It comprises general, Christian and Greco-Roman subjects. Concepts can be modified by keys to express, for instance, 'head of X'. So far, no formulation as a formal ontology has been undertaken. Van Gendt [76] could only partially represent ICONCLASS in SKOS. Even though it is a genuine aspect of cultural heritage, iconographic classification is not regarded as part of standard museum documentation.

CAMEO is a searchable information center developed by the Museum of Fine Arts, Boston [58]. The MATERIALS database contains chemical, physical, visual, and analytical information on over 10,000 historic and contemporary materials used in the production and conservation of artistic, architectural, archaeological, and anthropological materials. It offers only search by keywords and alphabetic order. The European funded Project CRISATEL developed a system and an ontology employing multiple generalization for art conservation comprising materials, techniques and methods of investigation and intervention for paintings, but the system has not been taken up by the community yet [20].

4.3 KOS of Particulars and Information Extraction

Understanding cultural heritage lives from contextual knowledge and concatenation of facts. Therefore it is most important to be precise about particular

[8] http://www.iconclass.nl/

persons, places, historical periods and objects, which appear as the major constituents that connect multiple facts. Relevant resources about particulars are organized as Knowledge Organisation Systems, sometimes also called "ontologies", even though the term does not apply to lists of particulars.

For instance, large reference lists of Persons are maintained by national libraries [10]. The Getty Research Institute maintains ULAN, the Union List of Artist Names [64], and TGN, the Thesaurus of Geographic Names [36], which lists a million of historical and current placenames organized in a hierarchy of geographic spatial inclusion. Also the Alexandria Gazetteer is an important resource of current placenames. In these resources, only the schema and the typologies can be regarded as a kind of ontology. The Alexandria Gazetteer contains an interesting list of feature types to classify places.

A large part of cultural heritage documentation, primary and secondary literature is in textual form. Even though databases have become standard tools for collection descriptions, curators prefer to express the relevant historical facts in free text. Therefore, automated information extraction (IE) becomes more and more important. Extracted information could be used to produce structured metadata and to instantiate ontology-based knowledge representation systems. Full text retrieval systems and text mining systems use to recognize concept names from ontologies. Ontologies should be tailored for this purpose, for instance be enriched with frequent synonyms. To our knowledge, there has been no such attempt for the more popular cultural heritage vocabularies.

So-called Named Entity Recognition of names of persons, things, places [57, 73], or recognition of date and time expressions [52] works reasonably well. Most systems use KOS or "gazetteers" to guess if a name is likely to be a person or a place name. Some languages, like Latin, have more distinct grammatical forms for location expressions, which makes the job to distinguish these categories easier [73]. Note that detecting a name does not mean that the individual has also been identified.

As the core ontologies presented in this chapter show, event information is particularly important for cultural heritage. Automatic event recognition could bring a break-through in the access to relevant historical knowledge. Automatic event recognition is the next step after recognizing named entities and dates. An event can normally be described by the kind of action, the participating things and people, date and place. Event recognition should be combined with NER. So far, there has been not too much work in this direction (for instance, [44,52,77]). An obstacle is the lack of formal ontologies relating characteristic action verbs, such as "printed", "discovered", "broke", "shot" with typical events, such as activities of creation, finding and destroying things, meetings, birth, death and killing.

5 Conclusions

Current ontologies for cultural heritage exhibit a focus on the material and physical aspects of the past. This is quite natural, since "heritage" in the narrower sense implies material evidence of the past. Information about events in the physical world is central to the understanding of heritage information and explicit formal representation of events a key element to integrate heritage information. Interesting is the convergence of core ontologies to very similar forms, which can be integrated, and their independence from a particular "cultural" view. The work of historians is more a detective work than that of a judge. This determines widely the character and focus of cultural heritage ontologies. Information is incomplete. More important than the conclusions is the careful collection of all evidence that could support the one or the other view about the past. In contrast to that, natural sciences would get rid of experimental data after a theory has been sufficiently supported by experiments.

Conclusions and judgment about the past are rather published in scholarly texts than encoded in data structures. This focus may be due to the characteristics of the reasoning in the sector, or just be enforced by the fact that IT methods have penetrated the sector from core documentation and management of physical collections. In the latter case, one may expect that cultural ontologies may in the future extend to other applications in the sector as well. May be formal ontologies dealing with the intellectual structuring of the sector, such as iconography, social interaction, and causation will find more attention in the future. Generally, we expect a greater diversity of conceptualization in the intellectual structure than in the description of material aspects, as represented by the CIDOC CRM.

Since many scholars question the utility of standardized terminology, the formalization of the major terminological systems in the sector is still poor, but this may be overcome by a gradual transfer of know-how and better appreciation of the specifics of cultural conceptualization by ontology engineers. The sector shows enough interest in using ontologies to solve the interoperability of data structures and engages in real implementations. Ontology languages seem to be sufficiently expressive for terminological problems. In the area of data structures semantics, reification problems (i.e. simultaneous use of ontologies and documentation of the discourse about them and documenting facts together with their observation), as well as partially categorical statements cannot sufficiently be described with current ontology languages.

In general, in the years of our collaboration with memory institutions and scholars we found that a major obstacle to introducing advanced computer science methods in the sector is a general underestimation of the complexity of cultural heritage conceptualization by the IT experts, which is equaled by the inability of domain experts to describe their conceptualizations in conscious, objective terms. Whoever wants to deal with the subject effectively must be prepared for a long knowledge engineering phase.

References

1. The CIDOC Conceptual Reference Model, Applications, http://cidoc.ics.forth.gr/uses_applications.html.
2. The CIDOC Conceptual Reference Model, References, http://cidoc.ics.forth.gr/references.html.
3. Conservation OnLine, Resources for Conservation Professionals http://palimpsest.stanford.edu/.
4. HEREIN Thesaurus: European Heritage Network, http://www.european-heritage.net/sdx/herein/thesaurus/introduction.xsp.
5. Simple Knowledge Organisation Systems (SKOS) - home page, http://www.w3.org/2004/02/skos.
6. STAR Project - Semantic Technologies for Archaeological Resources, http://hypermedia.research.glam.ac.uk/kos/star.
7. Successive drafts of the model and minutes of the Harmonisation Groups meetings are available from the CIDOC CRM Web site at http://cidoc.ics.forth.gr/frbr_inro.html and the FRBR Review Group Web site at http://www.ifla.org/VII/s13/wgfrbr/FRBR-CRMdialogue_wg.htm.
8. International Organization for Standardisation, MPEG-7 Overview ISO/IEC JTC1/SC29/WG11 N6828 http://www.chiariglione.org/mpeg/standards/mpeg-7/mpeg-7.htm, 2004.
9. T. Baker, M. Dekkers, R. Heery, M. Patel, and G. Salokhe. What Terms Does Your Metadata Use? Application Profiles as Machine-Understandable Narratives. *Journal of Digital Information 2(2)*, 2(No. 65), 2001.
10. R. Bennett, Chr. Hengel-Dittrich, E. O'neill, and B. Tillett. VIAF (Virtual International Authority File): Linking Die Deutsche Bibliothek and Library of Congress Name Authority Files. In *WLIC2006*, Seoul Korea, 2006.
11. P. Bonora, Ch. Ossicini, and G. Raffa. From Relational Metadata Standards to CRM Ontology: a Case Study in Performing Arts. In *Proc. of CIDOC2006*, Gothenburg, 2006.
12. J.M. Bower and M. et al Baca. *Union List of Artist Names - A User's Guide to the Authority Reference Tool, Version 1.0, Getty Art Information Program.* G.K. Hall, New York, 1994.
13. A. Cali. Reasoning in Data Integration Systems: Why LAV and GAV Are Siblings. In *Proc. of ISMIS 2003*, Lecture Notes in Computer Science 2871, pages 562–571. Springer, 2003.
14. D. Calvanese, G. Giacomo, M. Lenzerini, D. Nardi, and R. Rosati. Description Logic Framework for Information Integration. In *Proc. of the 6th International Conference on the Principles of Knowledge Representation and Reasoning (KR'98)*, pages 2–13., Trento, Italy, 1998.
15. M. Clauss. Epigraphic Data Bank http://www.manfredclauss.de/gb/index.html. Clauss/Slaby, 2003.
16. N. Crofts, M. Doerr, T. Gill, S. Stead, and M. Stiff. Definition of the CIDOC Conceptual Reference Model. Version 4.2 http://cidoc.ics.forth.gr/docs/cidoc_crm_version_4.2.doc, 2005.
17. A. D'Andrea and G. Ferrandino. Shared Iconographical Representations with Ontological Models. In *Proc. of Computer Applications and Quantitative Methods in Archaeology Conference, CAA2007*, Berlin, Germany, 2-6 April 2007. CAA.

18. M. Doerr. Mapping of the Dublin Core Metadata Element Set to the CIDOC CRM. Technical Report TR-274, FORTH-ICS, July 2000.
19. M. Doerr. The CIDOC CRM - An Ontological Approach to Semantic Interoperability of Metadata. *AI Magazine*, 24(3), 2003.
20. M. Doerr. Modelling Learning Subjects as Relationships. *Intuitive Human Interface*, LNAI 3359:200–214, 2004.
21. M. Doerr, J. Hunter, and C. Lagoze. Towards a Core Ontology for Information Integration. *Journal of Digital information*, 4(1)(169), 2003.
22. M. Doerr and D. Kalomoirakis. A Metastructure for Archeological Terminology. In *Proc. of the Computer Applications and Quantitative Methods in Archaeology Conference (CAA 2000)*, Ljubljana, Slovenia, April 18-21 2000.
23. M. Doerr, G. Markakis, M. Theodoridou, and M. Tsikritzis. DIATHESIS: OCR based semantic annotation of newspapers. In *Proc. of the third SEEDI International Conference: Digitization of cultural and scientific heritage*, Cetinje, Montenegro, September 13-15 2007.
24. M. Doerr and A. Sarris, editors. *CAA2002, The Digital Heritage of Archaeology, Proceedings of the 30th Conference*, Heraklion, Crete, April 2003. Hellenic Ministry of Culture.
25. M. Doerr, K. Schaller, and M. Theodoridou. Integration of Complementary Archaeological Sources. In *Computer Applications and Quantitative Methods in Archaeology Conference*, Prato, Italy, April 2004.
26. Ch. Fellbaum, editor. *WordNet: An Electronic Lexical Database.* MIT Press, 1998.
27. J.-Cl. Gardin. The structure of archaeological theories. *Studies in Modern Archaeology Vol. 3: Mathematics and Information Science in Archaeology: A Flexible Framework*, 3:7–25, 1990.
28. P. Gorgels, L. Aroyo, Y. Wang, R. Brussee, l. Rutledge, and N. Stash. Personalized Museum Experience: The Rijksmuseum Use Case. In *Proc. Museums and the Web 2007 Conference*, Toronto, Canada, 2007. Archives & Museums Informatics.
29. A. Grant, editor. *SPECTRUM - The Museum Documentation Standard.* The Museum Documentation Association, 1994.
30. A. Grant, J. Nieuwenhuis, and T. Petersen, editors. *International guidelines for museum object information: the CIDOC information categories.* 1995.
31. M. Greengrass. E-science Challenges in the World of Historical Studies, 2006.
32. T.-R. Gruber. Toward Principles for the Design of Ontologies Used for Knowledge Sharing. *International Journal of Human-Computer Studies*, 43((5-6)):907–928, 1995. Special issue on Formal Ontology in Conceptual Analysis and Knowledge Representation.
33. N. Guarino. Formal Ontology in Information Systems. In *FOIS'98*, pages 3–15, 1998.
34. M. Guercio, J. Barthelemy, and A. Bonardi. Authenticity Issue in Performing Arts using Live Electronics. In *Proceedings of the 4th Sound and Music Computing Conference (SMC07)*, pages 226–229, Lefkada, Greece, 2007.
35. G. Guizzardi, H. Herre, and G. Wagner. On the General Ontological Foundations of Conceptual Modeling. In *Proc. Conceptual Modeling - ER 2002: 21st International Conference on Conceptual Modeling*, volume 15, pages 65–78, Tampere, Finland, 2002. Springer-Verlag Berlin Heidelberg.
36. P. Harpring. Proper Words in Proper Places: The Thesaurus of Geographic Names. *MDA Information, 2(3), 5-12.*, 2(3):5 12, 1997.

37. R. Heery and M. Patel. Application Profiles: Mixing and Matching Metadata Schemas. *Ariadne*, (Issue 25), 2000.
38. S. Hermon and F. Niccolucci. A Fuzzy Logic Approach to Typology Archaeological Research. In *CAA 2002. The Digital Heritage in Archaeology. Computer Applications and Quantitative Methods in Archaeology*, pages 307–310, Heraklion, Crete, April, 2002 2002. Proceedings of the 30th Conference, CAA.
39. I. Hodder. *The Archaeological Process: An Introduction.* Blackwell Publishers, 1999.
40. J. Hunter. Combining the CIDOC/CRM and MPEG-7 to Describe Multimedia in Museums. *Museums and the Web*, 2002.
41. J. Hunter. *The Role of Information Technologies in Indigenous Knowledge Management*, chapter Chapter 9. Indigenous knowledge and libraries. Australian Academic & Researchers Libraries., 2005.
42. E. Hyvonen, E. Makela, M. Salminen, A. Valo, K. Viljanen, S. Saarela, M. Junnila, and S. Kettula. Museumfinland Finnish Museums on the Semantic Web. *Journal of Web Semantics*, 3((2)):224–241, 2005.
43. C. Kakali, I. Lourdi, Th. Stasinopoulou, L. Bountouri, C. Papatheodorou, M. Doerr, and M. Gergatsoulis. Integrating Dublin Core Metadata for Cultural Heritage Collections Using Ontologies. In *Proceeding of 10th International Conference on Asian Digital Libraries*, Hanoi, Vietnam, 10-13 December 2007. To appear.
44. M. Keynes. Ontology-driven Event Recognition on Stories. Technical report, KMI-TR-135, Knowledge Media Institute, 2003.
45. C. Lagoze. Business Unusual: How "event awareness" may breathe life into the catalog. In *Bicentennial Conference on Bibliographic Control in the New Millennium: Confronting the Challenges of Networked Resources and the Web*, Washington, DC., 2000.
46. C. Lagoze and J. Hunter. The ABC Ontology and Model. *Journal of Digital Information 2(2)*, (Article No. 77), 2001.
47. P. LeBoeuf. Frbr and Further. *Cataloging & Classification Quarterly*, 32((4)): 15–22, 2001.
48. P. LeBoeuf. *Functional Requirements for Bibliographic Records (Frbr): Hype or Cure-All.* Haworth Press, Inc, 2005.
49. P. LeBoeuf, Ch. Lahanier, G. Aitken, P. Sinclair, P. Lewis, and K. Martinez. Integrating Museum & Bibliographic Information: The SCULPTEUR Project. In *Proc. ICHIM 2005 Conference*, Paris, 2005.
50. J. Lehmann, S. Borgo, C. Masolo, and A. Gangemi. Causality and Causation in DOLCE. In Vieu L. (eds) Varzi, A.C., editor, *Formal Ontology in Information Systems, Proceedings of the International Conference FOIS 2004*, pages 273–284, Torino, November 4-6 2004. IOS Press Amsterdam.
51. A. Magkanaraki, S. Alexaki, V. Christophides, and D. Plexousakis. Benchmarking RDF schemas for the Semantic Web. In *Proc. First International Semantic Web Conference on the Semantic Web*, number 12, pages 132–146, Sardinia, Italy, 2002. Springer Berlin/Heidelberg.
52. J. Makkonen and H. Ahonen-Myka. Utilizing Temporal Information in Topic Detection and Tracking. In *Proc. of the 7th European Conference on Research and Advanced Technology for Digital Libraries (ECDL 2003)*, pages 393–404, Trondheim, Norway, 2003. Springer.
53. A. Maple. Faceted Access: A Review of the Literature, 1995.

54. C. Masolo, S. Borgo, A. Gangemi, N. Guarino, and A. Oltramari. Ontology Library (final). IST Project 2001-33052 WonderWeb: Ontology Infrastructure for the Semantic Web. Deliverable D18., 2003.
55. J. McCarthy. *Programs with Common Sense*, pages 403–418. Semantic Information Processing. MIT Press, Minsky, M. (ed). edition, 1969.
56. C. Meghini, M. Doerr, and N. Spyratos. Sharing Co-reference Knowledge for Data Integration. *DELOS Network of Excellence*, 2007. In Second DELOS Conference - Working Notes. Pisa, Italy.
57. A. Mikheev, M. Moens, and C. Grover. Named Entity Recognition without Gazetteers. In *EACL*, pages 1–8, 1999.
58. Boston Museum of Fine Arts. CAMEO: Conservation & Art Material Encyclopedia `http://cameo.mfa.org/`.
59. J. Mylopoulos, A. Borgida, M. Jarke, and M. Koubarakis. Telos: Representing Knowledge about Information Systems. *ACM Transactions on Information Systems*, 8((4)):325–362, 1990.
60. International Council of Museums ICOM. ICOM Definition of a Museum `http://icom.museum/statutes.html#2`, 2001.
61. International Council of Museums ICOM. ICOM Code of Ethics for Museums `http://icom.museum/ethics.html`, 2006.
62. M. Patel, T. Koch, M. Doerr, and Chr. Tsinaraki. D5.3.1: Semantic Interoperability in Digital Library Systems. Deliverable, Project no.507618, DELOS, A Network of Excellence on Digital Libraries, June 2005.
63. JL. Peterson. *Petri net theory and the modeling of systems.* Prentice-Hall, 1981.
64. Oxford University Press, editor. *Art & Architecture Thesaurus.* Getty Trust Publications, 1994.
65. SR. Ranganathan. *A descriptive account of Colon Classification.* Bangalore: Sarada Ranganathan Endowment for Library Science., 1965.
66. PA. Reed. CIDOC Relational Data Model. A Guide. ICOM/CIDOC, 1995.
67. V. Roux and P. Blasco. Logicisme et format scd: dune epistemologie pratique a de nouvelles pratiques editoriales hermes. 2004. CNRS - editions.
68. K.G Saur, editor. *IFLA Study Group on the functional requirements for bibliographic records. Functional requirements for bibliographic records: final report,* `http://www.ifla.org/VII/s13/frbr/frbr.htm`, volume vol. 19 of *New Series.* UBCIM Publications, Munich, 1998.
69. E. Schulten, H. Akkermans, G. Botquin, M. Doerr, N. Guarino, N. Lopes, and N. Sadeh. Call for Participants: The E-CommerceProduct Classification Challenge. *IEEE Intelligent System*, 16((4)):86–c3., 2001.
70. The SHIC Working Party. *Social, Historical and Industrial Classification (2nd edition).* The Museum Documentation Association, Cambridge, UK, 1993.
71. B. Smith. Ontology. *The Blackwell Guide to the Philosophy of Computing and Information*, pages 155–166, 2003. Floridi, L. (ed). Oxford: Blackwell.
72. B. Smith. Against Idiosyncrasy in Ontology Development. In B. Bennett and Fellbaum C., editors, *Formal Ontology in Information Systems (FOIS 2006)*, pages 15–26, Amsterdam, 2006. IOS Press.
73. DA. Smith and G. Crane. Disambiguating Geographic Names in a Historical Digital Library. In *Proc. of the 5th European Conference on Research and Advanced Technology for Digital Libraries (ECDL 2001)*, pages 127–136, Darmstadt, Germany, 2001. Springer.

74. D. Tudhope and C. Binding. A Case Study of a Faceted Approach to Knowledge Organisation and Retrieval. *DigiCULT, Resource Discovery Technologies for the Heritage Sector*, (Thematic Issue 6):2833, 2004.
75. D. Tudhope, C. Binding, D. Blocks, and D. Cunliffe. Query Expansion via Conceptual Distance in Thesaurus Indexed Collections. *Journal of Documentation*, 62((4)):509–533, 2006. Emerald Group Publishing Limited.
76. M. vanGendt, A. Isaac, L. vanderMeij, and S. Schloback. Semantic Web Techniques for Multiple Views on Heterogeneous Collections: A Case Study. In *Proc. of the 10th European Conference on Research and Advanced Technology for Digital Libraries (ECDL 2006)*, pages 426–437, Alicante, Spain, 2006. Springer.
77. M. Vargas-Vera and D. Celjuska. Event Recognition on News Stories and Semi-Automatic Population of an Ontology. *Web Intelligence*, pages 615–618, 2004.
78. SM. Vinas. *Contemporary Theory of Conservation.* Elsevier Oxford, 2005.

Part IV

Infrastructures for Ontologies

RDF Storage and Retrieval Systems

Alice Hertel, Jeen Broekstra, and Heiner Stuckenschmidt

[1] Fraunhofer Institute for Information and Data Processing, Fraunhoferstr. 1, 76131 Karlsruhe, Germany, alice.hertel@iitb.fraunhofer.de
[2] Technische Universiteit Eindhoven, P.O. Box 513, 5600 MB Eindhoven, The Netherlands, j.broekstra@tue.nl
[3] University of Mannheim, Schloss, 68131 Mannheim, Germany, heiner@informatik.uni-mannheim.de

Summary. Ontologies are often used to improve data access. For this purpose, existing data has to be linked to an ontology and appropriate access mechanisms have to be provided. In this chapter, we review RDF storage and retrieval technologies as a common approach for accessing ontology-based data. We discuss different storage models, typical functionalities of RDF middleware such as data model support and reasoning capabilities and RDF query languages with a special focus on SPARQL as an emerging standard. We also discuss some trends such as support for expressive ontology and rule languages.

1 Introduction

It is widely acknowledged that information access can benefit from the use of ontologies. For this purpose, available data has to be linked to concepts and relations in the corresponding ontology and access mechanisms have to be provided that support the integrated model consisting of ontology and data. The most common approach for linking data to ontologies is via an RDF representation of available data that describes the data as instances of the corresponding ontology that is represented in terms of an RDF Schema (compare chapter "Resource Description Framework"). Due to the practical relevance of data access based on RDF and RDF Schema, a lot of effort has been spent on the development of corresponding storage and retrieval infrastructures.

In this chapter, we summarize the state of the art with respect to existing storage and retrieval technologies for RDF data. In particular, we first review the general architecture of RDF infrastructures that normally consist of a storage and a middleware layer. We discuss important aspects of these layers covering different storage formats for RDF data, common middleware functionalities such as RDF Schema reasoning and basic operations for data access and manipulation. Throughout the chapter, we discuss these aspects on

S. Staab and R. Studer (eds.), *Handbook on Ontologies*, International Handbooks on Information Systems, DOI 10.1007/978-3-540-92673-3,

a general level and only point to particular systems to provide examples of concrete implementations. We further discuss RDF query languages as the most common interface for interacting with ontology-based RDF data and present the SPARQL language in more detail. In Sect. 7 we also provide a very brief overview of existing approaches to extend RDF storage and retrieval systems to support more complex ontology languages than RDF Schema. We close with a discussion of current trends and speculate about future developments.

2 Architecture of RDF Stores

An RDF store allows storage of RDF data and schema information, and provides methods to access that information. Thus, the two primary components of an RDF store are a repository and a middleware that builds on top of that repository. The middleware can be further divided into components as the access methods can be categorized into methods for adding, deleting, querying and exporting data. To describe the different components in detail, we assume a layered architecture as proposed in [1] and regard the layers from the bottom up.

Different repositories are imaginable, e.g. main memory, files or databases, but the access methods should remain the same. Thus, it is reasonable to encapsulate the access to the repository in an own layer, which provides well-defined interfaces to the upper layers and can be exchanged if another repository is used. The inference support also resides in this layer as close to the repository as possible. Sesame [1] implements such a layer and calls it the Storage And Inference Layer (SAIL).

The above mentioned access methods are located on a higher level and address the interfaces of the SAIL (or directly address the repository if there is no SAIL implementation). According to the different requirements of each access method they can be realized in different modules: The *admin module* provides the functionality for adding new data to and deleting data from the RDF store. Especially when loading data from files this requires parsing and validating RDF, so an RDF parser and an RDF validator are usually part of the admin module. The *query module* handles queries to the RDF store. As these queries can be formulated in any kind of RDF query language, several query modules may be necessary, each implementing a parser and handler for one query language. Finally, the *export module* allows a dump of the RDF store into files for data exchange with other systems.

These modules can be accessed locally or remotely, e.g. using SOAP or RMI. This is why the highest layer in the middleware contains protocol handlers that can manage different access modes. Figure 1 shows the generic architecture as proposed in [1]. This architecture is not only valid for Sesame – other RDF store implementations have a similar modular structure reflecting the different aspects of an RDF store.

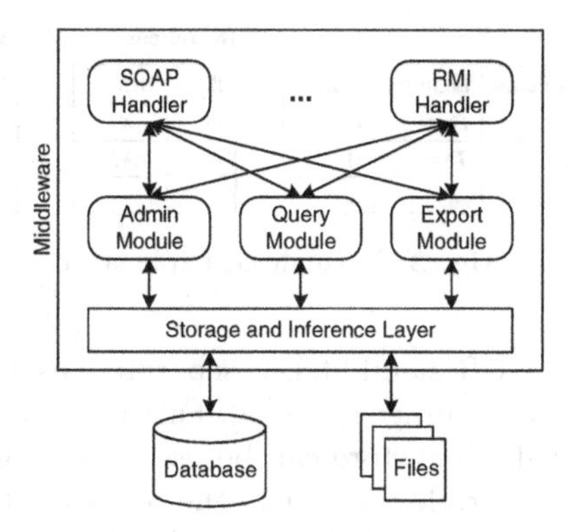

Fig. 1. Generic architecture of an RDF store (Sesame)

3 Storing RDF Data

RDF schemas and instances can be efficiently accessed and manipulated in main memory. For persistent storage the data can be serialized to files, but for large amounts of data the use of a database management system is more reasonable. Examining currently existing RDF stores we found that they are using relational and object-relational database management systems (RDBMS and ORDBMS).

Storing RDF data in a relational database requires an appropriate table design. There are different approaches that can be classified in generic schemas, i.e. schemas that do not depend on the ontology, and ontology specific schemas. In the following we describe the most important table designs showing their advantages and shortcomings.

3.1 Generic Schemas

The most simple generic schema is the triple store with only one table required in the database. The table contains three columns named `Subject`, `Predicate` and `Object`, thus reflecting the triple nature of RDF statements. This corresponds to the *vertical representation* for storing objects in a table in [2].

The greatest advantage of this schema is that no restructuring is required if the ontology changes. Adding new classes and properties to the ontology can be realized by a simple `INSERT` command in the table. On the other hand, performing a query means searching the whole database and queries involving joins become very expensive. Another aspect is that the class hierarchy cannot be modeled in this schema, what makes queries for all instances of a class rather complex.

The triple store can be used in its pure form [3], but most existing systems add several modifications to improve performance or maintainability. A common approach, the so-called *normalized triple store*, is adding two further

Triples:

Subject	Predicate	IsLiteral	Object
r1	*r2*	*False*	*r3*
r1	*r4*	*True*	*l1*
...	...	...	...

Resources:

ID	URI
r1	*...#1*
r2	*...#2*
...	...

Literals:

ID	Value
l1	*Value1*
...	...
...	...

Fig. 2. Normalized triple store

tables to store resource URIs and literals separately as shown in Fig. 2, which requires significantly less storage space [4]. Furthermore, a hybrid of the simple and the normalized triple store can be used, allowing to store the values themselves either in the triple table or in the resources table [5].

In a further refinement, the `Triples` table can be split horizontally into several tables, each modeling an RDF(S) property:

- `SubConcept` for the `rdfs:subClassOf` property, storing the class hierarchy
- `SubProperty` for the `rdfs:subPropertyOf` property, storing the property hierarchy
- `PropertyDomain` for the `rdfs:domain` property, storing the domains and cardinalities of properties
- `PropertyRange` for the `rdfs:range` property, storing the ranges of properties
- `ConceptInstances` for the `rdf:type` property, storing class instances
- `PropertyInstances` for the `rdf:type` property, storing property instances
- `AttributeInstances` for the `rdf:type` property, storing instances of properties with literal values

These tables only need two columns for `Subject` and `Object`. The table names implicitly contain the predicates. This schema separates the ontology schema from its instances, explicitly models class and property hierarchies and distinguishes between class-valued and literal-valued properties [1,6].

3.2 Ontology Specific Schemas

Ontology specific schemas are changing when the ontology changes, i.e. when classes or properties are added or removed. The basic schema consists of one table with one column for the instance ID, one for the class name and one for each property in the ontology. Thus, one row in the table corresponds to one instance. This schema is corresponding to the *horizontal representation* in [2] and obviously has several drawbacks: large number of columns, high sparsity, inability to handle multi-valued properties and the need to add columns to the table when adding new properties to the ontology, just to name a few.

Horizontally splitting this schema results in the so called one-table-per-class schema: one table for each class in the ontology is created. A class table provides columns for all properties whose domain contains this class. This is tending to the classic entity-relationship-model in database design and benefits queries about all attributes and properties of an instance.

However, in this form the schema still lacks the ability to handle multi-valued properties, and properties that do not define an explicit domain must then be included in each table. Furthermore, adding new properties to the ontology again requires restructuring existing tables.

Another approach is vertically splitting the schema, what results in the one-table-per-property schema, also called the *decomposition storage model*. In this schema one table for each property is created with only two columns for subject and object. RDF(S) properties are also stored in such tables, e.g. the table for `rdf:type` contains the relationships between instances and their classes.

This approach is reflecting the particular aspect of RDF that properties are not defined inside a class. However, complex queries considering many properties have to perform many joins, and queries for all instances of a class are similarly expensive as in the generic triple schema.

In practice, a hybrid schema combining the table-per-class and table-per-property schemas is used to benefit from the advantages of both of them. This schema contains one table for each class, only storing there a unique ID for the specific instance. This replaces the modeling of the `rdf:type` property. For all other properties tables are created as described in the table-per-property-approach (Fig. 3) [7]. Thus, changes to the ontology do not require changing existing tables, as adding a new class or property results in creating a new table in the database.

A possible modification of this schema is separating the ontology from the instances. In this case, only instances are stored in the tables described above. Information about the ontology schema is stored separately in four additional tables `Class`, `Property`, `SubClass` and `SubProperty` [8]. These tables can be further refined storing only the property ID in the `Property` table and the domain and range of the property in own tables `Domain` and `Range` [1]. This approach is similar to the refined generic schema, where the ontology is stored the same way and only the storage of instances is different.

To reduce the number of tables, single-valued properties with a literal as range can be stored in the class tables. Adding new attributes would then require to change existing tables. Another variation is to store all class instances in one table called `Instances`. This is especially useful for ontologies where there is a large number of classes with only few or no instances [8].

ClassA:

ID
...#1
...

Property1:

Subject	Object
...#1	*...#3*
...	...

ClassB:

ID
...#3
...

Fig. 3. Hybrid schema

3.3 Further Issues

There are further issues that may require an extension of the triple-based schemas and thus are affecting the design of the database tables:

- Storing multiple ontologies in one database
- Storing statements from multiple documents in one database

Both points are concerning the aspect of provenance, which means keeping track of the source an RDF statement is coming from.

When storing multiple ontologies in one database it should be considered that classes, and consequently the corresponding tables, can have the same name. Therefore, either the tables have to be named with a prefix referring to the source ontology [7] or this reference is stored in an additional attribute for every statement. A similar situation arises for storing multiple documents in one database. Especially, when there are contradicting statements it is important to know the source of each statement. Again, an additional attribute denoting the source document helps solving the problem [7].

The concept of named graphs [9] is including both issues. The main idea is that each document or ontology is modeled as a graph with a distinct name, mostly a URI. This name is stored as an additional attribute, thus extending RDF statements from triples to so-called quads. For the database schemas described above this means adding a fourth column to the tables and potentially storing the names of all graphs in a further table.

3.4 Object-Oriented Features

Current ORDBMS provide the *subtable* facility which allows for a better modeling of the subclass and subproperty relationships. The table of a subclass is then created as a subtable of the superclass table. Consequently, querying for all instances of a class does not require searching for all triples with the `rdfs:subClassOf` property or looking up a `SubClass` table. However, this feature should be used carefully, as a new subtable can only be added at the bottom of the hierarchy. Otherwise, the complete table hierarchy needs to be rebuilt [1,8].

Oracle[1] offers another object-relational feature: an own datatype to store RDF based on a graph data model. RDF triples can be persisted, indexed and queried, similar to other object-relational data types.

Although the RDF model has several object-oriented characteristics and most RDF stores are internally working with an object model, approaches to store RDF data and schema information using object database management systems (ODBMS) are rarely known. (Object-)Relational databases are still predominant, when large amounts of data have to be persisted on a server,

[1] See `http://download-east.oracle.com/otndocs/tech/semantic_web/pdf/rdfrm.pdf`

and object databases did not and will most probably not replace them. However, new developments of ODBMS may show some advantages over RDBMS in certain applications, e.g. for embeddable persistence solutions in mobile devices. This is why storing ontologies in an ODBMS is worth a closer look.

4 RDF Middleware

What we call RDF middleware is the layer implementing the access to the physical RDF data store. Besides an inference mechanism, the access layer should provide functions for creating, querying and deleting data in the store. While adding data requires parsing and ideally a validation of the incoming RDF sentences, querying the RDF store needs the implementation of some kind of query language as well as an interpretation and a translation of this query language into calls to the physical RDF storage. Another important feature of this layer is the possibility to export data to a file for exchange with other systems.

4.1 Inference for RDF

Inference for RDF is specified by the RDF(S) entailment rules described in [12]. The practice-relevant rules can be roughly divided into the two following groups:

- Inferring the transitive closure for the properties `rdfs:subClassOf` and `rdfs:subPropertyOf`
- Inferring class memberships analysing the use of properties and their domains and/or ranges

One approach is to compute the transitive closure using a recursive algorithm and to store it in database views. This algorithm constructs a view for each class, starting with the class table and adding the views of all of its subclasses examining the statements with the `rdfs:subClassOf` relationship in the database. Analogously, a view for each property is constructed from the `rdfs:subPropertyOf` relationships. A similar algorithm can be used to infer class membership from the properties of an instance [7].

An alternative is to use a production rule system that generates new facts from existing ones by forward chaining or applies backward chaining on a query presented to the system. This brings up an important aspect of the inference, namely the time, when the inference is executed. There are two possibilities:

- Inference in advance (*eager evaluation*)
- Inference at query runtime (*lazy evaluation*)

The eager evaluation is computing the deductive closure in advance, so the time to evaluate a query is reduced [1]. However, it also may cause a dramatic increase of the amount of stored data, potentially generating entailments that are rarely matching queries. In contrast, the lazy evaluation only performs evaluation of entailments matched by a given query, so no unused entailments need to be generated and stored. This significantly increases the query processing time. A compromise is to use both methods: those entailment rules that generate fewer entailments are evaluated in advance, while those requiring more storage space and less evaluation time are evaluated at query time [4].

Although we describe the inference mechanism as part of the middleware, the algorithms can also be defined as stored procedures in the database, leaving the inference task to the database management system. This depends on the capabilities of the DBMS used for storing the data.

4.2 Querying Data

For formulating a query to the RDF store there are several approaches:

- Implementing a proprietary query API
- Implementing a query language

Proprietary query APIs are defining their own query format. E.g. DLDB [7] is using conjunctive queries composed of atoms whose structure is based on First Order Logic. Constructing an SQL query is done through a translation algorithm that substitutes predicates and variables by table and column names. Another possibility is to create an own query language, e.g. KAON Query [6] or SeRQL [1].

Most RDF stores are using one of the common RDF query languages like RQL, RDQL or SPARQL [1, 4, 8]. This means implementing a parser that analyses the syntax of this query language. A potential intermediate step is to translate the parsed query into relational calculus [4], a graph [8] or the object model [1] to capture the query semantics. After that, the SQL query sentence is formed and sent to the database.

The syntax of the created SQL query usually depends on the underlying DBMS. This is why the implementation of an additional intermediate layer is reasonable that abstracts from the actual storage mechanism offering storage and retrieval functions. An example for that is the Storage And Inference Layer (SAIL) in Sesame. The layer can be exchanged according to the used DBMS and can even be placed on top of another SAIL to offer further functionality like caching recent query results [1].

An important aspect for accessing data is *query optimization*. It can be left to the database system, considering the sophisticated evaluation and optimization mechanisms of modern RDBMS. So, the query must be translated to SQL as completely as possible. This is the approach used by most RDF stores. Another approach is optimizing the query in the middleware itself, which is particularly interesting if the query engine should be independent of the underlying storage like in Sesame. Here, the query is translated into a set of SQL queries and joins or other operations are performed in the query engine. This not only requires an optimization strategy but also implies a transaction management, because one RQL query can result in multiple SQL queries and the state of the database must not change until all these queries are executed [1].

4.3 Adding, Deleting and Exporting Data

Adding data to the RDF store can be realized by creating new concepts, properties or instances in main memory using the API and then calling a function to store them into the knowledge base [5, 6]. Another possibility is reading RDF data from a file or an online source, which is implemented by all RDF stores as it is important for loading an ontology. Reading RDF data requires an RDF parser for reading in the statements and mapping them on the object model or directly on the database schema. Most systems use a parser that reads the RDF/XML notation, e.g. the ARP (Another RDF Parser), which is part of the Jena toolkit, or the Raptor RDF parser. Optionally, an RDF validator can be used to check the incoming data for correctness and for compliance with already loaded schemas [8]. In this case, the schemas should be loaded before the instances.

Delete operations in RDF stores have to be handled very carefully. While completely clearing the store is a quite simple function, deleting single statements can entail the deletion of other related statements. This not only requires recomputing the deductive closure for the RDF store, but also a mechanism for truth maintenance. Hence, deletions become quite costly [1].

To exchange data with other systems an export mechanism is required. Most RDF stores implement such an export function which allows to serialize the ontology and instance data from the RDF store into a file. The common formats for serializing RDF are N-Triples, N3 notation and RDF/XML notation.

5 RDF Query Languages

As mentioned in the preceding section, the use of query languages is the most common way of interacting with an RDF store. Many query languages already exist that could, in principle, be used to interact with RDF data. The most obvious example is SQL, the standard query language for relational databases. In this section, we will explore what properties a query language for semistructured data, and in particular for RDF, should have, and what the difference is with existing approaches such as SQL. We will then discuss several proposals for query languages. In particular, we will describe the SPARQL query language in more detail.

5.1 General Properties of Query Languages

We can identify several general properties with which one can characterize query languages. Here, we name six such properties:

- *Expressiveness:* Expressiveness indicates how powerful queries can be formulated in a given language. Ideally, a query language should be expressive enough to allow the retrieval of any arbitrary combination of values from the queried model, that is, be *complete* with respect to its datamodel. Usually, expressiveness is restricted to maintain other properties such as safety and to allow an efficient (and optimizable) execution of queries.
- *Closure:* The closure property requires that the results of an operation are again elements of the data model. This means that if a query language operates on a graph data model, the query results would again have to be graphs.
- *Adequacy:* A query language is called adequate if it uses all concepts of the underlying data model. This property therefore complements the closure property: For the closure, a query result must not be outside the data model, for adequacy the entire data model needs to be exploited.
- *Orthogonality:* The orthogonality of a query language requires that all operations may be used independently of the usage context.
- *Safety:* A query language is considered safe, if every query that is syntactically correct returns a finite set of results (on a finite data set). Typical concepts that cause query languages to be unsafe are recursion and negation.

5.2 Path Expressions

One of the main distinguishing features of query languages for semi-structured data is their ability to reach to arbitrary depths in the data graph. To do this, these languages all use the notion of path expressions. A path expression is a simple query, the result of which, for a given data graph, is a set of nodes. For example, consider the following bit of XML:

```
<?xml version="1.0"?>
<body>
  This page is written by
  <author>Jeen Broekstra</author>.
  <location>
    His tel.nr. at work is <tel>3686</tel>,
    his number at home is <tel>555722</tel>, and his
    room number is <room>HG7.76</room>.
  </location>
</body>
```

The result of the path expression `body.location.tel` would be the set of nodes with the associated values "3,686", "555,722".

Many useful regular expressions can be used in path expressions to facilitate more complex expressions than just specification of the complete path. For example, a regular expression `location|name` specifies either a `location` node or a `name` node. Another very useful pattern is the wildcard, which matches any node label. Using the symbol to express this, `body.tel` matches any path consisting of a `body` node followed by any node, followed by a `tel` node. Also, closure operations, like arbitrary repeats of a regular expression can be used. For example, `body*.tel` specifies the set of `tel` nodes that occur at arbitrary depth within the `body` node. At another level of abstraction, regular expressions can also be used to express matches on the actual string format of labels. For example the regular expression `body."[aA]uthor"` matches any `author` node within the `body`, possibly with the first letter capitalized.

Path expressions, although they are an essential feature of query languages for semistructured data, can only return a subset of nodes in the database. They can not construct new nodes, perform joins, or test values stored in the database. In other words: path expressions are necessary but not sufficient for a good query language on semistructured data. A query language that lacks path expressions cannot be considered adequate, nor sufficiently expressive for querying semistructured data.

5.3 Why not just SQL?

For strictly relational data (as opposed to semistructured data), SQL is by far the most widely supported query language, including support for large data storage, efficient indexing schemes, query optimizers, etc. It would therefore be attractive if we could use this robust and widely available technology for our purposes of querying semistructured data. Unfortunately, this can only be done at the cost of a very large gap between the data model in the repository (e.g. RDF) and the data-model on which the query language is based (the relational model).

To exemplify this, let us look at how the scenario would look for an XML implementation in a relational database: as a first step, we would have to encode the XML data model in the relational model. This would be possible by assigning each node in an XML tree a unique identifier, with each entry in the relational database linking such a node with all its descendants and attributes. The problems start when we want to use this as the basis for querying the XML structure: each XML query should be compiled into an SQL query on the underlying relational tables. Typically, a single XML query (such as: "return all descendants of a given node") must be compiled into a complicated set of SQL queries. It is not even clear whether a finite set of SQL queries could be generated for every reasonable XML query.

Although perhaps attractive as a short term solution, we feel that in the long run this is not an appropriate solution. Rather, techniques for large data storage, indexing schemes, query optimizers, etc., should be provided for the native data model (be it XML or RDF), instead of relying on these techniques for a completely different data model.

5.4 Querying RDF

RDF documents and RDF schemata can be considered at three different levels of abstraction:

1. At the *syntactic level* they are XML documents.
2. At the *structure level* they consist of a set of RDF triples.
3. At the *semantic level* they constitute one or more graphs with partially predefined semantics.

We can query these documents at each of these three levels. We will briefly consider the pros and cons of doing so for each level in the next few sections.

Querying at the Syntactic Level

As we have seen previously, RDF models can be written down in XML notation. It would therefore seem reasonable to assume that we can query RDF using an XML query language (e.g. XQuery[2]). However, this approach disregards the fact that RDF is not just an XML notation but has its own data model that is different from the XML tree structure: whereas XML is an ordered, node-labeled tree structure, RDF is an unordered, node- and edge-labeled graph structure. XML querying techniques have no functionality for dealing with differentiating between node and edge labels or with the absence of order or a tree root. Moreover, relationships in the RDF data that are not immediately apparent from the XML tree structure become very hard to query in this approach.

Querying at the Structure Level

When we abstract from the syntax any RDF document represents a set of triples, each triple representing a statement of the form subject-predicate-object. A number of query languages have been proposed and implemented that regard RDF documents as sets of such triples, and that allow to query such a triple set in various ways.

However, querying at this level means that we now interpret any RDF model *only* as a set of triples, including those elements which have been given special semantics in RDF Schema. For example, the fact that `rdfs:subClassOf` is a transitive relation is ignored at this level.

Querying at the Semantic Level

When we consider RDF models at the semantic level we query the full knowledge of everything that the RDF model entails, and not just those facts that happen to be represented explicitly.

[2] See `http://www.w3.org/TR/xquery/`

There are at least two options to achieve this goal:

1. Compute and store the deductive closure of a graph as a basis for querying.
2. Let a query processor infer new statements as needed per query.

While the choice of an RDF query language is, in principle, independent of the choice made in this respect, the fact remains that most RDF query languages have been designed to query a simple triple store and have no specific functionality or semantics to discriminate between data and schema information.

5.5 SPARQL

The SPARQL Query Language [23] is a W3C Candidate Recommendation for querying RDF, and as such is fast becoming the standard query language for this purpose. In September 2006, almost all major RDF query tools have begun implementing support for the SPARQL query language. Even though other query languages (e.g. SeRQL [1], RQL [24], RDQL [25]) have existed longer and have a more mature implementation base and a more expressive feature set, they typically are supported by only one or two tools, hindering interoperability. Several surveys and comparative analyses of these different query languages have been published, a fairly comprehensive one can be found in [27]. In this chapter, we will concentrate on the SPARQL query language, giving a brief introduction into its basic usage, highlighting some interesting features. For a formal analysis of the semantics and complexity of the SPARQL language, we recommend reading [26].

Basic Queries

The SPARQL query language is based on matching graph patterns. The simplest graph pattern is the triple pattern, which is like an RDF triple, but with the possibility of a variable instead of an RDF term in the subject, predicate or object positions. Combining triple patterns gives a basic graph pattern, where an exact match to a graph is needed to fulfill a pattern.

As a simple example, consider the following query:

```
PREFIX rdf: <http://www.w3.org/1999/02/22-rdf-syntax-ns#>
PREFIX rdfs: <http://www.w3.org/2000/01/rdf-schema#>
SELECT ?c
WHERE
{
    ?c rdf:type rdfs:Class .
}
```

The above query retrieves all triple patterns where the property is `rdf:type` and the object is `rdfs:Class`. In other words, this query, when executed, will retrieve all classes.

Note that like the namespace mechanism we have previously seen for writing down RDF in XML, SPARQL allows us to define prefixes for namespaces and use these in the query pattern, to make queries shorter and easier to read. In the rest of this chapter, we will omit the declaration of the "rdf" and "rdfs" prefixes, for brevity.

To get all instances of a particular class, for example the FOAF vocabulary class "Person", we write:

```
PREFIX foaf: <http://xmlns.com/foaf/0.1/>
SELECT ?i
WHERE
{
  ?i rdf:type foaf:Person .
}
```

SPARQL makes no explicit commitment to support RDFS semantics. Therefore, the result of this query depends on whether or not the system answering the query supports RDFS semantics. If it does, then the result of this query will include all instances of the subclasses of `Person` as well. If it does not support RDFS semantics, then it will only retrieve those instances that are explicitly of type "Person".

Using `Select–From–Where`

As in SQL, SPARQL queries have a SELECT–FROM–WHERE structure:

`SELECT` specifies the *projection*: the number and order of retrieved data.
`FROM` is used to specify the source being queried. This clause is optional; when not specified we can simply assume we are querying the knowledge base of a particular system.
`WHERE` imposes constraints on possible solutions in the form of graph pattern templates and boolean constraints.

For example, to retrieve all e-mail addresses of persons, we can write

```
SELECT ?x ?y
WHERE
{
  ?x foaf:mbox ?y .
}
```

Here `?x` and `?y` are variables, and `?x foaf:mbox ?y` represents a resource-property-value triple pattern.

We can create more elaborate graph patterns to get more complex information from our queries. For example, to retrieve all persons with name "Bob" and their phone numbers, we can write

```
SELECT ?x ?y
WHERE
{
    ?x foaf:name "Bob";
       foaf:mbox ?y .
}
```

Here `?x foaf:name "Bob"` collects all resources which have a name "Bob", as discussed, and binds the result to the variable `?x`. The second pattern collects all triples with predicate `mbox`. There is an *implicit join* here, in that we restrict the second pattern only to those triples, the subject of which is in the variable `?x`. Note that in this case we use a bit of syntax-shortcut as well: we use a semi-column to indicate that the following triple pattern shares its subject with the previous one, so the above query is equivalent to writing down:

```
SELECT ?x ?y
WHERE
{
    ?x foaf:name "Bob" .
    ?x foaf:mbox ?y .
}
```

We demonstrate an *explicit join* by a query that retrieves the name of all persons known by the person with name "Bob".

```
SELECT ?n
WHERE
{
   ?x rdf:type foaf:Person ;
      foaf:name ?n .
   ?c foaf:name "Bob" ;
      foaf:knows ?y .
   FILTER (?x = ?y).
}
```

In SPARQL, we use a FILTER condition to indicate a boolean constraint. In this case, the constraint is the explicit join of the variables `?x` and `?y` by using an equality (=) operator.

Optional Patterns

The graph patterns we have seen so far are mandatory patterns: either the knowledge base matches the complete pattern, in which case an answer is returned, or it does not, in which case the query does not produce a result. However, in many cases we may wish to be more flexible. Consider, for example, the following bit of RDF:

```
<foaf:Person rdf:about="#bob">
 <foaf:name>Bob</foaf:name>
</foaf:Person>

<foaf:Person rdf:about="#alice">
 <foaf:name>Alice</foaf:name>
 <foaf:mbox>alice@example.org</foaf:mbox>
</foaf:Person>
```

As you can see, this fragment contains information on two people. For one person it only lists the name, for the other it also lists the e-mail address. Now, we want to query for all people and their e-mail addresses:

```
SELECT ?name ?email
WHERE
{ ?x rdf:type foaf:Person ;
      foaf:name ?name ;
      foaf:mbox ?email .
}
```

The result of this query would be:

?name	?email
Alice	alice@example.org

So, despite the fact that Bob is listed as a person, the query does not return him: the query pattern does not match because he has no e-mail address.

As a solution we can adapt the query to use an optional pattern:

```
SELECT ?name ?email
WHERE
{ ?x rdf:type foaf:Person ;
       foaf:name ?name .
   OPTIONAL { ?x  foaf:mbox ?email }
}
```

The meaning is roughly "give us all the names of persons, and *if known* also their e-mail address" and the result looks like this:

?name	?email
Bob	
Alice	alice@example.org

This covers the basics of the SPARQL query language. For a full overview of the SPARQL language and an explanation of more advanced features, such as named graphs, we recommend reading the SPARQL specification at `http://www.w3.org/TR/rdf-sparql-query/`.

6 Scalability of RDF Stores

In terms of data storage and retrieval, scalability and performance is a very important issue. The performance of an RDF store depends on various factors: the underlying database system, the database representation of the RDF schema and instances, the efficiency of the query engine, and the performance of the inference engine. A detailed overview of the scalability and performance of different RDF stores would be out of scope of this chapter, but we can mention some interesting points.

Theoharis et al. [10] benchmarked different database representations and provide detailed results for the approaches described in Sect. 3. In this evaluation the ontology specific schema in its hybrid form performs better in terms of query execution times of taxonomic queries than the generic schemas. Although only one sort of queries has been evaluated this shows the weakness of the generic schemas. However, there is always a trade-off between the query execution times and the overhead for ontology evolution and table management: ontology specific schemas suffer from potentially large numbers of tables, and from the need to change the database schema when adding or deleting a class or property in the ontology.

An elaborate method and toolset to evaluate Semantic Web repositories as a whole is the Lehigh University Benchmark (LUBM) [11]. Although it is focussing on OWL applications, the LUBM can be applied to most of the RDF stores mentioned above, but there are only few evaluations available. LUBM provides means to generate a test dataset, several test queries, support for different degrees of reasoning as well as multiple performance metrics for load time, repository size, query response time, and query completeness and soundness.

The W3C maintains a web site recording the size of the largest deployed installations of triple stores.[3] End of February 2008 the site reports a number of systems that have been tested with about one billion statements. The largest data set is reported by the YARS2 System that is claimed to be able to store 7 billion triples generated using the LUBM benchmark.

7 Beyond RDF Schema

While the development of storage and retrieval systems for semantic data so far has been focussed on supporting RDF and RDF Schema there is also an interest in extending available infrastructures to more expressive languages. In particular, supporting more expressive ontology languages such as OWL-Lite and OWL-DL as well as expressive rule languages is a subject of active work. Other activities include the extension of representation and query languages with advanced features such as time [20], preference and uncertainty (e.g. [18, 19, 21]). In the following, we focus on the first kind of activities.

The most straightforward extension of existing RDF infrastructures is a support for ontologies encoded in OWL. As OWL can be serialized in RDF, the corresponding models can be stored in any RDF repository without changing the systems. The structural complexity of the OWL encoding in RDF, especially the high number of blank nodes, however, makes the access to these models rather cumbersome. In order to overcome these problems, many RDF stores use dedicated APIs as part of the middleware layer to support the storage, retrieval and manipulation of OWL ontologies. While some systems such as Jena use their own ontology API, other systems like KAON adopted the proposal for a standardized OWL API described in [22].

Naturally, extensions to more expressive languages do not only aim at providing support at the syntactic level, but also with respect to the semantics of the corresponding languages. As mentioned above, most RDF stores support RDF Schema reasoning on the basis of a specialized set of deduction rules. A common way of extending this fixed schema is to provide support for user defined rule sets. These rule sets can be used for defining parts of the semantics of OWL [16]. Examples of systems supporting OWL-Lite reasoning on the basis of custom rule sets are Sesame, Jena and OWLIM [14]. Besides this, customized rule sets can also be used for capturing domain specific knowledge [15] and for defining efficient subsets of the RDF Schema Semantics for particular applications [17].

[3] `http://esw.w3.org/topic/LargeTripleStores`

An alternative way of supporting OWL semantics is to provide an interface to dedicated Description Logic reasoners (e.g. Racer, FaCT or Pellet) either via specialized data structures or on the basis of the standardized DIG API (`http://dig.sourceforge.net/`). Systems differ in the amount of derivable knowledge that is actually integrated into the RDF model for query answering. The BOR reasoner (`http://www.ontotext.com/bor/`) for example computes the subsumption hierarchy of an OWL ontology and stores the derived subClass relations in the RDF model for further processing. Furthermore, there are some RDF compatible systems that implement expressive rule languages such as KAON2 which implements disjunctive datalog [13] or OntoBroker that implements F-Logic (cf. chapter "Ontologies in F-Logic").

8 Conclusion

After reviewing a number of existing RDF storage and retrieval systems, we can draw some conclusions about the state of the art and general trends in the fields. On the general level, we can say that there is strong convergence of technologies which is documented by the mergence of SPARQL as a standard query language but also in terms of features that are common to different systems. For instance, we can observe that most RDF stores are not really specialized database systems for RDF data but rather an intelligent middleware that wraps existing database technology. Besides providing special support for the graph data model that is characteristic for RDF data, the main functionality provided by this middleware is support for ontological reasoning. An observation that can be made in connection with these two main functions is the fact that almost all systems rely on relational databases that provide very limited support with respect to data model and reasoning. There are very little approaches that try to delegate some of these aspects to the storage model as well by using deductive or object oriented database technologies.

With respect to further development of RDF technologies, we can identify two trends. The first one that was already mentioned in Sect. 7 is the extension of existing systems to more expressive representation languages. In this context, rule languages (compare chapter "Ontologies and Rules") are the most promising candidates because it has been shown that rule-based reasoning has the potential to scale to very large data sets whereas ontological reasoning based on description logics shows serious limitations when large numbers of instances are involved. The other major direction of development concerns the scalability of RDF infrastructures to internet scale. In this context, approaches for distributed RDF processing are becoming more and more important. Both aspects, expressive representation languages and distribution are essential with respect to realizing the vision of the semantic web and are therefore important steps towards real semantic web applications.

References

1. Broekstra J (2005) Storage, querying and inferencing for Semantic Web languages. PhD Thesis, Vrije Universiteit, Amsterdam.
2. Agrawal R, Somani A, Xu Y (2001) Storage and querying of e-commerce data. In: Proceedings of the 27th Conference on Very Large Data Bases, VLDB 2001, Roma, Italy.
3. Oldakowski R, Bizer C, Westphal D (2005) RAP: RDF API for PHP. In: Proceedings of Workshop on Scripting for the Semantic Web, SFSW 2005, at 2nd European Semantic Web Conference, ESWC 2005, Heraklion, Greece.
4. Harris S, Gibbins N (2003) 3store: Efficient bulk RDF storage. In: Proceedings of the 1st International Workshop on Practical and Scalable Semantic Systems, PSSS 2003, Sanibel Island, FL, USA.
5. Jena2 database interface – database layout. http://jena.sourceforge.net/DB/layout.html.
6. Gabel T, Sure Y, Voelker J (2004) KAON – An overview. Insititute AIFB, University of Karlsruhe. http://kaon.semanticweb.org/main_kaonOverview.pdf.
7. Pan Z, Heflin J (2004) DLDB: Extending relational databases to support Semantic Web queries. Technical Report LU-CSE-04-006, Department of Computer Science and Engineering, Lehigh University.
8. Alexaki S, Christophides V, Karvounarakis G, Plexousakis D, Tolle K (2001) The ICS-FORTH RDFSuite: Managing voluminous RDF description bases. In: Proceedings of the 2nd International Workshop on the Semantic Web, Hongkong.
9. Caroll J, Bizer C, Hayes P, Stickler P (2004) Semantic Web publishing using named graphs. In: Proceedings of Workshop on Trust, Security, and Reputation on the Semantic Web, at the 3rd International Semantic Web Conference, ISWC 2004, Hiroshima, Japan.
10. Theoharis Y, Christophides V, Karvounarakis G (2005) Benchmarking database representations of RDF/S stores. In: Proceedings of the 4th International Semantic Web Conference, ISWC 2005, Galway, Ireland.
11. Guo Y, Pan Z, Heflin J (2005) LUBM: A benchmark for OWL knowledge base systems. Journal of Web Semantics 3(2):158–182.
12. RDF semantics – W3C recommendation. http://www.w3.org/TR/rdf-mt.
13. Hustadt U, Motik B, Sattler U (2007) Reasoning in description logics by a reduction to disjunctive datalog. Journal of Automated Reasoning 39(3):351–384.
14. Kiryakov A, Ognyanov D, Manov D (2005) OWLIM: A pragmatic semantic repository for OWL. In: Proceedings of the International Workshop on Scalable Semantic Web Knowledge Base Systems, SSWS 2005, WISE 2005, New York City, NY, USA.
15. ter Horst H (2005) Combining RDF and part of OWL with rules: Semantics, decidability, complexity. In: Proceedings of the 4th International Semantic Web Conference, ISWC 2005, Galway, Ireland.
16. ter Horst H (2005) Completeness, decidability and complexity of entailment for RDF Schema and a semantic extension involving the OWL vocabulary. Journal of Web Semantics 3:79–15.
17. Munoz J, Perez C, Gutierrez C (2007) Minimal deductive systems for RDF. In: Proceedings of the 4th European Semantic Web Conference, ESWC 2007, Innsbruck, Austria.

18. Bernstein A, Kiefer C (2005) iRDQL – Imprecise queries using similarity joins for retrieval in ontologies. In: Proceedings of the 4th International Semantic Web Conference, ISWC 2005, Galway, Ireland.
19. Siberski W, Pan J, Thaden U (2006) Querying the Semantic Web with preferences. In: Proceedings of the 5th International Semantic Web Conference, ISWC 2006, Athens, GA, USA.
20. Gutierrez C, Hurtado C, Vaisman A (2007) Introducing time into RDF. IEEE Transactions on Knowledge and Data Engineering, Special Issue on Knowledge and Data Engineering in the Semantic Web Era 19:207–218.
21. Hurtado C, Poulovassilis A, Wood P (2006) A relaxed approach to RDF querying. In: Proceedings of the 5th International Semantic Web Conference, ISWC 2006, Athens, GA, USA.
22. Bechhofer S, Lord P, Volz R (2003) Cooking the Semantic Web with the OWL API. In: Proceedings of the 2nd International Semantic Web Conference, ISWC 2003, Sanibel Island, FL, USA.
23. Prud'hommeaux E, Seaborne A (2006) SPARQL query language for RDF. W3C Candidate Recommendation. http://www.w3.org/TR/rdf-sparql-query.
24. Karvounarakis G, Christophides V, Plexousakis D, Alexaki S (2000) Querying community web portals. Techreport, Institute of Computer Science, FORTH, Heraklion, Greece. http://www.ics.forth.gr/proj/isst/RDF/RQL/rql.pdf.
25. Seaborne A (2004) RDQL – A query language for RDF. W3C Member Submission. http://www.w3.org/Submission/2004/SUBM-RDQL-20040109.
26. Perez J, Arenas M, Gutierrez C (2006) The semantics and complexity of SPARQL. In: Proceedings of the 5th International Semantic Web Conference, ISWC 2006, Athens, Georgia, USA.
27. Haase P, Broekstra J, Eberhart A, Volz R (2004) A comparison of RDF query languages. In: Proceedings of the 3rd International Semantic Web Conference, ISWC 2004, Hiroshima, Japan.

Tableau-Based Reasoning

Ralf Möller[1] and Volker Haarslev[2]

[1] Hamburg University of Technology, Germany, r.f.moeller@tu-harburg.de
[2] Concordia University, Montreal, Canada, haarslev@cse.concordia.ca

Summary. Tableau-based methods for satisfiability checking build the backbone of major contemporary ontology reasoning systems. The main idea of tableau-based methods for satisfiability checking is to systematically construct a representation for a model of the input formulae. If all representations that are considered by the procedure turn out to contain an obvious contradiction, a model representation cannot be found and it is concluded that the set of formulae is unsatisfiable.

In this chapter, tableau-based reasoning methods are formally introduced. We start with a nondeterministic basic version which subsequently will be extended with optimization techniques in order to demonstrate how practical systems can be built. We also demonstrate how computed tableau structures can be exploited for other inference problems in an ontology reasoning system.

1 Introduction

As part of the infrastructure for working with ontologies, reasoning systems are required. Reasoning is used at ontology development or maintenance time as well as at the time ontologies are used for solving application problems. In this section we will review so-called tableau-based decision procedures for inference problems arising in both contexts. We start with the satisfiability problem for a set of logical formulae. Speaking about ontologies, we focus on description logics, which provide the basis for standardized practical ontology languages. In this context, the set of formulae mentioned above is usually divided into a Tbox and an Abox for the intensional and extensional part of the ontology, respectively (see below for details). We are aware of the fact that ontology processing systems based on description logics also support some form of rules as well as means for specifying constraints among attributes of different individuals [5]. For introductory purposes, here we focus on satisfiability checking in basic description logics, however.

The main idea of tableau-based methods for satisfiability checking is to systematically construct a representation for a model of the input formulae. If all representations that are considered by the procedure turn out to contain

S. Staab and R. Studer (eds.), *Handbook on Ontologies*, International Handbooks on Information Systems, DOI 10.1007/978-3-540-92673-3,

an obvious contradiction (clash), a model representation cannot be found, and it is concluded that the set of formulae is unsatisfiable. In early publications on tableau-based proof procedures, in particular for first-order logics, the notation for the model representations was done using tables (tableaux in French). In recent approaches these tables are better described as graph structures. The name tableau is retained for historical reasons, however.

Initially, tableau-based methods for description logics have been developed for decidability proofs, and due to this fact, they are highly nondeterministic for expressive description logics. It turned out, however, that they can indeed be efficiently implemented using appropriate search strategies and index structures such that for typical-case inputs, acceptable runtimes can be expected even though the worst-case complexity is high. In practical systems, tableau structures are efficiently maintained during branch and bound (or backtracking) with the result that tableau-based methods have been successfully employed in ontology reasoning systems such as FaCT++, Pellet, or RacerPro (cf. [26] for an overview about description logic systems).

Although, in practical contexts, tableau-based methods are often applied in a refutation-based way (i.e., they are used to show unsatisfiability of a set of formulae), the graph structures computed for solving the ontology satisfiability problem can be reused for efficiently implementing higher-level reasoning services such as instance retrieval requests. In other words, in practical systems, tableau-based methods are not just used for satisfiability checking but are also used to compute index structures for subsequent calls to other reasoning services.

In this chapter, tableau-based reasoning methods are formally introduced. We start with a nondeterministic basic version which subsequently will be extended with optimization techniques in order to demonstrate how practical systems can be built. We also demonstrate how computed tableau structures can be exploited in an ontology reasoning system. In order to make this chapter self-contained, we shortly introduce the syntax and semantics of the description logic $\mathcal{ALC}$ and introduce Tboxes and Aboxes.

An overview on tableau algorithms for description logics can also be found in [7] as well as in [6]. In this chapter, we also consider optimization issues, and the presentation is oriented towards implementation aspects in order to complement the presentations in [6, 7].

1.1 Syntax and Semantics of $\mathcal{ALC}$

For a given application problem one chooses a set of elementary descriptions (or *atomic descriptions*) for *concepts* and *roles* representing unary and binary predicates, respectively. A set of *individuals* is fixed to denote specific objects of a certain domain. We use letters A and R for atomic concepts and roles, respectively. In addition, let $\{i, j, \ldots\}$ be the set of individuals. In $\mathcal{ALC}$ (Attributive Language with full Complement), descriptions for *complex concepts* C or D can be inductively built using the following grammar:

$$\begin{array}{rl} C, D \longrightarrow A & \mid \text{atomic concept} \\ C \sqcap D & \mid \text{conjunction} \\ C \sqcup D & \mid \text{disjunction} \\ \neg C & \mid \text{negated concept} \\ \exists R.C & \mid \text{existential quantification} \\ \forall R.C & \mid \text{value restriction} \end{array}$$

We introduce the concept descriptions $\top$ and $\bot$ as abbreviations for $A \sqcup \neg A$ and $A \sqcap \neg A$, respectively. Concept descriptions may be written in parentheses in order to avoid scoping ambiguities.

For defining the semantics of concept and role descriptions we consider *interpretations* $\mathcal{I}$ that consist of a non-empty set $\Delta^{\mathcal{I}}$, the domain, and an interpretation function $\cdot^{\mathcal{I}}$, which assigns to every atomic concept A a set $A^{\mathcal{I}} \subseteq \Delta^{\mathcal{I}}$ and to every atomic role R a set $R^{\mathcal{I}} \subseteq \Delta^{\mathcal{I}} \times \Delta^{\mathcal{I}}$. For complex concept descriptions the interpretation function is extended as follows:

$$\begin{aligned} (C \sqcap D)^{\mathcal{I}} &= C^{\mathcal{I}} \cap D^{\mathcal{I}} \\ (C \sqcup D)^{\mathcal{I}} &= C^{\mathcal{I}} \cup D^{\mathcal{I}} \\ (\neg C)^{\mathcal{I}} &= \Delta^{\mathcal{I}} \backslash C^{\mathcal{I}} \\ (\exists R.C)^{\mathcal{I}} &= \{x \mid \exists y.(x, y) \in R^{\mathcal{I}} \text{ and } y \in C^{\mathcal{I}}\} \\ (\forall R.C)^{\mathcal{I}} &= \{x \mid \forall y. \text{ if } (x, y) \in R^{\mathcal{I}} \text{ then } y \in C^{\mathcal{I}}\} \end{aligned}$$

The semantics of description logics is based on the notion of satisfiability. An interpretation $\mathcal{I} = (\Delta^{\mathcal{I}}, \cdot^{\mathcal{I}})$ *satisfies* a concept description C if $C^{\mathcal{I}} \neq \emptyset$. In this case, $\mathcal{I}$ is called a *model* for C.

A *Tbox* is a set of so-called *generalized concept inclusions* $C \sqsubseteq D$. For brevity the elements of a Tbox are called *GCIs*. An interpretation $\mathcal{I}$ satisfies a GCI $C \sqsubseteq D$ if $C^{\mathcal{I}} \subseteq D^{\mathcal{I}}$. An interpretation is a *model* of a Tbox if it satisfies all GCIs in the TBox. A concept description C *is subsumed by* a concept description D w.r.t. a Tbox if the GCI $C \sqsubseteq D$ is satisfied in all models of the Tbox. In this case, we also say that D *subsumes* C.

An *Abox* is a set of *assertions* of the form $C(i)$ or $R(i, j)$ where C is a concept description, R is a role description, and i, j are individuals. A concept assertion $C(i)$ is satisfied w.r.t. a Tbox $\mathcal{T}$ if for all models $\mathcal{I}$ of $\mathcal{T}$ it holds that $i^{\mathcal{I}} \in C^{\mathcal{I}}$. A role assertion $R(i, j)$ is satisfied w.r.t. a Tbox $\mathcal{T}$ if $(i^{\mathcal{I}}, j^{\mathcal{I}}) \in R^{\mathcal{I}}$ for all models $\mathcal{I}$ of $\mathcal{T}$. An interpretation satisfying all assertions in an Abox $\mathcal{A}$ is called a model for $\mathcal{A}$. An Abox $\mathcal{A}$ is called *consistent* if such a model exists, it is called *inconsistent* otherwise.

1.2 Decision Problems and Their Reductions

The definitions given in the previous section can be paraphrased as decision problems.

The *concept satisfiability* problem is to check whether a model for a concept exists. The *Tbox satisfiability problem* is to check whether a model for the Tbox exists. The *concept subsumption* problem (w.r.t. a Tbox) is to check whether $C^\mathcal{I} \subseteq D^\mathcal{I}$ holds (in all models of the Tbox).

The *Abox consistency problem* for an Abox $\mathcal{A}$ (w.r.t. a Tbox) is the problem to determine whether there exists a model of $\mathcal{A}$ (that is also a model of the Tbox). Another problem is to test whether an individual i is an instance of a concept description C w.r.t. a Tbox and an Abox (*instance test* or *instance problem*). The *instance retrieval* problem w.r.t. a query concept description C is to find all individuals i mentioned in the assertions of an Abox such that i is an instance of C.

The latter problem is a retrieval problem but, in theory, it can be reduced to several instance problems. Furthermore, the satisfiability problem for a concept C can be reduced to the consistency problem of the Abox $\{C(i)\}$. In order to solve the instance problem for an individual i and a concept description C one can check if the Abox $\{\neg C(i)\}$ is inconsistent. The concept subsumption problem can be reduced to an Abox consistency problem as well. If the Abox $\{C(i), \neg D(i)\}$ is not consistent, C is subsumed by D [6].

Thus, in theory, all problems introduced above can be reduced to the Abox consistency problem. Note that in practical systems, specific algorithms might be used to decide a certain problem.

2 Deciding the Consistency Problem for $\mathcal{ALC}$ Aboxes

A decision procedure for the $\mathcal{ALC}$ Abox consistency problem is described in this section using a so-called tableau-based algorithm. In order to simplify the presentation, in this section, we do not consider Abox consistency with respect to Tboxes. For Tboxes, among other extensions, additional machinery is required to ensure termination (see Sect. 3 for details).

As indicated in the introduction, the main idea of the Abox consistency algorithm is to systematically generate a representation for a model. In this process which searches for a model, some representations are generated which contain an obvious contradiction (*clash*), i.e., for an individual i we have $C(i)$ and $\neg C(i)$ in an Abox. The assertions $C(i)$ and $\neg C(i)$ are called the *culprits* for the clash. In case of a clash, the generated representations turn out to not to describe a model.

From a theoretical point of view, the algorithm described below is sound and complete but nondeterministic. In a practical implementation, indeterminism must be handled with systematic search techniques, and various heuristics have been described in the literature to guide the search process. If we see a tableau-based algorithm not as a theoretical vehicle for proving decidability of a logic, but as a practical way to solve the Abox consistency problem, then it becomes clear that it is important to be able to detect clashes as early as possible while the model representations are built. A representation with a clash no longer needs to be considered. Thus, we should be able to identify clashes not only for assertions with atomic concepts but also for assertions with complex concepts. Therefore, in contrast to other presentations of tableau algorithms we will not transform a concept into a form that makes the presentation (and analysis) of the tableau algorithm easier (negation normal form), but directly use a form that is efficient for detecting clashes in typical-case inputs (encoded normal form). The $\mathcal{ALC}$ Abox consistency algorithm described below checks whether there exists a model for the input Abox.

The algorithm operates on a set of Aboxes $\mathfrak{A}$. Each Abox represents an alternative to be investigated in the exhaustive model generation process. We also call such an internal Abox, i.e., an element of $\mathfrak{A}$, a *tableau* (but also use the term Abox in the following). Initially, the algorithm starts with a set $\mathfrak{A} = \{\mathcal{A}\}$ containing the input Abox $\mathcal{A}$. A set of rules is applied to an Abox from this set until no more rule applications are possible or no more rule applications are needed to determine the result. The rules are introduced below. If a rule is applied to an Abox, the Abox is replaced by one or more Aboxes. If it is replaced by more than one Abox (the so-called *successor Aboxes*), we say that the rule introduces a so-called *choice point.* A rule introducing a choice point is called a *nondeterministic* rule. All other rules are called *deterministic.* In any case, the Abox to which a rule is applied is replaced with new Aboxes that are "copies" of the original one plus some additional assertions.

If a tableau (Abox) is found to contain a clash, the tableau is called *closed*, otherwise it is called *open.* A tableau to which no rule can be applied is called *complete.* For a complete tableau the synonym *completion* is also used. If there exists a completion, i.e., an open tableau to which no more rules can be applied, the algorithm returns "yes" (indicating consistency). If all tableaux that could be generated by applying rules are closed, the algorithm returns "no" (inconsistency).

2.1 Concept Normalization and Encoding

In order to speed up the clash test, concepts are normalized using several transformation steps. First, double negations are eliminated, i.e., $\neg\neg C$ is replaced with C. Then, maximal sequences of conjunctions (possibly with nested parentheses) are flattened and represented with an n-ary conjunction term $\bigwedge\{C_1, C_2, \ldots, C_n\}$ (written as a prefix operator to the set of arguments). Corresponding representations $\bigvee\{C1, C2, \ldots, C_n\}$ are built for disjunctions. The interpretation function is extended in the obvious way

$$(\bigwedge\{C_1, C_2, \ldots, C_n\})^\mathcal{I} = (C_1)^\mathcal{I} \cap (C_1)^\mathcal{I} \cap \cdots \cap (C_n)^\mathcal{I},$$

$$(\bigvee\{C_1, C_2, \ldots, C_n\})^\mathcal{I} = (C_1)^\mathcal{I} \cup (C_1)^\mathcal{I} \cup \cdots \cup (C_n)^\mathcal{I}.$$

If there are two concepts C and $\neg C$ mentioned in a conjunction (disjunction) or the concept $\bot(\top)$ appears, the whole term $\bigwedge\{C_1, C_2, \ldots, C_n\}$ $(\bigvee\{C_1, C_2, \ldots, C_n\})$ is replaced with $\bot$ $(\top)$.

Afterwards, in an encoding process every concept description C and its negation $\neg C$ is inductively associated with a unique identifier. For instance, one could use numbers as unique identifiers and store concepts as records in an array, or it is possible use pointers to records (or objects) as unique identifiers. The fact that conjunctions (or disjunctions) are represented as sets enables the assignment of the same unique identifier to syntactically different but semantically equivalent conjunctive and disjunctive concept descriptions. The assignment of unique identifier to a concept is known as *encoding* a concept. If we use a concept description in the following text, we assume that we refer to its unique identifier.

The function $neg(C)$ is used to find the negation of a concept. The implementation of this function should require constant time (i.e., be as efficient as possible).[1]

Next, an internal representation of the input Abox with encoded concepts is built in a preprocessing step. We assume that normalized concepts are used from now on. For readability issues, however, in the presentation below, we still use concept descriptions as introduced above.

2.2 Tableau Rules

The tableau rules are applied to an Abox $\mathcal{A}$ as part of the set of Aboxes $\mathfrak{A}$ on which the algorithm operates. A rule can be *applied* whenever the precondition is satisfied and $\mathcal{A}$ is not in the set of closed Aboxes (this is an implicit condition). Initially, the set of closed Aboxes is empty. Saying that $\mathcal{A}$ is *replaced* by an Abox or a sequence of Aboxes we mean that $\mathcal{A}$ is removed from $\mathfrak{A}$ and the Aboxes generated by the rule are added to $\mathfrak{A}$. If a rule is applied to an assertion, we say the assertion is *expanded*:

- *Conjunction rule:* If $(\bigwedge\{C_1, \ldots, C_n\})(x) \in \mathcal{A}$ and $\{C_1(x), \ldots, C_n(x)\} \not\subseteq \mathcal{A}$, then replace $\mathcal{A}$ with $\mathcal{A} \cup \{C_1(x), \ldots, C_n(x)\}$.
- *Disjunction rule:* If $(\bigvee\{C_1, \ldots, C_n\})(x) \in \mathcal{A}$ and for all $i \in \{1, \ldots, n\}$ it holds that $C_i(x) \notin \mathcal{A}$, then replace $\mathcal{A}$ with a sequence of Aboxes $A_1, \ldots, A_n$ where $\mathcal{A}_1 = \mathcal{A} \cup \{C_1(x)\}, \ldots, \mathcal{A}_n = \mathcal{A} \cup \{C_n(x)\}$.
- *Existential quantification rule:* If $(\exists R.C)(x) \in \mathcal{A}$ but there is no individual name y such that $\{C(y), R(x, y)\} \subseteq \mathcal{A}$, then replace $\mathcal{A}$ with $\mathcal{A} \cup \{C(z), R(x, z)\}$ such that z is a fresh individual (i.e., an individual not occurring in $\mathcal{A}$).
- *Value restriction rule:* If $\{(\forall R.C)(x), R(x, y)\} \subseteq \mathcal{A}$ but $C(y) \notin \mathcal{A}$, then replace $\mathcal{A}$ with $\mathcal{A} \cup \{C(y)\}$.

[1] For instance, if numbers are chosen for the unique identifier, the unique identifier of the negation of a (non-negated) concept with number n could be $n + 1$. The encoding process must assign numbers accordingly. If (pointers to) objects are used for representing concepts, a field with a pointer to the negated concept provides for a fast implementation of *neg* at the cost of memory requirements probably being a little bit higher.

- *Negation rule:* If $\neg C(x) \in \mathcal{A}$ but $neg(C)(x) \notin \mathcal{A}$, then replace $\mathcal{A}$ with $(\mathcal{A} \cup \{ neg(C)(x)\}$.
- *Clash rule:* If $\{C(x), neg(C)(x)\} \subseteq \mathcal{A}$, then add $\mathcal{A}$ to the set of closed Aboxes.

The algorithm runs in a loop and applies a rule if its precondition is satisfied. A precondition of a rule is satisfied if there exists a substitution for the variables x or y with individuals such that the condition is satisfied. As indicated before, if the precondition is satisfied, a rule is applied to an Abox in $\mathfrak{A}$. Applying a rule means to execute the then-part applying the variable substitution computed from the if-part of the rule. The loop ends if a completion is found or if no rule is applicable.

The algorithm returns "yes" if there exists a completion and "no" otherwise. In principle, the rules defined above can be applied in any order. Later we will see that a rule application strategy might impose restrictions on the order of rule applications. Restrictions are introduced to find completions "early", i.e., the strategy is used for optimization purposes. If the expressivity of the language is extended, a particular rule application strategy might also be necessary to ensure termination or soundness and completeness.

A few additional definitions are appropriate for the analysis of the algorithm in the next subsection. If an Abox is replaced with one new Abox, the new Abox contains strictly more assertions. We call this process *and-branching*. Applying a nondeterministic rule (for the time being, the disjunction rule) might introduce several new Aboxes. We call this process *or-branching*.

The individuals mentioned in the original Abox are called *old* individuals, all other individuals are called *fresh*. A sequence of role assertions $R_1(x_1, x_2), R_2(x_2, x_3), \ldots, R_{n-1}(x_{n-1}, x_n), R_n(x_n, x_{n+1})$ is called a *path* (of length n) from x_1 to x_{n+1}. In a path of length 1, x_2 is called the (direct) *successor* and x_1 is called *predecessor* of x_2 (for a role R). The individuals x_i with $i \in \{2, n+1\}$ are called indirect successors of x_1.

Formal Properties

The formal properties of the algorithm are analyzed in three steps. We first show termination, and afterwards we prove soundness and completeness.

The procedure *terminates:* First, no rule can be applied twice to the same Abox with the same bindings for the variables x, y due to the preconditions (no infinite and-branching). Second, although new individuals are introduced by applying the existential quantification rule, the quantification concept is of a smaller size than the original concept. Hence, there can be no infinite applications of the existential quantification rule. The length of the longest Abox is bounded by the size of the input Abox. Third, no rule deletes an assertion, and therefore, Aboxes can only grow (i.e., no so-called yo-yo effects can occur [24, p. 547]).

The algorithm is sound: If the algorithm returns "yes" there exists a model satisfying all assertions of the input Abox. This is shown as follows. If the algorithm returns "yes" there exists a completion. From the completion $\mathcal{A}$ a so-called *canonical model* $\mathcal{I}_\mathcal{A} = (\Delta^\mathcal{I}_\mathcal{A}, \cdot^\mathcal{I}_\mathcal{A})$ can be constructed (cf. [6]):

1. Let $\Delta^\mathcal{I}_\mathcal{A}$ be the set of all individuals mentioned in $\mathcal{A}$.
2. For all atomic concept descriptions A let $A^\mathcal{I}_\mathcal{A} = \{x \mid A(x) \in \mathcal{A}\}$.
3. For all role descriptions R let $R^\mathcal{I}_\mathcal{A} = \{(x, y) \mid R(x, y) \in \mathcal{A}\}$.

By definition, all role assertions are satisfied by $\mathcal{I}_\mathcal{A}$. Now, using induction on the structure of concepts, it is easy to show that $\mathcal{I}_\mathcal{A}$ also satisfies all concept assertions in $\mathcal{A}$ (see [6] for details).

The algorithm is complete: If there exists a model for the input Abox, then the algorithms returns "yes". Or, by contraposition, it holds that if the algorithm returns "no", then there does not exist a model. If the algorithm returns "no", all tableaux are closed, i.e., there is a clash in each tableau. Under the assumption that there exists a model for an Abox to which a rule is applied, it is shown that a model for at least one of the generated Aboxes can be constructed by examining every rule (and hence, no alternative to be investigated is forgotten, for details see [6] again). Now since there is a clash in every tableau in case the algorithm returns "no", there cannot exist a model for the input Abox.

The $\mathcal{ALC}$ Abox consistency problem is PSPACE-complete, cf. [29]. The algorithm needs an exponential number of steps in the worst case. As we will see in the next subsection, there exists a rule application strategy such that intermediate tableaux can be discarded such that the algorithm runs in polynomial space in order to be worst-case-optimal.

2.3 Towards an Optimized Implementation

The tableau rules refer to assertions for specific individuals or check for a clash w.r.t. a specific individual. Thus, rather than using an arbitrary set data structure for representing a tableau, in a concrete implementation of the tableau algorithm, the set of assertions in an Abox is partitioned w.r.t. the individuals the assertions refer to (for $C(i)$ and $R(i, j)$ the assertion refers to i). We call such a partition an *individual partition* P_i. The access to the partition of an individual i should require almost constant time. If there is an assertion $R(i, j) \in P_i$, then there will also be a partition P_j for j (possibly empty). We say P_j *depends on* P_i.

Furthermore, looking at the preconditions of the rules, it is revealed that for each individual, the preconditions refer to specific concept constructors (conjunctions, disjunctions, existential quantifications, or value restrictions). Thus, for each individual partition, the set of conjunctions, disjunctions, existential quantifications, and value restrictions must be efficiently identifiable. For the latter two subsets, a further index over different roles might be considered.

The selection of a particular rule to apply is nondeterministic in the algorithm above. Various kinds of heuristics have been investigated to reduce the number of rule applications for typical-case inputs. First, in a practical implementation, best results have been achieved if the clash rule is applied with highest priority. Since no rules are applied to closed tableaux by definition, the number of applicability tests for rules is reduced if clashes are found early. The overhead for the clash rule must be kept at a minimum, however. Usually, in concrete implementations, the clash rule is (implicitly) applied whenever a concept assertion $C(x)$ is to be added to an Abox. Note that for $\mathcal{ALC}$, role assertions are not directly involved in a clash test – a condition that is no longer true for more expressive logics. Checking whether the assertion $neg(C)(x)$ is already an element of the tableau to which $C(x)$ is to be added is a frequently executed operation in a practical implementation, and has to be implemented very efficiently. As part of this so-called *clash test*, in a practical implementation it might become apparent that $C(x)$ is also already contained in the Abox. So, there is no maintenance effort for the Abox to which $C(x)$ is added.

The conjunction rule is applied with second-highest priority. Although the number of assertions to be handled in a partition is increased, the chance that a clash is detected early is also increased. Since in many contexts the Abox will indeed be consistent, conjunctions have to be "expanded" anyway. So, it is a good heuristic to prefer the conjunction rule over other rules.

In order to reduce the number of Aboxes to be handled as parts of $\mathfrak{A}$, in practical systems, deterministic rules are preferred over nondeterministic ones. In order to reduce memory requirements (and to meet the complexity class of the Abox consistency problem) the so-called *trace technique* has been developed. Employing the trace technique, the disjunction rule is applied before the deterministic value restriction rule. Then, for each existential quantification assertion $(\exists R.C)(x)$, it is ensured that all potentially applicable value restrictions are indeed available in the Abox. Thus, the existential quantification rule can be combined with the value restriction rule. Rather than only adding a concept assertion $C(y)$ based on the quantification concept as indicated in the existential quantification rule for a role R, additionally for every value restriction $(\forall R.D_i)(x)$ the assertion $D_i(y)$ is added, with y being the fresh individual introduced by the exists quantification rule. Then all assertions $\{C(y), D_1(y), \ldots, D_n(y)\}$ can be treated in isolation. If they turn out not to lead to a clash, the assertions, and all those derived from them, can be removed (and $(\exists R.C)(x)$ must somehow be marked to avoid repetitive rule applications).

In a practical implementation, the trace technique might not be adopted for various reasons. For instance, the removal of assertions might interfere with the idea to reuse of previous computation results, in particular if Tboxes are involved (see below). Or the strategy is to avoid the expansion of disjunctions but check the satisfiability of existential quantifications first.

2.4 Dealing with Indeterminism in a Tableau Algorithm

In the description above, a nondeterministic rule (in $\mathcal{ALC}$ only the disjunction rule) generates a sequence of new Aboxes. If Aboxes are created in a naive way, this can hardly be efficient. Thus, a practical implementation must find a way to implement a structure-sharing strategy for copies of Aboxes in order to avoid structures to be copied repeatedly. Copying complete structures is memory-extensive as well as time-consuming. Even with a structure-sharing approach, the naive generation of successor Aboxes should be avoided due to the memory-management overhead involved.

Obviously, the disjunction rule does not need to generate successors that immediately lead to a clash. If the disjunction rule would be applicable to some disjunct C_i in an assertion $(\bigvee\{C_1, \ldots, C_n\})(x)$ and $neg(C_i)(x) \in \mathcal{A}$ the corresponding successor Abox does not need to be generated (it will be closed according to the clash rule immediately). The detection of those situations requires some additional machinery in the implementation (boolean constraint propagation, BCP) [9]. Boolean constraint propagation is implemented in all major contemporary tableau-based reasoners. The challenge is to most efficiently determine which disjunctions can be virtually "shrunk" or "eliminated" in this way. Note that a disjunction becomes deterministic if only one disjunct "remains" after BCP, and it can be treated as a conjunction in this case (also w.r.t. priorities in rule applications). It is easy to see that termination and correctness is still fulfilled if boolean constraint propagation is employed.

Further optimizations that also have no impact on the correctness of the algorithm but provide for improved performance for typical-case inputs are possible. For instance, the disjunction rule requires as a precondition that the disjunct $C_i(x)$ is not already in $\mathcal{A}$. Thus, looking for a completion, in a concrete implementation it is advantageous to first apply the disjunction rule to those concept assertions $(\bigvee\{C_1, \ldots, C_n\})(x)$ with disjuncts C_i such that C_i is also mentioned in many other disjunctive assertions for the individual x. The application of the disjunction rule for the other disjunctions for x involving C_i is then "avoided" (due to the precondition of the disjunction rule). Efficiently finding those concepts C_i such that the number of occurrences in all disjunctions applying to an individual x is maximized (or large) is non-trivial, however. There is a tradeoff between the time spent in search for occurrences of a concept assertion $C_i(x)$, management of index structures for speeding up this search process, and the gain of this in terms of or-branching reduction.

Reusing previous results can help finding clashes early. If $\mathcal{A}$ is an Abox, then all Aboxes derived from $\mathcal{A}$ by applying a tableau rule are called *sibling* Aboxes. Information acquired for one successor Abox $\mathcal{A}'$ of $\mathcal{A}$ can be propagated to sibling Aboxes of $\mathcal{A}'$. Let us consider the successor Aboxes of an application of the disjunction rule to an Abox $\mathcal{A}$ again. If it turns out that one of the successor Aboxes $\mathcal{A}'$ with $C_i(x)$ being added contains a clash, then, $neg(C_i)(x)$ can be added to all (open) sibling Aboxes of $\mathcal{A}'$. Again, if

$neg(C_i)(x)$ is explicitly present in a sibling Abox, an application of the disjunction rule to the sibling Abox might be prevented and a clash might be revealed earlier. On the negative side it has to be mentioned that applying rules to $neg(C_i)(x)$ in a tableau also causes some overhead. In a practical implementation one might avoid the applications of the rules to assertions added this way (without impact on soundness and completeness).

Up to now, we have considered ways to find a completion earlier (i.e., we attempt to reduce or-branching). This heuristic is useful because the algorithm terminates if a completion is found. However, it is also possible to find ways to close tableaux early. Consider the following example Abox, which is obviously inconsistent (adapted from [17])

$$\mathcal{A}=\{(\bigvee\{C_1,D_1\})(i),\ldots,(\bigvee\{C_n,D_n\})(i),(\exists R.(\bigwedge\{A,B\}))(i),(\forall R.\neg A)(i)\}.$$

We assume that the tableau algorithm applies the disjunction rule to the first disjunction in $\mathcal{A}$. We get two new Aboxes $\mathcal{A}_1$ and $\mathcal{A}_2$. Both Aboxes are supersets of $\mathcal{A}$. In Fig. 1, Aboxes are indicated as circles. A set inclusion relation is indicated with a solid arrow (pointing from the superset to the subset). The assertions being added to an Abox w.r.t. its predecessor Abox are written next to the circle used for indicating the Abox. Aboxes that initially are not closed are presented with a bold outline. As indicated in Fig. 1, we assume that tableaux are represented using a kind of trie data structure.

In Fig. 1 we assume further rule applications to the Aboxes $\mathcal{A}_1, \mathcal{A}_3, \ldots, \mathcal{A}_{2n-1}$, and then $\mathcal{A}_{2n+1}, \mathcal{A}_{2n+2}$ and finally $\mathcal{A}_{2n+3}$. Initially, $\mathcal{A}_{2n+3}$ is assumed to be open. Now, a clash is found and $\mathcal{A}_{2n+3}$ is marked as closed by the clash rule (due to $\neg A(j)$ and $A(j)$ being an element of $\mathcal{A}_{2n+3}$).

The dashed lines (curved) indicate the dependencies of the assertions that are added to the respective Aboxes (not all dependencies are shown for

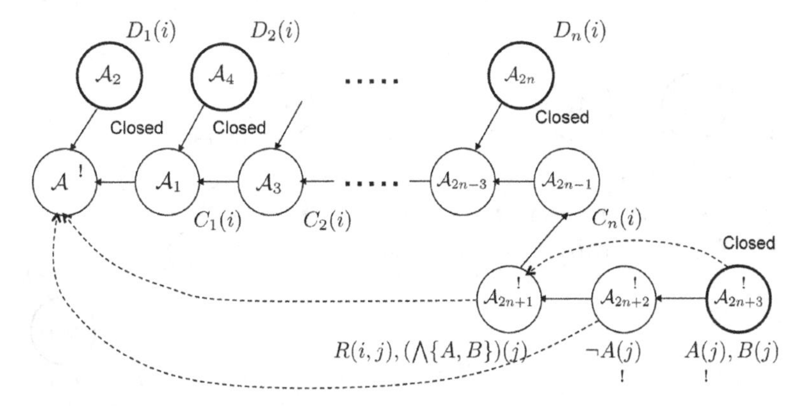

Fig. 1. A clash occurs in tableau, and exploring other open tableaux will not resolve it. The corresponding tableau can be closed even without a clash (see text)

readability reasons). Looking at Fig. 1 it should be apparent that the Aboxes $\mathcal{A}_2, \mathcal{A}_4, \ldots, \mathcal{A}_{2n}$, which are still open, will inevitably lead to the same clash. The idea to avoid repeatedly detecting the same clash over and over again is to find a way to close the Aboxes $\mathcal{A}_2, \mathcal{A}_4, \ldots, \mathcal{A}_{2n}$ in advance.

This can be achieved as follows. The clash occurs in $\mathcal{A}_{2n+3}$. The culprit in $\mathcal{A}_{2n+3}$ is $A(j)$. The other culprit $\neg A(j)$ is in $\mathcal{A}_{2n+2}$. Culprit assertions are indicated with an exclamation mark. Starting with the culprits and following the dashed lines the Aboxes are marked with exclamation markers. See Fig. 1 for the final marking in our example. Then, starting from the Abox with the "rightmost" culprit (in our case $\mathcal{A}_{2n+3}$) and following the solid lines in the direction of the arrows, the reachable Aboxes are visited. If an Abox $\mathcal{A}'$ marked with "!" pointing to an Abox $\mathcal{B}$ (with more than one incoming link) is reached, and the leaves of the other Aboxes pointing to the predecessor of $\mathcal{A}'$ are not closed, then the process stops. Then, all leaf Aboxes reachable in the inverse direction of the solid lines from the Abox $\mathcal{B}$ are marked as closed.

In the example shown in Fig. 1 the process stops at $\mathcal{B} = \mathcal{A}$. The closed Aboxes found by following the solid lines in the inverse direction are indicated with a corresponding label "Closed" in Fig. 1. Hence, futile rule applications to $\mathcal{A}_2, \mathcal{A}_4, \ldots, \mathcal{A}_{2n}$ are avoided.

A slightly modified example illustrates that the process does not necessarily close all Aboxes but only those which, due to the given culprits, do not lead to completions. The example is given as follows:

$$\mathcal{A} = \{(\bigvee\{C_1, D_1\})(i), \ldots, (\bigvee\{C_n, D_n\})(i), (\exists R.(\bigwedge\{\neg C_2, D\}))(i)\}.$$

In Fig. 2, the situation is shown after a few rule applications. Culprit markers indicate the assertions on which the clash depends. Walking from the rightmost culprit along the solid lines stops at $\mathcal{B} = \mathcal{A}_3$. All leaves reachable from $\mathcal{A}_3$ in the inverse direction of the solid lines are closed. In this example, $\mathcal{A}_2$

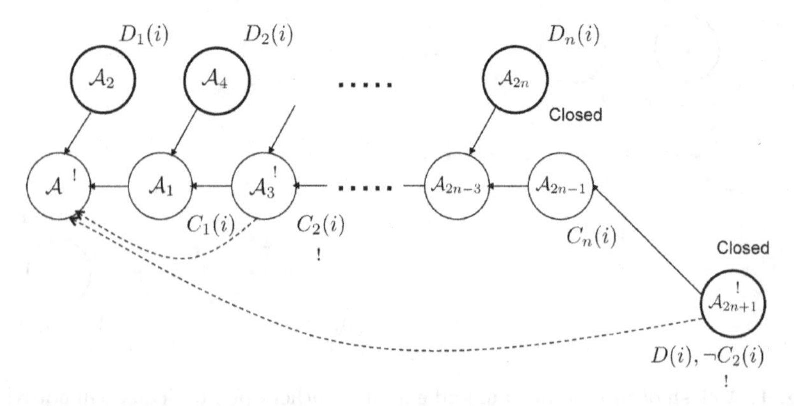

Fig. 2. Examining the dependencies reveals that $\mathcal{A}_4$ must not be automatically closed to retain completeness (see text)

and $\mathcal{A}_4$ remain open, which is necessary not to miss a possible completion to be constructed. If all successor Aboxes of an Abox are marked with a clash, their assertions are also seen as clash culprits and exclamation markers are propagated via the curved dependency links as introduced before [9].

In the literature, the technique described here is known as *backjumping*, which is a restricted form of dependency-directed backtracking [17]. We have presented it here using mathematical structures as part of the branch-and-bound strategy of a tableau algorithm. Note that Aboxes in the trie might indeed be reused, and thus, full dependency-directed backtracking might be achieved (with previous computations being maximally reused).

Usually, a partition for an individual is seen as a graph node with role assertions "pointing" to other graph nodes, and concept assertions defining the so-called "label" of a "node" (see below for a more formal introduction of a label). Hence, from an implementation point of view a tableau is seen as a graph with nodes and edges, both associated with a *label.* For nodes, the label is a set of concepts, and for edges it is a set of roles. To every path in the Abox trie from the root to a leaf (see Fig. 1 or Fig. 2) there corresponds a particular graph. For understanding tableau algorithms one can switch between the graph view and the tableau (or Abox) view. The graph corresponds to a model (see the construction of the canonical model mentioned above).

3 Dealing with Tboxes

The tableau algorithm introduced above must be extended to work with non-empty Tboxes. In theory, it is possible to transform all GCIs of the Tbox into a single GCI of the form $\top \sqsubseteq M$. The transformation is very simple. Instead of writing a GCI as $C_i \sqsubseteq D_i$ one could write $\top \sqsubseteq \neg C_i \sqcup D_i$, and thus, M is the conjunction $\bigwedge_i \{M_i\}$ of all $M_i = \neg C_i \sqcup D_i$ stemming from the GCIs in the Tbox (M and M_i are called *global constraints*). With the Tbox transformed into $\top \sqsubseteq \bigwedge\{M_1, \ldots, M_n\}$ we can see that the restriction M on the righthand side applies to all domain objects. The transformation is called *internalization* in the literature [8]. The tableau algorithm is extended with a new rule which adds $M(x)$ if there is an individual x mentioned in a tableau in which $M(x)$ is not already present:

- *GCI rule:* If $C(x) \in \mathcal{A}$ or $R(x, y) \in \mathcal{A}$ or $R(y, x) \in \mathcal{A}$ and $M(x) \notin \mathcal{A}$, then replace $\mathcal{A}$ with $\mathcal{A} \cup \{M(x)\}$.

A problem with this rule is that the algorithm then does not terminate. M might contain existential quantifications which cause new individuals y to be created, for which $M(y)$ is added and so on (infinite and-branching occurs). Some form of blocking must be enforced (see, e.g., [8]).

The tableau algorithm can be slightly changed to exploit these insights. We give a definition for the label of a partition. The *label* of a partition for an individual x is defined as $\{C \mid C(x) \in P_x\}$. Now, if there exists an individual partition P_k and there is no rule applicable to P_k nor to all partitions that depend on P_k, then no rule need to be applied to an assertion in a partition P_l if $label(P_l) \subseteq label(P_k)$ and k, l are fresh variables (otherwise, partitions for old individuals might block each other). We say P_l is *blocked* by the *witness* P_k. The witness must be a fresh individual. Note that if there is a clash detected for k the whole tableau is closed and no rule is applied to l anyway. The condition $label(P_l) \subseteq label(P_k)$ might become false if, due to other rule applications, new assertions are added for the individuals k or l in P_k and P_l, respectively. So, blocking conditions must be dynamically checked.

In order to show soundness in case of blocked partitions, one constructs a canonical interpretation for an individual i for which there exists a blocked partition P_i by defining tuples $(i, x) \in R^{\mathcal{I}}$ for every assertion $R(w, x)$ related to the witness w of i (for details see, e.g., [8]).[2]

The precondition that no rule is applicable to P_k and all partitions that depend on P_k could be dropped. But then one must make sure that P_l is a strict subset of P_k. Otherwise, to both partitions no rule would be applied and the algorithm would become incomplete. For more expressive logics, more expressive blocking conditions have to be defined (e.g., [21]). In the literature it has been shown that the extended algorithm is sound and complete for arbitrary $\mathcal{ALC}$ Tboxes.

One drawback from a practical point of view is that now a possibly large set of disjunctions is introduced for every individual mentioned in a tableau, since $M_i = \neg C_i \sqcup D_i$. Keeping in mind that, e.g., boolean constraint propagation is employed to deal with disjunctions in a practical system, it becomes clear that disjunctions always involve "heavy-weight" methods in a practical implementation of the tableau algorithm. For specific forms of GCIs, the disjunctions do not have to be explicitly generated, however. This is explained in the next subsection.

3.1 Lazy Unfolding

Let us assume, there is a global constraint of the form $L_i = \neg A \sqcup C$ in M such that A is an atomic concept description. Rather than adding this global constraint to every individual x, the idea is to only add $C(x)$ if adding $\neg A(x)$ would lead to a clash. For those global constraints, one can implicitly assume that individuals are instances of $\neg A$ "if not stated otherwise" (see the construction of the canonical interpretation). The global constraint L_i can be handled "in a lazy way" by a new rule which "unfolds" an assertion $A(x)$ in a tableau (cf. [3]).

[2] We use a way to construct the canonical interpretation that already considers additional concept constructors such as, say, number restrictions. In case of $\mathcal{ALC}$ it would be possible to map i to its witness w in the canonical interpretation.

We need some definitions for specifying the exact conditions under which soundness and completeness can be guaranteed. An atomic concept description A directly refers to an atomic description B if there exists a GCI $A \sqsubseteq C$ such that B is mentioned in C but not in the scope of an existential quantification or value restriction. A refers to B if it directly refers to B or there exists an atomic concept description B' such that A refers to B' and B' refers to B.

A global constraint of the form $L_i = \neg A \sqcup C$ need not be handled as a disjunction if A is an atomic concept description and C does not refer to A. Let us assume that global constraints that satisfy the conditions introduced above for L_i are collected not in M but into a set L. There must be no other global constraint in L with a disjunct $\neg A$ or disjunct A. Then, the following rule is used to deal with global constraints in L [3,21]:

- *Lazy unfolding rule 1:* If $A(x) \in \mathcal{A}$ and $(\neg A \sqcup C) \in L$ and $C(x) \notin \mathcal{A}$, then replace $\mathcal{A}$ with $\mathcal{A} \cup \{C(x)\}$.

In other words, only if there is an assertion $A(x)$ in a tableau, then $C(x)$ must be added because assuming $\neg A(x)$ would lead to a clash.

In case we also collect assertions of the form $A \sqcup C$ into the set of concepts L (and not into M) another rule must be added. Corresponding restrictions as for $\neg A \sqcup C$ apply:

- *Lazy unfolding rule 2:* If $\neg A(x) \in \mathcal{A}$, $(A \sqcup C) \in L$ and $C(x) \notin \mathcal{A}$, then replace $\mathcal{A}$ with $\mathcal{A} \cup \{C(x)\}$.

Lazy unfolding exploits the fact that one can safely assume that a domain object which is not explicitly enforced to be in $A^{\mathcal{I}}$ (or $(\neg A)^{\mathcal{I}}$ in the second case) is an element of $(\neg A)^{\mathcal{I}}$ (or $A^{\mathcal{I}}$ in the second case). See [22] for details.

3.2 GCI Absorption

Global constraints in L are handled more effectively. If, initially, the global constraints are not of the form that they can be put into L but must be stored in M, the goal is to transform them in a way that a maximum number of global constraints can be put into L, and possibly none must be kept in M, without changing the semantics of the Tbox. This transformation process is known as GCI absorption (see [15,16,22] for details).

In some cases, still some global constraints remain in M even if GCIs are transformed as describe above, unfortunately. For instance, this happens if there are two GCIs of the form $A \sqsubseteq (\exists R.A) \sqcap C$ and $\exists R.A \sqsubseteq A$ in a Tbox. The latter kind of GCI is only relevant for an individual x if there exists an assertion $R(x, y)$ in tableau.

For $\exists R\top. \sqsubseteq C$ an effective treatment is easily possible. The same holds for range restrictions $\top \sqsubseteq \forall R.C$. Let *domain*$(R)$ and *range*(R) denote sets of concepts (initially empty). We assume that all $(\forall R.\bot) \sqcup C$ are removed from M, and for each $(\forall R.\bot) \sqcup C$ removed, C is added to *domain*(R). In addition, all $\bot \sqcup \forall R.C$ are removed from M, and for each $\bot \cup \forall R.C$ removed, there is C added to $range(R)$:

- *Domain restriction rule:* If $R(x, y) \in \mathcal{A}$ and $C \in domain(R)$, then replace $\mathcal{A}$ with $\mathcal{A} \cup \{C(x)\}$.
- *Range restriction rule:* If $R(x, y) \in \mathcal{A}$ and $C \in range(R)$, then replace $\mathcal{A}$ with $\mathcal{A} \cup \{C(y)\}$.

See [10] for a description of the initial idea and [31] for an analysis of this technique. However, for $\exists R.A \sqsubseteq A$ a disjunction has to be added (to any individual for which there exists a role successor), which still could cause a combinatorial explosion if the wrong choice was made in a practical system. For instance, this could happen for the following Abox:

$$\mathcal{A} = \{\neg A(x_0), R(x_0, x_1), R(x_1, x_2), \ldots, R(x_{n-1}, x_n), A(x_n)\}.$$

For all $x_i, i \in \{0, \ldots, n-1\}$, a disjunction $(\forall R.\neg A) \sqcup A$ would be asserted, and choice points are set up. We assume that rules are first applied to the tableaux created for $\forall R.\neg A$. After several rule applications, a clash w.r.t. $\neg A(x_n)$ and $A(x_n)$ would be detected. The situation could be even worse, if there was a GCI $B \sqcap \exists R.A \sqsubseteq A$ with B being an atomic concept for which there exists a GCI $B \sqsubseteq D$. Thus, there is no way to absorb $B \sqcap \exists R.A \sqsubseteq A$ into L using the above-mentioned techniques.

In [23], a new transformation called *binary absorption* has been introduced to tackle this problem. Applying this transformation requires a new rule to be added to the tableau algorithm. This is discussed in the next subsection.

3.3 Binary Absorption

A GCI $B \sqcap \exists R.A \sqsubseteq A$ should not be transformed into a disjunction to be placed in M. It can be transformed into

$$\exists R^{-1}.\top \sqsubseteq A_1\ ,$$

$$A_1 \sqcap A \sqsubseteq \forall R^{-1}.A_2\ ,$$

$$A_2 \sqcap B \sqsubseteq A\ ,$$

where R^{-1} denotes the inverse of role R. The idea is to introduce a marker $A_1(y)$ for every y for which there is an assertion $R(x, y)$. The first GCI of the list above can be handled effectively by absorbing it into $range(R)$ as described before. A_2 is a marker indicating an instance of $\exists R.A$ (see the second GCI). The third GCI now enforces y to be an instance of A in the tableau. In order to deal with a conjunction of two atomic concepts on the left-hand side of a GCI the lazy unfolding rules can be extended as follows [23]:

- *Lazy unfolding rule 3*: If $\{A_1(x), A_2(x)\} \subseteq \mathcal{A}$, $(\neg(A_1 \sqcap A_2) \sqcup C) \in L$ and $C(x) \notin \mathcal{A}$, then replace $\mathcal{A}$ with $\mathcal{A} \cup \{C(x)\}$.

No disjunctions of this particular type have to be handled after the transformation is applied. Soundness and completeness of this approach have been shown in [23]. A disadvantage is that the tableau algorithm now must also handle inverse roles (denoted as R^{-1} in the GCIs above).

4 Tableau Structures for Subsumption Problems

Tableau-based reasoning can be exploited for solving other reasoning problems as well. First, we consider the subsumption problem, then we turn to the instance problem and the retrieval problem.

Subsumption problems occur very frequently if the so-called taxonomy of a Tbox is computed. This is usually done at ontology development time in order to check for modeling errors (unsatisfiable atomic concept descriptions, unwanted subsumption relationships, etc.). The taxonomy of a Tbox is a graph where the nodes are the atomic concept descriptions mentioned in the Tbox (including $\top$ and $\bot$), and the edges indicate whether a node is a most-specific subsumer of another node.

For expressive languages such as $\mathcal{ALC}$ the subsumption problem $A \sqsubseteq_? B$ can be reduced to the Abox consistency problem $\{(A \sqcap \neg B)(i)\}$ for some individual i. If the Abox is inconsistent the subsumption relation holds, otherwise it does not hold. For practical Tboxes, GCIs usually involve conjunctions on the right-hand side. Thus, if B is negated, disjunctions have to be handled by the tableau algorithm, which might lead to "unfocused" applications of the rules. For computing the taxonomy, many similar subsumption problems of the form $A_i \sqsubseteq_? B$ have to be solved, and hence, many Abox consistency problems $\{(A_i \sqcap \neg B)(i)\}$ are the consequence. In almost all cases the subsumption relation between A and B does not hold, and hence, the Abox is likely to be shown to be consistent. Quite some number of applications of tableau rules might be required, however (with large or-branching, and for realistic ontology considerable and-branching as well).

Therefore, in [15] the following technique was developed. Since $\neg B$ is used many times, its satisfiability is tested in isolation. Usually, $\neg B$ is satisfiable in practical contexts (otherwise, B would a synonym to $\top$). The test whether $\{\neg B(i)\}$ is consistent leads to a consistent Abox such that a label $\mathcal{L}_1$ is defined for i. We call this label a *pseudo model* (for an atomic concept description). The same is done now for A_i (let the label be called $\mathcal{L}_i$).

In [15] a process called *model merging* is defined. The idea of this process is to show non-subsumption by comparing the labels. Four conditions must be satisfied in order to conclude non-subsumption:

- For every $A \in \mathcal{L}_1$ there does not exist an $\neg A \in \mathcal{L}_i$.
- For every $\neg A \in \mathcal{L}_1$ there does not exist an $A \in \mathcal{L}_i$.
- For every $\exists R.C \in \mathcal{L}_1$ there does not exist an $\forall R.D \in \mathcal{L}_i$.
- For every $\forall R.C \in \mathcal{L}_1$ there does not exist an $\exists R.D \in \mathcal{L}_i$.

If non-subsumption cannot be concluded, the "full" test whether $\{(C \sqcup \neg D)(i)\}$ is consistent is performed. However, practical experiments have shown that this is not often required [15, 16], so there is hardly any overhead introduced by the model merging process. There is almost no search involved in comparing the labels in the way defined above if the assertions in the labels are indexed appropriately.

5 Conclusion

The tableau algorithm introduced in this section can be extended to deal with additional concept and role constructors. With the addition of new constructors, the rule application strategy becomes important for termination and correctness, not only for optimization. Furthermore, the blocking condition might become more complex. For instance, the following constructs have been investigated in the literature and tableau algorithms have been specified:

- Concrete domains (with feature composition) [4]
- Qualifying number restrictions [14]
- Number restrictions plus role conjunctions and GCIs [8]
- Transitive roles [27]
- Transitive roles, role hierarchies, GCIs, and features [16]
- Transitive roles, role hierarchies, GCIs, plus number restrictions and Aboxes [12]
- Transitive roles, role hierarchies, GCIs, number restrictions and Aboxes plus concrete domains without feature composition [11]
- Transitive roles, role hierarchies, GCIs, Aboxes, plus qualifying number restrictions and inverse roles [21]
- Nominals [1, 20, 28, 30]
- Role axioms [18]
- Concrete domains with role composition for description logics with GCIs [25]

For almost all of the language features in this list, efficient implementations based on tableau algorithms are available.

Tableau-based reasoning methods are very effective for concept satisfiability checking [19] as well as for Tbox-based reasoning tasks [13, 16, 32]. Even for some specific Tboxes for which it was assumed that resolution-based reasoning methods show better behavior, new techniques such as binary absorption have shown that tableau-based methods can exploit similar structures. Tableau-based Tbox reasoners such as FaCT++, Pellet or RacerPro are the fastest systems for expressive description logics for a wide range of expressive Tboxes that regularly occur in practice.

Acknowledgments

We would like to thank Sebastian Wandelt and Michael Wessel for comments on a draft of this chapter.

References

1. Franz Baader, Martin Buchheit, and Bernhard Hollunder. Cardinality restrictions on concepts. *Artificial Intelligence*, 88(1–2):195–213, 1996.
2. Franz Baader, Diego Calvanese, Deborah McGuinness, Daniele Nardi, and Peter F. Patel-Schneider, editors. *The Description Logic Handbook: Theory, Implementation, and Applications.* Cambridge University Press, Cambridge, 2003.
3. Franz Baader, Enrico Franconi, Bernhard Hollunder, Bernhard Nebel, and Hans-Jürgen Profitlich. An empirical analysis of optimization techniques for terminological representation systems or: Making KRIS get a move on. *Applied Artificial Intelligence. Special Issue on Knowledge Base Management*, 4: 109–132, 1994.
4. Franz Baader and Philipp Hanschke. A schema for integrating concrete domains into concept languages. In *Proc. of the 12th Int. Joint Conf. on Artificial Intelligence (IJCAI'91)*, pages 452–457, 1991.
5. Franz Baader, Ralf Küsters, and Frank Wolter. Extensions to description logics. In *[2]*, pages 219–261. 2003.
6. Franz Baader and Werner Nutt. Basic description logics. In *[2]*, pages 43–95. 2003.
7. Franz Baader and Ulrike Sattler. An overview of tableau algorithms for description logics. *Studia Logica*, 69:5–40, 2001.
8. Martin Buchheit, Francesco M. Donini, and Andrea Schaerf. Decidable reasoning in terminological knowledge representation systems. *Journal of Artificial Intelligence Research*, 1:109–138, 1993.
9. J. W. Freeman. *Improvements to Propositional Satisfiability Search Algorithms.* PhD thesis, Department of Computer and Information Science, University of Pennsylvania, 1995.
10. V. Haarslev and R. Möller. Practical reasoning in racer with a concrete domain for linear inequations. In *Proceedings of the International Workshop on Description Logics (DL-2002), Toulouse, France, April 19-21*, pages 91–98, 2002.
11. V. Haarslev, R. Möller, and M. Wessel. The description logic alcnhr+ extended with concrete domains. Technical Report FBI-HH-M-290/00, University of Hamburg, Computer Science Department, 2000.
12. Volker Haarslev and Ralf Möller. Expressive abox reasoning with number restrictions, role hierarchies, and transitively closed roles. In *Proc. of the 7th Int. Conf. on the Principles of Knowledge Representation and Reasoning (KR 2000)*, pages 273–284. Morgan Kaufmann, 2000.
13. Volker Haarslev and Ralf Möller. High performance reasoning with very large knowledge bases: A practical case study. In *Proc. of the 17th Int. Joint Conf. on Artificial Intelligence (IJCAI 2001)*, pages 161–168, 2001.
14. Bernhard Hollunder and Franz Baader. Qualifying number restrictions in concept languages. In *Proc. of the 2nd Int. Conf. on the Principles of Knowledge Representation and Reasoning (KR'91)*, pages 335–346, 1991.
15. Ian Horrocks. Optimisation techniques for expressive description logics. Technical Report UMCS-97-2-1, University of Manchester, Department of Computer Science, 1997.
16. Ian Horrocks. Using an expressive description logic: FaCT or fiction? In *Proc. of the 6th Int. Conf. on the Principles of Knowledge Representation and Reasoning (KR'98)*, pages 636–647, 1998.

17. Ian Horrocks. Implementation and optimization techniques. In *[2]*, pages 306–346. 2003.
18. Ian Horrocks, Oliver Kutz, and Ulrike Sattler. The even more irresistible $\mathcal{SROIQ}$. In *Proc. of the 10th Int. Conf. on Principles of Knowledge Representation and Reasoning (KR 2006)*, pages 57–67. AAAI Press, 2006.
19. Ian Horrocks and Peter F. Patel-Schneider. Optimizing description logic subsumption. *Journal of Logic and Computation*, 9(3):267–293, 1999.
20. Ian Horrocks and Ulrike Sattler. A tableaux decision procedure for $\mathcal{SHOIQ}$. In *Proc. of the 19th Int. Joint Conf. on Artificial Intelligence (IJCAI 2005)*, pages 448–453, 2005.
21. Ian Horrocks, Ulrike Sattler, and Stephan Tobies. Reasoning with individuals for the description logic $\mathcal{SHIQ}$. In David McAllester, editor, *Proc. of the 17th Int. Conf. on Automated Deduction (CADE 2000)*, volume 1831 of *Lecture Notes in Computer Science*, pages 482–496. Springer, 2000.
22. Ian Horrocks and Stephan Tobies. Reasoning with axioms: Theory and practice. In *Proc. of the 7th Int. Conf. on the Principles of Knowledge Representation and Reasoning (KR 2000)*, pages 285–296, 2000.
23. Alexander. K. Hudek and Grant Weddell. Binary absorption in tableaux-based reasoning for description logics. In *Proc. of the 2006 Description Logic Workshop (DL 2006)*. CEUR Electronic Workshop Proceedings, `http://ceur-ws.org/Vol-189/`, 2006.
24. C. Lutz. PSPACE reasoning with the description logic $\mathcal{ALCF}(\mathcal{D})$. *Logic Journal of the IGPL*, 10(5):535–568, 2002.
25. C. Lutz and M. Milicic. A tableau algorithm for description logics with concrete domains and general tboxes. *Journal of Automated Reasoning*, 2006. To appear.
26. Ralf Möller and Volker Haarslev. Description logic systems. In *[2]*, pages 282–305. 2003.
27. Ulrike Sattler. A concept language extended with different kinds of transitive roles. In Günter Görz and Steffen Hölldobler, editors, *Proc. of the 20th German Annual Conf. on Artificial Intelligence (KI'96)*, volume 1137 of *Lecture Notes in Artificial Intelligence*, pages 333–345. Springer, 1996.
28. Andrea Schaerf. Reasoning with individuals in concept languages. In *Proc. of the 3rd Conf. of the Ital. Assoc. for Artificial Intelligence (AIIA'93)*, Lecture Notes in Artificial Intelligence. Springer, 1993.
29. Manfred Schmidt-Schauß and Gert Smolka. Attributive concept descriptions with complements. *Artificial Intelligence*, 48(1):1–26, 1991.
30. Stephan Tobies. The complexity of reasoning with cardinality restrictions and nominals in expressive description logics. *Journal of Artificial Intelligence Research*, 12:199–217, 2000.
31. Dmitry Tsarkov and Ian Horrocks. Efficient reasoning with range and domain constraints. In *Proc. of the 2004 Description Logic Workshop (DL 2004)*, pages 41–50, 2004.
32. Dmitry Tsarkov, Ian Horrocks, and Peter F. Patel-Schneider. Optimizing terminological reasoning for expressive description logics. *Journal of Automated Reasoning*, 39(3):277–316, 2007.

Resolution-Based Reasoning for Ontologies

Boris Motik

School of Computer Science, University of Manchester, Manchester, UK,
bmotik@cs.man.ac.uk

Summary. We overview the algorithms for reasoning with description logic (DL) ontologies based on resolution. These algorithms often have worst-case optimal complexity, and, by relying on vast experience in building resolution theorem provers, they can be implemented efficiently. Furthermore, we present a resolution-based algorithm that reduces a DL knowledge base into a disjunctive datalog program, while preserving the set of entailed facts. This reduction enables the application of optimization techniques from deductive databases, such as magic sets, to reasoning in DLs. This approach has proven itself in practice on ontologies with relatively small and simple TBoxes, but large ABoxes.

1 Introduction

Tableau algorithms, introduced in chapter "Tableau-Based Reasoning", are nowadays the state-of-the-art for reasoning with description logic (DL) ontologies. This is mainly due to optimizations of the original algorithm that heuristically guide the search for a model. DLs such as the ones underlying the Web Ontology Language (OWL) (see chapter "Web Ontology Language OWL") are, however, complex logics, so no one reasoning method can be identified as the best. Rather, comparing different methods and identifying which ones are suitable for which types of problems can give us crucial insights into building practical reasoning systems. Therefore, alternatives to tableau calculi have been explored in the past.

Resolution and its refinements [4] are nowadays the most widely used calculi for general-purpose first-order theorem proving. They have been implemented in a number of practical systems, of which Vampire [28] is one of the most successful one. The general applicability of resolution is partly due to the powerful *redundancy elimination rules*, which can drastically reduce the search space.

Since resolution has been quite successful as a general theorem proving technique, it is natural to apply it to ontology reasoning. Decision procedures

S. Staab and R. Studer (eds.), *Handbook on Ontologies*, International Handbooks on Information Systems, DOI 10.1007/978-3-540-92673-3,

for various DLs have been developed in the past. It turns out that, even for relatively complex DLs, resolution-based algorithms can be derived easily and are quite elegant. While tableau algorithms need sophisticated blocking techniques to ensure termination [8], resolution-based algorithms terminate automatically as a side-effect of the resolution calculus. Furthermore, many resolution-based procedures are worst-case optimal [13, 25].

In this chapter, we outline the principles underlying most known resolution-based procedures for DLs. After introducing the basic notions in Sect. 2, we present a decision procedure for the DL $\mathcal{ALCHI}$ in Sect. 3. This DL provides many features characteristic of the DL languages, such as full Boolean connectives, (restricted) existential and universal quantification, inverse roles, and role hierarchies. Furthermore, the resolution decision procedure for this DL conveys the basic principles without overloading the presentation with technical detail. We also overview the problems involved in extending the algorithm to more expressive DLs.

Deductive databases have been successfully applied to answering queries over large data sets, so it is natural to apply them to DL reasoning with large ABoxes. To enable this, in Sect. 4 we present a technique that reduces an $\mathcal{ALCHI}$ knowledge base to a disjunctive datalog program without affecting the set of entailed ground facts. Thus, one can answer DL queries using the resulting disjunctive program, and, in doing so, one can apply known optimization techniques such as magic sets [6]. This transformation can be derived easily from the basic resolution-based decision algorithm.

The techniques presented in this chapter have been implemented in the DL reasoner KAON2.[1] Practical experience has shown that the reduction-based techniques work quite well for ontologies with relatively small and simple TBoxes, but large and complex ABoxes [23].

2 Preliminaries

2.1 The Description Logic $\mathcal{ALCHI}$

Description logics have been introduced in detail in chapter "Description Logics", but, to make this chapter self-contained, we present the definition of the DL $\mathcal{ALCHI}$. For a set of role names N_R, a *role* is either some $R \in N_R$ or an *inverse role* R^- for $R \in N_R$. An *RBox* $\mathcal{R}$ is a finite set of role inclusion axioms $R \sqsubseteq S$. For a set of *concept names* N_C, the set of *concepts* is the smallest set containing $\top$, $\bot$, A, $\neg C$, $C \sqcap D$, $C \sqcup D$, $\exists R.C$, $\forall R.C$, where A is a concept name, C and D are concepts, and R is a role. A TBox $\mathcal{T}$ is a finite set of *concept inclusion axioms* $C \sqsubseteq D$, where C and D are concepts. For a set of *individuals* N_I, an ABox $\mathcal{A}$ is a finite set of assertions of the form $C(a)$, $R(a, b)$, and $\neg R(a, b)$, where C is a concept, R is a role, and a and b are

[1] `http://kaon2.semanticweb.org/`

Table 1. Semantics of $\mathcal{ALCHI}$ by mapping to FOL

Mapping roles to FOL			
$\pi_{xy}(R)$	$= R(x,y)$	$\pi_{yx}(R)$	$= R(y,x)$
$\pi_{xy}(R^-)$	$= R(y,x)$	$\pi_{yx}(R^-)$	$= R(x,y)$
Mapping concepts to FOL			
$\pi_x(\top)$	$= \top$	$\pi_y(\top)$	$= \top$
$\pi_x(\bot)$	$= \bot$	$\pi_y(\bot)$	$= \bot$
$\pi_x(A)$	$= A(x)$	$\pi_y(A)$	$= A(y)$
$\pi_x(\neg C)$	$= \neg\pi_x(C)$	$\pi_y(\neg C)$	$= \neg\pi_y(C)$
$\pi_x(C \sqcap D)$	$= \pi_x(C) \wedge \pi_x(D)$	$\pi_y(C \sqcap D)$	$= \pi_y(C) \wedge \pi_y(D)$
$\pi_x(C \sqcup D)$	$= \pi_x(C) \vee \pi_x(D)$	$\pi_y(C \sqcup D)$	$= \pi_y(C) \vee \pi_y(D)$
$\pi_x(\exists R.C)$	$= \exists y : \pi_{xy}(R) \wedge \pi_y(C)$	$\pi_y(\exists R.C)$	$= \exists x : \pi_{yx}(R) \wedge \pi_x(C)$
$\pi_x(\forall R.C)$	$= \forall y : \pi_{xy}(R) \rightarrow \pi_y(C)$	$\pi_y(\forall R.C)$	$= \forall x : \pi_{yx}(R) \rightarrow \pi_x(C)$
Mapping axioms to FOL			
$\pi(C \sqsubseteq D)$	$= \forall x : \pi_x(C) \rightarrow \pi_x(D)$		
$\pi(R \sqsubseteq S)$	$= \forall x, y : \pi_{xy}(R) \rightarrow \pi_{xy}(S)$		
$\pi(C(a))$	$= \pi_x(C)\{x \mapsto a\}$		
$\pi((\neg)R(a,b))$	$= (\neg)\pi_{xy}(R)\{x \mapsto a, y \mapsto b\}$		
$\pi(\mathcal{K})$	$= \bigwedge_{\alpha \in \mathcal{T} \cup \mathcal{R} \cup \mathcal{A}} \pi(\alpha)$		

individuals. An $\mathcal{ALCHI}$ knowledge base $\mathcal{K}$ is a triple $(\mathcal{R}, \mathcal{T}, \mathcal{A})$. With $|\mathcal{K}|$ we denote the number of symbols needed to encode $\mathcal{K}$. We say that $\mathcal{K}$ is *extensionally reduced* if, in all ABox assertions $C(a)$, the concept C is a concept name or the negation of a concept name. Any $\mathcal{K}$ can be made extensionally reduced by replacing each assertion $C(a)$ where C is not of the appropriate form with an assertion $A_C(a)$ and an axiom $A_C \sqsubseteq C$, for A_C a new concept name.

In chapter "Description Logics", DLs are given a direct model-theoretic semantics. In this chapter, however, we use an equivalent semantics based on translation into first-order logic. In particular, we translate an $\mathcal{ALCHI}$ knowledge base $\mathcal{K}$ into a first-order formula $\pi(\mathcal{K})$, where π is the operator defined in Table 1. It is well-known that these two semantics are equivalent [7].

The basic inference problem for $\mathcal{ALCHI}$ is *checking satisfiability* of $\mathcal{K}$– that is, checking whether $\pi(\mathcal{K})$ is a satisfiable first-order formula. As discussed in chapter "Description Logics", other inference problems can be reduced to knowledge base satisfiability.

The *negation-normal form* $\mathsf{nnf}(C)$ of a concept C is the concept equivalent to C in which negation occurs only in front of concept names. The concept $\mathsf{nnf}(C)$ can be computed in time polynomial in the size of C by well-known transformations [2].

2.2 The Ordered Resolution Calculus

We use the well-known definitions of constants, variables, function symbols, terms, predicates, and formulae of first-order logic [4]. An *atom* A is a formula of the form $P(t_1, \ldots, t_n)$, where P is a predicate and t_i are terms. A *literal* L

is a positive atom A or a negative atom $\neg A$. A *clause* is a multiset of literals and is written as $L_1 \vee \cdots \vee L_n$. The *empty clause* is written as $\square$. Terms and formulae that do not contain variables are called *ground*. We say that formulae φ and ψ are *equisatisfiable* if φ is satisfiable if and only if ψ is satisfiable.

A *substitution* is mapping of variables to terms that is not identity on a finite number of variables; we often write it as $\{x_1 \mapsto t_1, \ldots, x_n \mapsto t_n\}$. An application of a substitution σ to a term t (formula φ) is written $t\sigma$ ($\varphi\sigma$) and it is the term (formula) obtained by replacing each free occurrence of a variable x with $x\sigma$. A substitution σ is a *unifier* of terms s and t if $s\sigma = t\sigma$. A unifier σ of s and t is called a *most general unifier* if, for each unifier η of s and t, a substitution ξ exists such that $x\eta = (x\sigma)\xi$ for every variable x. If a most general unifier σ of s and t exists, it is unique up to variable renaming [3], so we write $\sigma = \mathsf{MGU}(s,t)$.

The *skolemization* of a formula φ, written $\mathsf{sk}(\varphi)$, is obtained from φ by successively replacing each subformula $\exists x : \psi$ occurring positively or a subformula $\forall x : \psi$ occurring negatively in φ with a formula $\psi\{x \mapsto f(x_1, \ldots, x_n)\}$, where f is a new function symbol and $x_1, \ldots, x_n$ are the free variables of ψ different from x. It is well-known that φ and $\mathsf{sk}(\varphi)$ are equisatisfiable [26]. Finally, $\mathsf{Cls}(\varphi)$ is the set of clauses that is equisatisfiable with φ and is obtained from transforming $\mathsf{sk}(\varphi)$ into conjunctive normal form using the well-known transformations.

Ordered resolution [4] is a calculus that can be used to prove that a formula φ is unsatisfiable. Ordered resolution is a clausal calculus, so it cannot be applied to φ directly. First, one must compute $\mathsf{Cls}(\varphi)$. Next, one must fix the calculus' parameters. The first parameter is an *admissible* ordering on literals $\succ$ – that is, an ordering that is (1) well-founded, stable under substitutions (i.e., $L_1 \succ L_2$ implies $L_1\sigma \succ L_2\sigma$ for all literals L_1 and L_2 and each substitution σ), and total on ground literals; (2) $\neg A \succ A$ for all ground atoms A; and (3) $B \succ A$ implies $B \succ \neg A$ for all atoms A and B. A literal L is maximal w.r.t. a clause C if there is no literal $L' \in C$ such that $L' \succ L$, and L is strictly maximal w.r.t. C if there is no $L' \in C$ such that $L' \succeq L$. The second parameter is a *selection function*, which assigns to each clause C a possibly empty subset of negative literals of C.

An *inference rule* is a template that specifies how a conclusion is derived given a set of premises; an *inference* is an application of an inference rule to concrete premises. With $\mathcal{R}$ we denote the ordered resolution calculus, consisting of the following inference rules, where the clauses $C \vee A \vee B$ and $D \vee \neg B$ are called the *main premises*, $C \vee A$ is called the *side premise*, and $C\sigma \vee A\sigma$ and $C\sigma \vee D\sigma$ are called *conclusions* (as usual in resolution theorem proving, we make a technical assumption that the premises do not have variables in common):

Positive factoring: $$\frac{C \vee A \vee B}{C\sigma \vee A\sigma}$$

where (*i*) $\sigma = \mathsf{MGU}(A, B)$, (*ii*) $A\sigma$ is maximal with respect to $C\sigma \vee B\sigma$ and no literal is selected in $C\sigma \vee A\sigma \vee B\sigma$.

Ordered resolution: $$\frac{C \vee A \quad D \vee \neg B}{C\sigma \vee D\sigma}$$

where (i) $\sigma = \mathsf{MGU}(A, B)$, (ii) $A\sigma$ is strictly maximal with respect to $C\sigma$ and no literal is selected in $C\sigma \vee A\sigma$, (iii) $\neg B\sigma$ is either selected in $D\sigma \vee \neg B\sigma$, or it is maximal with respect to $D\sigma$ and no literal is selected in $D\sigma \vee \neg B\sigma$.

Ordered resolution is compatible with powerful *redundancy elimination techniques*, which allow deleting certain clauses during the theorem proving process without loss of completeness [4]. If a clause C is redundant in some set of clauses N, then C can be safely removed from N.

If a clause C is a tautology, then it is redundant in any set of clauses N. A sound and complete tautology check would itself require theorem proving, and would therefore be difficult to realize. Therefore, one usually only checks for *syntactic tautologies* – that is, clauses containing the literals A and $\neg A$. A clause C *subsumes* a clause D if there is a substitution σ such that $C\sigma \subseteq D$ and $|C| < |D|$. If a clause C is subsumed by a clause from a set of clauses N, then C is redundant in N.

A *derivation* by $\mathcal{R}$ from a set of clauses N is a sequence of sets of clauses $N_0, N_1, \ldots$ such that $N_0 = N$ and, for $i \geq 0$, either (1) $N_{i+1} = N_i \cup \{C\}$ where C is the conclusion of an inference by $\mathcal{R}$ from premises in N_i, or (2) $N_{i+1} = N_i \setminus \{C\}$ where C is redundant in N_i. Each derivation must be *fair* [4]; intuitively, this means that each applicable inference is performed after a finite number of steps. Ordered resolution is sound and complete [4]: if $\square \in N_i$ where N_i is derived by $\mathcal{R}$ from a set of clauses N_0, then N_0 is unsatisfiable; conversely, if N_0 is unsatisfiable, then, for each fair derivation by $\mathcal{R}$ from N_0, an integer i exists such that $\square \in N_i$. The process of computing a derivation by $\mathcal{R}$ from N_0 is called a *saturation* of N_0 by $\mathcal{R}$.

2.3 Disjunctive Datalog

We recapitulate the basic notions of disjunctive datalog [11]. A *datalog term* is a constant or a variable, and a *datalog atom* has the form $A(t_1, \ldots, t_n)$ or $t_1 \approx t_2$, where t_i are datalog terms. A *disjunctive datalog program with equality* P is a finite set of rules of the form $A_1 \vee \cdots \vee A_n \leftarrow B_1, \ldots, B_m$ where A_i and B_j are datalog atoms. The literals A_i are called *head literals*, whereas the literals B_i are called *body literals*. Each rule is required to be *safe* – that is, each variable occurring in the rule must occur in at least one body atom. A *fact* is a rule with $m = 0$. For the semantics, we take a rule to be equivalent to a clause $A_1 \vee \cdots \vee A_n \vee \neg B_1 \vee \cdots \vee \neg B_m$. We consider only Herbrand models, and say that a model M of P is *minimal* if there is no model M' of P such that $M' \subsetneq M$. A ground literal A is a *cautious answer* of P (written $P \models_c A$) if A is true in all minimal models of P. First-order entailment coincides with cautious entailment for positive ground atoms.

3 Deciding Satisfiability of $\mathcal{ALCHI}$ by Resolution

The fundamental principles for deciding a first-order fragment $\mathcal{L}$ by resolution have been established by Joyner [17]. First, one selects a sound and complete clausal calculus $\mathcal{C}$. Second, one identifies the set of clauses $\mathcal{N}_\mathcal{L}$ such that (1) $\mathcal{N}_\mathcal{L}$ is finite for a finite signature and (2) the translation of each formula $\varphi \in \mathcal{L}$ into clauses produces only clauses from $\mathcal{N}_\mathcal{L}$. Third, one demonstrates that $\mathcal{N}_\mathcal{L}$ is *closed* under $\mathcal{C}$; that is, one shows that applying an inference of $\mathcal{C}$ to clauses from $\mathcal{N}_\mathcal{L}$ produces a clause in $\mathcal{N}_\mathcal{L}$. This is sufficient to obtain a refutation decision procedure for $\mathcal{L}$: given any formula $\varphi \in \mathcal{L}$, a saturation by $\mathcal{C}$ of the clauses corresponding to φ will, in the worst case, derive all clauses of $\mathcal{N}_\mathcal{L}$. In this section, we apply these principles to obtain a procedure for checking satisfiability of an $\mathcal{ALCHI}$ knowledge base $\mathcal{K}$.

3.1 Translating the Knowledge Base into Clauses

The first step in deciding satisfiability of $\mathcal{K}$ is to transform $\mathcal{K}$ into an equisatisfiable set of clauses $\Xi(\mathcal{K})$. A straightforward way of doing so is to compute $\mathsf{Cls}(\pi(\mathcal{K}))$. Such an approach, however, has two important drawbacks. First, the size of the resulting clause set could be exponential in the size of $\pi(\mathcal{K})$, due to nesting of $\sqcap$ and $\sqcup$. Second, we should exploit the structure of the formula $\pi(\mathcal{K})$ in our algorithm, but $\mathsf{Cls}(\pi(\mathcal{K}))$ does not reflect this structure. To avoid these problems, we preprocess $\mathcal{K}$ using the *structural transformation* [26,27].

Definition 1. *For an $\mathcal{ALCHI}$ knowledge base $\mathcal{K}$, the knowledge base $\Theta(\mathcal{K})$ is computed as shown in Table 2.*

Intuitively, this transformation replaces complex concepts with simpler ones. The knowledge base $\Theta(\mathcal{K})$ does not contain $\sqcap$, so it can be translated into clauses without an exponential blowup.

Table 2. Structural transformation of $\mathcal{K}$

$\Theta(\mathcal{K})$	$= \bigcup_{\alpha \in \mathcal{R} \cup \mathcal{A}} \Theta(\alpha) \cup \bigcup_{C_1 \sqsubseteq C_2 \in \mathcal{T}} \Theta(\top \sqsubseteq \mathsf{nnf}(\neg C_1 \sqcup C_2))$
$\Theta(A \sqsubseteq B)$	$= \{A \sqsubseteq B\}$
$\Theta(A \sqsubseteq \neg B)$	$= \{A \sqsubseteq \neg B\}$
$\Theta(A \sqsubseteq C_1 \sqcap C_2)$	$= \Theta(A \sqsubseteq C_1) \cup \Theta(A \sqsubseteq C_2)$
$\Theta(A \sqsubseteq C_1 \sqcup C_2)$	$= \{A \sqsubseteq Q_{C_1} \sqcup Q_{C_2}\} \cup \Theta(Q_{C_1} \sqsubseteq C_1) \cup \Theta(Q_{C_2} \sqsubseteq C_2)$
$\Theta(A \sqsubseteq \exists R.C)$	$= \{A \sqsubseteq \exists R.Q_C\} \cup \Theta(Q_C \sqsubseteq C)$
$\Theta(A \sqsubseteq \forall R.C)$	$= \{A \sqsubseteq \forall R.Q_C\} \cup \Theta(Q_C \sqsubseteq C)$
$\Theta(R \sqsubseteq S)$	$= \{R \sqsubseteq S\}$
$\Theta(C(a))$	$= \{Q_C(a)\} \cup \Theta(Q_C \sqsubseteq C)$
$\Theta((\neg)R(a,b))$	$= \{(\neg)R(a,b)\}$

Note: A and B are concept names or $\top$; C, C_1, and C_2 are arbitrary concepts; R and S are roles; and Q_X is a new concept name not occurring in $\mathcal{K}$ that is unique for X.

Lemma 1. *An $\mathcal{ALCHI}$ knowledge base $\mathcal{K}$ and $\Theta(\mathcal{K})$ are equisatisfiable.*

Proof. Consider a single application of Θ. It is obvious that the axioms obtained after the transformation imply the axiom before the transformation, which proves the ($\Leftarrow$) direction. For the ($\Rightarrow$) direction, simply observe that each interpretation I of $\mathcal{K}$ can be extended to an interpretation I' of $\Theta(\mathcal{K})$ by interpreting each newly introduced concept Q_X as X. □

To obtain a set of clauses corresponding to $\mathcal{K}$, we translate $\Theta(\mathcal{K})$ into first-order logic using the operator π from Table 1, skolemize it, and transform the result into conjunctive normal form. This is captured by the following definition:

Definition 2. *For an $\mathcal{ALCHI}$ knowledge base $\mathcal{K}$, let $\Xi(\mathcal{K}) = \mathsf{Cls}(\pi(\Theta(\mathcal{K})))$.*

We now show that clausification does not affect the satisfiability of a knowledge base, and that it produces clauses of a certain syntactic structure:

Lemma 2. *The following claims hold for each $\mathcal{ALCHI}$ knowledge base $\mathcal{K}$:*

1. *$\mathcal{K}$ is satisfiable if and only if $\Xi(\mathcal{K})$ is satisfiable.*
2. *$\Xi(\mathcal{K})$ can be computed in time polynomial in $|\mathcal{K}|$.*
3. *Each clause in $\Xi(\mathcal{K})$ is of the form as shown in Table 3.*

Proof. (1) Equisatisfiability of $\mathcal{K}$ and $\Xi(\mathcal{K})$ is a direct consequence of Lemma 1. (2) The number of recursive invocations of Θ and the number of new concepts Q_X are linear in $|\mathcal{K}|$. Hence, $|\Theta(\mathcal{K})|$ is linear in $|\mathcal{K}|$, so $|\Xi(\mathcal{K})|$ is polynomial in $|\mathcal{K}|$. (3) It is easy to see that $\Theta(\mathcal{K})$ contains only axioms from the left-hand side of Table 3, which are translated into clauses as shown on the right-hand side of the table. □

Table 3. Clause types after clausification

Axiom	Clause
$R \sqsubseteq S$	$\neg R(x,y) \vee S(x,y)$
$R^- \sqsubseteq S^-$	$\neg R(y,x) \vee S(y,x)$
$R \sqsubseteq S^-$	$\neg R(x,y) \vee S(y,x)$
$R^- \sqsubseteq S$	$\neg R(y,x) \vee S(x,y)$
$A \sqsubseteq \bigsqcup (\neg) B_i$	$\neg A(x) \vee \bigvee (\neg) B_i(x)$
$A \sqsubseteq \exists R.B$	$\neg A(x) \vee R(x, f(x))$ $\neg A(x) \vee B(f(x))$
$A \sqsubseteq \exists R^-.B$	$\neg A(x) \vee R(f(x), x)$ $\neg A(x) \vee B(f(x))$
$A \sqsubseteq \forall R.B$	$\neg A(x) \vee \neg R(x,y) \vee B(y)$
$A \sqsubseteq \forall R^-.B$	$\neg A(x) \vee \neg R(y,x) \vee B(y)$
$A(c)$	$A(c)$
$(\neg)R(c,d)$	$(\neg)R(c,d)$
$(\neg)R^-(c,d)$	$(\neg)R(d,c)$

Note: The function symbol f is different for each axiom.

3.2 Saturation by Ordered Resolution

Since ordered resolution ($\mathcal{R}$) is a sound and complete calculus, we can use it to check satisfiability of $\Xi(\mathcal{K})$. To obtain a decision procedure, we just need to ensure that each saturation of $\Xi(\mathcal{K})$ by $\mathcal{R}$ terminates; that is, we must ensure that we can derive only finitely many clauses from $\Xi(\mathcal{K})$ by applying the rules of $\mathcal{R}$. There are two main reasons why we might derive an infinite number of clauses.

First, we might derive clauses with ever deeper terms. This is shown by the following example, in which the selected literals are underlined:

$$\dfrac{\dfrac{\underline{C(a)} \quad \underline{\neg C(x)} \vee C(f(x))}{\underline{C(f(a))}} \qquad \underline{\neg C(x)} \vee C(f(x))}{C(f(f(a)))}$$

Second, we might derive clauses with an unbounded number of variables. For example, the following inference increases the number of variables by one, and repeating it for the conclusion produces clauses with an arbitrary number of variables:

$$\dfrac{\neg C(x) \vee \neg R(x, y) \vee \underline{C(y)} \qquad \underline{\neg C(y)} \vee \neg R(y, z) \vee C(z)}{\neg C(x) \vee \neg R(x, y) \vee \neg R(y, z) \vee C(z)}$$

The inferences that ordered resolution performs on a given set of premises are determined by the parameters of the calculus – the literal ordering and the selection function. By choosing these parameters appropriately, we can restrict the resolution inferences in a way that allows us to establish a bound on the term depth and on the number of variables. In the first example, if we ensure that $C(f(x)) \succ \neg C(x)$, then the second premise can participate in an inference only on literal $C(f(x))$; since $C(f(x))$ and $C(a)$ do not unify, no inference of $\mathcal{R}$ is applicable to $\underline{C(a)}$ and $\neg C(x) \vee \underline{C(f(x))}$. In the second example, the undesirable inference can be prevented if we select $\neg R(x, y)$.

The following definition fixes the parameters for $\mathcal{R}$ that, as we shall see shortly, restrict the inferences on $\Xi(\mathcal{K})$ in a way which ensures termination.

Definition 3. *Let $\mathcal{R}_{DL}$ denote the calculus $\mathcal{R}$ parameterized as follows:*

- *The literal ordering is any admissible ordering $\succ$ such that, for all function symbols f and predicates R, C, and D, we have $R(x, f(x)) \succ \neg C(x)$ and $D(f(x)) \succ \neg C(x)$.*
- *The selection function selects every negative binary literal in each clause.*

An ordering compatible with Definition 3 can be obtained by instantiating a *lexicographic path ordering* [10]; see [22, Sect. 4.4] for details.

Table 4. Types of $\mathcal{ALCHI}$-clauses

1	$\neg R(x,y) \vee S(x,y)$
2	$\neg R(x,y) \vee S(y,x)$
3	$\mathbf{P}(x) \vee R(x,f(x))$
4	$\mathbf{P}(x) \vee R(f(x),x)$
5	$\mathbf{P_1}(x) \vee \mathbf{P_2}(f(x))$
6	$\mathbf{P_1}(x) \vee \neg R(x,y) \vee \mathbf{P_2}(y)$
7	$\mathbf{P(a)}$
8	$(\neg)R(a,b)$

Note: $\mathbf{P}(t)$ is a possibly empty disjunction of the form $(\neg)P_1(t) \vee \cdots \vee (\neg)P_n(t)$ for t a term of the form x, $f(x)$, or a; $\mathbf{P(a)}$ is a possibly empty disjunction of the form $P_1(a_1) \vee \cdots \vee P_m(a_m)$; and the empty clause $\square$ is of type 5.

It is easy to see that an application of $\mathcal{R}_{DL}$ to clauses from Table 3 can produce clauses of the form not shown in the table. Therefore, we generalize Table 3 to *$\mathcal{ALCHI}$-clauses*, shown in Table 4. It is easy to see that $\Xi(\mathcal{K})$ contains only $\mathcal{ALCHI}$-clauses. As we show next, when applied to $\mathcal{ALCHI}$-clauses, each $\mathcal{R}_{DL}$ inference produces an $\mathcal{ALCHI}$-clause.

Lemma 3. *Each $\mathcal{R}_{DL}$ inference, when applied to $\mathcal{ALCHI}$-clauses, produces an $\mathcal{ALCHI}$-clause.*

Proof. We summarize all possible $\mathcal{R}_{DL}$ inferences on all types of $\mathcal{ALCHI}$-clauses in Table 5. For the sake of brevity, we omit inferences in which participating literals are complemented. The notation $n + m = k$ above each inference means that the inference premises are of types n and m, and the conclusion is of type k. Due to the requirement on the literal ordering $\succ$, a literal of the form $(\neg)A(x)$ occurring in a clause C can participate in an inference only if C does not contain a literal of the form $(\neg)B(f(x))$ or $R(x,f(x))$. Furthermore, a ground literal $A(a)$ does not unify with a literal $A(f(x))$, and $R(a,b)$ does not unify with $R(x,f(x))$. Hence, ground clauses can participate only in inferences with clauses not containing terms of the form $f(x)$. One can easily see that the conclusion is always an $\mathcal{ALCHI}$ clause. $\square$

The following lemma shows that the number of $\mathcal{ALCHI}$-clauses is finite for a finite knowledge base $\mathcal{K}$. In fact, the bound on the number of derivable clauses can be used to estimate the complexity of the algorithm.

Lemma 4. *For an $\mathcal{ALCHI}$ knowledge base $\mathcal{K}$, the longest $\mathcal{ALCHI}$-clause over the signature of $\Xi(\mathcal{K})$ is polynomial in $|\mathcal{K}|$, and the number of such clauses different up to variable renaming is exponential in $|\mathcal{K}|$.*

Proof. The number c of unary predicates in the signature of $\Xi(\mathcal{K})$ is linear in $|\mathcal{K}|$, since each concept introduced by Θ corresponds to one nonliteral subconcept of C. Similarly, the number f of unary function symbols in the signature

Table 5. Possible inferences by $\mathcal{R}_{DL}$ on $\mathcal{ALCHI}$-clauses

1 + 3 = 3:

$$\frac{\underline{\neg R(x,y)} \vee S(x,y) \quad \mathbf{P}(x) \vee \underline{R(x,f(x))}}{\mathbf{P}(x) \vee S(x,f(x))}$$

2 + 3 = 4:

$$\frac{\underline{\neg R(x,y)} \vee S(y,x) \quad \mathbf{P}(x) \vee \underline{R(x,f(x))}}{\mathbf{P}(x) \vee S(f(x),x)}$$

1 + 4 = 4:

$$\frac{\underline{\neg R(x,y)} \vee S(x,y) \quad \mathbf{P}(x) \vee \underline{R(f(x),x)}}{\mathbf{P}(x) \vee S(f(x),x)}$$

2 + 4 = 3:

$$\frac{\underline{\neg R(x,y)} \vee S(y,x) \quad \mathbf{P}(x) \vee \underline{R(f(x),x)}}{\mathbf{P}(x) \vee S(x,f(x))}$$

6 + 3 = 5:

$$\frac{\mathbf{P_1}(x) \vee \underline{\neg R(x,y)} \vee \mathbf{P_2}(y) \quad \mathbf{P}(x) \vee \underline{R(x,f(x))}}{\mathbf{P}(x) \vee \mathbf{P_1}(x) \vee \mathbf{P_2}(f(x))}$$

6 + 4 = 5:

$$\frac{\mathbf{P_1}(x) \vee \underline{\neg R(x,y)} \vee \mathbf{P_2}(y) \quad \mathbf{P}(x) \vee \underline{R(f(x),x)}}{\mathbf{P}(x) \vee \mathbf{P_1}(f(x)) \vee \mathbf{P_2}(x)}$$

5 + 5 = 5:

$$\frac{\mathbf{P_1}(x) \vee \mathbf{P_2}(f(x)) \vee \underline{\neg A(f(x))} \quad \underline{A(x)} \vee \mathbf{P_3}(x)}{\mathbf{P_1}(x) \vee \mathbf{P_2}(f(x)) \vee \mathbf{P_3}(f(x))}$$

5 + 5 = 5:

$$\frac{\mathbf{P_1}(x) \vee \underline{\neg A(x)} \quad \underline{A(x)} \vee \mathbf{P_2}(x)}{\mathbf{P_1}(x) \vee \mathbf{P_2}(x)}$$

5 + 5 = 5:

$$\frac{\mathbf{P_1}(x) \vee \mathbf{P_2}(f(x)) \vee \underline{\neg A(f(x))} \quad \underline{A(f(x))} \vee \mathbf{P_3}(f(x)) \vee \mathbf{P_4}(x)}{\mathbf{P_1}(x) \vee \mathbf{P_2}(f(x)) \vee \mathbf{P_3}(f(x)) \vee \mathbf{P_4}(x)}$$

---

7 + 5 = 7:

$$\frac{\mathbf{P_1}(\mathbf{a}) \vee \underline{\neg A(b)} \quad \underline{A(x)} \vee \mathbf{P_2}(x)}{\mathbf{P_1}(\mathbf{a}) \vee \mathbf{P_2}(b)}$$

7 + 7 = 7:

$$\frac{\mathbf{P_1}(\mathbf{a}) \vee \underline{\neg A(b)} \quad \underline{A(b)} \vee \mathbf{P_2}(\mathbf{c})}{\mathbf{P_1}(\mathbf{a}) \vee \mathbf{P_2}(\mathbf{c})}$$

8 + 1 = 8:

$$\frac{\underline{R(a,b)} \quad \underline{\neg R(x,y)} \vee S(x,y)}{S(a,b)}$$

8 + 2 = 8:

$$\frac{\underline{R(a,b)} \quad \underline{\neg R(x,y)} \vee S(y,x)}{S(b,a)}$$

8 + 6 = 7:

$$\frac{\underline{R(a,b)} \quad \mathbf{P_1}(x) \vee \underline{\neg R(x,y)} \vee \mathbf{P_2}(y)}{\mathbf{P_1}(a) \vee \mathbf{P_2}(b)}$$

8 + 8 = 5:

$$\frac{\underline{R(a,b)} \quad \underline{\neg R(a,b)}}{\square}$$

of $\Xi(\mathcal{K})$ is linear in $|\mathcal{K}|$, since each function symbol is introduced by skolemizing one concept of the form $\exists R.C$. Consider now the longest $\mathcal{ALCHI}$-clause Cl_6 of type 6. Such a clause contains a possibly negated literal $A(x)$ for each unary predicate A, and a possibly negated literal $A(f(x))$ for each pair of unary predicate and function symbols, yielding at most $\ell = 2c + 2cf$ literals, which is polynomial in $|\mathcal{K}|$. Each $\mathcal{ALCHI}$-clause of type 2 is a subset of Cl_6, so there are 2^ℓ such clauses; that is, the number of clauses is exponential in $|\mathcal{K}|$. For other $\mathcal{ALCHI}$-clause types, the bounds on the length and on the number of clauses can be derived in an analogous way. □

We now state the main result of this section:

Theorem 1. *For an $\mathcal{ALCHI}$ knowledge base $\mathcal{K}$, saturating $\Xi(\mathcal{K})$ by $\mathcal{R}_{DL}$ decides satisfiability of $\mathcal{K}$ and runs in time that is at most exponential in $|\mathcal{K}|$.*

Proof. By Lemma 4, the number of clauses derivable by $\mathcal{R}_{DL}$ from $\Xi(\mathcal{K})$ is exponential in $|\mathcal{K}|$. Each inference can be performed in time polynomial in the size of clauses. Hence, the saturation terminates after performing at most an exponential number of steps. Since $\mathcal{R}_{DL}$ is sound and complete, it decides satisfiability of $\Xi(\mathcal{K})$, and by Lemma 2 of $\mathcal{K}$ as well, in time that is exponential in $|\mathcal{K}|$. □

3.3 An Example

We now present a simple example. Let $\mathcal{K}$ be the following knowledge base:

$$\exists S.A \sqsubseteq \exists R.B \tag{1}$$

$$B \sqsubseteq C \tag{2}$$

$$\exists R.C \sqsubseteq D \tag{3}$$

$$S(a, b) \tag{4}$$

$$A(b) \tag{5}$$

Let us assume that we want to check whether $\mathcal{K} \models D(a)$; as shown in chapter "Description Logics", this so if and only if $\mathcal{K} \cup \{\neg D(a)\}$ is unsatisfiable. Hence, let $\mathcal{K}'$ be the knowledge base $\mathcal{K}$ extended with the assertion $\neg D(a)$.

To check satisfiability of $\mathcal{K}'$ using resolution, we first apply structural transformation. For (1), we obtain the following:

$$\Theta(\top \sqsubseteq \forall S.\neg A \sqcup \exists R.B) = \{\top \sqsubseteq Q_1 \sqcup Q_2\} \cup \Theta(Q_1 \sqsubseteq \forall S.\neg A) \cup \Theta(Q_2 \sqsubseteq \exists R.B)$$

By Definition (1), we should introduce a new name for the concepts $\neg A$ and B; however, both $Q_1 \sqsubseteq \forall S.\neg A$ and $Q_2 \sqsubseteq \exists R.B$ can be translated into $\mathcal{ALCHI}$-clauses in a straightforward way. Hence, we do not further apply Θ, and neither we do so for (2) and (3). We obtain the set $\Xi(\mathcal{K}')$ as follows (the meaning of underlining will be explained shortly):

$$\begin{array}{rcll}
\top \sqsubseteq Q_1 \sqcup Q_2 & \rightsquigarrow & Q_1(x) \vee \underline{Q_2(x)} & (6)\\
Q_1 \sqsubseteq \forall S.\neg A & \rightsquigarrow & \neg Q_1(x) \vee \underline{\neg S(x,y)} \vee \neg A(y) & (7)\\
Q_2 \sqsubseteq \exists R.B & \rightsquigarrow & \neg Q_2(x) \vee \underline{R(x,f(x))} & (8)\\
Q_2 \sqsubseteq \exists R.B & \rightsquigarrow & \neg Q_2(x) \vee \underline{B(f(x))} & (9)\\
B \sqsubseteq C & \rightsquigarrow & \neg B(x) \vee \underline{C(x)} & (10)\\
\exists R.C \sqsubseteq D & \rightsquigarrow & D(x) \vee \underline{\neg R(x,y)} \vee \neg C(y) & (11)\\
S(a,b) & \rightsquigarrow & \underline{S(a,b)} & (12)\\
A(b) & \rightsquigarrow & \underline{A(b)} & (13)\\
\neg D(a) & \rightsquigarrow & \underline{\neg D(a)} & (14)
\end{array}$$

To saturate $\Xi(\mathcal{K}')$ by $\mathcal{R}_{DL}$, we use a literal ordering $\succ$ compatible with Definition 3, where we break ties by comparing predicates alphabetically. The literals that are either selected or maximal are underlined. We now saturate $\Xi(\mathcal{K}')$; R(xx+yy) means that a clause was obtained by resolving (xx) and (yy).

$$\begin{array}{rll}
D(x) \vee \neg Q_2(x) \vee \underline{\neg C(f(x))} & \text{R}(8{+}11) & (15)\\
D(x) \vee \neg Q_2(x) \vee \underline{\neg B(f(x))} & \text{R}(15{+}10) & (16)\\
D(x) \vee \underline{\neg Q_2(x)} & \text{R}(16{+}9) & (17)\\
D(x) \vee \underline{Q_1(x)} & \text{R}(17{+}6) & (18)\\
\underline{\neg Q_1(a)} \vee \neg A(b) & \text{R}(7{+}12) & (19)\\
\underline{D(a)} \vee \neg A(b) & \text{R}(18{+}19) & (20)\\
\underline{\neg A(b)} & \text{R}(14{+}20) & (21)\\
\square & \text{R}(13{+}21) & (22)
\end{array}$$

We derived the empty clause, so the set of clauses $\Xi(\mathcal{K}')$ is unsatisfiable, and so is $\mathcal{K}'$, which implies $\mathcal{K} \models D(a)$.

3.4 Extending the Algorithm to the More Expressive DLs

We now overview the problems encountered in extending this basic algorithm to more expressive DLs and point to the relevant literature for the solutions.

Boolean Role Expressions

The DL $\mathcal{ALCB}$ [25] is obtained from $\mathcal{ALCHI}$ by allowing for concepts $\forall E.C$ and $\exists E.C$ and axioms $E_1 \sqsubseteq E_2$, where $E_{(i)}$ are *Boolean role expressions* R, $\neg E$, $E_1 \sqcup E_2$, and $E_1 \sqcap E_2$. As shown in [25], $\mathcal{ALCB}$ can easily be decided by extending the algorithm from this section. The main difference is that translating an $\mathcal{ALCB}$ knowledge base to clauses can produce clauses of the following form:

$$\neg R_1(x,y) \vee \cdots \vee \neg R_n(x,y) \vee S_1(x,y) \vee \cdots \vee S_m(x,y) \tag{23}$$

If $n = 0$, such clauses can cause termination problems. For example, resolving the clauses (24) and (25) produces the clause (26):

$$\underline{R(x,y)} \tag{24}$$
$$A(x) \vee \underline{\neg R(x,y)} \vee B(y) \tag{25}$$
$$A(x) \vee \underline{B(y)} \tag{26}$$

The clause (26) contains two clauses of type 6 that do not share a variable. Resolving such clauses with other clauses of that form can easily produce clauses with an arbitrary number of variables. For example, resolving (26) with (27) produces (28), which contains more variables than either of the premises:

$$\underline{\neg B(y)} \vee C(y) \vee D(z) \tag{27}$$
$$A(x) \vee C(y) \vee D(z) \tag{28}$$

This problem, however, can be solved in a simple way: since $A(x)$ and $B(y)$ are variable-disjoint, similarly as in the DPLL procedure [9], we can *split* the clause (26) into $A(x)$ or $B(y)$ – that is, we can guess which subclause is true. This reduces (26) to a clause of type 6, which does not cause termination problems. Splitting makes the procedure nondeterministic: deriving the empty clause under one of the guesses does not mean that the original clause set is unsatisfiable; rather, we must derive the empty clause under all possible guesses. Hence, such an algorithm runs in NExpTime. This is worst-case optimal, since $\mathcal{ALCB}$ is an NExpTime-complete logic [21].

Transitivity Axioms

Many DLs allow roles to be declared as transitive [12]. Translation of transitivity axioms produces clauses of the following form:

$$\neg R(x,y) \vee \underline{\neg R(y,z)} \vee R(x,z) \tag{29}$$

Such clauses are difficult for resolution. For example, if we also have the clause (30), then it can be resolved with (29) to produce (31):

$$A(x') \vee \underline{R(x',f(x'))} \tag{30}$$
$$\underline{\neg R(x,x')} \vee A(x') \vee R(x,f(x')) \tag{31}$$

Clause (31) is similar to (30), but it contains two variables; hence, further resolution inferences with (31) might produce clauses with even more variables.

To prevent the increase in the number of variables, one might select the negative literal in (31). While this prevents the introduction of arbitrarily many variables, it allows the derivation of arbitrarily deep terms; for example, a resolution of (30) and (31) produces the following clause:

$$A(x) \vee R(x, f(f(x))) \tag{32}$$

There are several ways to address this problem. In [18], resolution has been extended with simplification rules that transform clauses of the form (31) and (32) into simpler clauses without affecting satisfiability.

Another solution is to replace transitivity axioms with new concept inclusion axioms that capture the effects of the transitivity axioms. Roughly speaking, a transitivity axiom $\mathsf{Trans}(S)$ is replaced with axioms $\forall R.C \sqsubseteq \forall S.(\forall S.C)$, for each R with $S \sqsubseteq^* R$ and C a "relevant" concept from $\mathcal{K}$; for more details, please see [22, Sect. 5.2]. Similar encodings have been considered in modal logic [29] and in DLs with role conjunctions [30].

Number Restrictions

As explained in chapter "Description Logics", many DLs provide for number restrictions $\geqslant n\, R.C$ and $\leqslant n\, R.C$. The algorithm from this section can be extended to such concepts by using the well-known translation of number restrictions into first-order logic:

$$\geqslant n\, R.C \;\rightsquigarrow\; \exists y_1, \ldots, y_n : \bigwedge_{1 \leq i \leq n+1} [R(x, y_i) \wedge C(y_i)] \wedge \bigwedge_{1 \leq i < j \leq n} y_i \not\approx y_j$$

$$\leqslant n\, R.C \;\rightsquigarrow\; \forall y_1, \ldots, y_{n+1} : \bigwedge_{1 \leq i \leq n+1} [R(x, y_i) \wedge C(y_i)] \rightarrow \bigvee_{1 \leq i < j \leq n+1} y_i \approx y_j$$

These translations employs the equality predicate $\approx$. Ordered resolution alone is not an efficient calculus for theorem proving with equality. Therefore, deciding DLs with number restrictions typically requires the application of a calculus optimized for theorem proving with equality. *Basic superposition* [5, 24] is one such calculus, which introduces new rules that take into account the semantics of equality.

In [13], a decision procedure for the DL $\mathcal{SHIQ}^-$ (a DL obtained from $\mathcal{SHIQ}$ by imposing certain restrictions on the usage of number restrictions) based on basic superposition. In [14], this algorithm has been generalized to $\mathcal{SHIQ}$ by extending basic superposition with a *decomposition* inference rule, which simplifies certain clauses. All these procedures are worst-case optimal (i.e., they run in ExpTime) for unary coding of numbers. It is known that $\mathcal{SHIQ}$ is ExpTime-complete even for binary coding of numbers [30]; however, the assumption of unary number coding is standard in practical DL reasoning systems.

Nominals

Another common construct considered in DLs are nominals. Although such a result has not been published, it would be straightforward to extend the algorithms from [13,14] to handle the DL $\mathcal{SHOQ}$. The combination of inverse roles and nominals, however, is rather difficult to handle. Intuitively, such a logic does not have the tree-model property. Still, in [19], basic superposition has been extended with decomposition and novel *nominal generation* rule to obtain a decision procedure for $\mathcal{SHOIQ}$. The resulting decision procedure is, however, not optimal: it runs in triple exponential time, whereas $\mathcal{SHOIQ}$ is NExpTime-complete [30].

4 Reasoning by Reduction to Logic Programming

We now present an algorithm for reducing an $\mathcal{ALCHI}$ knowledge base to a disjunctive datalog program that entails the same set of ground atoms. As discussed in [23], such a reasoning technique is particularly suitable for knowledge bases that have a rather small and simple TBox but a large ABox.

4.1 The Main Difficulty

For an $\mathcal{ALCHI}$ knowledge base $\mathcal{K}$, our goal is to derive a disjunctive datalog program $\mathsf{DD}(\mathcal{K})$ such that $\mathcal{K} \models \alpha$ if and only if $\mathsf{DD}(\mathcal{K}) \models \alpha$ for α of the form $A(a)$ or $R(a,b)$. Thus, we can use $\mathsf{DD}(\mathcal{K})$ instead of $\mathcal{K}$ for query answering, and in doing so, we can apply all optimization techniques known from deductive databases, such as magic sets [6] or join-order optimizations [1].

As shown in Table 1 and in [7], there is a close correspondence between description logics and first-order logic. Consider the following knowledge base:

$$\mathcal{K} = \{A \sqsubseteq \exists R.A, \exists R.\exists R.A \sqsubseteq B, A(a)\} \tag{33}$$

A naïve attempt to reduce $\mathcal{K}$ into disjunctive datalog is to translate $\mathcal{K}$ into a first-order formula $\pi(\mathcal{K})$, skolemize it, translate it into conjunctive normal form, and rewrite the obtained set of clauses into rules. For $\mathcal{K}$, such an approach produces the following logic program $\mathsf{LP}(\mathcal{K})$:

$$R(x, f(x)) \leftarrow A(x) \tag{34}$$

$$A(f(x)) \leftarrow A(x) \tag{35}$$

$$B(x) \leftarrow R(x,y), R(y,z), A(z) \tag{36}$$

$$A(a) \tag{37}$$

Clearly, $\mathcal{K}$ and $\mathsf{LP}(\mathcal{K})$ entail the same set of ground facts. The program $\mathsf{LP}(\mathcal{K})$, however, contains a function symbol in a recursive rule (35). This raises the issue of how to answer queries in $\mathsf{LP}(\mathcal{K})$. Namely, well-known query evaluation techniques will not terminate on $\mathsf{LP}(\mathcal{K})$; for example, using bottom-up saturation, we shall derive $A(f(a))$, $R(a, f(a))$, $A(f(f(a)))$, $R(f(a), f(f(a)))$, $B(a)$, and so on. Obviously, such an algorithm will continue deriving ever deeper facts, and will therefore never terminate. Note that we need all previously derived facts to derive $B(a)$ from $\mathsf{LP}(\mathcal{K})$, and that we do not know a priori when all relevant ground facts have been derived, so that we might stop the saturation.

This problem could be solved by employing an appropriate cycle detection mechanism. In [16], such an approach has been used to derive a decision procedure for the DL $\mathcal{ALC}$ based on hyperresolution. Using specialized algorithms for evaluating queries in $\mathsf{LP}(\mathcal{K})$ takes us, however, away from our original goal of applying deductive database optimization techniques to description logics. In a way, such an algorithm could be viewed as an alternative notation for the tableau calculus, for which it is unclear how to apply optimization techniques such as magic sets.

To avoid potential problems with termination, our goal is to derive a true disjunctive datalog program $\mathsf{DD}(\mathcal{K})$ without function symbols. For such a program, queries can be evaluated using any standard technique; furthermore, all existing optimization techniques known from deductive databases can be applied directly. Hence, the main problem that we deal with is the elimination of function symbols from $\mathsf{LP}(\mathcal{K})$.

4.2 The Translation Algorithm

From Table 5, we see that (1) a ground clause cannot participate in an inference with a nonground clause containing a function symbol, and (2) if one premise in an inference by $\mathcal{R}_{DL}$ is ground, the conclusion is ground as well. Hence, we can perform all inferences among nonground clauses first, after which we can simply delete all nonground clauses containing function symbols. The remaining clause set consists of clauses without function symbols, which can easily be translated into a disjunctive datalog program, by moving positive literals into rule heads and negative literals into rule bodies. A minor problem arises if the resulting rules contain unsafe variables. We deal with such clauses using a simple trick: we introduce a new predicate HU and add an assertion $HU(a)$ for each individual a; next, we append $HU(x)$ to the body of each rule in which x is an unsafe variable.

Definition 4. *Let $\mathcal{K} = (\mathcal{R}, \mathcal{T}, \mathcal{A})$ be an extensionally reduced $\mathcal{ALCHI}$ knowledge base. Then, $\Gamma(\mathcal{T} \cup \mathcal{R})$ is the set of clauses obtained by*

- *saturating $\Xi(\mathcal{T} \cup \mathcal{R})$ by $\mathcal{R}_{DL}$, and then*
- *deleting all clauses containing function symbols.*

The disjunctive datalog program $\mathsf{DD}(\mathcal{K})$ is obtained from $\Gamma(\mathcal{T} \cup \mathcal{R}) \cup \Xi(\mathcal{A})$ using the following transformations:

- *Each clause of the form $A_1 \vee \cdots \vee A_n \vee \neg B_1 \vee \cdots \vee \neg B_m$ is rewritten into a rule $A_1 \vee \cdots \vee A_n \leftarrow B_1, \ldots, B_m$.*
- *If a variable x occurs in some rule only in the head, then the literal $HU(x)$ is added to the rule body.*

- *The fact $HU(a)$ is added to the program for each constant a occurring in $\mathcal{K}$.*

If $\mathcal{K}$ is not extensionally reduced, then $\mathsf{DD}(\mathcal{K}) = \mathsf{DD}(\mathcal{K}')$, where $\mathcal{K}'$ is an extensionally reduced knowledge base obtained from $\mathcal{K}$ as explained in Sect. 2.1.

We now state the properties of $\mathsf{DD}(\mathcal{K})$:

Theorem 2. *The following claims hold for each $\mathcal{ALCHI}$ knowledge base $\mathcal{K}$:*

1. *$\mathcal{K}$ is satisfiable if and only if $\mathsf{DD}(\mathcal{K})$ is satisfiable.*
2. *$\mathcal{K} \models \alpha$ if and only if $\mathsf{DD}(\mathcal{K}) \models_c \alpha$, where α is of the form $A(a)$ or $R(a,b)$ for A a concept name and R a role.*
3. *$\mathcal{K} \models C(a)$ for a complex concept C iff $\mathsf{DD}(\mathcal{K} \cup \{C \sqsubseteq Q\}) \models_c Q(a)$ for Q a new concept name.*
4. *The number of literals in each rule in $\mathsf{DD}(\mathcal{K})$ is at most polynomial, the number of rules in $\mathsf{DD}(\mathcal{K})$ is at most exponential, and $\mathsf{DD}(\mathcal{K})$ can be computed in time exponential in $|\mathcal{K}|$.*

Proof. (1) Table 5 shows that each inference with at least one ground premise (these are the inferences below the dashed line) always produces a ground conclusion. Hence, in saturating $\Xi(\mathcal{K})$ by $\mathcal{R}_{DL}$, we can perform all inferences among nonground clauses first. Furthermore, Table 5 also shows that ground clauses can participate in inferences only with clauses not containing function symbols. Hence, after performing all inferences among nonground clauses of $\Xi(\mathcal{K})$, we can delete all clauses with terms of the form $f(x)$.

By Definition 2, $\Xi(\mathcal{T} \cup \mathcal{R})$ is exactly the set of nonground clauses of $\Xi(\mathcal{K})$, so $\Gamma(\mathcal{T} \cup \mathcal{R})$ is exactly the set of clauses obtained by saturating the nonground part of $\Xi(\mathcal{K})$ and deleting the clauses containing function symbols. Furthermore, it is easy to see that $\Gamma(\mathcal{T} \cup \mathcal{R}) \cup \Xi(\mathcal{A})$ is satisfiable if and only if $\mathsf{DD}(\mathcal{K})$ is satisfiable. Namely, both clause sets are function-free and they differ only in that the unsafe variables in the latter set are bound using the predicate HU which enumerates the entire Herbrand universe.

(2) Simply observe that $\mathcal{K} \models \alpha$ if and only if $\mathcal{K} \cup \{\neg\alpha\}$ is unsatisfiable. The latter is the case if and only if $\mathsf{DD}(\mathcal{K} \cup \{\leftarrow \alpha\}) = \mathsf{DD}(\mathcal{K}) \cup \{\leftarrow \alpha\}$ is unsatisfiable, which is the case if and only if $\mathsf{DD}(\mathcal{K}) \models_c \alpha$.

(3) Follows in the same manner as (2).

(4) Follows immediately from Lemma 4. □

4.3 An Example

We now continue the example from Sect. 3.3 and compute a disjunctive datalog program $\mathsf{DD}(\mathcal{K})$. The first step in the algorithm is to compute $\Xi(\mathcal{T} \cup \mathcal{R})$; clearly, it consists of the clauses (6)–(11).

The next step is to compute $\Gamma(\mathcal{T} \cup \mathcal{R})$ by saturating $\Xi(\mathcal{T} \cup \mathcal{R})$ by $\mathcal{R}_{DL}$. This was already done in Sect. 3.3: the saturated set contains the clauses (6)–(11) and, additionally, (15)–(18).

The next step is to remove all clauses containing function symbols. Therefore, we remove the clauses (8), (9), (15), (16). The final step is to compute $\mathsf{DD}(\mathcal{K})$ by moving all negative literals into the body and the positive literals into the head. The clauses (6) and (18) are unsafe, so we additionally add the literals $HU(x)$ to the body of the rules.

$$Q_1(x) \vee Q_2(x) \leftarrow HU(x) \tag{38}$$
$$\leftarrow Q_1(x), S(x,y), A(y) \tag{39}$$
$$C(x) \leftarrow B(x) \tag{40}$$
$$D(x) \leftarrow R(x,y), C(y) \tag{41}$$
$$D(x) \leftarrow Q_2(x) \tag{42}$$
$$D(x) \vee Q_1(x) \leftarrow HU(x) \tag{43}$$

Finally, we add to $\mathsf{DD}(\mathcal{K})$ the ABox and the facts involving HU:

$$S(a,b) \tag{44}$$
$$A(b) \tag{45}$$
$$HU(a) \tag{46}$$
$$HU(b) \tag{47}$$

It is straightforward to verify that $\mathsf{DD}(\mathcal{K}) \models D(a)$, in accordance with Theorem 2.

It is instructive to compare the algorithm from this section with tableaux algorithms from chapter "Tableau-Based Reasoning". Tableau algorithms introduce new individuals in order to satisfy the existential quantifiers. In contrast, the programs obtained by the reduction do not represent such individuals at all. In our example, $\mathsf{DD}(\mathcal{K})$ is function-free, so the universe of the program is restricted to the constants explicitly mentioned in it. Thus, the models of $\mathcal{K}$ and $\mathsf{DD}(\mathcal{K})$ coincide only on positive ground facts, and are unrelated for the facts involving unnamed objects.

To understand why the saturation of the TBox and RBox by $\mathcal{R}_{DL}$ is necessary, consider the role of each rule in $\mathsf{DD}(\mathcal{K})$. While the axiom (2) in $\mathcal{K}$ is applicable to all individuals in a model, the rule (40) is applicable only to named individuals. The relationship between (3) and (41) is analogous. To compensate for the fact that (40) and (41) derive consequences only about named individuals, $\mathsf{DD}(\mathcal{K})$ contains the rule (42), which is produced by the saturation of $\Xi(\mathcal{T} \cup \mathcal{R})$ by $\mathcal{R}_{DL}$. This rule acts as a shortcut: instead of introducing for each x in Q_2 an R-successor y in B by (8), propagating y to C by (10), and then concluding that x is in D by (11), the rule (42) derives that all instances of Q_2 are instances of D in one step. This ensures that $\mathsf{DD}(\mathcal{K})$ and $\mathcal{K}$ entail the same set of ground facts.

4.4 Discussion

By Theorem 2, the program $\mathsf{DD}(\mathcal{K})$ is independent of the query, as long as the query is a concept name or a role. Hence, $\mathsf{DD}(\mathcal{K})$ can be computed once, and can be used to answer any query involving only concept names. If the query involves a complex concept C (even if C is a negated concept name), then query answering can be reduced to entailment of positive ground facts, by introducing a new name Q and by adding the axiom $C \sqsubseteq Q$ to the TBox. Obviously, $\mathsf{DD}(\mathcal{K} \cup \{C \sqsubseteq Q\})$ may depend on C. Namely, by saturating $\Gamma(\mathcal{T} \cup \mathcal{R})$, the reduction algorithm derives all nonground consequences of $\mathcal{K}$, and a complex query concept can introduce new nonground consequences, which should be taken into account in the reduction.

Theorem 2 allows $|\mathsf{DD}(\mathcal{K})|$ to be exponential in $|\mathcal{K}|$, which may seem discouraging. Note, however, that the number of rules depends on $|\mathcal{T} \cup \mathcal{R}|$ and not on $|\mathcal{A}|$. This is important for *data complexity* [31] – the complexity under the assumption that the TBox and RBox are fixed. Under such an assumption, $|\mathsf{DD}(\mathcal{K})|$ becomes polynomial in $|\mathcal{A}|$, which has been used in [15] to show that checking satisfiability of $\mathcal{SHIQ}$ knowledge bases is NP-complete for data complexity. Also, a *Horn fragment* of $\mathcal{SHIQ}$ has been identified that does not provide for disjunctive reasoning but exhibits polynomial data complexity. To deal with the exponential blowup in the number of rules, an optimization has been presented in [13, 22] that allows many rules to be removed from $\mathsf{DD}(\mathcal{K})$ without invalidating Theorem 2. Practical experience has shown that the number of remaining rules is typically twice the number of axioms in $\mathcal{K}$ [23].

4.5 Adding Number Restrictions

The reduction algorithm presented in [13, 22] differs from this one mainly in that it can handle knowledge bases with number restrictions. We now outline the differences between this algorithm and the one presented in this section. Namely, if $\mathcal{K}$ is an $\mathcal{ALCHI}$ knowledge base, all functional terms encountered in a saturation of $\Xi(\mathcal{K})$ by $\mathcal{R}_{DL}$ are nonground (see Table 5). This is no longer the case if $\mathcal{K}$ is an $\mathcal{ALCHIQ}$ knowledge base. Namely, the translation of number restrictions can produce clauses such as (48). To see why such clauses case problems, let us assume that some other axioms produce the clauses (49)–(50).

$$\underline{\neg R(x, y_1)} \vee \underline{\neg R(x, y_2)} \vee y_1 \approx y_2 \tag{48}$$

$$\neg C(x) \vee \underline{R(x, f(x))} \tag{49}$$

$$\underline{R(a, b)} \tag{50}$$

By resolving (48) with (49) and (50), we obtain the following clause:

$$\neg C(a) \vee f(a) \approx b \tag{51}$$

This clause differs from clauses of type 7 from Table 4 in that it contains a ground functional term. The functional terms from clauses such as (51) can participate in further inferences, so we cannot just remove all clauses with function terms.

The solution is to represent ground functional terms in clauses of the form (51) using new constants. Thus, the clause (51) is encoded as the following clause, where a_f is a new constant unique for a pair of a and f:

$$\neg C(a) \vee a_f \approx b \tag{52}$$

After saturation of TBox and RBox, the nonground clauses from the saturated set are transformed in a certain way that reflects such an encoding of the ground clauses. It is important to understand that the constants such as a_f have no deeper semantic meaning; they are just a proof-theoretic aid that allows the simulation of inferences of basic superposition in disjunctive datalog.

5 Conclusion

This chapter overviews the algorithms for reasoning in description logics by resolution. These algorithms are interesting because they are worst-case optimal, but are also suitable for practical implementation [23]. Furthermore, such algorithms can be used to reduce a DL knowledge base to a disjunctive datalog program. This allows the application of known reasoning algorithms from deductive databases to reasoning with large ABoxes. Practical experience has shown that such algorithms are quite suitable for ontologies with relatively small and simple TBoxes but large ABoxes.

A challenge for future research is to obtain a more elegant and perhaps worst-case optimal algorithm for reasoning with nominals. Namely, reasoning with nominals requires reasoning about the cardinality of sets, which is known to be difficult for resolution. Another challenge is to provide methods for dealing with transitivity and general role inclusion axioms, such as the ones available in the DL $\mathcal{SROIQ}$ [20].

References

1. S. Abiteboul, R. Hull, and V. Vianu. *Foundations of Databases*. Addison Wesley, 1995.
2. F. Baader, D. Calvanese, D. McGuinness, D. Nardi, and P. F. Patel-Schneider, editors. *The Description Logic Handbook: Theory, Implementation and Applications*. Cambridge University Press, January 2003.
3. F. Baader and W. Snyder. Unification Theory. In A. Robinson and A. Voronkov, editors, *Handbook of Automated Reasoning*, volume I, chapter 8, pages 445–532. 2001.

4. L. Bachmair and H. Ganzinger. Resolution Theorem Proving. In A. Robinson and A. Voronkov, editors, *Handbook of Automated Reasoning*, volume I, chapter 2, pages 19–99. 2001.
5. L. Bachmair, H. Ganzinger, C. Lynch, and W. Snyder. Basic Paramodulation. *Information and Computation*, 121(2):172–192, 1995.
6. C. Beeri and R. Ramakrishnan. On the power of magic. In *Proc. PODS '87*, pages 269–283, San Diego, CA, USA, 1987.
7. A. Borgida. On the Relative Expressiveness of Description Logics and Predicate Logics. *Artificial Intelligence*, 82(1–2):353–367, 1996.
8. M. Buchheit, F. M. Donini, and A. Schaerf. Decidable Reasoning in Terminological Knowledge Representation Systems. *Journal of Artificial Intelligence Research*, 1:109–138, 1993.
9. M. Davis, G. Logemann, and D. Loveland. A Machine Program for Theorem-Proving. *Communications of the ACM*, 5(7):394–397, 1962.
10. N. Dershowitz and D. A. Plaisted. Rewriting. In A. Robinson and A. Voronkov, editors, *Handbook of Automated Reasoning*, volume I, chapter 9, pages 535–610. 2001.
11. T. Eiter, G. Gottlob, and H. Mannila. Disjunctive Datalog. *ACM Transactions on Database Systems*, 22(3):364–418, 1997.
12. I. Horrocks and U. Sattler. A Description Logic with Transitive and Inverse Roles and Role Hierarchies. *Journal of Logic and Computation*, 9(3):385–410, 1999.
13. U. Hustadt, B. Motik, and U. Sattler. Reducing $\mathcal{SHIQ}^-$ Description Logic to Disjunctive Datalog Programs. In *Proc. KR 2004*, pages 152–162, Whistler, Canada, 2004.
14. U. Hustadt, B. Motik, and U. Sattler. A Decomposition Rule for Decision Procedures by Resolution-based Calculi. In *Proc. LPAR 2004*, pages 21–35, Uruguay, 2005.
15. U. Hustadt, B. Motik, and U. Sattler. Data Complexity of Reasoning in Very Expressive Description Logics. In *Proc. IJCAI 2005*, pages 466–471, Edinburgh, UK, 2005.
16. U. Hustadt and R. A. Schmidt. On the Relation of Resolution and Tableaux Proof Systems for Description Logics. In *Proc. IJCAI '99*, pages 202–207, Stockhom, Sweden, 1999.
17. W. H. Joyner. Resolution Strategies as Decision Procedures. *Journal of the ACM*, 23(3):398–417, 1976.
18. Y. Kazakov and H. de Nivelle. A Resolution Decision Procedure for the Guarded Fragment with Transitive Guards. In *Proc. IJCAR 2004*, pages 122–136, Cork, Ireland, 2004.
19. Y. Kazakov and B. Motik. A Resolution-Based Decision Procedure for $\mathcal{SHOIQ}$. In *Proc. IJCAR 2006*, pages 662–667, Seattle, WA, USA, 2006.
20. O. Kutz, I. Horrocks, and U. Sattler. The Even More Irresistible SROIQ. In *Proc. KR 2006*, pages 68–78, Lake District, UK, 2006.
21. C. Lutz and U. Sattler. The Complexity of Reasoning with Boolean Modal Logics. In *Proc. AiML 2000*, pages 329–348, Leipzig, Germany, 2001.
22. B. Motik. *Reasoning in Description Logics using Resolution and Deductive Databases.* PhD thesis, Univesität Karlsruhe (TH), Karlsruhe, Germany, January 2006.

23. B. Motik and U. Sattler. A Comparison of Reasoning Techniques for Querying Large Description Logic ABoxes. In *Proc. LPAR 2006*, pages 227–241, Cambodia, 2006.
24. R. Nieuwenhuis and A. Rubio. Theorem Proving with Ordering and Equality Constrained Clauses. *Journal of Symbolic Computation*, 19(4):312–351, 1995.
25. H. De Nivelle, R. A. Schmidt, and U. Hustadt. Resolution-Based Methods for Modal Logics. *Logic Journal of the IGPL*, 8(3):265–292, 2000.
26. A. Nonnengart and C. Weidenbach. Computing Small Clause Normal Forms. In A. Robinson and A. Voronkov, editors, *Handbook of Automated Reasoning*, volume I, chapter 6, pages 335–367. 2001.
27. D. A. Plaisted and S. Greenbaum. A Structure-Preserving Clause Form Translation. *Journal of Symbolic Logic and Computation*, 2(3):293–304, 1986.
28. A. Riazanov and A. Voronkov. The design and implementation of VAMPIRE. *AI Communications*, 15(2–3):91–110, 2002.
29. R. A. Schmidt and U. Hustadt. A Principle for Incorporating Axioms into the First-Order Translation of Modal Formulae. In *Proc. CADE-19*, pages 412–426, USA, 2003.
30. S. Tobies. *Complexity Results and Practical Algorithms for Logics in Knowledge Representation.* PhD thesis, RWTH Aachen, Germany, 2001.
31. M. Vardi. The Complexity of Relational Query Languages. In *Proc. STOC '82*, pages 137–146, San Francisco, CA, USA, 1982.

Ontology Repositories

Jens Hartmann[1], Raúl Palma[2], and Asunción Gómez-Pérez[2]

[1] Center for Computing Technologies (TZI), University of Bremen, Germany, jh@tzi.de

[2] Ontology Engineering Group, Laboratorio de Inteligencia Artificial, Facultad de Informática, Universidad Politécnica de Madrid, Spain, palma@fi.upm.es, asun@fi.upm.es

Summary. The growing use and application of ontologies in the last years has led to an increased interest of researchers and practitioners in the development of ontologies, either from scratch or by reusing existing ones. Reusing existing ontologies instead of creating new ones from scratch has many benefits: It lowers the time and cost of development, avoids duplicate efforts, ensures interoperability, etc. In fact, ontology reuse is one of the key enablers for the realization of the Semantic Web. However, currently, ontologies are mostly developed from scratch, due to several reasons. First, ontologies are usually tailored to work for specific applications, restricting its potential reusability. Second, developers usually follow a monolithic approach when developing ontologies, usually covering different domains, hampering the reusability of relevant parts for other applications. Third, ontologies are rather difficult to find due to the lack of standards for documenting them and appropriate tools supporting intelligent ontology discovery and selection by end users. In this chapter, we define a generic ontology repository framework that enables the implementation of fully-fledged ontology repositories providing the technological support to the aforementioned issues. We distinguish between the ontology repository itself and the software to manage the repository, and describe their main aspects and services. Finally, we present two exemplary systems based on this framework.

1 Introduction

Knowledge reuse and access is one of the leading motivations for the Semantic Web. Driven by those intensions an increasing amount of ontologies can be found nowadays on the Web distributed among personal or institutional web pages. One of the key problems the ontology engineering community has to face at the moment is that most ontologies are built from scratch – rather than reusing existing ones – leading to high engineering efforts and costs. One of the main reasons is that most existing ontologies are build having a specific application scenario in mind, making them similar to custom software. This leads to ontologies that are tailored to work with specific applications, but are

S. Staab and R. Studer (eds.), *Handbook on Ontologies*, International Handbooks on Information Systems, DOI 10.1007/978-3-540-92673-3,

not knowledge representation artifacts in the traditional sense. When designing these ontologies, engineers focus on expected behavior in the application rather than on reuse and interoperability with other ontologies. Another problem is that ontologies trying to cover domains in the knowledge representation sense are often too big to be reused efficiently. These ontologies try to capture the complete domain knowledge whilst ontology engineers normally only need to reuse certain parts for their ontology. Nevertheless, having modular ontologies is not enough to facilitate the reusability of ontologies if developers are not able to find them efficiently. We need an appropriate infrastructure that enables an intelligent ontology discovery and selection by end users.

Currently, initial collections of ontologies have been created during the last years, e.g., the DAML Library.[1] Apparently these resources are mostly created by hand and annotated manually. Many of these ontologies fail to follow consistent representation and storage conventions, so humans and machines are hampered by finding and reusing them. The process of identifying and accessing ontological resources, which can be summarized as *ontology retrieval*, is consequently affected by non-existing mechanisms and standards for storing and representing ontologies.

These circumstances point out the strong requirement for novel methods facilitating an efficient access and reuse of ontologies within a large scalable and reliable infrastructure, so called *ontology repositories*. The storage of knowledge encoded by ontologies can only be part of the solution. Crucial for ontology repositories is that additional knowledge about ontologies, so called *meta knowledge*, is managed together in such a repository.

We propose a Generic Ontology Repository Framework (GORF) including specific module support, tailored to exactly those requirements. We base our framework on experiences gained by realizing our ontology repository ONTHOLOGY[2] and previous work on the ontology metadata vocabulary OMV.[3] After their deployment, it became evident that having a central place to find ontologies and ontology modules alone does not solve the problem of quality assurance or knowing which module is the most suited for a specific task. We therefore integrate an Open Rating System (ORS) into our repository to assist in the retrieval process and prove quality assurance through feedback from the community itself. To fill the initial void of modules, existing ontologies can be partitioned (depending on their knowledge representation formalism, e.g., using the approach and tool mentioned in [6]) to create relevant modules.

In the following we discuss essential aspects of ontology repositories. To begin with, in Sect. 2 we describe the historical development from data to ontology repositories. In Sect. 3 we will describe the generic architecture of an ontology repository and corresponding management systems. Core elements

[1] Cf. `http://www.daml.org/ontologies/`

[2] Cf. `http://www.onthology.org/`

[3] Cf. `http://omv.ontoware.org/`

and services of an Ontology Repository are discussed in Sect. 4. Further we illustrate management systems for ontology repositories in Sect. 5 and exemplify a centralized vs. a decentralized solution in Sect. 6. We conclude in Sect. 7 and comment on further steps.

2 From Data Repositories to Ontology Repositories

In this section we present an overview of how repositories have evolved throughout time from general purpose data repositories to specialized ontology repositories.

In literature exist many different meanings and definitions to what a data repository is, and in general to what a repository is. Hence we will first discuss what we understand by a data repository, instead of giving another definition. We consider a data repository as a collection of digital data that is available to one or more entities (e.g., users, systems) for a variety of purposes (e.g., learning, administrative processes, research) and that has the characteristics proposed by Heery and Anderson [17]:

- Content is deposited in a repository, whether by the content creator, owner or third party.
- The repository architecture manages content as well as metadata.
- The repository offers a minimum set of basic services, e.g., put, get, search, access control.
- The repository must be sustainable and trusted, well-supported and well-managed.

The term data library is usually used in the literature to refer to subject specific datasets (e.g., climate data library, time series data library, geospatial data library). Moreover, a data library tends to house local data collections and provides access to them through various means. Thus, in general a data library usually provides access to the complete dataset instead of providing the basic services (e.g., search, put, get) a data repository offers.

Around the middle 1990s the term digital library (previously also known as electronic library or virtual library) was first made popular by the NSF/DARPA/NASA Digital Libraries Initiative. According to [1] a digital library is a managed collection of information, with associated services, where the information is stored in digital formats and accessible over a network. The information stored can be very diverse and used by many different users. In general a digital library is considered similar to a traditional library (i.e., it used by users to find information that others have created, and use it for study, reference, or entertainment) but it takes advantage of the new technologies to deliver the information to users.

Data warehouses [18] became popular during the late 1980s and early 1990s. The purpose of a data warehouse is to perform analysis of the stored data for management's decision making. Data is entered into this repository periodically, usually in an append-only manner. A data repository however, does not necessarily have that analysis functionality provided by a data warehouse.

Similarly to data repository, it is also possible to find many different meaning and definitions to what is a knowledge base. Yet, in general, a knowledge base is a central repository of knowledge artifacts. Usually a knowledge base may use an ontology to formally represent its content and its classification scheme, but it may also include unstructured or unformalized information expressed in natural language or procedural code. Also, in contrast to a data repository, usually the purpose of the knowledge base is to allow automated deductive reasoning over the stored knowledge (i.e., decide how to act by running formal reasoning procedures over the base of knowledge).

It is not surprising that some years ago, the ontology and semantic web community became interested in using repositories to hold semantic content (e.g., ontologies). Within the last years, ontologies have seen an enormous development and application in many domains, especially in the context of the semantic web. Academia and industry are developing and using ontologies to provide new technologies and support daily operations. Therefore, currently there exists a large amount of ontologies developed by many different parties which makes necessary the means to share and reuse them.

Initial efforts to collect the base of existing ontologies proposed the creation of library systems (i.e., known as Ontology library systems) that offered various functions for managing, adapting and standardizing groups of ontologies [8]. These systems defined an important environment in grouping and reorganizing ontologies for further re-use, integration, maintenance, mapping and versioning. They defined an evaluation model based on the functionality the library system provided. Examples of library systems are: WebOnto, Ontolingua, DAML Ontology Library System, SchemaWeb, etc.

Currently, efforts are put in the creation of ontology repositories. An ontology repository is similar to what Ding et al defined as an ontology library system [8], but they also have some differences. In the remaining of this chapter we will propose an widely-accepted definition of these terms.

3 Generic Ontology Repository Framework

Ontology reuse is still rarely encountered today. This is partly due to the problem of finding suitable ontologies to reuse, and the way most ontologies are created, namely without reusability in mind. Also, most of the established ontologies containing domain knowledge are simply too big to be easily reused, and no quality information is available on web ontologies.

We argue that ontology engineers can adopt from software engineers a way how ontologies could be designed, namely modular. This way, small, reusable components (ontology modules) are produced during creation. To manage and provide access to ontologies we propose an *Generic Ontology Repository Framework GORF* with specific module and rating support. So not only ontologies or modules can be found in a single place, one can also see reviews about their quality or their usefulness in different scenarios. This way ontology engineers have a one-stop-shop for reusable knowledge artifacts.

The term *ontology repository* can be seen as evolved term coming from the classical understanding of data repositories [18]. In the remaining we rely on the following understanding of an ontology repository and corresponding management systems.

Definition 1 (Ontology Repository and Management System). *An Ontology Repository (OR) is a structured collection of ontologies (schema and instances), modules and additional meta knowledge by using an Ontology Metadata Vocabulary. References and relations between ontologies and their modules build the semantic model of an ontology repository. Access to resources is realized through semantically-enabled interfaces applicable for humans and machines. Therefore a repository provides a formal query language.*

Software to manage an ontology repository is known as Ontology Repository Management System (ORMS). An ORMS is a system to store, organize, modify and extract knowledge from an Ontology Repository.

The main driving motivation creating ontology repositories is to support knowledge access and reuse for humans and machines. Hence ontology repositories on the one hand act as a storage facility and on the other provide access to knowledge through defined interfaces and policies. To achieve these goals, comprehensive facets must be considered by an ontology repository when handling ontologies. In general, these facets can be separated into access-related and storage-related aspects. A general requirement is that ontology repositories can support the entire ontology lifetime, i.e., ranging from the ontology engineering process to the desired application within specialized tools or tasks. Additionally, long term knowledge conservation is one of the crucial ontology repository tasks.

On a technical level, practical realizations of ontology repositories might differ in their concrete implementation. In contrast to that, relevant components or services on a conceptual level are reusable among different technical solutions. Consequentially, we now present a conceptual framework for ontology repositories. Based on different ontology repository implementations and realizations in the past [16], we identified a set of relevant components and services which are embedded into a scalable and reliable framework.

To be more specific the Generic Ontology Repository Framework (GORF) extends conceptually the SEAL (SEmantic portAL) [14] framework and preliminary work on ontology repositories, as described in [16]. As a result, the remaining framework GORF facilitates semantic-driven access and reuse of ontologies, thereby maintaining the vision of the semantic web. Furthermore, the framework remains scalable and can be distributed and interconnected to other repositories – like the one used in KMI's Watson[4] – using, e.g., web services. We assume that there will be technically heterogeneous solutions for semantic applications and especially for ontology repositories on the WWW. To ensure an easy and efficient knowledge exchange between applications, knowledge workers, and repositories, GORF acts as a framework identifying required components and services on a conceptual level, as shown in Fig. 1.

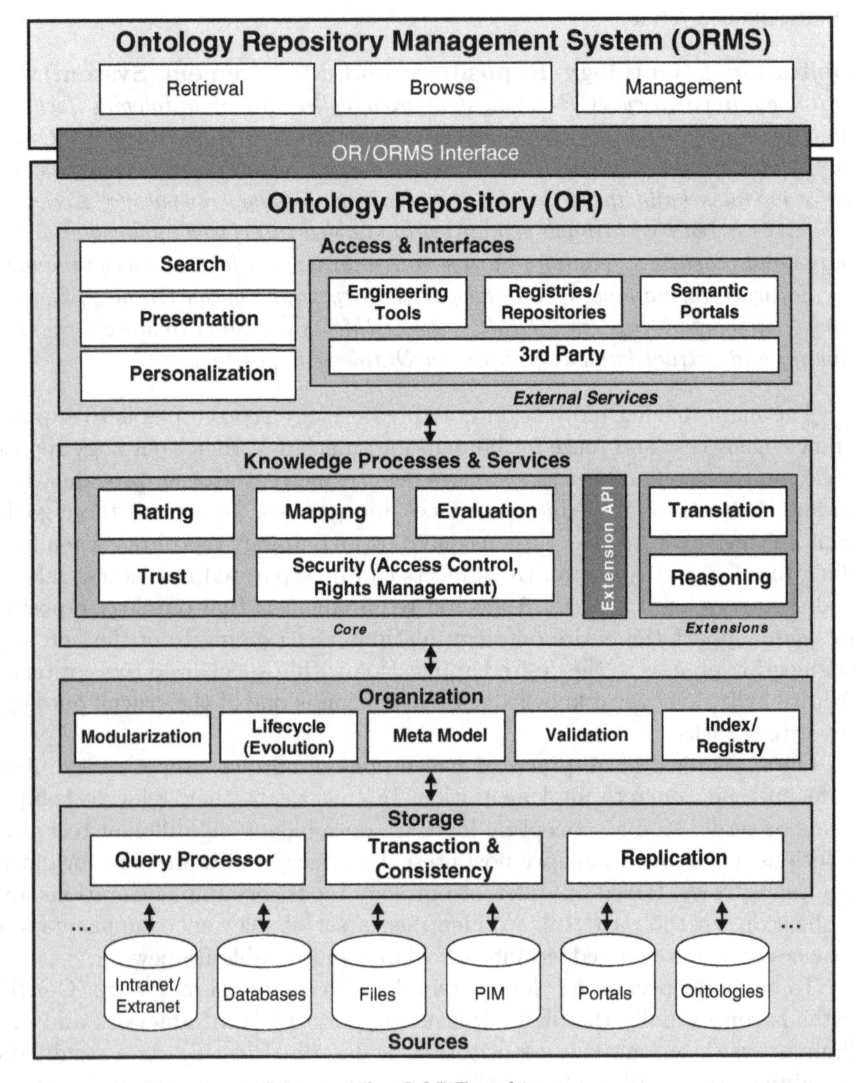

Fig. 1. The GORF architecture

[4] Cf. http://watson.kmi.open.ac.uk/

GORF distinguishes between the ontology repository itself and software to manage the repository. The latter one is called Ontology Repository Management System (ORMS). We claim that such knowledge intensive applications are also build using semantic technologies, as shown in [12]. In the following we briefly discuss the main aspects and services of an Ontology Repository and an ORMS.

4 Ontology Repositories

The framework for ontology repositories includes five conceptual layers. These layers can be seen as knowledge workflows, from the bottom to the top layer. In the following, we discuss each layer briefly.

4.1 Knowledge Access

Maintaining the vision of the Semantic Web [3], the architecture provides interfaces for humans and machines. Presenting knowledge to users involves a sophisticated visualization of knowledge for users with different interests and experiences. Thus, the framework must provide adaptable views on the stored knowledge, as shown in [14]. Although an ontology represents a commonly shared conceptualization of a domain, users typically have their own personal views and may request different visualizations. Therefore, an ontology repository should provide personalization services, which can become key success factors.

To sum up, the following aspects show up as elementary access functions:

- *Presentation and visualization:* The access layer generates flexible graphical user interfaces for users in different formats, e.g., HTML output. Underlying templates define where, how, and when the framework presents knowledge to the user. Furthermore, the framework is able to dynamically generate ontology browsing interfaces and navigation bars.
  So, the presented framework for ontology repositories must provide several views of the stored knowledge. The presentation layer generates graphical user interfaces for users, e.g., by producing HTML output. Underlying templates define where, how, and when the framework presents knowledge to the user. Portals based on SEAL dynamically generate the ontology's browsing interface and navigation bar. To support users interacting with the portal, we developed a context-sensitive help system that provides useful tips and explanations based on the current context.
- *Searching and querying a repository:* Our framework offers several search and query functionalities to the user. These include both standard full-text search forms and complex query forms, such as allowing a query for specific concepts or using simple query logic for a set of queries.

- *Personalization:* Although an ontology represents a commonly shared conceptualization of a domain, users typically have their own personal views and may request different visualizations. So, a semantic portal, especially a community one, should provide personalization services, which can become key enablers for successful repositories.

Besides functionalities for accessing knowledge and rendering its presentation, the access component defines the interface used by the ORMS to manage the OR.

4.2 Knowledge Processes and Services

Crucial to ontology repositories are processes and services for handling the stored knowledge within the repository:

- *Rating:* GORF will support an Topic-Specific Trust Open Rating System (TS-ORS) that can provide means to ensure the quality of ontologies and modules in the repository. Open Rating Systems (ORS) [10] have become increasingly popular over the last years. Nowadays, businesses have realized their value in customer satisfaction and information, and a variety of websites employ them. Examples are Epinions,[5] where products can be reviewed, Slashdot,[6] where articles and news can be reviewed, Amazon[7] (in the user review section), and iTunes[8] where users can review music. The general idea of ORS is to give everyone a voice, not only the so-called experts. One nice aspect about not excluding anyone is that ORS scale even when the rate of new content is growing steadily. Normally, in ORS, a meta-rating approach is used. This means that reviews are rated useful or not by other users. Based on these meta-reviews, a Web of Trust can be computed [11] and reviews and products can be ranked. One of the problems of the original algorithms and models [10, 11] proposed for ORS was that they were inflexible in the way objects could be reviewed (only complete objects) and how trust could be expressed (only globally). This is problematic because most people are only experts in certain fields and not in all. Another problem is that content in the system could only be rated as a whole, not specific properties separately. When applying ORS to review complicated content like ontologies [23], this is not sufficient. Users have to be able to review only those parts of the content they understand or have expertise on reviewing. To solve this problem, Lewen et al introduced topic-specific trust in ORS [20] and extended the underlying ORS model, making it a TS-ORS.

[5] Cf. http://www.epinions.com
[6] Cf. http://www.slashdot.org
[7] Cf. http://www.amazon.com
[8] Cf. http://www.apple.com/itunes

- *Mapping:* Mapping – frequently also called alignment – of ontologies is a core task to achieve interoperability and therefore a foundational service of an ontology repository. Because most ontologies reflect a subject-oriented view of the world, different people will model knowledge differently. Being able to link these different representations is important for the success of the Semantic Web. Thus, an ontology repository needs to support mapping mechanisms.
- *Evaluation:* In contrast to ontology rating that is a subjective assessing of an ontology, ontology evaluation can be seen as an assessment of the quality and the adequacy of an ontology or parts of it regarding a specific aim, goal or context. So far, several methods for evaluating ontologies have been proposed. An overview can be found in [29]. Selected evaluation strategies can be implemented in an *evaluation component* and applied in a large repository.
- *Trust:* Trust is an ongoing and currently not fully solved research problem in the area of the Semantic Web. However, we see trust management as an important functionality for managing knowledge in ontology repositories. The ORS partly addresses this issue in GORF.
- *Security:* Due to intellectual property rights, commercial licenses, patents or copyrights, not all knowledge artifacts may be accessible by the public. Therefore clear access control and right management functionalities are required. While knowledge access might be restricted, meta knowledge like OMV [15] remains accessible and processable. As a result, commercially used knowledge artifacts can be identified in a repository while the access is secured by specialized services like payment systems.

Additional processes or services might be added or attached through an *extension component.* For example, reasoning or validation services might be included here.

4.3 Knowledge Organization

The developed framework GORF can handle massive amounts of knowledge stored in a repository. Additionally, the technology allows for having multiple portals as access points on top of one ontology repository. Using the knowledge representation mechanism, we developed several continuative knowledge-organization methods to provide fast and effective access to knowledge:

- *Modularization:* The introduced framework aims to use ontology modules as key elements to be stored in a repository – alongside existing ontologies – for the reasons motivated beforehand. The core idea behind ontology modularization is the identification of reusable knowledge artifacts which are adaptable to different tasks and remain domain-independent. In contrast to an ontology, which aims at providing a domain-specific conceptualization in one construct for a set of tasks, a module represents a shared,

domain-independent conceptualization which is adaptive to and intended for re-occurring tasks and applications. In general, ontology modules are comparable to software libraries in the software engineering domain.
A key pre-condition for modularization is an expressive modeling language for ontologies, e.g., [5]. Currently first approaches can be found in literature for the modularization of ontologies and connecting these modules [30]. These approaches are, however, mostly driven by the underlying logic – except the ones working on structural information, like graph-based modularization approaches (for an overview we refer to [7]).
The main principle of modularization is described in an abstract way whereby we distinguish the process of modularization on existing ontologies and modularization whilst an ontology is being created. The first task, modularization on existing ontologies, can be considered as an ontology re-engineering task and the latter one represents an ontology engineering task. We assume that in most cases the effort required to identify useful modules in large and possibly unknown ontologies is too expensive for manual modularization. So automatic or at least semi-automatic mechanisms are required. First steps in this direction can be found in [6], where the ontology engineering environment itself provides means to extract modules out of existing ontologies, keeping the logical entailments intact. So the main challenge in extracting modules from existing ontologies is to identify modules that are best suited for reuse and maintenance issues.
When new ontologies are created, they can be designed with a module-based approach in mind. Software engineers are used to program following the object-oriented modeling paradigm [26], meaning they encapsulate required behavior in smaller blocks according to functionality. If ontology engineers would follow the same principle, the task of ontology engineering could become less cumbersome and less costly, due to an increase in reusable knowledge components, namely ontology modules.
At a certain point ontology modules are likely to require connections to other modules or ontologies to provide the functionality required for the given application. Techniques for linking ontologies and modules range from simple use of the *owl:imports* statement to fully fledged linking mechanisms [2, 4, 9]. These techniques differ in the assumptions they make regarding the source and target ontologies or modules. Some approaches require that the local domains and terminologies are disjoint [19], others require the use of special semantics. For an evaluation of their properties and usefulness in a distributed scenario as well as a more detailed introduction of their internal workings, see [30].
Which formalism the ontology engineer uses at the end to connect ontologies or modules is dependent on requirements of the specific application [21]. For our purposes it is just important to note that different formalisms exist, and modular ontologies can be created using different modules.

- *Lifecycle:* The support for evolution of knowledge, in particular for ontologies, is a major requirement for ontology repositories. In contrast to static content, ontologies are changing and demand mechanisms for updating and evolving knowledge over time. Stojanovic [27] defines an ontology evolution process model based on a requirement analysis. The approach relies on a declarative specification of change requests and evolution strategies. The approach has been practically realized within the KAON [22] ontology engineering framework. Such evolution mechanisms can be included into GORF through so called *lifecycle components.*
- *Meta model:* Ontologies are commonly used as a shared means of communication between computers and between humans and computers. Ergo ontologies should be represented, described, exchanged, shared and accessed based on open standards such as the W3C standardized web ontology language OWL. However, most ontologies today exist without any additional information about authorship, domain of interest and other metadata about ontologies. Therefore, searching and identifying existing ontologies which are potentially reusable because they are for example applied in similar domains, used within similar applications, or have similar properties, is a rather hard and tedious task.
  We argue that meta knowledge (seen as meta model) in the sense of machine processable information for the Web[9] helps to improve accessibility and reuse ontologies. Further, it can provide other useful resource information to support maintenance. We claim that metadata does not only help when it is applied (or attached) to documents, but also to ontologies. As a consequence, ontologies which are annotated by metadata require an infrastructure including metadata support – like the registry component in the GORF. Metadata simply consisting of attribute-value pairs is not sufficient for efficient knowledge access and reuse. Therefore first metadata vocabularies for ontologies have been developed [13, 15] and successfully applied in numerous applications [16]. In a more language centered way an approach relating UML with ontologies has been developed [5].
- *Validation:* Integrating different knowledge sources requires capable validation services. Validation components mainly analyze and validate the syntax of ontologies.
- *Registry and indices:* Knowledge artifacts may exist as entities, modules, schemes, or ontologies as a whole. Those artifacts are indexed by a registry component (OMR – Ontology Metadata Registry).
  The OMR provides services for storage, cataloging, discovery, management, and retrieval of ontology metadata definitions. The OMR provides the means to support advanced semantic searches of ontologies based on their characteristics. In general the OMR can be a GORF component or an independent system.

[9] Cf. `http://www.w3.org/Metadata/`

Analogically to the Dublin Core Metadata Initiative's (DCMI) Metadata Registry,[10] the OMR is designed to promote the discovery and reuse of existing ontologies. It provides users, and applications, with an authoritative source of information about the characteristics of ontologies, thus simplifying the discovery process.
Registry services and indices accelerate access to a repository, especially for search. Generally, indices are useful for concepts, relations, and full-text search capability. Our facilities offer repository administrators the possibility to freely define further indices.

Summarizing, the knowledge organization layer provides efficient methods for handling and organizing knowledge in repositories.

4.4 Knowledge Storage

To support the envisioned large, scalable application scenario of GORF, we use a highly scalable storage mechanism. Distributed repositories are set up in a cluster for handling several requests. Therefore the storage layer includes components for querying, transactions and replication of knowledge within such repositories.

- *Query processor:* The query processing component handles queries for single knowledge artifacts in a repository. As query language we prefer standardized languages like the well-known SPARQL[11] query language:
- *Transaction and consistency:* Performing and handling access to knowledge simultaneously requires sophisticated mechanisms preventing inconsistencies. Thus, transaction and consistency components analyze and check queries against the underlying knowledge storages.
- *Replication:* Being designed for scenarios with a high number of users and queries, GORF provides adaptable knowledge replication mechanisms. Those mechanisms replicate knowledge storages and additionally distribute them among pre-defined spaces.

The knowledge storage layer provides mechanisms for handling and accessing distributed knowledge sources, which are described below.

4.5 Knowledge Sources

The presented approach provides a sophisticated framework for integrating knowledge from different sources like *files, data bases, ontologies* or other *semantic portals.* The framework is capable of using existing sources along with their attached infrastructure. Therefore, our framework can rest atop

[10] `http://dublincore.org/dcregistry/`
[11] Cf. `http://www.w3.org/TR/rdf-sparql-query/`

existing technologies and act as a kind of semantic layer for these technologies to use the developed integration mechanisms.

Knowledge sources are typically distributed and heterogeneous, and tend to change during semantic interrelation, aggravating the task of integrating information into one common knowledge repository. The layer comprises two modules. The generic knowledge integration module shares and integrates knowledge from previously unknown sources. The interconnected-integration module handles sources that are closely interconnected technically and semantically. This module mainly integrates content such as other portals and semantic metadata.

5 Ontology Repository Management Systems

An *Ontology Repository Management System (ORMS)* is a semantically-enabled software to *store, organize, modify* and *extract* knowledge from an Ontology Repository. Two systems, namely Oyster[12] and ONTHOLOGY[13] [16] are already available to the end user.

In general, the main tasks of an ORMS are providing access to knowledge resources, supporting retrieval and allocating sufficient management mechanisms.

Retrieval

Ontology retrieval for humans and machines is a key functionality of an ontology repository management system. The retrieval component provides mechanisms to manage search and discovery functions of an ontology repository. For example consider the allocation of indices or the provision of metadata.

Browsing and Navigation

Semantically-driven navigation through knowledge stocks enables users to identify new and potentially useful knowledge artifacts within a repository. The navigation through repositories can be guided by specialized ontologies for semantic navigation, as introduced in [14]. The *browsing and navigation* component therefore allows mainly the selection of such navigation ontologies. Based on a usage analysis [28], a repository manager is able to evolve deployed navigation ontologies.

Management

An ontology repository is administrated through a management component which contains all administrative functionalities required to store, organize

[12] Cf. `http://oyster.ontoware.org/`

[13] Cf. `http://www.onthology.org/`

and maintain the knowledge within a repository. In general, manageable components in a repository provide interfaces to the *management component*. The main task hereby is to collect all manageable functionalities and to enable a standardized and centralized access to all relevant administrative functionalities. The entire business logic is implemented within each repository component itself. Thus, the management component itself does not contain real business logic. As a result, components within GORF are easily interchangeable and the whole framework remains flexible and scalable.

To sum up, an Ontology Repository Management System (ORMS) is a powerful tool to manage ontology repositories, even several distributed ones together. This way, established workflows and processes can be easily interchanged among other repositories reducing maintenance efforts and increasing the usability of such repositories.

6 Centralized Vs. Decentralized Systems

We now present exemplary running systems based on GORF. In detail, we present two complementary applications, namely the decentralized P2P system Oyster and the centralized ontology portal ONTHOLOGY. In general, the two tools differ in their usage perspective and are appropriate for different tasks. However, as we will see, only the combined application of both tools will offer users the full potential of ontology management.

6.1 Centralized Systems

Ensuring a scalable and reliable access to ontologies, optimization techniques are required. One well-known approach is a hybrid storage mechanism from the data warehouse area which materializes content to provide faster access. We present the conceptual design of a centralized ontology portal and its implementation, so-called ONTHOLOGY standing for "anthology of ontologies."

Scope

Centralized systems allow to reflect long-term community processes in which some ontologies become well accepted for a domain or community and others become less important. Such well accepted ontologies and in particular their metadata need to be stored in a central metadata portal which can be accessed easily by a large number of users whereby the management procedures are well defined. Hence, a main goal of a centralized metadata portal is to act as large evidence storage of metadata resp. their related ontologies to facilitate access, reuse and sharing as required for the Semantic Web.

Actors

We identified several different user roles for ONTHOLOGY: The *visitor* is an anonymous user, he is allowed to browse the public content of the portal. A visitor can become a *user* by completing an application form on the website. In order to avoid unnecessary administrative work, a user is added automatically to the membership database. Users can customize their portal, e.g., the content of their start-page or their bookmarks. If a user wants to submit metadata to the portal, this submission has to be reviewed before it is published. ONTHOLOGY establishes a *review process* in order to ensure a certain level of quality. *Reviewers* check the new submissions before they are published. The *technical administrator* is responsible for any other task mainly the maintenance of the portal.

Functionalities

Functionalities of ONTHOLOGY can be separated into two groups based on the usage. Indeed, *basic functionalities* which are provided to every user who accesses the portal and *sophisticated functionalities* for reviewers and administrators. The main operations a user can perform on the repository are (1) *Search*, (2) *Submit* and (3) *Export*.

The search and export can be performed by any visitor without being registered to the repository. Since providing new metadata is based on a certain community confidence, a visitor has to register at the portal to be able to submit data.

Architecture

ONTHOLOGY consists of an ontology repository and an ORMS. Exemplary, Sesame[14] or KAON[15] can be used as back-end metadata storage solution for an ontology-based representation. Furthermore, *access* and in particular the *management* of the repository must be guaranteed, too. Therefore, ONTHOLOGY is based on the proposed framework GORF. It supports queries to multiple sources, but beyond that also intensive use of the schema information itself to allow for automatic generation of navigational views such as navigation hierarchies that appear as `has- part`-trees or `has- subtopic` trees in the ontology. In addition to that mixed ontology and content-based presentation is supported. Further information can be found at [13, 15].

6.2 Decentralized Systems

In this section we describe the distributed ontology registry (Oyster).

---

[14] `http://www.openrdf.org/`

[15] `http://kaon.semanticweb.org/`

Oyster[24] is a Peer-to-Peer application that exploits semantic web techniques in order to provide a solution for exchanging and re-using. In order to achieve this goal, Oyster implements the proposal for a metadata standard OMV[13] as the way to describe ontologies.

Oyster Design

The Oyster system[16] was designed using a service-oriented approach, and it provides a well defined API. Accessing the registry functionalities can be done using directly the API within any application, invoking the web service provided or using the included java-based GUI as a client for the distributed registry. As part of the design, Oyster identifies an ontology metadata entry by the URI of the ontology it describes, therefore two ontology metadata entries are considered the same when the URI of both ontologies are the same. However, due to the distributed nature and potentially large size of the Peer-to-Peer network, two ontology metadata entries might refer to the same ontology but have different URI, in which case they are considered duplicates.

In Oyster, ontologies are used extensively in order to provide its main functions described in the following:

Creating and importing metadata: Oyster enables users to create metadata about ontologies manually, as well as to import ontology files and to automatically extract the ontology metadata available, letting the user fill in missing values. The ontology metadata entries are aligned and formally represented according to two ontologies: (1) the OMV ontology, (2) a topic hierarchy (i.e., the DMOZ topic hierarchy), which describes specific categories of subjects to define the domain of the ontology.

Formulating queries: A user can search for ontologies using simple keyword searches, or using more advanced, semantic searches. Here, queries are formulated in terms of these two ontologies. This means queries can refer to fields like name, acronym, ontology language, etc., or queries may refer to specific topic terms.

Routing queries: Users may query a single specific peer (e.g., their own computer, because they can have many ontologies stored locally and finding the right one for a specific task can be time consuming, or users may want to query another peer in particular because this peer is a known big provider of information), or a specific set of peers (e.g., all the members of a specific organization), or the entire network of peers (e.g., when users have no idea where to search), in which case queries are routed automatically in the network.

Processing results: Finally, results matching a query are presented in a result list. The answer of a query might be very large, and contain many duplicates due to the distributed nature and potentially large size of the P2P network. Such duplicates might not be exact copies because of the semi structured nature of the metadata, so the ontologies are used again to measure

[16] For a complete information and for downloading Oyster system we refer the reader to `http://ontoware.org/projects/Oyster/`

the semantic similarity between different answers and to remove apparent duplicates. As proposed by the ontology metadata standard, all the different realizations of an ontology (ontology documents) can be grouped by the same ontology base to give a more organized view of the results.

Oyster Architecture

The high-level design of the architecture of a single Oyster node in the Peer-to-Peer system is shown in Fig. 2. In the following, we discuss the individual components of the system architecture.

The *Local Repository* of a node contains the metadata about ontologies that it provides to the network. It supports query formulation and processing and provides the information for peer selection. In Oyster, the Local Repository is based on KAON2 and it supports SPARQL as its query language.

The *Knowledge Integrator* component is responsible for the extraction and integration of knowledge sources (i.e., ontologies) into the Local Repository. Oyster supports automatic extraction of metadata for OWL, DAML+OIL, and RDF-S ontology languages. This component is also in charge of how duplicate query results are detected and merged.

The *Query Manager* is the component responsible for the coordination of the process of distributing queries. It receives queries from the user interface, API or from other peers. Either way it tries to answer the query or distribute it further according to the content of the query. The decision to which peers a query should be sent is based on the scope of the query (i.e., a specific set of peers or entire network) and optionally on the knowledge about the expertise of other peers.

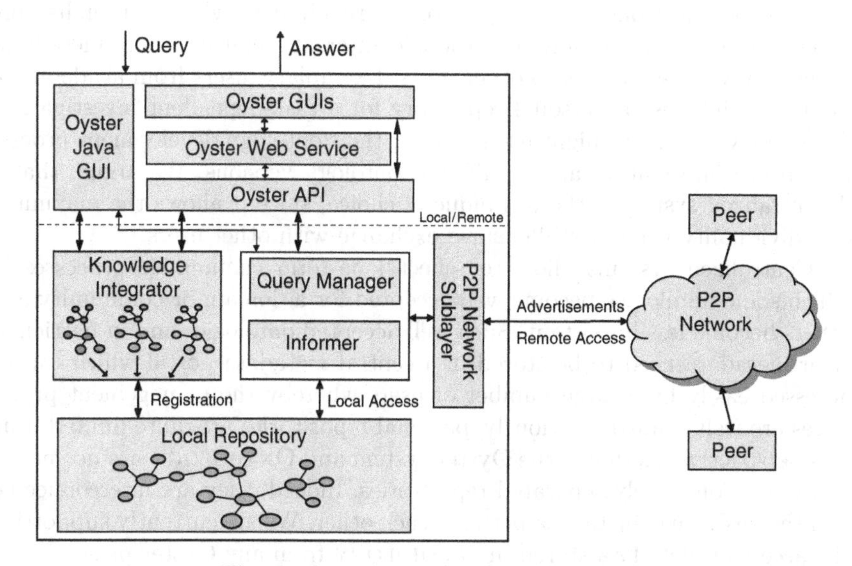

Fig. 2. Overview of Oyster architecture

The *Informer* component is in charge of proactively advertising the available knowledge of a Peer in the Peer-to-Peer network and to discover peers along with their expertise. This is realized by sending advertisements about the expertise of a peer. In Oyster, these expertise descriptions contain a set of topics (i.e., ontology domains) that the peer is an expert in. Peers may accept these advertisements, thus creating a semantic link to the other peer. These semantic links form a semantic topology, which is the basis for intelligent query routing.

The *Peer-to-Peer network sub-layer* is the component responsible for the network communication between peers. It provides communication services for the data exchange with remote nodes, i.e., to propagate advertisement messages and to realize the access to remote repositories. In Oyster, we rely on an RMI-based implementation, however, other communication protocols would be possible as well.

The API, WS and GUI components provide alternative ways for accessing Oyster functionalities, i.e., the API defines a set of methods that expose all of the functionalities, the web service encapsulates the API exposing a reduced set of functionalities and the GUIs provide ready-to-use clients for the Oyster network.

Additional registry functionalities can be provided by engineering components. Some of these components are described in [25].

6.3 Discussion

Both presented applications are covering a variety of different tasks. Indeed, for a user who wants to store metadata individually similar to managing his personal favorite song list, a repository is required to which a user has full access and can perform any operation (e.g., create, edit or delete metadata) without any consequences to other users. Exemplary, users from academia or industry might use a personal repository for a task-dependent investigation, or ontology engineers might use it during their ontology development process to capture information about different ontology versions. We argue, that a decentralized system is the technique of choice, since it allows the maximum of individuality while it still ensures exchange with other users.

Centralized systems allow to reflect long-term community processes in which some ontologies become well accepted for a domain or community and others become less important. Such well accepted ontologies and in particular their metadata need to be stored in a central metadata portal which can be accessed easily by a large number of users whereby the management procedures are well defined. Obviously, personal repositories are quite limited from this perspective. Actually, the Oyster system and ONTHOLOGY are not necessarily two completely separated repositories. Indeed, they are interconnected and they exchange metadata between each other. We are currently supporting the access of metadata stored in ONTHOLOGY from any Oyster peer.

The benefit of connecting both systems lies mainly in the simple use of existing ontology metadata information within Oyster. So, while users are applying or even developing their own ontologies they can manage their own metadata along with other existing metadata in one application (in Oyster). If some metadata entries from Oyster have reached a certain confidence, an import into ONTHOLOGY can be performed easily. In combination, both systems ensure efficient and effective ontology metadata management for various use cases.

7 Conclusions

Ontology repositories will be a crucial cornerstone facilitating efficient knowledge access and reuse especially in the context of the Semantic Web. We have presented our Generic Ontology Repository Framework GORF including rating and module support. We expect that there will be a shift in ontology engineering towards developing ontologies in a modular way. We are optimistic that then the critical mass of ontology modules in our repository can be reached, and ontology engineers will start reusing them and providing new ones.

Already existing realizations like ONTHOLOGY and Oyster illustrate the benefits of such systems. We assume that ontology repositories will play an important role in realizing the Semantic Web vision.

Acknowledgement

Research reported in this chapter was partially supported by two European Projects: The Network of Excellence KnowledgeWeb (FP6-507482) and the NeOn Project (FP6-027595).

References

1. William Y. Arms. *Digital Libraries*. MIT, Cambridge, MA, 2001.
2. Jie Bao, Doina Caragea, and Vasant Honavar. Towards collaborative environments for ontology construction and sharing. In *International Symposium on Collaborative Technologies and Systems (CTS 2006)*, pages 99–108. IEEE Press, 2006.
3. T. Berners-Lee, J. Hendler, and O. Lassila. The Semantic Web. *Scientific American Magazine*, 284(5):34–43, 2001.
4. Alexander Borgida and Luciano Serafini. Distributed description logics: Directed domain correspondences in federated information sources. In *OTM Federated Conference CoopIS/DOA/ODBASE*, pages 36–53, 2002.
5. Saartje Brockmans, Robert M. Colomb, Elisa F. Kendall, Evan Wallace, Christopher Welty, Guo Tong Xie, and Peter Haase. A model driven approach for building OWL DL and OWL full ontologies. In Isabel Cruz et al., editor,

The Semantic Web – ISWC 2006: 5th International Semantic Web Conference, volume 4273 of *LNCS*, pages 187–200, Athens, GA, USA, Nov 2006. Springer.
6. Bernardo Cuenca Grau, Ian Horrocks, Yevgeny Kazakov, and Ulrike Sattler. Just the right amount: Extracting modules from ontologies. In *Proc. of the Sixteenth International World Wide Web Conference (WWW 2007)*, 2007.
7. M. d'Aquin, M. Sabou, and E. Motta. Modularization: A key for the dynamic selection of relevant knowledge components. In *1st International Workshop on Modular Ontologies (WoMo 2006), co-located with ISWC*, 2006.
8. Y. Ding and D. Fensel. Ontology library systems: The key to successful ontology reuse, 2001. In *Proc. 1st Int Semantic Web Working Symposium (SWWS'01)*, 2001.
9. Bernardo Cuenca Grau, Bijan Parsia, and Evren Sirin. Working with multiple ontologies on the semantic web. In *International Semantic Web Conference*, pages 620–634, 2004.
10. Ramanathan Guha. Open rating systems. Technical report, Stanford University, CA, USA, 2003.
11. Ramanathan Guha, Ravi Kumar, Prabhakar Raghavan, and Andrew Tomkins. Propagation of trust and distrust. In *Proc. of the Thirteenth International World Wide Web Conference*, pages 403–412, New York, NY, May 2004. ACM Press.
12. Jens Hartmann. *Ontology-based modeling and realization of knowledge management systems*. PhD thesis, University of Karlsruhe (TH), Institute AIFB, Karlsruhe, 2007.
13. Jens Hartmann and Raul Palma. OMV – Ontology Metadata Vocabulary for the Semantic Web, 2006. v. 2.0, available at `http://omv.ontoware.org/`.
14. Jens Hartmann and York Sure. An infrastructure for scalable, reliable semantic portals. *IEEE Intelligent Systems*, 19(3):58–65, 2004.
15. Jens Hartmann, York Sure, Peter Haase, Raul Palma, and Mari Carmen Suárez-Figueroa. OMV – Ontology metadata vocabulary. In Chris Welty, editor, *ISWC 2005 – In Ontology Patterns for the Semantic Web*, Nov 2005.
16. Jens Hartmann, York Sure, Raul Palma, Peter Haase, Mari Carmen Suárez-Figueroa, Rudi Studer, and Asunción Gómez-Pérez. Ontology metadata vocabulary and applications. In Robert Meersman, editor, *International Conference on Ontologies, Databases and Applications of Semantics. In Workshop on Web Semantics (SWWS)*, Oct 2005.
17. Rachel Heery and Sheila Anderson. Digital repositories review, February 2005.
18. William H. Inmon and W. H. Inmon. *Building the Data Warehouse, 3rd Edition.* Wiley, New York, 2002.
19. Oliver Kutz, Carsten Lutz, Frank Wolter, and Michael Zakharyaschev. E-connections of description logics. In *Description Logics Workshop, CEUR-WS Vol. 81*, 2003.
20. Holger Lewen, Kaustubh Supekar, Natalya F. Noy, and Mark A. Musen. Topic-specific trust and open rating systems: an approach for ontology evaluation. In *Proceedings of the 4th International Workshop on Evaluation of Ontologies for the Web (EON2006) at the 15th International World Wide Web Conference (WWW 2006)*, Edinburgh, UK, May 2006.
21. Frank Loebe. Requirements for logical modules. In *1st International Workshop on Modular Ontologies (WoMo 2006), co-located with ISWC*, 2006.
22. Alexander Maedche, Boris Motik, and Ljiljana Stojanovic. Managing multiple and distributed ontologies in the Semantic Web. *VLDB Journal*, 12(4):286–302, 2003.

23. Natalya F. Noy, Ramanathan Guha, and Mark A. Musen. User ratings of ontologies: Who will rate the raters? In *Proc. of the AAAI 2005 Spring Symposium on Knowledge Collection from Volunteer Contributors*, Stanford, CA, 2005.
24. R. Palma and P. Haase. Oyster – sharing and re-using ontologies in a peer-to-peer community. In *International Semantic Web Conference*, pages 1059–1062, 2005.
25. Raul Palma, Peter Haase, Yimin Wang, and Mathieu d'Aquin. D1.3.1 propagation models and strategies. Technical Report D1.3.1, Universidad Politécnica de Madrid, Nov 2007.
26. J. Rumbaugh, M. Blaha, W. Premerlani, F. Eddy, and W. Lorensen. *Object-Oriented Modeling and Design.* Prentice-Hall, Upper Saddle River, NJ, 1991.
27. Ljiljana Stojanovic. *Methods and tools for ontology evolution.* PhD thesis, Universität Karlsruhe (TH), Universität Karlsruhe (TH), Institut AIFB, Karlsruhe, 2004.
28. Nenad Stojanovic, Jens Hartmann, and Jorge Gonzalez. The OntoManager – a system for usage-based ontology management. In *Proceedings of FGML Workshop. Special Interest Group of German Information Society (FGML – Fachgruppe Maschinelles Lernen der GI e.V.)*, 2003.
29. Denny Vrandecic and York Sure. How to design better ontology metrics. In Wolfgang May and Michael Kifer, editors, *Proceedings of the 4th European Semantic Web Conference (ESWC'07)*, Innsbruck, Austria, Jun 2007. Springer.
30. Yimin Wang, Jie Bao, Peter Haase, and Guilin Qi. Evaluating formalisms for modular ontologies in distributed information systems. In Massimo Marchiori and Jeff Z. Pan, editors, *Proceedings of The First International Conference on Web Reasoning and Rule Systems (RR2007)*, LNCS 4524, pages 178–182, Innsbruck, Austria, Jun 2007. Springer.

Ontology Mapping

Natalya F. Noy

Stanford University, Stanford, CA, 94305, USA, noy@stanford.edu

1 Why Is Ontology Mapping Difficult?

A quick scan through ontologies mentioned in this book, would indicate that many ontologies in use today overlap in content. Even for such, seemingly uncontroversial, domains, as anatomy, there are several ontologies representing them. Consider, for instance, the ontology repository at the National Center for Biomedical Ontologies [32].[1] Even among the small number of well-accepted and widely used ontologies there, several contain representation of human anatomy: the Foundational Model of Anatomy, the National Cancer Institute Thesaurus, the GALEN ontology. This situation is not surprising as different applications require different views on and different representations of the domain.

However, if we want to have the applications using different ontologies to "talk" to one another, or if we want to integrate data that is annotated with or structured according to different ontologies, we must first find the *correspondences* between concepts in these ontologies. The process of finding such correspondences is called *ontology mapping*. Ontology mapping (also referred to as ontology matching, or ontology alignment) is one of the most active areas of ontology research. Creating high-quality ontology mappings automatically is the holy grail of the Semantic Web research. Ontologies have gained popularity in the AI community as a means for establishing explicit formal vocabulary to share between applications. Therefore, one can say that one of the goals of using ontologies is not to have the problem of heterogeneity at all. It is of course unrealistic to hope that there will be an agreement on one or even a small set of ontologies. While having some common ground either within an application area or for some high-level general concepts could alleviate the problem of semantic heterogeneity, we will still need to map between ontologies, whether they extend the same foundational ontology or are developed independently.

[1] biointology.org

S. Staab and R. Studer (eds.), *Handbook on Ontologies*, International Handbooks on Information Systems, DOI 10.1007/978-3-540-92673-3,

We define an *ontology mapping* as a set of correspondences between components of two ontologies. These correspondences can be equivalence relationships, they can be subclass or superclass relationships, transformation rules, and so on. The process of finding ontology mapping is often referred to as ontology matching.

So, what are the types of differences between ontologies? In part summarizing earlier surveys, Klein [25] categorizes different types of mismatches between ontologies. The first class of mismatches are mismatches at the *language level* – mismatches in expressiveness and semantics of ontology language. The languages can differ in their syntax, but, more important, constructs available in one language (e.g., stating that classes are disjoint) are not available in another. Even semantics of the same language primitives could be different (e.g., whether declaration of multiple ranges of a property have union or intersection semantics). The *normalization* process therefore often precedes ontology-matching [24] and translates source ontologies to the same language, resolving these differences. It is important to note that in recent years, with the acceptance of RDFS and OWL for representing ontologies, the problem of resolving language-level mismatches became far less important.

However, even for ontologies expressed in the same language, possible *ontology-level* mismatches abound. A partial list of ontology-level mismatches includes using the same linguistic terms to describe different concepts; using different terms to describe the same concept; using different modeling paradigms (e.g., using interval logic or points for temporal representation); using different modeling conventions and levels of granularity; having ontologies with differing coverage of the domain, and so on.

Let us start with an example to illustrate the problem. We will use this example throughout the chapter. Suppose we have two airlines and ontologies describing their two respective reservation systems. Figure 1 presents small portions of these ontologies. Both ontologies have a class `Reservation` which represents each reservation record. For simplicity, we assume that each flight is a direct flight with no stops or connections.

In both ontologies, this class has a number of properties describing the reservation. In the ontology for the first airline (the top figure, white rectangles representing classes), there is a reservation number (string property `reservationNumber`), the date the reservation was made (`reservationDate`), the price of the ticket, the string representing the airports where the flight departs from (`from`) and where it lands (`to`). The records representing the time and date of the departure and arrival (instances of the `TimeAndDate` class) are values for the `departure` and `arrival` properties. There is a reference to the aircraft (property `aircraft` pointing to a class `PlaneModel`) and a property where all passengers are listed (`passengers`). Each passenger record is an instance of class `Passenger`, or, more specifically, one of its subclasses, `Child` or `Adult`.

The second ontology (the bottom figure, gray rectangles representing classes) has a similar structure: each reservation also has a number

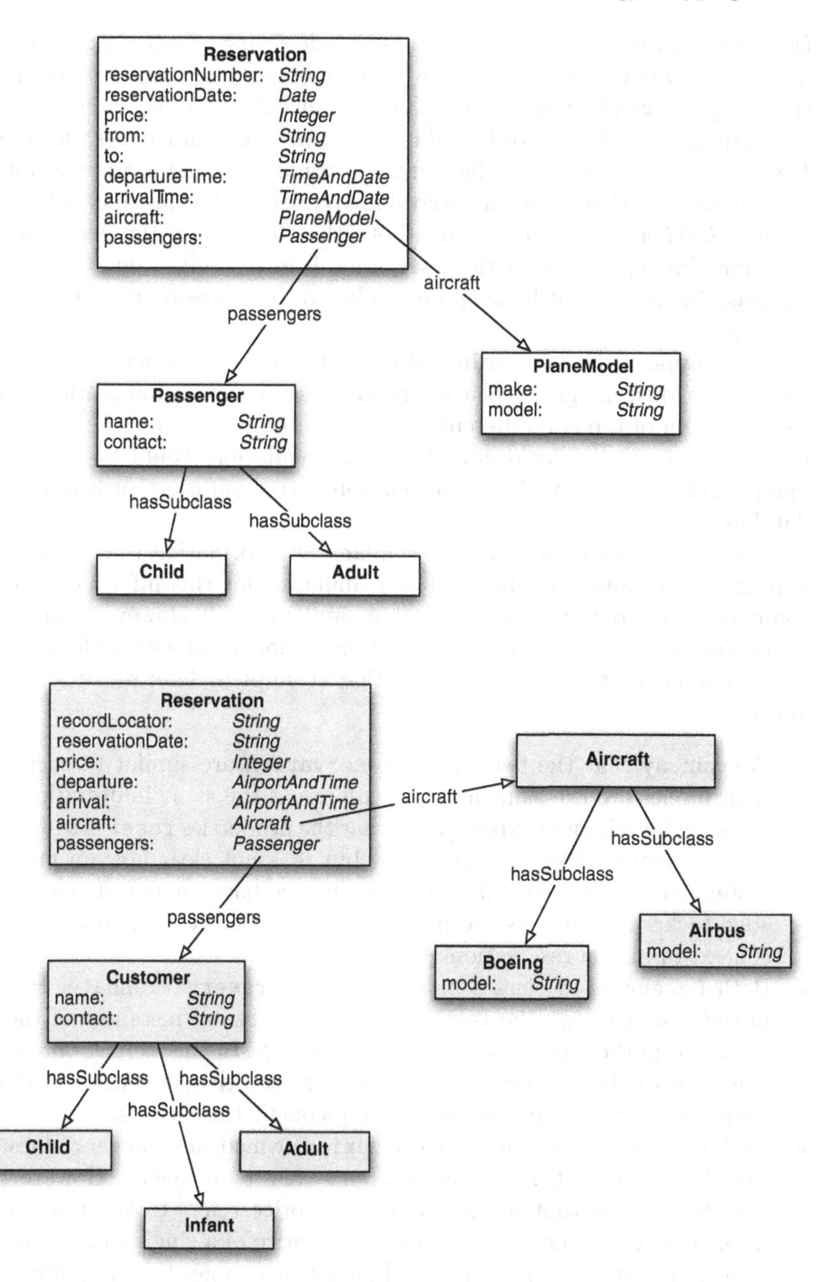

Fig. 1. Example of two ontologies representing airline reservations

(recordLocator) and the date it was made (reservationDate). There is information about the price of the ticket (property price) and the aircraft (property aircraft with values that are instances of the class Aircraft, or, more specifically, one of its subclasses; two of the subclasses, Boeing and Airbus are presented in the figure; there could be many more, one for each aircraft maker). Departure and arrival are represented as instances of the class AirportAndTime that encapsulates both the airport and the departure date and time. Passenger list for the reservation is also a collection of instances of the class Customer that looks quite similar to the Passenger class in the first ontology.

Now suppose the two airlines decided to merge. The merge means that the airlines must integrate their reservation systems. This integration requires reconciliation of the two different ontologies used to describe the reservations in the two airlines. The result of this reconciliation would be an ontology mapping that would enable transformation of reservation records into a single database.

The two ontologies look rather similar, the information that they capture is roughly the same and the level of granularity for this information is very comparable; many terms are identical or similar as well. However, after careful examination, we can see that the mapping is not at all straightforward even for a human expert, let alone for tools that attempt to determine the mapping automatically:

- We can say that the two classes Reservation are similar to each other: their names are the same and they represent the same information.
- It is easy for a human expert to see that the properties reservationNumber and recordLocator are equivalent, but it is not clear how an automatic system can identify this fact; as humans, we have enough domain knowledge to know that these terms usually refer to the same concept in the context of airline reservations.
- Both Reservation classes have a property reservationDate. However, in the first ontology the values of this property are instances of the class Date and in the second they are simply strings. In the merged ontology we will have to choose one representation or the other and to convert dates from one format to another when we reconcile the records.
- Both ontologies have the property price, which has integer values. One would think that this property is very easy to reconcile. However, it is easy to imagine that in one system the price refers to the price for the whole reservation and in the other it is a price of a single ticket and needs to be multiplied by the number of passengers to get the price for the full reservation. Note that just by looking at the ontology, we simply cannot tell what the price refers to in either case. We need additional information, for example in the form of documentation to determine whether the price properties in the two ontologies are equivalent.

- In the first ontology, the information about the departure and arrival location and the time and date of arrival is represented directly in the reservation record. In the second ontology, the departure and arrival information (the airport and the date and time) are encapsulated as instances of the class `AirportAndTime`.
- Classes `PlaneModel` and `Aircraft` are equivalent – they both represent various aircraft makes and models – but have different names.
- The classes `Passenger` and `Customer` have different names but are equivalent in this context. Note that in general, for any two ontologies, classes named *Passenger* and *Customer* may not be equivalent, but they are in this context and given the structure of these two ontologies.
- In the first ontology, there are two subclasses of the class `Passenger`: `Child` and `Adult`. In the second ontology, the breakdown is different, with the class `Infant` representing children under 2. Thus the class `Child` in the first ontology is the union of the classes `Child` and `Infant` from the second; the two classes `Child` are not themselves equivalent.
- Finally, while in the first ontology the various makes of an aircraft are represented as a string property `make` of the class `PlaneModel`, in the second ontology this distinction is made in different subclasses, such as `Boeing` and `Airbus`.

As you can see, even such small example of two ontology snippets with very similar domain coverage and granularity, poses many difficulties in both determining the correspondences between classes and properties and finding them automatically.

The goal of this chapter is to discuss the major thrusts of approaches to semantic integration produced by various projects in the ontology community and the user-centered tools that support the ontology mapping in practice. We do not attempt to provide a comprehensive review of the state of the art in ontology mapping. We refer the reader to an excellent and thorough review by Euzenat and Shvaiko [34] for that purpose.

We discuss four dimensions of ontology-mapping research in this chapter:

Mapping discovery: Given two ontologies, how do we find similarities between them, determine which concepts and properties represent similar notions, and so on.

Interactive specification of mappings: tools for enabling users and ontology developers to define the compare ontologies interactively, define the mappings, perhaps with the semiautomated help from the tool itself.

Declarative formal representations of mappings: Given two ontologies, how do we represent the mappings between them to enable reasoning with mappings.

Reasoning with mappings: Once the mappings are defined, what do we do with them, what types of reasoning are involved?

In the rest of this chapter, we explore these dimensions.

2 Discovering Mappings

Many researchers agree that one of the major bottlenecks in semantic integration is mapping discovery. There are simply too many ontologies and database schemas available and they are too large to have manual definition of correspondences as the primary source of mapping discovery. Furthermore, in the world where software agents will roam the (semantic) web, they will need to map structures they know about to new structures they come across on-the-fly. Hence, the task of finding mappings (semi-) automatically has been an active area of research in both database and ontology communities [31, 34].

We identify two major architectures for mapping discovery between ontologies. In the first approach, we create a mapping between two ontologies, O_1 and O_2, by using another (or several other ontologies) as an intermediary: we map both O_1 and O_2 to this third ontology or terminology and use this set of two mappings to infer the mapping between O_1 and O_2. In the second approach, the two ontologies are mapped directly to each other, by using heuristic-based techniques, machine learning, graph comparison, or the use of background knowledge.

2.1 Using a Shared Ontology

Recall that the goal of ontologies is to facilitate knowledge sharing. As a result, ontologies are often developed with the explicit goal of providing the basis for future semantic integration. Here, the vision is that a general upper ontology is agreed upon by developers of different applications, who then extend this general ontology with concepts and properties specific to their applications. A number of very general ontologies formalizing notions such as processes and events, time and space, physical objects, and so on, are being developed and some of them are becoming accepted standards (chapter "Foundational Choices in DOLCE"). The explicit goal of these ontologies is to have domain-specific ontologies extend them, thus providing the grounding in common vocabulary for these ontologies. Two of the ontologies that are built specifically with the purpose of being formal foundational ontologies are the Suggested Upper Merged Ontology (SUMO) [27] and DOLCE [15]. SUMO is an effort by the IEEE Standard Upper Ontology Working Group aimed at developing "a standard upper ontology that will promote data interoperability, information search and retrieval, automated inferencing, and natural language processing." The SUMO ontology defines such high-level concepts as `Object`, `ContinousObject`, `Process`, `Quantity`, `Relation`, and so on, providing axioms in first-order logic that describe properties of these concepts and relations among them. Similarly, the DOLCE ontology is a formal foundational ontology developed as a foundational ontology in the WonderWeb project, which comprises a large number of European research groups. The goal of DOLCE is to provide a common reference framework for WonderWeb ontologies to facilitate sharing of information among them. In its representation, DOLCE aims at capturing "ontological categories underlying natural language and human common-sense."

While many researchers hope that domain- and application-specific ontologies will reuse the foundational ontologies, like SUMO and DOLCE, and that such reuse will indeed facilitate semantic interoperation between applications based on these ontologies, we do not yet have enough experience reports with such approaches to claim it a success. There are reports on both the successes [30] and difficulties [35] of such reuse.

There are also implemented semantic-integration tools that exploit the idea that if two ontologies extend the same reference ontology in a consistent way, then finding correspondences between their concepts is easier. For example, the Process Specification Language (PSL) [19], developed at the National Institute for Standards and Technology, is an ontology that is endorsed as an International Standard within the International Organization of Standardization (ISO) (see also chapter "Using the PSL Ontology"). PSL was designed to "facilitate correct and complete exchange of process information among manufacturing systems such as scheduling, process modeling, [and] process planning" [20]. The designers of PSL have developed it as an interlingua for ontologies representing these different process. All theories within the PSL ontology have been verified with respect to the intended semantics of their terminology. Grüninger and Kopena [20] developed an integration architecture with the PSL ontology at the center and mappings between ontologies for specific manufacturing processes and the PSL ontology. The mappings are defined semiautomatically by presenting ontology developers with a set of questions (in natural language) helping them to map terms in their process-specific ontology to the terms in PSL. The system then generates two-way mappings between the task-specific ontology, such as scheduling and the PSL interlingua. Note that the generation of these mappings is defined formally and is not based on heuristics. These mappings can be composed to provide mappings between any task-specific ontologies.

Finally, the third approach is to use a reference ontology or terminology as background knowledge that helps in aligning the ontologies that need to be aligned. In this model, the source ontologies are first mapped to a reference ontology that serves one of the two purposes: (1) providing wider domain coverage to bridge the coverage gap of the source ontologies or (2) providing the additional structure and semantic richness that the source ontologies lack. For example, the developers of S-Match [16] used WordNet as the terminology that provided wide coverage to bridge the source ontologies. Zhang and Bodenreider used the relationships in UMLS to verify their automatically generated mappings between two anatomy ontologies [36]. In both cases, the researchers used simple lexical-matching techniques to map concepts from the source ontologies to the shared ontology. Aleksovski and colleagues mapped two simple terminologies that had very little structure to a rich background medical ontology; the background ontology then provided the additional information that was needed to map the two source terminologies [1]. The researchers mapped the source ontologies to the shared ontology manually and then used the rich information in the shared ontology to infer the mappings between the sources.

2.2 Structure-Based, Machine-Learning, and Other Approaches

It is certainly helpful to have ontologies that we need to match to refer to the same foundational ontology or to conform to the same reference ontology. However, we often do not have this "luxury" and need to create mappings between ontologies that perhaps use the same specification language but do not have any vocabulary beyond the specification language in common. Most researchers agree that automatic mapping between ontologies in this context is beyond our grasp at the moment, but many techniques have produced good results.

Ontologies are often richly structured, with many links between definitions of classes and properties. Thus, many approaches exploit this richness by comparing various elements of the structure of the ontologies to be mapped. Consider again the example in Fig. 1. A simple lexical mapping can identify the correspondence of the two classes `Reservation`. While lexical mapping will not identify the equivalence of the `Passenger` and `Customer` class in this context, we can infer this equivalence by looking at the structure of the two ontologies: the two properties `passengers` have the same labels, the equivalent classes as their domains (`Reservation`) and have `Passenger` and `Customer` as their respective ranges. These are the types of features that tools that analyze ontology structure look at. For example, COMA++ [6] represents both ontologies as graphs, treating all classes and properties as nodes and then compares the structure of the two graphs. QOM [9] uses a range of ontology features such as concept labels, domains and ranges of properties, relations between classes, and so on to find how similar concepts in source ontologies are. Similarly, Euzenat and Valtchev [12] exploit the full gamut of features of the OWL ontologies to compute a weighted combination of similarities in OWL concept definitions: their labels, domains and ranges of properties, restrictions on properties (such as cardinality restrictions), types of concepts, subclasses and superclasses, and so on. PROMPT [28] is an interactive ontology-mapping tool that guides users through the process and suggesting which classes and properties can be merged. It records the mappings identified both by the system and by the user during merging to create a declarative mapping specification between source ontologies. To make suggestions, PROMPT also uses a mixture of lexical and structural features, as well as input from the user during an interactive merging session to find the mappings. Another algorithm in the toolset – ANCHORPROMPT [28] – treats an ontology as a graph with classes as nodes and slots as links. The algorithm analyzes the paths in the subgraph limited by the anchors and determines which classes frequently appear in similar positions on similar paths. These classes are likely to represent semantically similar concepts.

GLUE [7] is an example of a system that employs machine-learning techniques to find mappings. GLUE uses multiple learners exploiting information in concept instances and taxonomic structure of ontologies. GLUE uses a probabilistic model to combine results of different learners. The learners that GLUE uses currently relies on ontologies having instances and they work much better if many slot values have text in them rather than references to other instances. For example, GLUE might identify the equivalence of the classes `Passenger` and `Customer` in Fig. 1 by classifying their respective instances and learning that they are similar.

Sabou and colleagues [33] have used the ontologies on the Semantic Web itself to find mappings. In this work, in order to find correspondences between terms in two ontologies, the authors find the same terms in ontologies in a semantic web ontology repository. If there is an ontology where the two terms are linked directly (say, through a subclass-of) relationship, then the authors infer that the original terms are related as well.

Many researchers have now shown that the real power of these various methods for discovering mappings lies in their combination. The tools that have been showing the most success in performance recently [13] are the ones that combine several very different approaches and manage to figure out the right way to combine them in terms of relative weights of different components. For instance, FOAM [10] uses machine-learning techniques to determine the relative weights of different matchers. Falcon-AO [23] uses three different matchers as input to the component that integrates them: one matchers looks at the similarity of concept definitions, treating them as lexical entities and comparing their vector-space models; the second matcher uses simple lexical techniques to find both similarities and differences in concept labels; the third uses a graph comparison algorithm to compare graph representations of the two ontologies. A central controller then dynamically determines the relative weight of each component based on its perceived success with the specific source ontologies.

Researchers have also recognized that finding simple one-to-one mappings, essentially representing equivalent or similar concepts, is not sufficient. Two of the key aspects that researchers also look at are *complex mappings* and *approximate mappings*.

Complex mappings express specialization or generalization relationship between concepts, or perhaps even contain an expression linking entities together. For instance, the class `Child` in the first ontology in Fig. 1 is a union of the classes `Infant` and `Child` in the second ontology. S-Match [17] is one example of a system that identifies complex mappings. S-Match starts by grounding their source ontologies in WordNet terms but then run a SAT prover on the mappings to determine other types of mappings (such as generalization, specialization or disjointness): the authors reformulate the matching problem as that of propositional satisfiability.

In many cases, exact mappings either cannot be derived or simply do not exist: the entities from different ontologies may be related, but, have, for example, largely overlapping meaning. Consider, for instance, different frequent-flyer programs for airlines. The frequent-flyer program in an ontology describing one airline would be very similar but not exactly the same as a frequent-flyer program for another airline. Gligorov and colleagues [18] formalized approximate mapping by decomposing a concept definition into a set of sub-definition and then determining how many of these sub-definitions map another concept that they are trying to match.

3 Interactive Tools for Specifying Mappings

As the results of the OAEI tests show, even the best automatic tools for discovering mappings still leave a lot of work for the user to do: to verify the mappings and to add the ones that the algorithms missed.

However, many of the ontology-mapping tools focus only on the algorithm and provide only rudimentary user interface. The most common interface is the use of command line to provide the file names or URIs for the source ontologies and to get a text listing of the mappings.

Clio [21] and COMA++ [6] are among the few exceptions – tools that support graphical user interfaces. The number of visual paradigms that the tools use to display the mappings is quite limited though. Clio was developed by IBM for generating mappings between relational and XML schemas. Clio can infer correspondences in the source and target schemas and it also allows users to draw correspondences between parts of the schemas. Once the user verifies the correspondences suggested by Clio, Clio generates queries to drive the translation from the source schema to the target schema. COMA++ automatically generates mappings between the source and target schemas, and draws lines between matching terms. Users can also define their own term matches. Both tools draw mappings between the source and target schemas as arrows and support in-tool navigation of the source ontologies.

In Sect. 2, we focused on the tools to discover mappings and in Sect. 4 on various representations. We will now discuss an interactive tool with a graphical user interface that allows developers to leverage the best components of other tools easily by integrating them in a single plugin framework.

In the Protégé group, we have developed a suite of tools to support users in the process of ontology mappings. While in the early days of the PROMPT suite, we focused on algorithms [29], we have later come to realize the dearth of the user-oriented tools in the field of ontology mapping. Most researchers focus on the algorithms and leave the fine-tuning of the mapping as an "exercise to the reader."

We believe that in order for the ontology-mapping tools to "step out" of the research labs and to be adopted in real-world and industrial setting, both the performance of automatic ontology-mapping algorithms and the quality of cognitive support in ontology-mapping tools must improve. Recognizing that in many cases, researchers must focus on one or the other of these tasks, we have developed a plugin framework that covers the full spectrum of ontology mapping, from specifying algorithms for initial comparison to executing the mappings. We have developed a reference implementation for each of the steps, including a number of cognitive aids. Developers can plug in their own components and have the plugins developed by others (including our team) fill in the missing pieces to have a comprehensive end-user tool.

The PROMPT plugin framework [14] allows developers to replace any of the components that we have just described with their own. The plugin framework works by providing Java interfaces for various types of plugins (comparison algorithm, visualization components, etc.). A plugin developer chooses the interface they wish to implement, and then supplies the appropriate method bodies in order to perform the operations they wish to execute. More specifically, we view the ontology-mapping process as a sequence of the following components (Fig. 2):

Perform initial comparison of the ontologies: an algorithm compares two ontologies and produces a list of candidate mappings.

Present candidate mappings to the user enabling him to analyze the results. This step includes components for cognitive support (various visualizations of the source and target ontologies, options to filter content presented in the display, etc.) and interactive comparison algorithms that are invoked either explicitly by the user or as a result of mappings being verified.

Fine tune and save the mappings in a declarative mapping format.

Execute mappings to transform instances from source to target or to perform other operations.

In the current implementation, developers can replace components of any of the steps in this list, and our plan is to make all of the steps replaceable.

For example, the integration of FOAM and PROMPT is one of the PROMPT plugins available as part of PROMPT distribution. A developer of an algorithm plugin can specify not only how to invoke the algorithm, but also how the configuration screen presented to the user should look like.

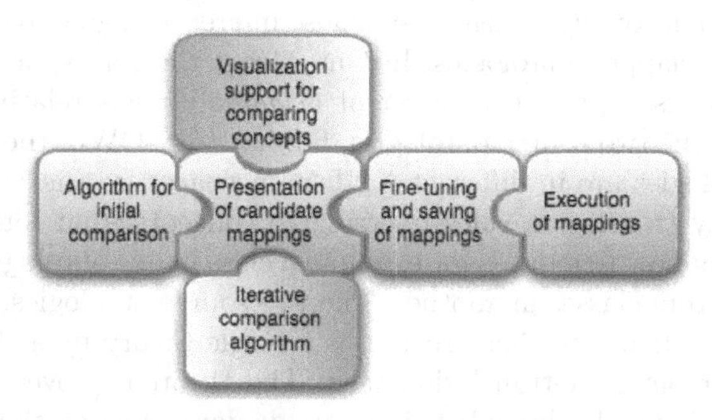

Fig. 2. Configurable steps in the PROMPT plugin framework. Developers can replace any of the components in the figure with their own implementation

4 Representations of Mappings

While developing tools for automatic and semiautomatic ontology matching is a large thrust of semantic-integration research in the ontology community, it is definitely not the only one. The high expressive power of ontology languages provides the opportunity for representing mappings themselves in expressive terms. We will discuss several representations of mappings here: using the ontology language itself to express mappings; defining bridging axioms in first-order logic to represent transformations; representing mappings as instances in an ontology of mappings; and using views to describe mappings from a global ontology to local ontologies.

We can use the constructs provided by the OWL language itself to express mappings between concepts in different ontologies (chapter "Web Ontology Language: OWL"). To express equivalence between classes, properties and individuals, we can use the following three OWL constructs, respectively: `owl:equivalentClass`, `owl:equivalentProperty`, and `owl:sameAs`. For instance, we can say that a class `Boeing` in one ontology (Fig. 1) is equivalent to the class `PlaneModel` where the property `make hasValue` "Boeing." Similarly, such RDFS and OWL constructs as `rdfs:subclassOf` can express generalization and specialization relations. Because these constructs can link any two arbitrary class expressions, OWL itself is an expressive ontology-mapping language. Mappings expressed as OWL constructs become part of the ontology itself, and there is no clear separation between the mappings and the ontology definitions.

Sometimes, however, we want to be more precise about the nature of the mapping and to separate the definition of the mappings from the definition of ontology concepts. The C-OWL language, for example, takes such approach [2]. It keeps ontologies "contextualized," that is not integrated with other ontologies. C-OWL defines explicit mapping rules to express correspondences between concepts in the ontologies. Semantically, these rules express how to translate instances from the source ontology to the target ontology.

In the OntoMerge system [8] developed for semantic integration on the Semantic Web, authors use a general-purpose inference engine to enable translation between mapped ontologies. In OntoMerge the correspondence between two ontologies is expressed as a set of *bridging axioms* relating classes and properties of the two source ontologies. Similar to C-OWL, the vocabulary of the two ontologies are in different XML namespaces, so the bridging axioms are essentially translation rules referring to concepts from source ontologies and specifying how to express for example a class in one ontology by collecting information from classes in another. The two source ontologies, together with the bridging axioms are then treated as a single theory by a theorem prover optimized for ontology-translation task. The theorem prover runs either in forward-chaining or backward-chaining mode depending on the task at hand.

Several researchers use ontologies themselves to represent mappings declaratively, as instances in an ontology. The *mapping ontology* by Crubézy and colleagues [5] or the *Semantic Bridge Ontology* of the MAFRA framework [26], for instance, define the structure of specific mappings and the transformation functions to transfer instances from one ontology to another. This ontology can then be used by tools to perform the transformations. Such an ontology usually provides different ways of linking concepts from the source ontology to the target ontology, transformation rules to specify how values should be changed, and conditions and effects of such rules. Then a mapping between two ontologies constitutes a set of instances of classes in the mapping ontology and can be used by applications to translate data from the source ontology to the target. The mapping ontology mentioned above [5], for example, provides declarative means for defining many-to-one or many-to-many aggregation relationships between concepts in the source and target ontologies, as well as one-to-many concept-decomposition relations. It allows specification of recursive mappings, complex mappings between that collect information from several related concepts, and other mechanisms.

Finally, researchers also used views to define mappings between ontologies, similar to defining mappings in information integration, both in global-as-view (GAV) and local-as-view (LAV) setting. The OIS framework [3] is a good example of such approach. In OIS, a global ontology is used to provide access to local ontologies. Both global and local ontologies are defined using Description Logics. The mappings are defined as views over either the global or the local ontologies. In other words, a predicate from one ontology is defined as a query (and DL expression) over predicates in another ontology.

5 We Have the Mappings: Now What?

Naturally, defining the mappings between ontologies, either automatically, semiautomatically, or interactively, is not a goal in itself. The resulting mappings are used for various integration tasks: data transformation, query answering, or web-service composition, to name a few.

Given that ontologies are often used for reasoning, it is only natural that many of these integration tasks involve reasoning over the source ontologies and the mappings. For example, the OntoMerge system mentioned earlier [8] uses reasoning to perform several tasks related to ontology translation. The first task is translating instances that conform to one ontology (the source) to instances conforming to another ontology (the target), given the mapping between the source and target. To perform this task, OntoMerge first creates a merged ontology that includes the source, the target, and the mapping and performs inference on this merged ontology. Afterwards, OntoMerge performs a projection step, where it retains only the new conclusions reached that exclusively reference the target vocabulary.

For the second task that OntoMerge deals with – generating ontology extensions – consider for example, two ontologies describing Web services:

OWL-S[2] and WSDL.[3] Suppose we have defined a mapping between these two ontologies. Suppose also that we have an ontology describing ticket-purchasing web services – a domain-specific extension of the OWL-S ontology (chapter "Semantic Web Services"). This ticket-purchasing ontology creates subclasses of some of the classes in OWL-S, fills in some of the property values, and so on. In other words, it *extends* the OWL-S ontology. If we have a mapping between OWL-S and WSDL, OntoMerge can automatically generate a WSDL description of ticket-purchasing – an extension of the WSDL ontology. Note that this case is different from data translation since we are dealing with subontologies rather than instances conforming to ontologies. In both of these tasks, OntoMerge uses forward-chaining reasoner to perform the translation.

In the OIS framework [3], ontologies are expressed in Description Logics and therefore it is natural that DL reasoners are used to answer queries in the data-integration framework. The authors address the general task of answering queries posed in terms of the global ontology using the data in the local ontologies. However, while even in expressive Description Logics, computing certain answers to queries is decidable, it may often be intractable. In recent research, the authors have explored less expressive subsets of Description Logics [4], making this type of query answering tractable.

6 The State of the Art

The best way to assess the state of the art in creating ontology mappings automatically is to look at the results of the Ontology Alignment Evaluation Initiative (OAEI).[4] OAEI is a principled formal effort to compare the performance of different ontology-mapping algorithms on the same set of data. The OAEI is run annually since 2004 and more systems participate in the evaluation each year. The total of 18 systems participated in 2007.

The OAEI organizers published the ontologies to be compared and the tool developers apply their tools to find correspondences between these ontologies. The organizers then compare the results to a set of reference alignments to determine precision and recall of the performance of individual tools. The ontologies range in the size and complexity, from relatively small ontologies that are designed primarily to understand which features of ontologies tools take into account, to large "real-world" ontologies representing such complex domains as anatomy.

The performance of the tools varies depending on the setting. For the simplest test cases, the best tools have close to perfect recall and precision. For the large more complex test cases such as anatomy ontologies, for example, the

[2] `http://www.daml.org/services/owl-s/1.0/`

[3] `http://www.w3.org/TR/wsdl`

[4] `http://oaei.ontologymatching.org`

F-measure (the harmonic means of precision and recall) for the best performing generic domain independent tools (such as Falcon-AO [23] and ASMOV [22]) is 0.73–0.75. The tools that take into account domain knowledge (e.g., the tools specifically designed for alignment of ontologies in biomedical domain, such as AOAS [36]), reach the precision of 92% and recall of 80% (F-measure 0.86). More important, the precision and recall measures for the best systems keep improving every year [11].

The data provided by the OAEI initiative has affected quite dramatically the field of ontology-mapping algorithms by providing a well-documented and well-studied set of reference alignments that tool-developers can use to assess their methods: It is impossible to publish a paper about an ontology-mapping algorithm today without providing the results of how the algorithm performs on the OAEI data set. It is important to note, however, that OAEI does not evaluate all types of ontology-mapping systems: For instance, this type of evaluation is not well suited for interactive ontology-mapping tools, such as PROMPT. At the same time, the OAEI results show that user interaction will still be required in the foreseeable future if an application needs precise and complete mappings.

Finally, in addition to improving precision and recall of ontology-mapping algorithms, there are a number of other fundamental questions that researchers are addressing: How do we explain the mappings produced by the tools to the users? What are the best ways to support interactive ontology mapping? Are imperfect or inconsistent alignment useful in some settings or must alignments always be 100% precise to be useful? What are the settings that con tolerate approximate mappings? What are the levels of precision and recall that make such imprecise mappings useful? How do we maintain mappings between ontologies as the ontologies evolve? When ontologies change, many of the mappings remain valid, but some are probably invalidated? How do we know which ones? Can we design a "tollbox" of mapping approaches that we can custom-tailor to fit a given problem? Can generic domain-independent ontology-mapping methods perform acceptably well, or do we need domain-specific approaches? Researchers are actively investigating these and many other intriguing challenges in ontology mapping and we can fully expect this research area to continue to provide new advances and to make automatic semantic integration – that holy grail of the Semantic Web – more attainable.

Acknowledgments

Some portions of this chapter appeared as an article in SIGMOD Record in 2004. I would like to thank AnHai Doan, Michael Grüninger, and Yannis Kalfoglou for thoughtful and helpful comments on that article. Sean Falconer implemented many components of the PROMPT plugin framework. Most important, I am extremely grateful to Mark Musen for the many years of guidance and support in this work.

References

1. Z. Aleksovski, M. Klein, W. ten Kate, and F. van Harmelen. Matching unstructured vocabularies using a background ontology. In *15th International Conference on Knowledge Engineering and Knowledge Management (EKAW'06)*, 2006.
2. P. Bouquet, F. Giunchiglia, F. van Harmelen, L. Serafini, and H. Stuckenschmidt. C-OWL: Contextualizing ontologies. In *International Semantic Web Conference (ISWC)*, Sanibel Isalnd, Florida, 2003.
3. D. Calvanese, G. Giacomo, and M. Lenzerini. Ontology of integration and integration of ontologies. In *Description Logic Workshop (DL 2001)*, pages 10–19, 2001.
4. D. Calvanese, G. De Giacomo, M. Lenzerini, R. Rosati, and G. Vetere. DL-lite: Practical Reasoning for Rich DLs. In *International Workshop on Description Logics (DL2004)*, Whistler, Canada, 2004.
5. M. Crubézy and M. A. Musen. Ontologies in support of problem solving. In S. Staab and R. Studer, editors, *Handbook on Ontologies*, pages 321–342. Springer, Berlin, 2003.
6. H. Do. *Schema Matching and Mapping-based Data Integration.* PhD thesis, Department of Computer Science, Universitt Leipzig, 2006.
7. A. Doan, J. Madhavan, P. Domingos, and A. Halevy. Learning to map between ontologies on the semantic web. In *The Eleventh International WWW Conference*, Hawaii, US, 2002.
8. D. Dou, D. McDermott, and P. Qi. Ontology translation on the semantic web. In *International Conference on Ontologies, Databases and Applications of Semantics*, 2003.
9. M. Ehrig and S. Staab. QOM – Quick Ontology Mapping. In *3rd International Semantic Web Conference (ISWC2004)*, Hiroshima, Japan, 2004.
10. M. Ehrig and Y. Sure. FOAM – framework for ontology alignment and mapping; results of the ontology alignment initiative. In *Proceedings of the Workshop on Integrating Ontologies*, volume 156, pages 72–76, October 2005.
11. J. Euzenat, A. Isaac, C. Meilicke, P. Shvaiko, H. Stuckenschmidt, O. Šváb, V. Svátek, W. van Hage, and M. Yatskevich. Results of the ontology alignment evaluation initiative 2007. In *2nd International Workshop on Ontology Matching (OM-2007) at ISWC 2007*, 2007.
12. J. Euzenat and P. Valtchev. Similarity-based ontology alignment in OWL-Lite. In *The 16th European Conference on Artificial Intelligence (ECAI-04)*, Valencia, Spain, 2004.
13. J. Euzenat, M. Mochol, P. Shvaiko, H. Stuckenschmidt, Ondrej Sváb, Vojtech Sváte, Willem Robert van Hage, and Mikalai Yatskevich. Results of the ontology alignment evaluation initiative 2006. In *International Workshop on Ontology Matching at ISWC-2006*, Athens, GA, 2006.
14. S. Falconer, N. Noy, and M. A. Storey. Towards understanding the needs of cognitive support for ontology mapping. In *International Workshop on Ontology Matching at ISWC-2006*, Athens, GA, 2006.
15. A. Gangemi, N. Guarino, C. Masolo, and A. Oltramari. Sweetening wordnet with DOLCE. *AI Magazine*, 24(3):13–24, 2003.
16. F. Giunchiglia, P. Shvaiko, and M. Yatskevich. S-match: an algorithm and an implementation of semantic matching. In *European Conference on Semantic Web (ESWC 2004)*, pages 61–75, 2004.

17. F. Giunchiglia, P. Shvaiko, and M. Yatskevich. Semantic matching. In *1st European semantic web symposium (ESWS'04)*, pages 61–75, Heraklion, Greece, 2004.
18. R. Gligorov, Z. Aleksovski, W. ten Kate, and F. van Harmelen. Using Google distance to weight approximate ontology matches. In *Seventeenth World Wide Web Conference WWWW-07*, Banff, Canada, 2007.
19. M. Grüninger. A guide to the ontology of the process specification language. In S. Staab and R. Studer, editors, *Handbook on Ontologies*. Springer, Berlin, 2003.
20. M. Grüninger and J. Kopena. Semantic integration through invariants. In A. Doan, A. Halevy, and N. Noy, editors, *Workshop on Semantic Integration at ISWC-2003*, Sanibel Island, FL, 2003.
21. M. Hernandez, R. Miller, L. Haas, L. Yan, C. T. Howard Ho, and X. Tian. Clio: A semi-automatic tool for schema mapping. In *SIGMOD Record*, 2001.
22. Y. Jean-Mary and M. Kabuka. ASMOV: ontology alignment with semantic validation. In *JointSWDB-ODBIS Workshop on Semantics, Ontologies, Databases*, Vienna, Austria, 2007.
23. N. Jian, W. Hu, G. Cheng, and Y. Qu. Falcon-ao: Aligning ontologies with falcon. In *K-Cap Workshop on Integrating Ontologies*, 2005.
24. Y. Kalfoglou and M. Schorlemmer. Ontology mapping: the state of the art. *The Knowledge Engineering Review*, 18(1):1–31, 2003.
25. M. Klein. Combining and relating ontologies: an analysis of problems and solutions. In *IJCAI-2001 Workshop on Ontologies and Information Sharing*, pages 53–62, Seattle, WA, 2001.
26. A. Maedche, B. Motik, N. Silva, and R. Volz. MAFRA - a mapping framework for distributed ontologies. In *13th European Conference on Knowledge Engineering and Knowledge Management EKAW*, Madrid, Spain, 2002.
27. I. Niles and A. Pease. Towards a standard upper ontology. In *The 2nd International Conference on Formal Ontology in Information Systems (FOIS-2001)*, Ogunquit, Maine, 2001.
28. N. F. Noy and M. A. Musen. The PROMPT suite: Interactive tools for ontology merging and mapping. *International Journal of Human-Computer Studies*, 59(6):983–1024, 2003.
29. N. F. Noy. Tools for mapping and merging ontologies. In S. Staab and R. Studer, editors, *Handbook on Ontologies*, pages 365–384. Springer, Berlin, 2003.
30. S. Polyak, J. Lee, M. Grüninger, and C. Menzel. Applying the process interchange format(PIF) to a supply chain process interoperability scenario. In *Workshop on Applications of Ontologies and Problem Solving Methods, ECAI'98*, Brighton, England, 1998.
31. E. Rahm and P. A. Bernstein. A survey of approaches to automatic schema matching. *VLDB Journal*, 10(4), 2001.
32. D. L. Rubin, S. E. Lewis, C. J. Mungall, S. Misra, M. Westerfield, M. Ashburner, I. Sim, C. G. Chute, H. Solbrig, M. A. Storey, B. Smith, J. Day-Richter, N. F. Noy, and M. A. Musen. The national center for biomedical ontology: Advancing biomedicinethrough structured organization of scientific knowledge. *OMICS: A Journal of Integrative Biology*, 10(2), 2006.
33. M. Sabou, J. Gracia, S. Angeletou, M. dAquin, and E. Motta. Evaluating the semantic web: A task-based approach. In K. Aberer, K. S. Choi, and N. Noy, editors, *6th International Semantic Web Conference (ISWC 2007)*, Busan, Korea, 2007. Springer.

34. P Shvaiko and J Euzenat. A survey of schema-based matching approaches. *Journal on Data Semantics*, 4:146–171, 2005.
35. A. Valente, T. Russ, R. MacGrecor, and W. Swartout. Building and (re)using an ontology for air campaign planning. *IEEE Intelligent Systems*, 14(1):27–36, 1999.
36. S. Zhang and O. Bodenreider. Alignment of multiple ontologies of anatomy: Deriving indirect mappings from direct mappings to a reference. In *AMIA Annual Symposium*, pages 864–868, 2005.

Part V

Ontology-Based Infrastructure and Methods

Ontologies and Software Engineering

Dragan Gašević[1], Nima Kaviani[2], and Milan Milanović[3]

[1] Athabasca University, Canada, dgasevic@acm.org
[2] University of British Columbia, Canada, nimak@ece.ubc.ca
[3] University of Belgrade, Serbia, milan@milanovic.org

Summary. The chapter analyzes the state of the art in the use of ontologies for various software engineering tasks. The chapter starts from defining software engineering as an application context for ontologies. Next, it introduces a framework that identifies places in software lifecycle where ontologies can contribute to improve the current state of software engineering.

1 Introduction

Fast growth of communication and mobile technologies, constant demands for new services, and increased number of computer users, are some of the key reasons of the constantly increasing need for more software. This naturally requires effective methods for engineering software that will be able to respond adequately to the needs for which the software was built, and yet to allow for higher levels of productivity of software engineers. However, today's state of the art and practice demonstrates that both perspectives are still suffering from serious problems. On one hand, the Standish Group published its well-known "Chaos Report" in 1994 in which it was noted that only 16% of software projects were successful, 31% were failures, and some 53% were challenged. The 2006 report demonstrates a bit better situation where 35% of software projects were successful, 19% were failures, and 46% were challenged [9]. On the other hand, productivity methods are struggling with new challenges such as better methods for software maintenance (e.g., tracing place in the code when adding new or updating present functionalities to the software [69]) or facilitating collaboration of software teams (e.g., mutual understanding between different parties collaborating in requirement engineering, especially in the context of global software development [18]).

While software is a technical category designed to perform specific tasks by using computer hardware, it is also a social category which nowadays is used in almost every aspect of human's life. In fact, software is a knowledge repository where knowledge is largely related to the application domain, and not

S. Staab and R. Studer (eds.), *Handbook on Ontologies,* International Handbooks on Information Systems, DOI 10.1007/978-3-540-92673-3,

to software as an entity [4]. So, we need to be able to share and interoperate (application) knowledge stored in software with the knowledge about all relevant aspects surrounding and influencing software (e.g., domain knowledge, new requirements, policies, and contexts, in which people use and interact with software) in order to get software to the more advanced levels. This knowledge sharing and management requires the use of explicit definition of knowledge, as it is a basic need for machines to be able to interpret knowledge. This is why the software engineering community has recognized ontologies as a promising way to address current software engineering problems [14, 32].

Researchers have so far proposed many different synergies between software engineering and ontologies. For example, ontologies are proposed to be used in requirement engineering [47], software modeling [45], model transformations [42], software maintenance [43], software comprehension [70], software methodologies [30], and software community of practice [1]. Moreover, software engineering technologies are proposed for modeling and reasoning over ontologies. These synergies between ontologies and software engineering have also attracted attention of standardization bodies and have some on-going activities. Ontology-Driven Architecture (ODA) is an effort of the W3C's Software Engineering Best Practices Working Group that tries to develop best practices for using ontologies in software engineering [66]. Probably, the most important result so far is the Ontology Definition Metamodel (ODM) that is proposed to be the Object Management Group (OMG)'s standard [54]. The ODM standard allows for integrating ontology languages (i.e., ontologies) into the software development process based on model-driven engineering principles [7]. Although all of these different efforts demonstrate many benefits to different aspects of software and ontology engineering or give a nice description of the state of the art in the area [14, 32], none of them analyze and evaluate applications of ontologies in different aspects of software engineering by following a comprehensive software life cycle framework.

In this chapter, we start from defining software engineering as an application context for ontologies, and proceed to defining a framework that identifies places in software life cycle where ontologies can contribute to improve the current state of software engineering. We consequently have organized the structure of this chapter to use this framework for analyzing the use of ontologies in different phases of software life cycle. Note that the chapter does not discusses Semantic Web rules (see chapter "Ontologies and Rules") or upper layers of the Semantic Web cake, but fully focuses on ontologies in software engineering.

2 Software Engineering

The goal of this section is to define software engineering, describe some typical software life cycle phases, artifacts used and produced in them, participants, their interactions, and relevant domain and application knowledge. Based on this discussion, we define a unified framework for the use of ontologies in software engineering to which we are going to refer in the rest of the chapter.

The most commonly used definition of software engineering is the one given in the IEEE Standard Glossary for Software Engineering [38], where software engineering is defined as "the application of a systematic, disciplined, quantifiable approach to the development, operation, and maintenance of software, that is, the application of engineering to software." It is obvious that this definition has a very strong foundation on the life cycle of software, i.e., how it is built (i.e., development); how it is used (i.e., operation); and how it is updated, and renewed (i.e., maintenance). Therefore, it is natural to discuss about software engineering by focusing on software life cycle phases. While different methodologies (e.g., Rational Unified Process (RUP) or adaptive methodologies such as agile development) consider different phases for software life cycle, we use the phases of software life cycle as defined in [61] given the dominant use of object-oriented paradigm, while the definition of all stages are based on [38]. In Fig. 1, we give an overview of all software life cycle phases with their parallel activities; used and produced artifacts; types of interactions and collaborations; and participants and their roles. Each of the software life cycle phases can be defined as follows [38]:

- *Analysis phase* determines what has to be done in a software system. After determining what kind of software is needed to be developed, the *requirements phase* is the first and the most important step. In this phase, the requirements for a software product are defined and documented. This is usually done in collaboration with end-users and domain experts, where

	Process/Methodology				
	Analysis	Design	Implementation & Integration	Maintenance	Retirement
	Documenting				
	Testing				
Artifacts	• Documents • Models • Mock-ups	• Documents • Models • Components • Mock-ups	• Documents • Models • Source code • Components • Databases • Unit tests • Test data • ...	• Documents • Models • Source code • Components • Databases • Unit tests • Test data • ...	• Documents
Interactions/ Collaborations	• Wikis • Chats, Emails, Forums • CVS • Specialized tools	• Wikis • Chats, Emails, Forums • CVS • Specialized tools	• Wikis • Chats, Emails, Forums • CVS • Specialized tools	• Wikis • Chats, Emails, Forums • CVS • Bug reports • Execution logs	
Participants	• Domain experts • End-users • System analysts	• Analysts • Designers • Domain experts	• Developers • Designers • Domain experts (testing) • End-users (testing)	• Maintainers • End-users • Domain experts (testing)	• Maintainers

Fig. 1. Software development life cycle: phases, artifacts, interaction and collaboration, and participants

the critical point is to establish common understanding of the domain under study. Once requirements are defined, they are formally specified in the form of a legal document. Typically, this document includes functional requirements, performance requirements, interface requirements, design requirements, and development standards; to eliminate all ambiguousness, incompleteness, and contradictions. Modeling approaches are recommended at this stage (e.g., RUP recommends using UML use cases and class diagrams), while some researchers recommend using even some more formal approaches (e.g., Petri nets [39]).

- *Design phase* defines detailed designs for application domain, architecture, software components, interfaces, and data. Since all the design should be verified and validated to satisfy requirements, usually this phase regards the use of modeling (e.g., UML). The more formal designs are defined, the less potential errors will be, and the more potentials will exist for automatic software implementation (e.g., code generation). Therefore, the software engineering community puts a lot of attention to the discipline called model-driven engineering (MDE) to enable model-driven development (MDD) of software products [24]. Moreover, model transformations (model-to-model; model-to-text; and text-to-model) are the key concepts of MDD which allow for round trip engineering (i.e., forward and reverse engineering) of software.
- *Implementation phase* creates a software product from the design documentation and models. This phase also debugs and documents the software product. This phase assumes the use of programming languages to encode specified designs, and testing techniques (e.g., unit testing) to eliminate any potential bugs. Besides eliminating software bugs, it is also important to be able to check whether implementations are fully valid w.r.t. the models (aka., model-based testing [3]).
- *Integration phase* is the process of combining software components (e.g., Web services), hardware components, or both, into an overall system. This phase is usually done in parallel with the implementation phase. Besides importance of a high-quality and up-to-date documentation, this stage also requires testing, such as acceptance testing (by end-users) and integration testing (i.e., checking the integration with other components).
- *Maintenance phase* is the process of modifying a software system or component after delivery to correct faults, improve performance or other attributes, or adapt to a changed environment, i.e., any change after acceptance of the software by the client. This phase highly depends on the quality of documentation in order to trace parts of software to be changed. Of course, this phase also assumes documenting all changes as well as testing software for its compliance to the initial and newly-defined requirements.
- *Retirement phase* is the period of time in the software life cycle during which support for a software product is stopped. This may happen in cases where a drastic change in design, implementation, or documentation has occurred. This phase also has to be well-documented to explain why a software product is retired.

However, the current software practice suffers from a lack of traceability of all artifacts and elements produced/used in different stages of the life cycle (e.g., requirement documents and source code), that can substantially affect software development and especially software maintenance [69]. As already pointed out, software is a socio-technical category which necessitates keeping track of all relevant human–human and human–software interactions (e.g., chat discussions that may explain why some design decision were made) [1].

It is also very important to mention that every software product strongly relies on the application-specific domain knowledge, standards, and policies related to the software system under study. In addition, every software development process follows some methodologies,[1] and it is useful to relate methodology tasks and activities with the software artifacts produced/used in different life cycle phases. Moreover, each task in the software development life cycle is important to be assigned to a person (e.g., software programmer) that has competencies needed. Very often, such knowledge is not represented explicitly, and thus it is very hard to establish traceability links between such knowledge and produced software artifacts and interactions used in all phases of software life cycle. Such knowledge can further stimulate social interactions and locate peers that can help in dealing with some specific software development issues. In the rest of the chapter, we present how ontologies can assist in establishing the missing semantic links in the above software life cycle phases.

3 Analysis

According to the Standish Group report from 1994,[2] the main reasons for software project failures are issues caused by the poor or inappropriate software analysis. The three reasons for software success are user involvement, executive management support, and a clear statement of requirements, while the main reasons for challenged and failed software projects are the lack of user input, incomplete requirements and specifications; and changing requirements and specifications. All these reasons stress the need for mutual understanding between requirement engineers and end-users and the importance of the preciseness of the requirement specification.

3.1 Ontologies as Requirement Engineering Products

The above arguments motivated researchers to look at ontologies as a solution to improve the state of the art in this area. Breitman and Leite argue that ontologies should be sub-products of the requirement engineering phase [10]. It follows the idea of Hendler that on the web we can have many

[1] Today's methodologies follow incremental and iterative software development.

[2] http://www.spinroot.com/spin/Doc/course/Standish_Survey.htm

application-oriented ontologies that should be interconnected to facilitate knowledge sharing between different applications [34]. Thus, their requirement engineering process has a particular sub-process for ontology construction. This process is inspired by the layered ontology engineering approach [49], where the main source for creating ontologies is the language's extended lexicon. The lexicon is built by eliciting the important terms from the relevant source documents, and mapping the terms to the appropriate constructs (e.g., classes) of the ontology language used in the application under study. Looking further to some other ontology development technologies, we can also find out that requirement engineering and ontology engineering are even sharing some common methodologies. For example, the DOGMA ontology engineering framework [65] uses a scenario-based approach to engineer ontologies for application domains. The most important thing from both cases is that the ontology is a product of the analysis phase, which means that all parties involved in this process should agree upon the ontology developed. This, in fact, should eliminate the lack of misunderstanding of the users' needs and should further be propagated to the design phase (e.g., by transforming such an ontology to models – cf. Sect. 4). Another benefit is that all documents (e.g., stories) that are used for requirement acquisition could be semantically annotated with the ontologies created from them to represent intelligent content [13]. If such ontologies are further used in the design phase (e.g., models), we can then have traceability between these two software development stages (i.e., analysis and design) and establish mapping relations with other ontologies to provide traceability with other potentially relevant sources of knowledge.

The use of upper-level ontologies is also well-known in software engineering when developing domain models that are usually part of the requirement specification. Typically, an upper-level ontology (e.g., Bunge–Wand–Weber [BWW] model) is used as a definition of the background theory (or the perspective to the world) based on which the domain model is built. Current software development methodologies (e.g., RUP) suggest UML-based domain models as the results of the analysis phase. The current experience demonstrates that if one wants to make such domain models valid w.r.t. the upper level ontology, then a modeling language should be constrained in order to allow the use of only those models that are compliant to the upper ontology. For example, Evermann and Wand [23] constrain the specification of the UML (i.e., UML metamodel) by using the Object Constraint Language (OCL), so that every UML model is fully compliant with the BWW model.

3.2 Requirement Engineering Approaches

Requirement engineering phase assumes the use of many different sources, which are not only end-users and domain experts, but also policies and standards. Requirement engineering also implies the use of different methodologies such as goal-driven, viewpoints-oriented, and scenario-based approaches, or their combinations [47]. None of these approaches usually allow for using different approaches collaboratively, since they are mainly constrained by the tools they use. Recognizing this problem, Lee and Gandhi proposed an ontology-based framework, aka Onto-ActRE, which promotes cohesiveness between the artifacts generated from different modeling techniques and creates a shared understanding from multiple dimensions [47]. The central point of this solution is a Problem Domain Ontology that integrates (1) goal-driven scenario composition, (2) requirements domain model, (3) viewpoints hierarchy, and (4) other domain specific taxonomies. Leveraging PDO represented in OWL and the Jena Semantic Web framework, they developed the GENeric Object Model (GenOM) tool that, for example, allows requirement engineers to utilize the requirements domain model along with the goals from the goal hierarchy and the associated stakeholders in a viewpoints hierarchy. Although not suggested by Lee and Gandhi, the requirements domain model can be obtained from a domain ontology developed by some of the approaches discussed earlier.

3.3 Requirement Engineering Collaboration

Collaboration appears to be the crucial activity in successful requirements engineering, especially in the current global software development landscape. The main challenges to be addressed are [16]: (1) knowledge acquisition and sharing, (2) effective communication and coordination, and (3) aligning RE processes and tools. We have already commented on how ontologies can address (1), but there is a need to combine it with (2) to facilitate efficient collaboration and coordination of involved parties. The use of Wikis appears to be a promising solution to this task. Wikis demonstrate the use of ontologies to define the structure (e.g., concepts such as Use Case, and Actor) and types of documents used in the requirements engineering phase based on the story telling approach [17]. Software engineering can benefit form semantic Wikis as frameworks for (application) ontology engineering by using collective intelligence [64]. Collaborative results produced in semantic Wikis can directly be translated to models used in the design phase (cf. Sect. 4 for details).

Not only are Wikis means of collaboration in requirements engineering, but stakeholders also communicate by other communication channels and tools such as chats and discussion forums [63]. It would definitely be an important research challenge to leverage ontologies for managing knowledge contained in all these channels (e.g., semantically annotating discussion messages [67] to represent contextual knowledge about why and how some decisions were made). Finally, for a successful collaboration of distributed stakeholders, it is also important that they fully understand the different cultural, geographical, and organizational boundaries. For example, a common problem in collaboration could be a misunderstanding of different requirements engineering tools, methodologies they are based on, and levels of details of the requirement specification requested [18]. While ontologies like the Problem Domain Ontology can certainly harmonize different ontology engineering approaches, there are some open opportunities for applications of ontologies such as to describe requirement engineering tools and methodologies (and connect them with general software engineering development methodologies [30]) or to harmonize communication among stakeholders with different cultural and technical origins.

3.4 Requirements Verification

Testing of identified and specified requirements is a critical activity of the analysis phase, as it is very important to make sure that all involved stakeholders with different backgrounds and levels of knowledge agree upon the requirements specification. Probably, the most effective way is to use formal model-based animations (e.g., UML use-cases and classes) that present defined requirements. However, as UML does not have formally defined semantics, it is very hard to run simulations that formally analyze the models defined [26]. Although development of methods for formal analysis of models is set as one of the main challenges in the area of model-driven engineering [27], there are already some proven formal languages that have successfully been used for verification of requirement specifications. For example, Jorgensen and Bossen suggest the use of Petri nets for defining executable use-cases [39]; demonstrating potentials of Petri net analysis for requirement engineering.

However, Petri nets are a formalism for modeling processes rather than for modeling a structure (e.g., domain model) of a system under study. The question is then how to combine domain ontologies developed in some of the above-mentioned ways and process formalisms such as Petri nets? Brockmans et al. proposed a mechanism for semantic annotation of Petri nets by using concepts from domain ontologies [12]. Taking a similar approach, [28] demonstrates that ontology alignment techniques can assist in the automatic business process integration. This example can stimulate some other applications of ontologies to semantically enrich requirement engineering and even improve traceability of all artifacts produced in this phase to be used in other software life cycle phases. For example, one could trace requirement document from Petri net models by using ontology concepts annotating Petri net elements. Moreover, these semantic links between requirement documents and Petri net models via the domain ontology could further increase the degree of "intelligence" of content [13]. This ontology annotation of Petri nets can also serve as an interesting direction for further integration of ontologies and models to develop mechanisms to semantically annotate models used in the design phase (e.g., [46] investigates workflow and composition languages). Such semantically annotated artifacts will further be interlinked with the artifact of the implementation and integration phases (e.g., Semantic Web services – chapter "Semantic Web Services").

4 Design

As already mentioned, the design phase assumes a comprehensive definition of the software system under study. As a result, this phase heavily relies on the use of modeling principles and best software practices such as software patterns. Due to the importance of modeling in this phase, in this section, we first introduce model-driven engineering (MDE) as a software engineering discipline that promotes software development fully based on modeling principles. Then, we discuss how MDE helps to integrate ontologies into software design, and finally, we conclude this section by discussing how ontologies can be applied to improve the use of design patterns.

4.1 Model-Driven Engineering

Model Driven Engineering (MDE) is a new software engineering discipline in which the process heavily relies on the use of models [7]. Models are the central MDE concepts and are specified by using modeling languages (e.g., UML or ODM), while modeling languages are defined by metamodels. A metamodel is a model of a modeling language. That is, a metamodel makes statements about what can be expressed in the valid models of a certain modeling language [62]. The core idea of MDE is to increase the productivity of software developers by increasing level of abstraction when developing some software. Once models have been developed, they can be translated to different platform specific implementations (e.g., Java or C#). The OMG's Model Driven Architecture (MDA) is a possible architecture for MDE [50].

MDA consists of three layers, namely: M1 (model) for defining models of systems under study; M2 (metamodel) for defining modeling languages (e.g., UML and Common Warehouse Metamodel [CWM]); and M3 (metametamodel) where only one metamodeling language is defined (i.e., MOF) [53]. The relations between different MDA layers can be considered as instance-of or conformant-to, which means that a model is an instance of a metamodel, and a metamodel is an instance of a metametamodel. Besides MOF, MDA also includes the Object Constraint Language (OCL) to define (more formal) constraints over MOF-defined MDA layers, so that more precise model definitions can formally be verified. OCL is also defined by a MOF-based metamodel resided on the M2 level. Another MDE architecture is Eclipse Modeling Framework (EMF), which is different from MDA just in using Ecore on the M3 layer instead of MOF (cf. Sect. 4.5).

4.2 MDE and Ontologies

Cranefield was the first to explore the synergy of software modeling languages and ontologies [15]. He started from the assumption that there are similarities between the standard concepts of UML and those of ontologies (e.g., classes, relations, and inheritance). Having this in mind, he proposed the use of UML for modeling ontologies due to the wide-acceptance of UML by software engineers and many already-developed UML models, which would facilitate adoption of ontologies by software practitioners. Moreover, software engineers can also benefit from the use of ontology reasoning services (e.g., consistency checking) to reason over UML models. In this way, one can connect software design and ontology development. This motivated several other researchers to look into the problem of similarities and differences between ontology and software modeling languages (mainly UML). The details about findings could be found in [29].

The above-mentioned activities initiated a standardization process at the Object Management Group (OMG) to issue a request for proposals for the Ontology Definition Metamodel (ODM) in 2003. The aim was to define a MOF-based metamodel for the OWL (cf. chapter "Resource Description Framework") and RDF(S) (cf. chapter "Web Ontology Language: OWL") ontology languages (i.e., ODM), corresponding ontology UML profile (to use standard UML tools for modeling ontologies), and transformations between ODM and other relevant ontology and modeling languages. These activities resulted in the OMG's ODM specification [54] that defines MOF-based metamodels for Semantic Web ontology languages, RDF(S) and OWL as well as metamodels of Common Logic, Entity-Relationship models, and Topic Maps. The ODM specification also specifies model transformations (by using the OMG's Query/View/Transformations (QVT) standard transformation language [55]) of the ODM and RDFS metamodels with the metamodels of the following languages: UML, Common Logics, Topic Maps, and Entity-Relationship. IBM's tool Integrated Ontology Development Toolkit (IODT) is the most complete implementation of ODM [57].

Note also that application (and domain) ontologies can also be used in the design of software architectures. Grønmo et al. demonstrated how MDE principles can be used to model Semantic Web services (i.e., OWL-S) by extending UML activity diagrams [31]. This approach reminds of the approach for semantic annotation of Petri nets [12]. This demonstrates the importance of further exploration of how ontologies can be integrated into custom modeling languages (e.g., Business Process Modeling Notation – BPMN). This effort can have several contributions to software engineering such as (1) improved traceability of software models when maintaining software and (2) improved software integration capacity, especially, in the context of service-oriented architectures.

4.3 Software Models and Business Vocabularies

Not always should domain ontologies be defined in the analysis phase, but the requirements specification can be only in the form of documents written in natural language. This implies that we should define our domain models (i.e., ontologies) in the design phase from scratch. So, for this task it will be useful to have an automatic or a semi-automatic approach to produce ontologies for requirement documents (see chapter "Ontology Engineering Environments"). Moreover, we should also be able to update textual requirement documents automatically with the changes of ontologies.

Semantics of Business Vocabulary and Business Rules (SBVR) is a promising solution to the above problem [56]. SBVR is the OMG specification that defines a metamodel for capturing expressions in a controlled natural language and representing them in formal logic structures. The SBVR metamodel is compatible with Common Logic. Given that the ODM specification defines the mappings between the Common Logics metamodel and the metamodels of both OWL and RDF(S), the ODM specification provides a bridge to transform SBVR to OWL, RDF(S), UML, Topic Maps, Entity-Relationship models, and Description Logics.

4.4 Ontologies and Model Reasoning

Software modeling tools are usually very intuitive and allow software designers to use a visual notation of modeling languages (e.g., UML). However, today's software modeling tools lack the support for formal validation of software models, and discovering some potentially hidden implications of such models (e.g., inconsistencies and redundancies), which may impact the overall quality of software designs [27]. Trying to address these issues, Berardi et al. explored the use of description logics to enable reasoning over UML class models [6]. The main finding of their research is that UML class diagrams are EXPTIME-hard, even under restrictive assumptions including only binary associations, only minimal multiplicity constraints, and generalizations (between classes and associations) with disjointness and completeness constraints. They also demonstrated how reasoning over UML class models can become EXPTIME-complete by disabling the arbitrary use of first order OCL predicates, but still allowing disjointness constraints on the generalization hierarchies. A practical contribution is a reasoner that allows for reasoning over UML class models. There are similar on-going research activities in the MDE community to provide formal semantics for UML [27]. The ontology community also considers the use of some UML features (e.g., composite structures) in the future OWL extensions (e.g., OWL 1.1) [58].

4.5 Ontologies and Model Transformations

Model transformation plays an important role and represents the central operation for handling models in MDE [7]. Model transformation is the process of producing one model from another model of the same system [50]. Model transformations are usually defined between different modeling languages that are defined by different metamodels, and hence the process of transformation is usually called metamodel-driven model transformation. The OMG adopted the MOF2 Query/View/Transformation (QVT) specification [55] to address this need. One of the most commonly used QVT implementations is ATLAS Transformation Language (ATL) [5], which is the official Eclipse recommendation for model-to-model transformations, and yet is an open-source solution. Based on the previous discussions on the similarities between ontologies and models, there have been several approaches that propose the use of ontology alignment techniques to attack the problem of model transformation. The ModelCVS system addresses this problem by transforming (i.e., lifting) metamodels into ontologies (i.e., transforming Ecore to ODM) [42]. Then, such obtained ontologies are further refactored to represent explicitly some hidden concepts that are usually not precisely represented in metamodels, but should be placed in ontologies. Finally, ontology matching algorithms are executed over such ontologies (e.g., COMA++), and discovered mapping relations are encoded into the ATL transformations. The ontology-based model transformation (ontMT) approach in an attempt that semantically annotates metamodels with the concepts from a reference ontology of a domain [60]. ontMT makes use of such semantic annotations to reason over concepts of the metamodels being mapped and generates model transformations (i.e., ATL) from inferred mapping relations.

Both of these applications of ontologies to model transformations have been recognized as valuable contributions to the MDE area. However, there are many important research questions that should be solved such as: combinations of both approaches to make the process of model transformation more effective; applying ontologies and ontology matching at the model (M1) level of MDA, for example, to improve software refactoring; and applications of ontology matching to contribute round-trip engineering (i.e., code generation and reverse engineering) by complementing the efforts for model-to-text and text-to-model transformations [40].

4.6 Ontologies and Software Patterns

Using the experience form the urban architecture, software engineering adopted the concept of software patterns as an attempt to describe successful solutions to common software problems. The pattern, in short, is a thing, which happens in the world, and the rule which tells us how and when to create it. A pattern language is a network of multiple patterns, with links between related patterns. The most known type of software patterns are design patterns which nowadays are used in almost all applications. Patterns are, in fact, shared knowledge of software engineering, and represent a way for common understanding of software designs. Patterns are described in literary forms, such as sonnets. This works fine if patterns are intended to be understood by software engineers, but if they need to be interpreted by tools, there is a need for a formal representation of patterns [20]. For example, the BORE tool leverages the ontologies to encode the pattern language for usability design patterns [36]. BORE does not automate user interface design, as for effective user-interface design, talent and creativity of the software designer is very important. However, the design pattern ontology helps designers improve their knowledge about patterns and share the design experience with other designers easier. As suggested for using semantic annotations of models with ontologies, semantic annotations of design patterns and artifacts can also improve the maintenance, so that one can trace the knowledge on which the design was based [20].

5 Implementation and Integration

The design of software products should specify how the system should finally be implemented and integrated with other software systems, so that the software product eventually accomplishes the requirements initially set. This phase usually looks at lower computer-specific details and is done by using programming languages. Although the goal of MDE is to allow for automatically generating as much implementation code as possible along with many promising results, the current state of the art indicates that many implementation details should still be done manually. This section explores the potentials of using ontologies in the implementation and integration phases.

5.1 Implementation

In this section, we distinguish between three different approaches to the use of ontologies in software implementation.

First, as already indicated, some approaches claim that ontologies could be used in the same manner as models in MDE. Thus, we should be able to generate the implementation of a software system from an ontology, possibly the domain ontology that we created in the analysis phase and refined in the design phase. Following this approach, Cranefield created transformations of UML models to Java code (e.g., classes) besides the RDF(S) ontologies [15], and thus provided a complementary Java implementation for an RDF(S) ontology. However, this approach did not provide mechanisms for preservation of the semantic definitions in ontologies (e.g., OWL restrictions). RDFReactor is a more recent approach that allows for mapping RDF(S) ontologies to Java. Although it does not support OWL, it improves the previous work by eliminating some non-safe type usages (e.g., Java.util.List for properties) by the use of domain specific classes generated from the ontology and leverages the use of static semantic analysis used by compilers of programming languages.

Second, given the AI origins of ontologies, ontologies can also be used in the implementation of software systems in a more declarative way, but yet to use conventional object-oriented programming languages (e.g., Java). HP's Jena Semantic Web framework offers a Java API for handling RDF(S) and OWL ontologies. Examples of alternatives for Jena are the Protégé OWL API and Protégé-Frames API [45]. In this case, we can say that ontologies are not used only for code-generation (like it is the case with MDE and approaches such as [41] and [68]), but ontologies are also a part of the run-time software behavior. A good aspect of implementations based on generic ontology APIs is that they are more dynamic in terms of allowing for on-the-fly ontology changes and updates. However, in these implementations, software developers can not resolve run-time issues by using static semantic analysis. Such run-time issues can hardly be handled with standard exception mechanisms of programming language [45]. In addition, it is not possible to benefit from widely-adopted techniques for software testing, such as JUnit. This group of implementations can also benefit from the ODM specification, as the OWL and RDF(S) metamodels can also be programmatically managed by using model handlers (i.e., their APIs) such as EMF and Java Metadata Interface (JMI, `http://java.sun.com/products/jmi/`).

Third, ontologies can be used as a part of the implementation logic in software systems that are implemented by using rule-based languages (e.g., Jess or JBoss Rules). This is the most flexible software implementation approach, as it not only allows for dynamically changing ontologies, but also rules. Then, an inference engine is responsible for execution of rules. Given that most of rule languages define rules over vocabularies and ontologies, this implementation technique can nicely be applied to ontologies [33]. However, rule-based languages are not the widely adopted implementation approach in software engineering and this approach is mainly used for implementation of smaller specialized components with high degree of dynamicity (e.g., e-Negotiations).

Besides the above-mentioned approaches, the use of ontology-based semantic annotations can additionally improve software development life cycle. For example, Java annotation mechanism can be used to semantically interconnect parts of Java code and ontology conceptualization. Not does this can only be useful for JavaBeans[3] to perform some advanced reasoning (e.g., consistency checking), but it can also produce some benefits for the overall software maintenance (e.g., license ontologies can be useful to apply different license policies to different parts of source code). Finally, the potential text mining and ontology-based analysis of the code can be interesting to provide (semi-) automatic approaches to verify some implementation requirements and their designs, similar to the use of ontologies for detection of design errors [37].

5.2 Integration

The most important contribution of ontologies to software integration is semantic Web services. Semantic Web services, as the augmentation of Web service descriptions through Semantic Web annotations, facilitate the higher automation of service discovery, composition, invocation, and monitoring on the Web. In this section, we focus on a relevant topic: *semantic middleware*.

The concept of middleware is applied to managing heterogeneity of various software components and technologies used in a distribute software system. However, it is very important to have environments for developing such

[3] `http://blogs.sun.com/bblfish/entry/java_annotations_the_semantic_web`

middleware-based distributed systems. Application servers are component-based middleware platforms that provide functionalities for developing distributed systems which can use the components developed the developers or third parties. The current management of the functionalities of application severs is based on the use of administrative tools and XML configuration. While this brings a lot of flexibility, there are still many complexity management issues for developers and administrators. These issues are chiefly cased by the lack of an explicit representation of the data in configuration files, or having no commitment to any abstract model that can improve the interpretation of data when developing and analyzing distributed systems [51].

Studying the above issues, Oberle [51] identified the typical challenges at development time as: component dependencies and versioning, licensing, capability descriptions, service classification and discovery, semantics of parameters, and automatic generation of component and service metadata. In addition, typical run time use-cases, requiring more advanced complexity management approaches are: access rights management, error handling, transactional settings, and secure communication. Trying to provide a more generic solution that can be independent of a particular application domain as much as possible, Oberle et al. proposed a stack of the ontologies based on the *DOLCE* (Descriptive Ontology for Linguistic and Cognitive Engineering) generic ontology and its sub-module for descriptions and situations that define patterns for (re)structuring domain ontologies. At the top, the core ontologies of components (typical concepts characterizing components in application servers) and of services (typical concepts characterizing services) are defined (see `http://cos.ontoware.org` and chapter "An Ontology for Software"). These two ontologies are then specialized in domain ontologies by adding concepts specific for a domain of discourse. These ontologies are leveraged in KAON SERVER, a semantic application server that is implemented as an extension of the open-source JBoss application server. Thanks to the ontological description of components, KAON SERVER can perform more advanced analysis of the components used in a distributed system by making use of ontology reasoning and query languages, and thus helps developers and administrators with more contextualized feedback (e.g., who can access a particular component).

The organization of ontologies on which KAON SERVER is based, indicates why it is important to ground domain ontologies in upper-level ontologies (e.g., DOLCE) in the early software life cycle development phases (e.g., analysis and design). There are many potential benefits for this approach. For example, if our requirement domain models are based on upper-level ontologies, requirement engineers will be able to search for suitable components in the analysis phase. Moreover, the implementation of such systems can later be capable of more flexible integrations with software systems. Indeed, a similar approach for integration of business processes based on the use of Semantic Web services have already been proposed [22], However, the future research should define methodologies that can guide the use of generic ontologies used and refined in all software life cycle phases (see chapters "Foundational Choices in DOLCE" and "An Ontology for Software" for more information).

6 Maintenance

Any change ever since the client accepts the software, is related to the maintenance software development life cycle phase. When developing software, software engineers need a lot of knowledge about application domain, technologies used, algorithms applied, software testing, and past and new requirements. However, this knowledge is usually not recorded and for software maintainers (which are not necessarily the original software developers) it is very hard to fully understand the system being maintained. It is then not surprising why software maintainers spend 40–60% of their time just to understand the system being maintained [59]. The current software development practice tries to address this problem by requesting software developers and maintainers to document as much of this knowledge as possible. However, documenting software is usually not enough, as software maintainers need an easy assess to the knowledge relevant to the given context of software maintenance. Traceability links between various software artifacts are needed, to make this process more efficient. This is why some researchers argue that software maintenance is a knowledge management task where ontologies play a critical role [19].

To enable the support for managing knowledge of software maintenance, Anquetil et al. developed a comprehensive ontology for software maintenance consisting of five sub-ontologies [2]: the software system ontology with concepts such as software system, users, and documentation; the computer science skills ontology with concepts such as computer science technologies and modeling languages; the modification process ontology with concepts such as modification request and maintenance activity; the organizational structure ontology with concepts such as organizational unit and directive; and the application domain ontology that associates domain concepts with tasks to be performed. Applying Post-Mortem Analysis (a method to elicit knowledge in software engineering), they developed a methodology that allows for explicit representation of knowledge of different stages of the ISO/IEC 14764 maintenance process (i.e., after modification analysis, after implementation of the modification, and at the end of the project) by using their software maintenance ontology. However, this approach does not consider the problem of establishing traceability links with the already developed artifacts in the previous phases. To do so, the knowledge management process requires (semi-) automatic approaches to capture the knowledge encoded in legacy systems.

Witte et al. address the above problems by developing two ontologies, namely, the source code ontology (i.e., an ontology of major concepts of object-oriented programming languages) and documentation ontology (i.e., an ontology of different concepts that may appear in documents related to programming, such as programming languages and data structures) [70]. This experiment demonstrated that these two ontologies allow for establishing traceability links between software documentation (i.e., JavaDoc) and source code. Moreover, such links can help in the validation of, for example, documentation, by checking whether relations described in the documentation (e.g., between a class and a method) actually exist, and whether the documentation reflects the state of the implementation in the source code. This approach allows for even more advanced software maintenance use cases, including, identification of security concerns in source code (e.g., checking whether public and non-final attributes can be updated outside of the class they belong to) and architectural recovery and restructuring (e.g., checking whether documented architecture such as layered architecture is actually implemented).

Other authors demonstrate that it is possible to perform even more advanced software analysis by using ontologies [43]. Besides using software ontology model (FAMIX-based and language-independent ontology of object-oriented code), this approach uses a bug ontology (inspired by Bugzilla) and a version ontology. These ontologies are first populated by parsing a source code extracted from a CVS. Then, the ontologies are queried by using iSPARQL, an extension of SPARQL that adopts the concept of virtual triples. The three ontologies and iSPARQL can assist software maintainers in use cases such as: code evolution visualization (e.g., how a class evolved in different revisions); detection of code smells (e.g., long parameter list); application of code metrics (e.g., big classes with many methods and attributes and their correlation with bug reports); and ontology reasoning (e.g., methods that are not invoked, aka orphan methods). This project also reports on the scalability issues of today's Semantic Web technologies (e.g., reasoners) which can be another stimulus for the great interest of software engineering in the future research on integrating searching and reasoning approaches on the Web [25].

For software maintainers it can also be important to know what designs are implemented in the maintained source code [20]. An ontology of design patterns can be used to analyze source code and discover design patterns implemented. In addition, this ontology can assist in providing common understanding between software developers and software maintainers.

As we initially indicated, software is a social category, and so is software maintenance. Thus, it is also important to allow for capturing other relevant knowledge related to software maintenance (e.g., exchange of experiences on discussion forums). Capturing such type of knowledge facilitates communication between developers and helps with locating the developers with the most suitable skills, experiences, and reputations. The Dhruv system addresses this problem and facilitates connecting three different types of knowledge, i.e., content, interaction, and community. There are certainly a lot of potentials to experiment with the use of ontologies for social networking (e.g., FOAF) to build networks of software developers. Additionally, this can also be applied to analyze the trustworthiness of software based on the level of its developer's reputation. A good example in the line of this research direction is the Baetle ontology (`http://code.google.com/p/baetle/`) that combines a bug ontology with several other ones (e.g., atom, workflow, and description of a project's ontologies). In addition, software maintenance can benefit from the use of domain ontologies that were built during the software development (as described in the previous sections) or extracted from already developed artifacts [8]. Moreover, there is a potential to use ontologies to semantically annotate the logs of software behaviors, which can be useful for software maintenance (e.g., to synthesize models of behavior [44] and compare them with models defined in the analysis and design phases [11]).

7 Conclusions

To the best of our knowledge, there has been no approach addressing the issues of the retirement phase. Although retirement usually means the end of the use of a software product, it could be very important for software developers to be able to create repositories of retired software, as each retired software system contains a lot of knowledge encoded in its implementation [48]. Thus, we need methods to extract knowledge out of retired software systems, especially those that are implemented using legacy technologies. Some of the ontology-based approaches to software maintenance [20, 43] could be used as good directions for (re-)using knowledge from retired software systems. Ontologies could also play an important role in the on-going effort of the OMG for Architecture-Driven Modernization (ADM) and their metamodel for Knowledge Discovery Metamodel (KDM) [52].

To sum up the current state of the use of ontologies, we refer back to Sect. 2 and analyze the orthogonal dimensions to the software life cycle phases (i.e., documenting, testing, artifacts, interaction/collaboration, and participants). Documentation is probably the most commonly analyzed application of ontologies with ontologies used in all software life cycle phases. Domain, upper-level, and document structure ontologies are chiefly used to improve documentation. Still, the documentation activity could additionally benefit from ontologies by developing intelligent tools for software annotation that will for example have features for checking validity of documentation statements w.r.t. the software artifacts [70]. Ontologies also help to have clear semantic relations between different software artifacts and documentations (e.g., models and documents), and thus building software documentations as intelligent contents [13]. This research also indicates that there is a need for developing standard ontologies of documentation structure and types.

Using ontologies for software testing is probably the least explored aspect of software engineering. In fact, we have seen that (upper-level) ontologies are only used to validate requirements and detect design errors [37]. Given a lot of attention to model-based software testing, ontologies are definitely a promising technology (as discussed in Sect. 6) to even outperform MDE-based approaches (e.g., UML) thanks to their strong formal and reasoning foundation. Therefore, further research topics such as semantic annotation of logs of software behaviors for intelligent monitoring, semantic annotations of unit and integration tests, ontology-based reverse engineering, and ontology-based software metrics can bring many potential benefits to software engineering.

The use of ontologies for various software artifacts is probably one of the areas that has attracted a lot of attention so far. Domain and upper-level ontologies, ontologies for documentation, source code, bugs, ontology-based models, model transformations, requirements, and design patterns, are just some examples that are used for important software engineering tasks such as adding more semantics to the artifacts, improving traceability links, consistency checking of models, generating model transformations, and software metrics. While all these attempts are well-recognized by both the Semantic Web and software engineering communities, further exploration of semantic annotation mechanisms of software models and implementation code, integration of ontologies and metamodeling architectures, and a comprehensive traceability model of software artifacts, are some of the biggest challenges concerning the aspects of software knowledge artifacts.

Interaction and collaboration are fundamental requirements for successful software engineering. The current efforts already demonstrate some interesting results for some of the software life cycle phases (e.g., facilitating mutual understanding of stakeholders and semantic Wikis for requirement acquisition). However, social aspects of design, implementation, integration, and maintenance phases are almost the dark side of ontologies [35]. Investigating the use of collaborative tagging and folksonomies to improve collaborative experience when designing, implementing and integrating; leveraging social networking ontologies (e.g., FOAF) for annotating software artifacts; and multi-cultural understanding; are some of the potential applications where ontologies can improve interaction capturing and facilitate better collaboration in software engineering [21]. Ontologies can also be a suitable technology for integration of software development environments and collaborative tools (e.g., adding chats like GTalk chats in Gmail). In addition, competence ontologies can help locate software engineers with competencies needed for particular projects, which is one of the most common issues in today's software knowledge management, especially in the domain of global software development [18].

Another important area is to describe software processes and methodologies. Not only do ontologies of methodologies have potentials to be related with modeling languages [30], but they can actually be used to semantically interlink, for example, particular project tasks and activities with all different artifacts produced/used, participants responsible, and interactions done.

References

1. Anupriya Ankolekar, Katia Sycara, James Herbsleb, Robert Kraut, and Chris Welty. Supporting Online Problem-solving Communities with the Semantic Web. In *Proc. of the 15th Int'l Conf. on WWW*, pages 575–584, 2006.

2. Nicolas Anquetil, Káthia M. de Oliveira, Kleiber D. de Sousa, and Márcio G. Batista Dias. Software Maintenance seen as a Knowledge Management Issue. *Inf. Softw. Technol.*, 49(5):515–529, 2007.
3. L. Apfelbaum. Model Based Testing. In *Soft. Quality Week 1997 Conf.*, 1997.
4. Phillip G. Armour. Software: hard data. *Commun. ACM*, 49(9):15–17, 2006.
5. ATLAS Transf. Language, 2006. `http://www.sciences.univ-nantes.fr/lina/atl`.
6. Daniela Berardi, Diego Calvanese, and Giuseppe De Giacomo. Reasoning on UML class diagrams. *Artif. Intell.*, 168(1):70–118, 2005.
7. Jean Bézivin. On the unification power of models. *Software and System Modeling*, 4(2):171–188, 2005.
8. Kalina Bontcheva and Marta Sabou. On Self-Validating Rule Bases. In *Proc. of the 2nd Int'l WSh on Semantic Web Enabled Software Eng.*, 2006.
9. Grady Booch. The Irrelevance of Architecture. *IEEE Soft.*, 24(3):10–11, 2007.
10. Karin Breitman and Julio Cesar Sampaio do Prado Leite. Ontology as a Requirements Engineering Product. In *11th IEEE Int'l Requirements Eng. Conf.*, pages 309–319, 2003.
11. Saartje Brockmans, Robert M. Colomb, Elisa F. Kendall, Evan Wallace, Christopher Welty, Guo Tong Xie, and Peter Haase. A Model Driven Approach for Building OWL DL and OWL Full Ontologies. In *Proc. of the 5th Int'l Semantic Web Conf.*, pages 187–200, 2006.
12. Saartje Brockmans, Marc Ehrig, Agnes Koschmider, Andreas Oberweis, and Rudi Studer. Semantic Alignment of Business Processes. In *Proc. of the 8th Int'l Conf. on Enterprise Info. Sys.*, pages 191–196, 2006.
13. Tobias Bürger. Putting Business Intelligence into Documents. In *Proc. of the WSh. on Semantic Business Process and Product Lifecycle Management*, 2007.
14. Calero Coral, Ruiz Francisco, and Piattini Mario. *Ontologies for Software Engineering and Software Technology.* Springer, Berlin, Heidelberg, 2006.
15. Stephen Cranefield. UML and the Semantic Web. In *Proceedings of the Semantic Web Working Symposium*, pages 113–130, 2001.
16. Daniela Damian. Stakeholders in Global Requirements Engineering: Lessons Learned from Practice. *IEEE Software*, 24(2):21–27, 2007.
17. Björn Decker, Eric Ras, Jörg Rech, Pascal Jaubert, and Marco Rieth. Wiki-Based Stakeholder Participation in Requirements Engineering. *IEEE Software*, 24(2):28–35, 2007.
18. K. C. Desouza, Y. Awazu, and P. Baloh. Managing Knowledge in Global Software Development Efforts: Issues and Practices. *IEEE Soft.*, 23(5):30–37, 2006.
19. Márcio Dias, Nicolas Anquetil, and Kàthia de Oliveira. Organizing the Knowledge Used in Software Maintenance. *J. UCS*, 9(7):641–658, 2003.
20. Jens Dietrich and Chris Elgar. Towards a web of patterns. *Web Semantics: Science, Services and Agents on the World Wide Web*, 5(5):108–116, 2007.
21. Jens Dietrich and Nathan Jones. Using Social Networking and Semantic Web Technology in Software Engineering–Use Cases, Patterns, and a Case Study. In *Proce. of the 2007 Australian Software Eng. Conf.*, volume 0, pages 129–136, Los Alamitos, CA, USA, 2007. IEEE Computer Society.
22. Stefan Dietze, Alessio Gugliotta, and John Domingue. A Semantic Web Services-based Infrastructure for Context-Adaptive Process Support. In *Proc. of the International Conf. on Web Services*, pages 537–543, 2007.
23. Joerg Evermann and Yair Wand. Toward Formalizing Domain Modeling Semantics in Language Syntax. *IEEE Trans. Software Eng.*, 31(1):21–37, 2005.

24. Jean-Marie Favre. Foundations of Meta-Pyramids: Languages vs. Metamodels - Episode II: Story of Thotus the Baboon.
25. D. Fensel and F. van Harmelen. Unifying Reasoning and Search to Web Scale. *IEEE Internet Computing*, 11(2):96–95, 2007.
26. R. B. France, S. Ghosh, T. Dinh-Trong, and A. Solberg. Model-driven development using UML 2.0: promises and pitfalls. *Computer*, 39(2):59–66, 2006.
27. Robert France and Bernhard Rumpe. Model-driven Development of Complex Software: A Research Roadmap. In *Proc. of 28th Int'l Conf. on Software Eng. - Future of Software Engineering*, pages 37–54, 2007.
28. Dragan Gašević and Vladan Devedzić. Petri net ontology. *Knowl.-Based Syst.*, 19(4):220–234, 2006.
29. Dragan Gašević, Dragan Djurić, and Vladan Devedzić. *Model Driven Architecture and Ontology Development.* Springer, 2006.
30. César Gonzàlez-Pèrez and Brian Henderson-Sellers. *An Ontology for Software Development Methodologies and Endeavours*, volume Ontologies for Software Engineering and Software Technology, pages 123–151. Springer, 2006.
31. Roy Grønmo, Michael C. Jaeger, and Hjørdis Hoff. Transformations Between UML and OWL-S. In *Proc. of the 1st European Conference Model Driven Architecture - Foundations and Applications*, pages 269–283, 2005.
32. Hans-Jörg Happel and Stefan Seedorf. Applications of Ontologies in Software Engineering. In *Proc. of the Int'l WSh. on Semantic Web Enabled Software Engineering*, 2006.
33. Marek Hatala, Ron Wakkary, and Leila Kalantari. Rules and ontologies in support of real-time ubiquitous application. *J. Web Sem.*, 3(1):5–22, 2005.
34. J. Hendler. Agents and the Semantic Web. *IEEE Int. Sys.*, 16(2):30–37, 2001.
35. J. Hendler. The Dark Side of the Semantic Web. *IEEE Int. Sys.*, 22(1):2–4, 2007.
36. Scott Henninger and Padmapriya Ashokkumar. An Ontology-Based Infrastructure for Usability Design Patterns. In *Proc. of the Int'l WSh. on Semantic Web Enabled Software Engineering*, pages 41–55, 2005.
37. Allyson Hoss. *Ontology-Based Methodology for Error Detection in Software Design.* PhD thesis, Louisiana State University Graduate School, 2006.
38. IEEE Standard Glossary of Software Engineering Terminology-Description, 1990. `http://ieeexplore.ieee.org/servlet/opac?punumber=2238`.
39. J.B. Jorgensen and C. Bossen. Executable use cases: requirements for a pervasive health care system. *IEEE Software*, 21(2):34–41, Mar-Apr 2007.
40. Frédéric Jouault, Jean Bézivin, and Ivan Kurtev. TCS:: a DSL for the specification of textual concrete syntaxes in model engineering. In *Proc. of the 5th Int'l Conf. on Generative Prog. and Component Eng.*, pages 249–254, 2006.
41. Aditya Kalyanpur, Daniel Jiménez Pastor, Steve Battle, and Julian A. Padget. Automatic Mapping of OWL Ontologies into Java. In *Proc. of the 16th Int'l Conf. on Software Eng. and Knowledge Eng.*, pages 98–103, 2004.
42. Gerti Kappel, Elisabeth Kapsammer, Horst Kargl, Gerhard Kramler, Thomas Reiter, Werner Retschitzegger, Wieland Schwinger, and Manuel Wimmer. Lifting Metamodels to Ontologies: A Step to the Semantic Integration of Modeling Languages. In *Proc. of the ACM/IEEE 9th Int'l Conf. on Model Driven Eng. Languages and Sys.*, pages 528–542, 2006.
43. Christoph Kiefer, Abraham Bernstein, and Jonas Tappolet. Analyzing Software with iSPARQL. In *Proc. the 3rd ESWC Int'l WSh. on Semantic Web Enabled Software Eng.*, 2007.

44. Ekkart Kindler, Vladimir Rubin, and Wilhelm Schäfer. Process Mining and Petri Net Synthesis. In *Proc. of Business Process Management WSh.*, pages 105–116, 2006.
45. Holger Knublauch. Ontology-Driven Software Development in the Context of the Semantic Web: An Example Scenario with Protege/OWL. In *Proc. of 1st Int'l WSh on the Model-Driven Semantic Web*, 2004.
46. Florian Lautenbacher and Bernhard Bauer. A Survey on Workflow Annotation and Composition Approaches. In *Proc. of the Wsh. on Semantic Business Process and Product Lifecycle Management*, 2007.
47. Seok Won Lee and Robin A. Gandhi. Ontology-based Active Requirements Engineering Framework. In *Proc. of the 12th Asia-Pacific Software Eng. Conf.*, pages 481–490, 2005.
48. Raghavendra Rao Loka. Software Development: What Is the Problem? *IEEE Computer*, 40(2):112–111, Feb 2007.
49. Alexander Maedche. *Ontology Learning for the Semantic Web*. Kluwer Academic Publishing, Boston, 2002.
50. J. Miller and J. Mukerji. MDA Guide Version 1.0.1. Technical report, Object Management Group (OMG), 2003.
51. Daniel Oberle. *The Semantic management of middleware*. Springer, 2006.
52. OMG KDM. Architecture-Driven Modernization (ADM): Knowledge Discovery Meta-Model, 2007. `http://www.omg.org/cgi-bin/apps/doc?ptc/07-03-15.pdf`.
53. OMG MOF. Meta Object Facility (MOF) Core, v2.0, 2006. `http://www.omg.org/cgi-bin/doc?formal/2006-01-01`.
54. OMG ODM. Ontology Definition Metamodel, 2006. `http://www.omg.org/cgi-bin/doc?ad/06-05-01.pdf`.
55. OMG QVT. MOF QVT Final Adopted Specification, 2005. `http://www.omg.org/docs/ptc/05-11-01.pdf`.
56. OMG SBVR. Semantics of Business Vocabulary and Business Rules, 2005. `http://www.omg.org/docs/bei/05-08-01.pdf`.
57. Yue Pan, GuoTong Xie, Li Ma, Yang Yang, Zhaoming Qiu, and Juhnyoung Lee. Model-Driven Ontology Engineering. *Journal of Data Semantics VII*, pages 57–78, 2006.
58. P. F. Patel-Schneider and I. Horrocks. OWL 1.1 Web Ontology Language: Overview, 2006. `http://www.w3.org/Submission/owl11-overview/`.
59. S.L. Pfleeger. *Software engineering: theory and practice*. Prentice-Hall, 1998.
60. Stephan Roser and Bernhard Bauer. An Approach to Automatically Generated Model Transformations Using Ontology Engineering Space. In *Proc. of the 2nd Int'l WSh. on Semantic Web Enabled Software Eng.*, 2006.
61. Stephen R. Schach. *Object-Oriented and Classical Software Engineering*. McGraw-Hill, 2006.
62. E. Seidewitz. What models mean. *IEEE Software*, 20(5):26–32, 2003.
63. Vibha Sinha, Bikram Sengupta, and Satish Chandra. Enabling Collaboration in Distributed Requirements Management. *IEEE Software*, 23(5):52–61, 2006.
64. Katharina Siorpaes and Martin Hepp. myOntology: The Marriage of Ontology Engineering and Collective Intelligence. In *Proc. of the Wsh. on Bridging the Gep between Semantic Web and Web 2.0*, pages 127–138, 2007.
65. P. Spyns, Y. Tang, and R. Meersman. A model theory inspired collaborative ontology engineering methodology. *Journal of Applied Ontology*. submitted.

66. P. Tetlow, J. Z. Pan, D. Oberle, E. Wallace, M. Uschold, and E. Kendall. Ontology Driven Architectures and Potential Uses of the Semantic Web in Systems and Software Engineering. W3C Working Draft, 2006.
67. V. Uren, P. Cimiano, J. Iria, S. Handschuh, M. Vargas-Vera, E. Motta, and F. Ciravegna. Semantic Annotation for Knowledge Management: Requirements and a Survey of the state of the art. *J. of Web Semantics*, 4(1):14–28, 2006.
68. Max Völkel and York Sure. RDFReactor – From Ontologies to Programmatic Data Access. In *Poster Proc. of the 4th Int'l Semantic Web Conf.*, 2005.
69. Christopher A. Welty. Software Engineering. In Franz Baader, Diego Calvanese, Deborah L. McGuinness, Daniele Nardi, and Peter F. Patel-Schneider, editors, *Description Logic Handbook*, pages 373–387. Cambridge University Press, 2003.
70. René Witte, Yonggang Zhang, and Juergen Rilling. Empowering Software Maintainers with Semantic Web Technologies. In *Proc. of the 4th European Semantic Web Conference*, pages 37–52. Springer, 2007.

Semantic Web Services

Jos de Bruijn[1], Mick Kerrigan[2], Michal Zaremba[2], and Dieter Fensel[2]

[1] Faculty of Computer Science, Free University of Bozen-Bolzano, Italy,
debruijn@inf.unibz.it
[2] Semantic Technology Institute (STI), University of Innsbruck, Austria,
mick.kerrigan@sti2.at, michal.zaremba@sti2.at, dieter.fensel@sti2.at

Summary. Semantic Web services are a prominent application area for ontologies, and Semantic Web technologies in general. Using Semantic technologies such as ontologies for describing Web services enables automating tasks such as discovering, combining, and executing services. In this chapter we survey the aspects relevant to the description of Semantic Web services through an overview of the Web Service Modeling Ontology (WSMO), which provides a conceptual model for describing services.

We then survey in more detail the various uses of ontologies in Web service descriptions. Finally, we describe other prominent Web service description frameworks and contrast them with WSMO, in particular WSDL-S, OWL-S, and SWSF.

1 Introduction

The Semantic Web [6, 25] aims at making the vast amount of information on the Web accessible to machines through the annotation of Web content using machine-understandable formats such as RDF[1] and to enable comprehension and integration of this information through the use of ontologies. However, these annotations refer only to static knowledge; additionally, ontologies are – generally speaking – static domain descriptions. Web services [3] are concerned with providing functionality over the Web, and are thus more than pieces of static information. An example of such functionality is the sale of books over the Web; see, for example, the Amazon Web services.[2] Web service technologies such as SOAP[3] and WSDL[4] provide means for the structured XML-based annotation of, and interaction with, Web services. However, the description of the functionality of services using these technologies is limited to natural language text and a description of the structure of inputs and outputs. These

[1] See http://www.w3.org/RDF and chapter "Resource Description Framework" of this handbook.
[2] http://aws.amazon.com/
[3] http://www.w3.org/TR/soap/
[4] http://www.w3.org/TR/wsdl

S. Staab and R. Studer (eds.), *Handbook on Ontologies,* International Handbooks on Information Systems, DOI 10.1007/978-3-540-92673-3,

limitations make it hard to understand the functionality of a service, let alone automatically discover, combine, and execute Web services. Consequently, the location, selection, combination, and usage of Web services requires considerable human effort.

Semantic Web services [18] aim to combine Semantic Web and Web service technologies to overcome these limitations and enable automation of the mentioned Web service usage tasks (i.e., discovery, selection, composition, execution, etc.). The use of Semantic Web technologies – especially ontologies – for the description of Web services has a number of benefits. First of all, ontologies are specified using formal languages that have associated reasoning methods, which is an important prerequisite for the automation of Web service usage tasks. Second, ontologies (are intended to) reflect a common understanding of a particular domain, shared among a potentially large group of stakeholders. The use of ontologies as a common vocabulary for Web service descriptions has the potential to increase the understanding and reusability of Web service descriptions.

In order for the vision of Semantic Web services to be realized, it is necessary to identify all the aspects related to the description of Web services in a single conceptual framework. The Web Service Modeling Ontology WSMO [10] is such a conceptual model; it identifies the functional and non-functional aspects of Web service offerings, as well as user requests, to enable their description in an adequate manner. In this chapter we give an overview of the modeling aspects of Web services as identified by WSMO, and describe the typical current and envisioned role of ontologies in Web service description and usage. In this presentation we are mainly concerned with the description of single Web services, rather than the composition (e.g., [7, 20, 26]) and choreography of multiple services.[5]

This chapter is further structured as follows. Firstly in Sect. 2 we explain the Web Service Modeling Ontology in more detail specifically focusing on the elements that make up a WSMO Web service description. We then proceed to describe a number of uses of ontologies in Web service descriptions, firstly reviewing the concept of modeling Goals and Web services as concepts of an ontology and then presenting the use of ontologies as the terminology for a Web service description, in Sect. 3. In Sect. 4 we briefly review two prominent other approaches to Semantic Web services, namely SAWSDL and OWL-S, and describe how the mentioned uses of ontologies in Web service description can be realized in these approaches. Finally we provide conclusions in Sect. 5.

2 WSMO: An Ontology for Modeling Web Services

The Web Service Modeling Ontology WSMO [10] aims to describe all aspects related to services that are accessible through a Web service interface. Ultimately the goal is to enable the total or partial automation of tasks that

[5] See, e.g., `http://www.w3.org/TR/ws-cdl-10/`

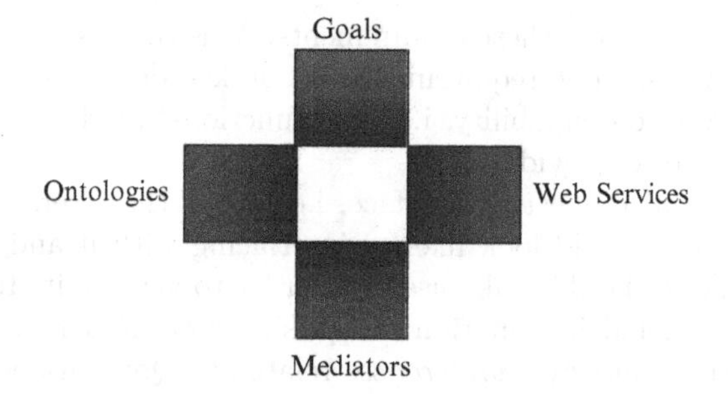

Fig. 1. WSMO top level elements

usual occur in the process of using Web services. Such tasks include discovering services that can fulfill a piece of functionality on behalf of the end user, select the most appropriate service when more than one is available, compose multiple services to perform complex tasks, resolve heterogeneity issues on both the data and process levels, and ultimately invoke Web services on the end user's behalf.

Figure 1 shows the four top-level elements of WSMO:

- *Ontologies* provide formal and explicit specifications of the vocabulary used by the other modeling elements in WSMO. The use of shared ontologies specified in formal languages increases interoperability and allows for automated processing of the descriptions. See chapters "What Is an Ontology", "Description Logics", "Ontologies in F-Logic", "Resource Description Framework", and "Web Ontology Language: OWL" in this handbook for descriptions of the nature of ontologies and the formal and shared languages used for their specification.
- A Web service is a piece of functionality accessible over the Web. A WSMO *Web service* is made up of three parts, namely:
  - The *capability*, which describes the functionality offered by the service.
  - The *interface*,[6] which describes (a) how to interact with the service, through its *choreography* and (b) how the service makes use of other services in order to provide its functionality, through its *orchestration*
  - The *non-functional information*, comprising meta-data (e.g., Dublin Core [30]) and Quality of Service (QoS) related parameters [23,28].
- The way in which service requesters use Web services may be very different from what was envisaged by the provider of the service. Thus it is important that requirements of the requester are given the same importance as the description of services. Thus WSMO provides *goals* as a mechanism for describing the requirements a given service requester has when searching

[6] Note that WSMO allows including multiple interfaces in a Web service description, thereby facilitating interaction with the service in different ways.

for services that meet these requirements. As is the case for the description of Web services, these requirements are broken down into:
  - The requested capability, i.e., the functionality the requester expects the service to provide.
  - An optional requested interface, i.e., what the interaction pattern of the service should look like for interfacing with it and which services this service should make use of in order to achieve its functionality.
  - Non-functional information comprising metadata related to the goal description and *user preferences* related to QoS parameters.
- The open and distributed nature of the Web requires resources to be decoupled. In other words, WSMO descriptions are created in (relative) isolation from one another and thus the potential for heterogeneity problems between resources is high. Such heterogeneity issues can exist between the formats of the data exchanged between service requesters and providers, the process is used for invoking them and the protocols used in communication. WSMO *Mediators* are responsible for overcoming these heterogeneity problems; WSMO emphasizes the centrality of mediation by making mediators a first class component of the WSMO model. An example of a WSMO mediator for resolving data heterogeneity is a mediator that performs transformation of instant information from one ontology to another through the use of ontology mappings [19]. More information on ontology mappings can be found in chapter "Ontology Mappping".

Ontologies and ontology languages have been explained in detail throughout this book (chapters "What Is an Ontology" "Description Logis", "Ontologies in F-Logic", "Resource Description Framework", and "Web Ontology Language: OWL"). We focus here on the structure of Web service and goal descriptions and how they relate to each other. When clear from the context, we refer to WSMO Web service and goal descriptions simply as a services and goals, respectively. Recall that Web services define the information needed for a machine to interpret the usability of a Web service to fulfill a requester's requirements, which are encoded as a goal. Figure 2 presents the elements of a Web service description, namely non-functional properties, a capability, a choreography and an orchestration. The term interface is used to describe the combination of the choreography and orchestration of a service. The structure of a goal is the same as that of a Web service and automating a given task in the process of using Web services is essentially the interaction of a given part of the goal description with a given part of one or more Web service descriptions. Therefore below we describe the elements that make up goals and Web services by describing how they interact with one another in the process of automatically finding and using Web services.

To perform Web service discovery, in other words to automatically find services that can fulfill the user's requirements, the capability of the goal is compared with the capabilities of known services. A capability is a description of the functionality provided by a service (or requested by a requester) and is

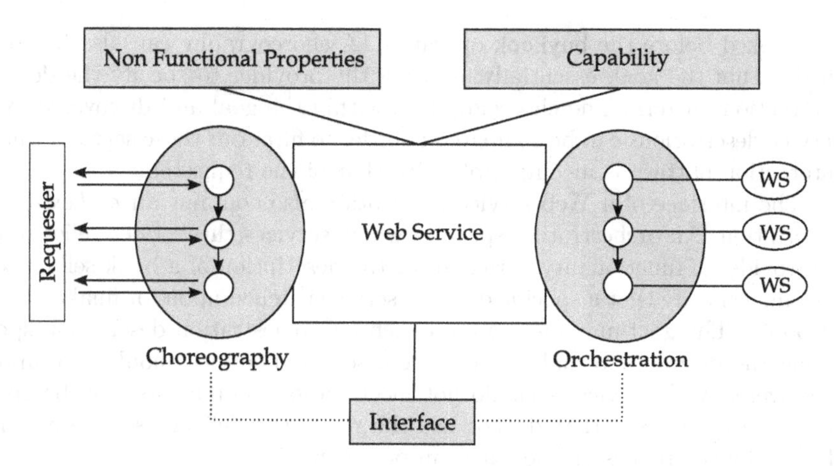

Fig. 2. Elements of a Web service description

described in terms of conditions on the state of the world that must exist for execution of the service to be possible and conditions on the state of the world that are guaranteed to hold after execution of the service. WSMO makes a distinction between the state of the information space, i.e., the inputs and outputs of the service, and the state of the world.

Based on these considerations a capability description comprises four main elements. *Preconditions* describe conditions on the state of the information space prior to execution. Therefore, preconditions specify requirements on the inputs the service, e.g., typing. There may exist additional conditions that must hold in the real world in order for the service to successfully execute. These conditions, called *Assumptions*, are not necessarily checked by the service before execution but are crucial to the successful execution of the service (e.g., the balance on a credit card must be sufficient to conclude a purchase). *Postconditions* describe conditions on the state of the information space after execution has occurred, thus describing properties of the outputs of the service, as well as the relationship between the inputs and the outputs. Many services will have real world effects, for example when purchasing a book using a book selling service a physical book will be delivered to the requester. *Effects* are conditions that are guaranteed to hold in the real world after execution.

The process of discovering services by comparing the capabilities of goal and Web service descriptions may yield a number of services that are capable of achieving the user's goals. However, compatibility of the capabilities of a given goal and Web service does not mean that a given Web service is desirable for the requester. The *interface* of a Web service specifies how to interact with the service in terms of a *choreography*, this choreography essentially provides information about the relationships between different operations on the Web service, for example the `login` operation of a book selling service must be invoked before the buyBook operation. A choreography can also be specified within the goal, essentially allowing the provider to specify the desired interaction pattern. The choreographies within the goal and discovered Web service descriptions can be compared in order to filter out those services whose interaction pattern is incompatible with that of the requester.

The interface of a Web service description also contains an *orchestration* description. An orchestration specifies which services this service relies upon to provide its functionality, for example the description of a book selling service may specify that a specific delivery service is relied upon for final delivery of books. The goal may also contain such an orchestration description specifying the desired external services the discovered service should rely upon. Discovered Web services that do not meet these requirements may be eliminated, e.g., services that do not use the requested delivery service are not desired by the requester and thus can be ignored.

After discovering those services whose functionally meets the requester's requirements and filtering out those that do not match in terms of their interaction pattern or the services upon which they rely there may still be multiple services that can achieve the user's goal. In this case the most desirable a Web service must be selected from the list. To perform this selection the non-functional properties of the discovered Web services are compared against the requested non-functional properties within the goal. Non-functional properties, as their name suggests, are used to capture non-functional aspects of a given Web service. These non-functional properties typically provide a description of the Quality of Service of the service, e.g., reliability, scalability, and security. By comparing the requested non-functional properties of the goal to those of the discovered services we can eliminate those services that do not meet the minimum requirements laid out by the goal and rank the remaining services to find the service that best fits the requester's non-functional requirements. Having selected the right service for the requester, based on functional, interface and non-functional parameters, automatic invocation of the selected service is possible using the choreography description of the service.

In the remainder of this chapter we are concerned with the functional description of goals and Web services, i.e., the description of capabilities.

3 Ontologies in Web Service Descriptions

Current Web service standards lack the necessary means to enable automation of the service usage process; they do not allow specifying the semantics of services in a machine-processable manner. In this section we describe how ontologies and formal Semantic Web languages may be used for describing user requirements and Web service functionality.

Most Web service usage tasks require descriptions of there functionality and/or its interfaces. Web service discovery requires a description of the desired (goal), as well as the provided (service) functionality, and a means to compare them. Web service composition requires a description of the functionality of all services taking part in the composition, as well as their interaction, in order to verify whether the considered combination realizes the requested functionality. Invocation of services requires a description of the choreography of the service, in order to know how to invoke it, and to know which data, should be sent to the service, in which format, and which output may be expected. Non-functional descriptions of services (e.g., cost, security) play an important role in the selection of services according to user preferences. Ontologies play an important role in the description of the functional and non-functional aspects, as well as the interfaces, of services.

In this section we review three typical uses of ontologies in goal and Web service descriptions:

1. Modeling goals and Web services as concepts in a task ontology
2. Modeling inputs and outputs as ontology concepts
3. Using an ontology to provide the basic vocabulary for the rich functional description of goals and services

These paradigms are of increasing complexity, and allow increasingly detailed description of goals and services. The first and third paradigm are only concerned with the description of the functionality, the second paradigm is mostly concerned with (a simplified view of) the interface of the service.

In this presentation we limit ourselves to the use of ontologies in the functional description of goals and services and do not address the use of ontologies in non-functional or behavioral descriptions. For more details about the use of ontologies for non-functional description of services we refer the reader to [16, 22, 28]; for more details about the use of ontologies for behavioral description of services we refer the reader to [20].

3.1 Goals and Web Services as Task Ontology Concepts

One could view the tasks that may be performed by a Web service as the domain of interest of a particular *task ontology*. The concepts in this ontology represent the tasks that may be performed by a service, and that may be requested in a goal. Furthermore, such tasks can be organized in a task hierarchy, and further relations between tasks (such as simple kinds of composition) may be expressed in the ontology, depending on the expressiveness of the ontology language.

Figure 3 depicts an example task hierarchy for the domain of book sales. In the figure, arrow represents the `subClassOf` relationship. A task A being a subclass of another task B indicates that A is more specialized. For example, *BookSelling* is a specialization of *Selling*. Furthermore, the figure depicts a disjointness relation between the tasks *Shipping* and *Selling*, indicating that the tasks have nothing in common.

If the functionality of a goal is expressed as a concept in the same, or a related, task ontology, then Web service discovery can be reduced to simply

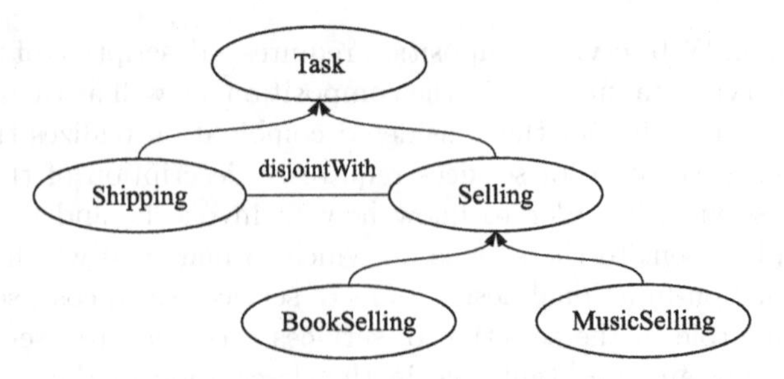

Fig. 3. Example task hierarchy for the bookselling domain

checking the relationship between the concept representing the goal and the concept representing the Web service.

In particular, if a description logic-based language (e.g., OWL DL [8, 11]; see also chapter "Description Logics") is used for the description of this ontology, then description logic reasoning services ([4]; see also the chapter "Tableau-Based Reasoning" and "Resolution-Based Reasoning for Ontologies") can be used for checking equivalence or subsumption of such concepts. The matching of goals and Web services using description logic reasoning was explored in detail by Li and Horrocks [14], reusing and refining matching notions introduced in [24], [21], and [29].

In description logics concepts represent sets, and sub-concept and concept equivalence relations correspond to set inclusion and set equality, respectively. One can thus understand a Web service description $\mathcal{WS}$ as a set of *elements* that can be delivered through execution of the Web service. In this setting, the goal description $\mathcal{G}$ represents all the elements desired by the user. As pointed out in [12], it is up to the person modeling the goals and Web services to decide what these elements are meant to represent.

Given an ontology $\mathcal{O}$, a concept $\mathcal{WS}$ (representing the Web service), and a concept $\mathcal{G}$ (representing the goal), the following matching notions are distinguished:

Exact match The goal $\mathcal{G}$ is equivalent to the Web service $\mathcal{WS}$, given the ontology $\mathcal{O}$, denoted $\mathcal{G} \equiv_{\mathcal{O}} \mathcal{WS}$.

PlugIn match The goal $\mathcal{G}$ is a sub-concept of the Web service $\mathcal{WS}$, given the ontology $\mathcal{O}$, and thus the Web service can be "plugged in" place of the goal, denoted $\mathcal{G} \sqsubseteq_{\mathcal{O}} \mathcal{WS}$.

Subsume match The Web service $\mathcal{WS}$ is a sub-concept of the goal $\mathcal{G}$, given the ontology $\mathcal{O}$, denoted $\mathcal{WS} \sqsubseteq_{\mathcal{O}} \mathcal{G}$.

Intersection match The intersection of the Web service $\mathcal{WS}$ and the goal $\mathcal{G}$, given the ontology $\mathcal{O}$, is not empty: $\mathcal{WS} \sqcap \mathcal{G} \not\sqsubseteq_{\mathcal{O}} \bot$.

Disjoint The intersection of the Web service $\mathcal{WS}$ and the goal $\mathcal{G}$, given the ontology $\mathcal{O}$, is empty, denoted $\mathcal{WS} \sqcap \mathcal{G} \sqsubseteq_{\mathcal{O}} \bot$.

So, an exact match can be understood as stating that all and only those elements that fulfill the user desires $\mathcal{G}$ are provided by the Web service $\mathcal{WS}$;

a plug-in match means that all elements requested in $\mathcal{G}$ are provided by the service $\mathcal{WS}$; a subsume match means that all elements provided by $\mathcal{WS}$ are requested in $\mathcal{G}$, but there might be some elements that are not provided by $\mathcal{WS}$; an intersection match means that some elements requested in $\mathcal{G}$ are (potentially) provided by $\mathcal{WS}$; finally, a disjoint match means that none of the elements requested in $\mathcal{G}$ is provided by $\mathcal{WS}$.

The mentioned notions of matching were further refined by Keller et al. [12]:

- A Match means that the Web service $\mathcal{WS}$ provides all elements requested by the goal $\mathcal{G}$: the Exact match and PlugIn match are both considered a Match.
- A Partial match means that some of the elements requested in $\mathcal{G}$ can be provided by $\mathcal{WS}$: the Subsume match and Intersection match are both considered a Partial match.
- A Non-match means that the Web service does not provide any of the requested elements: the Disjoint match is a Non-match.

Example 1. Consider a goal $\mathcal{G}_1$ corresponding to a request for a service that sells books and ships them. So, the goal is to find a service whose functionality is the union of bookselling and shipping: $\mathcal{G}_1$ = `unionOf(BookSelling Shipping)` (i.e., $\mathcal{G}_1 \equiv \mathit{BookSelling} \sqcup \mathit{Shipping}$). Imagine now a service $\mathcal{S}$ that sells books and music: $\mathcal{S}$ = `unionOf(BookSelling MusicSelling)` (i.e., $\mathcal{S} \equiv \mathit{BookSelling} \sqcup \mathit{MusicSelling}$). It is easy to see that there is an intersection, and thus a Partial match between $\mathcal{G}_1$ and $\mathcal{S}$: the service $\mathcal{S}$ provides bookselling, but not shipping.

Consider now a goal $\mathcal{G}_2$ corresponding to a request for a service that provides shipping: $\mathcal{G}_2$ = `Shipping`. Clearly, $\mathcal{G}_2$ and $\mathcal{S}$ are disjoint (by the disjointness of `Shipping` and `Selling`, depicted in Fig. 3), so there is a Non-match between $\mathcal{G}_2$ and $\mathcal{S}$.

Considering the structure of WSMO capability descriptions (precondition, postcondition, etc.), there does not appear to be a means for referencing concepts in a task ontology. However, a capability description is a description of the functionality of the service. Likewise, concepts in a task ontology represent all elements that are requested by or can be delivered through a goal or Web service, respectively. Therefore, such concepts are in fact descriptions of service functionality, and are thus capability descriptions that can be used as such in WSMO Web service and goal descriptions.

3.2 Inputs and Outputs as Ontology Concepts

Most classical approaches to Web service description (e.g., WSDL) model Web services in terms of the structure of the input and output messages of the service. This is analogous to the way the interfaces of functions and methods are typically described in popular programming languages such as C++ and

Java; the description of input and output messages of the service corresponds to the signature of a function or method. This description of the signature of a Web service tells the user the format of the messages to be sent to the service, and those returned by the service. However, the structure of a message describes only the format of the message, not its semantics (intention).

The situation can be improved by using a shared ontology for the description of inputs and outputs in goal and the Web service descriptions. In the goal description, the input corresponds to the information the user is able or willing to provide, and the output corresponds to the desired output of the service. In the service description, the input corresponds to the information the provider requires before the service can be executed, and the output corresponds to the information that is produced by the service after successful execution.

This scenario is similar to the one described in the previous section. However, in contrast to the previous scenario, the service description does not correspond to the functionality of the service, but to the type of information the service takes as input, and the type of information of its output. In case the service is an information-providing service, the functionality of the service can actually be described in terms of inputs and outputs.[7] However, if the service is a world-altering service (e.g., a shipping service), then input and output do not capture the functionality of a service; in fact, such a service might not have any output of interest; the added value of the service lies beyond the informational world (e.g., physical delivery of a product to a home).

When describing inputs and outputs using description logic concepts, matching notions similar to the ones in the previous section can be defined. There are, however, two complicating factors: we need to distinguish between input and output and the service may require several inputs and have several outputs, as illustrated by the following example.

Example 2. Consider the simple domain ontology for the book and music selling domain in Fig. 4. In the figure, the arrow stands for the `subClassOf` relationship.

Consider now the service $\mathcal{S}$ from Example 1. As input, the service expects instances of the concepts `Product` (the item to be purchased) and `CCInfo` (the credit card information of the customer); as output, the ser-

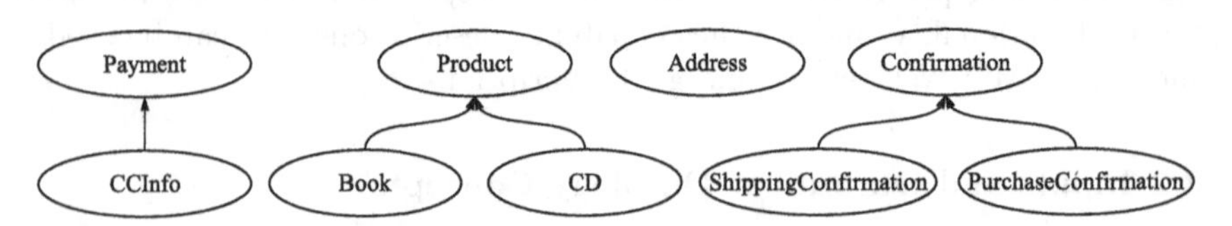

Fig. 4. Example ontology for the bookselling domain

[7] For a complete definition of the functionality of information-providing services, the relation between the inputs and outputs should be specified as well.

vice provides an instance of the concept PurchaseConfirmation. The goal $\mathcal{G}_1$ includes willingness of the requester to provide instances of Book, CCInfo, and Address as inputs of the service, and requires instances of the concepts PurchaseConfirmation and ShippingConfirmation as outputs.

Clearly, the requester can provide all inputs required by the service. However, the service can only provide some of the outputs required by the requester.

We now extend the matching notions of the previous section to matching notions for signatures:

- A Signature full match means that all inputs requested by the service are provided by the goal, and all outputs requested by the goal are provided by the service.
- A Signature output match means that all outputs requested by the goal are provided by the service.
- A Partial signature match means that some outputs requested by the goal are provided by the service.
- A Signature non-match means that none of the outputs requested by the goal are provided by the service.

We now define these notions formally. Recall the description logic notation introduced in chapter "Description Logics".

We represent the individual inputs and outputs of a service $\mathcal{S}$ as description logic concepts $\mathcal{S}_{I1}, \ldots, \mathcal{S}_{Im}$ and $\mathcal{S}_{O1}, \ldots, \mathcal{S}_{On}$, respectively, where $\mathcal{S}_{Ii}$ and $\mathcal{S}_{Oj}$ $(1 \leq i \leq m,\ 1 \leq j \leq n)$ range over the individual inputs and outputs of the service. Likewise for the goal $\mathcal{G}$.

We now define the overall Web service and goal inputs and outputs $(\mathcal{S}_I, \mathcal{S}_O, \mathcal{G}_I, \mathcal{G}_O)$ as description logic concepts, based on the individual inputs and outputs:

$$\begin{aligned}
\mathcal{S}_I &\equiv \exists hasInput.\mathcal{S}_{I1} \sqcap \cdots \sqcap \exists hasInput.\mathcal{S}_{Im},\\
\mathcal{S}_O &\equiv \exists hasOutput.\mathcal{S}_{O1} \sqcap \cdots \sqcap \exists hasOutput.\mathcal{S}_{On},\\
\mathcal{G}_I &\equiv \exists hasInput.\mathcal{G}_{I1} \sqcap \cdots \sqcap \exists hasInput.\mathcal{G}_{Ik}, \text{ and}\\
\mathcal{G}_O &\equiv \exists hasOutput.\mathcal{G}_{O1} \sqcap \cdots \sqcap \exists hasOutput.\mathcal{G}_{Ol}.
\end{aligned}$$

- There is a Signature full match if $\mathcal{G}_I \sqsubseteq_O \mathcal{S}_I$ and $\mathcal{S}_O \sqsubseteq_O \mathcal{G}_O$.
- There is a Signature output match if $\mathcal{S}_O \sqsubseteq_O \mathcal{G}_O$.
- There is a Partial signature match if $\mathcal{S}_O \sqcap \mathcal{G}_O \not\sqsubseteq_O \bot$.
- There is a Signature non-match if $\mathcal{S}_O \sqcap \mathcal{G}_O \sqsubseteq_O \bot$.

From the definition we can see that whenever there is a Signature full match, there is a Signature output match, and whenever there is a Signature output match, there is a Partial signature match. Conversely, if there is a Signature non-match, then there is no Partial signature match; if there is no Partial signature match, there is no Signature output match; and if there is no Signature output match, then there is no Signature full match.

Example 3. Consider the Web service $\mathcal{S}$ and the goal $\mathcal{G}_1$ from Example 2. The inputs of $\mathcal{S}$ are `Product` and `CCInfo`; the output is `PurchaseConfirmation`. The input and output concepts of $\mathcal{S}$ are defined as follows.

$$\mathcal{S}_I \equiv \exists hasInput.\texttt{Product} \sqcap \exists hasInput.\texttt{CCInfo}$$
$$\mathcal{S}_O \equiv \exists hasOutput.\texttt{PurchaseConfirmation}$$

The inputs and outputs of $\mathcal{G}_1$ are `Book`, `CCInfo` and `Address`, and `Purchase-Confirmation` and `ShippingConfirmation`, respectively. The input and output concepts of $\mathcal{G}_1$ are defined as follows.

$$\mathcal{G}_{1I} \equiv \exists hasInput.\texttt{Book} \sqcap \exists hasInput.\texttt{CCInfo} \sqcap \exists hasInput.\texttt{Address}$$
$$\mathcal{G}_{1O} \equiv \begin{array}{l} \exists hasOutput.\texttt{PurchaseConfirmation} \sqcap \\ \exists hasOutput.\texttt{ShippingConfirmation} \end{array}$$

We can now easily verify that $\mathcal{S}_O \not\sqsubseteq_O \mathcal{G}_{1O}$, so there is no Signature output match, and consequently no Signature full match. Nevertheless, $\mathcal{S}_O \sqcap \mathcal{G}_{1O} \not\sqsubseteq_O \perp$ holds, so there is a Partial signature match between $\mathcal{G}_1$ and $\mathcal{S}$.

A similar model for describing inputs and outputs as ontology concepts, and its use in the context of Web service discovery, was introduced by Sycara et al. [27]. There are two main distinctions between the approach described in this section and the approach by Sycara et al.:

- Sycara et al. use one concept to represent all inputs, and one concept to represent all outputs, whereas we consider an arbitrary number of inputs and outputs, and create the input and output concepts $\mathcal{S}_I$ and $\mathcal{S}_O$ using the relations *hasInput* and *hasOutput* for the purpose of matching goals and services.
- Where our notions of matching are mostly concerned with the outputs of the service, Sycara et al. distinguish between input and output matching. Our conjecture is that, for the task of Web service discovery, the service requester is mostly interested in Web services that provide the desired outputs, rather than services that accept its knowledge as input. Generally speaking, a service requester will have far more knowledge than is required as inputs for a single Web service invocation, and we do not expect that all knowledge of the requester will be explicitly described in a goal, nor do we expect that the requester is able to guess exactly which imports will be required by the Web services that are potentially of interest when creating the goal description.

In WSMO, the *precondition* and *postcondition* of a goal or Web service capability define conditions on the input and output of a service. In fact, the concepts representing the inputs to the service are conditions on the input: the input is required to be a member of this particular concept; similarly, the concepts representing the outputs are conditions on the output of the service. Therefore, the input concepts are part of a Web service precondition, and the output concepts are part of a Web service postcondition.

3.3 Ontologies as Terminologies for Web Service Description

The approach of describing the functionality of Web services in terms of inputs and outputs, introduced in the previous section, has a number of limitations, for example:

- It is not possible to describe the relationship between the inputs and outputs; output concepts provide a very limited notion of *postcondition*.
- This approach to description only deals with inputs and outputs, and thus a partial description of the pre- and postconditions in the capability of a goal or Web service. Consequently, the description only deals with the information space, and not *assumptions* on the state of the world, and *effects* in the real world of the execution of the service.

Consider the description of the service $\mathcal{S}$ in Example 2. The service has an input of type `Product` and an output of type `PurchaseConfirmation`. However, it is not entirely clear from the description to which product this purchase confirmation refers. One could assume that it refers to the input product; one could also imagine that the purchase is of a similar product, in case the requested product is not available. Generally speaking, the output of the service is related to the input, and it is usually beneficial – and indeed necessary – to specify this relationship, especially in a setting, as with Semantic Web services, where service finding and usage are meant to be automated.

Concerning assumptions and effects, consider the descriptions of the service $\mathcal{S}$ and the goal $\mathcal{G}_1$ in Example 2. One would expect a number of assumptions to be part of the description of $\mathcal{S}$, e.g., the requested product is in stock, and the balance on the credit card is sufficient, and one would expect certain effects to be part of the description of $\mathcal{G}_1$, e.g., the product is delivered to the address provided as input to the service.

As pointed out above, the description of inputs and outputs as ontology concepts is not sufficient to describe the functionality of world-altering services. In fact, it also has limitations when considering information-providing services, because it is not possible to describe the *relationship* between the input and output when viewing them merely as concepts of an ontology. In the remainder of this section we consider more expressive ways of describing goal and Web service capabilities, in which it is possible to describe the relationships between inputs, outputs, assumptions, and effects.

The execution of a Web service alters the state of the world. Therefore, we need a notion of *state*. We principally distinguish between two states: the *pre-state* is the state before and the *post-state* is the state after execution of a service. Additionally, WSMO distinguishes between the state of the information space and the state of the real world. Therefore, when considering a single service, we are concerned with four states, namely: (1) the pre-state of the information space, (2) the post-state of the information space, (3) the pre-state of the real world, and (4) the post-state of the real world. The precondition and assumption are conditions on the pre-states of the information space and real world, respectively. The postcondition and effect are conditions on the post-states of the information space and real-world, respectively, and describe the relationship between the respective pre- and post-states. Specifically, the information space consists of the inputs and outputs of the service, and thus the precondition consists of conditions on the inputs, and the postcondition consists of conditions on the outputs and relationships between inputs and outputs.

Even though this type of description is the most expressive we consider in this chapter, relatively little research has been done into the use of such expressive Web service description, when compared with the more simple kind of description based on task ontologies and interface description that we described above. A notable exception is [20], in which the authors used the situation calculus [17] to capture the semantics of Web service descriptions and used Petri Nets for various Web service related tasks such as composition and verification of services. Also, Keller et al. [13] propose a state-based semantic framework to formalize and reason with goal and Web service capabilities.

For both approaches it is the case that if the elements of the capability are specified using first-order logic formulas, which is by definition the case for the situation calculus, then typical reasoning tasks can be reduced to corresponding reasoning tasks in first-order logic. *Realizability*, which corresponds to checking whether a Web service description can in theory be realized, i.e., there may be a Web service that *realizes* the description, can be reduced to satisfiability checking in first-order logic, and *functional refinement*, which corresponds to checking whether the capability of one goal or Web service is a refinement of the capability of another, can be reduced to checking entailment in first-order logic.

For the purposes of this presentation we do not give formal definitions of either of these tasks; we illustrate the notion of functional refinement using an example. For convenience, we only consider preconditions and postconditions in the examples; however, they may be straightforwardly extended to include also assumptions and effects, which are treated analogously. We use first-order logic formulas for the description of the capabilities in the examples. Note that the free variables in the precondition ϕ^{pre} and postcondition ϕ^{post} are shared between the conditions. Intuitively, a capability corresponds to an implication $(\forall)\phi^{pre} \Rightarrow \phi^{post}$, where $(\forall)$ denotes universal closure, and $\Rightarrow$ denotes state change (as opposed to material implication in classical logic): for all inputs (variable assignments) holds that if the pre-state is such that ϕ^{pre} is true, then the state will change to a post-state, in which ϕ^{post} is true. Classes in an ontology correspond to unary predicates.

Example 4. We use the bookselling ontology in Fig. 4. We assume that this ontology is our background knowledge and is true in both the pre- and post-state; it is thus implicitly part of both the pre- and postcondition.

Consider a book and CD selling and shipping Web service $\mathcal{S}$ with a precondition $\mathcal{S}^{pre}$ and a postcondition $\mathcal{S}^{post}$. With the precondition $\mathcal{S}^{pre}$ we want to specify that there should be an input that identifies a book or a CD and there should be one which is a member of the class `Address`. The individual inputs are denoted by the variables x and y, respectively:

$$\mathcal{S}^{pre} \equiv (Book(x) \vee CD(x)) \wedge Address(y).$$

With the postcondition $\mathcal{S}^{post}$ we want to specify that there is a confirmation of purchase, denoted by the variable z, and that the product x is shipped to y:

$$\mathcal{S}^{post} \equiv \exists z(PurchaseConfirmation(z) \wedge confirms(z,x)) \wedge isShippedTo(x,y).$$

Consider now a goal $\mathcal{G}$ representing requests for a bookselling and shipping service, with the precondition $\mathcal{G}^{pre}$ that specifies the willingness to provide a book and address:

$$\mathcal{G}^{pre} \equiv Book(x) \wedge Address(y).$$

The postcondition $\mathcal{G}^{post}$ specifies the requirement that the book is shipped to the provided address:

$$\mathcal{G}^{post} \equiv shippedTo(x,y).$$

One can now verify that $\mathcal{S}$ is a functional refinement of $\mathcal{G}$: any state that is compliant with (i.e., that satisfies) the precondition of the goal $\mathcal{G}^{pre}$ is also compliant with the precondition of the Web service $\mathcal{S}^{pre}$; and any post-state compliant with the postcondition of the Web service $\mathcal{S}^{post}$ is also compliant with the postcondition of the goal $\mathcal{G}^{post}$. So, any execution of the Web service $\mathcal{S}$ with inputs satisfying the precondition of the goal (e.g., the inputs provided by the user) completely fulfills the user's requirements, and thus there is a match between the goal and the service.

4 Other Frameworks for Semantic Web Service Description

In this chapter we have described how ontologies can be used in the description of Web services, in the context of WSMO, a framework for the description of semantic Web services. Now, there are several other frameworks for describing semantic Web services that also allow using ontologies in Web service descriptions. In this section we briefly review the most prominent other frameworks for semantic Web service description. The frameworks we consider are WSDL-S [1], SAWSDL [9], OWL-S [15], and SWSF [5].

4.1 WSDL-S / SAWSDL

WSDL-S was proposed as a member submission to the W3C in November 2005 between the LSDIS Laboratory and IBM [1]. After submitting WSDL-S to W3C, the proposal has been superseded by SAWSDL [9], which is a

W3C Recommendation. SAWSDL is a restricted and homogenized version of WSDL-S in which annotations such as preconditions and effects are not explicitly specified, as there is no current agreement about their usefulness or their meaning in the broader Semantic Web services community.

In contrast to WSMO, SAWSDL is a lightweight approach to Web service description. It extends WSDL and allows associating semantic annotations with Web services, building on pre-existing standards. Using the extensibility of SAWSDL, semantic annotations in the form of URI references to external models, such as WSMO or OWL-S (presented in the next subsection) can be added to the interface, operation and message constructs. SAWSDL is independent from the language used for defining the semantic models and explicitly regards the possibility of using WSMO, OWL-S and UML as potential candidates, as illustrated in the SAWSDL usage guide [2]. As such, SAWSDL is complementary to WSMO.

SAWSDL extends WSDL with a set of attributes and elements that may be used to associate semantic annotations with WSDL descriptions. For the annotation of individual Web services, a bottom-up approach is followed, meaning that WSDL message types, used for Web service inputs and outputs, are mapped to the concepts in domain-specific ontologies (see also Sect. 3.2). Additionally, WSDL operations, which are description of Web service functionality, may be mapped to ontological concepts in a task ontology (see also Sect. 3.1). The user goals may be represented using service templates based on concepts from domain ontologies.

4.2 OWL-S

OWL-S (formerly known as DAML-S) [15] is an OWL ontology that facilitates the description, discovery, invocation, composition and monitoring of services. The ontology consists of three main elements: *Service Profile*, *Service Grounding* and *Service Model* (see Fig. 5). These concepts are used to describe

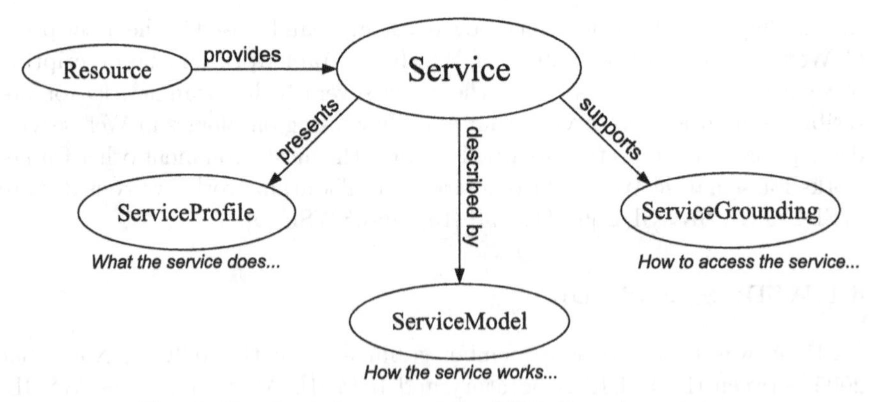

Fig. 5. Top-level elements of the OWL-S service ontology

(1) what the service provides to prospective clients, (2) how the service can be used, and (3) how interaction with the service can take place, respectively.

The Service Profile contains a number of non-functional properties of the Web service, including: the service name, a textual description of the service, contact information of the service responsible, an external categorization of the service, and, finally, an expandable list of non-predefined properties.

The functional characterization of Web services is expressed in terms of the information transformation and the state change produced by the execution of the service. The state change, modeled by preconditions and effects, refer to the change on the state of the world as a consequence of executing the service, and information transformation, modeled by inputs and outputs, which refer to what information is required and what information is produced (generally depending on the information provided as input) by the service. The schema to describe IOPEs (inputs, outputs, preconditions, and effects) instances is defined in the *Service Model*, not in the *Service Profile*. Therefore, these instances are described in the *Service Model* and referenced from the *Service Profile*. Such IOPEs may be used for expressive Web service descriptions, as illustrated in Sect. 3.3, although outputs cannot be used for expressing the relationship between the inputs and outputs. Then, when leaving out the preconditions and effects from the IOPE, one obtains a description of inputs and outputs as in Sect. 3.2. Additionally, service models may refer to a *service category*, which could be a concept in a task ontology (cf. Sect. 3.1).

Both OWL-S and WSMO aim (and claim) to provide the necessary means for creating semantic descriptions for Web services, i.e., to enable the vision of the Semantic Web services. Although both approaches have identical aims, there are certain differences between the two. One striking difference is that, in contrast to OWL-S, WSMO defines the concept of mediator as a first-class citizen of the framework. The other major difference is the structure of Web service descriptions. OWL-S defines an ontology comprising the main elements of a service, where the service element serves as an organizational point of reference for declaring Web services. The description of an individual Web service is an instance of the Service concept. The structure of WSMO descriptions is somewhat different as it offers four main top-level elements (*ontology*, *Web service*, *goal* and *mediator*) that can refer to each other either by using mediators or by importing ontologies (i.e., ontologies that contain the vocabulary to be used in the semantic descriptions).

4.3 SWSF

The Semantic Web Services Framework (SWSF) [5] is a specification produced by the SWSL Committee[8] of the Semantic Web Service Initiative (SWSI). SWSF has its own conceptual model, called Semantic Web Service Ontology (SWSO) and language, called Semantic Web Service Language (SWSL).

[8] `http://www.daml.org/services/swsl/`

SWSO has been influenced by OWL-S and adopts its three ontologies, i.e., service profile, model and grounding. The difference and the key contribution of SWSO is its rich behavioral process model, which is based on the Process Specification Language PSL.[9] Therefore, SWSO can support more powerful descriptions and reasoning over Web service descriptions. SWSL has two subsets, SWSL-FOL and SWSL-Rules, which are based on first-order-logic and logic programming, respectively.

5 Conclusions

Existing E-Business solutions typically require the implementation of costly and custom infrastructures by each of the business partners involved. Web service technologies are a milestone on the path towards flexible interoperability among distributed and independent software systems. However, while Web services provide a uniform infrastructure for the provision of services over the Web, they deliver only syntactical descriptions that are hardly amenable to automation. The process of dynamically creating ad-hoc interactions between companies, as envision by Web services, remains unattainable. Semantic Web services, as presented in this chapter, are an application of Semantic Web technologies, and ontologies in particular, to Web service description. Such semantic descriptions enable machine processing of an automated reasoning about Web service functionality, as well as the mechanisms used to invoke them and the data used as inputs and outputs.

We have seen how ontologies can be used for the formal description of both user requests and Web service functionality. Specifically, the three ways of using ontologies in Web service descriptions we addressed in this chapter were: describing goals and Web services as concepts in a *task ontology*, describing *inputs and outputs* using concepts in a domain ontology, and using ontologies as terminologies for expressive *state-based Web service descriptions*. We have also described how the first and second approach can be formalized using Description Logics (such as OWL DL) and how Description Logic reasoning can be used for the task of Web service discovery.

Comparing the approaches to describing Web services, there is a difference in the detail and preciseness of the descriptions both between and within the approaches. At the one end of the spectrum there are the lightweight task ontologies (such as the one depicted in Fig. 3) whose concepts are used for rather coarse-grained Web service descriptions. At the other end, there are the detailed state-based (pre- and postcondition) descriptions that use heavyweight background ontologies. In between, there are the more detailed, more heavyweight task ontologies, containing a more precise axiomatization of the domain, state-based descriptions with more lightweight background ontologies, etc.

---

[9] `http://www.mel.nist.gov/psl/`

As discussed in detail in chapter "Exploring the Economical Aspects of Ontology Engineering", creating ontologies, and especially heavyweight ontologies, have an (often considerable) cost. The engineering of Web service descriptions based on such ontologies brings additional cost, which can be considered relatively low in case they are based on task ontologies, but will be high for the case of detailed state-based descriptions. The existence of ontologies on the (Semantic) Web, which may be reused in Web service descriptions, would reduce the cost of describing goals and Web services. Nonetheless, authors of such descriptions will need to make a trade-off between the detail of the descriptions – descriptions with higher detail will lead to more accurate results in Web service discovery and a higher degree of automation in selection and execution – and the effort required to create the descriptions.

References

1. Rama Akkiraju, Joel Farrell, John Miller, Meenakshi Nagarajan, Marc-Thomas Schmidt, Amit Sheth, and Kunal Verma. Web service semantics – WSDL-S. W3C Member Submission 7 November 2005.
2. Rama Akkiraju and Brahmananda Sapkota. Semantic annotations for WSDL and XML schema – usage guide. Technical Report 28 August 2007, W3C, 2007.
3. Gustavo Alonso, Fabio Casati, Harumi Kuno, and Vijay Machiraju. *Web Services*. Springer, Berlin Heidelberg, 2004.
4. Franz Baader, Diego Calvanese, Deborah L. McGuinness, Daniele Nardi, and Peter F. Patel-Schneider, editors. *The Description Logic Handbook*. Cambridge University Press, Cambridge, 2003.
5. Steve Battle et al. Semantic web service framework. W3C Member Submission 9 September 2005.
6. Tim Berners-Lee, James Hendler, and Ora Lassila. The semantic web. *Scientific American*, 284(5):34–43, May 2001.
7. Piergiorgio Bertoli, Jörg Hoffmann, Freddy Lécué, and Marco Pistore. Integrating discovery and automated composition: from semantic requirements to executable code. In *Proceedings of the 2007 IEEE International Conference on Web Services (ICWS 2007)*, pages 815–822. IEEE Computer Society, 2007.
8. Mike Dean and Guus Schreiber. OWL web ontology language reference. Recommendation 10 February 2004, W3C, 2004.
9. Joel Farrell and Holger Lausen. Semantic annotations for WSDL and XML schema. Recommendation 28 August 2007, W3C, 2007.
10. Dieter Fensel, Holger Lausen, Axel Polleres, Jos de Bruijn, Michael Stollberg, Dumitru Roman, and John Domingue. *Enabling Semantic Web Services – The Web Service Modeling Ontology*. Springer, Berlin Heidelberg, 2006.
11. Ian Horrocks and Peter Patel-Schneider. Reducing OWL entailment to description logic satisfiability. *Journal of Web Semantics*, 1(4):345–357, 2004.
12. Uwe Keller, Rubén Lara, Holger Lausen, Axel Polleres, and Dieter Fensel. Automatic location of services. In *Proceedings of the 2nd European Semantic Web Conference (ESWC2005)*, pages 1–16. Springer, 2005.
13. Uwe Keller, Holger Lausen, and Michael Stollberg. On the semantics of functional descriptions of web services. In *Proceedings of the 3rd European Semantic Web Conference (ESWC2006)*, Budva, Montenegro, June 2006. Springer.

14. Lei Li and Ian Horrocks. A software framework for matchmaking based on semantic web technology. In *Proceedings of the 12th International World Wide Web Conference (WWW2003)*, pages 331–339, 2003.
15. Dean Martin et al. OWL-S: Semantic markup for web services. W3C Member Submission 22 November 2004.
16. E. Michael Maximilien and Munindar P. Singh. A framework and ontology for dynamic web services selection. *IEEE Internet Computing*, 8(5):84–93, 2004.
17. John McCarthy and Patrick Hayes. Some philosophical problems from the standpoint of artificial intelligence. In B. Meltzer and D. Michie, editors, *Machine Intelligence*, volume 4, pages 463–502. Edinburgh University Press, Edinburgh, 1969.
18. Sheila McIlraith, Tran Cao Son, and Honglei Zeng. Semantic web services. *IEEE Intelligent Systems, Special Issue on the Semantic Web*, 16(2):46–53, 2001.
19. Adrian Mocan and Emilia Cimpian. An ontology-based data mediation framework for semantic environments. *International Journal on Semantic Web and Information Systems (IJSWIS)*, 3(2):66–95, 2007.
20. Srini Narayanan and Sheila A. McIlraith. Analysis and simulation of web services. *Computer Networks*, 42(5):675–693, 2003.
21. Tommaso Di Noia, Eugenio Di Sciascio, Francesco M. Donini, and Marina Mongiello. A system for principled matchmaking in an electronic marketplace. In *Proceedings of the 12th International World Wide Web Conference (WWW2003)*, pages 321–330, 2003.
22. Justin O'Sullivan, David Edmond, and Arthur H. M. ter Hofstede. The price of services. In *Proceedings of the 3rd International Conference on Service-High-End Computing (ICSOC2005)*, pages 564–569, 2005.
23. Justin O'Sullivan, David Edmond, and Arthur H.M. ter Hofstede. Formal description of non-functional service properties. Technical report, Queensland University of Technology, Brisbane, 2005.
24. Massimo Paolucci, Takahiro Kawamura, Terry R. Payne, and Katya Sycara. Semantic matching of web services capabilities. In *Proceedings of the 1st International Semantic Web Conference (ISWC2002)*, Sardinia, Italy, 2002.
25. Nigel Shadbolt, Tim Berners-Lee, and Wendy Hall. The semantic web revisited. *IEEE Intelligent Systems*, 21(3):96–101, 2006.
26. Evren Sirin, Bijan Parsia, Dan Wu, James A. Hendler, and Dana S. Nau. HTN planning for web service composition using SHOP2. *Journal of Web Semantics*, 1(4):377–396, 2004.
27. Katia Sycara, Massimo Paolucci, Anupriya Ankolekar, and Naveen Srinivasan. Automated discovery, interaction and composition of semantic web services. *Journal of Web Semantics*, 1(1):27–46, 2003.
28. Ioan Toma, Douglas Foxvog, and Michael C. Jaeger. Modeling QoS characteristics in WSMO. In *Proceedings of the 1st workshop on Middleware for Service Oriented Computing (MW4SOC 2006)*, pages 42–47, Melbourne, Australia, 2006.
29. David Trastour, Claudio Bartolini, and Chris Preist. Semantic web support for the business-to-business e-commerce pre-contractual lifecycle. *Computer Networks*, 42(5):661–673, 2003.
30. Stuart Weibel, John Kunze, Carl Lagoze, and Misha Wolf. Dublin core metadata for resource discovery. RFC 2413, IETF, 1998.

Ontologies for Machine Learning

Stephan Blöhdorn[1] and Andreas Hotho[2]

[1] Knowledge Management Research Group, Institute AIFB, University of Karlsruhe, 76128 Karlsruhe, Germany, bloehdorn@aifb.uni-karlsruhe.de
[2] Knowledge and Data Engineering Group, Department of Mathematics and Computer Science, University of Kassel, 34121 Kassel, Germany, hotho@cs.uni-kassel.de

Summary. The growing amounts of ontologies and semantically annotated data has led to considerable interest in mining these richly structured data sources. While research has actively addressed the issue of inducing semantic structures from conventional types of data, approaches for mining semantically annotated data still constitute an emerging field of research. Approaches in this direction either investigate how semantic structures can help to advance classical Machine Learning tasks or how semantic structures can themselves become the objects of interest. In this chapter, we review some of the main topics at the intersection of Machine Learning and Semantic Web research.

1 Introduction

Recent efforts of research and industry in the area of the Semantic Web (SW) and ontologies together with the standardization of the Resource Description Framework (RDF) and the Web Ontology Language (OWL) [38] have led to an increasing amount of available ontologies, taxonomies and knowledge structures of various types as well as a rising number of semantic annotations. As of March 2009, the statistics of the SW Search Engine SWOOGLE[1] count a total of 1,615,237 publicly available "error-free pure SW Documents". As typed graphs, supplemented by the formal semantics of the employed ontology languages and the associated possibilities for logical reasoning, these data sources also exhibit an unconventional structure as compared to traditional data sources like single database tables or textual data.

In the last years, research has actively addressed the problem of learning knowledge structures *for* the Semantic Web in the context of both *Ontology Learning* (the topic of chapter "Ontology Learning" in this volume) and *Information Extraction* (the topic of chapter "Information Extraction").

Along another line, research has recently started to investigate how existing SW data sources can be mined and analysed by inductive learning

[1] http://swoogle.umbc.edu/

S. Staab and R. Studer (eds.), *Handbook on Ontologies*, International Handbooks on Information Systems, DOI 10.1007/978-3-540-92673-3,

techniques. Two communities contribute to this trend. On the one hand, the Semantic Web community has started to incorporate concepts of inductive learning into their research work, an area which is also referred to as *Semantic Web Mining* [79]. On the other hand, the Machine Learning (ML) community increasingly investigates how semantic structures can help to perform classical ML tasks or how richly structured data sources, including SW-type data, can become the subject of learning techniques. Some of these approaches do not (yet) explicitly build on SW technology but rather on own formalisms and/or representations. However, these formalisms mostly have counterparts in the Semantic Web or could be used in an analogous manner. We strongly believe that with the increased availability of Semantic Web data, the issue of mining the knowledge inclosed therein – i.e. mining *from* the Semantic Web as opposed to mining *for* the Semantic Web – will become a substantial and important research field.

In this chapter, we review various attempts to combine ML techniques and ontologies, semantically annotated data or both. This is an exciting and rapidly expanding but also highly scattered research area. Our exposition includes both, references to explicit *Semantic Web Mining* research and references to activities in the ML community that show links to SW research efforts. In order to structure the different contributing research fields, the next section starts with an overview over the whole research area and the content of this chapter.

2 The Machine Learning and Semantic Web Research Landscape

The use of ontologies and comparable declarative knowledge representation paradigms within ML tasks is an emerging field of research which draws from contributions from various communities and is shaped by a large number of diverse paradigms. In this section, we aim at structuring this research field along two major dimensions.

On the one hand, approaches can be organised according to the *type of the objects of interest* to be analysed by the learning techniques:

"Ordinary" data: In this setting, the objects of interest are arbitrary data items which have already been the subject of investigation in conventional ML settings. However, their content and conventional representation can be mapped to entities found in ontological structure. As an example, consider textual data which on the one hand has a classical representation for ML in terms of the Bag-of-Words (BOW) model but whose content can on the other hand be described further by means of lexical ontologies such as WordNet.

Ontology Entities: In this setting, the objects of interest are parts of an ontological structure themselves. It covers all cases where ontological entities become the focus of the mining activities. This class could be further divided according to the entity type, i.e. whether entities reside on the schema or on the instance level of the ontology.

Ontologies: Finally, this group of approaches covers all cases, where sets of ontological axioms, i.e. whole ontologies or parts of ontologies are the object of ML interest.

On the other hand, the contributions in the field can be roughly organised according to the *structural component* of the ML technique which is modified primarily:

Feature representation: Techniques of this class use knowledge from the ontology to modify classical feature representations, e.g. by adding features which can be deduced from the ontology.

Similarities and distances: While techniques of the previous class explicitly transform the data representation, techniques in this class achieve a similar effect implicitly. This is done by distorting the pairwise instance similarities or distances which form the input to many ML algorithms by means of calculations which take the structure and knowledge of the ontology into account.

Model class: For this class of approaches, the knowledge about the dependencies of entities in the ontology becomes part of the overall ML model, e.g. in the form of constraints on the solution space or in the form of probabilistic dependencies between instances and features.

Algorithm: For this class of techniques, the knowledge encoded in the ontology enters the overall machine learning technique only on the algorithmic level.

Table 1 visualizes the approaches that will be covered in this chapter along these dimensions. We will take up this classification in the respective sections. Clearly, this structure covers only some of the relevant aspects and the distinction between two classes along these dimensions may sometimes not be as crisp as our exposition suggests. At the same time, this chapter does not claim to provide an exhaustive review of all relevant approaches. On the contrary, the choice of techniques to be covered is certainly biased by the research interests of the authors. However, the analysis provides some intuition on how the overall research landscape is structured.

The topics of this chapter are arranged as follows: In Sect. 3, we sketch a number of approaches to exploit background knowledge encoded in ontologies within *Text Mining* applications. In this context, ontologies support the generation of informative features for the use with classical Text Mining (TM) algorithms. This section is more comprehensive and detailed than the other sections, which, as a tribute to the limited space, can only sketch some of the main ideas and point to further sources of information for the interested reader. In Sect. 4, we review a number of approaches from the area of *Similarity Measures* that make use of semantic knowledge representation mechanisms.

Table 1. Landscape of relevant approaches at the intersection of ML and SW covered in this chapter

		Objects of interest		
		"Ordinary" data (e.g. texts)	Ontology entities	Ontologies
Structural component	Feature representation	Ontologies for text mining *Sect. 3*	Link-based object classification *Sect. 5*	Graph classification *Sect. 5*
	Similarities and distances	Semantic smoothing kernels *Sect. 4*	Similarities and kernel functions for ontological entities *Sect. 4*	Graph kernels Graph matching *Sect. 5*
	Model	⟵ Statistical Relational Learning ⟶ *Sect. 6*		
	Algorithms	–	Social network analysis of ontologies *Sect. 5* Inductive Logic Programming *Sect. 6*	–

In particular we also cover an exciting field of modern ML research, *Kernel Methods*, that make use of a specific class of such similarity measures. In Sect. 5, we survey a field of ML research called *Link Mining* which addresses various learning tasks on data that exhibits a link structure. Then we sketch how this work can be adapted to ontological data. In Sect. 6, we introduce the field of *Statistical Relational Learning* that naturally lends itself to application to SW data and sketch its relations to the somewhat more traditional field of *Inductive Logic Programming*. We conclude with a short summary in Sect. 7.

3 Ontologies for Text Mining

The term *Text Mining (TM)* was first phrased by Feldman and Dagan [26] in 1995 to describe a new field of data analysis. Text Mining comprises various facets but in general it refers to the application of methods from Machine Learning to textual data. An overview over the topic is given by Hotho et al. [43]. Some subfields of Text Mining go beyond the detection of patterns from texts as wholes but rather focus on the extraction of factual knowledge from them – a field which is known as Information Extraction (IE) covered in chapter "Information Extraction" in this volume. Text Mining is distinguished primarily by special preprocessing methods to prepare the textual data for the analysis by Machine Learning techniques. It is not surprising that Text Mining as a field naturally overlaps with other computer science disciplines that deal with the processing of natural language such as Information Retrieval (IR) [2], Web Mining [14] as well as Natural Language Processing (NLP) [59].

From the data mining perspective, Text Mining mainly targets three different application areas. On the one hand, *Text Clustering* is of interest to allow for a better way to explore huge text collections and to add structure for navigation. On the other hand, *Text Classification* aims at learning models that enable the assignment of thematic categories to unseen texts, e.g. to support news providers by classifying their incoming news, but also for spam detection. The survey by Sebastiani [75] provides a good overview of the topic. Finally, the *visualization* of large text corpora for fast and simple exploratory inspection is a nontrivial task which is often necessary to get first insights into huge and otherwise hardly usable textual resources [27, 86].

Ontologies represent additional background knowledge which can be exploited to better solve typical Text Mining tasks [8]. In the context of our classification of approaches to Machine Learning with ontologies in Sect. 2, the techniques we report on fall into the class of approaches that deal with arbitrary instances outside the ontology structure. At the same time, all of these approaches mainly aim at a modified feature representation paradigm. In Sect. 4, we will shortly look at an alternative technique, namely the modification of the underlying similarity measures by means of an appropriate kernel function.

3.1 Preprocessing

In this part, we shortly sketch both the conventional and the ontology-enhanced representation of text data for ML settings.

Text Representation

In Text Mining, documents are typically represented as so called *Bag-of-Words* vectors as originally proposed by Salton [72] for IR. This means that one counts words of documents independently of their order resulting in a vector representation. Every document is represented as a vector $\mathbf{d}$ which consist of the frequencies of every word in the document. For example, in Fig. 1 the word "oil" appears twice in the text. The dimension of the resulting vector space is given by the number of distinct words of the corpus. Figure 1 illustrates this situation for a document from the well-known Reuters-21578 corpus [55]. On the left side the original document is depicted which results in the bag of words vector next to the document.

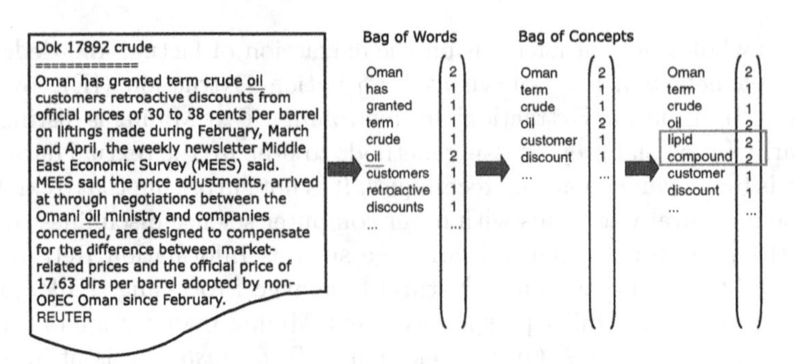

Fig. 1. Bag of words example

Non-descriptive words, so called *stopwords*, are often removed from this representation based on stopword lists.[2] One linguistic property of words in text is their morphological variability. For text mining, it is necessary to have the same word in the same spelling to be able to compute a meaningful similarity. To address this issue, words are usually not used in their inflected form, but only as their stem. For example, the stemming algorithm introduced by Porter [67] is the most often used heuristics for the English language.

Term weighting techniques, such as *TFIDF* weight tf, the frequency of a word in a document, with *idf*, a factor that discounts its importance when it appears in many documents in the corpus. It is defined as:

$$\text{tfidf}(d,t) := \log(tf(d,t)+1) * \log\left(\frac{|D|}{df(t)}\right),$$

where $df(t)$ is the document frequency of term t that counts in how many documents term t appears. See Amati et al. [1] for a discussion of such measures. Detection of multi-word expressions, names and abbreviations are examples of further preprocessing steps which may or may not be of interest for a particular application. For general information on preprocessing, the interested reader is referred to the surveys by Sebastiani [75] and Hotho et al. [43].

Incorporating Background Knowledge from Ontologies

The background knowledge we will exploit further on is encoded in an ontology. The ontological background knowledge is incorporated into the vector space model by applying additional preprocessing steps. After deriving the typical bag of words representation, the vector dimensions are mapped to concepts of a given ontology or knowledge base.

Enriching the term vectors with explicit concepts from the ontology has two benefits. First it resolves synonyms; and second it introduces more general concepts which help to identify related topics and provides some kind of

[2] The stopword list of the SMART project which is available at `ftp://ftp.cs.cornell.edu/pub/smart/english.stop` is commonly used for English.

connection between documents which addresses the same or a very similar topic with different words. For instance, a document about beef may not be related to a document about pork by the cluster algorithm if there are only "beef" and "pork" in the term vector. But if the more general concept "meat" is added to both documents, their semantic relationship is revealed. We have investigated the influence of three different strategies for adding or/and replacing terms by concepts on the clustering/classification performance [41]. By mapping words to concepts we compute a new concept vector. The first strategy uses all available information by performing the mining on both vectors together. The second strategy removes all words of the word vector which could be mapped on a concept. The last strategy bases the analysis only on the concept vector. A mapped vector for our example document of Fig. 1 is given next to the bag of words vector. Only concepts which are found in the given ontology (in this case Wordnet was used) are present in the vector.

For our purpose, a knowledge base needs a lexical component to allow for an appropriate mapping of the words of the text documents to the concepts of the ontology. Obviously, this mapping yields new challenges like the handling of the emerging concept vector and the detection of the meaning of a word to find the right concept. In this context, an important problem is to find the *right* concept for a word in a given context which have more than one meaning. This is the word sense disambiguation problem [45]. A perfectly mapped resource seems to be helpful for clustering text [18], but the current state of the art does not reach this level. First studies show the need of word sense disambiguation for clustering [41], but a detailed analysis on the clustering performance of correctly disambiguated word senses does not exist yet.

The mapping of words to concepts solves also the synonymy problem. Adding additional hypernyms/super concepts allows for relating very similar topics which are the content of different documents but which a user would expect in the same cluster. By changing the document representation in a way that different words of the vector are mapped to the same (super) concept, to represent the same or a very similar topic by a common representation, the clustering algorithm should be better able to group such documents together. By adding more super-concepts we start to add noise and in result the performance will drop because topics become related which have not so much in common. The rightmost vector in Fig. 1 illustrates the extension of the concept "oil" by the two super-concepts "lipid" and "compound". In this small example, the super-concepts have the same count as oil but this could change if other sub-concepts of, e.g. lipid would be present in the document.

3.2 Approaches for Different Learning Tasks

We now focus on some of our own results that use ontologies to improve clustering and classification tasks [41, 9, 7].

Text Clustering

Text document clustering methods can be used to find groups of documents with similar content. The result of a clustering is typically a partition of the set of documents. Each cluster consists of a set of documents. Usually the quality of a clustering is considered better if the contents of the documents within one cluster are more similar and between the clusters more dissimilar. Most clustering methods group the documents only by considering their distribution in document space (for example, in the vector space model for text documents). A good survey can be found in [4] and a discussion of the performance of different Text Clustering approaches in [78].

We illustrate the integration of background knowledge into the text clustering process by results of Hotho et al. [41] and Hotho [42] using a variant of the popular k-means clustering algorithm. In these experiments, we applied the usual preprocessing steps on the Reuters-21578 corpus [55], the FAODOC corpus and a small Java corpus. As a lexical ontology, WordNet [61], a lexical ontology of the English language, was used. It provides not only a morphological component which significantly improves the preprocessing but also contains synonymy, hypernym/super-concept and frequency information about polysemous words. The main outcome of our experiment was the following: TFIDF weighting improves the text clustering performance significantly and is also helpful to integrate the background knowledge as it gives a good weight to the concepts. Word sense disambiguation is necessary during the mapping of words to concepts. There are indications that the "add strategy", which uses both words and concepts equally, outperforms all other integration strategies. The integration of super-concepts into the concept vector additional improves the performance of the text clustering approach.

Not only the performance of Text Clustering can be improved by using background knowledge. The integration of super-concepts provides also a very good basis for clustering visualization. Hotho et al. [44] use Formal Concept Analysis (covered in chapter "Formal Concept Analysis" of this volume) to compute the visualization. The resulting concept lattice makes the exploration of a new corpus easier then inspecting unrelated clusters as it provides a good overview over the different topics of the corpus by relating clusters to each other. High level concepts from the ontology are used to describe the commonalities of different clusters. The structure of the lattice helps also to drill down to very specific clusters while maintaining a clear relation to a major topic.

To date, the work on integrating background knowledge into text clustering is quite heterogeneous. Green [32] uses WordNet to construct chains of related synsets from the occurrence of terms for document representation and subsequent clustering. Green does not evaluate performance and scalability of his approach as compared to standard BOW-based clustering of documents. Also Kushal Dave [50] has explored WordNet. He did not perform word sense disambiguation and only found that WordNet synsets decreased clustering performance in all his experiments.

Text Classification

The automatic process of learning a model, based on a given set of training examples, which is then able to predict the class label of a new text document is known as Text Classification. Early methods that produced good results were Rocchio, k-Nearest Neigbhour (kNN) or neural networks in the middle of the 1990s. Meanwhile, more advanced Machine Learning approaches like Support Vector Machines (SVMs) or Boosting show very impressive Text Classification performance. A good survey is presented by Sebastiani [75]. As an example, Fig. 2 illustrates the general classification model of linear classifiers like the Perceptron or Support Vector Machines.

In this section, we report the main idea of integrating formally represented knowledge into the learning step with the goal to improve the prediction performance. We follow the presentation of our work in [9], where we showed how background knowledge in form of simple ontologies can improve Text Classification results by directly addressing the problems of multi-word expressions, synonymous words, polysemous words, and the lack of generalization. We used a hybrid approach for document representation based on the common term stem representation which is enhanced with concepts extracted from the used ontologies as introduced above. For the actual classification, we suggested to use the AdaBoost algorithm using decision stumps as base classifiers which has been proved to produce accurate classification results in many experimental evaluations and seems to be well suited to integrate different types of features. Evaluation experiments on three text corpora, namely the Reuters-21578, OHSUMED and FAODOC collections showed that our approach lead to improvements in all cases. We also showed that in most cases the improvement can be traced back to two distinct effects, one being situated mainly on

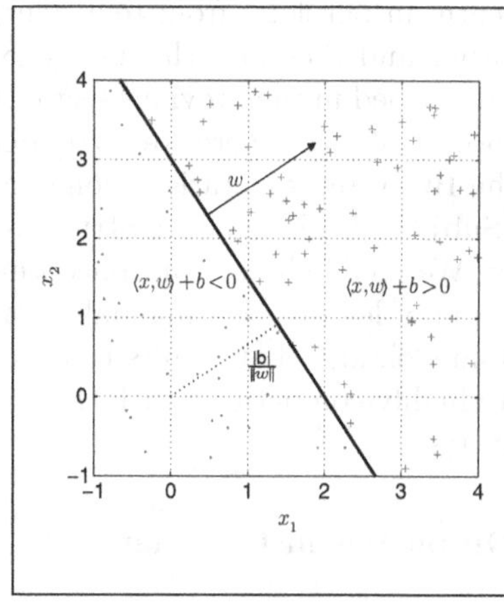

Linear classifiers are defined by discriminant functions of the form:

$$f(x) = \text{sign}(\langle x, w\rangle + b).$$

Given a set of input examples, training algorithms for linear classifiers try to estimate optimal parameter w (normal vector) b (bias). The resulting decision boundary is given by a *hyperplane* (here a straight line in 2D). Each side of the hyperplane corresponds to a particular classification decision. Popular training algorithms for linear classifiers are the *Perceptron* [71] or *Support Vector Machines (SVMs)* [13, 76].

Fig. 2. Excursus: Linear Classification

the lexical level (e.g. detection of multi-word expressions) and the generalization on the conceptual level (resolving synonyms and adding super-concepts).

Other results from similar settings including background knowledge are reported in [74] and [84]. Ureña Lóez et al. [82] and de Buenaga Rodrguez et al. [18] show a successful integration of WordNet for a document categorization task, but the result is based on *manually* designed synset vectors. They use the Reuters corpus for evaluation and improve the classification results of the Rocchio and Widrow-Hoff algorithms by 20 percent points. This result can be seen as an upper bound as the manual word sense disambiguation can be considered as perfect.

3.3 Related Approaches

A number of other approaches have varied the basic settings we have reported in the previous section. Beside the usual preprocessing of text, several other approaches like classification on n-grams or smoothing with Latent Semantic Indexing (LSI) were investigated to improve the performance in selected Text Mining settings. In the following, we report on two particularly interesting directions.

Text Mining with Automatically Learned Ontologies

So far, the ontological structures employed for the classification and clustering task are created manually by knowledge engineers which requires a high initial modelling effort. Research on Ontology Learning (covered in chapter "Ontology Learning" of this volume) has started to address this problem by developing methods for the automatic construction of conceptual structures out of large text corpora mostly in an unsupervised process. To reduce the modelling effort, the next step is to first learn an ontology from text which perfectly matches with the topics of the corpus and then add this newly extracted knowledge to the mining process as described in the previous sections. This approach was undertaken by Bloehdorn et al. [7], where we compared results both (1) to the baseline given by the BOW representation alone and (2) to results based on the MeSH (Medical Subject Headings) Tree Structures as a manually engineered medical ontology. We could show that conceptual feature representations based on a combination of learned and manually constructed ontologies outperformed the BOW model, and that results based on the automatically constructed ontologies are highly competitive with those of the manually engineered MeSH Tree Structures.

Using Background Knowledge from Ontologies in Information Retrieval

Several researchers have reported positive results concerning query expansion in the context of IR applications. In early work on the topic, Salton and Lesk

[73], found that expansion with synonyms improved performance, while using broader or narrower terms produced too inconsistent results for being actually useful. Wang et al. [85] report that a variety of lexical–semantic relations improved retrieval performance. A comprehensive study by Voorhees [83] indicated that query expansion is especially useful when queries are relatively short. Gonzalo et al. [31] compare indexing id disambiguated words and indexing of WordNet synsets and show that the first variant improves upon the plain term vector model but performs worse than the second variant.

4 Similarities and Kernel Functions for Knowledge Structures

Various Machine Learning algorithms can be designed in such a way that the only required input is a matrix of pairwise similarities or distances among the input items. The definition of appropriate similarity and dissimilarity measures is a topic that plays a key role in different areas of Artificial Intelligence. This chapter deals with a selection of approaches for defining appropriate similarity or dissimilarity measures in the context of Machine Learning with Ontologies. In the context of the classification provided in Table 1, the group of techniques reported here corresponds to the second row.

In the simplest case, algorithms can work with an arbitrary similarity function which is usually only required to be positive, reflexive and symmetric. Algorithms that work with distances rather than similarities often require that the distance (i.e. dissimilarity) measures comply with the requirements of a metric. Examples of algorithms that pose only minor requirements on the properties of the employed measures are the k-Nearest Neighbour (kNN) algorithm for classification or agglomerative clustering techniques. Classical similarity measures defined on feature vectors are the inner product or the cosine, and the corresponding canonical dissimilarity measure is the Euclidean distance as e.g. required for k-means clustering.

Most naturally, there are strong relations between feature representations and (dis-)similarity measures: Changes to the feature representation may imply different measures and changes to the measures may implicitly correspond to a modified feature representation. A particularly interesting class of similarity measures are *kernel functions*. Kernel functions compute the similarities of data instances in such a way that the result is equivalent to an inner product in some (possibly unknown) vector space. Formally, any function $\kappa : X \times X \to \mathbb{R}$ that for all $x, z \in X$ satisfies $\kappa(x, z) = \langle\phi(x), \phi(z)\rangle$ is a valid kernel, whereby X is some input domain under consideration and ϕ is a mapping from X to a feature space F. It can be shown that the class of such functions can be characterised as the class of functions which are positive semi-definite. The reason for the large interest in kernel methods is the fact that the correspondence to a vector space makes it possible to use kernel functions together with many Machine Learning algorithms whose models are tied to a geometric interpretation within the corresponding vector space without representing objects explicitly in this space. An example would be the notion of a separating hyperplane in the case of classification with linear classifiers as described in Fig. 2, e.g. in the case of classification with SVMs. These models can usually be better analysed in terms of their statistical behaviour and usually lead to superior generalization performance. At the same time, the explicit construction of a feature vectors can be avoided and would often be practically impossible due to the large number of dimensions involved. Kernel methods thus form an interesting group of methods for dealing with complex and structured data, such as data embedded in an ontology structure [33]. For more details on the theory of kernel functions, the interested reader is referred to the introductory article by Müller et al. [64] or to the book by Shawe-Taylor and Cristianini [76].

4.1 Ontology-Based Kernel Functions for Semantic Smoothing

Semantic Smoothing Kernels are a technique for incorporating ontological background knowledge into a kernel function for vector representations of textual data. This kernels implicitly mimic parts of the effects of the explicit feature transformations we have investigated in the previous section. Semantic smoothing kernels were initially proposed by Siolas and d'Alche Buc [77] and subsequently revisited in [3,6,60]. These kernels still work on standard vectorial data but try to incorporate information about the semantic dependencies between the dimensions of the vector space. This approach is interesting in those cases where the dimensions of the vector space can not be regarded as mutually orthogonal dimensions as, e.g. in the standard bag-of-words representation used in text mining settings. In their basic form, these kernels are of the type $k(x, y) = x'\mathbf{Q}y$, whereby $\mathbf{Q}$ is a symmetric and positive semi-definite smoothing matrix that encodes the similarity of the respective attributes. Specifically, an off-diagonal entry $\mathbf{Q}_{ij}$ specifies the similarity between two features i an j, e.g. two similar terms like "beef" and "pork" as discussed earlier. In most approaches, the design of the smoothing matrix $\mathbf{Q}$ is guided by the arrangement of the entities corresponding to the vector dimensions (e.g. terms) in an ontological structure. The positive semi-definiteness of the smoothing matrix $\mathbf{Q}$ implies a possible decomposition as $\mathbf{Q} = \mathbf{P}\mathbf{P}'$, thereby revealing the underlying feature mapping as a linear transformation into a different space, usually a concept space similar to the one introduced in Sect. 3.

Advanced approaches have exploited the property that kernels can be combined and embedded in one another. As an example, Bloehdorn and Moschitti [10,11] have combined ontology-based semantic smoothing kernels with *Tree Kernels* [15] defined on the syntactic structure of sentences, yielding powerful combined kernels.

4.2 Similarities and Dissimilarities for Ontology Entities

A general framework for similarity measures in ontologies is proposed by Ehrig et al. [24]. The framework is based on distinguishing several layers of similarity, namely the data layer which takes into account similarities of the involved data types, the ontology layer which corresponds to measures that use the ontology structure and the context layer which takes into account how ontology entities are used in some external context.

Bernstein et al. [5] as well as Hefke et al. [37] provide software frameworks for computing semantic similarity measures according to a wide range of similarity measures. As an exemplary application, Kiefer et al. [47] use the similarity framework reported by Bernstein et al. [5] to mine software repositories.

D'Amato et al. [16, 17] provide a set of similarity and dissimilarity measures for concept descriptions in the description logic $\mathcal{ALC}$. The measures are based both on the syntax and on the semantics of the descriptions including extensions for involving individuals and for evaluating their dissimilarity.

4.3 Kernel Functions for Ontology Entities

Most of the work on kernels for structured data is rooted in the influential work on *convolution kernels* by Haussler [36]. The relevance of this kind of research towards ontologies and metadata becomes obvious when looking at learning problems where the instances are described in the knowledge representation language of interest. Such an approach is taken by Frasconi et al. [28], where objects are described using a simple knowledge representation language consisting of a type hierarchy, relations between objects and attributes. The kernel, which is set up in a similar way as the convolution kernel and is motivated by mereotopological considerations, is capable of improving classification accuracy on a biomedical dataset compared to state-of-the art Inductive Logic Programming (ILP) learning algorithms.

Gärtner et al. [35] have proposed a logic-based kernel on instances represented as (closed) terms in the typed higher-order logic of Lloyd [58]. Neglecting the technicalities of the approach, this logic essentially allows the construction of complex types such as sets or lists out of other types, including standard types as, e.g. natural numbers. The work presents a kernel defined on terms in the associated logic including a proof that the kernel is positive definite on all basic terms. This work can be seen as the first principled framework for defining kernel functions on data items represented in a declarative knowledge representation formalism.

A first endeavour to investigate the use of kernels for actual Semantic Web data is presented by Bloehdorn and Sure [12]. The paper introduces a framework for kernel computations on instance data that makes use of various layers of instance similarity. These layers exploit the similarity of the type structure, the similarity of data property extensions as well as common object properties. Experiments on a classic Semantic Web dataset, the SWRC ontology [80], show that the design of effective kernels within this framework requires only little conceptual overhead.

In a different spirit, Raedt and Passerini [68] have proposed a kernel on PROLOG proof trees. In this setting, the individuals are described in the context of global ontological background knowledge in first-order logic. In contrast to the approaches mentioned previously, the ontological descriptions are not investigated and incorporated into the kernel computation directly, but rather indirectly by means of traces left in a PROLOG reasoner. Specifically, the idea of this kernel is then to measure the similarity of two individuals by means of the similarity of the proof trees of a special logic program, called the *visitor program*. The proof trees are compared using standard tree kernels as mentioned above while the visitor program is designed to probe certain characteristics of the individuals that may be of interest for the domain and learning task at hand. Experiments using this kernel, for example on Bongard scenes and Protein Fold classification, show promising results. In contrast to other approaches, this kernel allows to exploit background knowledge in a principally different way whereby the problem of kernel design is shifted from the explicit design of kernels to operate on the instance data to the design of appropriate visitor programs.

On the schema-level, Fanizzi and d'Amato [25] propose a declarative kernel for concept descriptions in the description logic $\mathcal{ALC}$. The kernel is structurally based on the convolution kernel but takes into account the semantics of the overall logic by describing classes in terms of their (known) instantiations.

5 Link Mining

In many ways, ontological data can be seen as a collection of linked resources. Typically, as in the case of RDF or OWL, they form heterogeneous networks with many resource and link types, whereby the type information itself is again arranged in a linked structure, e.g. a subsumption hierarchy. But also datasets that are not explicitly encoded according to SW standards may exhibit a certain level of semantics that is present within a link structure. As an example, consider bibliographic data linking publications, authors, and venues.

Link mining refers to data mining techniques that explicitly consider such links when building predictive or descriptive models of linked data. A good survey over the field is given by Getoor and Diehl [30]. Typical link mining tasks include group detection, classification of links and resources, prediction of (missing) links and subgraph discovery. Link mining constitutes a research field taking influences from different communities such as social network analysis, hypertext mining, graph mining and web analysis. The term *link mining* was chosen to put a special emphasis on the links as main subject of analysis. In the following, we review some link mining subfields that relate to ontologies and the Semantic Web. In the classification of techniques we have proposed above, the techniques in this section mainly relate to learning from ontological entities, represented as nodes or edges in the ontology graph, or from whole ontologies as such, i.e. whole graphs or parts of it.

5.1 Semantic Network Analysis of Ontologies

Since the last decade, the social network analysis community has started to discover the Internet and the Web as fruitful application domains for their techniques (e.g. analysing the link structure of the Internet [48]). The use of network analysis techniques for the analysis of Semantic Web type data has been undertaken by Hoser et al. [40]. The paper illustrates the benefits of applying such techniques to ontologies and the Semantic Web. In particular, it discusses how different notions of centrality describe the core content and structure of an ontology. From the rather simple notion of degree centrality over betweenness centrality to the more complex eigenvector centrality based on Hermitian matrices [39], the paper illustrates the insights these measures can provide on ontologies.

Ding et al. [20] has developed an algorithm for ranking the importance of Semantic Web objects at three levels of granularity: documents, terms, and RDF graphs of the semantic web search engine Swoogle. The proposed OntoRank approach applies mainly the idea of PageRank [65] on Semantic Web documents and the links between them.

5.2 Link-Based Object Classification

This group of techniques refers to algorithms that directly exploit the link structure in a graph (e.g. in an ontology) to classify objects (e.g. ontological entities). A good overview of this kind of approaches is given in [30, Sect. 4] for a wide range of methods and possible applications. Referring back to the field of kernel methods introduced in Sect. 4, the *diffusion kernel* presented of Kondor and Lafferty [49] constitutes an example of such a technique. In this case the diffusion kernel tries to exploit local relations which imply global information by using exponentials of the matrix power series on the kernel matrix.

5.3 Subgraph Detection and Graph Matching

This topic builds upon subgraph mining techniques to find frequent or informative substructures in graph instances. An example of the utilization of such algorithms on Semantic Web data is given by Ramakrishnan et al. [69], which describes methods for discovering interesting subgraphs in an RDF graph based on semantic information associated with edges. The main focus is to relate entities of the graph with other entities given the underlying RDF graph structure.

Graph matching refers to techniques that try to detect similar substructures in pairs of graphs. Seen in the context of the Semantic Web, this field immediately evokes thoughts about the field of *Ontology Mapping*. Work in this area includes Doan et al. [21] who presents GLUE, a system which is able to use different sources to learn the similarities for the mapping process.

Interesting recent work in this direction that explicitly takes into account the interaction between mining the structure of the ontologies to be aligned and checking the semantics of the resulting mappings is reported by Udrea et al. [81]. The paper presents the *Integrated Learning In Alignment of Data and Schema (ILIADS)* algorithm that is based on an interleaving of a hierarchical clustering algorithm with an incremental logical inference algorithm. Clustering entities creates new relationships among the entities; these new relationships may have logical consequences in OWL Lite.

5.4 Graph Classification

Unlike link-based object classification, which attempts to mine the nodes in a graph, graph classification is a learning problem in which the goal is to classify an entire graph as a positive or negative example in a classification setting. A typical approach to address this problem is to discover features on the input graphs, thereby building on subgraph mining techniques to find frequent or informative substructures in the graph instances. The detected substructures are then used for transforming the overall graphs into vectorial data, and then traditional classifiers are used for classifying the instances. However, finding all frequent substructures is usually computationally prohibitive.

Again, kernel methods have been designed to efficiently work on graph data, but these approaches have usually been restricted to specific kinds of graphs (e.g. trees). Kernels for arbitrary graphs have proved to be more difficult to design. Approaches in this direction are reported in [34].

6 Statistical Relational Learning

In this section we start with a short review of Inductive Logic Programming approaches and relate them to Semantic Web paradigms. Then we focus on the upcoming new research area Statistical Relational Learning which combines logic and probabilistic learning approaches. We will give some references to this area and discuss connections to Semantic Web. In the context of our overview of the research landscape, these approaches can deal with various types of input items but do so mainly on the level of modifying the learning model or modifying the actual learning algorithms.

6.1 ILP and the Semantic Web

Inductive Logic Programming (ILP), a term phrased by Muggleton [63], is a research area formed at the intersection of Machine Learning and Logic Programming. Good introductions are given by Lavrač and Džeroski [52], Muggleton and Raedt [62] as well as by Dzeroski [23]. ILP uses logic as the uniform representation for examples, background knowledge and hypotheses, typically in the context of classification tasks. Given a first-order logic encoding of the background knowledge and a dataset represented as a set of logical axioms, ILP systems try to derive an logic program which explains all the positive but none of the negative examples in classification setting. ILP systems are centred around techniques for refining and generalizing hypotheses that do not yet fully explain the data and are thus based on a search process through a partially ordered space of inductive hypotheses. A unifying theory of ILP is built up around lattice-based concepts such as refinement, least general generalization, inverse resolution and most specific corrections. Successful applications areas for ILP systems include the learning of structure-activity rules for drug design, prediction of protein structure and fault diagnosis rules for technical systems.

As Semantic Web standards are largely built upon the foundations of first-order logic, the application of ILP techniques to learn on Semantic Web data is a natural step. Lisi [56] shows that current ILP systems could serve the learning purpose if they were more compliant with the standards of representation for ontologies and rules in the Semantic Web and/or inter-operable with well-established Ontological Engineering tools that support these standards.

ILP systems tuned towards learning in description logics (which form the basis of most Semantic Web endeavours, particularly OWL) are presented by Lisi and Esposito [57] and Lehmann and Hitzler [54]. A crucial issue to apply ILP techniques successfully is the definition of generality orders for inductive hypotheses. This issue is investigated in detail by Lehmann and Hitzler [53]. The paper presents a study on desirable properties of the employed refinement operators in description logics and shows that ideal refinement operators do not exist, as an indication of the hardness inherent in learning in description logics. The authors also show how the set of desirable properties can be constrained to make learning feasible.

6.2 Combination of Probabilistic and Logic-Based Learning Approaches

Statistical Relational Learning (SRL) is a relatively young research area which focuses on the combination of probabilistic and logic models with the goal to be better able to describe real world phenomena. Traditional statistical machine learning is able to capture uncertainty, but only within one relation – whereas traditional ILP and relational learning approaches are able to work on multiple relations, but cannot handle noise. The combination of both approaches tries to overcome these limitations, which is a critical point when working with heterogeneous and richly interlinked Semantic Web data. Methods developed in this area are applied to richly structured data which is available for, e.g. hypertext classification, topic prediction of bibliographic entries, or in any kind of social networks. Other applications areas for SRL includes communication data, customer networks, collaborative filtering, trust networks, biological data, sensor networks, and natural language data. The book by Getoor [29] provides a good introduction to this area.

There are four distinct areas which are the starting points of SRL research: (1) ILP, (2) statistical learning as well as (3) probabilistic and (4) logical inference. Researchers from these areas extended in the last years well known approaches to bridge the gap between logical and probabilistic approaches. One example is the extension of the popular propositional rule learning algorithm CN2 to ICL (Inductive Constraint Logic) in [51]. ICL is now able to work on data represented in first-order logic while CN2 works one relation only. The resulting combined approach shows the strength of newly emerging SRL field.

6.3 Statistical Relational Learning Challenges and Applications on the Semantic Web

There exist many application areas for SRL. We focus here on applications with relation to the Semantic Web. The emergent field of Statistical Relational Learning offers a variety of methods to overcome existing Semantic Web problems; E.g. the logic used in the Semantic Web was not designed to deal with uncertainty but SRL has made first steps to combine logic and uncertainty. On the other hand these problems cause new challenges for the Statistical Relational Learning community as new kind of data arises. In the rest of this section we will shortly review first solutions combining both areas.

Statistical relational learning techniques are well suited to knowledge-intensive learning, because they allow input knowledge to be expressed in a rich relational language, while being able to handle noise in this input. In general, many different types of knowledge can potentially be integrated into SRL. First steps towards an automatic knowledge integration are reported by Domingos et al. [22]. In an endeavour to design such a system that is able to support the building of large knowledge bases by mass collaboration, [70] have designed an architecture for incorporating knowledge from a large number of sources into a learner. The learner uses SRL techniques to handle inconsistency among different sources and high variability in source quality. The system uses a Bayesian Logic Programs representation [46] to extract a Bayesian network which in turn is used to answer given queries. The approach was successfully applied in a printer troubleshooting domain (cf. Domingos et al. [22]).

Many good examples of applications of SRL techniques to the Semantic Web come from the area of ontology mapping. Given initial mappings between knowledge structures from different sources, one can learn generalizations of them using SRL techniques like the content learner and the name learner, both utilizing the well know naive Bayes classifier on a bag of tokens. Amongst others, Doan et al. [21] as well as Dhamankar et al. [19] have reported such experiments for XML data and for SW ontologies.

Popescul et al. [66] present an extension of a typical text classification approach which is related to the work of Sect. 3. It uses a richer set of features from a relational database like the concepts in the text mining setting which allows for an improvement of the learned predictive model over a typical propositional one. Features include not only typical textual ones, but also exploit the semantic links between documents based on citing and author information.

7 Conclusion and Outlook

In this paper, we have reviewed contributions from different communities to the emerging field of *Semantic Web Mining* by discussing Machine Learning techniques that directly or indirectly use ontologies or related declarative knowledge representation paradigms. At this point, the work in this area is highly scattered among various subfields of Machine Learning theory and practice and among current Semantic Web research efforts. All these areas continue to evolve and are likely to undergo various transformations in the years to come.

We have introduced Inductive Logic Programming in Sect. 6.1 as the predecessor of today's Semantic Web mining efforts which continually evolves to accommodate more and more of the current Semantic Web development. From a theoretical perspective, the kernel methods paradigm and the techniques from the field of statistical relational learning we presented in Sects. 4 and 6, respectively, are likely to have the highest impact on modern Semantic Web mining efforts. From a practical perspective, the fields of text mining, presented in Sect. 3, and link mining, presented in Sect. 5 already show ontology-enhanced applications actually working or – in the case of link mining – point to practical approaches that show substantial potential to be applied and transferred to the field of the Semantic Web.

References

1. G. Amati, C. Carpineto, and G. Romano. Fub at trec-10 web track: A probabilistic framework for topic relevance term weighting. In *The Tenth Text Retrieval Conference (TREC 2001)*. National Institute of Standards and Technology (NIST), online publication, 2001.
2. Ricardo Baeza-Yates and Berthier Ribeiro-Neto. *Modern Information Retrieval*. Addison Wesley, May 1999.
3. Roberto Basili, Marco Cammisa, and Alessandro Moschitti. A semantic kernel to classify texts with very few training examples. In Stephan Bloehdorn, Andreas Hotho, and Wray Buntine, editors, *Proceedings of the Workshop on Learning in Web Search at the 22nd International Conference on Machine Learning (ICML 2005), August 7–11, 2005, Bonn, Germany*, pages 10–17, 2005. Published online in August 2005 at `http://cosco.hiit.fi/search/learninginsearch05/ICML_W4.pdf`.

4. Pavel Berkhin. Survey of clustering data mining techniques. Technical report, Accrue Software, San Jose, CA, 2002.
5. Abraham Bernstein, Esther Kaufmann, Christoph Kiefer, and Christoph Bürki. SimPack: A generic Java library for similiarity measures in ontologies. Technical report, Department of Informatics, University of Zurich, Zurich, Switzerland, 2005.
6. Stephan Bloehdorn, Roberto Basili, Marco Cammisa, and Alessandro Moschitti. Semantic kernels for text classification based on topological measures of feature similarity. In *Proceedings of the 6th IEEE International Conference on Data Mining (ICDM 2006), 18–22 December 2006, Hong Kong, China*. IEEE Computer Society, Washington, DC, USA, 2006.
7. Stephan Bloehdorn, Philipp Cimiano, and Andreas Hotho. Learning ontologies to improve text clustering and classification. In Myra Spiliopoulou, Rudolf Kruse, Andreas Nürnberger, Christian Borgelt, and Wolfgang Gaul, editors, *From Data and Information Analysis to Knowledge Engineering: Proceedings of the 29th Annual Conference of the German Classification Society (GfKl 2005), March 9–11, 2005, Magdeburg, Germany*, volume 30 of *Studies in Classification, Data Analysis, and Knowledge Organization*, pages 334–341. Springer, Berlin–Heidelberg, Germany, 2006.
8. Stephan Bloehdorn, Philipp Cimiano, Andreas Hotho, and Steffen Staab. An ontology-based framework for text mining. *LDV Forum - GLDV Journal for Computational Linguistics and Language Technology*, 20(1):87–112, 2005.
9. Stephan Bloehdorn and Andreas Hotho. Text classification by boosting weak learners based on terms and concepts. In *Proceedings of the 4th IEEE International Conference on Data Mining (ICDM 2004), 1–4 November 2004, Brighton, UK*, pages 331–334. IEEE Computer Society, Washington, DC, USA, 2004.
10. Stephan Bloehdorn and Alessandro Moschitti. Combined syntactic and semantic kernels for text classification. In Gianni Amati, Claudio Carpineto, and Gianni Romano, editors, *Advances in Information Retrieval – Proceedings of the 29th European Conference on Information Retrieval (ECIR 2007), 2–5 April 2007, Rome, Italy*, volume 4425 of *Lecture Notes in Computer Science*, pages 307–318, Berlin–Heidelberg, Germany, 2007. Springer, Berlin–Heidelberg, Germany.
11. Stephan Bloehdorn and Alessandro Moschitti. Structure and semantics for expressive text kernels. In Mario J. Silva, Alberto H. F. Laender, Ricardo A. Baeza-Yates, Deborah L. McGuinness, Bjoern Olstad, Oystein Haug Olsen, and Andre O. Falcao, editors, *Proceedings of the 16th ACM Conference on Information and Knowledge Management (CIKM 2007), November 6–9, 2007, Lisbon, Portugal*, pages 861–864. ACM Press, New York, NY, USA, 2007.
12. Stephan Bloehdorn and York Sure. Kernel methods for mining instance data in ontologies. In Karl Aberer, Key-Sun Choi, and Natasha Noy, editors, *The Semantic Web – Proceedings of the 6th International Semantic Web Conference and the 2nd Asian Semantic Web Conference (ISWC 2007 + ASWC 2007), November 11–15, 2007, Busan, Korea*, number 4825 in Lecture Notes in Computer Science, pages 58–71. Springer, Berlin–Heidelberg, Germany, 2007.
13. Bernhard E. Boser, Isabelle M. Guyon, and Vladimir N. Vapnik. A training algorithm for optimal margin classifiers. In David Haussler, editor, *Proceedings of the Fifth Annual Workshop on Computational Learning Theory (COLT '92), July 27–29, 1992, Pittsburgh, PA, USA*, pages 144–152. ACM Press, New York, NY, USA, 1992.

14. Soumen Chakrabarti. *Mining the Web: Discovering Knowledge from Hypertext Data.* Morgan-Kauffman Publishers, San Francisco, CA, USA, 2002.
15. Michael Collins and Nigel Duffy. Convolution kernels for natural language. In Thomas G. Dietterich, Suzanna Becker, and Zoubin Ghahramani, editors, *Advances in Neural Information Processing Systems 14 – Proceedings of the 2001 Neural Information Processing Systems Conference (NIPS 2001), December 3–8, 2001, Vancouver, British Columbia, Canada*, pages 625–632. MIT Press, Cambridge, MA, USA, 2002.
16. C. d'Amato, N. Fanizzi, and F. Esposito. A semantic similarity measure for expressive description logics. In A. Pettorossi, editor, *Proceedings of Convegno Italiano di Logica Computazionale (CILC05) 21–22 June 2005, Rome, Italy*, 2005.
17. Claudia d'Amato, Nicola Fanizzi, and Floriana Esposito. A dissimilarity measure for ALC concept descriptions. In Hisham M. Haddad, editor, *Proceedings of the 2006 ACM Symposium on Applied Computing (SAC 2006), April 23–27, 2006, Dijon, France*, pages 1695–1699. ACM, New York, NY, USA, 2006.
18. M. de Buenaga Rodriguez, J. M. Gomez Hidalgo, and B. Díaz-Agudo. Using WordNet to complement training information in text categorization. In *Recent Advances in Natural Language Processing II*, volume 189. John Benjamins, 2000.
19. Robin Dhamankar, Yoonkyong Lee, AnHai Doan, Alon Y. Halevy, and Pedro Domingos. imap: Discovering complex mappings between database schemas. In Gerhard Weikum, Arnd Christian König, and Stefan Deßloch, editors, *SIGMOD Conference*, pages 383–394. ACM, 2004.
20. Li Ding, Rong Pan, Timothy W. Finin, Anupam Joshi, Yun Peng, and Pranam Kolari. Finding and ranking knowledge on the semantic web. In Yolanda Gil, Enrico Motta, V. Richard Benjamins, and Mark A. Musen, editors, *Proc. of the International Semantic Web Conference*, volume 3729 of *Lecture Notes in Computer Science*, pages 156–170. Springer, 2005.
21. AnHai Doan, Jayant Madhavan, Pedro Domingos, and Alon Halevy. Learning to map between ontologies on the semantic web. In *Proceedings to the Eleventh International World Wide*, Honolulu, Hawaii, USA, May 2002.
22. P. Domingos, Y. Abe, C. Anderson, A. Doan, D. Fox, A. Halevy, G. Hulten, H. Kautz, T. Lau, L. Liao, J. Madhavan, Mausam, D. Patterson, M. Richardson, S. Sanghai, D. Weld, and S. Wolfman. Research on statistical relational learning at the university of washington. In *Proceedings of the IJCAI-2003 Workshop on Learning Statistical Models from Relational Data*, 2003.
23. Saso Dzeroski. Multi-relational data mining: an introduction. *SIGKDD Explorations*, 5(1):1–16, 2003.
24. Marc Ehrig, Peter Haase, Nenad Stojanovic, and Mark Hefke. Similarity for ontologies – a comprehensive framework. In D. Bartman, F. Rajola, J. Kallinikos, D. Avison, R. Winter, P. Ein-Dor, J. Becker, F. Bodendorf, and C. Weinhardt, editors, *Information Systems in a Rapidly Changing Economy: Proceedings of the 13th European Conference on Information Systems (ECIS 2005), May 26–28, 2005, Regensburg, Germany*. Association for Information Systems, 2005. Published online at `http://is2.lse.ac.uk/asp/aspecis/`.
25. Nicola Fanizzi and Claudia d'Amato. A declarative kernel for ALC concept descriptions. In Floriana Esposito, Zbigniew W. Ras, Donato Malerba, and Giovanni Semeraro, editors, *Foundations of Intelligent Systems, 16th International Symposium*, volume 4203 of *Lecture Notes in Computer Science*, pages 322–331. Springer, 2006.

26. R. Feldman and I. Dagan. Knowledge discovery in textual databases (kdt). In *Proceedings of the First International Conference on Knowledge Discovery (KDD)*, pages 112–117, 1995.
27. Blaz Fortuna, Marko Grobelnik, and Dunja Mladenic. Visualization of text document corpus. *Informatica (Slovenia)*, 29(4):497–504, 2005.
28. P. Frasconi, A. Passerini, S. Muggleton, and H. Lodhi. Declarative kernels. Technical Report RT 2/2004, Dipartimento di Sistemi e Informatica, Universit'a di Firenze, 2004.
29. Ben Taskar Lise Getoor, editor. *Introduction to Statistical Relational Learning (Adaptive Computation and Machine Learning)*. 2007.
30. Lise Getoor and Christopher P. Diehl. Link mining: a survey. *SIGKDD Explor. Newsl.*, 7(2):3–12, 2005.
31. J. Gonzalo, F. Verdejo, I. Chugur, and J. Cigarrán. Indexing with WordNet synsets can improve text retrieval. In *Proceedings ACL/COLING Workshop on Usage of WordNet for Natural Language Processing*, 1998.
32. Stephen J. Green. Building hypertext links by computing semantic similarity. *IEEE Transactions on Knowledge and Data Engineering (TKDE)*, 11(5):713–730, 1999.
33. Thomas Gärtner. A survey of kernels for structured data. *SIGKDD Explorations*, 5(1):49–58, 2003.
34. Thomas Gärtner. *Kernels for Structured Data*. PhD thesis, University of Bonn, Germany, 2005.
35. Thomas Gärtner, John W. Lloyd, and Peter A. Flach. Kernels and distances for structured data. *Machine Learning*, 57(3):205–232, 2004.
36. D. Haussler. Convolution kernels on discrete structures. Technical Report Technical Report UCS-CRL-99-10, UC Santa Cruz, 1999.
37. Mark Hefke, Valentin Zacharias, Andreas Abecker, Qingli Wang, Ernst Biesalski, and Marco Breiter. An extendable java framework for instance similarities in ontologies. In Yannis Manolopoulos, Joaquim Filipe, Panos Constantopoulos, and José Cordeiro, editors, *ICEIS 2006 - Proceedings of the Eighth International Conference on Enterprise Information Systems: Databases and Information Systems Integration, Paphos, Cyprus, May 23–27, 2006*, pages 263–269, 2006.
38. I. Horrocks, P. F. Patel-Schneider, and F. van Harmelen. From SHIQ and RDF to OWL: The Making of a Web Ontology Language. *Journal of Web Semantics*, 1(1), 2003.
39. Bettina Hoser. *Analysis of Asymmetric Communication Patterns in Computer Mediated Communication Environments*. Universittsverlag Karlsruhe, 2005.
40. Bettina Hoser, Andreas Hotho, Robert Jäschke, Christoph Schmitz, and Gerd Stumme. Semantic network analysis of ontologies. In York Sure and John Domingue, editors, *The Semantic Web: Research and Applications: Proceedings of the 3rd European Semantic Web Conference (ESWC 2006), Budva, Montenegro, June 11–14, 2006*, volume 4011/2006 of *Lecture Notes in Computer Science*, pages 514–529, Heidelberg, 2006. Springer.
41. A. Hotho, S. Staab, and G. Stumme. Ontologies improve text document clustering. In *Proceedings of the International Conference on Data Mining – ICDM-2003*. IEEE Press, 2003.
42. Andreas Hotho. *Clustern mit Hintergrundwissen*, volume 286 of *Diski*. Akademische Verlagsgesellschaft Aka GmbH, Berlin, 2004.

43. Andreas Hotho, Andreas Nürnberger, and Gerhard Paaß. A brief survey of text mining. *LDV Forum - GLDV Journal for Computational Linguistics and Language Technology*, 20(1):19–62, MAY 2005.
44. Andreas Hotho, Steffen Staab, and Gerd Stumme. Explaining text clustering results using semantic structures. In *Principles of Data Mining and Knowledge Discovery, 7th European Conference, PKDD 2003, Dubrovnik, Croatia, September 22–26, 2003*, LNCS, pages 217–228. Springer, SEP 2003.
45. N. Ide and J. Véronis. Introduction to the Special Issue on Word Sense Disambiguation: The State of the Art. *Computational Linguistics*, 24(1):1–40, 1998.
46. K. Kersting and L. De Raedt. Bayesian logic programs. *Arxiv preprint cs.AI/0111058*, 2001.
47. Christoph Kiefer, Abraham Bernstein, and Jonas Tappolet. Mining software repositories with isparol and a software evolution ontology. *msr*, 0:10, 2007.
48. Jon M. Kleinberg. Authoritative sources in a hyperlinked environment. *Journal of the ACM*, 46(5):604–632, 1999.
49. Risi Imre Kondor and John D. Lafferty. Diffusion kernels on graphs and other discrete input spaces. In *Proceedings of the Nineteenth International Conference on Machine Learning (ICML 2002)*, pages 315–322, San Francisco, CA, USA, 2002. Morgan Kaufmann Publishers Inc.
50. David M. Pennock Kushal Dave, Steve Lawrence. Mining the peanut gallery: opinion extraction and semantic classification of product reviews. In *Proceedings of the Twelfth International World Wide Web Conference, WWW2003*, pages 519–528. ACM, 2003.
51. Wim Van Laer and Luc De Raedt. How to upgrade propositional learners to first order logic: A case study. In Georgios Paliouras, Vangelis Karkaletsis, and Constantine D. Spyropoulos, editors, *Machine Learning and Its Applications*, volume 2049 of *Lecture Notes in Computer Science*, pages 102–126. Springer, 2001.
52. N. Lavrač and S. Džeroski. *Inductive Logic Programming: Techniques and Applications*. 1994.
53. Jens Lehmann and Pascal Hitzler. Foundations of refinement operators for description logics. In *Proceedings of the 17th International Conference on Inductive Logic Programming (ILP)*, 2007.
54. Jens Lehmann and Pascal Hitzler. A refinement operator based learning algorithm for the $\mathcal{ALC}$ description logic. In *Proceedings of the 17th International Conference on Inductive Logic Programming (ILP)*, 2007.
55. D.D. Lewis. Reuters-21578 text categorization test collection, 1997.
56. Francesca Lisi. A methodology for building semantic web mining systems. *Foundations of Intelligent Systems*, pages 306–311, 2006.
57. Francesca A. Lisi and Floriana Esposito. Mining the semantic web: A logic-based methodology. In Mohand-Said Hacid, Neil V. Murray, Zbigniew W. Ras, and Shusaku ER Tsumoto, editors, *Foundations of Intelligent Systems – Proceedings of the 15th International Symposium, ISMIS 2005, Saratoga Springs, NY, USA, May 25–28, 2005*, Lecture Notes in Computer Science, pages 102–111, Heidelberg, 2005. Springer.
58. J.W. Lloyd. *Logic for Learning: Learning Comprehensible Theories from Structured Data.* Springer, Berlin–Heidelberg, Germany, 2003.
59. Christopher D. Manning and Hinrich Schtze. *Foundations of Statistical Natural Language Processing.* The MIT Press, June 1999.

60. Dimitrios Mavroeidis, George Tsatsaronis, Michalis Vazirgiannis, Martin Theobald, and Gerhard Weikum. Word sense disambiguation for exploiting hierarchical thesauri in text classification. In Alípio Jorge, Luís Torgo, Pavel Brazdil, Rui Camacho, and João Gama, editors, *Knowledge Discovery in Databases: Proceedings of the 9th European Conference on Principles and Practice of Knowledge Discovery in Databases (PKDD 2005), Porto, Portugal, October 3–7, 2005*, pages 181–192. Springer, 2005.
61. G. A. Miller. WordNet: a Lexical Database for English. *Communications of the ACM*, 38(11):39–41, 1995.
62. S.H. Muggleton and L. De Raedt. Inductive logic programming: Theory and methods. *Journal of Logic Programming*, 19,20:629–679, 1994.
63. Stephen Muggleton. Inductive logic programming. *New Generation Computing*, 8(4):295–318, 1991.
64. K.-R. Mller, S. Mika, G. Rätsch, S. Tsuda, and B Schölkopf. An introduction to kernel-based learning algorithms. *IEEE Transactions on Neural Networks*, 12(2):181–202, 2001.
65. L. Page, S. Brin, R. Motwani, and T. Winograd. The pagerank citation ranking: Bringing order to the web. In *Proceedings of the 7th International World Wide Web Conference*, pages 161–172, Brisbane, Australia, 1998.
66. Alexandrin Popescul, Lyle H. Ungar, Steve Lawrence, and David M. Pennock. Statistical relational learning for document mining. In *ICDM*, pages 275–282. IEEE Computer Society, 2003.
67. M. F. Porter. An algorithm for suffix stripping. *Program*, 14(3):130–137, 1980.
68. Luc De Raedt and Andrea Passerini. Kernels on prolog proof trees: Statistical learning in the ILP setting. *Journal of Machine Learning Research*, 7(Feb): 307–342, 2006.
69. Cartic Ramakrishnan, William H. Milnor, Matthew Perry, and Amit P. Sheth. Discovering informative connection subgraphs in multi-relational graphs. *SIGKDD Explor. Newsl.*, 7(2):56–63, 2005.
70. Matthew Richardson and Pedro Domingos. Building large knowledge bases by mass collaboration. In *K-CAP '03: Proceedings of the 2nd international conference on Knowledge capture*, pages 129–137, New York, NY, USA, 2003. ACM Press.
71. F. Rosenblatt. The perceptron: A probabilistic model for information storage and organization in the brain. *Psychological Review*, 65:386–408, 1958.
72. G. Salton. *Automatic Text Processing: The Transformation, Analysis and Retrieval of Information by Computer.* Addison-Wesley, 1989.
73. G. Salton and M.E. Lesk. Computer evaluation of indexing and text processing. In G. Salton, editor, *The SMART Retrieval System: Experiments in Automatic*, pages 143–180. Prentice-Hall, 1971.
74. S. Scott and S. Matwin. Text Classification Using WordNet Hypernyms. In *Proceedings of the COLING/ACL Workshop on Usage of WordNet in Natural Language Processing Systems.* Association for Computational Linguistics, Montral, Canada, 1998.
75. F. Sebastiani. Machine learning in automated text categorization. *ACM Computing Surveys*, 34(1):1–47, 2002.
76. John Shawe-Taylor and Nello Cristianini. *Kernel Methods for Pattern Analysis.* Cambridge University Press, Cambridge, UK, June 2004.

77. Georges Siolas and Florence d'Alche Buc. Support vector machines based on a semantic kernel for text categorization. In *IEEE-INNS-ENNS International Joint Conference on Neural Networks (IJCNN)*, volume 5, pages 205–209, 2000.
78. M. Steinbach, G. Karypis, and V. Kumar. A comparison of document clustering techniques. In *KDD Workshop on Text Mining*, 2000.
79. Gerd Stumme, Andreas Hotho, and Bettina Berendt. Semantic web mining - state of the art and future directions. *Journal of Web Semantics*, 4(2):124–143, 2006.
80. York Sure, Stephan Bloehdorn, Peter Haase, Jens Hartmann, and Daniel Oberle. The SWRC ontology - Semantic Web for research communities. In Carlos Bento, Amilcar Cardoso, and Gael Dias, editors, *Progress in Artificial Intelligence – Proceedings of the 12th Portuguese Conference on Artificial Intelligence (EPIA 2005), December 5–8, 2005, Covilh, Portugal*, volume 3803 of *Lecture Notes in Computer Science*, pages 218–231. Springer, Berlin–Heidelberg, Germany, DEC 2005.
81. Octavian Udrea, Lise Getoor, and Renee Miller. Leveraging data and structure in ontology integration. In *Proceedings of ACM-SIGMOD 2007 International Conference on Management*, 2007.
82. L. A. Ureña Lóez, M. de Buenaga Rodríguez, and J. M. Gómez Hidalgo. Integrating linguistic resources in tc through wsd. *Computers and the Humanities*, 35(2):215–230, 2001.
83. E. M. Voorhees. Query expansion using lexical-semantic relations. In *Proceedings of the 17th annual international ACM SIGIR conference on Research and development in information retrieval*, pages 61–69, 1994.
84. B. B. Wang, R. I. Mckay, H. A. Abbass, and M. Barlow. A comparative study for domain ontology guided feature extraction. In *Proceedings of the 26th Australian Computer Science Conference (ACSC-2003)*, pages 69–78, Adelaide, Australia, 2003. Australian Computer Society, Inc.
85. Y.-C. Wang, J. Vandendorpe, and M. Evens. Relational thesauri in information retrieval. *Journal of the American Society for Information Science*, 36(1):15–27, 1985.
86. James A. Wise. The ecological approach to text visualization. *Journal of the American Society for Information Science*, 50(13):1224–1233, 1999.

Information Extraction

Claire Nédellec[1], Adeline Nazarenko[2], and Robert Bossy[1]

[1] INRA, Name.Lastname@jouy.inra.fr
[2] Université Paris-Nord, Name.Lastname@lipn.univ-paris13.fr

Summary. Information Extraction (IE) addresses the intelligent access to document contents by automatically extracting information relevant to a given task. This chapter focuses on how ontologies can be exploited to interpret the textual document content for IE purposes. It makes a state of the art of IE systems from the point of view of IE as a knowledge-based NLP process. It reviews the different steps of NLP necessary for IE tasks: named entity recognition, term analysis, semantic typing and identification specific relations. It stresses on the importance of ontological knowledge for performing each step and presents corpus-based methods for the acquisition of the required knowledge.

This chapter shows that IE is an ontology-based activity and argues that future effort in IE should focus on formalizing and reinforcing the relation between the text extraction and the ontology model. The discussion gives authors' insights on the integration of ontological knowledge in IE systems from a formal and pragmatic point of view.

Examples in this chapter are taken from IE tasks for biology since this domain attracts a large community of IE specialists and provides a large number of ontological resources.

1 Introduction

As the volume of textual information is exponentially increasing, it is more than ever a key issue for knowledge management to develop intelligent tools and methods to give access to document content and extract relevant information. Information Extraction (IE) is one of the main research fields that attempt to fulfill this need. It aims at automatically extracting well-defined and domain specific data from free or semi-structured textual documents. The extraction of instances of appointments from on-line news is a typical example. IE interprets "Yesterday, Mr. Smith as been appointed as Chief Executive Officer of AAACompany Inc." into the knowledge structure: *Appointment (Smith, AAACompany, CEO, Yesterday)* where the arguments respectively play the role of person, company, title and date of the appointment. Once

S. Staab and R. Studer (eds.), *Handbook on Ontologies,* International Handbooks on Information Systems, DOI 10.1007/978-3-540-92673-3,

formalised in such a way, the content of the document may support formal calculus or logical inference as needed by knowledge management applications.

The relation between IE and ontologies can be considered in two non-independent manners. As IE can extract ontological information from documents, it is exploited by ontology learning and population methods for enriching ontologies. This issue is specifically discussed in chapter "Ontology Learning". Conversely, this chapter focuses on how ontologies can be exploited to interpret the textual document content for IE purposes. We will show here that IE is an ontology-based activity and we will argue that future effort in IE should focus on formalising and reinforcing the relation between the text extraction and the ontology model.

Examples from the biology domain will illustrate the presentation of IE concepts. Biology is a relevant application domain because of the importance of text-mining for the biology community, the availability of structured resources such as document collections and nomenclatures, the clear expression of application requirements and finally, the amount of evaluation material (e.g. Genia [44], BioCreative [24, 45], LLL [34], TREC [21]). This paper first introduces Information Extraction (Sect. 2), then an example of a knowledge-based IE system is presented in Sect. 3. On the basis of that example, we assert the fact that IE is an ontology-based process. This statement is developed in the following sections that detail the role of the various knowledge resources in IE (Sects. 4–7). The last section (Sect. 8) discusses open and challenging issues for IE.

2 What Is IE?

2.1 Definition

The IE field was initiated by the DARPA MUC program (Message Understanding Conference) in 1987 [16]. MUC has originally defined IE as the task of (1) extracting specific, well-defined types of information from the text of homogeneous sets of documents in restricted domains and (2) filling pre-defined form slots or templates with the extracted information.

A typical IE task is illustrated by Fig. 1 in functional genomics, a sub-field of biology. IE process recognises two names, *GerE* and *cotA*, as protein and gene names respectively. It also recognises a genic interaction relation between them and fills the form accordingly.

In the simplest case, extracted textual fragments fill the form slots and no more text pre-processing is required. However IE cannot be reduced to

```
Sentence: ''GerE stimulates the expression of cotA.''
Genic interaction form
   Agent: Protein(GerE)
   Target: Gene(cotA)
```

Fig. 1. IE example

simple keyword filtering. Any fragment must be interpreted with respect to its context and its expected role in the form. In the example above, *GerE* must be understood as a *genic interaction protein agent* and background knowledge about molecular biology is necessary to carry out this interpretation.

IE systems were initially designed as shallow text-understanding systems , which relied on targeted and local techniques of text exploration rather than in-depth semantic analysis of the text. Then the limitations of the first IE systems called for new approaches more deeply and more formally relying on text analysis and ontological knowledge.

2.2 IE Overall Process

Operationally, IE relies on document pre-processing and extraction rules (typically regular expressions or patterns) to identify and interpret the target text. The extraction rules specify the conditions the preprocessed text must match and how the relevant textual fragments can be interpreted to fill the forms. Figure 2 gives an example of a rule that can extract the genic interaction information of Fig. 1.

The rule assumes that gene and protein names occurring on both sides of an *interaction_verb* denote a genic interaction between the corresponding protein and gene.

A typical IE system includes three processing steps [22]:

1. *Text analysis:* From text segmentation into sentences and words in the simplest case to full linguistic analysis. In the example from Figs. 1 and 2, the linguistic analysis should segment the text into words, identify the gene and protein names as well as the interaction verb and derive the successor relation from the word order. Section 2.3 details these Natural Language Processing (NLP) steps.
2. *Rule selection:* Information extraction rules are associated with triggers, usually keywords. The presence of trigger items activates the checking of the conditional parts of the corresponding rules. For instance, the rule of Fig. 2 could be triggered by the occurrence of gene and protein names.
3. *Rule application:* Once a rule has been triggered, all contextual conditions of the rules are checked and the form is filled according to the conclusions of the matching rules. The result may be a filled form as in Fig. 1 or an annotated text.

The rules are usually declarative but they may be expressed in different ways. The rule example of Fig. 2 is represented in a first-order logic formalism. The simplest rules extract simple slot values (i.e. dates, person names) whereas

```
genic_interaction(X,Y) :-
   protein_name(X), gene_name(Y), interaction_verb(Z),
   successor(Z,X), successor(Z,Y)
```

Fig. 2. Extraction rule example

more complex ones extract several related values at the same time. This is referred to as multi-slot extraction, which requires a relational formalism [14].

Forms of increasing complexity are taken into consideration [46]:

- Entity form filling requires to identify items in the text that represent domain referential entities (e.g. protein and gene names).
- Domain event form filling requires to extract events that represent actual relations between entities (e.g. the agent role of a protein in a genic interaction).
- Merging forms issued from different parts of the text that provide information about a same entity or event.
- Scenario forms relate several event and entity forms that, considered together, describe a temporal or logical sequence of actions and events.

2.3 Text Processing

From the very beginning, the main issue in IE appeared to be the design of efficient extraction rules able to extract all relevant information pieces and only the relevant ones. The difficulty comes from the intrinsic richness and complexity of natural language where a given word or phrase may have different meanings (polysemy) and several formulations may express the same information (paraphrases). If the rules rely on surface clues (i.e. the presence of a given specific lexical item, the word distance or order), a whole set of very specific rules must be designed for each new IE application.

If the text is pre-analysed, information extraction rules can be expressed in a more abstract and powerful way. In that case, the rules apply on the result of the pre-analysis, which is a normalised representation of the text. For instance, the successor relations of the rule of Fig. 2 are replaced in the rule of Fig. 3 by the subject and object syntactic dependencies. The rule is then more general and easier to interpret in terms of domain knowledge. Syntactic dependencies are independent of the word order and reflect the agent and target semantic roles more accurately. The same rule applies to sentences in passive voice such as *CotA is activated by GerE.* Usual linguistic analysis steps include morphology, syntactic and semantic analysis. The *morphology analysis* focuses on the form of textual units, usually referred as words. It includes the *segmentation* of the character stream into a sequence of words based on character separators (e.g. spaces, punctuation signs). In specific cases one must rely on linguistic hints: some poly-lexical units such as "*Bacillus subtilis*" in

```
genic_interaction(X,Y) :-
   protein_name(X), gene_name(Y),
   interaction_verb(Z)
   subject(Z,X), object(Z,Y)
```

Fig. 3. Abstract extraction rule example

biology can be considered as single words, while a single token can be viewed as the contraction of several words. The *lemmatisation* associates a normalised form (the lemma) to each word (infinitive form for verbs, singular form for nouns and pronouns) by removing marks that bear flexional features. The *morphological tagging* associates morphological features (tense, number, gender, presence of non-alphabetical characters and case) to words. The *syntactic analysis* performs two dependent tasks. The *part-of-speech* (POS) tagging assigns a syntactic category to words (e.g. noun, verb, adverb).

The *parsing* identifies the sentence structure by grouping words into phrases. Depending on the parser, syntactic dependencies between words or phrases (e.g. subject-verb dependency) may also be computed.

The *semantic analysis* builds a formal representation of the text meaning. In IE, the semantic analysis is traditionally restricted to (1) the identification of the *semantic textual units* (named entities and terms) that refer to the relevant domain objects, (2) the *semantic typing* that associates concepts to those semantic units, and (3) the tagging of *domain specific relations* between them.

The text analysis process relies on linguistic and domain knowledge. The most traditional *lexical* resource is the *named entity dictionary*, a nomenclature of the names of domain entities, such as genes and proteins in biology, but other resources can also be exploited. We will show in the following that IE is an ontology-driven approach to text analysis that heavily relies on lexical and ontological resources.

2.4 IE as a Text-Ontology Mapping

The overall process of IE aims at mapping text to ontology. IE selects and interprets relevant pieces of the input text in terms of form slot values. The form slot values are derived from the semantic analysis while the form itself represents an ontological knowledge structure.

This mapping can be formalised into the annotation of the text by the ontology, as shown in Fig. 4. The text fragments are tagged with ontological concepts and relations according to the IE goals: the ontological concepts *protein*, *negative interaction* and *gene* are linked to the semantic units *GerE*, *inhibits* and *sigK* respectively. The ontological relations *agent(protein, interaction)* and *target(gene, interaction)* are instantiated by *agent(GerE, inhibits)* and *target(sigK, inhibits)*.

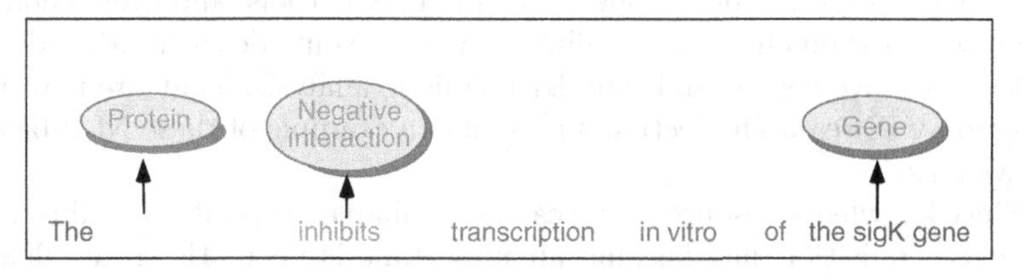

Fig. 4. Example of semantic annotation

Most approaches rely on the assumption that semantic units are denoted by noun phrases, that relations are denoted by predicates (verbs or verb nominalisations) and that predicate arguments can be identified through neighbourhood relations or syntactic dependency paths. However in the general case, linking a text to an ontology is not as straightforward as these methods assume [8,35]. The text and the conceptual model are fundamentally different and cannot be directly mapped to each other and that calls for intermediate levels of knowledge.

The *lexical knowledge* plays the role of *mediator*. Various types of lexical resources can be exploited in IE, from the named entity dictionaries to the domain terminologies or ontological thesauri. The lexical mediation between the text and the ontology is complex to formalise. First, there is no one-to-one relation neither between text fragments and lexicon entries nor between those entries and ontological entities due to linguistic phenomena, like variation, polysemy and ellipsis. We will show in the following sections that the lexica are associated to sets of rules that govern the recognition or disambiguation of the lexicon items. They contribute to building links between the text and the lexicon on the one hand, and between the lexicon and the ontology on the other hand. The various types of knowledge are traditionally considered as distinct resources although they partially overlap, maintaining the coherence between them remains an open question.

The importance of lexical resources has raised the problem of their acquisition. Indeed the development of applications in specific domains generally requires the adaptation of available knowledge resources needed for the various linguistic processing. Thus the issues of the re-usability, the acquisition and the formalisation of knowledge become central.

2.5 IE State of the Art

In the 1990s, IE quickly became operational for extracting simple information pieces from short and homogeneous documents such as conference announcements. But extracting relational information (e.g. gene–protein interaction) from free texts (e.g. abstracts of scientific papers) remained challenging.

The IE field then evolved since the beginning of the 2000s toward semantic processing, knowledge acquisition and ontologies. This has led to the development of a new generation of IE systems (e.g. [3,41]).

Two major phenomena have made this progress possible:

- An increasing number of operational linguistic tools, and even whole integrated NLP pipelines, are available for people outside the NLP field. These tools achieve deeper and sounder linguistic analysis. They are now widely used by IE research. Section 3 presents an example of these NLP-based IE systems.
- Since knowledge resources are scarcely available in specific domains, knowledge acquisition has become an important issue in IE since 1998 [30].

Corpus-based Machine Learning (ML) was soon recognised by the IE field as a relevant alternative to costly manual knowledge acquisition and adaptation in particular for the acquisition of information extraction rules and ontologies [13]. In fact, various kinds of knowledge (named entities, terms, types, semantic relations and properties) can be acquired with specifically designed learning methods and training corpora (see Sects. 4–7).

Current IE systems therefore evolved to sophisticated platforms that combine various NLP and ML steps, [4, 17]. For example, tools are available to extract named entities, even in domains where many unknown variant forms are frequent like in biology. Extraction of relational information has also become operational [34, 39]. However, these systems have often been specifically designed for a given application. The NLP and ML processes and the underlying data model used to integrate them are chosen on an ad hoc basis, which hinders the genericity and the re-usability of the systems. An open challenge is the design of a formalised and integrated approach of IE where the whole process is properly decomposed into elementary tasks and where the role of the various knowledge resources is made explicit. We argue that a precise decomposition of processes and resources is necessary to achieve a generic IE architecture that would be reusable and tunable for many IE tasks.

3 IE as a Knowledge-Based NLP Process

This section describes IE as a knowledge-based NLP process in more detail by outlining a generic IE architecture.

3.1 Architecture of a Linguistic-Enabled IE System

We illustrate the role of text and ontology processing in IE by the Alvis semantic analysis pipeline.[1] Alvis provides a software framework to develop domain specific distributed semantic search engines. The semantic analysis is based on the NLP platform Ogmios that generates ontological-based representations of textual documents [32]. It is suitable for developing various textual-based applications, including IE as well as IR, QA and more generally any application relying on semantic annotation of documents. Comparable architectures have been proposed to manage text processing over the last decade: GATE [7], KIM [36] or UIMA[15], to mention only generic ones. As other platforms, Ogmios is configurable and designed to integrate various existing NLP tools in an operational way. Considerable attention has been paid to scalability and efficiency, so that Alvis is able to process large and heterogeneous collections of documents [19].

Figure 5 outlines Alvis pipeline architecture. The different NLP steps are operated by distinct modules denoted by the boxes on the top layer, each one

[1] `http://cosco.hiit.fi/search/alvis.html`

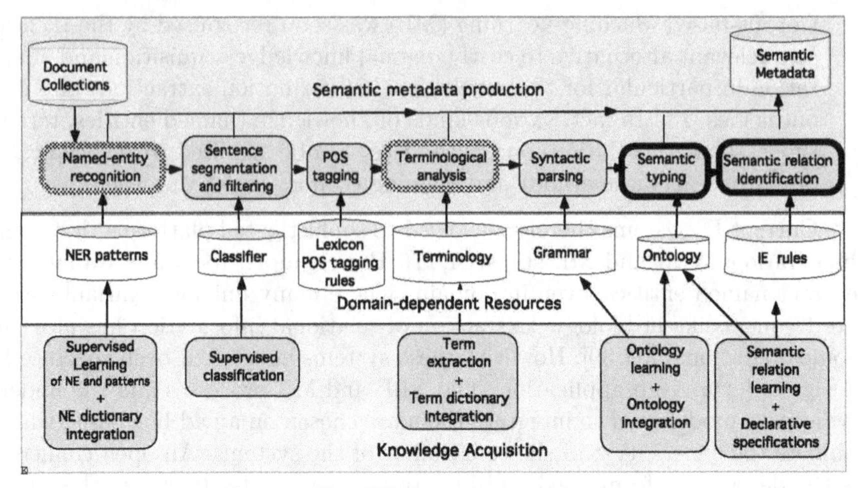

Fig. 5. Alvis NLP pipeline architecture

carrying out a specific process. Each module relies on the information produced by previous components and produces information that contributes to the interpretation of the document. The information is represented as annotations recorded in an XML stand-off format [31].

3.2 Semantic-Based Text Analysis

The linguistic steps that were presented in Sect. 2.3 are implemented in Alvis NLP pipeline. We illustrate in Table 1 the linguistic data produced by the linguistic analysis of the example of Fig. 4 represented in a logic-based language.

Relevant *named entities* or *terms* are first identified as *semantic units* by the dotted line-framed components of Fig. 5. Pre-processing steps of segmentation into documents, paragraphs, sentences and words, morphological analysis and syntactic category tagging are required for semantic unit recognition. Once the semantic units are identified, they are typed with fine-grained concepts and associated by domain-specific relations (bold line-framed boxes) from the ontology. This latter task requires syntactic dependency or neighbourhood relation analysis.

This process is knowledge intensive. The components use linguistic resources as figured by the middle layer boxes. They are typically domain-dependent and application-driven. The clear distinction between the process and the knowledge bases (KB) reduces the adaptation to new domains to the revision of the following knowledge bases: named entity dictionaries, terminologies, ontologies and IE rules. Two specialised versions of the Ogmios platform have been deployed to develop IR applications for scientific papers

Table 1. Example of linguistic analysis result

```
Words:
   word(the), word(GerE), word(protein),word(inhibit), ...

Named entities:
   entity(GerE), entity(sigK), entity(sigma K)

Syntactic categories:
   cat(the, determinant), cat(GerE, noun), cat(inhibit, verb), ...

Terminology:
   term(GerE protein), term(in vitro) term(sigK gene)

Syntactic dependencies:
   subject(GerE, inhibit), object(transcription, inhibit), ...
```

in biology[2] and patents in agro-biotechnologies.[3] Their development did not require any adaptation of the pipeline components themselves except for syntactic parsing.[4]

3.3 Coupling Semantic Annotation and Knowledge Acquisition

Acquisition methods are closely integrated into the Alvis pipeline as figured by the bottom layer of Fig. 5.

The Alvis pipeline is self-feeding since the training data needed for the acquisition of the resource of a given component is derived from the documents enriched with annotations of preceding components. For instance, the acquisition of a terminology requires a training corpus of segmented POS-tagged text that is achieved by the three first components of the pipeline. The pipeline is therefore exploited in two different modes.

In *production mode* the pipeline is applied to a corpus in order to feed an external application, typically IE or IR with annotated documents. The components and the KB are stable and their choice is driven toward this application. In that mode, the pipeline is a critical element of an external service and usually processes massive amounts of data, so it must be reliable, stable, fast and scalable.

In *acquisition mode*, relevant components of the pipeline are applied to a corpus in order to build training examples for an ML algorithm that aims at acquiring the KB of other components. As the amount of documents processed in acquisition mode is typically smaller than in production mode scalability and speed performance are less vital. However, the flexibility and the

---

[2] `http://search.cpan.org/~thhamon/Alvis-NLPPlatform-0.3/`

[3] `https://www.epipagri.org/index.cgi?rm=mode\_sr`

[4] Link Grammar Parser has been tuned to parse biological texts [38].

modularity of the pipeline are critical, since KB acquisition requires several experiments and fine tuning of the intermediate representations of the training corpus.

The following sections explain the role of the various types of knowledge in IE by detailing how they are exploited in the text annotation process of the Alvis IE system and how they are acquired.

4 Handling Named Entities

In its usual meaning, the term *named entity* (NE) designates proper nouns but it is also often used for other types of invariant terms (e.g. dates and chemical formulae). The proper names are *rigid designators* that designate a referential entity in an unambiguous way [25]. More generally, the named entities are linguistic expressions that denote ontological objects in documents. They are important to identify because they act as referential anchors from an informational point of view. Their role in IE and more generally in text understanding is widely acknowledged [35]. For instance, it is easier to guess what a document is about if one knows that it mentions *Hiroshima* and *1945*, which are both named entities. NE are also exploited as extraction rule triggers as they are often quite easy to identify.

Named entity recognition (NER) identifies the named entities in texts and associates a canonical form and a semantic category to them. A canonical form is a unique representative of a set of forms denoting the same entity. The semantic type is a rough ontological knowledge about the NE. NER relies on named entity dictionaries often tuned for a specific domain and a specific type of documents.

In the general language, variant and ambiguous NE frequently occur and even more in sub-languages of technical and scientific domains. For instance, *Paris* is ambiguous since it alternatively refers to entities belonging to different semantic categories: either a person or a place. The gene name *cat* may also refer either to a protein or to the mammalian. Many different name variation types can be observed: acronyms (*chloramphenicol acetyltransferase / CAT*), abbreviations (*Bacillus subtilis / B. subtilis*), ellipses (*EPO mimetic peptide / EPO*), typographic variations (*sigma K / sigma(K) / sigma-K*), synonymy due to renaming (*SpoIIIG / sigma G*). Each type of variation is handled differently [37,43].

4.1 Named Entity Recognition

Because of the ambiguities and variations, named entity tagging cannot be achieved by simple dictionary matching; it also involves matching of the context of the candidate NE by NER rules. On one hand, disambiguation rules specify in which context a given NE candidate should be interpreted as belonging to a given category. On the other hand, variation rules enrich the dictionary with lists of synonyms or are applied on the fly to recognise variant forms.

Active domains constantly produce documents containing new concepts and new NE, thus dictionaries are quite hard to keep fully up to date. Additional recognition rules are able to palliate the NE dictionaries incompleteness. These rules exploit the morphology of the candidate NE and various contextual clues in documents.

The respective roles of the dictionary and the rules are illustrated by one experiment we did in biology [33] where the NER performance increased from 75% of recall and precision with simple dictionary mapping to 93% with disambiguation and recognition rules.

In Alvis, NE tagging is a two-step process. The first one only involves matching dictionary entries on the text to identify known NE. This tagging is used afterwards for word and sentence segmentation to avoid the interpretation of abbreviation dots as sentence separators (e.g. *Bacillus sp. BT1*). The second NER step is achieved after documents have been segmented and lemmatised and word morphology have been analysed. The conditions of the NER rules are checked for each candidate phrase, then NE are disambiguated and associated with their semantic type. Dictionary-based annotations are removed when they correspond to ambiguous NE and new NE annotation are added.

4.2 Corpus-Based Acquisition of Named Entities

Supervised ML methods can be applied to automatically acquire disambiguation and recognition rules in order to improve existing NE dictionaries. The reference training corpus is pre-annotated using existing NE dictionaries, then manually reviewed by human experts. Negative examples are automatically generated under the closed world assumption. The main features computed to describe the training examples of NE are usually typographic (length, case, presence of symbols and digits). Non-typographic features are based on neighbourhood words. In Alvis this acquisition is performed in two steps: feature selection, then induction of a decision tree by C4.5 from the Weka library [33]. Variation rules acquisition is done in a similar way: annotation of synonym pairs in a training corpus and application of supervised learning.

The Alvis experiment on biology has shown that the quality of the manual annotation is critical for NER rule learnability. NE and non-NE should be clearly distinguished in the training corpus and the frontier should be strictly defined [10] in annotation guidelines in order to achieve a high annotation quality. The fuzzy frontier between entities denoted by proper nouns and by terms is an important source of errors: some technical terms are so detailed that they definitely designate an entity in the context of the document although their morphology is not a proper noun one (see for instance the term *spore coat protein A* which is synonym of the NE *CotA*). The recognition of these terms as entities raises the question of how detailed must be a term to be considered as a named entity, which is often not easy to answer. A second related problem is the undetermined generality level of the objects to be recognised (instances vs. concepts). A proper noun may define a family of instances or a general concept and not a single instance (e.g. *ABC transporters*). The well-known problem of name boundaries is a third source of errors. NE often occur with their synonyms in an apposition or adjective role, as in "monoclonal antibodies (mAb)", or with roles and properties, like in "mouse synaptophysin gene". It is important to distinguish NE from their roles, properties and alternate names because it facilitates manual annotation and considerably increases machine learning performance as demonstrated by our experiments [33].

5 Term Analysis

Less widely acknowledged than NE recognition, term recognition is nevertheless crucial for further linguistic and semantic processing because terms often denote ontological concepts. Terminological analysis is a traditional step in sub-language analysis. It helps to identify the most relevant semantic units and it reduces the wording diversity.

5.1 Term Tagging

In Alvis, term tagging consists in the projection of the terminology on the corpus. The text fragments that correspond to a given term are tagged with a canonical form, but no semantic category. Only flexion variations are possible at that stage (e.g. plural transformation). The simpler the tagging process is, the richer the resource must be. This calls for powerful acquisition methods.

5.2 The Role of Terminologies

A terminology is a knowledge source that describes the *specific* vocabulary of a given domain. It is composed of a list of terms, single or multi-word lexical units. For instance, the well-known medical terminological resource MeSH thesaurus (Medical Subject Headings)[5] contains the terms: *amino acid*, *protein* and *DNA-binding protein*. Simple term lists are not sufficient for most terminological applications because term surface forms may vary incredibly following morpho-syntactic rules. For instance, two terms are morpho-syntactic variants of each other if one is an inflected or derived form of the other (*Aortic stenosis / Stenosis of the aorta*) or if one can be altered to the other, via a regular set of transformation rules, such as permutation (*Aortic Subvalvular Stenosis / Subvalvular Aortic Stenosis*) [12]. Morpho-syntactic variations

[5] `http://www.nlm.nih.gov/mesh/`

apart flexions are pre-computed in the terminologies, that explicitly list term variants. The term variation problem is not handled in the same manner as for named entities because the morpho-syntactic variation rules are not reliable enough for recognising new terms on the fly as they can be for NE.

For a given domain, there are as many terminologies as application goals. Although specific, terms of scientific and technical terminologies hardly ever match the actual document text. For instance we observed that MeSH and Gene Ontology (GO) lexicons, although useful and widely recognised biomedical resources, have a very poor coverage on our corpus of 16,000 sentences of PubMed paper abstracts: less than 1% of the GO and MeSH terms. The reason is that existing terminologies have often been designed for other purposes than automatic text analysis and do not reflect writing usages in corpora.

5.3 Term Acquisition

Terminological knowledge acquisition tools have been proposed since the 1990s [9]. Term identification methods generally exploit linguistic information like chunk boundaries (e.g. punctuation), morpho-syntactic patterns (*noun noun* or *adjective noun*) and more often statistic criteria to filter incidental terms. *YaTeA*, the Alvis term extractor, performs the acquisition of term candidates from corpora on the basis of POS tags and endogenous disambiguation [6].

The results of term extractors remain noisy, however. Expert knowledge is necessary to filter out irrelevant terms and to validate the most relevant ones. For instance, YaTeA extracts the two terms "heterologous polypeptide" and "suitable polypeptide" from biological documents. Both terms match the same morpho-syntactic pattern *adj noun*, but the second one must be filtered out by manual validation, because the adjective "suitable" does not convey additional relevant information to "polypeptide".

The automatic acquisition of term variants greatly increases existing terminological resource coverage on the corpora. In the same way as for candidate terms, candidate variants must be validated. The term list is then organised into synonym classes of term variants (similar to WordNet synsets) and the most representative of them is chosen as the canonical form. We have integrated a separate term variation computing tool, FASTR [23] into Alvis. In our biology experiments, FASTR increases the terminology size from 7,000 to 10,000 valid terms, gathered into 5,272 classes.

6 Semantic Typing with Conceptual Hierarchies

Once the semantic units (named entities and terms) have been identified, they must be related to the concepts of an ontology by semantic tagging and the concepts play the role of semantic types. Compared to NE broad typing, finer-grained ontological categories are considered in this task. In the case where concepts are organised into generality hierarchies, semantic tagging selects the generality level relevant to a given application. The tagging should both highlight contrasts among critical objects (e.g. protein and genes in genomics) and attenuate or remove unessential differences (e.g. rhetoric or stylistic considerations in scientific documents, *the result indicates* or *the result shows*).

6.1 The Lexicon-Ontology Mapping

In the simplest case, ontology concept labels can be mapped to the text semantic units (e.g. *protein* as a concept maps to *protein* as a word) or through a one-to-one relation with a term or named entity lexicon entry. However, this process is not straightforward because some semantic units are ambiguous and can be assigned different ontological types. A typical example is *star* that denotes both an *astronomical object* and a *famous actor*. Contextual disambiguation rules associate the concepts of the ontology to the relevant lexical knowledge, in a similar way to NE type disambiguation rules. Various strategies involve various degrees of linguistic analysis and ontological inference in order to build the relevant context.

Available ontologies are scarcely used for automatic text analysis. They are usually designed for domain modelling and inference, without the task of text analysis in mind so they are hardly usable for that purpose. In the best case, ontologies are used for manual text indexing such as MeSH indexing of MedLine abstracts or gene annotation by Gene Ontology entries [5]. For instance, the GO label, "negative regulation of translation in response to oxidative stress" is very explicit, specific and useful for manual annotation of biological processes. However, it never occurs literally in scientific documents where the expressions which mean the same, "important antioxidant involved in the stress response" and "negative role for these stress response factors in this translational control" are preferred. This observation does not stand only for biology, but more generally for technical and scientific domains that tend to use complex vocabulary.

6.2 Semantic Type Disambiguation

Disambiguation rules mainly rely on two types of contextual information: sets of neighbour words or syntactic dependencies. In the first case, each alternative meaning of an ambiguous term is attached to a set of usual neighbour words. An occurrence of this term is then interpreted according to the closest set. For instance, for disambiguating the word *tiger* in Flickr legends of photos as being a name of *mammal* or of a *Mac OS version*, word sets such as (*mac*, *apple*, *OSX*, *computer / animal*, *cat*, *zoo*, *Sumatra*) can be mapped to its context [27]. This strategy fails when fine-grained tagging is required or when the alternative meanings are too close.

Finer disambiguation is achieved by taking into account the syntactic relations that connect the ambiguous term to its context. In the Alvis pipeline, disambiguation rules take advantage of the results of a syntactic dependency parsing that must match constraints defined along with ontology nodes.

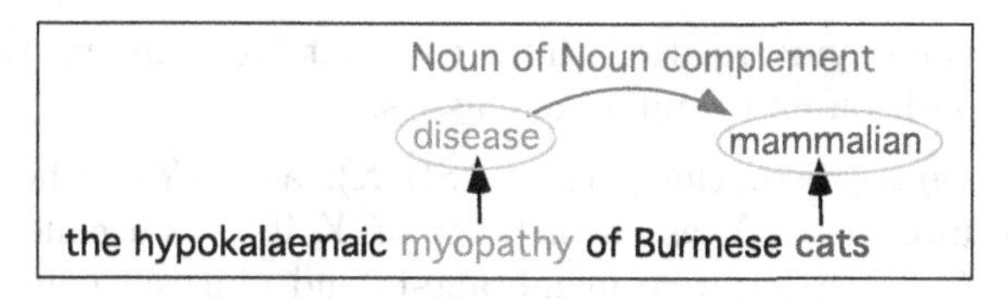

Fig. 6. Word sense disambiguation of CAT

For instance, the word *cat* has many different meanings in biology, among which, a mammalian species or a gene name, though both senses are not found in the same contexts. Given the relevant ontology, in the sentence of Fig. 6, *myopathy* would be first semantically annotated as *disease*. Then a *Noun of Noun complement* dependency is computed between *myopathy* and *cat*. *cat* can be correctly assigned to *mammalian* by verifying the constraint Noun_of_Noun(diseases, mammalian) that states that mammalians can have diseases while this constraint does not apply to genes.

This strategy greatly improves the quality of the disambiguation compared to simple neighbourhood-based strategies. However syntactic parsing is hardly applicable to very large datasets for computational performance reasons. It is appropriate for rather small specific collections.

6.3 Acquisition of Conceptual Hierarchies

Corpus-based learning methods assist the acquisition of ontological hierarchies and disambiguation rules. Two main classes of acquisition methods can be applied: *distributional semantics* and *lexico-syntactic patterns* (see chapter "Ontology Learning").

Distributional semantics identifies sets of terms frequently occurring in similar contexts in the training corpora. The definitions of context are the same as used in disambiguation: either word windows or syntactic dependencies. Various distance metrics have been proposed, all of which are based on co-occurrence frequency measures. Sets of close terms are supposed to be semantic classes and the generality relation is derived from set inclusions. The learning result must be manually validated; it happens that the distance does not denote a semantic proximity but a weaker relation. Linguistic phenomena like metonymy and ellipsis are typical sources of erroneous classes. Distributional semantics is considered robust on large corpora such as Web collections, but machine learning is more efficient when applied to homogeneous corpora with a limited vocabulary, reduced polysemy and limited syntactic variability. In the case of heterogeneous corpora, syntactic context is preferred over neighbourhood because the generated classes are of higher quality [18]. Indeed syntactic dependencies constitute a more homogeneous feature set and the shared syntactic contexts of a resulting class can be easily converted into semantic annotation disambiguation rules.

Research on lexico-syntactic patterns is largely inspired by traditional terminological methods [42], popularised by Hearst's work on pattern design for identifying hyperonymy relations from free text [20]. Among the many patterns, the apposition and copula are classics:

- *Indefinite apposition:* the pattern "SU(X), a SU(Y)" where *SU* means semantic unit, gives X as an instance of Y, if Y is a concept. From the sentence "csbB, a putative membrane-bound glucosyl transferase", *csbB* is interpreted as an instance of *transferase* because *csbB* is a named entity and *transferase* is defined as a concept.
- *Copula construction:* "SU(X) be one of SU(Y)" or "SU(X), e.g. SU(Y)". The fact that the NE *abrB* is an instance of *gene* concept is extracted from "to repress certain genes, e.g. abrB".

The quality of the relations depends on the patterns. Pattern matches may be rare and precise (e.g. X *also known as* Y) as well as frequent and weak (e.g. X *is a* Y may denote a property of X instead of a specialisation relation between X and Y). Dedicated corpora such as textbooks, dictionaries or on-line encyclopedia are more productive although smaller than large Web document sets. The patterns may be automatically learnt from training examples annotated by hand or by bootstrapping learning from known pairs [1].

Pattern-based approaches are less productive than distributional semantics approaches, because of the low number of patterns matches in the corpus while distributional semantics potentially relate all significant words and phrases of the domain. However, the type of the relation is better specified and easier to interpret.

Corpus-based learning is an efficient and operational way to assist the acquisition of lexicon-based ontologies. The synonymy and hyperonymy links extracted from text represent important lexical knowledge. But their modelling into the ontology strongly requires human interpretation and validation. One has to decide what should be considered as a property or a class (e.g. is *four wheels vehicle* a property or a type of *vehicle*?). The distinction between instances and abstract concepts cannot be automated. The independent knowledge bits must be properly integrated (e.g. if X *is_a* Y and X *is_a* Z, what can be said about the relationship between X and Y?). Moreover the methods rely on the assumption that the learnt concepts are represented by explicit semantic units and that the formulation variations can be handled at the lexicon level. They are not applicable if the link between the text and the ontology is more complex and requires the application of inference rules at the ontology level.

7 Identification of Ontological Specific Relations

Information extraction of events consists in identifying domain specific ontological relations in documents between instances of concepts represented by semantic units. The domain specific relations are defined in the ontology and reflected by the IE template slots (see for instance, the slot *Interaction_Agent* in Fig. 1 and the *Agent* relation in Fig. 4).

The recognition of the relation instances in the text consists in first checking the actual occurrence of candidate arguments in the text: there should be semantic units in the text with the same semantic types as the relation arguments (e.g. protein and gene in the interaction relation) The second step checks the presence of a relation between them by using IE rules as described in Sect. 2.2. Thus it does not consist in just tagging the text with entity couples that are known to hold a given relation.

7.1 Designing Relation Extraction Rules

In complex cases, the arguments are not well-defined semantic units in a way that contextual explanations and evidence can be easily provided [11]. The definition of the argument results from a complex interpretation, to the point that no argument can be declaratively defined although the relation is observed. In the same way, relations themselves may not be supported by local and delimited lexical or textual fragments such as verbs (e.g. "stimulates" in Fig. 1). In both cases, it is out of the scope of current IE methods to produce a consistent semantic abstraction of the text on which pure semantic rules could apply and the interpretation could be fully formalised.

Then conditions of the IE rules usually include clues difficult to interpret in terms of ontological knowledge. For example, neighbourhood does not necessarily denote semantics but may capture some shallow knowledge that is useful in certain limited contexts. Rules often combine various matching conditions that pertain to different levels of text annotation (e.g. mixing conceptual, typographic, positional and syntactic criteria as in the examples of Figs. 2 and 3).

The design of efficient IE rules becomes a complex problem that remains open after many years of active research. Manual design is tedious, rarely comprehensive and unreliable (Sect. 7.1). Acquiring extraction rules by Machine Learning from training corpora saves expert time but was limited to rather simple cases until recently. Learning relational extraction rules remains challenging (Sect. 7.2) but the availability of new text analysis tools promises a lot of progress. The recent progress in performance and availability of syntactic dependency parsers had also a very positive effect on the system abstraction ability. When syntactic parsing conditions are combined with ontology-based semantic types, it may be easier to relate the rule conditions to the ontological definition of the objects [34] as illustrated in the example of Figs. 2 and 3.

7.2 IE Rule Learning

Learning IE rules for identifying specific domain relations is done by supervised learning applied on a training corpus where the target information was manually tagged. The abstraction degree of the learnt rules strictly depends on the representation of the training examples. Their features are derived from the linguistic analysis of the training corpus. The number of errors in abstract features like syntactic dependencies tends to be higher than in low level information such as word segmentation. The Machine Learning methods (e.g. ILP) applied to complex representations such as relational representation are also more sensitive to the errors occurring in the example description as opposed to statistics-based methods.

Pre-processing the training examples by feature selection or more complex inference may reduce the number of errors, while preserving discriminant features. This is the track followed by LP-Propal method based on the Propal algorithm [2]. LP-Propal takes as input the corpus after full processing by the linguistic pipeline of Fig. 5. Then, given a declarative list of linguistic properties, LP-Propal selects the relevant features for the training example representation [28]. For instance, the term *expression* can be neglected in biology, when it occurs in "A activates the expression of B", because "A activates B" is fully equivalent with respect to the IE task. This sentence simplification reduces data sparseness and improves the homogeneity of the training corpus.

The application of LP-Propal to one of the LLL[6] challenge dataset on genic interaction extraction yields 89.3% recall and 89.6% precision [29], which is very promising with respect to previous LLL results [34] and comparable to BioCreative results [24]. The positive role of the syntactic parsing has been experimentally measured by applying LP-Propal to the same dataset with the neighbourhood relation instead of the syntactic dependencies. It yields a poor precision (22.8%) and recall (34.7%), which confirms the importance of a deep linguistic analysis for IE.

8 Discussion

As mentioned above, IE has made significant progress and powerful IE systems are now operational. The previous sections have described on which principles a generic and modular IE system should be founded. This last section focuses on the key issues that remain to be solved in order to fully ground IE on ontologies.

8.1 Beyond the Development of NLP Toolboxes

By acknowledging the needs for domain-specific applications, the IE field has been exploring horizons outside the frame of MUC, which was rather generalist. This called for a more sophisticated linguistic analysis to take into account the diversity of sub-language formulations and to improve the richness and reliability of the extracted information. The IE performances greatly improve in consequence as shown in Sect. 7.2. The NLP underlying analysis is more expensive in term of computational time, but the IE is also more robust.

[6] Learning Language in Logic.

The availability of NLP toolboxes and pipelines helped IE system designers to achieve these results by exploiting and integrating various natural language processes into a unique IE system. An important effort in software integration was necessary, because NLP tools are usually developed independently by different teams and may have partially overlapping roles. For instance syntactic taggers, such as the popular TreeTagger, perform their own word segmentation and lemmatisation. The integration of a POS tagger with a third-party segmenter raises complex token alignment problems. The integration of each processing step in the NLP pipeline raises similar questions that should be properly solved for avoiding concurrent annotations and inconsistencies.

However, focus has been put on software integration rather than on knowledge integration and several problems remain to be addressed. More fundamentally, IE approaches correspond to a relatively narrow form of text-understanding:

- The analysis is mostly limited to the scope of sentences. IE does not take the whole document discourse into consideration to the exception of anaphora resolution that extends the analysis to neighbour sentences.
- Sophisticated ontology-based inference models beyond generality tree climbing are rarely involved. The conditions of the extraction rules are usually considered as independent.

8.2 Lexical Knowledge as a Mediator Between Text and Ontology

We have argued in Sect. 2.3 that for text interpretation the lexical knowledge plays a necessary role of *mediator* between the text and the conceptual model.

We have shown that a lexical base is composed of a lexicon and a set of rules. Their relative importance varies from one source to the other. The terminology represents the simplest case where the variants are listed in the lexicon and no rule is used. The domain specific relations represent an opposite case where the lexicon is quasi-absent, all the knowledge being embodied in the rules.

To be fully operational, maintainable and reusable, this complex knowledge structure should be properly represented in expressive knowledge representation languages. Lexicon and ontology representations have drawn a lot of attention the last years [8,11,26,40], while their link with the various contextual rules was less comprehensively studied. Integrating both knowledge types in an operational IE system remains challenging.

8.3 Toward Formalised and Integrated Knowledge Resources

With the progress of formalisation, IE research cannot longer consider ontologies as organised vocabulary or hierarchies of terms as thoroughly demonstrated in chapter “Ontology and the Lexicon”.

While formal languages for ontology representation have made great advances, there are few formal or operational proposals to tie ontologies to linguistic knowledge. This gap severely hinders the progress of IE and more generally of all textual content analysis technologies (e.g. IR, Q/A, summarising). As illustrated in Sect. 3, sophisticated and operational IE pipelines are available for developing new applications. However the cost of maintaining and reconfiguring them exponentially increases with the complexity of the linguistic knowledge. The field would gain a lot in moving the focus from software integration to knowledge integration.

Another open question comes from the partial overlap between the various types of knowledge, which are traditionally considered as distinct resources. For instance, it is sometimes difficult to distinguish named entities and terms. From an ontological point of view, they have different status. NE correspond to instance labels while terms correspond to concepts and concept labels. NE rather appear in the leaves of the ontology, while terms appear in internal nodes. The distinction is also useful from a pragmatic operational point of view but it is not sound from a linguistic point of view. In the same manner, NE dictionaries and ontologies often overlap, because NE dictionaries include NE semantic types that should be related to the ontology. Developing a coherent set of knowledge source or integrating these various knowledge sources into a single knowledge base (KB) requires that the specific scope of each one is clearly defined.

A third problem concerns the integration the learnt lexical knowledge into the available knowledge bases. This question is particularly critical for ontologies as reflected by the ontology population and ontology alignment issues.

From a research point of view, the IE field has quickly evolved towards the integration of research results from natural language processing, knowledge acquisition and ontology domains. The results on ontology formalisation and the development of new representation languages has a very positive effect on IE modelling effort while linguistic processing and knowledge acquisition methods increase the operationality of IE systems.

References

1. E. Agichtein and L. Gravano. Snowball: Extracting relations from large plain-text collections. In *Proceedings of the Fifth ACM International Conference on Digital Libraries*, 2000.
2. E. Alphonse and C. Rouveirol. Lazy propositionalization for relational learning. In W. Horn, editor, *Proc. of the 14th European Conference on Artificial Intelligence (ECAI'2000)*, pages 256–260. IOS Press, 2000.
3. S. Ananiadou and J. McNaught. *Text Mining for Biology and Biomedicine.* Artech House Books, 2006.
4. Rie Kubota Ando. Biocreative ii gene mention tagging system at ibm watson. In L. Hirschmann, M. Krallinger, and A. Valencia, editors, *Proceedings of the Second BioCreative Challenge Evaluation Workshop.* CNIO, 2007.

5. A. R. Aronson, O. Bodenreider, Chang H. F., S. M. Humphrey, Mork J. G., S. J. Nelson, T. J. Rindflesch, and W. J. Wilbur. The nlm indexing initiative. In *Proceedings of the AMIA Symp.*, pages 17–2, 2000.
6. S. Aubin and T. Hamon. Improving term extraction with terminological resources. In T. Salakoski, F. Ginter, S. Pyysalo, and T. Pahikkala, editors, *Advances in Natural Language Processing (Proceedings of the 5th International Conference on NLP (FinTAL'06*, LNAI 4139, pages 380–387. Springer, 2006.
7. K. Bontcheva, V. Tablan, D. Maynard, and H. Cunningham. Evolving GATE to meet new challenges. *Natural Language Engineering*, 2004.
8. P. Buitelaar, M. Sintek, and M. Kiesel. A lexicon model for multilingual/multimedia ontologies. In *The Semantic Web: Research and Applications; Proceedings of the 3rd European Semantic Web Conference (ESWC06)*, Lecture Notes in Computer Science, Vol. 4011. Springer, 2006.
9. M. T. Cabré, R. Estopà, and J. Vivaldi. Automatic term detection: a review of current systems. In Didier Bourgault, Christian Jacquemin, and Marie-Claude L'Homme, editors, *Recent Advances in Computational Terminology*, volume 2 of *Natural Langage Processing*, pages 53–87. John Benjamins, Amsterdam, 2001.
10. N. Chinchor and P. Robinson. Muc-7 named entity task definition (version 3.5). In *Message Understanding Conference Proceedings, MUC-7*. NIST, 1998.
11. P. Cimiano, P. Haase, M. Herold, M. Mantel, and P. Buitelaar. Lexonto: A model for ontology lexicons for ontology-based nlp. In Paul Buitelaar, Key-Sun Choi, Aldo Gangemi, and Chu-Ren Huang, editors, *Proceedings of the OntoLex07 Workshop held in conjunction with the 6th International Semantic Web Conference (ISWC07) "From Text to Knowledge: The Lexicon/Ontology Interface"*, Busan (South Korea), November 2007.
12. B. Daille. Variations and application-oriented terminology engineering. *Terminology*, 11(1):181–197, 2005.
13. Riloff E. Automatically constructing a dictionary for information extraction tasks. In *Proceedings of AAAI93*, pages 811–816, 1993.
14. Ciravegna F. Learning to tag for information extraction from text. In *Proceedings of the ECAI-2000 Workshop on Machine Learning for Information Extraction*, 2000.
15. David Ferrucci and Adam Lally. Uima: an architectural approach to unstructured information processing in the corporate research environment. *Nat. Lang. Eng.*, 10(3-4):327–348, 2004.
16. R. Grishman and B. Sundheim. Message understanding conference-6: A brief history. In *16th International Conference on Computational Linguistics*, Copenhagen, Denmark, 1996.
17. Zhou GuoDong and Su Jian. Exploring deep knowledge resources in biomedical name recognition. In Nigel Collier, Patrick Ruch, and Adeline Nazarenko, editors, *COLING 2004 International Joint workshop on Natural Language Processing in Biomedicine and its Applications (NLPBA/BioNLP) 2004*, pages 99–102, Geneva, Switzerland, August 28th and 29th 2004. COLING.
18. B. Habert, E. Naulleau, and A. Nazarenko. Symbolic word clustering for medium-size corpora. In *Proceedings of the 16th International Conference on Computational Linguistics*, volume 1, pages 490–495, Copenhagen, Denmark, 1996.
19. Thierry Hamon, Adeline Nazarenko, Thierry Poibeau, Sophie Aubin, and Julien Derivière. A Robust Linguistic Platform for Efficient and Domain specific Web

Content Analysis. In *Proceedings of the 8th Conference RIAO'07 (Large-Scale Semantic Access to Content)*, Pittsburgh, USA, May 2007.
20. M. A. Hearst. Automatic acquisition of hyponyms from large text corpora. In *Proceedings of the 15th International conference on Computational Linguistics*, volume 2, pages 539–545, Nantes, 1992.
21. W. Hersh, A. Cohen, L. Ruslen, and P. Roberts. Trec 2007 genomics track overview. In *TREC 2007 Proceedings*, 2007.
22. J. R. Hobbs, D. Appelt, J. Bear, D. Israel, M. Kameyama, M. Stickel, and M. Tyson. Fastus: A cascaded finite-state transducer for extraction information from natural language text. In E Roche and Y Schabes, editors, *Finite-State Language Processing*, chapter 13, pages 383–406. MIT Press, 1997.
23. Christian Jacquemin. A symbolic and surgical acquisition of terms through variation. In S. Wermter, E. Riloff, and G. Scheler, editors, *Connectionist, Statistical and Symbolic Approaches to Learning for Natural Language Processing*, pages 425–438. Springer-Verlag, 1996.
24. M. Krallinger. The interaction-pair and interaction method sub-task evaluation. In *proceedings of the BioCreAtIvE II Workshop*, at CNIO, Madrid, Spain, 2007.
25. S. Kripke. Naming and necessity. In G. Harman D. Davidson, editor, *Semantics of Natural Language*. Reidel, Dordrecht, 1972.
26. B. Lauser and M. Sini. From agrovoc to the agricultural ontology service/concept server: an owl model for creating ontologies in the agricultural domain. In *Proceedings of the 2006 international conference on Dublin Core and Metadata Applications (DCMI'06): "Metadata for knowledge and learning"*, pages 76–88. Dublin Core Metadata Initiative, 2006.
27. K. Lerman, A. Plangprasopchok, and C. Wong. Personalizing image search results on flickr. Technical report, arXiv, 2007.
28. A.-P. Manine and C. Nédellec. Alvis deliverable d6.4.b: Acquisition of relation extraction rules by machine learning. Technical report, Institut National de la Recherche Agronomique, http://genome.jouy.inra.fr/bibliome/docs/D6.4b.pdf, March 2007.
29. A.-P. Manine, E. Alphonse and P. Bessières. Information extraction as an ontology population task and its application to genic interactions. *ICTAI '08: Proceedings of the 2008 20th IEEE International Conference on Tools with Artificial Intelligence*, http://dx.doi.org/10.1109/ICTAI.2008.117, pages 74–81. IEEE Computer Society, Washington, DC, USA, 2008.
30. S. Miller, M. Crystal, H. Fox, L. Ramshaw, R. Schwartz, R. Stone, R. Weischedel, and the Annotation Group. Algorithms that learn to extract information–BBN: Description of the SIFT system as used for MUC. In *Proceedings of the Seventh Message Understanding Conference (MUC-7)*, 1998.
31. A. Nazarenko, E. Alphonse, J. Derivière, T. Hamon, G. Vauvert, and D. Weissenbacher. The ALVIS format for linguistically annotated documents. In *Proceedings of the 5th international conference on Language Resources and Evaluation, LREC 2006*, pages 1782–1786. ELDA, 2006.
32. A. Nazarenko, C. Nédellec, E. Alphonse, S. Aubin, T. Hamon, and A.-P. Manine. Semantic annotation in the Alvis project. In W. Buntine and H. Tirri, editors, *Proceedings of the International Workshop on Intelligent Information Access*, pages 40–54, Helsinki, Finlande, 2006.
33. C. Nédellec, P. Bessières, R. Bossy, A. Kotoujansky, and A.-P. Manine. Annotation guidelines for machine learning-based named entity recognition in

microbiology. In M. Hilario and C. Nedellec, editors, *Proceedings of the Data and text mining in integrative biology workshop, associé ECML/PKDD*, pages 40–54, Berlin, Allemagne, 2006.

34. Claire Nedellec. Learning language in logic - genic interaction extraction challenge. In Cussens J. and Nedellec C., editors, *Proceedings of the Learning Language in Logic (LLL05) workshop joint to ICML'05*, pages 40–54, 2005.
35. S. Nirenburg and V. Raskin. *Ontological semantics*. MIT Press, 2004.
36. B. Popov, A. Kiryakov, D. Ognyanoff, D. Manov, and A. Kirilov. Kim - a semantic platform for information extraction and retrieval. *Nat. Lang. Eng.*, 10 (3-4):375–392, 2004.
37. J. Pustejovsky, J. Castano, B. Cochran, M. Kotecki, M. Morrell, and A. Rumshisky. Linguistic knowledge extraction from medline: Automatic construction of an acronym database. In *Proceedings of the 10th World Congress on Health and Medical Informatics (Medinfo 2001)*, 2001.
38. S. Pyysalo, T. Salakoski, S. Aubin, and A. Nazarenko. Lexical adaptation of link grammar to the biomedical sublanguage: a comparative evaluation of three approaches. *BMC Bioinformatics*, 7(Suppl 3), 2006.
39. Saetre R., Yoshida K., Yakushiji A., Miyao Y., Matsubayashi Y., and Ohta T. Akane system: Protein-protein interaction pairs in biocreative2 challenge, ppi-ips subtask. In L. Hirschmann, M. Krallinger, and A. Valencia, editors, *Proceedings of the Second BioCreative Challenge Evaluation Workshop.* CNIO, 2007.
40. A. Reymonet, J. Thomas, and N. Aussenac-Gilles. Modelling ontological and terminological resources in owl dl. In Paul Buitelaar, Key-Sun Choi, Aldo Gangemi, and Chu-Ren Huang, editors, *Proceedings of the OntoLex07 Workshop held in conjunction with the 6th International Semantic Web Conference (ISWC07) "From Text to Knowledge: The Lexicon/Ontology Interface"*, Busan (South Korea), November 2007.
41. F. Rinaldi, G. Schneider, K. Kaljurand, M. Hess, and M. Romacker. An environment for relation mining over richly annotated corpora: the case of genia. *BMC Bioinformatics*, 7(Suppl 3), 2006.
42. Juan C. Sager. *A Practical Course in Terminology Processing.* John Benjamins Publishing Company, 1990.
43. A. S. Schwartz and M. A. Hearst. Introduction to the bio-entity recognition task at jnlpba. In *Proceedings of the Pacific Symposium on Biocomputing (PSB 2003).* International Conference on Computational Linguistics (COLING'04), 2003.
44. Ohta T., Tateisi Y., Mima H., and Tsujii J. Genia corpus: an annotated research abstract corpus in molecular biology domain. In *Proceedings of the Human Language Technology Conference*, 2002.
45. J. Wilbur, L. Smith, and L. Tanabe. Biocreative ii: Gene mention task. In *proceedings of the BioCreAtIvE II Workshop*, at CNIO, Madrid, Spain, 2007.
46. Y. Wilks. Information extraction as a core language technology. In M. T. Pazienza, editor, *Information Extraction.* Springer, Berlin, 1997.

Browsing and Navigation in Semantically Rich Spaces: Experiences with Magpie Applications

Martin Dzbor, Enrico Motta, and Laurian Gridinoc

Knowledge Media Institute, The Open University, UK,
M.Dzbor@open.ac.uk, E.Motta@open.ac.uk, L.Gridinoc@open.ac.uk

Summary. Semantic Web is a medium for knowledge exchange, where knowledge produced by one agent is consumed by another agent who may extend or modify it. Semantic Web also affords novel opportunities for acquiring knowledge – including approaches favoring automated selection, reuse and integration of external, just-in-time gathered semantic resources. As semantic resources are no longer specifically developed for a single purpose, their re-contextualization within other web resources (e.g., web pages) is becoming a more pressing challenge. In this chapter, we look at the case when external semantic resources discovered in the web-sized corpus are re-contextualized to enhance the user experience of an arbitrary web content visited by a particular user. We first review different approaches showcasing different facets of semantic browsing and define the notion of 'semantic browsing' in general terms. Next, we share our experiences with Magpie, an in-house semantic web browsing framework, and illustrate new functional features such a semantically-enriched browsing tool may offer on the example of introducing additional user interaction modalities and developing a capability to work with multiple background knowledge models simultaneously. In the discussion we re-visit the defining tenets of 'semantic browsing' and look at how the reuse of just-in-time discovered and applied semantic resources really addresses the issue of enabling the user to re-contextualize semantic data for the purposes of text analysis, data interpretation, relationship discovery, and knowledge validation.

1 Introduction

The Web is often seen as one of the fundamental inventions of the twentieth century, which helped to shape the notion of the networked resources, networked economy, and ultimately networked world. The Web matured and became a fairly user-oriented information space – mainly as an effect of emerging interactive applications collectively known as "Web 2.0". Although Social Web and Web 2.0 are not subjects of this chapter (for interesting insights, see, e.g., [1]), it is useful to note the growth in the volume of structured data these applications produce, and also the population of users creating, browsing, or otherwise involved in resource networking is reaching a billion. In its

S. Staab and R. Studer (eds.), *Handbook on Ontologies*, International Handbooks on Information Systems, DOI 10.1007/978-3-540-92673-3,

own right, services like MySpace or Second Life have active user bases equivalent to well-sized countries, and majority of these users interact with the Web through browsing online web pages, publishing new information, new connections and introducing ever more links into this vast information space.

Despite this vast user base, the Web (and also its recent social enrichment) has some limitations – most of them related to the issues of conveying knowledge (rather than merely data) and interpreting it (rather than merely retrieving). To address this limitation the vision of Semantic Web [2] as a "Web of Data" has emerged since early 2000s. In this vision, web resources are annotated with *semantic markup*, using knowledge representation languages such as RDF(S) [3] or OWL [4]. One rationale for the semantic markup considers the nature of connections one can express. Unlike standard HTML, semantic markup languages allow expressing not only ad-hoc, generic links between the resources, but also formal statements *about* the interesting properties of the web resources, about external entities, and their conceptual, i.e., *named relationships*. To enable meaningful knowledge sharing and inter-operability, such markup needs to be based on some ontologies [5] – knowledge-level models that capture some shared and agreed on understanding of the parties that want to collaborate or otherwise make use of knowledge.

Once such conceptual commitments are established, one gains an opportunity to request much richer information from the Web (here "the Web" is used as a large-scale data repository). For instance, instead of merely finding a list of *typically* collocated key words, thanks to semantic markup and conceptual models, one would be able to tell that "carbohydrates" are a specific group of "organic compounds", and as such they share some generic characteristics of all organic materials, but at the same time "sugars" or "sugar acids" are narrower and conceptually more specific terms. Admittedly, this is a fairly trivial enrichment, but this neighbourhood of conceptually related terms may be used, e.g., to expand the user's original search query. In a different domain, instead of merely retrieving a list of articles containing term "user modelling" among keywords, we can combine the conceptual annotations and data from several ontological models, so as to obtain, for example, a list of leading experts publishing on that topic or a list of publishing outlets where such topic *may* be appearing.

However, there is a certain three-way tension between (1) the dependence of the Semantic Web on semantic annotations (done on a large scale), (2) the cost and complexity of providing these semantic annotations, and (3) the cognitive complexity for human user to interact with the semantic annotations. In this chapter we briefly touch on the third aspect of this tension; with an emphasis given to the user aspects. We consider how a user can interact with semantic mark-up – by means of turning it into web-browseable resources and by means of combining it with standard web pages or other (textual) documents: If a user cannot or does not know how to access knowledge that takes form of those rich conceptual connections that form Semantic Web, as described above, then the knowledge has very little value.

In this chapter, we touch on the issue of semantic navigation on three levels. First, we look at recent approaches to semantic browsing and navigation in general, trying to identify four families of user interaction styles. Second, we sum up our experiences with a tool from one of the reviewed families – Magpie, and highlight what functional features are valuable from the end user's viewpoint. Finally, we devote the rest of the chapter to exploring the future of semantic navigation on the Web.

2 Existing Semantic Web Browsing Applications

We start by looking at characteristics and evolution of Semantic Web browsing tools in general. In terms of desired functionality, Quan and Karger [6] suggested that the primary purpose of a browser for the Semantic Web is "to separate the content – the proper purview of the publisher serving the information, from the presentation – an issue in which the end user or their local application should have substantial say". Let us therefore briefly look at how their requirement was addressed in early Semantic Web browsing tools.

2.1 Early Semantic Web Browsing Approaches

First prototypes of tools that claimed to support some aspects of Semantic Web browsing appeared around year 2002–2003. One common trait of these early tools was a close relationship with the Web. Indeed, in the absence of the key ingredient – semantic markup – these tools looked to the available web pages and featured a range of entity recognition algorithms [7]. For example, Knowledge and Information Management (KIM) [8] was a platform for automatic semantic annotation, web page indexing and retrieval. It could recognize named entities (such as job titles or geographic names) in text, and use the findings to assign ontological definitions to the entities in the text and thus to capture semantic relationships between terms mentioned. KIM extended the GATE platform [9], which was built into KIM proxy, and thus enabled on-the-fly annotation of web pages.

Another system, this time getting inspired by the hypertext and the web browsing paradigm, COHSE [10] implemented semantic markup using so-called open hypermedia approach. Early versions of COHSE recognized entities in a web page, semantically marked these entities, and enabled the user to click on any of these dynamic hyperlinks to navigate to other web pages tagged with the same term. Thus, an idea here was to add to the author-defined hyperlinks also dynamic links, which may reflect user's interest in certain terms. Similarly as KIM, the markup capability was embedded in a COHSE proxy, but unlike KIM, COHSE enabled some tuning of this proxy, e.g., by picking up a different "bag of terms" to drive entity recognition.

2.2 Recent Advances in Semantic Web Browsing

More recent approaches to facilitating access to semantic markup explored a wide range of other metaphors reused from other domains. In particular, style sheets became one of such approaches that inspired several tools supporting Semantic Web browsing. For example, PiggyBank [11] and Exhibit [12] are two champions of the idea to let users publish structured data on standard web servers with no installation, database administration, and little programming. The structure of semantic markup gets exploited and mapped onto some specific (user selected) visual and/or interactive widgets. Thus, unlike tools mentioned earlier in Sect. 2.1, the emphasis shifted from acquiring conceptual data to exposing it.

Another user interaction metaphor that got exploited when the amount of semantic data reached larger proportions was faceted navigation [13]. The key principle of this metaphor was that large data collections (e.g., libraries or galleries) have many dimensions according to which data can be viewed, searched or navigated. Thus, faceted navigation is a user interaction style whereby users filter an appropriate set of data records by progressively, step-by-step selecting from valid dimensions of a particular classification. The idea was originally used to expose standard databases – e.g., the Flamenco demonstrator [13]. Recent achievements in *semantic* faceted browsing include, e.g., Browse RDF [14], a generic RDF browser, or `/facet` [15], an RDF browser used in a manner similar to Flamenco, but for the Dutch cultural heritage data that were annotated semantically rather than merely stored in a database.

Some other ideas that got explored included the metaphor of "semantic overlays", or semantic layering, as it was introduced by the authors of Magpie [16–18]. The idea was visually similar to KIM and COHSE from Sect. 2.1, but rather than linking to other web pages annotated with a similar term, a semantic menu was suggested as a container for several semantically annotated relationships or properties, which the user could invoke. Thus, the approach supported *named relationships* on top of standard, anonymous hyperlinks; later, the relationships were generalized to cover semantic services (as a procedural way to uncover or compute a particular relationship). AKTive Document [19] later extended Magpie's viewing metaphor to the world of text and data authoring.

Further details on different metaphors and tools are provided in the subsequent sections – here we tried to sketch how the idea of accessing semantic markup started and evolved. Also, a differently scoped review and human–computer interaction focused analysis of tools for navigating the ontologies and other semantic content can be found in our other publications [20, 21].

2.3 What Is Semantic Web Browsing?

We have mentioned Quan and Karger's view on this topic earlier, but the requirement to separate content from its presentation is more of a philosophy than an actual definition. If we wanted to define what Semantic Web browsing is about, we suggest the following:

Semantic Web browsing (or navigation) is a family of user interaction styles that rely on techniques for rendering all information that can be found in semantic markup data stores about a specific resource for the purpose of exposing the information space(s) or context(s) around a specific resource. In addition to this generic and abstract definition, a few specific points apply:

- Data is usually expressed in a standard, web-compatible formalism, such as RDF(S) or OWL (the same can be extended to techniques that are usually expressed in formalisms like SOAP or WSDL).
- Data is usually meant as a combination of schema/ontology-level *model* and assertions about specific *facts*.
- Information space is usually equivalent to *Data – Link – Data* structures that can be dereferenced to hyperlinks comprehensible to standard web servers.

Based on our experiences and feedback we obtained with the demonstrators of Magpie technology, we suggested in [20] some additional criteria. From the four criteria mentioned in [20], the *navigation using the markup committed to ontologies* is covered by the above definition, but we argued for an additional pragmatic requirement, namely the capability to expose data and navigate across *multiple ontologies*.

Before continuing let us briefly say what applications do not satisfy the definition above. First, tools like IsaViz or CropCircles are schema visualization techniques, not semantic browsers per se: (1) they mainly visualize the schema but not the data, and (2) they consider hierarchical links (as in "a category of all lions is classified under category of mammals"). Second, tools like Flickr[1] are *tag (cloud) browsers*, not semantic browsers per se: (1) the links between data items are established based on co-occurrence and are thus anonymous, with no committed meaning and (2) the focus is on the data, schemas are largely not considered.

Third, there are other approaches to accessing Semantic Web data that may feed into browsing, but are, in principle, different user interaction paradigms. For example, a data store can be queried (with a user formulating a query in SPARQL, RDQL, etc.), and appropriate data rendered (e.g., with a style sheet). This is a more controlled user interaction, whereas browsing has an element of serendipity (cf. with searching and browsing on the Web [22,23]). Obviously, as with the standard Web, results of querying can be browsed to access further information, which may not have been retrieved in the original query.

[1] Flickr (`http://www.flickr.com`) is a registered trademark of Yahoo!

The research addressing the above criteria is ongoing and new tools are produced. In terms of user interaction, while there are XML style sheet formalisms applicable also to Semantic Web languages, more generic challenge of how a human user can (and wants to) interact with the Semantic Web content is only now getting a more significant attention. The issue of using multiple ontological frames to open up the interpretation choices is even less developed. Nevertheless, let us consider a few specific user interaction metaphors that satisfy the above definition and can be seen as browsers for semantically marked up data.

In this overview, we consider three dimensions of data presentation, as defined in a broader framework for classifying semantic search tools:[2] (1) data selection, (2) data organization, and (3) user feedback. These dimensions form the basis for describing four distinct families of tools as follows:

Navigation in graph structures: the focus is on the organization of nested data in a form of trees or graphs with expansible and clickable nodes.

Faceted navigation: primary feature here is the opportunity for a continuous query refinement and step-by-step formulation of the user need.

Navigation with templates: primary feature of this family is (1) strong focus on selecting data properties, and (2) using a rich repertoire of visualization metaphors to present nested data records.

Navigation with semantic overlays: primary feature here is (1) data selection being dynamic and relying on a loaded schema, and (2) data organization done in a form of embedding data into a plain text (e.g., web page).

2.4 Navigating in Semantic Graphs

Tabulator [24] is one possible form of *a* browser for semantically linked data. It started as a project and tool to demonstrate the serendipitous re-use and to address the "explore vs. analyze" tension in user interface design in an open-world of interlinked semantic data. Tabulator is a generic browser for linked data, without the expectation of providing domain-specific interfaces. However, it permits domain-specific functionality (such as calendar, money or address book management) to be loaded transparently from the web.

Unlike in tools discussed in Sects. 2.6 or 2.7, where the web documents are primary and the semantic layers or annotations secondary, for the Tabulator the logical, semantic graph is the primary source of data; the web documents are optional and secondary. Hence, the user can explore the graph of data as a conjunction of all the graph documents that have been read (in a particular browsing session). While this approach may not allow directly browsing the documents, it allows the user to check the provenance or source of any piece of information included in the browseable graphs.

[2] For background and the schema see section "Example" in `http://swuiwiki.webscience.org/swuiwiki/index.php?title=Semantic_Search_Survey`

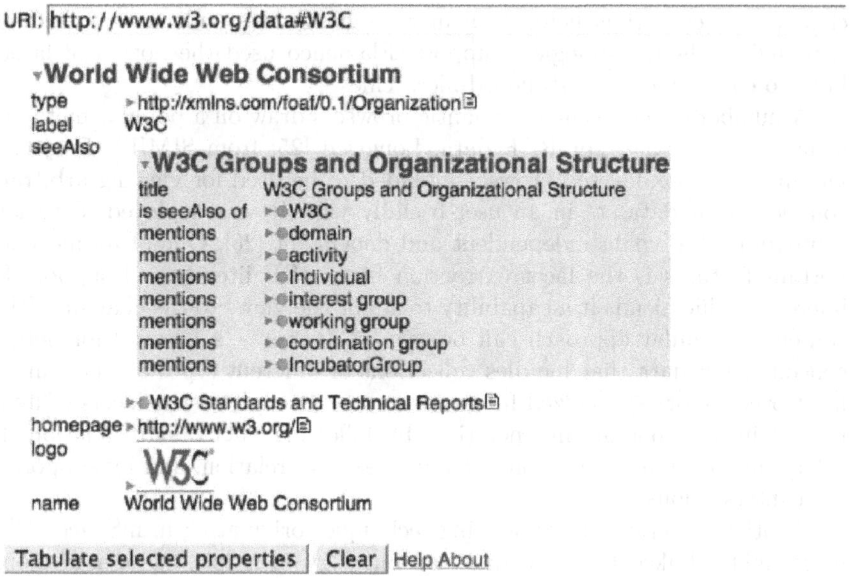

URI: http://www.w3.org/data#W3C

World Wide Web Consortium
type http://xmlns.com/foaf/0.1/Organization
label W3C
seeAlso
W3C Groups and Organizational Structure
title W3C Groups and Organizational Structure
is seeAlso of W3C
mentions domain
mentions activity
mentions Individual
mentions interest group
mentions working group
mentions coordination group
mentions IncubatorGroup
W3C Standards and Technical Reports
homepage http://www.w3.org/
logo W3C
name World Wide Web Consortium

Tabulate selected properties | Clear | Help About

Fig. 1. An outline view of a simple set of triples related to organizational knowledge

Tabulator (Fig. 1) operates in two modes: exploration and analysis. In the exploration (open world) mode it allows the user to interact with a large data graph composed of several partial graphs. This is achieved without the user having to provide all the data – Tabulator implicitly follows links that may contain RDF data about relevant nodes. Linked data is typically presented using a graph metaphor; i.e., as nodes and arcs, which in Tabulator are called outline views. When Tabulator is in the analysis mode, the user may select some nodes or arcs to define patterns of a query, which is then executed by the tool against the available data graphs. Query results may be displayed in different views and may be mashed together.

One interesting proposal coming from the Tabulator project is about the requirement to include some form of user interface "tips" in the ontologies that can be interpreted by generic applications, such as Tabulator, effectively choose appropriate and most useful user interaction components to data from unfamiliar domains.

2.5 Faceted Navigation

Large datasets (e.g., libraries or museums) have many dimensions along which they can be browsed, searched or navigated. One interaction strategy for such data – in addition to simple searching and browsing – is *faceted* browsing, where users filter an item sub-set by progressively selecting from valid dimensions of an appropriate classification. On a non-semantic level, the strategy was piloted in Flamenco [13] that used metadata to guide users through the choices of views, thus helping them to organize the underlying collection. Although without ontological support, Flamenco used the notion of lateral links to complement the standard hierarchies.

A number of data-centric semantic browsers draw on a popular metaphor of faceted browsing. For RDF data, Longwell [25] from SIMILE Project is an out-of-the-box faceted browser intended to be used for viewing arbitrary, complex RDF datasets in an user-friendly way. It was deployed in various contexts both domain independent and dependent [26]. One of its most important features is the facet extraction from RDF literals and support for inference, which lends it a capability to adapt the views to the changing RDF content. A similar approach can be seen in `/facet` – a browser for heterogeneous RDF data that handles collections of different types of items unlike most other, more specialized faceted browsers. It explores the facets of items related in a taxonomic manner (i.e., by following `subClassOf` relation, by collapsing facets related through `subPropertyOf` relation, and by support of their intersections.

Another generic suite of faceting techniques originating in mSpace [27] is more tightly linked to ontologies – an ontology in mSpace acts as the only source of facet classification. The user is then left with selecting suitable browsing pathways (i.e., sequences of facet dimensions that fit their preferences). In mSpace, Semantic Web is seen as a rich, multidimensional *hypertext system*. mSpace thus extends the faceted browsing paradigm and its functional operations like slicing, sorting, swapping, adding, or subtracting to semantic data sets – one of its many demonstrators is in Fig. 2. It supports direct manipulation of the ontology content and the selection of instances associated with the current configuration of facet sequences. Its logic also provides for automatic reasoning to ensure that only meaningful attribute ordering/selection occurs. In addition to traditional facets, mSpace adds a few experimental semantic add-ons – e.g., its numeric volume indicators used as predictors of what to expect *after* choosing given facet, or their early adoption of geographic maps as a visualization medium.

2.6 Navigation Using Styles and Templates

This approach to browsing emerged from the fact that much information present on the Web is already stored in relational form, in the database-driven web sites. Therefore, another way to resolve the aforementioned knowledge acquisition bottleneck is to take advantage of the structural clues of this structured web content to re-create the original information stored in the databases backing this content.

Thresher [28] is a system build on top of the Haystack platform [29], which allows non-technical users, rather than content providers, to "unwrap" the semantic structures buried inside human-readable Web. It provides users with an interface to "demonstrate" the extraction of semantic meanings, and through

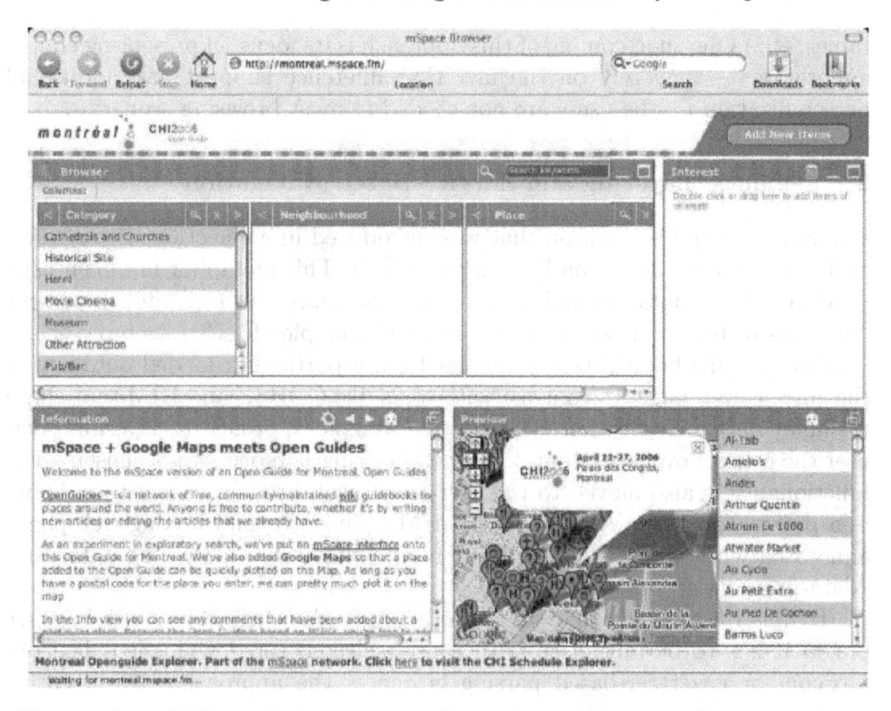

Fig. 2. A multi-faceted view on an mSpace demonstrator used as a conference delegate management system

such demonstrations it learns mappings between the regularities in the document structure and the semantics. Thresher then automatically applies the mappings to similar documents. Thus, extraction of semantic data is separated from its presentation (see Quan and Karger's criterion in Sect. 2), and is accessible for further reuse via the Haystack platform.

A notable trend related to templates is the push away from the heavy clients towards lightweight clients – often in a form of plugins or bookmarklets. A lightweight equivalent of Haystack is PiggyBank [11], which runs as a browser plugin, allowing the user to collect and browse found semantic information. Solvent, another web browser plugin, performs Thresher's role for PiggyBank, allowing the user to visually annotate the document with common vocabularies (e.g., Doublin Core) and generate "scrapers" that extract, convert and store in PiggyBank the "unwrapped" semantic information.

Once information has been extracted, one way to reuse it is to re-publish it back to the Web. Here, Exhibit, another tool from the same family as the above tools, is a JavaScript-based approach that exposes structured data to the Web using styles and templates. Rather than directly showing graph structures, Exhibit, PiggyBank and others emphasize the need to make the semantic content human-friendly – hence, templates and styles serve to "prettify" the graphs and show them in a variety of familiar metaphors (e.g., timelines, maps, tables, etc.) One shortcoming of this approach is its focus on presenting simple annotations – they rely on the fact that inference support and additional personalization by the users are not needed in most browsing scenarios.

2.7 Semantic Layering and Service-Based Navigation

Semantic layering is a notion that was introduced in connection with semantic browsing and navigation by Magpie [16,17]. This metaphor has been used previously by annotating and entity discovery algorithms (e.g., [9]) to present outcomes of text analysis. In the context of Magpie (Fig. 3), its browser extension (plugin) highlights the entities from a particular, loaded ontology in the current web page. In contrast with tools like COHSE and KIM that share a similar data presentation strategy, the Magpie approach to layering gives user the control over what ontology (i.e., layering perspective) is chosen – the functionality is also moved to the user-facing web browser rather than being in a proxy. Tools like ViEWS, KIM, or Magpie go beyond linking web pages annotated with same terms, and put more emphasis on accessing data (and "data-properties" clusters).

To illustrate the notion of semantic layering take Magpie, as a web browser plugin it has to be initialized with a user-selected (or downloaded) lexicon. Lexicon- or gazetteer-based parsing is one of the approaches to an entity

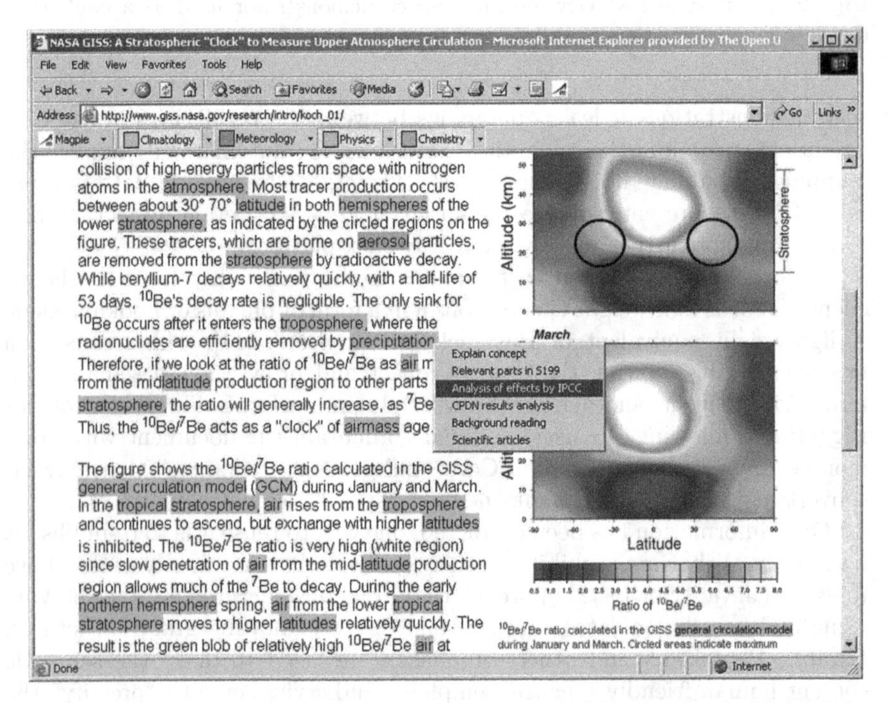

Fig. 3. A web page related to climate science with Magpie plug-in highlighting concepts relevant from the perspective of climatology course

recognition task [7]; in case of Magpie it helps keeping the time overheads of calculating a semantic layer (even for data sets with thousands of data nodes and lengthy web pages) comparable to those for network latency. Once semantic layer has been matched to the web page, the user can (de-)activate the view of different categories. This activation *presents a semantic layer* over the original document. This approach to visualizing semantic data as layers puts the users in control of what knowledge is visible at any time, which in turn reduces the problem of overwhelming the users with too much information.

Annotated and highlighted concepts become hotspots that allow the user to request a menu with a set of actions (formally coded as web services) for a relevant item. Here it suffices to say that web service choices depend on the ontological classification of a particular concept in the selected ontology and on what services are available for a given ontology. Magpie plugin is wrapping a user's click (i.e., the request for a particular service) into a URI, which is then unwrapped to communicate with the actual web service using SOAP over HTTP. The results from the individual web service may be constructed based on data retrieval (Fig. 4a), knowledge-level inference (Fig. 4b), statistical correlation, or their combination (Fig. 4c).

One feature of this approach to semantic navigation is the notion of ontological perspective and its selection by the user. It can be seen in other tools of this family; e.g., VIeWs [30] enables visitors of an information portal to choose between several, ontology-grounded perspectives, which, in turn, inform what knowledge will be made available to them. VIeWs emphasizes knowledge customization for different audiences – e.g., tourists vs. business visitors to a region. In chronological overview, we also mentioned AKTive Document [19] as a way to use the layering metaphor to create, share, import and reuse annotations from multiple sources, e.g., of other team members.

3 Semantic Web Browsing: Experiences with Magpie

In the previous publications (e.g., [18]) we used Magpie as a dynamic educational tool for undergraduate students at The Open University(UK). In this scenario, Magpie facilitated to the students a course-specific perspective on scientific texts, analyses and publications. A screenshot of the climate science demonstrator was shown in Fig. 3. Since Magpie is a generic framework, other demonstrators were built with different ontologies (or ontology mashups) and with different sets of semantic web services and inference shortcuts. In one such example Magpie reuses a 50,000 term large thesaurus of terminology related to agriculture from the Food and Agriculture Organization of the United Nations.[3] Web services developed for FAO include term translations (in conceptual and natural languages, shown in Fig. 4a) and the conceptual navigation through semantically close entities (entities in Agrovoc are added in evolutionary rather than systemic or taxonomic manner).

[3] Further details about this thesaurus are at `http://www.fao.org/agrovoc`

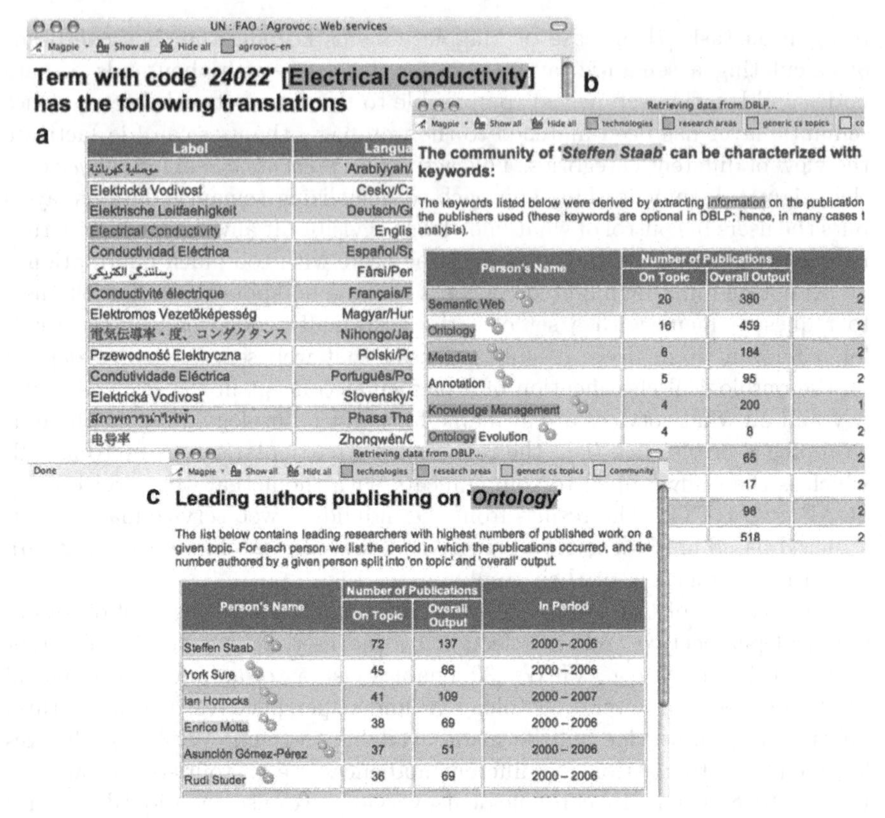

Fig. 4. Some Magpie services: (**a**) data retrieval for Agrovoc term translation, (**b**) inferencing the characterization of a person's, and (**c**) inference combined with statistical analysis providing a list of experts for a given topic

Since most of our evaluation work has been done with Magpie, we summarize the experiences and performance of this tool, which we then propose to address in coming research activities.

3.1 Positive Experiences with Magpie

As shown in previous publications, Magpie is a generic and flexible semantic browser – in terms of supporting any ontological viewpoint the user is willing to commit to and interpret or annotate web pages. In Fig. 3 Magpie presents climatology as a science closely related to physics and other base sciences. In another demo (services of which are shown in Fig. 4b,c) ontological viewpoints reflect two styles of making sense of an ill-defined area, and use them in the domain of training young researchers.

One feature that came out of evaluations as positive was the opportunity for the user to access data even in situations, which would otherwise require the user to formulate a complicated and lengthy query. Second aspect that affected users' performance in our tests was the actual support for navigation; i.e., semantic data was not only retrieved, the framework enabled composing partial data retrieval services into a more complex service that showed the user how to apply analytic or synthetic compositions to obtain more valuable information. When compared with established techniques, such as Google Scholar in the domain of academic support, the value of semantically enriched platform showed in identifying similar researchers or topics, in identifying groups of researchers formed around a theme rather than explicitly joining any specific mailing list, discussion board, or working group.

Another positive aspect is the capability to interact with the user via semantic web services – these can even be derived automatically based on a given ontology. Services can be obviously composed, and thus a more natural and richer user experience can be achieved. For example, the content of service response in Fig. 4b comprises inferences of several independent web services (e.g., community of practice in terms of people and in terms of topics addressed). Hence, a user's single click gives access to a more comprehensive result, which would normally require more complex, manual composition of partial results, their interpretation, etc. The degree of sophistication of the web services is independent of the Magpie architecture.

Semantic shortcutting is also useful – merely highlighting concepts from a user-selected lexicon (which may be built by combining chunks from several ontologies) in a text gives an indication of its relevance. Combining this with services acting as an inference shortcut, even fairly sophisticated data relationships can be accessed with a single click. While this might not be useful for every user, in analytic and synthetic tasks (such as compilation of expertise sources on a given topic), Magpie shortcuts can cut the processing time from hours to a few seconds (for other analyses see also [31]).

3.2 Shortcomings of the Magpie Approach

Although the "single ontology = single interpretative perspective" paradigm used by Magpie reduces the size of the problem space, this reduction is not always helpful. Although it focuses the user's attention (as intended), it also unduly restricts the breadth of the acquired knowledge (this was clearly not intended). For instance, during a study session a student may come across a few similar but semantically not entirely identical study materials. This means that at each page, the student would benefit from minor tuning of the used ontology, glossary and/or service menu. These tunings reflect slight shifts within a broader problem space, which is a fairly common tactic we use everyday to deal with the open situations. Thus, Magpie's design actually features a gap between the inherent notion of a single, formal, sound but 'semantically closed ontology guaranteeing a certain precision within the domain, and the desire to open up the interaction by supporting multiple services, as well as multiple ontologies (at the same time, without explicit user's reloading step).

Another shortcoming of the "layering" in Magpie was the limitation of categories fitting the screen estate – hence, most Magpie applications were limited to between four and seven top-level categories (buttons) and six to eight web services forming a menu. Since, screen limits are unlikely to change, new approaches to presenting semantic data need to be explored; in particular, when one extends the "single ontology" perspective to multiple ontologies, more content becomes available, more services may be found and invoked. Yet, exploration without guidance and tracking may quickly degrade to chaotic and blind clicking.

4 Future of Semantic Browsing

As we highlighted in Sect. 2, out of the four criteria for an application enabling the user to browse the Web using the semantic links, the least advanced is the second – the capability to apply multiple ontological perspectives in multiple user contexts. Therefore, we first touch on the issue of acquiring ontologies from an open, distributed environment of the Web. Then we suggest how multiple ontologies may be interacted with on the level of user interfaces.

4.1 Finding Distributed Ontologies

Coping with multiple ontologies on the user level depends, to some extent, on an infrastructure supporting quick and efficient selection of ontologies. However, as the number of ontologies and semantically marked up data is growing at a rapid pace, it outpaced our understanding of the quality of this generated and designed content in the distributed Semantic Web resources. Our recent advances in infrastructure known as Watson [32] enable access to networked ontologies while *enriching* them with information about their quality and dependencies on other ontologies. This gives us insights into the nature of Semantic Web content, so that the new generation of Semantic Web applications (e.g., more flexible semantic web browsers) access the needed content in a more efficient way.

Watson offers a scalable infrastructure for discovering and selecting ontologies distributed over the Web. It is a stand-alone (i.e., semantic browsing independent) infrastructure with several benefits over similar tools. For instance, Swoogle [33] – a well-known ontology search engine, has a broad coverage of semantic content, but suffers from its index not reflecting any deeper sense of "ontology quality" and also, index is driven by term occurrence, little semantic structure is taken into account.[4] Swoogle's approach to semantic resources is akin to Google's PageRank; i.e., reflecting ontology

[4] Swoogle's "web" view leads to *semantic* duplicates or near-duplicates, i.e., files with different URLs but semantically equivalent. Not taking semantic duplication into account may skew the performance of applications like semantic browsers.

popularity rather than more practical aspects such as domain coverage or similarly.

For the purpose of browsing on the open Web, the added value of infrastructures like Watson is in the fact that *semantic and qualitative analysis* of the harvested content is done independently of the semantic browser – on the infrastructural level. In addition to basic analytic information (e.g., data format or expressiveness), one can learn from Watson about the *topological and networked relationships* among ontologies and semantic data sets. All this helps us to acquire a heterogenous volume of semantic content applicable to any given web page; next, we briefly describe how the support for multiple ontological frames can be realized on the user level.

4.2 Semantic Browsing Using Multiple Ontologies

Our new semantic browser ("PowerMagpie") relies on the generic Watson framework introduced in Sect. 4.1. Watson extends our *semantic layering* technique (Sect. 2.7) by feeding multiple ontologies to it. Thus, PowerMagpie may make different use of the retrieved ontologies, based on their quality, topic coverage, or expressiveness, rather than merely finding *any* semantic content containing a given keyword.

Unlike the previous versions of Magpie that were restricted to semantic layering based on categories specific to a single ontology, the new framework is more flexible. Apart from offering multiple ontologies for any web page, PowerMagpie can discover additional semantically related content, which is not directly or indirectly referred by the user-selected ontology. This capability uses the fact that each ontology models a certain aspect of the world, from a particular, non-exhaustive perspective. Hence, it makes sense to view one ontology in the context, i.e., in a relation with other ontologies on the Web.

The strategy of finding semantic similarities is common, e.g., in query expansion, but not in search engines. The majority of search engines bases the similarity on the lexical proximity of resources, which, in turn, draws upon the underlying search index. When such a similarity-computing service is implemented outside the search engine scope – i.e., it cannot exploit the document index to explore the resource neighbourhood – a dynamic, document-specific descriptive vector of terms needs to be computed. In our framework, this capability is referred to as *document fingerprint*, and it is somewhat resembling a summary of the document, a set of key defining concepts. The fingerprint terms are submitted to the Watson Semantic Web Gateway (Sect. 4.1). The key idea of interfacing Watson rather than generic engines, such as Google, is to *reuse* already formalized and represented conceptual commitments captured in numerous ontologies that Watson harvested on the Web.

The technique is inherently iterative: the web browser plugin starts with an initial document fingerprint and tests its conceptual fitness against the existing ontologies. From the most relevant ontologies one can calculate semantic neighbours of the matched concepts, which, when returned to the PowerMagpie plugin serve as candidate fingerprint extensions. The plugin attempts to find matches to these fingerprint extensions, thus disambiguating between the different perspectives in which a document might be interpreted.

Different ontologies not only facilitate different navigational paths for the document interpretation, they also offer opportunities for an implicit annotation of the page and for an implicit ontology population. In many semantic web browsers in the past, annotations were merely visual and transient. Now, with discovering new ontologies it makes sense to store the annotations locally. One formalism that has been recently agreed upon to facilitate this reuse is RDFa [34], which supports either ad-hoc or ontology-driven annotation in an easily parsed and reusable style – e.g., for the purposes of social semantic annotations or tagging applications.

4.3 Functional Overview

PowerMagpie is implemented as a web browser bookmarklet,[5] which allows for a simple installation, invocation, and "at-the-glass"[6] integration with existing semantic extensions (e.g., the aforementioned RDFa visualizers). The prototype comprises two complementary components: the *User Interface* and the *Back-end Service.* The former extends a web browser and acts as the key element for user interaction. The back-end, on the other hand, facilitates ontology discovery, selection and matching services. The functions performed by the prototype include the following:

Term selection and ranking exploits the structure of a web page. For example, assigning the term appearing in the title or in a heading more weight than to a term randomly found in a document paragraph. This traditional filtering technique is extended by a calculation of weights from the popularity and frequency of these terms and of lexically similar terms in the actual ontology index maintained by Watson and Yahoo search engine.

Web page processing is very simple: upon invocation, the document object model (DOM) of the page is serialized into XML and shared with the back-end. The back-end carries out term extraction using TF/IDF[7] weighing [35] against the existing Watson indexes. During this process the actual lexicon is compiled from found entities and is checked against Yahoo for yielding additional lexical signatures of the document.

[5] A bookmarklet is a small application stored as a bookmark URL in a web browser or as a hyperlink on a web page.

[6] This is a common strategy in portal development, where small sub-applications can be quickly linked to create more complex applications, and our formalism of choice, RDFa, allows this kind of integration client-side.

[7] TF/IDF or "term frequency/inverse document frequency" is a well-known method for weighing terms in a text corpus.

Ontology selection draws on the generic Watson Semantic Web Gateway, which pre-computes ontological indexes and thus simplifies and accelerates the selection process. There are several challenges on this level, due to the requirement on the web browser to respond to the user's requests in real time. Hence, ontology selection and processing must be also done in real time.

Semantic matches are then returned to the web browser, where every matched entity is associated with a location in the web page and is expressed as an XPath expression [36]. This uniquely identifies concept occurrences in the text, and is a precursor for semantic disambiguation.

Semantic layering is basically a visualization of matches in the web browser. For example, one can see the matches on the level of entire ontologies or on the level of concepts shared by the discovered ontologies but conceptualized differently. These different visual views on the semantic content then create a dedicated semantic layer (or a skin) over the web page, and three types of visualized content are shown in Fig. 5a,b.

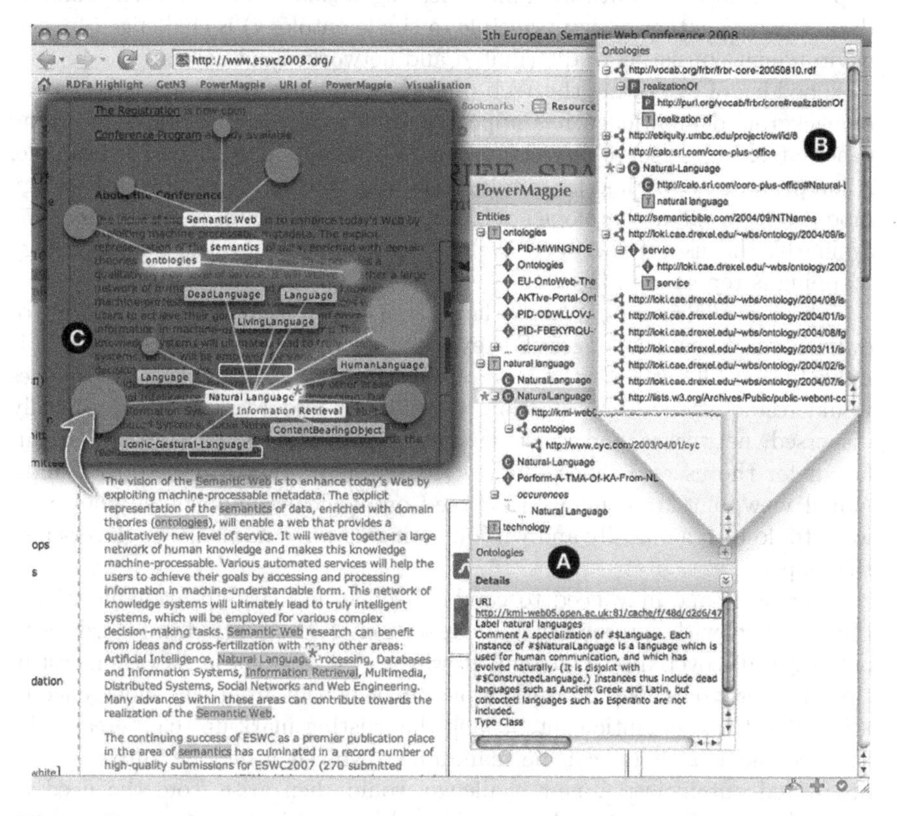

Fig. 5. "PowerMagpie" prototype of a Semantic Web browser with skins implementing three distinct views of the concept "Natural Language": a conceptual view (**a**), an ontological view (**b**) and a navigational graph visually superimposed on text (**c**)

Semantic browsing reflects further exploration of connected concepts that is permitted by the discovered ontologies. A prototype visualization[8] of ontology coverage and relationships is shown in Fig. 5c. The highlighted ontology-concept link represents the user's annotation of the ontology choice for that particular web page.

5 Discussion

The Semantic Web is gaining momentum and more semantic data is available online. This has an impact on the application development strategies. The original Magpie as described in the previous edition of this book [16] came out in the era before the aforementioned momentum in the Semantic Web became visible, so its assumption of no or little semantic mark-up available is now obsolete. The momentum implies that the new generation of Semantic Web application needs to work with more heterogeneous and distributed semantic data. Hence, another design principle (a single ontology) is challenged by this environment consisting of distributed and networked ontologies.

The idea of exploiting the Web (and the Semantic Web) as a large source of background knowledge has appeared in several recent works concerning generic tasks (e.g., sense disambiguation or ontology matching). For example, Alani proposed a method for ontology learning that relies on reusing ontology modules from online ontologies relevant to keywords from a user query [37]. Similarly, the use of the open Web as background knowledge for ontology mapping is reported in [38].

The use of Semantic Web at large as a resource in its own right introduces several new challenges. For example, in the open Semantic Web, it is unlikely that all ontologies and various lexicons derived from those ontologies would reside at the same location. Ontological resources are geographically dispersed, networked and richly interlinked. Given this, it is no longer sufficient for the user to choose ontology. Users may want to create their individual viewpoint from many networked ontological components. They may want to do it dynamically and without bringing any knowledge engineers into the loop.

Moreover, one may need to combine semantic mark-up available within the web page with external semantic assertions coming, e.g., from third-party ontologies discovered by Watson Gatewayor similar engines. As we suggest in this chapter, tools like "PowerMagpie" may benefit from this emergence of a large body of semantic content [39]: the existing mark-up (e.g., created by the document authors) can be maintained/evolved using the automatically discovered ontologies. A new challenge would then arise from the need to reconcile the differences and to combine these multiple sources in a manner

[8] A joint work with Takayuki Goto, Knowledge-as-Media Research Group, National Institute of Informatics and University of Tokyo.

that is transparent and useful for the end user. In our prototypes this has been piloted using the metaphor of "skinning" the web page with a set of views each supporting a particular form of semantic navigation.

5.1 Benefits of Using Multiple Ontologies

The key issue with older version of Magpie (and also other semantic web browsing tools) is the requirement upon the user to select, load and activate an ontology that would drive the application. For instance, in Magpie this has been achieved by means of ontology-derived lexicons, from which the application toolbar and semantic menus were created. The shortcoming is that such an approach assumes the user knows which ontology to use and where to load it from. Obviously, loading an inappropriate ontology would yield false positives (e.g., identifying terms in text such as "Cork" being an instance of "Tropical Wood" rather than "City" and a part of "Ireland").

Moreover, the user rarely has suitable means to assess the fitness of a particular ontology to the semantic annotation and interpretation of any given web page. Hence, the approach of combining Watson ontology discovery, analysis and access engine (Sect. 4.1) with an iterative matching between the document and the ontologies (Sect. 4.2) automates these key decision tasks for the user. The automation is achieved by means of:

- *Ranking terms* in the resource with an intention to produce an initial document fingerprint with semantic commitments
- *Selecting ontologies* by matching the set of key descriptive terms identified in the previous step to the index of harvested semantic content
- *Ordering discovered ontologies* to assist the user with assessing their fitness in terms of domain coverage, richness, and expressiveness
- *Creating dynamic semantic layers* based on serializing selected semantic content from the discovered ontologies and visualizing it in text

Semantic browsing is promoted in this chapter as a process of constructing and using semantic layers that are expressive and flexible in nature. Rather than using solely instances for semantic browsing (as, e.g., in faceted browsers and in semantic layering approaches), the proposed approach reminds the skins that visually amend user interfaces of many software applications. The "skinning" approach to semantic browsing embeds semantics *onto* any background text and thus supports advanced semantic analyses; for example:

- On the level of ontologies it supports the identification of different perspectives and their role in facilitating different routes in semantic browsing.
- On the level of conceptual entities the technique enables the user to compare what are the different meanings (and implications) of particular commitments in different ontologies.
- On the level of concept links the technique shifts the user's attention away from singular entities and presents ontological relationships among the entities in the web document.

- On the level of instances it supports ontology population and extension by mashing up and merging conceptual commitments from several sources into one (possibly persistent) skin representation.

5.2 Advances in Semantic Browsing

The idea of interlinking semantic annotation, semantic browsing and semantic services is gaining popularity. Annotation is no longer a separate objective in its own right; new annotation tools aim to offer additional services, e.g., validation or consistency checking. A major challenge in the domain of semantic browsing stems from the need to make the association between semantic services and semantic mark-up more open and more flexible. Furthermore, a good motivator with usable semantic browsing techniques are needed to make semantic browsing a mainstream user activity.

A potentially interesting input is likely to come from the deployment of modular ontologies and specialized services as opposed to monolithic ontologies with tightly integrated web services. The modular approach allows some of the services to be involved in evolving general knowledge captured on the Semantic Web. Some methods may use statistical techniques, whereas other services may rely more on social trust. It seems that knowledge evolution may provide a good test case and a motivator for semantic browsing tools – whether it is a formal evolution of knowledge within specific ontologies or repositories or a social, user-driven evolution based on the annotations and tags of users relying on similar ontologies in their browsing.

Another motivator for semantic browsing tools might stem from a rise of new techniques for information extraction, text analysis, knowledge validation or relationship discovery. While the low-level techniques rapidly change and become outdated, semantic browsers with sufficiently flexible architectures may benefit from those changes – a web browser is a very low-cost tool to be upgraded by ordinary users. Thus, the visual components of a semantic browser may hide the flux of underlying technologies – one would be able to use the latest *knowledge technologies* without any major re-design of the existing *user interaction techniques*. Semantic browsers may thus become a bridge to enable a shift from closed, single perspective application development to a smarter, on-demand knowledge construction.

Acknowledgements

The authors were supported by the climate*prediction*.net, Dot.Kom, KnowledgeWeb, Advanced Knowledge Technologies (AKT), OpenKnowledge and NeOn projects. climate*prediction*.net was sponsored by the UK Natural Environment Research Council and UK Department of Trade e-Science Initiative. Dot.Kom (Designing Adaptive Information Extraction from Text for Knowledge Management) by the IST Framework 5 (grant IST-2001-34038).

KnowledgeWeb was an IST Framework 6 Network of Excellence (grant FP6-507482), AKT an Interdisciplinary Research Collaboration (IRC) sponsored by the UK Engineering and Physical Sciences Research Council by grant GR/N15764/01, and OpenKnowledge and NeOn are Framework 6 research projects partly supported by grants IST-2005-027253 and IST-2005-027595.

References

1. P. Brusilovsky. Social Information Access: The Other Side of the Social Web. In Viliam Geffert, editor, *In 34th SofSem Conference on Current Trends in Theory and Practice of Computer Science*, pages 5–22. Springer, 2008.
2. T. Berners-Lee, J. Hendler, and O. Lassila. The Semantic Web. *Scientific American*, 279(5):34–43, 2001.
3. D. Brickley and R. Guha. Resource Description Framework (RDF) Schema Specification. W3C Recommendation, World Wide Web Consortium, 2000.
4. F. van Harmelen, J. Hendler, I. Horrocks, D.L. McGuinness, P.F. Patel-Schneider, and L.A. Stein. OWL web ontology language reference. W3C Recommendation, World Wide Web Consortium, 2002.
5. T.R. Gruber. Towards principles for the design of ontologies used for knowledge sharing. *Intl. J. of Human-Computer Studies*, 43(5/6):907–928, 1993.
6. D. Quan and D.R. Karger. How to make a semantic web browser. In *Proc. of the 13th Intl. Conf. on World Wide Web*, pages 255–265. ACM Press, New York, USA, 2004.
7. L. Hirschman and N. Chinchor. Named Entity Task Definition. In *Proc. of the 7th Message Understanding Conf. (MUC-7)*, 1997.
8. B. Popov, A. Kiryakov, A. Kirilov, D. Manov, D. Ognyanoff, and M. Goranov. KIM – Semantic Annotation Platform. In *Proc. of the 2nd Intl. Semantic Web Conf.*, pages 834–849. Springer, Florida, USA, 2003.
9. H. Cunningham, D. Maynard, K. Bontcheva, and V. Tablan. GATE: A Framework and Graphical Development Environment for Robust NLP Tools and Applications. In *40th Anniversary Meeting of the Association for Computational Linguistics (ACL)*. Pennsylvania, USA, 2002.
10. L. Carr, S. Bechhofer, C. Goble, and W. Hall. Conceptual Linking: Ontology-based Open Hypermedia. In *Proc. of the 10th Intl. WWW Conf.* ACM Press, Hong-Kong, 2001.
11. D. Huynh, S. Mazzocchi, and D.R. Karger. Piggy Bank: Experience the Semantic Web Inside Your Web Browser. *J. of Web Semantics*, 5(1):16–27, 2007.
12. D. Huynh, D. Karger, and R. Miller. Exhibit: lightweight structured data publishing. In *In Proc. of the 16th Intl. World Wide Web Conf.*, pages 737–746. ACM Press, Canada, 2007.
13. P. Yee, K. Swearingen, K. Li, and M. Hearst. Faceted Metadata for Image Search and Browsing. In *Proc. of the ACM Conf. on Computer-Human Interaction (CHI)*, 2003.
14. E. Oren, R. Delbru, and S. Decker. Extending faceted navigation for RDF data. In *Proc. of the 5th Intl. Semantic Web Conf.*, pages 559–572. Georgia, US, 2006.
15. M. Hildebrand, J. van Ossenbruggen, and L. Hardman. /facet: A browser for heterogeneous semantic web repositories. In *Proc. of the 5th Intl. Semantic Web Conf.*, pages 272–285. Georgia, US, 2006.

16. J. Domingue, M. Dzbor, and E. Motta. Semantic Layering with Magpie. In S. Staab and R. Studer, editors, *Handbook on Ontologies in Information Systems*. Springer, 2003.
17. M. Dzbor, J. Domingue, and E. Motta. Magpie: Towards a Semantic Web Browser. In *Proc. of the 2nd Intl. Semantic Web Conf.*, pages 690–705. Springer, Florida, USA, 2003.
18. M. Dzbor, J. Domingue, and E. Motta. Opening Up Magpie via Semantic Services. In *Proc. of the 3rd Intl. Semantic Web Conf.*, pages 635–649. Springer, Japan, 2004.
19. V. Lanfranchi, F. Ciravegna, and D. Petrelli. Semantic Web-based Document: Editing and Browsing in AktiveDoc. In *Proc. of the 2nd European Semantic Web Conf.*, pages 623–632. Springer, Greece, 2005.
20. M. Dzbor, E. Motta, and J. Domingue. Magpie: Experiences in supporting Semantic Web browsing. *J. of Web Semantics*, 5(3):204–222, 2007.
21. M. Dzbor and E. Motta. Engineering and Customizing Ontologies: The Human-Computer Challenge in Ontology Engineering. In M. Hepp, P. De Leenheer, A. de Moor, and Y. Sure, editors, *Ontology Management: Semantic Web, Semantic Web Services, and Business Applications*, volume 7 of *Semantic Web and Beyond*, chapter 10, page 350. Springer, Heidelberg (Germany), 2007.
22. A. Broder. A taxonomy of web search. *ACM SIGIR Forum*, 36:3–10, 2002.
23. R. Guha, R. McCool, and E. Miller. Semantic Search. In *Proc. of the 12th Intl. Conf. on World Wide Web*, pages 700–709. ACM Press, 2003.
24. T. Berners-Lee. Tabulator: Exploring and analyzing linked data on the Semantic Web. In *Proc. of the 3rd Semantic Web User Interaction Wksp (at ISWC)*, page `http://swui.semanticweb.org/swui06`. Georgia, US, 2006.
25. E. Pietriga, Ch. Bizer, D. Karger, and R. Lee. Fresnel: A browser-independent presentation vocabulary for RDF. In *Proc. of the 5th Intl. Semantic Web Conf.*, pages 158–171. Georgia, US, 2006.
26. S. Mazzocchi, S. Garland, and R. Lee. SIMILE: Practical Metadata for the Semantic Web. In *XML.com Online*. O'Reilly (`http://www.xml.com/pub/a/2005/01/26/simile.html`), 2005.
27. M.C. Schraefel, D.A. Smith, A. Owens, A. Russel, C. Harris, and M.L. Wilson. The evolving mSpace platform: leveraging the Semantic Web on the Trail of the Memex. In *Proc. of the Intl. Conf. on Hypertext*. ACM Press, Austria, 2005.
28. A. Hogue and D.R. Karger. Thresher: automating the unwrapping of semantic content from the World Wide Web. In *Proc. of the 14th Intl. Conf. on World Wide Web*, pages 86–95. ACM Press, Japan, 2005.
29. D. Quan, D. Huynh, and D.R. Karger. Haystack: A Platform for Authoring End User Semantic Web Applications. In *Proc. of the 2nd Intl. Semantic Web Conf.*, pages 738–753. Florida, USA, 2003.
30. P. Buitelaar and T. Eigner. Semantic Navigation with VIeWs. In *UserSWeb: Wksp. on User Aspects of the Semantic Web*, 2005.
31. M. Dzbor and D. Rajpathak. Report on the status and evaluation of ASPL-v2. Deliverable D3.3.7, KnowledgeWeb Project, 2007.
32. M. d'Aquin, M. Sabou, M. Dzbor, C. Baldassarre, L. Gridinoc, S. Angeletou, and E. Motta. WATSON: A Gateway for the Semantic Web. In *Posters of the 4th European Semantic Web Conf.*, 2007.
33. L. Ding, R. Pan, T. Finin, A. Joshi, Y. Pen, and P. Kolari. Finding and ranking knowledge on the Semantic Web. In *Proc. of the 4th Intl. Semantic Web Conf.*, pages 156–170. Springer, Ireland, 2005.

34. B. Adida and M. Birbeck. RDFa Primer 1.0: Embedding RDF in XHTML. W3C Working Draft, World Wide Web Consortium, 2007.
35. Ch. D. Manning, P. Raghavan, and H. Schütze. *Introduction to Information Retrieval.* Cambridge University Press, to appear in 2008.
36. J. Clark and S. DeRose. XML Path Language (XPath). W3C Recommendation, World Wide Web Consortium, 1999.
37. H. Alani. Ontology Construction from Online Ontologies. In *Proc. of the 15th Intl. Conf. on World Wide Web.* ACM Press, Scotland, 2006.
38. M. Sabou, M. d'Aquin, and E. Motta. Using the Semantic We as Background Knowledge for Ontology Mapping. In *Proc. of the Ontology Mapping Workshop (collocated with ISWC 2006).* Georgia, US, 2006.
39. M. Dzbor. Best of Both: Using Semantic Web Technologies to Enrich User Interaction with the Web. In Viliam Geffert, editor, *In 34th SofSem Conference on Current Trends in Theory and Practice of Computer Science*, pages 34–49. Springer, 2008.

36. [illegible] and [illegible] ... [illegible]. W3C Working Draft, World Wide Web Consortium, 2002.
37. [illegible] ... [illegible] and H. Schütze. *Introduction to Information Retrieval*. Cambridge University Press, to appear in 2008.
38. [illegible] ... Meta Language (XML). W3C Recommendation, World Wide Web Consortium, 1999.
39. [illegible] ... and Ontology. [illegible] ... World Wide Web. ACM Press, [illegible].
40. [illegible] and [illegible]. Using [illegible] ... Knowledge [illegible] ... Proceedings [illegible] ...
41. M. [illegible] ... [illegible] ... [illegible] ... [illegible].

Part VI

Ontology-Based Applications

Ontologies for Knowledge Management

Andreas Abecker[1] and Ludger van Elst[2]

[1] FZI Forschungszentrum Informatik, Department of Information Process Engineering (IPE), 76131 Karlsruhe, Germany, abecker@fzi.de
[2] German Research Center for Artificial Intelligence (DFKI GmbH), Knowledge Management Department, 67655 Kaiserslautern, Germany, elst@dfki.uni-kl.de

Summary. Within Computer Science and Artificial Intelligence, the term *ontologies* was coined in the *Knowledge Sharing and Reuse Effort*, for efficient engineering of (distributed, cooperating) knowledge-based systems. It is not surprising that it soon entered the Knowledge Management (KM) area: *Sharing* and *reuse* of personal, group, and organizational knowledge are among the central goals aimed at in most KM projects. In this chapter we introduce the main ideas of KM, as well as the role of and requirements for information technology (IT) in KM. We discuss the potential of ontologies as elements in IT support for KM. We characterize their current role in research and practice, derive a working focus for the near future, and conclude with an outlook on trends in KM software and their implications on ontologies.

1 Information Technology for Knowledge Management

Knowledge Management is an interdisciplinary topic with roots in Economics, Information Technology, Pedagogy, Psychology, and Organization Theory (cp. [20, 33]). We can define *Knowledge Management* as a:

- *Systematically managed* organizational activity
- Which views *implicit* and *explicit knowledge* as a key strategic resource of an organization, and thus
- Aims at improving the handling of knowledge *at the individual, team, organization, and inter-organizational level*
- *In order to achieve organizational goals* such as better innovation, higher quality, more cost-effectiveness, or shorter time-to-market
- By employing tools, techniques, and theories from manifold areas such as *IT, strategic planning, change management, business process management, innovation management, human resource management*, and others
- In order to achieve a planned impact on *people, processes, technology*, and *culture* in an organization

S. Staab and R. Studer (eds.), *Handbook on Ontologies*, International Handbooks on Information Systems, DOI 10.1007/978-3-540-92673-3,

From the very beginning of KM, two streams of research and applications could be identified, following the *process-centered* and the *product-centered view* on KM, respectively.[1]

(1) The *process-centered view* mainly understands KM as a social communication process. It assumes that the most important knowledge source of an organization are its employees, and that solving *wicked problems* [18] is merely a process of achieving social commitment than one of problem solving. Consequently, knowledge exists, is created, and is further developed in the interaction among people and tasks – such that the focus of IT should be to enable, to facilitate, and to support communication and collaboration.

Technical solutions in this area comprise, e.g., yellow page and expert-finder systems for determining the right communication partner, Computer-Supported Collaborative Work (CSCW) systems for effective collaboration between geographically separated people, or Skill Management systems for the systematic and planned acquisition and development of human skills.

In this view, organizational measures play a particularly important role, e.g., the installation of expert networks, the running of training courses, the facilitation of virtual teams and communities of practice, and all kinds of cultural KM support.

(2) The *product-centered view* assumes that knowledge can exist outside of people and can be treated as an object within IT systems. It focuses on knowledge documents, their creation, storage, and reuse in computer-based *organizational memories* (OMs). It is based on the idea of explicating, documenting, and formalizing knowledge in order to have it as a tangible resource, and on the idea of supporting the user's individual knowledge development/usage by presenting the right information sources at the appropriate time.

The transition from intangible (implicit and tacit) to tangible (explicit) knowledge in the form of standardized processes and templates, of FAQs, lessons learned, best practices documents, etc., allows a company to enhance its structural capital to some extent – maybe at the price of reducing creativity and flexibility. Basic techniques for this approach come from Document Management, Knowledge-Based and Information Systems.

In this view, organizational measures aim at fostering the use and improving the value of information systems by bonuses, or by installing organizational roles/processes for high-quality knowledge-content management.

Figure 1 illustrates the typical software support for both approaches to KM. By analyzing the landscape of IT support for KM (cp. [42]), one may identify the following four types of KM applications:

---

[1] The same dichotomy was also called *personalization vs. codification strategy*, *organic vs. mechanistic* approach, or *community model vs. cognitive model* view, see [1,44].

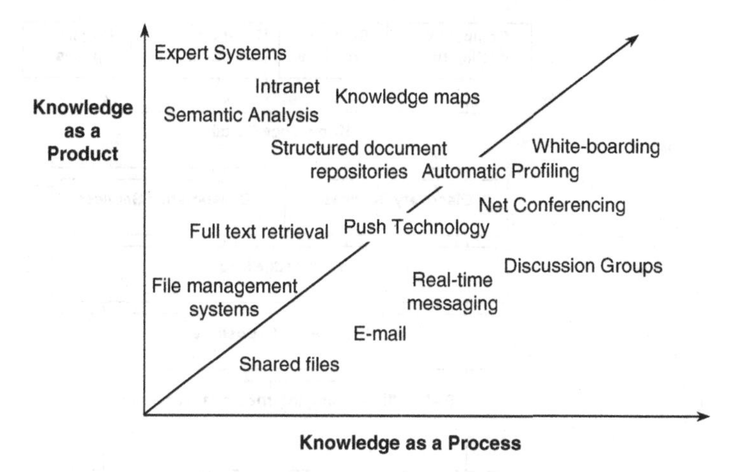

Fig. 1. Software support for the product and the process approach to KM

Type 1 Applications: Conventional Software Basis

Type 1a Applications: Standard Software Applications. Especially the early success stories of KM were – although IT-enabled and heavily IT-dependent – not building upon any *new* IT solution; they just employed conventional technology like databases or discussion boards. This should not be underestimated in practice. However, it is not interesting in the context of this book.

Type 1b Applications: Integrated Standard Software. The first generation of dedicated KM tools is characterized by the deep integration of manifold aspects of KM support in *one* software suite, hence incorporating both product and process aspects of KM. Typical representatives are the big KM tool suites successful in the market. Tools like Livelink, Lotus Notes, Autonomy's or Inxight's products, usually combine (1) many types of synchronous and asynchronous communication and coordination for collaboration and process management, with (2) functionalities for document and web-content management, many individual and organization-wide information management functions (push and pull services, Intranet portal functionalities, etc.) and (3) advanced document classification and information retrieval (IR, enterprise and Internet search) technology.

Figure 2 depicts an abstract, comprehensive KM tool architecture which:

- Incorporates data and information from manifold sources
- Organizes it according to a common corporate knowledge map
- Provides collaboration and discovery services working upon these organizational knowledge sources
- Feeds these services through a common knowledge portal into operational business processes and into KM processes

Type 1 applications often do not maintain a knowledge-rich, explicit ontological basis; nevertheless, the box "Knowledge Map" in the middle of the

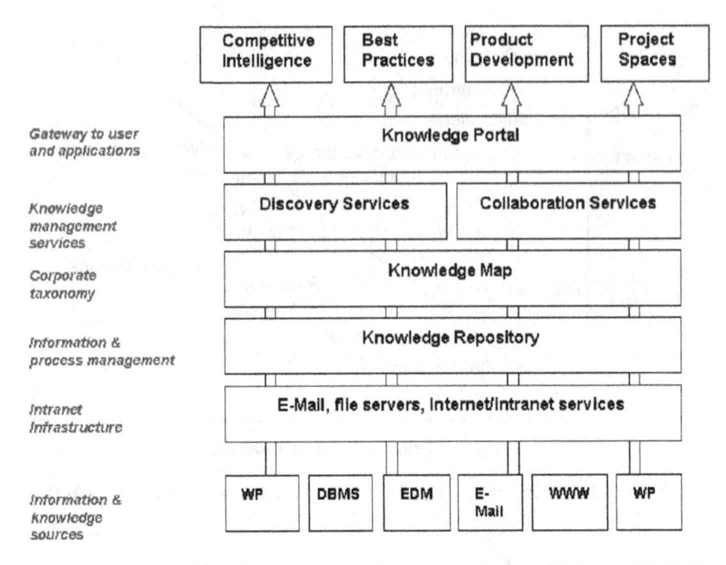

Fig. 2. Abstract KM system architecture, adapted from OVUM

picture points out the central role of a shared language to connect people to people, people to information, and information to information, in an organization. This is the target area for more "heavy-weight," knowledge-based approaches in order to improve KM systems and services by ontologies.

Type 2 Applications: Intelligent Software Basis

Type 2a Applications: Intelligence-Enhanced Solutions. While Type 1 applications are based on "conventional" IT, we here subsume applications based on Artificial Intelligence methods, including ontologies as a core enabler. Figure 3 gives some examples arranged according to their role in different KM core processes. Type 1 represents the current, product-based, state of practice in (mainly, big) industries and administrations, whereas Type 2 comprises many mature prototypes, as well as niche-products offered by small or medium-sized, leading-edge companies, with operational solutions mostly deployed for early adopters.[2] Type 2a applications are today's most important field of ontology usage in KM. Ontology-supported tools and functionalities comprise:

- Intelligent Search and Retrieval in Intranet and Internet
- Information Gathering, Information Extraction and Information Integration with ontologies as target data structure
- Semantic Community Web Portals
- Expert Systems and Intelligent Advisor Systems

[2] For concrete software products it is sometimes difficult to clearly distinguish Type 1 and Type 2. But, for the purpose of this book, it is sufficient to clarify whether an application is based on explicit, formal ontologies, or not.

	Share knowledge	Distribute knowledge	Capture & codify knowledge	Create knowledge
Traditional systems	E-Mail, Group Collaboration, Discussion Groups, P2P technology, Intranet Portals	Word processing, DTP, Document Management	*All systems that codify knowledge are knowledge-based*	Brainstorming software, mindmapping, statistical analysis
Knowledge -based systems	Ontology-based portals	Expert Systems, Lessons-Learned + Best Practice Systems	Knowledge acquisition and coding tools	Knowledge discovery and data mining systems, Creativity systems

Fig. 3. Traditional vs. knowledge-based KM technology (i.e., type 1 vs. type 2 applications), adapted from [77]

We will discuss these fields in more detail in Sect. 3 below.

Type 2b Applications: Enhanced Solutions Integrated. KM is by definition *boundary spanning*, i.e., bridging the gaps between departments and organizations, between people and information, and between different kinds of software services; so, it is both a challenge and a chance of KM software design to try capturing "the whole picture," i.e., integrate the product and the process view, and cover the whole architecture sketched in Fig. 2. Because they aim at exactly this goal, Type 1b applications are the first tools which deserve the dedicated name "KM software." Now, the interesting, but hardly ever posed, question for Type 2 applications is how different knowledge-based functions in a comprehensive KM application can exploit synergies. We will sketch some approaches in Sect. 4 below.

While the first generation KM success stories typically were built on type 1a, or, seldom, type 2a applications – like a Lessons Learned database, an Expert System, or a Yellow Page system – the big commercial KM toolboxes are comprehensive type 1b approaches which integrate manifold complementary functionalities. Seen from the IT – or, AI – point of view, the interesting questions are to which extent 1a services can be improved towards ontology-based 2a approaches, and what possibilities arise when thinking about integrated type 2b applications. Before we come to these questions, we will shortly discuss some general requirements for KM software and their implications for the use of ontologies.

2 Requirements for KM Software and Ontologies

We discuss three major KM requirements and their implications for the ontology topic, namely *(1)* minimalization of upfront knowledge engineering, *(2)* integration of KM support with everyday work procedures and tools, as well as *(3)* integration of heterogeneous kinds of information.

R1: Minimalization of Upfront Knowledge Engineering

Since KM is considered an additional organizational task, orthogonal to the "productive" work, expensive start-up activities would be a major barrier for successful KM initiatives. On the other hand, it seems clear that no ontology-based approach can be introduced without an explicit commitment of all people involved and without their contributions to ontology engineering. Hence, all topics dealing with a smooth and cost-efficient introduction of ontology-based applications are particularly important for ontology-based KM:

- *Method-Driven Ontology Engineering.* There are many far-developed ontology modeling and management tools.[3] Knowledge Management provides significant difficulties for such a framework: The ontology shall (often) be built and maintained for a community-spanning use, seen from different perspectives, in an evolving domain, by non-Knowledge Engineers. So, we expect (in the ideal case) not only incorporation of legacy structures and text analysis results, plus technical solutions for the distributed creation and use of an ontology, but also a convincing methodological approach (see chapter "Ontology Engineering Methodology" of this book) built into the ontology tool suite for guiding and supporting the user. We expect community concepts like distributed discussion support, versioning concepts, help for managing the informal-formal transition in a group discussion process for ontology building,[4] and, in the optimum solution, an integrated support for all steps of a *business-oriented* KM methodology. There are several approaches in this spirit, for instance:
  - In the DECOR, the PROMOTE, and the SPEDE approaches in the realm of Business-Process Oriented Knowledge Management, top-level business-process analysis is intertwined with (and, actually, provides the overall frame for) knowledge acquisition for ontology engineering [1, 19, 36].
  - In the OPAL ontology-modeling framework, business-specific terms are offered to the "not-ontology-expert" modeler as modeling templates [49].
- *Ontology learning.* At least the "first cut" ontologies to start with in an organization should avoid the typical "cold start" problems by building as much as possible upon structures already explicit in the organization or hidden in the organization's text documents. The use of machine learning and text analysis algorithms for ontology structuring and population is discussed in this book in chapters "Ontology Learning" and "Information Extraction."
- *Ontology reuse.* Regarding *cross-organizational ontology reuse*, not only technical and methodological provisions must be made (cp. chapters

[3] Like, for instance, the NeOn toolkit (`http://www.neon-toolkit.org/`), KAON (`http://kaon.semanticweb.org`), or Protégé (`http://protege.stanford.edu`), cp. [29].

[4] Cp. Tadzebao, WebODE, or DILIGENT [29, 75].

"Ontology Engineering Methodology" and "Ontology Mapping" in this book), it is also a "political" and economic challenge to bring together a significant number of companies – typically competitors – to embark upon creating a shared ontology of their application domain. So, the main success stories are driven by public efforts and/or research scientists, like medicine, biology/lifesciences (cp. chapter "Application of Ontologies in Bioinformatics" in this book), and genetics [51,58]. Other areas with good chances for reuse concern broadly shared concepts in product and business modeling required for e-commerce and product-data sharing, as well as areas dominated by few global players, like the insurance sector (cp. `http://www.acord.org`). In such sectors, ontologies or, at least shared data models, have been motivated by data-exchange and resource-integration scenarios; but these sectors could profit even more from comprehensive KM. Regarding *organization-internal reuse*, the problem is obviously much smaller such that a company strongly dedicated to ontology-based applications might profit much from internally reusing ontologies and ontology modules – provided the organization structure and processes foster and support that.

R2: Integration of KM Support with Everyday Work Procedures/Tools

In order to achieve a good user acceptance and to realize a maximum effect on knowledge workers' task performance, it is useful to integrate KM software as seamlessly as possible with the tools already in use for daily work. Several research prototypes and industrial case studies address this goal by coupling knowledge storage and retrieval with workflow enactment which controls the operational business process [1,36]. Here, the ontology is the "glue" between operative tasks and KM tasks, on one hand (describing task-specific knowledge needs expressed in terms of an application domain ontology), and the Organizational Memory System, on the other hand (annotating knowledge resources semantically with ontology-based metadata).

The authors of [34] go a step further: they propose an ontological foundation of all business modeling based upon (1) a static ontology (the things in the world their attributes and relationships); (2) a dynamic ontology (states, state transitions, and processes); (3) a social ontology (agents, positions, roles, organization forms); and (4) an intentional ontology (believes, goals, etc., of agents). Such a comprehensive semantics-based business model could be the basis for powerful KM services and systems. The EULE system presented by [54] formally represents and enacts even more task knowledge:

- Process aspects (temporal and causal relationships)
- Normative aspects (deontic knowledge)
- Terminological aspects (concepts and their relationships)

are modeled in an insurance application, in order to realize partially automatic problem-solving and knowledge-based information retrieval. Such heavy-weight approaches are promising in domains (1) which are strongly regulated by law, norms, and regulations, such that (2) a high degree of formalization can be achieved, and (3) where once-formalized knowledge can be employed in manifold different applications (cp. [13,71]). Another application domain with similar characteristics are medical guidelines [52].

R3: Dealing with Heterogeneous Kinds of Information

Looking for a *practical* definition of "knowledge" (in contrast to data and information), it seems important that knowledge is always oriented towards *action* – this aspect is already treated above; other aspects concern the fact that knowledge is strongly related to *context* and that it has a *network* character – showing how pieces belong together. Technically, this leads to the requirement that KM applications often have to process data, information, and information sources created for capturing knowledge (like lessons-learned entries or best-practice documents) in a highly integrated manner. As a solution approach, such knowledge documents are annotated with metadata which can be processed automatically and set into relation with application data. Hence, a KM application should be built upon an *Information Ontology* [1] which defines:

- Which *types of documents* occur.
- What *metadata attributes* they have and which ontologies determine the *value ranges* of these attributes, where:
  - This may differ from document type to document type: a lesson learned may have a pointer to the project it was created in and the question how successful this project was; whereas a technical report may have an attribute for the location of the hardcopy of the document in the library, or links to experts for the technology described.
  - Such metadata attributes may also be application specific; e.g., in an *e-Learning* application (which can be seen as a specific KM task) it might be important to specify how difficult to understand a document is and which prior knowledge is required; whereas in a *Knowledge Trading* scenario [4] attributes for pricing models, IPR issues and contract models might be required.
- What *relationships between documents* are represented; linking logically related documents is a powerful mechanism for representing context; for example, in the EULE system knowledge with different degree of formalization is linked together (e.g., formal inference rules and their textual explanations [54]); discourse representation and group decision support systems implement in a somehow "hardwired" manner an information ontology by providing different kinds of message types (e.g., issues, arguments, questions) and relationships (e.g., explains, corroborates, contradicts) for documenting argumentation structures in meetings or discussion processes [46]; in an e-learning system specific relations may describe that some lessons require other lessons as (mandatory or useful) prior knowledge, or that an example illustrates some definition.

These ideas have been applied in the field of *Experience Management* in Software Engineering where sophisticated domain-specific information ontologies have been developed in order to identify those facets of a software-development experience which are important for assessing its later reusability in another situation [74]. Other areas which employ case-based reasoning approaches for reuse of, e.g., technical designs, go into similar directions (cp. [56]). Another area which can be seen a part KM or, at least, pretty close to KM, is eLearning – where also many sophisticated information ontologies exist already and are used mainly for personalization purposes (see, e.g., [31]).

Of course, real-world KM applications (and their ontology aspects) have not only to meet the requirements described above, but also hold a *rigorous cost-benefit analysis*. A detailed analysis of an expected ontology life cycle can be a powerful guide to achieve an optimal level of formalization in terms of costs and benefits. Likewise an explicit handling of an ontology's sharing scope helps minimizing negotiation costs as well as the complexity of revision processes in case of ontology evolution (these dimensions and their trade-offs are theoretically discussed in [69]). After these design considerations, we show some practical examples in the following sections.

3 Ontologies in Intelligence-Enhanced Applications

O'Leary characterizes the role of IT in KM as "converting and connecting" [47], with the following specific functions:

- Conversion of data and text into knowledge
- Conversion of individual and group's knowledge into accessible knowledge
- Connection of people and knowledge to other people and other knowledge
- Communication of information between users
- Collaboration between different groups
- Creation of new knowledge that would be useful to the organization

Typical ontology-based KM applications to support these functions are:

(1) Knowledge Portals for Communities of Practice. A *Community of Practice* (CoP) is a, typically, informal, self-organizing group of individuals with an interest in a particular practice, for example the group of people in a company who do the same (or partially overlapping) jobs [6]. The CoP might be contained within an organization, or spread across several. The CoP members have in common a desire to develop their competence, either for pleasure or pride in their ability, or for improving their work efficiency. CoP members typically exchange "war stories," insights or advice on specific problems, or tasks connected with their common practice. A CoP can act as a part of the organizational memory, transfer best practice, provide mechanisms for situated learning, and act as a focus for innovation.

Knowledge Portals, or Community Portals act as an information intermediary which structures all aspects relevant to a given, specific topic, in order to allow a community of users to flexibly and easily access a huge amount of information in different formats (today, usually text documents) related to this topic, to exchange information and communicate about the topic in quest, and to maintain and extend the content base accessed via this Internet (or, Intranet) portal [61]. Normally, such a portal comprises browsing and searching mechanisms for documents, as well as community services like on-line forums, mailing lists and news articles. Examples comprise the OntoWeb Semantic Web community portal [60], or the RiboWeb portal for molecular biology [7].

(2) Organizational Memories. An Organizational Memory Information System OMIS, or, for short, Organizational Memory OM [1, 23] – or, as a specialization, a Project Memory [28] – is a computer system within an organization which continuously gathers and actualizes knowledge and information (from within and from outside the organization) and provides it to the end user in a context-dependent and task-specific, manner, thus offering proactive assistance to a knowledge worker dealing with knowledge-intensive tasks. The OMIS collects manifold types of information, such as best practice and lessons learned documents, news articles, document templates, company regulations and manuals, CAD drawings, minutes of meetings, etc. Typical OMIS functionalities comprise integration of knowledge with different degree of formalization, intelligent problem-solving assistance by automatic generation of partial problem solutions [54], and context-aware, task-specific retrieval of information [1].

(3) Lessons Learned Archives. A Lesson Learned (LL) is a piece of knowledge gained through experience, which if shared, would benefit the work of others.[5] It is typically generated from a customer project in a debriefing step, or created by an innovation or adverse experience which lead to some shareable insight to promote repeated application, or avoid reoccurrence, respectively. Lessons Learned systems are typically used in Consulting firms [48], large technology companies, or in big government institutions, like military. Technologically, the challenge in LL systems lies in finding (and filling) an appropriate metadata schema (or, information ontology) which allows to precisely assess the potential value of a given LL as a reuse candidate in a new situation [76]. As a related problem, the question of matchmaking arises (compare stored LL metadata with characteristics of the current situation to estimate whether the application of the LL will be useful) which is today often addressed by CBR (Case-Based Reasoning) methods. The authors of [72] distinguish four types of LL systems according to the way the systems capture their input *(passive vs. (semi-)automatic)* and according to the way the LLs are published to the users *(push vs. pull)*.

[5] See `http://www.aic.nrl.navy.mil/~aha/lessons/`, cp. [76].

(4) Expert Finder and Skill Management Systems. Since tacit and not (yet) explicated knowledge is at least as important for KM as explicit, documented knowledge, the "classical" means for connecting people to people – yellow page systems, simple expert directories, and personal web pages – are typical "quick win" applications for KM (cp. [9]).
More advanced approaches for expert finders (e.g., for project-team configuration, specific technical questions, or strategic knowledge-development plans in the organization) try to avoid the manual creation and continuous maintenance of skill profiles; instead, existing explicit information is analysed, like documents created by a person, documented trainings and formal qualifications, project membership, collaboration or co-authorship relations, information flows, etc. [78]. Further improvements comprise sophisticated matching functions for skill profiles (compare an employee's skills with a job's required skill profile) [11, 64]. In the ideal case, such functionalities are integrated into the personal and organizational skill and Human Resource management functions for planning, monitoring, staffing, etc. [11]. Further extensions comprise additional value-adding services like automatic scheduling of appointments for knowledge exchange between users, provision of extra information during interactions, negotiation support for knowledge exchange planning, etc.
Ontologies are normally used to structure the area of competencies, sometimes also to structure the environment in which competencies were acquired, used and further developed, i.e., projects, publications, etc. Most important, ontologies provide the background knowledge for knowledge-based matching functions and complex similarity measures for comparing skill profiles. Examples for ontology-based skill management include the SwissLife case study from the insurance sector [39], the OntoProper case in the software sector [64], or the DaimlerChrylser case [11].

3.1 Usage and Benefits of Ontologies

In the above mentioned, major knowledge-based KM applications, ontologies are mainly used for the following three general purposes (see also [21]):

O1: Ontologies Support Knowledge Search, Retrieval, and Personalization

The most important application of ontologies in KM – besides browsing interfaces in Knowledge Portals – is certainly to improve search and retrieval of documents by exploiting ontological background knowledge about the application domain.

In [43], the basic ideas are described how to use taxonomies for increasing recall of information retrieval (IR) when browsing and querying. Normally, in the case of an empty or small answer set, taxonomic knowledge is used for extending the query by sub-concepts or super-concepts.

In the Electronic Fault Recording system for structured documentation and retrieval of maintenance experiences for a complex and large mechanical device,[6] the retrieval of potentially useful documented experience is not only supported by a detailed machine-model in terms of is-a and has-part relations; similarity of potentially useful situations can also be assessed using modeled links for hydraulic and electrical connections between machine parts, as well as analogy relationships between similarly constructed machines [10].

The authors of [40] propose a declarative search-heuristics language in an ontology-based skill management prototype: Potentially useful information is inferred via graph-traversal following domain-specific links in the knowledge-base, e.g., about project-team membership. Similarly, rules about relationship instantiations are used in [64] for skill inferencing.

In general, the more specifically a domain is described, the more powerful inferences for query expansion and query reformulation are possible; however, detailed models are expensive to acquire and maintain, such that we have the typical KM trade-off asking for economic rationality when deciding between "high-tech" and "low-tech" approaches.

While the approaches above usually increase recall of IR, precision is not so often treated explicitly. The KonArc prototype (for storage and retrieval of experience in a database for software-solution designs) used domain-specific information about incompatibilities of search constraints (e.g., between operating systems and specific software packages) for early detecting empty answers sets (and also explaining the contradictions to the user) [59]. For yellow pages and product catalogues, it is shown in [30] that an ontology coupled with a linguistic knowledge-base can increase *both* recall and precision, because it supports query disambiguation for polysemous query terms.

The so-far discussed approaches all describe information *pull* situations; of course, ontologies are also a means to provide the vocabulary for expressing personal interest profiles for information *push services* which automatically deliver knowledge and information for categories a user is interested in – be it in personalized knowledge portals that are offered by many KM tool-suite vendors, in KM-oriented RSS feeds, or in mobile KM scenarios which need a proactive knowledge supply. An example is the myPlanet system which creates personalized news with the help of an ontology-based user profile [35].

In general, the issues of sophisticated, ontology-based representations and processing of (life and work) *context*, *user profiles*, and *user activities* in order to realize high-precision retrieval, proactive, context-dependent knowledge supply, personalization of retrieval and presentation, collaborative retrieval, usage mining, proactive knowledge collection, group-knowledge sharing, etc., are still active and promising research topics in the intersection of KM and ontology research.

---

[6] Log entries of maintenance experience comprise fault events, maintenance measures, repair actions, etc.

O2: Ontologies Serve as the Basis for Information Gathering, Integration, and Organization

KM deals with knowledge resources of different degree of formality, often informal text documents. On the other hand, the more formally represented information we have, the more and better formal inferences are possible – for query answering and passage retrieval, for derivation of new knowledge, or for comparing and integrating facts and documents from different sources. More formalized information (i.e., facts related to a predefined schema) allows, e.g., to partially automate problem solving or to integrate IR results into operative business applications. The basis for such inferences are the *information ontology* structuring the metadata of informal knowledge sources, and the *domain ontologies* structuring the content area of documents and providing background knowledge for inferences. This background knowledge may comprise information *search knowledge* as well as domain-specific *application knowledge. Information Extraction* (IE) algorithms (see chapter "Information Extraction" in this book) for (semi-)automatically annotating metadata to documents and *Text Categorization* techniques [57] for finding semantic content indexes map informal sources to values of formal metadata attributes.

For realizing *Business Intelligence* applications in a KM context, domain ontologies provide the target data structures for gathering information from different sources in the Internet or a corporate Intranet. For example, in [47] a Price Waterhouse Coopers application is described that fills information frames about management changes in companies by analyzing a stream of business news articles. Similar applications are reported for filtering specific events out of news articles about economy or politics, for analyses in the military sector, and for fact extraction from personal web pages or publication web pages (see [38] for a technology survey). The authors of [5] describe an ontology-based application which creates narrative biographical sketches of artists based upon information automatically gathered, extracted and integrated from Web pages.

The Ontobroker [22] showed early how formal inferences can support retrieval and analysis of distributed information in the WWW. Prototype systems like PowerAqua, AquaLog or ORAKEL show how advanced ontology-based Semantic Web methods, natural-language processing, and machine learning can be combined in systems that lift document retrieval to real *question answering* from distributed resources (see, for instance, [12,41]). In [68], an overview of the state of the art is given regarding *semantic document annotation and metadata extraction* – which is an indispensable prerequisite to make full use of Semantic Web retrieval and processing technologies; the authors also discuss KM requirements and unsolved issues in this respect.

O3: Ontologies Support Knowledge Visualization

Different aspects of visualization for information search have been discussed in the literature on Human–Computer Interaction (HCI) and in the Digital

Library community (see, e.g., [37]). With the advent of the Internet society, such methods gain growing interest (cp. [16]) for surveying and analyzing big amounts of information with complex interconnections. A survey about applicable visualization methods is given by [32], including (1) basic *graph layout approaches* (like H-tree layouts, balloon views, radial views, tree-maps, cone trees, hyperbolic views, etc.); (2) *navigation and interaction* techniques (such as zoom-and-pan, focus+context techniques like fisheye distortion, and approaches to incremental exploration); and (3) *clustering* for grouping data based on a chosen semantics and reducing the number of shown nodes or the complexity of the created view by methods like ghosting, hiding, or grouping.

Such methods can be used for inspecting the metadata and content descriptions of knowledge stocks in order to create new knowledge by analysis and recombination of existing knowledge. In such cases visualization may help to illustrate structure (e.g., content density) and distribution of content in a document corpus, as well as relationships between specific metadata attributes (like time or geographic relationships regarding document content or document creation, as well as co-authorship relations between people). Visualization of content structures can even be useful for *intra*-document analysis for long documents like government reports, classical literature, socio-economic almanacs, etc. – in order to get a rough overview of topics discussed, of their textual manifestation, and their interrelationships, or in order to have a quick, topics-based access to document parts. Visualization is also valuable for finding useful knowledge items in vaguely specified search situations where (partially) exploring the information space is a part of problem-solving and helps clarifying the problem specification and/or its solution space.

In the meanwhile, visualization for topic-oriented document access went into commercial practice.[7] A number of commercial companies offer tools for knowledge and information visualization, for instance:

- USU AG (`http://www.usu.de`) or intelligent views GmbH (`http://www.i-views.de/`), among others, use a semantic network interface for browsing, navigating, and exploring the major topics and topic interrelationships in a collection of text documents, in combination with a search engine or for enterprise knowledge portals.
- ADUNA (`http://www.aduna-software.com/`) offers a visualization of hierarchically classified objects which can be used to show instantiated

[7] One enabling factor for commercially successful visualization suites for knowledge organization and access, may have been the IEEE Topic Map standard, see `http://www.topicmaps.org/`. Topic maps are often seen a competitor to ontologies because they serve partially similar purposes, but have different roots, some incompatible basic design decisions, a different research community. However, they have partially similar goals and application areas and complementary strengths to the mainstream ontology approaches – in particular, the design for human understanding and manipulation – such that the authors see them allies in the long term, rather than competitors.

taxonomies or ontologies – this is used, e.g., to display how search results of a desktop search can be grouped according to their relevance for certain keywords and keyword combinations [63].

- Ontopia or empolis provide generic topic map software for different purposes, including a topic map navigator (see also `http://www.topicmap.com/`).

Altogether, visual approaches can be a great support for understanding, searching, and investigating huge amounts of information and metadata. An overview of research and practice of visualization for the Semantic Web (e.g., RDF Graph visualization) can be found in [27]. However, in the authors' opinion, the economically valid use cases and scenarios should be better understood; analysing the quantifiable value-added of visual approaches as well as their critical success factors still seems to be a promising field for applied computer science.

Independent from the question which visualization approach is used (even with a simple, tree-structured browsing interface), KM usually deals with sharing complex knowledge content between people with quite different background and interests; this may often lead to the requirement that *multiple views* onto the same knowledge base should be provided. This is to some extent contradicting to the goal of creating a *widely shared* ontology for enabling communication between people; nevertheless, this requirement should not be neglected in practice – in particular, regarding the future trends of more distributed KM scenarios (see Sect. 5). Preliminary considerations about the technical support for such scenarios are presented in [59], based on the idea that specific, user-oriented GUI views can be created from special *presentation ontologies* created by selection and transformation operations from one (or more) system-internal ontologies.

3.2 Challenges for Ontologies in KM

Since Type 2A applications more or less represent the state-of-the-art in using ontologies for KM, we summarize some challenges which we see for the near future of research and technology transfer in this area:

Evaluation: It is already an indispensable need for KM applications to show their economic benefits to the project sponsors – which is not easy. In order to be successful, we have to find *success criteria* and develop *metrics* to assess whether ontology-based applications are more useful than solutions with "low tech" approaches. Although there exist already methodologies for ontology-based KM projects (for instance [65]), the aspects of benchmarking or quantitative performance criteria are rarely tackled.

Evolution: Since we are talking about long-living systems in dynamic environments, also ontological structures must be evolved cost-effectively to avoid decreasing system performance. A maintenance methodology for Case-Based Reasoning systems which might be transferable to KM, has been proposed by [55]. Its prerequisite are *quantitative quality and performance indicators* for the KM system. A well-structured analysis of the field of change discovery for ontologies distinguishing between structure-driven (obvious structural deficiencies of the ontology), data (instance) driven, and usage-driven indicators for required changes, was given by [62]. Concrete, usage-driven ontology evolution processes with the example of user interactions on an ontology-based e-business portal, were illustrated in [8].

Inference: As already argued, exploiting the power of inferences would show the usefulness of knowledge-rich approaches in contrast to, e.g., taxonomy-based ones. We should search for domains requiring powerful reasoning mechanisms and expressive domain descriptions. This may include aspects not yet fully adopted in ontology-based KM systems, such as the use of manifold link types [67], the representation of uncertainty and vagueness in domain modeling, or the definition of *similarity* on top of ontologies as it is demonstrated in CBR systems.

4 Ontologies Towards Enhanced Integrated Solutions

We mentioned already that exploiting synergy effects between different applications in the complex KM scenario can be an interesting source of innovation – for both new ideas and improved effectiveness of existing software functionalities. This area – especially with respect to ontologies – is not yet explored very well; but, we give some examples for work into this direction:

- We reported in [2] on performance improvements for document analysis (DA) and information extraction from paper documents by using *expectations* generated taking into account open workflow instances. The link between workflow system and DA is established by process, domain and DA ontologies and their mutual mappings. Similarly, task-specific IR is realized by coupling IR needs to workflow tasks.
- ONTOCOPI [6] is a tool for identifying potential members of a (hidden) Community of Practice by uncovering informal relationships between people trough traversal of instantiations of ontologically described formal relationships, like is-coauthor-of. *Recommender systems* learn about user preferences over time for realizing precise information push (cp. chapter "Ontology-Based Recommender System"). It is described in [45] how both systems can mutually benefit using the same ontological basis as the link between them.
- Typical software systems for supporting the process-view on KM comprise groupware (CSCW) and workflow systems. If, on the other hand, personal interest and skills are described formally on an ontological basis, groupware and CSCW support can be improved using this information. Examples for more intelligent CSCW support are more knowledgeable task

assignment to employees in a workflow application, more knowledgeable project staffing when configuring a new team, or better informed briefing of participants before a virtual meeting.

5 Future Trends

Comprehensive KM frameworks emphasize that Knowledge Management can take place at the individual, the group, organizational, and interorganizational level. The software functionalities discussed in Sects. 3 and 4 are mostly used to support the group and organizational level. Focussing on the personal and the interorganizational level, are logical next steps. Economically, the transition to interorganizational KM is driven by the movements towards the *Extended Enterprise* which tries to integrate logistics and production processes along the whole production chain (cp. [50]), and towards the *Virtual Enterprise* which is configured ad-hoc for specific projects from independent small units, in order to dynamically establish a temporary value-creation chain. One can easily see that such scenarios provide both more chances and more challenges to KM than traditional enterprise-internal scenarios.

Technically, the concepts of *Distributed Organizational Memory* (DOM) [3,69] and *Agent-Mediated Knowledge Management* (AMKM, [24,26,70]) have been introduced to deal with highly dynamic and highly distributed environments. Projects and systems like NAUTICUS, Jasper II, COMMA, FRODO, KDE, or EDAMOK identified KM-specific functionalities to be provided in an AMKM scenario by different kind of electronic agents or agent societies;[8] examples are collaborating agents for knowledge capture, retrieval, summarization, and user-profile refinement, or agent sub-societies to support annotation, ontology management and maintenance, metadata and user management, as well as matchmaking and retrieval. Such approaches often maintain process and role models as first-order citizens of their framework and often address issues such as sophisticated user-context acquisition and usage (typically, working context on the desktop) for offering high-precision, task-oriented KM services. In general, the issue of *context* in OMs is still a challenging research area [14].

A possible approach to realize AMKM or DOMs, is Peer-to-Peer technology (P2P, [15, 66]). For instance, in [14, 69] it is suggested to engineer social order/social mechanisms (like rights and obligations) into P2P KM systems for coordinating agents' activities. One essential problem behind is how to balance private issues and organizational issues in a complex and dynamic scenario.

Some other recent trends, only enumerated in a sketchy manner:

[8] Due to space limitations, we cannot include references for all the named systems; but they can be found in [70].

- The idea the *semantic desktop* is to use ontology-based, topic-oriented structuring mechanisms in the background for organizing and finding information from everyday-applications in the personal, private information space. The idea of the *social semantic desktop* transcends this from the personal towards the group information space. The NEPOMUK project investigates how such mechanisms can be used for personal and for community knowledge management (`http://nepomuk.semanticdesktop.org/`).
- *Folksonomies* exploit the power of large user communities with lightweight semantic technologies to achieve nevertheless a good quality of indexing for information retrieval. The transition between such lightweight *social software* approaches and more heavyweight, ontology-based approaches is an open question with a particular importance for KM because it addresses the trade-off between costs and quality.
- *Process knowledge* slowly becomes a topic of interest in KM, and in advanced, ontology-based information management projects. On one hand, business tasks and business processes are a source of context for knowledge creation and search; on the other hand, process and task execution knowledge itself may be a shareable, reusable asset in an organization; lastly, knowledge workers' productivity depends much on sensible task management support. Hence, manifold research topics can be found in this area and its combination with more traditional information management issues (see, for example, [17,53]).
- Last but not least, there is some technological as well as methodological convergence visible between several complementing and overlapping areas: In particular, skill and human resource management, e-Learning, and KM. Technically, first integrative works address ontological foundations of many affected fields (cp. the PALETTE project [73]) and more integrations of process-view and product-view tools for KM, like OM and groupware/social software (cp. [25]).

References

1. Abecker A (2004) Business-process oriented knowledge management – concepts, tools, and applications. PhD thesis, University of Karlsruhe.
2. Abecker A, Bernardi A, Maus H, Sintek M, Wenzel C (2000) Information supply for business processes – coupling workflow with document analysis and information retrieval. Knowledge-Based Systems 13(5):271–284.
3. Abecker A, Bernardi A, van Elst L (2003) Agent technology for distributed organizational memories. In: Proc ICEIS-03, vol. 2, pp. 3–10, 2003.
4. Abecker A, Reuschling C, Tabor S, Apostolou D, Maass W, Mentzas G (2003) Towards an information ontology for knowledge asset trading. In: Proc ICE-2003, pp. 187–194.
5. Alani H, Kim S, Millard DE, Weal MJ, Hall W, Lewis PH, Shadbolt NR (2003) Automatic ontology-based knowledge extraction from Web documents. IEEE Intelligent Systems 18(1):14–21.

6. Alani H, O'Hara K, Shadbolt N (2002) ONTOCOPI: methods and tools for identifying communities of practice. In: Intelligent Information Processing Conference, IFIP World Computer Congress (WCC), Montreal, Canada.
7. Altmann R, Bada M, Chai X, Carillo MW, Chen R, Abernethy N (1999) RiboWeb: an ontology-based system for collaborative molecular biology. IEEE Intelligent Systems 14(5):68–76.
8. Apostolou D, Mentzas G, Klein B, Abecker A, Maass W (2008) Interorganizational knowledge exchanges. IEEE Intelligent Systems 2008(July):65–74.
9. Becerra-Fernandez I (2001) Locating expertise at NASA. Knowledge Management Review 4(4):33–37.
10. Bernardi A, Sintek S, Abecker A (1998) Combining artificial intelligence, database technology, and hypermedia for intelligent fault recording. In: Proc ISOMA'98.
11. Biesalski E, Abecker A, Breiter M (forthcoming, 2008) Towards integrated, intelligent human resource management. Applied Ontology: An Interdisciplinary Journal of Ontological Analysis and Conceptual Modeling. IOS.
12. Bloehdorn S, Cimiano P, Duke A, Haase P, Heizmann J, Thurlow I, Völker J (2007) Ontology-based question answering for digital libraries. In: Proc 11th ECDL.
13. Boer A, Hoekstra R, Winkels R (2001) The CLIME ontology. In: Proc Second Int Workshop on Legal Ontologies, University of Amsterdam.
14. Bonifacio M, Bouquet P, Mameli G, Nori M (2002) KEx: a peer-to-peer solution for distributed knowledge management. In: Karagiannis D, Reimer U (eds) PAKM 2002. Springer, Berlin.
15. Bonifacio M, Giunchiglia F, Zaihrayeu I (2005) Peer-to-Peer knowledge management. In: Proc I-KNOW'05.
16. Börner K, Chen C (eds) (2001) Visual interfaces to digital libraries–Its past, present, and future. In: Workshop at 1st ACM+IEEE Joint Conf on Digital Libraries (JCDL-2001).
17. Carr L, Miles-Board T, Wills G, Woukeu A, Hall W (2004) Towards a knowledge-aware office environment. In: Karagiannis D, Reimer U (eds) PAKM 2004. Springer, Berlin.
18. Conklin E, Weil W (1997) Wicked problems: naming the pain in organizations. White Paper of Group Decision Support Systems Inc.
19. Cottam H (1999) Ontologies to assist process oriented knowledge acquisition. Technical Report SP142, SPEDE Project, University of Nottingham.
20. Davenport T, Prusak L (1997) Working Knowledge: How Organizations Manage What They Know. Harvard Business School Press, Boston.
21. Davies J, Fensel D, van Harmelen F (eds) (2002) Towards the Semantic Web: Ontology-Driven Knowledge Management. Wiley, London.
22. Decker S, Erdmann M, Fensel D, Studer R (1999) Ontobroker: ontology based access to distributed and semi-structured information. In: Meersman R, Tari Z, Stevens SM (eds) IFIP TC2/WG2.6 Eighth Working Conference on Database Semantics (DS-8). IFIP Conference Proceedings 138. Kluwer, Dordrecht.
23. Dieng-Kuntz R, Matta N (eds) (2002) Knowledge management and organizational memories. Kluwer, Dordrecht.
24. Dignum V (2004) A model for organizational interaction: based on agents, founded in Logic. PhD thesis, Universiteit Utrecht. SIKS dissertation series.

25. Falbo RA, Arantes DO, Natali ACC (2004) Integrating knowledge management and groupware in a software development environment. In: Karagiannis D, Reimer U (eds) PAKM 2004. Springer, Berlin.
26. Gandon F, Poggi A, Rimassa G, Turci P (2002) Multi-agent corporate memory management system. Journal of Applied Artificial Intelligence 16(9–10):699–720.
27. Geroimenko V, Chen C (eds) (2003) Visualizing the Semantic Web. Springer, Berlin.
28. Golebiowska J, Dieng-Kuntz R, Corby O, Mousseau D (2001) Building and exploiting ontologies for an automobile project memory. In: Gómez-Pérez A, et al (eds) IJCAI-01 Workshop on Ontologies and Information Sharing, CEUR-Proceedings 47.
29. Gómez-Pérez A, Corcho-Garcia O, Fernandez-Lopez M (2003) Ontological Engineering. Springer, Berlin.
30. Guarino N, Masolo C, Vetere G (1999) Ontoseek: content-based access to the web. IEEE Intelligent Systems, 14(3):70–80.
31. Henze N, Dolog P, Nejdl W (2004) Reasoning and ontologies for personalized e-Learning in the Semantic Web. Journal of Educational Technology and Society 7(4):82–97.
32. Herman I, Melancon G, Marshal MS (2000) Graph visualization and navigation in information visualization: a survey. IEEE Transactions on Visualization and Computer Graphics 6(1):24–43.
33. Holsapple CW (ed) (2003) Handbook on knowledge management, vols. 1 and 2. International Handbooks on Information Systems. Springer, Berlin.
34. Jurisica I, Mylopoulos J, Yu E (2004) Ontologies for knowledge management: an information systems perspective. Journal of Knowledge and Information Systems (2004)6:380–401.
35. Kalfoglou Y, Domingue J, Motta E, Vargas-Vera M, Buckingham-Shum S (2001) MyPlanet: an ontology-driven web-based personalised news service. In: Gómez-Pérez A, Gruninger M, Stuckenschmidt H, Uschold M (eds) IJCAI-01 Workshop on Ontologies and Information Sharing. CEUR-Proceedings.
36. Woitsch R, Karagiannis D (2005) Process oriented knowledge management: a service based approach. Journal of Universal Computer Science 11(4):565–588.
37. Kumar V, Furuta R, Allen RB (1997) Metadata visualization for digital libraries: interactive timeline editing and review. In: Proc Third ACM Conf on Digital Libraries.
38. Laender A, Ribeiro-Neto B, da Silva A, Teixeira J (2002) A brief survey of web data extraction tools. ACM SIGMOD Record 31(2).
39. Lau T, Sure Y (2002) Introducing ontology-based skills management at a large insurance company. In: Modellierung 2002.
40. Liao M, Hinkelmann K, Abecker A, Sintek M (1999) A competence knowledge base system for the organizational memory. In: Puppe F (ed) Proc XPS-99. Springer, Berlin.
41. Lopez V, Uren V, Motta E, Pasin M (2007) AquaLog: an ontology-driven question answering system for organizational semantic intranets. Journal of Web Semantics 5(2):72–105.
42. Maier R (2007) Knowledge Management Systems, 3rd edition. Springer, Berlin.
43. McGuiness D (1998) Ontological issues for knowledge-enhanced search. In: Guarino N (ed) Proc FOIS'98. IOS, Amsterdam.
44. Mentzas G, Apostolou D, Young R, Abecker A (2002) Knowledge asset networking. Advanced Information and Knowledge Processing. Springer, London.

45. Middleton S, Alani H, Shadbolt N, Roure DD (2002) Exploiting synergy between ontologies and recommender systems. In: Proc WWW2002, Semantic Web Workshop.
46. Mulholland P, Zdrahal Z, Domingue J, Hatala M (2000) Integrating working and learning: a document enrichment approach. Journal of Behaviour and Information Technology 19(3):171–180.
47. O'Leary D (1998) Knowledge management systems: converting and connecting. IEEE Intelligent Systems 13(3):30–33.
48. O'Leary D (1998) Using AI in knowledge management: knowledge bases and ontologies. IEEE Intelligent Systems 13(3):34–39.
49. Missikoff M, Schiappelli F (2005) A method for ontology modeling in the business domain. In: EMOI-INTEROP 2005.
50. Pedersen MK, Larsen MH (2000) Inter-organizational systems and distributed knowledge management in electronic commerce. In: Svensson L, Snis U, Sorensen C, Fägerlind H, Lindroth T, Magnusson M, Östlund C (eds) Proc IRIS 23.
51. Pisanelli DM (ed) (2004) Ontologies in medicine. Vol. 102 in Studies in Health Technology and Informatics. IOS, Amsterdam.
52. Pisanelli DM, Gangemi A, Steve G (2000) The role of ontologies for an effective and unambiguous dissemination of clinical guidelines. In: Dieng R, Corby O (eds) Proc 12th EKAW. Springer, Berlin.
53. Rath AS, Kröll M, Andrews K, Lindstaedt S, Granitzer M, Tochtermann K (2006) Synergizing standard and ad-hoc processes. In: Reimer U, Karagiannis D (eds) PAKM 2006. Springer, Berlin.
54. Reimer U, Margelisch A, Staudt M (2000) EULE: a knowledge-based system to support business processes. Knowledge-Based Systems 13(5):261–269.
55. Roth-Berghofer T (2002) Knowledge maintenance of case-based reasoning systems. The SIAM methodology. PhD thesis, University of Kaiserslautern.
56. Schaaf M, Maximini R, Bergmann R, Tautz C, Traphöner R (2002) Supporting electronic design reuse by integrating quality-criteria into CBR-based IP selection. In: Proc 6th ECCBR.
57. Sebastiani F (2002) Machine learning in automated text categorization. ACM Computing Surveys 34(1):1–47.
58. Sidhu AS, Dillon TS, Chang E, and Chen JY (eds) (2007) Special issue on ontologies for bioinformatics. International Journal of Bioinformatics Research and Applications (IJBRA) 3(3):261–428.
59. Sintek M, Tschaitschian B, Abecker A, Bernardi A, Müller H-J (2000) Using ontologies for advanced information access. In: Domingue J (ed) PAKeM 2000.
60. Spyns P, Oberle D, Volz R, Zheng J, Jarra M, Sure Y, Studer R, Meersman R (2002) Ontoweb – a semantic web community portal. In: Proc PAKM-2002.
61. Staab S, Maedche A (2000) Knowledge portals: ontologies at work. The AI Magazine 22(2):63–75.
62. Stojanovic L (2004) Methods and tools for ontology evolution. PhD thesis, University of Karlsruhe.
63. Stuckenschmidt H, van Harmelen F (2001) Knowledge-based validation, aggregation and visualization of meta-data: analyzing a web-based information system. In: Zhong N, Yao Y, Ohsuga S (eds) Proc Web Intelligence 2001. Springer, Berlin.
64. Sure Y, Maedche A, Staab S (2000) Leveraging Corporate Skill Knowledge - From ProPer to OntoProPer. In: Mahling D, Reimer U (eds) Proc PAKM-2000.

65. Sure Y, Staab S, Studer R (2002) Methodology for development and employment of ontology based knowledge management applications. ACM SIGMOD Record 31(4):18–23.
66. Susarla A, Liu D, Whinston A (2002) Peer-to-peer knowledge management. In: [33].
67. Tudhope D, Alani H, Jones C (2001) Augmenting thesaurus relationships: possibilities for retrieval. Journal of Digital Information 1(8):41.
68. Uren V, Cimiano P, Iria J, Handschuh S, Vargas-Vera M, Motta E, Ciravegna F (2006) Semantic annotation for knowledge management: requirements and a survey of the state of the art. Journal of Web Semantics 4(1):14–28.
69. van Elst L, Abecker A (2002) Ontologies for information management: balancing formality, stability, and sharing scope. Expert Systems with Applications 23(4):357–366.
70. van Elst L, Dignum V, Abecker A (eds) (2004) Agent-Mediated Knowledge Management. LNAI 2926. Springer, Berlin.
71. van Engers TM, Kordelaar PJ, den Hartog J, Glassée E (2002) POWER: programme for an ontology based working environment for modelling and use of regulations and legislation. In: DEXA-2002 Electronic Government Workshop.
72. van Heijst G, van der Spek R, Kruizinga E (1998) The lessons learned cycle. In: Borghoff UM, Pareschi R (eds) Information Technology for Knowledge Management. Springer, Berlin.
73. Vidou G, Dieng-Kuntz R, Ghali AE, Evangelou C, Giboin A, Tifous A, Jacquemart S (2006) Towards an ontology for knowledge management in communities of practice. In: Reimer U, Karagiannis D (eds) PAKM-2006. Springer, Berlin.
74. Gresse von Wangenheim C (2002) Operationalizing Reuse of Software Measurement Planning Knowledge. PhD thesis, University of Kaiserslautern.
75. Vrandecic D, Pinto HS, Sure Y, Tempich C (2005) The DILIGENT knowledge processes. Journal of Knowledge Management 9(5):85–96.
76. Weber R, Aha D, Becerra-Fernandez I (2001) Intelligent lessons learned systems. International Journal of Expert Systems Research and Applications 20(1):17–34.
77. Weber R, Kaplan R (2002) Knowledge-based knowledge management. In: Faucher C, Jain L, Ichalkaranje N (eds) Innovations in Knowledge Engineering. Physica, Heidelberg.
78. Yimam D (2000) Expert finding systems for organizations: domain analysis and the DEMOIR approach. In: Ackerman M, Cohen A, Pipek V, Wulf V (eds) Beyond Knowledge Management: Sharing Expertise. MIT, Cambridge.

Application of Ontologies in Bioinformatics

Robert Stevens[1] and Phillip Lord[2]

[1] School of Computer Science, University of Manchester, Oxford Road, Manchester M13 9PL, UK, robert.stevens@manchester.ac.uk
[2] School of Computing Science, Newcastle University, Claremont Road, Newcastle Upon Tyne NE1 7RU, UK, phillip.lord@newcastle.ac.uk

Summary. The use of ontologies has become a mainstream activity within bioinformatics. In a largely descriptive science such as biology, the need to have a common understanding of things described is obvious. The need to be able to apply computational methods to the large quantities of data being produced also suggests a computational requirement to standardise descriptions in biology.

As a mechanism for describing the categories of entities and their characteristics, ontologies offer many of the features that can support a descriptive science. The main use of ontologies in bioinformatics has been the delivery of controlled vocabularies. In this chapter we explore this use of ontology, but also other uses, especially those that have a deeper computational aspect. We take a broad view of ontology to include many ontology-like resources and classify the uses of ontology and ontology-like artifacts. We present a series of case studies and conclude by describing the current state and future directions for bio-ontologies.

1 Introduction

In this chapter, we explore the uses within bioinformatics of ontologies and other ontology-like artefacts, some of which were described in chapter "Ontologies for Formal Representation of Biological System". That chapter provided a motivation for the use of an ontology and described the range now available. In the first edition of this volume, we explored why bioinformaticians have become so interested in the development and use of ontologies [44]. This interest has become consolidated to the point that the development and use of ontologies has become mainstream. Yet, as we will see in this chapter, the different kinds of uses of ontologies within bioinformatics is quite narrow, while the potential uses are more wide ranging. We will see that the initial goal of ontology has been basic data integration for use by humans. Ontologies should offer the means to drive computational use of biological data and it is this aspect that we wish to investigate further in this chapter.

The production of data in biology has become industrialized; so, therefore, must its analysis. The lack of the laws or grand theories of physics means that

S. Staab and R. Studer (eds.), *Handbook on Ontologies*, International Handbooks on Information Systems, DOI 10.1007/978-3-540-92673-3,

much inference in bioinformatics is still reliant on the processing of factual data – the knowledge we have about the entities in the biological world. A common understanding of what is described by the data collected is obviously a great help in such an endeavor. The primary means of delivering such a common understanding in bioinformatics[1] is by talking about the same entities in the same way – controlling the vocabulary used for representing information in the data resources. Delivery of controlled vocabulary for a "de facto integration" [46] is still the primary use of bio-ontologies.

The need for a common reference for the functional attributes of gene products, by the genome projects for different organisms motivated the development of the Gene Ontology (GO) [46]. A common understanding requires agreement upon those categories, for example, of molecular functions that exist. The labels chosen for those categories provides a vocabulary (an ontology is not a vocabulary but can deliver one). The control arises from the commitment to use that ontology delivered vocabulary to describe the attributes of classes of gene products in cross-species and community-wide resources. As described in Sect. 3, this has great utility not only in querying resources, but also in their analysis.

Doing a Google search with "`define: ontology`" gives an answer with approximately 20 slightly different definitions. These do, however, cluster into two distinct definitions:

1. A discipline of philosophy concerned with the description of that which exists
2. A shared understanding of what a community understands about a domain that allows machine reasoning

In essence, these are both concerned with descriptions of the "things" in the world or the description of those entities as they appear within information. The emphasis of the second, however, is that of the *shared* use of the description and its use by computers. As we will see, defining what it is to be a member of a class, then agreeing the label for that class assists both human and computers in data processing. In a knowledge-based discipline, such as bioinformatics, having a machine-processable form of knowledge to allow a wide range of scientific inferences is vital. We claim, however, that description for the sake of description, without including the computer is potentially highly restrictive.

An ontology, according to the philosophers who coined the term, is a description of the categories and membership criteria of those things which exist. Computer scientists have latterly taken this term and shifted its meaning somewhat [18]. An ontology still describes things, but the emphasis is on the *shared understanding* of conceptualizations. The goal of a computer science ontology is to enable machines to manipulate symbolic representations

[1] Here we take a broad definition of bioinformatics to mean the storage, management and analysis of biological data by computational means to answer biological questions.

of knowledge. Whether or not a broad or narrow view of ontology is taken, ontologies and ontology-like artefacts are all about description of the world or human understanding of the world. This can range from an attempt to record the true account of reality through thesauri, vocabularies, classification schema to glossaries. Irrespective of the representation and the level of reasoning supported, they all provide some description and/or definition of things in the world. Within bioinformatics it is possible to find many different kinds of knowledge artifact described as ontologies [10]. In this chapter we will take the broader view of ontology and explore how those descriptions are used within bioinformatics applications.

In Sect. 2 we classify the uses to which ontologies have been put within bioinformatics. Then in Sect. 3 we look at some case studies of these uses. In Sect. 4 we discuss the current state and future directions for ontologies within bioinformatics.

2 Classifying Uses of Bio-Ontologies

Ontologies, whether from the computer science or philosophical perspective, are all about description. The applications of ontologies within biology are therefore all rooted in description. Figure 1 shows a classification scheme (very deliberately not an ontology) for the uses of ontology and ontology-like artefacts within biology. Obviously describing the world is a use in itself

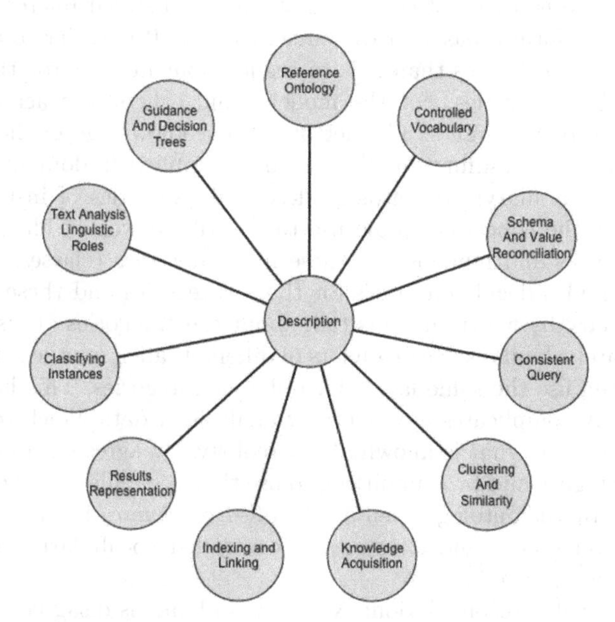

Fig. 1. A classification scheme for the uses of ontology and ontology-like artefacts within biology

and consequently all the uses are *narrower* uses of description. Potential uses of ontology have previously been categorized [48]. These categorizations are still within ours, but we wished to emphasize the role of description, the interrelatedness of these uses and also to present those uses at a finer granularity. We have already mentioned one of the principle uses of such descriptions in bioinformatics – using the labels of the concepts for the delivery of a controlled vocabulary.

Other uses of ontologies exploit the structure of the relationships between the concepts. Having annotated data with a controlled vocabulary, the structure of the ontology can be used to query instance data or navigate instance data. To move from a shared understanding which is fit for humans to use towards one exploitable by machines, it is necessary to introduce a more *strict semantics* (a precise description of the relationships between concepts) and is facilitated by a *richer expressivity* (the ability to express different kinds of relationships). Additional semantic strictness and expressivity does not necessarily enable new uses per se, but can allow more extensive uses in the same area.

The uses to which ontological description can be put include, but are not limited to:

Reference ontology: Defining the classes of entities within a domain, hopefully both logically and in a human orientated fashion can be of utility in its own right. Simply affording a community of discourse an encyclopædia of that which is known acts as a *reference* source for that domain. The Foundational Model of Anatomy [11, 38] can be seen in this light. Even when there is a lack of consensus about a self-styled reference ontology, it can still form a basis for discourse, as it is often easier to argue about definitions of entities than it is to argue about mere words that are used as labels for entities. For the modeler and others, the act of modeling itself can offer insights. The act of making knowledge explicit can force questioning of assumptions that are often implicit in domain discourse.

Controlled vocabulary: An ontology describes categories of instances in the world or the concepts people use to describe a world. There is a world of instances and humans put these into categories (classes, types, etc.). Humans also decide on labels for those categories and these provide the vocabulary by which humans talk about the categories of instances. Unfortunately, humans decide on lots of different labels for the same category and often use the same labels for different categories. This heterogeneity massively complicates any query or analysis of data which relies on manipulations of what is known about biology. By agreeing upon the labels for a category and by committing to use that vocabulary for the categories defined by the ontology then a *controlled vocabulary* has been developed. The development and examples of controlled vocabularies in biology is described in Sect. 3.

Schema and value reconciliation: Not only do humans disagree on the labels given to categories, but they also disagree on the categories themselves.

There are many legitimate ways to describe the world; models are, after all, virtually neither complete nor wholly correct. This can be due to different perspectives on the same issues, e.g., taking either a developmental or structural view of anatomy will give different categories [41]. Other descriptions will either be partial or skewed due to some application bias [21]. Many databases exist within bioinformatics that represent similar or overlapping extents [42]. Thus to get a complete coverage of a domain of interest these data need to be pooled. Unfortunately, differing conceptualizations for the data from different datasets mean that these datasets cannot be compared without *reconciliation.* A community agreement on the categories and their definitions, in the form of an ontology, can provide a common data model which can drive reconciliation of both the differing schema and values. These ontologies can define either an intentional definition of the constraints to which an instance must comply or provide a template for the attributes of the instances. This use of an ontology to specify a model to drive both schema and value reconciliation is common both within and without bioinformatics [26]. This use is explored in Sect. 3.3.

Consistent query: Obviously once there is a common conceptualization, a common set of labels for the concepts and the instances all comply with that ontology, then the querying and analysis of data can be greatly eased. Different ontological representations afford different kinds of query facility [55]. Simply using a controlled vocabulary allows better querying by exact matching. Using the taxonomic structure of an ontology allows queries to retrieve "instances of this class" that implies all the instances of the subclasses (as these are instances of the query class). Querying data in some way is a prime motivation of much work in bioinformatics, consequently *query* pervades Sect. 3.

Knowledge acquisition: Having described the classes of instances in a domain, a practitioner will often want to describe instances of those classes. Ontologies can either specify templates for the attributes that instances of a class must be given or describe what is known about an instance, explicitly stated or not [50]. As a result, ontologies can be used to generate forms by which instances are gathered or acquired [16]. Similarly, data can be transformed to comply with the ontology to generate a *knowledge base* (the combination of ontology and instances of the classes in the ontology). Several examples of these have been seen in bioinformatics [2, 24, 39] and were described in the first edition of this handbook [44]. The ontology then offers the means by which those instances can be queried and otherwise exploited in a sophisticated manner.

Clustering and similarity: Rather than straight-forward querying, an ontology can be used to *cluster* data items. For example, if the genes detected by a microarray chip are annotated with Gene Ontology terms, one can take differentially expressed genes and cluster them against the aspects of the Gene Ontology. For instance, the set of up-regulated genes on a chip could be clustered about the GO biological process ontology. Taking the lowest common subsumer – the most specific term that all the genes of interest share – provides the analyst with an idea of what might be happening in the condition under investigation [13, 46]. The degree of similarity shared by the members of a cluster remains a question. In bioinformatics, we are well used to the notion of sequence similarity and how it is to be interpreted [3]. Recently, as the amount of semantically annotated data has risen, the notion of semantic similarity has become prominent. Following the introduction of this notion into bioinformatics [33], the possibility has been realised for querying data at a semantic level in the style of "these two entities have an 42% functional similarity". The use of description to enable clustering and measures of semantic similarity will be explored in Sect. 3.2.

Indexing and linking: As already described, ontologies and ontology -like artefacts can provide structured, controlled vocabularies. These are often used to describe data objects. One consequence of this is to *index* those data. Just as with a traditional book index, this is a mechanism for quick retrieval. This has an obvious closeness to querying and searching. Perhaps the most prominent example of indexing the biomedical arena is the use of MeSH (Medical Subject Headings) [32] to index PubMed abstracts.

Results representation: One of the more recent uses of ontologies, is in the description of primary results before they are lodged in a publicly available resource. For example, the MGED Ontology [51] enables the description of microarray experiments and their results. This use of ontologies at the time of publication of the experiment differs from *post-hoc*, interpretative annotation that, for example, the Gene Ontology provides. Since the MGED ontology, this use has become more widespread in the proteomics community [22] and finally for all biomedicine with the Ontology for Bio-Medical Investigations (OBI), previously known as FUGO [52].

Classifying instances: An ontology describes the classes of instances in a domain. Definitions of those classes provide knowledge of how to recognize a domain instance as a member of a particular class. Given a set of facts about instances the ontology can be used to classify those instances to place them into categories or classes. This use is described in Sect. 3.4.

Text analysis / linguistic roles: Ontologies and ontology-like resources are widely used in text-mining applications [9]. Thesauri, such as WordNet [15], have uses in determining word types, synonyms, spelling variants, etc., that have obvious relevance to text-mining applications. Ontologies, as already described, provide vocabularies and these are very helpful in text-mining applications. Finally, the structure of ontologies can help in offering "possibilities" for associations between words or concepts and the classification of word types to broad ontological distinctions such as *Event*, *Role*, *Process*, etc. Text mining is very important in bioinformatics [4], but the role of ontology in this large area is beyond the scope of this chapter.

Guidance and decision trees: Ontologies, by capturing knowledge about a domain and encapsulating constraints about class membership, can offer guidance around a domain and support decision making processes. In query formulation, for instance, an ontology can inform an application or human operator information about what can be said about an entity [2,17]. So when querying about transcription complexes, for instance, an ontology might offer information about transcription factors, binding sites in promoters, polymerases, etc., but not about entities relevant to replication and other possibly irrelevant processes. The constraints in an ontology can reduce the space of possibilities, which is useful in a large and complex domain such as biology and bioinformatics. Similarly, given a set of facts about symptoms, an ontology can prompt a user to provide more discriminating facts to distinguish between classes [30].

There are a range of potential uses for bio-ontologies within bioinformatics. We have presented a simple classification scheme of their uses in order to help orientation and navigation within the field. All uses can be traced back to the *description* of entities in a domain which is an end in and of itself. Many of the uses are minor variations on major themes of controlled vocabulary, controlled structure and the querying that such knowledge models support. In the next section we take examples from biomedicine to illustrate this scheme.

3 Case Studies

3.1 Using Controlled Vocabulary

The single most common use of ontology in bioinformatics is to provide a *controlled vocabulary*, which is then used to provide annotation for database entries. The pre-eminent example for this is the Gene Ontology (GO) [46]. This project started around 1998 as a collaboration between a number of model organism databases [46]. GO was created to address the considerable difficulty of inter-operability between the different genome databases; to search sequence data cross-species was and still is straight-forward but, at this time, diversity in the nomenclature for genes and their products meant that similar searches over the knowledge of biology were difficult or impossible.

The Gene Ontology is focused on describing three features or *aspects* of biology: the *molecular function* defined as the biochemical (or molecular) activity of a gene; the *cellular component* defined the location in the cell that a gene product is active; and the *biological process* defined as the biological objective, or the series of events to which the molecular function contributes [47].[2]

[2] http://www.geneontology.org

Each ontology is structured as a set of terms linked together with two relationships, *is-a* and *part-of*. This representation can be represented and is frequently described as a Directed Acyclic Graph (DAG).[3] The simplicity of this representation has been one significant reason for the success of GO [5], as it has proven to be straight-forward for many biologists.

As well as the ontology, there are also a large number of annotations – database records describing various gene products (generally proteins or genes as proxies for proteins). At the time of writing, there were around 3,000,000 Uniprot [12] proteins, with some 19,000,000 annotations.[4] A protein sequence or its database entry may be annotated with one or more terms from the three aspects of GO. The exact relationship between the protein described by the database record and the GO term depends on the aspect of GO to which the term belongs. For example, for Molecular Function an annotation means that a given molecule of the protein has the *propensity* to act in the way defined by the GO Term [8], in a given set of contexts. It does not mean that all molecules of that protein either will or are always capable of behaving in this way.

For a GO annotation, the association between a term and the proteins is supplemented with "Evidence Codes"; this is a term from an additional controlled vocabulary that describe the kind of evidence that was used to suggest the association. These range from "TAS" or traceable author statement; in paraphrase this means that the evidence came From a statement in a review paper, rather than a primary research paper, which suggests that it is well enough acknowledged in the community. Other codes, such as IEP – Inferred from Expression Pattern – describe the kind of experimental evidence that has been used.

The success of the Gene Ontology has spawned a large number of tools for its use. Perhaps the best known of these are Amigo – a website which functions as a browser for the Gene Ontology, shown in Fig. 2. This tool is backed by the GODatabase, a relational schema and set of associated code that allows rapid search and query over the Gene Ontology.

Perhaps the most common use of GO or its annotations is for the analysis of microarray results.[5] Annotation of data in databases is still the largest activity in the bio-ontology sector. The Open Biomedical Ontologies (OBO) ontologies cover a range of topics within biology (see `http://obo.sf.net`). This reflects a basic need within the discipline to overcome the vast heterogeneity in their databases and permit broader analyses across their data. This technically simple application has had profound effects on bioinformatics.

---

[3] The less common term "poset" or partially ordered set is also used.

[4] `http://www.geneontology.org/GO.current.annotations.shtml`

[5] At least judged by the number of tools available; microarray tools form the largest subsection of related tools on the GO website (`http://www.geneontology.org/GO.tools.shtml`).

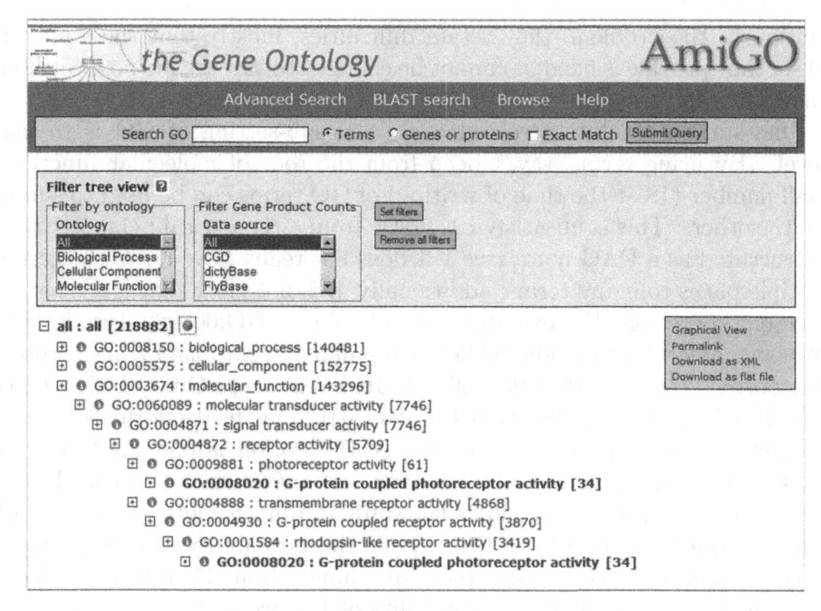

Fig. 2. Amigo

3.2 Statistical Uses of Ontologies

We now describe the application of statistics applied to and with the Gene Ontology.[6] In microarray experiments, more genes are found that show interesting expression patterns than can be reasonably examined by hand. A scientist will often wish to understand what kind of processes these genes display. There are a number of different tools available which perform this analysis, including GOMiner [56] and GOstat [7]. Of these, GOMiner uses Fisher's two-sided exact test to determine whether a "category" (that is, the proteins annotated with a given GO term or any of its children) is over-represented among the set of genes of interest (that are, for example, over- or under-expressed according to the microarray results).

The continued development of GO means that it now has around 24,000 terms. This large ontological structure has meant that GO has become difficult to present to users, particularly in the context of an expanded hierarchical viewer. The GO consortium's response to this was the introduction of "GO Slims" – defined subsets of GO. As well as a general purpose slim (the Generic GO slim), there are others tailored for specific purposes; for example the yeast and plant GO slim focuses on those terms which are important for the given organism; both contain "*cell wall*", (GO:0005616) for example, while only the plant slim contains "*thylakoid*", (GO:0009579). While these provide a partial

[6] In this section, we will talk exclusively about GO, as it is the ontology which has been statistically analysed most widely. Most of the techniques could also apply to other ontologies.

solution to the problem, they retain difficulties; mostly that the size of the subset they provide is fixed; it cannot be changed to suit the purpose (although a new GO Slim could be created).

One standard solution to automate the sub-setting of GO is to use a "level". By using terms, say 1 deep from the root of molecular function, a small number (18 at the time of writing) of GO terms can be used to summarize the others. This is unsatisfying, firstly from a theoretical perspective, GO is structured as a DAG not a tree – it does not really have levels as there are multiple paths to many terms; additionally, it is not clear that levels actually represent specificity. For example, "*ice binding*", (GO:0050825) is three levels below molecular function, while "*high-affinity tryptophan transmembrane transporter activity*", (GO:0005300) is 10 levels deep. One solution to this is provided by the GO Partition Database [1]. This uses information theory to determine how specific a GO term is; the notion is familiar from internet search engines – common words have low *information content* (IC) and are not useful for searching, unusual words are much more so. The IC of a term is given by $-\ln_2(pT_n)$ where pT_n is the probability of a term (or any of its children) occurring in any particular annotation. Therefore, "*molecular function*" (GO:0003674) has a probability of 1, since every time a (molecular function) term occurs, it must be a child of this GO term. A set of terms with similar information contents, therefore, implicitly defines a set of terms of similar specificity. One thing to note, is that this knowledge is based on a corpus – a body of GO annotations. Therefore, for example, the IC based on the Saccaromyces Gene Database (SGD) [14] annotation would differ somewhat from that of Uniprot. Choosing the correct corpus is, therefore, likely to be important in obtaining relevant results.

As well as summarizing GO, there are many applications that need a numerical measure of the *semantic similarity* between two GO terms or, in more common use, two entities annotated with one or more GO terms. Initial attempts to develop these measures came from WordNet [15], a electronic lexicon, and used variations on path distance between terms as their measure [25, 34]. We call these *structural measures*; related techniques have been applied directly to the Gene Ontology [23]. These techniques have some of the problems described earlier with ontology levels as a mechanism for specificity; different edges in GO, for example, do not necessarily have the same weight. Information Content based measures have also been applied; initially to WordNet [36], but later specifically to the Gene Ontology [33]. These measures use the IC ($-\ln_2(pT_n)$) of the common parents of the terms of interest. As GO is a DAG, this will normally be a set of parents, in which case the term with the lowest IC is used. These measures were strongly correlated with sequence similarity [27], and gene expression [49] and, therefore, do appear to be related to the underlying biology. Since this time, they have found a variety of applications, including the prediction of gene function [45], validating functional networks [37].

3.3 Schema Reconciliation

There is great heterogeneity in how bioinformatics data are organized – that is, the schema of the databases differ. Complete integration of resources necessarily involves reconciliation at the level of data and the schema in which the data are held. Ontologies can be used with great effect to provide a structure for how data should be organized. In schema reconciliation, the general idea is that the differing representations of the data are re-modeled to fit with the ontological description of the data. Again, this is the general idea of a *shared* description of an understanding of what exists in a domain and the community members *committing* to use that description.

One widely used and current examples of schema reconciliation in bioinformatics using ontology is the BioPAX project [29]. Biological pathway exchange or BioPAX has the general goal of enabling the exchange of the vast quantities of pathway data. The BioPAX consortium oversees the development of a rigorous, open-source standard for the representation of all forms of biological pathways. It does this by providing a common conceptual framework, a set of common terms and a common format for exchange and integration.

There are two top level classes in the BioPAX ontology: *entity* and *utilityClass*. Entities describe the biology while the utility classes record knowledge about the pathway data such as cross-references to other databases, evidence codes, and experimental conditions. *Pathways* are a subclass of *entity*, along with two sibling classes, *interaction* and *physicalEntity*. A *pathway* has components that are of the class *pathwayStep*, a utility class. Each *pathwayStep* contains a set of *stepInteractions* that describe the physical interactions, such as catalysis, modulation, biochemical reaction, complex assembly, and transport that make up that step in the pathway, or another pathway. A *pathway*, such as glycolysis (the conversion of glucose to pyruvate), MAPK (the intra-cellular transmission of growth factor signals), or apoptosis (biochemical events leading to a programmed cell death)is composed of instances of *interactions*. Interactions can occur between entities so that *interactions* of *interactions* and *interactions* of *pathway* can be represented.

Figure 3 shows how the interactions from one step in the glycolysis pathway are mapped to the *entity* class hierarchy as defined in the BioPAX ontology. A biochemical reaction (across figure at bottom) is mapped to the BioPAX root class *entity* (top of the figure). In this biochemical reaction, there are three instances of the class *physicalEntity*, of these, two are instances of the class *smallMolecule*, β-D-glucose-6-phosphate and D-fructose-6-phosphate, and one an instance of the class *protein*, phosphoglucose isomerase, (*enzyme* is not explicitly represented as a subclass of *protein* in BioPAX). This biochemical reaction converts β-D-glucose-6-phosphate into D-fructose-6-phosphate. The reaction itself is controlled by the enzyme phosphoglucose isomerase and is part of the glycolytic pathway.

Once the *physicalEntities* that participate in the reactions are identified, together with the interaction roles they play in the reaction, we can represent

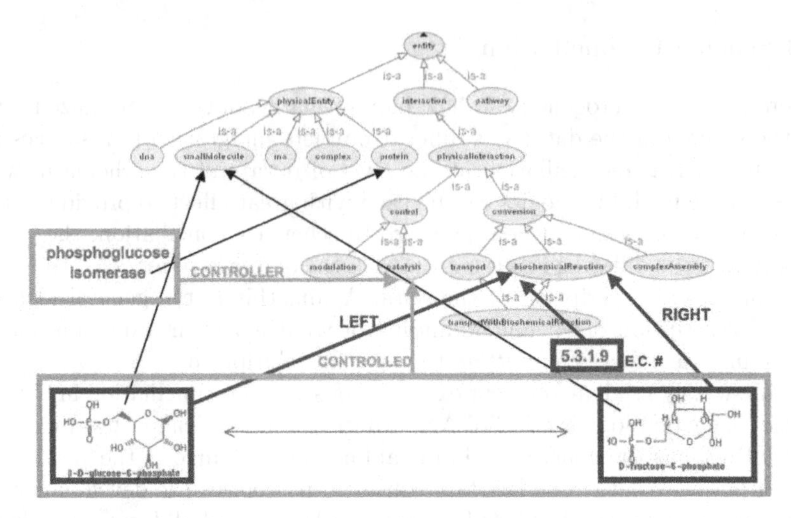

Fig. 3. Image showing the BioPAX ontology and a reaction from the glycolysis pathway

them in BioPAX. Thus in BioPAX, an instance is created of *biochemicalReaction* with the property *LEFT*[7] filled with β-D-glucose-6-phosphate, the property *RIGHT*[8] filled by D-fructose-6-phosphate, and `E.C.#`[9] property filled with *5.3.1.9*. An instance of the *catalysis* class is created with property *CONTROLLER* filled with phosphoglucose isomerase, and the property *CONTROLLED* with the reaction name *PGLUCISOM-RXN*.

Each physical interaction has participants which are instances of one of the subclasses of *PhysicalInteraction* or instances of the class *physicalEntityParticipant*. The class *physicalEntityParticipant* is also a *utilityClass* and is used to describe a physical entity in the context of an interaction. A Physical_Entity_Participant specifies the *physicalEntity* in the context of an interaction by adding the properties *CELLULAR-LOCATION* and *STOICHIOMETRIC-COEFFICIENT*.

For each database, committing to the BioPAX ontology, a converter is made that maps that resource to elements of the ontology. The BioPAX ontology is general enough in how it models the elements of pathways to capture a wide range of the existing resources. It does this by modeling at a high level of abstraction. BioPAX does not attempt to make a canonical view of bio-pathways – a standard view, for instance, of glycolysis. Rather it describes the elements of pathways, their steps, types of interaction and so on. This means the actual pathways, enzymes, small molecules, and so on, are instances of these classes. This works in practice, although it is not ontologically rigorous [28].

[7] The left-hand side of the reaction.
[8] The right-hand side of the reaction.
[9] The enzyme classification number.

The mapping in Fig. 3 is not actually formed for the values as described, but for the schema of the client resource. The BioPAX ontology, therefore, only reconciles at the level of schema [28]. The values or instances held within the *schema* are not reconciled. For example, all the metabolic pathway resources contain adenosine triphosphate (ATP) as part of their descriptions. ATP would be mapped to *Small_molecule* in BioPAX, but the vocabulary used to state that it is "ATP" can still vary. Another level of reconciliation is needed at the value or instance level to allow full inter-operability at this level [28]. This returns to the basic use of ontology within bioinformatics – that of providing a controlled vocabulary for describing entities within bioinformatics databases. Nevertheless, the BioPAX ontology, despite its flaws [28], does provide a significant step in reconciling an important collection of bioinformatics data.

It is clear that not every resource to be mapped into BioPAX will have a schema element equivalent to all classes in the ontology. Here, the constraint based nature of OWL can help. An OWL class describes what is known about instances of that class. Simply asserting an instance to be a member of that class implies that the restriction on that class apply to the instance. So, for example, BioPAX states that all *PhysicalEntity* have a *CellularLocation*, but if a client database does not give cellular location, it is simply assumed to exist. OWL's ability to describe under-specified knowledge in this way is of great utility for this kind of modeling as it means resources can be compliant without over-committing [43]. Ontologies describe the things described by the data, rather than just providing a description of the data. This extra level of abstraction affords a level of flexibility in mapping from data-oriented languages. The relative richness of the modeling permitted in ontologies, the constraints, allow precision and accuracy in the description of the entities being represented in the data and a consequently higher fidelity in the mapping.

Once data are mapped into a common schema, it provides another level of query. The BioPAX initiative is rare within bioinformatics for being schema reconciliation as an end in itself. Schema reconciliation as described for BioPAX is a common factor in many systems, but is rarely done purely for the sake of schema reconciliation. TAMBIS (see Sect. 3.4) uses an ontology as a common schema, integrating through queries diverse and distributed bioinformatics data resources. Here, the schema reconciliation is part of the query answering process. Similarly, in the work of the Health Care and Life Sciences working group of the W3C[10] an ontology is built to which instances are imported from client resources [40]. The individual mappings to the ontology bring all the data instances together in a common representation or description of the domain. Schema and data reconciliation are an inherent part of bioinformatics and ontologies are a standard technique for tackling the problem in computer science that is widely applied in bioinformatics.

[10] `http://www.w3.org/2001/sw/HCLS`

3.4 Classifying Instances

Ontologies are description and definitions of instances in the world and we have already seen their utility in de facto integration with increased recall and precision in queries across diverse resources. In this section, we take this theme further by looking at broader ways of using ontologies to query data in the form of instances described by an ontology and to enable the recognition of the types of instances present. These are both forms of classification:

1. An ontological class has an extent of instances. By creating a class, a set of instances is being described. A query also describes a set of instances. In this way, a query classifies instance – a query puts instances into a class.
2. A defined class captures the properties that are sufficient to recognize an instance as being a member of that class [35], that is, an instance is classified against the ontology.

In this section we will concentrate on the second form of classification. Our chapter in the first edition [44] described systems such as TAMBIS [17] and RiboWeb [2] that describe querying of data. In the case of TAMBIS, classes were dynamically constructed against an ontology and re-written to retrieve instances from external resources. With RiboWeb, the ontology was used to guide data acquisition and analysis to form a knowledge base that could then be queried.

In the second form of classification, we are moving much more towards ontology capturing knowledge for computational use. As described, the most straight-forward way of doing this is to recognise when an instance belongs to a particular class.

Bioinformatics is rich with tools designed to detect features on proteins and DNA sequences. From the features detected, a human bioinformatician is given clues by which data can be interpreted and classified. Typical of this is the classification of protein sequences by the presence of a certain configuration of features that, for example, suggest a certain catalytic or other behavior. Tools such as InterPro and InterProScan [31] provide a bioinformatician with a set of sequence features for a protein, but it is up to a human to interpret this information as to which class of protein sequences a particular protein sequence belongs. Ontologies offer a mechanism for capturing the knowledge by which humans recognize collections of features to draw conclusions.

We are beginning to see examples of this very general technique in the bioinformatics arena. For example, genome complements of protein phosphatases have been classified [54]. In this work, an ontology written in OWL describing a class of enzymes called protein phosphatases was constructed, with some fifty classes of protein phosphatase defined in terms of the sets of

```
Class: TyrosineRreceptorProteinPhosphataseSequence
   EquivalentTo: ProteinSequence That
        hasdomain SOME ProteinTyrosinePhosphataseDomain
        and hasdomain EXACTLY 1 TransmembraneDomain
Class: R2AProteinPhosphataseSequence
   EquivalentTO: ProteinSequence That
   hasDomain EXACTLY 2 ProteinTyrosinePhosphataseDomain
   and hasdomain EXACTLY 1 TransmembraneDomain
   and hasdomain EXACTLY 4 FibronectinDomains
   and hasDomain EXACTLY 1 ImmunoglobulinDomain
   and hasDomain EXACTLY 1 MAMDomain
   and hasDomain EXACTLY 1 Cadherin-LikeDomain
   and hasdomain ONLY (TyrosinePhosphataseDomain
        or TransmembraneDomain or
        FibronectinDomain or ImnunoglobulinDomain
        or Clathrin-LikeDomain or
        ManDomain)
```

Fig. 4. Definitions written in OWL for two classes of protein phosphatase. The definitions describe those protein features that are sufficient to recognise a particular class of phosphatase. Note that for *TyrosineRreceptorProteinPhosphataseSequence* only two features are asserted, but others may be present without affecting the classification. For the *R2AProteinPhosphataseSequence*, however, all features are specified; any others being present would mean that an instance is not a member of this class

sequence features[11] both necessary and sufficient to recognize membership of a particular class of protein phosphatase. This used a *qualified cardinality restriction*, which states the numbers of a particular sequence feature required to be present [53] (see Fig. 4 for two examples).

The proteins from a genome were analyzed with InterProScan to detect their sequence features. The scan results were processed and transformed to produce a collection of OWL individuals for each named protein, along with assertions as to which protein sequence features were present.

The instances are classified against the ontology, producing a catalog of the phosphatases present in the genome. Note that in Fig. 4, that the description of the R2A phosphatase is closed (the

```
hasDomain ONLY
```

clause) – it defines the class in terms of the sequence features that must be present and only those features that can be present. If any more features are present it cannot be a member of that class. Similarly, the definition of a receptor tyrosine phosphatase is open, it describes what features must be present,

[11] The sequence of amino acid residues in a protein determine how the protein "folds" into a three-dimensional shape. This shape determines the functionality of the protein. Biologists have determined some patterns of amino acid residue that indicate certain features of these "shapes" that are diagnostic for functionality.

but leaves it open as to which others can be present. The presence of the catalytic feature and the transmembrane feature are sufficient, irrespective of other features present, to recognize membership of this class. The classification takes advantage of this openness, with local closure to classify instances as far as possible. Any instance classifying part way down the hierarchy is, in effect, a protein phosphatase whose complete description is not yet in the ontology – that is, a putative new class of phosphatase.

FungalWeb [6] takes a related, but much broader approach to classifying instances. FungalWeb also brings in the factor of data integration across multiple resources via an ontology in order to classify instances of enzymes of interest from fungal biochemistry. The FungalWeb ontology draws together fungal species, genes, protein families, enzymes, the reactions they catalyze, functions, processes and commercial applications of those enzymes. The ontology is derived from database schema for bioinformatics databases; pre-existing ontologies; and de novo ontology development. Instance data are drawn from domain literature and client databases.

In FungalWeb, the instances are: The species of fungus; named proteins, as individuals, are classified by their reaction; chemical names represent individuals of enzyme substrate and product; and industrial applications of enzymes were modeled as individuals. Properties from the ontology allows these individuals to be related by assertions. An example set of individuals can be seen in Fig. 5, where the concept *Enzyme* is linked by *able to modify* to the concept *Substrate*, where the instance *Pectin* is specified. *Enzyme* is also linked to *Commercial Enzyme Product* by the role *usedInProduct* (which is negated). Lastly *Enzyme* is linked to the concept *Fungi* by the role *has been reported to be found in*. The tabulated results show two columns of Enzyme and Fungi instances arranged in nine rows. The instances represent all enzymes not used in commercial products, where the enzyme is known to act on the substrate pectin. The corresponding fungus known to produce such an enzyme (pectolase) is listed.

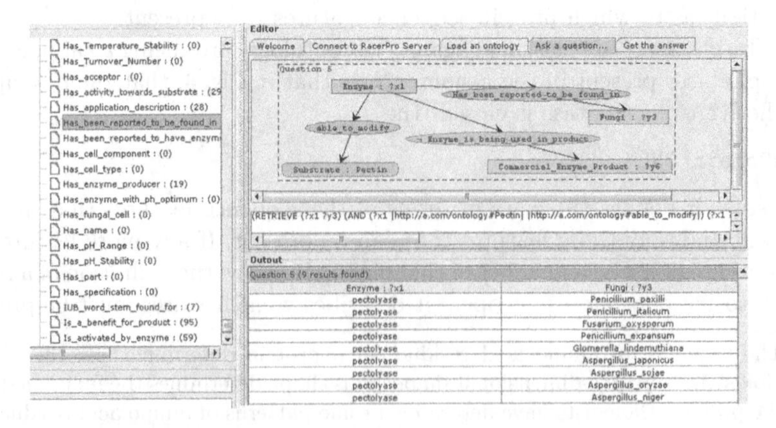

Fig. 5. A query from the FungalWeb OntoIQ tool for finding enzymes not currently used to act on pectin

FungalWeb used the NRQL language with the Racer reasoner [19] to classify instances in order to answer queries. An example query given (taken from [6]) is to find enzymes with oxygen as an electron acceptor that have an industrial use. The ontology has a class of *enzyme*, whose membership definition is to have oxygen as an electron acceptor. As with the phosphatase example earlier, instances can be classified according to such criteria. Similarly, a class of industrial process can be described as using a particular class of enzyme – in this case "coal liquefaction". FungalWeb also uses NRQL [20] to find, for example, enzymes not known to take part in a particular industrial process. The utility of NRQL in this case is to pose a locally closed query on the open world A-box.

Both these examples show ontologies written in OWL being used to reason over bioinformatics data in the form of OWL instances. In the case of the phosphatases, an ontology has been used to drive biological discoveries. OWL ontologies provide a method, through their necessary and sufficient conditions on classes, by which the features for recognition of class membership – domain knowledge – can be computationally encoded. Bioinformatics, through tools such as InterProScan, provide the computational means for recognizing features on data. If we have the means to encode first, those features in an ontology; second, the class definitions in terms of those features by which an individual can be recognized to be a member of a class; and third, the means by which features can be detected an encoded as OWL individuals; then we have a general mechanism for classifying data.

4 Discussion

We see many inter-related uses of ontological description of entities in the world. The overwhelming use of ontologies in bioinformatics is still the annotation of data to provide a common way of describing these data and then to enable the querying, clustering and further analysis of these data. This has been enormously powerful. It has enabled a large range and quantity of biological queries and insights to be gained. It has made the very expensively generated biological data much more useful.

Whilst an ontological description of a domain is useful as a way to capture knowledge and stimulate thinking about a domain, a more exciting prospect is the opportunity to capture domain knowledge such that we can make computational use of knowledge in a symbolic form. The mass annotation of biological data has begun this action with clustering for the analyses of experimental data. Statistical measures over the annotated data, exploiting the structure in which the ontology terms are held provides more sophisticated analysis of these data.

Drawing together diverse data into a common setting at both the level of schema and value, whether transiently or in a more sustained fashion, enables richer queries and analysis. The richness of and high-fidelity of ontological description makes it a good candidate for such reconciliation.

Within the life-sciences, real computational use of knowledge is still in its infancy. Bioinformatics still has much to gain from basic annotation of its data with names supplied by an ontology. So much is enabled by this simple device that more complex analyzes from reasoning over symbolic knowledge are not yet demanded by biologists themselves. We have seen the beginnings of such computational use and it will test the scalability of current Semantic Web technologies and languages. All the current activities, however, are laying the foundation for a much deeper exploitation of bioinformatics data through the application of ontologies.

References

1. Gil Alterovitz, Michael Xiang, Mamta Mohan, and Marco F. Ramoni. GO PaD: the gene ontology partition database. *Nucleic Acids Research*, 35(Suppl 1):D322–327, 2007.
2. R. Altman, M. Bada, X.J. Chai, M. Whirl Carillo, R.O. Chen, and N.F. Abernethy. RiboWeb: An ontology-based system for collaborative molecular biology. *IEEE Intelligent Systems*, 14(5):68–76, 1999.
3. S.F. Altschul, W. Gish, M. Miller, E.W. Myers, and D.J. Lipman. Basic local alignment search tool. *Journal of Molecular Biology*, 215:403–410, 1990.
4. S. Ananiadou and B. Stapley, editors. *Text mining for biology.* IOS Press, 2005.
5. Michael Bada, Robert Stevens, Carole Goble, Yolanda Gil, Michael Ashburner, Judith A. Blake, J. Michael Cherry, Midori Harris, and Suzanna Lewis. A Short Study on the Success of the Gene Ontology. *Web Semantics Science, Services and Agents on the World Wide Web*, 1(2):235–240, 2004.
6. Christopher J.O. Baker, Xiao Su, Volker Haarslev, and Greg Butler. Semantic web infrastructure for fungal enzyme biotechnologists. *Web Semantics: Science, Services and Agents on the World Wide Web*, 4(3):168–180, September 2006. Special issue on Semantic Web for Life Sciences.
7. Tim Beissbarth and Terence P. Speed. GOstat: find statistically overrepresented Gene Ontologies within a group of genes. *Bioinformatics*, 20(9):1464–1465, 2004.
8. Judith Blake, David Hill, and Barry Smith. Gene ontology annotations: What they mean and where they come from. In: 10th Annual Bio-Ontologies SIG, 2007. `http://bio-ontologies.org.uk`.
9. Olivier Bodenreider. Lexical, terminological and ontological resources for biological text mining. In: S. Ananiadou and B. Stapley, editors, *Text mining for biology.* IOS Press, 2005.
10. Olivier Bodenreider and Robert Stevens. Bio-ontologies: current trends and future directions. *Brief Bioinform*, 7(3):256–274, 2006.
11. James F. Brinkley, Dan Suciu, Landon T. Detwiler, John H. Gennari, and Cornelius Rosse. A framework for using reference ontologies as a foundation for the semantic web. In: *Proceedings, American Medical Informatics Association Fall Symposium*, pages 96–100, 2006.
12. UniProt Consortium. The universal protein resource (uniprot). *Nucleic Acids Research*, 35(Database issue):D193–D197, Jan 2007.

13. K.D. Dahlquist, N. Salomonis, K. Vranizan, S.C. Lawlor, and B.R. Conklin. GenMAPP, a new tool for viewing and analyzing microarray data on biological pathways. *Nature Genetics*, 31(1):19–20, 2002.
14. Selina S Dwight, Rama Balakrishnan, Karen R Christie, Maria C Costanzo, Kara Dolinski, Stacia R Engel, Becket Feierbach, Dianna G Fisk, Jodi Hirschman, Eurie L Hong, Laurie Issel-Tarver, Robert S Nash, Anand Sethuraman, Barry Starr, Chandra L Theesfeld, Rey Andrada, Gail Binkley, Qing Dong, Christopher Lane, Mark Schroeder, Shuai Weng, David Botstein, and J. Michael Cherry. Saccharomyces genome database: underlying principles and organisation. *Brief Bioinform*, 5(1):9–22, Mar 2004.
15. C. Fellbaum. *Wordnet: an electronic lexical database.* Mit Pr, 1998.
16. Kevin Garwood, Phillip Lord, Helen Parkinson, Norman W. Paton, and Carole Goble. Pedro ontology services: A framework for rapid ontology markup. In: A. Gómez-Pérez and J. Euzenat, editors, *European Semantic Web Conference*, pages 578–591. Springer, Berlin, 2005.
17. C.A. Goble, R. Stevens, G. Ng, S. Bechhofer, N.W. Paton, P.G. Baker, M. Peim, and A. Brass. Transparent Access to Multiple Bioinformatics Information Sources. *IBM Systems Journal Special issue on deep computing for the life sciences*, 40(2):532–552, 2001.
18. T. R. Gruber. Towards Principles for the Design of Ontologies Used for Knowledge Sharing. In: N. Guarino and R. Poli, editors, *Formal Ontology in Conceptual Analysis and Knowledge Representation*, Deventer, The Netherlands, 1993. Kluwer, Dordrecht.
19. V. Haarslev and R. Moller. RACER system description. *Proc. of the Int. Joint Conf. on Automated Reasoning (IJCAR 2001)*, 2083:701–705, 2001.
20. V. Haarslev, R. Möller, and M. Wessel. Querying the semantic web with racer + nrql. In: *Proceedings of the KI-2004 International Workshop on Applications of Description Logics (ADL'04), Ulm, Germany, September 24*, 2004.
21. V. Heijst, G Shreiber, and B. Wielinga. Using explicit ontologies in KBS. *International Journal of Human-Computer Studies*, 46(2/3):183–292, 1997.
22. Andrew R. Jones and Frank Gibson. An update of data standards for gel electrophoresis. *Practical Proteomics*, 7(S1):35–40, 2007
23. Cliff A Joslyn, Susan M Mniszewski, Andy Fulmer, and Gary Heaton. The gene ontology categorizer. *Bioinformatics*, 20 Suppl 1:i169–i177, Aug 2004.
24. P.D. Karp, M. Riley, M. Saier, I.T. Paulsen, S.M. Paley, and A. Pellegrini-Toole. The EcoCyc and MetaCyc Databases. *Nucleic Acids Research*, 28:56–59, 2000.
25. C. Leacock and M. Chodorow. Combining local context and WordNet similarity for word sense identification. *WordNet: An Electronic Lexical Database*, 49(2):265–283, 1998.
26. Maurizio Lenzerini. Data integration: a theoretical perspective. In: *PODS '02: Proceedings of the twenty-first ACM SIGMOD-SIGACT-SIGART symposium on Principles of database systems*, pages 233–246, New York, NY, USA, 2002. ACM Press.
27. P.W. Lord, R.D. Stevens, A. Brass, and C.A. Goble. Semantic similarity measures as tools for exploring the Gene Ontology. In: *Pacific Symposium on Biocomputing*, pages 601–612, 2003.
28. Joanne Luciano and Robert Stevens. e-science and biological pathway semantics. *BMC Bioinformatics*, 8(Suppl 3):S3, 2007.
29. J.S. Luciano. PAX of mind for pathway researchers. *Drug Discovery Today*, 10:937–42, 2005.

30. Robert Minchin, Fabio Porto, Christelle Vangenot, and Sven Hartmann. Symptoms ontology for mapping diagnostic knowledge systems. In: *CBMS '06: Proceedings of the 19th IEEE Symposium on Computer-Based Medical Systems*, pages 593–598, Washington, DC, USA, 2006. IEEE Computer Society.
31. Nicola J Mulder, Rolf Apweiler, Teresa K Attwood, Amos Bairoch, Alex Bateman, David Binns, Peer Bork, Virginie Buillard, Lorenzo Cerutti, Richard Copley, Emmanuel Courcelle, Ujjwal Das, Louise Daugherty, Mark Dibley, Robert Finn, Wolfgang Fleischmann, Julian Gough, Daniel Haft, Nicolas Hulo, Sarah Hunter, Daniel Kahn, Alexander Kanapin, Anish Kejariwal, Alberto Labarga, Petra S Langendijk-Genevaux, David Lonsdale, Rodrigo Lopez, Ivica Letunic, Martin Madera, John Maslen, Craig McAnulla, Jennifer McDowall, Jaina Mistry, Alex Mitchell, Anastasia N Nikolskaya, Sandra Orchard, Christine Orengo, Robert Petryszak, Jeremy D Selengut, Christian J A Sigrist, Paul D Thomas, Franck Valentin, Derek Wilson, Cathy H Wu, and Corin Yeats. New developments in the interpro database. *Nucleic Acids Res*, 35(Database issue):D224–D228, Jan 2007.
32. Stuart J. Nelson, Douglas Johnston, and Betsy L. Humphreys. Relationships in Medical Subject Headings. In: Rebecca Bean, Carol A.; Green, editor, *Relationships in the organization of knowledge*, pages 171–184. Kluwer, Dordrecht, 2001.
33. P.W. Lord, R.D. Stevens, A. Brass, and C.A. Goble. Investigating semantic similarity measures across the Gene Ontology: the relationship between sequence and annotation. *Bioinformatics*, 19(10):1275–83, 2003.
34. R. Rada, H. Mili, E. Bicknell, and M. Blettner. Development and application of a metric on semantic nets. *Systems, Man and Cybernetics, IEEE Transactions on*, 19(1):17–30, 1989.
35. Alan Rector, Nick Drummond, Matthew Horridge, Jeremy Rogers, Holger Knublauch, Robert Stevens, Hai Wang, and Chris Wroe. OWL pizzas: practical experience of teaching owl-dl: common errors and common patterns. In: *14th International Conference on Knowledge Engineering and Knowledge Management EKAW 2004*, pages 63–81, 2004.
36. P. Resnik. Using information content to evaluate semantic similarity in a taxonomy. *IJCAI*, pages 448–453, 1995.
37. Oleg Rokhlenko, Tomer Shlomi, Roded Sharan, Eytan Ruppin, and Ron Y. Pinter. Constraint-based functional similarity of metabolic genes: going beyond network topology. *Bioinformatics*, page btm319, 2007.
38. Cornelius Rosse and Jose L. V. Mejino. A reference ontology for bioinformatics: the foundational model of anatomy. *Journal of Biomedical Informatics*, 36:478–500, 2003.
39. Daniel L. Rubin, Farhad Shafa, Diane E. Oliver, Micheal Hewett, and Russ B. Altman. Representing genetic sequence data for pharmacogenomics: an evolutionary approach using ontological and relational models. In: Chris Sander, editor, *Proceedings of Tenth International Conference on Intelligent Systems for Molecular Biology*, volume 18, pages 207–215, 2002.
40. Alan Ruttenberg, Tim Clark, William Bug, Matthias Samwald, Olivier Bodenreider, Helen Chen, Donald Doherty, Kerstin Forsberg, Yong Gao, Vipul Kashyap, June Kinoshita, Joanne Luciano, M. Scott Marshall, Chimezie Ogbuji, Jonathan Rees, Susie Stephens, Gwen Wong, Elizabeth Wu, Davide Zaccagnini,

Tonya Hongsermeier, Eric Neumann, Ivan Herman, and Kei-Hoi Cheung. Advancing translational research with the Semantic Web. *BMC Bioinformatics*, 8, 2007.

41. Robert Stevens. Foreword. In: Richard Baldock Albert Burger, Duncan Davidson, editor, *Anatomy Ontologies for Bioinformatics, Principles and Practice*. Springer, Berlin, November 2008.
42. Robert Stevens, Phil Lord, and Duncan Hull. Using distributed data and tools in bioinformatics applications. In: Thomas Lengauer, editor, *Bioinformatics – From Genomes to Therapies; Volume 3*, pages 1627–1649. Wiley-VCH, New York, 2005.
43. Robert Stevens, Mikel Ega na Aranguren, Katy Wolstencroft, Ulrike Sattler, Nick Drummond, Matthew Horridge, and Alan Rector. Using owl to model biological knowledge. *International Journal of Human Computer Studies*, 65(7):583–594, 2007. Special issue on limitations of ontology.
44. Robert Stevens, Chris Wroe, Phillip Lord, and Carole Goble. Ontologies in bioinformatics. In: Stefan Staab and Rudi Studer, editors, *Handbook on Ontologies in Information Systems*, pages 635–657. Springer, Berlin, 2003.
45. Ying Tao, Lee Sam, Jianrong Li, Carol Friedman, and Yves A. Lussier. Information theory applied to the sparse gene ontology annotation network to predict novel gene function. *Bioinformatics*, 23(13):529–538, 2007.
46. The Gene Ontology Consortium. Gene ontology: tool for the unification of biology. *Nature Genetics*, 25:25–29, 2000.
47. The Gene Ontology Consortium. Creating the gene ontology resource: design and implementation. *Genome Research*, 11(8):1425–1433, 2001.
48. M. Uschold and R. Jasper. A framework for understanding and classifying ontology applications, 1999.
49. H. Wang, F. Azuaje, O. Bodenreider, and J. Dopazo. Gene expression correlation and gene ontology-based similarity: an assessment of quantitative relationships. *Computational Intelligence in Bioinformatics and Computational Biology, 2004. CIBCB'04. Proceedings of the 2004 IEEE Symposium on*, pages 25–31, 2004.
50. H. Wang, A. Rector, N. Drummond, M. Horridge, J. Seidenberg, N.F. Noy, M.A. Musen, T. Redmond ad D.L. Rubin, S. Tu, and T. Tudorache. Frames and owl side by side. 9th International Protg Conference. Stanford, CA., 2006.
51. Patricia L. Whetzel, Helen Parkinson, Helen C. Causton, Liju Fan, Jennifer Fostel, Gilberto Fragoso, Laurence Game, Mervi Heiskanen, Norman Morrison, Philippe Rocca-Serra, Susanna-Assunta Sansone, Chris Taylor, Joseph White, and Christian J. Stoeckert. The mged ontology: a resource for semantics-based description of microarray experiments. *Bioinformatics*, 22(7):866–873, 2006.
52. P.L Whetzel, R.R. Brinkman, H.C. Causton, L. Fan, D. Field, J. Fostel, G. Fragaso, T. Gray, M. Heiskanen, T. Hernandez-Boussard, N. Morrison, H. Parkinson, P. Rocca-Serra, S-A. Sansone, D. Schober, B. Smith, R. Stevens, C.J. Stoeckert, C. Taylor, J. White, and A. Wood. the fugo working group development of fugo: An ontology for functional genomics investigations. *OMICS: A journal of integrative biology*, 10:199–204, June 2006.
53. K. Wolstencroft, A. Brass, I. Horrocks, P. Lord, U. Sattler, D. Turi, and R. Stevens. A little Semantic Web goes a long way in biology. In: *Proc. of the 4th Int. Semantic Web Conf. (ISWC2005)*, volume 3729/2005 of *LNCS*, Springer, Berlin, 2005.
54. K. Wolstencroft, P. Lord, L. Tabernero, A. Brass, and R. Stevens. Protein classification using ontology classification. *Bioinformatics*, 22(14):e530–538, 2006.

55. Chris Wroe and Robert Stevens. Ontologies for molecular biology. In: Thomas Lengauer, editor, *Bioinformatics - From Genomes to Therapies*, pages 1061–1085. Wiley-VCH, New York, 2005.
56. B.R. Zeeberg, W. Feng, G. Wang, M.D. Wang, A.T. Fojo, M. Sunshine, S. Narasimhan, D.W. Kane, W.C. Reinhold, S. Lababidi, et al. GoMiner: a resource for biological interpretation of genomic and proteomic data. *Genome Biol*, 4(4):R28, 2003.

Semantic Portals for Cultural Heritage

Eero Hyvönen

Semantic Computing Research Group, Helsinki University of Technology (TKK) and University of Helsinki, eero.hyvonen@tkk.fi, http://www.seco.tkk.fi/

Summary. Cultural heritage is a promising application domain for semantic web technologies due the semantic richness and heterogeneity of cultural content, and the distributed ways in which the content is created in memory organizations and by citizens. This chapter overviews issues and research related to creating semantic portals for publishing cultural heritage collections and other content on the web.

1 Benefits of Cultural Semantic Portals

Cultural content on the web is available in various forms (documents, images, audio tracks, videos, collection items, learning objects, etc.), concern various topics (art, history, handicraft, etc.), is written in different languages, is targeted to both laymen and experts, and is provided by different independent memory organizations (museums, archives, and libraries) and individuals. The difficulty of finding and relating information in this kind of heterogenous content provision and data format environment creates an obstacle for end-users of cultural contents, and a challenge to organizations and communities producing the contents.

Portals try to ease these problems by collecting content of various publishers into a single site [50]. Portal types include *service portals* collecting a large set of services together (e.g., Yahoo! and other "start pages"), *community portals* [53] acting as virtual meeting places of communities, and *information portals* [43] acting as hubs of data. Much of the semantic web content will be published using *semantic information portals* [38, 43]. Such portals are based on semantic web standards[1] and machine "understandable" content, i.e., metadata, ontologies, and rules, in order to improve structure, extensibility, customization, usability, and sustainability of traditional portal designs.

[1] http://www.w3.org/2001/SW/

S. Staab and R. Studer (eds.), *Handbook on Ontologies,* International Handbooks on Information Systems, DOI 10.1007/978-3-540-92673-3,

Cultural heritage is a promising application domain for semantic portals [3,4,6,25,26,48,57]. They are useful from the end-users' view point in several ways:

- *Global view to heterogeneous, distributed contents.* The contents of different content providers can accessed through one service as a single, seamless, and homogenous repository [25]. Only a single user interface has to be learned.
- *Automatic content aggregation.* Satisfying an end-user's information need often requires *aggregation* of content from several information providers [26, 50], a task suitable for semantic web technologies. For example, when looking for data about an artist, relevant information may be provided by museum collections, libraries, archives, authority records, ontologies, and other sources.
- *Semantic search.* In traditional portals, search is usually based on free text search (e.g., Google), database queries, and/or a stable classification hierarchy (e.g., Yahoo! and dmoz.org). Semantic content makes it possible to provide the end-user with more "intelligent" facilities based on ontological concepts and structures, such as *semantic search* [10], *semantic autocompletion* [24], and *faceted search* [19,21,27,42,47].
- *Semantic browsing and recommendations.* Semantic content also facilitates semantic browsing [17] (cf. chapter "Browsing and Navigation in Semantically Rich Spaces: Experiences with Magic Applications") and recommendations [58] (cf. chapter "Ontology-based Recommender Systems"). Here semantic associations between search objects can be exposed to the end-user as recommendation links, possibly with explicit explanations.
- *Other intelligent services.* Also other kind of intelligent services can be created based on machine interpretable content, such as knowledge and association discovery [49], personalization [2, 4], and semantic visualizations based on, e.g., historical and contemporary maps and time lines [36].

Semantic portals are very attractive from the content publishers viewpoint, too:

- *Distributed content creation.* Portal content is usually created in a centralized fashion by using a content management system (CMS). This approach is costly and not feasible if content is created in a distributed fashion by independent publishers, e.g., by different museums and other memory organizations. Semantic technologies can be used for harvesting and aggregating distributed heterogenous content (semi-)automatically into global content portals [25].
- *Automated link maintenance.* The problems of maintaining links up-to-date is costly from the portal maintenance viewpoint. In semantic portals, links can be created and maintained automatically based on the metadata and ontologies.

- *Shared content publication channel.* In the cultural domain the publishers usually share the common goal of promoting cultural knowledge in public and among professionals. A semantic portal can provide the participating organizations with a shared, cost-effective publication channel [28].
- *Enriching each other's contents semantically.* Interlinking content between collaborating organizations enriches the contents of everybody "for free".
- *Reusing aggregated content.* The content aggregated into a semantic portal can be reused in different applications and cross-portal systems [59].

A cultural semantic information portal includes several major components. First, we need a *content model* for representing cultural metadata, ontologies, and rules. Second, a content creation system is needed for creating and harvesting content. Third, the portal publishes semantic services for (1) human end-users as *intelligent user interfaces* and possibly for (2) other portals and applications as *web services.* In the following these components are explained in more detail.

2 Content Models for Semantic Cultural Portals

The semantic web "layer cake model" makes the distinction between a syntactic data level based on XML[2], and semantic levels above it:

- *Metadata level.* The RDF data model[3] (cf. chapter "Resource Description Framework") is used for representing metadata about cultural resources.
- *Ontology level.* The RDF Schema and web ontology language OWL[4] (cf. chapter "Web Ontology Language: OWL") are used for representing ontologies [14] (cf. chapter "Ontologies for Cultural Heritage") that describe vocabularies and concepts concerning the real world and our conception of it.
- *Logic level.* Logic rules (cf. chapter "Ontology and Rules") can be used for deriving new facts and knowledge based on the metadata and ontologies.
- *Trust level.* At the highest conceptual level issues of trustworthiness of content, copyrights, etc., are of concern.

In the following, metadata, ontology, and logic layers are considered from the viewpoint of semantic cultural portals. Issues related to trust on the semantic web in the cultural heritage domain have thus far not been discussed much in the literature.

[2] http://www.w3.org/XML/
[3] http://www.w3.org/RDF/
[4] http://www.w3.org/2004/OWL/

2.1 Metadata Schemas

Cultural content in museum collections, libraries, and other content repositories is usually described using *metadata schemas* (also called *annotation schemas* or *annotation ontologies*). These templates specify a set of obligatory and optional elements, i.e., properties, by which the metadata for content items should be described. For example, the Dublin Core (DC) Metadata Element Set[5] lists 15 standardized[6] elements, such as *dc:title, dc:creator*, and *dc:subject*, with additional elements and element refinements. Encoding guidelines tell how to express the elements in RDF/XML and using HTML/XHTML meta and link elements. Qualifiers, such as encoding schemes, enumerated lists of values, and other processing clues are used to provide more detailed information about a resource. For example, "date" is a DC element that can further be specified as "date published" or "date last modified". The core elements can be extended in an interoperable way by using the "dumb-down" principle. It means that in any use of a qualified DC element, the qualifier may be dropped and the remaining value of the element should still be a term that is useful for discovery, although with less precision.

DC is used as a basis in more detailed cultural metadata schemas, such as the Visual Resource Association's (VRA) Core Categories.[7] Its element set provides a categorical organization for the description of works of visual culture as well as the images that document them. Most VRA elements are defined as subproperties of corresponding DC elements. An example of an instance of VRA metadata in the CHIP portal [2, 4] is given below in RDF Turtle notation.[8] The schema has properties such as *vra:type* (the type of the art-work as a reference to the VRA vocabulary), *vra:title* (literal title of the art-work), *vra:creator, vra:subject, vra:culture*, and *vra:material*. Element values with a namespace are references to underlying ontologies.

```
rijks:artefactSK-C-K
  vra:type vra:Work ;
  vra:title "The Night Watch" ;
  vra:date "1642" ;
  vra:creator: 500011051 ;          # Rembrandt
  vra:subject iconclass:45F31 ;     # Call to arms
  vra:culture tgn:7006952 ;         # Amsterdam
  vra:material aat:30015050 .       # Oil paint
```

A metadata schema makes it possible to specify relevant aspects of the search objects, such as the "author", "title", and "subject" of a document, and focus search according to these. Sharing a metadata schema between different content providers facilitates, for example, multi- or metasearch.[9] Here the user types in a query in a metaportal. The query is then distributed to underlying

[5] http://dublincore.org/documents/1998/09/dces/
[6] NISO Standard Z39.85-2001 and ISO Standard 15836-2003.
[7] http://www.vraweb.org/
[8] http://www.dajobe.org/2004/01/turtle/
[9] http://en.wikipedia.org/wiki/Metasearch_engine

systems and the results are aggregated for the end-user. Protocols such as Z39.52[10] and Search and Retrieve via URL (SRU)[11] of the Library of Congress can be used here. For example, the Australian Museums and Galleries Online[12] and Artefacts Canada[13] are multi-search engines over nation-wide distributed cultural collections.

Another approach to creating metaportals is to first harvest the content into a global database, and search the global repository. Protocols such as Open Access Initiative Protocol for Metadata Harvesting (OAI-PMH)[14] can be used for distributed content publishing and harvesting.

Schema definitions tackle the problems of *syntactic* and *semantic interoperability* of content objects. Syntactic interoperability can be obtained by harmonizing encoding conventions (e.g., a date format) and other structural forms for representing data (e.g., an XML schema). Semantic interoperability is obtained by shared conventions for interpreting the syntactic representations, e.g., that the property *dc:subject* describes the subject matter of a document as a set of keywords taken from a thesaurus. Making different metadata schemas semantically interoperable includes two subtasks. First, semantic interoperability of element values has to be addressed using (shared) vocabularies and ontologies, and second, if multiple metadata schemas are involved, interoperability problems between different schema elements has to be solved. In below, these two issues are discussed in more detail.

2.2 Vocabularies and Ontologies

Metadata schemas specify data formats but do not tell how to fill the element values in the formats. Additional standards and guidelines are necessary to guide the choice of terms or words (data values) as well as the selection, organization, and formatting of those words (data content). Data value standards have been traditionally specified by constructing controlled vocabularies and thesauri [1, 15]. Examples of cultural thesauri include the Thesaurus for Graphic Materials I (TGM I)[15] for indexing pictorial materials, ICONCLASS[16] for art, the Art and Architecture Thesaurus (AAT)[17] for fine art, architecture, decorative arts, archival materials, and material culture, the Union List of Artist Names (ULAN),[18] the Thesaurus of Geographic Names

[10] http://www.cni.org/pub/NISO/docs/Z39.50-brochure/
[11] http://www.loc.gov/standards/sru/
[12] http://www.collectionsaustralia.net/
[13] http://www.chin.gc.ca/
[14] http://www.openarchives.org/
[15] http://www.loc.gov/rr/print/tgm1/
[16] http://www.iconclass.nl/
[17] http://www.getty.edu/research/conducting_research/vocabularies/aat/
[18] http://www.getty.edu/vow/ULANSearchPage.jsp

(TGN),[19] the Library of Congress Authority Files,[20] and the terminologies and standards of the MDA (formerly Museum Documentation Association).[21] An example of a data content standard is the Cataloging Cultural Objects (CCO) guidelines.[22]

Many cultural thesauri have been transformed [55,56] into SKOS format[23] to be used in cultural semantic portals [48,57]. However, although a syntactic transformation into SKOS is useful, it is not always enough from a semantic viewpoint. The fundamental problem with traditional thesauri is that its semantic relations have been constructed mainly to help the indexer in finding indexing terms, and understanding the relations needs implicit human knowledge. Unless the meaning of the semantic relations of a thesaurus is made more explicit and accurate for the computer to interpret, the SKOS version is equally confusing to the computer as the original thesaurus, even if semantic web standards are used for representing it.

For example, there are many problems in utilizing the Broader Term (BT) relations of thesauri [30]: (1) BT relations do not necessarily structure the terms into a full-blown hierarchy that would be useful, e.g., in faceted search, but into a forest of small subhierarchies. (2) The semantics of the BT relation is ambiguous: it may mean either subclass-of-relation, part-of relation (of different kinds, cf. [13]), or instance-of relation. As a result, the BT relation cannot, e.g., be used for property inheritance. (3) The transitivity of the BT relation chains is not guaranteed from the instance-class-relation point of view. If x is an instance of class A whose broader term is B, then it is not necessarily the case that x is an instance of B, although this a basic assumption in RDFS and OWL. For example, assume that x is an instance of "make-up mirror", whose broader term is "mirror", and that its broader term is "furniture". When searching with the concept "furniture" one would expect that instances of furniture are retrieved, but in this case the result would include confusingly make-up mirrors, too, if transitivity is assumed. A solution to these fundamental problems is to actually refine and reorganize the semantic structures of a thesaurus into a light-weight ontology, e.g., along the lines proposed in [31].

Several domain ontologies are used in describing cultural metadata. This raises up the problem of making ontologies mutually interoperable. There are solution approaches for this, such as ontology mapping and alignment [18] (cf. chapter "Ontology Mapping"), sharing common foundational logical principles like in DOLCE[24] (cf. chapter "Foundational Choices in DOLCE"), and using shared horizontal top ontologies, such as the IEEE SUMO.[25] It is likely,

[19] `http://www.getty.edu/research/conducting_research/vocabularies/tgn/`
[20] `http://authorities.loc.gov/`
[21] `http://www.mda.org.uk/stand.htm`
[22] `http://www.vraweb.org/ccoweb/cco/index.html`
[23] `http://www.w3.org/2004/02/skos/`
[24] `http://www.loa-cnr.it/DOLCE.html`
[25] `http://suo.ieee.org/`

that in many cases several identifiers (URIs) will be in use for denoting a single concept even if this is not desirable in general. For example, registries of same geographical locations are maintained at different countries and by different service providers using their own identifiers. In such cases, dereferencing services will be needed to map resource identifiers denoting same concepts with each other.

2.3 Metadata Schema Interoperability

If a portal aggregates cultural contents described using different kind of schemas (e.g., for artifacts, music, maps, books, cultural sites), the schema element structures have to be made interoperable in one way or another, including the element values. If the element structures in the schemas refine each other, then using subproperties and the dumb-down principle of DC applications may be applied. In other cases, the metadata schemas can be made interoperable by transforming them into a shared underlying form.

An approach to this is the CIDOC Conceptual Reference Model (CIDOC CRM) [11] (cf. chapter "Ontologies for Cultural Heritage"), an annotation ontology standard[26] developed as an underlying schema into which other metadata schemas in the cultural domain can be transformed for interoperability. This model "provides definitions and a formal structure for describing the implicit and explicit concepts and relationships used in cultural heritage documentation".[27] The framework includes 81 classes, such as *crm:Man-Made Object*, *crm:Place*, and *crm:Time-Span*, and a large set of 132 properties relating the entities with each other, such as *crm:HasTime-Span* and *crm:IsIdentifiedBy*.

Another approach to semantic metadata schema interoperability has been developed in the CultureSampo portal[28] [26,45]. The cultural content types in this system include a wide variety of cultural objects, such as artifacts, paintings, photographs, videos, music, biographies, epics, cultural sites, and historical events. The original metadata from the content providing memory organizations use several schemas, including DC, ULAN, and CIDOC CRM. In the 2007 version of this portal [45], content integration was performed by transforming content into a light-weight knowledge base describing the domain world based on events and their thematic roles [52], such as agent, goal, and place. For example, the DC metadata of a painting tells that there has been a painting event with the value of *dc:creator* in the agent role. This event instance can be used for enriching the painter's biography, that is also represented in terms of underlying events, such as the painter "being born" at a certain place—another event that can be derived from the relational embedded meaning of the relation *ulan:birthPlace* used in ULAN. In contrast to CIDOC CRM, the events and thematic role values are based on large shared

[26] Since 2006, CIDOC CRM has been an official ISO standard 21127:2006.
[27] `http://cidoc.ics.forth.gr/`
[28] `http://www.kulttuurisampo.fi/`

domain ontologies of tens of thousands of concepts, and only few thematic and other relationships between them. The domain ontologies are used not only for explicating relational meaning of metadata schemas in an interoperable way, but also for making *element values* semantically interoperable, an issue not addressed by the CIDOC CRM standard. The homogenized event-based knowledge can be used, e.g., as a basis for semantic recommendations [46].

2.4 Logic Rules for Cultural Heritage

A collection of cultural metadata and related ontologies constitute a knowledge base. On the logical level, rules can be used for deriving new facts and knowledge based on the repository, i.e., for explicating the implicit content of the repository, and enriching the content semantically. Some examples illustrating different ways of using rules in semantic cultural portals and systems are given below:

- *Explicating content of metadata schemas.* Many metadata formats contain implicit knowledge embedded, e.g., in the relational meaning of the element names. In [45] rule sets for three cultural metadata schemas are presented for explicating such knowledge in terms of events.
- *Enriching semantic content.* Common sense rules may be used for enriching annotations, thus extending the machine's understanding about culture. In [27], for example, family relation rules (and others) we used to explicate implicit family relations, such as "grand father of", between persons in order to link photographs of relatives together while browsing the repository.
- *Semantic recommendations with explanations.* In [25] some 300 rules and associations, such as "doctoral hats are related to academic ceremonies" or "distaffs are related to spinning events", were used to represent simple common sense knowledge and associations between ontological concepts. A semantic recommendation service was then established that, based on additional logical rules, could (1) dynamically find out chained semantic associations between cultural objects based on ontologies and the common sense relations, and (2) at the same time construct literal explanations of why the association would be of interest. In [32] semantic process descriptions of cultural processes, such as traditional farming and fishing, were used as basis for relating cultural objects with each in meaningful ways.
- *Projecting search facets.* In faceted search, rules can be used for constructing facet hierarchies based on ontological structures, such as the subclass-of and part-of-relations. Furthermore, rules can be used to solve the problem of projecting search items to facet categories, which may be complicated [27,29,58]. From a software engineering viewpoint, using logic rules for projections separates facets from the annotation ontologies and annotations, which makes it possible to apply the same faceted search engine to knowledge bases based on different kind of ontologies and annotation schemas [39].

- *Association discovery.* Association discovery can be based on rules trying to find paths between resources in a knowledge base [26,48,49].

3 Cultural Content Creation

Several kinds of content need to be created for a semantic portal, including ontologies, terminologies, and semantic annotations. Also creating rules for, e.g., semantic recommendations can be seen as a form of content to be created. In below, ontology, terminology, and annotation creation are discussed in some more detail.

The core of a semantic cultural heritage portal is typically a set of domain ontologies that are used for annotating cultural contents. Many vocabularies and ontologies, such as AAT, are used for defining *universals*, i.e., general concepts, classes, or types of individuals, such as "chair" (artifact ontology), "wood" (material ontology), "painter" (actor types), or "city" (geographical concepts). In creating ontologies, it is advisable to try to reuse existing ontologies or transform existing thesauri into semantic web formats, as discussed earlier. The ontologies can also be created or enhanced manually using an ontology editor such as Protégé.[29]

Another basic type of ontologies are instance-rich ontologies or registries of individuals. Such ontologies include, for example, geo-ontologies, such as TGN, and actor ontologies (persons and organizations), such as ULAN. This kind of ontologies of individuals are based on a (usually small) ontology of classes (universals), such as "city" or "person", that is *populated* with individuals from, e.g., a database. This kind of instance ontologies can be used for annotating content (e.g., used as *dc:creator* values), but the instances may, at the same time, be used as a content type of its own value in the portal (e.g., a biography).

The terminology used in a portal is typically defined by associating ontological resources with preferable and alternative labels (e.g., using properties *rdfs:label* or *skos:altLabel*). Resource identifiers (URIs) of concepts, used by the machine, refer to concepts that are in principle language independent. However, labels used by humans can be multi-lingual, based on XML markup (e.g., *xml:lang*). This is essential when creating multilingual portals.

The content providers often use different literal terms to refer to the same resources when describing metadata in legacy systems. For example, literals "United States" and "US" may be used to refer to the same country. This problem of *synonymy* can be approached by using alternative labels. On the other hand, the same term may be used to refer to different concepts, such as river "bank" and financial "bank". In order to eliminate such *homonymy* in terminology, it is advisable that an ontology uses a unique labeling of terms for concepts (e.g., "bank (financial)"). However, this does not solve the problem disambiguating meanings of terms occurring in natural language descriptions.

[29] `http://protege.stanford.edu/`

Content in memory organizations is usually available as relational legacy databases, whose annotations are literal terms and free text descriptions. Such annotations are often intended for human usage, use various syntactic conventions, are often semantically ambiguous, and may contain syntactic typing errors. When transforming legacy metadata into semantic web formats, a key problem is how to map textual descriptions in the metadata with ontological concepts, e.g., how to determine that the string "bank" in a *dc:subject* description of a photograph refers to the concept "river bank" and not "financial bank". In below, the task of transforming literal element values used in legacy systems into ontological references needed on the semantic web is discussed. The semantic portal MuseumFinland [25] and its content creation model [28] is used as a concrete example of the more general problem.

Metadata in this system originates from different DC like metadata schemas used in three museums, represented in different kind of database tables using different cataloging database systems. These tables are transformed into an RDF repository in two steps depicted in Fig. 1: First, the heterogenous relational tables in each museum are harmonized by transforming them into an XML metadata schema format that is shared by the co-operating content providers. This transformation ensures syntactic interoperability among all data sources, and partial semantic interoperability in terms the meaning of the metadata schema elements, since a single element set is used. Second, semantic interoperability between metadata sources is obtained by transforming the XML descriptions into the final RDF metadata schema format used by the portal. During this XML-to-RDF transformation the essential task is to move from term space into concept space by changing literal terms, used at the XML level as element values, into corresponding concept URIs referring to seven underlying domain ontologies (e.g., Artifacts, Places). The URIs created in this phase connect metadata RDF with domain ontology RDF, resulting into a single large semantic RDF triple store used for querying and as a basis for logical reasoning.

A major problem in the RDF transformation above is how to disambiguate the meanings of homonyms (e.g., "bank") that may occur as keywords, free indexing terms, or in free text descriptions in different element values. Several methods can be applied here. For example, the type of the metadata element in which a homonymous expression is used, can often be used for semantic disambiguation effectively [28]. However, when dealing with the *dc:subject*

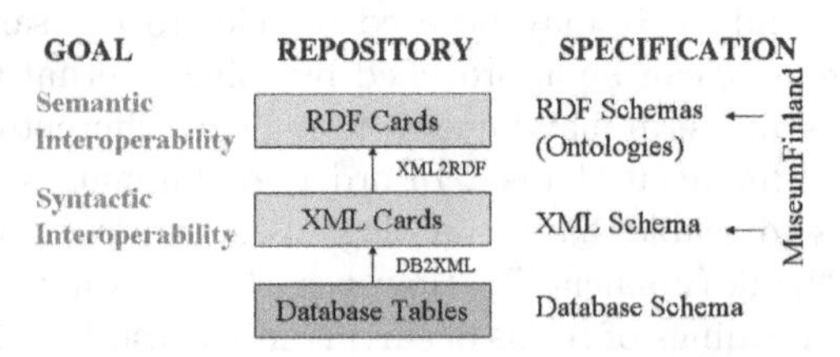

Fig. 1. Transforming legacy museum collection data from database tables into RDF

element (or similar ones) that can have values taken from different vocabularies, such contextual disambiguating information is not available, and human decision help is more often needed.

Another practical problem is spelling errors in metadata, and the variance of synonyms and correct syntactic encoding practices used at different organizations at different times, in different languages, and even by different catalogers. For example, the name of Ivan Ayvazovsky (Russian painter, 1817–1900) has 13 different labels in ULAN (Ajvazovskij, Aivazovski, Aiwasoffski, etc.), and the first, middle, and last names can be ordered and shortened in many different ways.

Still another problem of transforming literals into URIs is complicated free text descriptions that may be used as element values, such as the material description "cow leather with decorations". Free text descriptions in metadata are in general difficult to search for due their syntactic variance, and for the same reason, difficult to transform into URI references automatically. The problem can be approached by using in indexing controlled vocabularies or ontologies. However, even then the problem remains when dealing with *free indexing terms*. These terms are, by definition, legal keywords of a thesaurus that are not listed as entries. For example, plant and animal types as well as person and location names can be used as free indexing terms. When encountering such a term, it cannot usually be associated with the underlying ontologies without human help.

In a distributed content creation environment, free indexing concepts pose a challenge for ontology maintenance, too. In many cases new concepts should to populated into the ontologies and be shared, too. For example, when a painting of a new, formerly unknown artist is cataloged in a museum, the other catalogers and organizations should be made aware of her/him in order to prevent creation of multiple identifiers for the artist and later confusion of identities.

A solution approach to this is to connect annotation creation tools to centrally maintained *ontology library services* that provide the clients with up-to-date information about the vocabulary resources available, and facilitates creation and sharing of new resources collaboratively. An implementation of such a service is the ONKI Ontology Server[30] [31,59] that can be used for creating mash-up annotation applications in a way analogous to creating Google Maps mash-ups.

Sharing unique URIs for concepts is preferable on the semantic web, but in practice there will be multiple URIs referring to a single resource. Creation of multiple identifiers for free indexing concepts cannot be eliminated totally in practice, and multiple identifiers will be created purposefully, too. Global dereferencing services will be needed in the future telling, e.g., that the concept of "London" in UK refers to the same thing as "Londres" in France.

[30] `http://www.seco.tkk.fi/services/onki/`

After creating semantically interoperable RDF metadata, content harvesting and aggregation can be done either (1) off-line before starting the portal or (2) on-line dynamically when answering end-user queries. The on-line approach is more dynamic. However, from the viewpoint of creating intelligent end-user services, the off-line approach seems more promising: (1) By creating a global knowledge base first off-line, reasoning can be easily done at the global scale across local contents, which facilitates, e.g., generation of recommendation links between the content of different content providers. (2) Knowledge can be compiled and critical reasoning tasks performed off-line beforehand for faster response times. For example, the *rdf:type* instance-class-relations can be explicated as RDF-triples based on the transitive closures of the subclass-of hierarchies. (3) The portal is independent of the content providers' possibly unreliable web services when running the system.

4 Semantic Portal Services

The goal of semantic information portals for cultural heritage is to provide the end-user with intelligent services for finding and learning the right information based on her own preferences and the context of using the system. In the following, some possibilities of providing the end-users with intelligent services using semantically annotated metadata are shortly explored.

4.1 Semantic Search

In information retrieval [5] search is usually based on finding occurrences of words in documents. On the semantic web, search can be based on finding the concepts related to the documents at the metadata and ontology levels, in addition to the actual text or other features of the data. With concept-based methods document meanings and queries can be specified more accurately which usually leads to better recall and precision, especially if both the query and the underlying content descriptions are concept-based.

With non-textual cultural documents, such as paintings, photographs, and videos, metadata-based search techniques are a must in practice, although also content-based information retrieval methods [44] (CBIR) and multimedia information retrieval (MIR) [37] can be used as complementary techniques. Here the idea is to utilize actual document features (at the data level), such as color, texture, and shape in images, as a basis for information retrieval. For example, an image of Abraham Lincoln could be used as a query for finding other pictures of him, or a piece of music could be searched for by humming it. Bridging the "semantic gap" between low level image and multimedia features and semantic annotations is an important but challenging research theme [23]. Still another approach to do semantic search is to analyze and build search on the content using linguistic and/or statistical methods, without using annotated semantic metadata [8].

A key problem of semantic search is mapping the literal search words, used by humans, to underlying ontological concepts, used by the computer. Depending on the application, only meaningful queries expressed by terms that are relevant to the domain and content available, other queries result in frustrating "no hits" answers. A way to solve the problem is to provide the end-user with a vocabulary as a subject heading category tree, a facet, as in Yahoo! and dmoz.org. By selecting a category, related documents are retrieved. Faceted search [19, 21, 27, 42, 47] is a natural generalization of this, where the user can make several simultaneous selections from *multiple* orthogonal facets. They are exposed to the end-user in order to (1) provide her with the right query vocabulary, and (2) for presenting the repository contents and search results and the amounts of hits in facet categories. The result set can be presented to the end-user according to the facet hierarchies for better readability. This is in contrast with traditional search where results are typically presented as a list in decreasing relevence order. The number of hits resulting from a category selection is always shown to the user before the selection. This eliminates queries leading to "no hits" dead-ends, and guides the user in making next constraining selections on the facets.

Faceted search has been integrated with the idea of ontologies and the semantic web [27]. The facets can be constructed algorithmically from a set of underlying ontologies that are used as the basis for annotating search items. Furthermore, the mapping of search items onto search facets can be defined using logic rules. This facilitated more intelligent semantic search of indirectly related items. A method for ranking the search results in faceted search based on fuzzy logic is presented in [22], and [54] presents a card sorting approach for specifying and using end-user facets independently from the indexing ontologies.

The faceted search paradigm is based on *facet analysis* [41], a classification scheme introduced in information sciences by S.R. Ranganathan already in the 1930s. The idea of faceted search has been invented and developed independently by several research groups, and is also called view-based search [42] and dynamic taxonomies [47]. Several semantic cultural heritage portals make use of faceted search, such as [21, 25]. However, faceted search is not a panacea for all information retrieval tasks. Google-like keyword search interface is usually preferred if the user is capable of expressing her information need terms of accurate keywords [12].

4.2 Semantic Autocompletion

Keyword search can be integrated with semantic search by extending search to the labels of ontological resources or facet categories. For example, in [25] keyword search is integrated with faceted search in the following way: First, search keywords are matched against category names in the facets in addition to text fields in the metadata. The result set of hits is shown containing all objects in any of the categories matched in addition to all objects whose metadata directly contains the keyword. The hits are grouped by the categories found. Second, a new dynamic facet is created in the user interface for disambiguating the different possible ontological interpretations and roles of the keyword. This facet contains all facet categories whose name (or other property values) matches the keyword. They tell the end-user the different interpretations and roles of the keyword. For example, the keyword "Nokia" matches in the portal with the mobile phone company resource in the "Manufacturer" facet role, with the city of Nokia in the facet roles "Place of manufacturing" and "Place of usage", and with some other resources that have the string in their name. By selecting one of the interpretations, the user is able to disambiguate the meanings easily and constrain search further.

The idea of searching ontologies and facet categories for disambiguating intended meanings and roles has been generalized into the notion of *semantic autocompletion* [24]. The idea here is to generalize traditional text autocompletion by trying to guess, based on ontologies and reasoning, the search concept the user is trying to formulate after each input character in an input field. For example, the user may type in the query in French and the semantic autocompletion service finds the possible intended search concepts in English after each input character.

Autocompletion has become a popular way to find meaningful keywords in large search vocabularies after Google Suggest[31] was released. The idea is applied in several semantic cultural portals, such as [26, 33, 48, 57].

4.3 Semantic Browsing and Recommending

In addition to semantic search, semantic content facilitates *semantic browsing*. Faceted search is already a kind of combination of searching and browsing because search is based on selecting links on facets. However, in semantic browsing the general idea is not to constrain the result set but rather to expand it by trying to find objects of potential interest outside of the hit list. The idea is to support browsing documents through associative links that are created based on the underlying metadata and ontologies, not on hardwired anchor links encoded by humans in HTML pages.

A simple form of a semantic browser are RDF browsers and tabulators [7]. Their underlying idea has been explicated as the "linked data"[32] principle proposing that when an RDF resource (URI) is rendered in a browser, the attached RDF links to related resources should be shown. When one of these links is selected, the corresponding a new resource is rendered, and so on.

A more developed related idea is *recommendation systems* [9]. Here the logic of selecting and recommending of related resources can be based on also other principles than the underlying RDF graph. For example, collaborative filtering [20] is based on browsing statistics of other users. Also logic rules on

[31] `http://www.google.com/webhp?complete=1&hl=en`

[32] `http://www.w3.org/DesignIssues/LinkedData.html`

top of an RDF knowledge base can be used for creating semantic recommendation links [58] and, at the same time, *explanations* telling the end-user why the recommendation link was selected in this context. In [2, 4] explanations for recommended art works can be obtained based on a user profile of interest and features of the artworks. In [32] ontological models of narrative stories and processes in the society, such as fishing or slash farming, were used as a basis for creating recommendation links between cultural resources. Still another approach to create recommendation links with explanations is to use similarity measures of event-based annotations [46].

4.4 Relational Search

Semantic recommending is related to *relational search*, where the idea is to try to search and discover serendipitous semantic associations between different content items [26, 48, 49]. The idea is to make it possible for the end-user to formulate queries such as "How is X related to Y" by selecting the end-point resources, and the search result is a set of semantic connection paths between X and Y. For example, in Fig. 2 the user has specified two historical persons, the Finnish artist Akseli Gallen-Kallela (1865–1931) and the French emperor Napoleon I (1769–1821) in the CultureSampo portal [26]. The underlying knowledge base contains an ontologized version of the ULAN vocabulary in RDF with over 100,000 persons and organizations, and semantic autocompletion is used for finding the right query resources. The system has discovered an association chain between the persons based on "patronOf", "teacherOf", "knows", and "studentOf" properties.

4.5 Personalization and Context Awareness

In many occasions the functioning of a semantic portal should not be static but adapt dynamically according to the (1) personal interests of the end-user

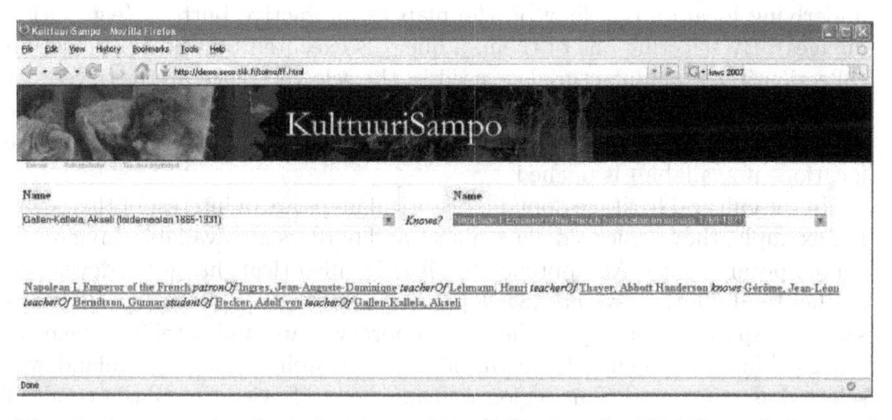

Fig. 2. An example of relational search in [26] using the ULAN vocabulary and database

and (2) the context of usage, such as time and location [51]. Visitors in semantic cultural portals, like in physical museums, are usually not interested in everything found in the underlying collections, and would like to get information at different levels of detail. An important aspect of a semantic cultural portal is then adaptation of the portal to different personal information needs and interests. An example of a personalized cultural semantic portal is [2,4], where user profiling and personalization is based on metadata obtained by asking the users about her interests by rating pieces of artworks.

An example of location-based adaptability is the mobile phone user interface of [25]. By pushing a special button on the interface, collection artifacts either manufactured or used nearby can be retrieved based on a geolocation service proving the coordinate information of the phone. It can be envisioned that this kind of location-based and navigational services will be available in future cultural portals based on phones supporting GPS positioning and radio-frequency identification (RFID) tags.

Also time is an important parameter for contextualizing portal services. For example, recommending the end-user to visit a site in the nature during winter may not be wise due to snow, or to direct her to a museum when it happens to be closed.

4.6 Visualization and Mash-Ups

Visualization is an important aspect of the semantic web dealing with semantically complicated and interlinked contents [16]. In the cultural heritage domain, maps, time lines, and methods for visualizing complicated and large semantic networks are of special interest.

Maps are useful in both searching content and in visualizing the results. A widely used approach to using maps in portals is to use mash-up map services. For example, [36] presents a mash-up combining Google Maps[33] and a semantic cultural portal [25]. The map interface is used for showing the places of the underlying location ontology on the map as interactive buttons (e.g., cities, villages). By selecting one of them, a query is executed by which all museum collection items manufactured or used in the selected place are retrieved. At the same time, additional search links to seven different traditional portals are shown. For example, by selecting the Wikipedia link, an article about the location (if available) is opened.

In the cultural heritage domain, historical maps are of interest of their own. For example, they depict old place names and borders not available anymore in contemporary maps. An approach to visualize historical changes is developed in the Temp-O-Map system [34, 36] that makes it possible to lay old maps semi-transparently on top of the contemporary maps and satellite images of Google Maps. To demonstrate the idea, the Karelia region of Finland was selected as a test case. This region was annexed to Soviet-Union as a result of

[33] `http://maps.google.com/`

the Second World War, after which most old Finnish place names in the region were changed into Russian ones making it difficult to the end-user to bridge the semantic gap between old and new names and locations. The system is connected into an ontology modeling over 1,000 regional changes of Finnish municipalities in 1860–2007. Historical municipalities of different time periods are available as facets for finding historical places on the maps. By selecting a category, the tool focuses the map view to the center point of the region [35].

Another important dimension for visualizing cultural content is time. A standard approach for temporal visualization is to project search objects on a time line, as in [26, 48]. A generic mash-up tool for creating time lines is the Simile time line.[34] A time line can be used both for querying and for visualizing search results.

4.7 Cross-Portal Reuse of Content

Portal contents can be reused in other web applications and portals due to semantic web standards. Reusing semantic content in this way is a kind of generalization of the idea of "multi-channel publication" of XML, where a single syntactic structure can be rendered in different ways. In a similar vein, semantic metadata can be reused without modifying it through *multi-application publication.*

One possibility to facilitate cross-portal reuse is to merge triple stores, and provide services to end-users based on the extended knowledge base. For example, the learning object video portal [33] is able to provide recommendation links to the cultural museum collection portal [25] in this way. Another way of reusing content is to keep the portals separate and publish their functionalities as web services to be used by other semantic portals [59]. Both traditional web services or light-weight mash-ups based on the REST principle can be used. Here portal functionalities can be used in other portals on the HTML user interface level with just a pair of additional Javascript code added on the HTML level. This approach is related to the idea of using Google AdSense[35] advertisements, but generalized on a semantic level and used for publishing portal services. For example, there is a semantic widget for reusing the semantic search functionality and contents of the portal [25] in external web pages [40]. If a page, for instance, contains information about skating, then the widget can query and show dynamically, using AJAX, images and semantic links to skates and related objects in the museum collection portal.

5 Conclusions

Cultural heritage provides a semantically rich application domain in which useful vocabularies and collection contents are available, and where the organizations are eager to make their content publicly accessible. A major

[34] `http://simile.mit.edu/timeline/`
[35] `http://www.google.com/adsense/`

application type in the area has been semantic portals, often aggregating content from different collections, thus providing cultural organizations with a shared cost-effective publication channel and the possibility of enriching collaboratively the contents of each other's collections. For the end-user, new kinds of intelligent semantic services and ways of visualizing content can be provided. It can be envisioned that in the near future ever larger cultural semantic portals crossing geographical, cultural, and linguistic barriers of content providers at different countries will be developed, such as Europeana[36]. Also more systems for enriching the collections by end-user created content and tagging in the spirit of Web 2.0 will be seen, such as Steve Museum[37] and Powerhouse Museum[38].

A major practical hinder for publishing cultural content on the semantic web is that current legacy cataloging system do not support creation of ontology-based annotations. If semantic annotations cannot be created in memory organization when cataloging content, then costly manual work is needed when transforming and disambiguating literal legacy metadata into ontological references in semantic portals. A solution approach to this fundamental problem is to provide ontologies as publicly available ontology services, and to reuse them—as well as semantically annotated portal contents—as ready-to-use functionalities (widgets) in legacy systems using mash-up techniques [31, 59].

Acknowledgements

This work was supported by the national semantic web ontology project FinnONTO[39] 2003–2010 in Finland, funded by the National Funding Agency for Technology and Innovation (Tekes) and a consortium of 38 public organizations and companies.

References

1. J. Aitchison, A. Gilchrist, and D. Bawden. *Thesaurus construction and use: a practical manual.* Europa, London, 2000.
2. L. Aroyo, R. Brussee, L. Rutledge, P. Gorgels, N. Stash, and Y. Wang. Personalized museum experience: The Rijksmuseum use case. In J. Trant and D. Bearman, editors, *Museums and the Web 2007: Proceedings*, pages 137–144, 2007.
3. L. Aroyo, E. Hyvönen, and J. van Ossenbruggen, editors. *Cultural Heritage on the Semantic Web. Workshop Proceedings. The 6th International Semantic Web Confereence and the 2nd Asian Semantic Web Conference, Busan, Korea.* 2007.

---

[36] http://www.europeana.eu/
[37] http://www.steve.com/
[38] http://www.powerhousemuseum.com/collection/database/
[39] http://www.seco.tkk.fi/projects/finnonto/

4. L. Aroyo, N. Stash, Y. Wang, P. Gorgels, and L. Rutledge. CHIP demonstrator: Semantics-driven recommendations and museum tour generation. In *Proceedings of ISWC 2007 + ASWC 2007, Busan, Korea*, pages 879–886. Springer, 2007.
5. R. Baeza-Yates and B. Ribeiro-Neto. *Modern Information Retrieval.* Addison-Wesley, 1999.
6. V. R. Benjamins, J. Contreras, M. Blázquez, J.M. Dodero, A. Garcia, E. Navas, F. Hernandez, and C. Wert. Cultural heritage and the semantic web. In *The Semantic Web: Research and Applications*, pages 433–444. Springer, 2004.
7. T. Berners-Lee, Y. Chen, L. Chilton, D. Connolly, J. Hollenbach R. Dhanaraj, A. Lerer, and D. Sheets. Tabulator: Exploring and analyzing linked data on the semantic web. In *The 3rd International Semantic Web User Interaction Workshop (SWUI 2006)*, Nov 2006.
8. W. Buntine, K. Valtonen, and M. Taylor. The ALVIS document model for a semantic search engine. In *The 2nd Annual European Semantic Web Conference, Demos and Posters, Heraklion, Crete*, May 2005.
9. R. Burke. Knowledge-based recommender systems. In A. Kent, editor, *Encyclopedia of Library and Information Systems*, volume 69. Marcel Dekker, New York, 2000.
10. S. Decker, M. Erdmann, D. Fensel, and R. Studer. Ontobroker: Ontology based access to distributed and semi-structured unformation. In R. Meersman et al., editor, *DS-8: Semantic Issues in Multimedia Systems*, pages 351–369. Kluwer, New York, 1999.
11. M. Doerr. The CIDOC CRM – an ontological approach to semantic interoperability of metadata. *AI Magazine*, 24(3):75–92, 2003.
12. J. English, M. Hearst, R. Sinha, K. Swearingen, and K.-P. Lee. Flexible search and navigation using faceted metadata. Technical report, University of Berkeley, School of Information Management and Systems, 2003.
13. C. Fellbaum, editor. *WordNet. An electronic lexical database.* MIT, Cambridge, MA, 2001.
14. D. Fensel. *Ontologies: Silver bullet for knowledge management and electronic commerce (2nd Edition).* Springer, Berlin, 2004.
15. D. J. Foskett. Thesaurus. In *Encyclopaedia of Library and Information Science*, volume 30, pages 416–462. Marcel Dekker, New York, 1980.
16. V. Geroimenko and C. Chen, editors. *Visualizing the Semantic Web: XML-based Internet and Information Visualization.* Springer, Berlin, 2002.
17. C. Goble, S. Bechhofer, L. Carr, D. De Roure, and W. Hall. Conceptual open hypermedia = the semantic web? In *Proceedings of the WWW2001, Semantic Web Workshop, Hongkong*, 2001.
18. A. Hameed, A. Preese, and D. Sleeman. Ontology reconciliation. In S. Staab and R. Studer, editors, *Handbook on ontologies*, pages 231–250. Springer, Berlin, 2004.
19. M. Hearst, A. Elliott, J. English, R. Sinha, K. Swearingen, and K.-P. Lee. Finding the flow in web site search. *CACM*, 45(9):42–49, 2002.
20. J. H. Herlocker, J. A. Konstan, and J. Riedl. Explaining collaborative filtering recommendations. In *Computer Supported Cooperative Work*, pages 241–250. ACM, 2000.
21. M. Hildebrand, J. van Ossenbruggen, and L. Hardman. /facet: A browser for heterogeneous semantic web repositories. In *Proceedings of the 5th International Semantic Web Conference (ISWC 2006)*. Springer, Nov 2006.

22. M. Holi and E. Hyvönen. Fuzzy view-based semantic search. In *Proceedings of the 1st Asian Semantic Web Conference (ASWC2006), Beijing, China.* Springer, 2006.
23. L. Hollink. *Semantic annotation for retrieval of visual resources.* PhD thesis, Free Univerisity of Amsterdam, 2006. SIKS Dissertation Series, No. 2006-24.
24. E. Hyvönen and E. Mäkelä. Semantic autocompletion. In *Proceedings of the first Asia Semantic Web Conference (ASWC 2006), Beijing.* Springer, 2006.
25. E. Hyvönen, E. Mäkela, M. Salminen, A. Valo, K. Viljanen, S. Saarela, M. Junnila, and S. Kettula. MuseumFinland—Finnish museums on the semantic web. *Journal of Web Semantics*, 3(2):224–241, 2005.
26. E. Hyvönen, T. Ruotsalo, T. Häggström, M. Salminen, M. Junnila, M. Virkkil, M. Haaramo, T. Kauppinen, E. Mäkelä, and K. Viljanen. CultureSampo–Finnish culture on the semantic web. The vision and first results. In *Semantic Web at Work—Proceedings of STeP 2006.* Finnish AI Society, Espoo, Finland, 2006. Also in: Klaus Robering (Ed.), Information Technology for the Virtual Museum. LIT Verlag, Berlin, 2008.
27. E. Hyvönen, S. Saarela, and K. Viljanen. Application of ontology techniques to view-based semantic search and browsing. In *The Semantic Web: Research and Applications. Proceedings of the First European Semantic Web Symposium (ESWS 2004).* Springer, 2004.
28. E. Hyvönen, M. Salminen, S. Kettula, and M. Junnila. A content creation process for the Semantic Web, May 2004. Proceeding of OntoLex 2004: Ontologies and Lexical Resources in Distributed Environments, Lisbon, Portugal.
29. E. Hyvönen, A. Valo, K. Viljanen, and M. Holi. A logic-based semantic web HTML generator — a poor man's publishing approach. In *Proceedings of WWW2004, New York, Alternate Track Papers and Posters*, May 2004.
30. E. Hyvönen, K. Viljanen, E. Mäkelä, T. Kauppinen, T. Ruotsalo, O. Valkeapää, K. Seppälä, O. Suominen, O. Alm, R. Lindroos, T. Känsälä, R. Henriksson, M. Frosterus, J. Tuominen, R. Sinkkilä, and J. Kurki. Elements of a national semantic web infrastructure — case study Finland on the semantic web (invited paper). In *Proceedings of the First International Semantic Computing Conference (IEEE ICSC 2007), Irvine, California*, Sept 2007. IEEE.
31. E. Hyvönen, K. Viljanen, J. Tuominen, and K. Seppälä. Building a national semantic web ontology and ontology service infrastructure — the FinnONTO approach. In *Proceeedings of the 5th European Semantic Web Conference (ESWC 2008).* Springer, 2008.
32. M. Junnila, E. Hyvönen, and M. Salminen. Describing and linking cultural semantic content by using situations and actions. In *Semantic Web at Work — Proceedings of STeP 2006.* Finnish AI Society, Espoo, Finland, Nov 2006. Also in: Klaus Robering (Ed.), Information Technology for the Virtual Museum. LIT Verlag, Berlin, 2008.
33. T. Känsälä and E. Hyvönen. A semantic view-based portal utilizing Learning Object Metadata, August 2006. 1st Asian Semantic Web Conference (ASWC2006), Proceedings of the Semantic Web Applications and Tools Workshop.
34. T. Kauppinen, C. Deichstetter, and E. Hyvönen. Temp-O-Map: Ontology-based search and visualization of spatio-temporal maps. In *Demo track at the European Semantic Web Conference ESWC 2007, Innsbruck, Austria*, June 4–5 2007. `http://www.eswc2007.org/demonstrations.cfm`.

35. T. Kauppinen, R. Henriksson, R. Sinkkilä, R. Lindroos, J. Väätäinen, and E. Hyvönen. *Ontology-based Disambiguation of Spatiotemporal Locations.* Proceedings of the 1st International Workshop on Identity and Reference on the Semantic Web (IRSW2008), CEUR Workshop Proceedings, Vol 422, `http://ceur-ws.org`, 2008.
36. T. Kauppinen, R. Henriksson, J. Väätäinen, C. Deichstetter, and E. Hyvönen. Ontology-based modeling and visualization of cultural spatio-temporal knowledge. In *Semantic Web at Work — Proceedings of STeP 2006.* Finnish AI Society, Espoo, Finland, Nov 2006.
37. Michael S. Lew, Nicu Sebe, Chabane Djeraba, and Ramesh Jain. Content-based multimedia information retrieval: state of the art and challenges. *ACM Transactions on Multimedia computing, communications, and applications*, pages 1–19, Feb 2006.
38. A. Maedche, S. Staab, N. Stojanovic, R. Struder, and Y. Sure. Semantic portal — the SEAL approach. Technical report, Institute AIFB, University of Karlsruhe, Germany, 2001.
39. E. Mäkelä, E. Hyvönen, and S. Saarela. Ontogator — a semantic view-based search engine service for web applications. In *Proceedings of the 5th International Semantic Web Conference (ISWC 2006)*, Nov 2006.
40. E. Mäkelä, K. Viljanen, O. Alm, J. Tuominen, O. Valkeapää, T. Kauppinen, J. Kurki, R. Sinkkilä, T. Känsälä, R. Lindroos, O. Suominen, T. Ruotsalo, and E. Hyvönen. In L. Nixon, R. Cuel, and C. Bergamini, editors, Proceedings of the Workshop on *First Industrial Results of Semantic Technologies, co-located with ISWC 2007 + ASWC 2007, Busan, Korea*, 2007. CEUR Workshop Proceedings, Vol. 293.
41. A. Maple. Faceted access: A review of the literature. Technical report, Working Group on Faceted Access to Music, Music Library Association, 1995.
42. A. S. Pollitt. The key role of classification and indexing in view-based searching. Technical report, University of Huddersfield, UK, 1998. `http://www.ifla.org/IV/ifla63/63polst.pdf`.
43. D. Reynolds, P. Shabajee, and S. Cayzer. Semantic Information Portals. In *Proceedings of the 13th International World Wide Web Conference, Alternate track papers and posters*, New York, NY, USA, May 2004. ACM.
44. Y. Rui, T. Huang, and S. Chang. Image retrieval: current techniques, promising directions and open issues. *Journal of Visual Communication and Image Representation*, 10(4):39–62, April 1999.
45. T. Ruotsalo and E. Hyvönen. An event-based method for making heterogeneous metadata schemas and annotations semantically interoperable. In *Proceedings of ISWC 2007 + ASWC 2007, Busan, Korea*, pages 409–422. Springer, 2007.
46. T. Ruotsalo and E. Hyvönen. A method for determining ontology-based semantic relevance. In *Proceedings of the International Conference on Database and Expert Systems Applications DEXA 2007, Regensburg, Germany.* Springer, 2007.
47. G. M. Sacco. Dynamic taxonomies: guided interactive diagnostic assistance. In N. Wickramasinghe, editor, *Encyclopedia of Healthcare Information Systems.* Idea Group, 2005.
48. G. Schreiber, A. Amin, M. van Assem, V. de Boer, L. Hardman, M. Hildebrand, L. Hollink, Z. Huang, J. van Kersen, M. de Niet, B. Omelayenko, J. van Ossenbruggen, R. Siebes, J. Taekema, J. Wielemaker, and B. Wielinga.

MultimediaN E-Culture demonstrator. In *Proceedings of the Fifth International Semantic Web Conference (ISWC 2006)*, pages 951–958. Springer, November 2006.
49. A. Sheth, B. Aleman-Meza, I. B. Arpinar, C. Bertram, Y. Warke, C. Ramakrishnan, C. Halaschek, K. Anyanwu, D. Avant, F. S. Arpinar, and K. Kochut. Semantic association identification and knowledge discovery for national security applications. *Journal of Database Management on Database Technology*, 16(1):33–53, Jan–Mar 2005.
50. T. Sidoroff and E. Hyvönen. Semantic e-goverment portals — a case study. In *Proceedings of the ISWC-2005 Workshop Semantic Web Case Studies and Best Practices for eBusiness SWCASE05*, CEUR Workshop Proceedings, Vol 155, `http://ceur-ws.org`, Nov 2005.
51. S. Sirmakessis, editor. *Adaptive and Personalized Semantic Web*. Springer, Berlin, 2006.
52. J. Sowa. *Knowledge Representation. Logical, Philosophical, and Computational Foundations.* Brooks/Cole, 2000.
53. S. Staab, J. Angele, S. Decker, M. Erdmann, A. Hotho, A. Maedche, H.-P. Schnurr, R. Studer, and Y. Sure. Semantic Community Web Portals. In *Proceedings of the 9th International World Wide Web Conference*, Amsterdam, The Netherlands, May 2000. Elsevier.
54. O. Suominen, K. Viljanen, and E. Hyvönen. User-centric faceted search for semantic portals. In *Proceedings of the 4th European Semantic Web Conference (ESWC 2007), Innsbruck, Austria*, pages 356–370. Springer, 2007.
55. M. van Assem, V. Malaise, A. Miles, and G. Schreiber. A method to convert thesauri to SKOS. In *Proceedings of the Third European Semantic Web Conference (ESWC 2006)*. Springer, 2006.
56. M. van Assem, M. R. Menken, G. Schreiber, J. Wielemaker, and B. Wielinga. A method for converting thesauri to RDF/OWL. In *Proceedings of 3rd International Semantic Web Conference (ISWC 2004), Hiroshima, Japan.* Springer, 2004.
57. J. van Ossenbruggen, A. Amin, L. Hardman, M. Hildebrand, M. van Assem, B. Omelayenko, G. Schreiber, A. Tordai, V. de Boer, B. Wielinga, J. Wielemaker, M. de Niet, J. Taekema, M.-F. van Orsouw, and A. Teesing. Searching and Annotating Virtual Heritage Collections with Semantic-Web Techniques. In *Proceedings of Museums and the Web 2007*, San Francisco, California, March 2007. Archives and Museum Informatics.
58. K. Viljanen, T. Känsälä, E. Hyvönen, and E. Mäkelä. ONTODELLA – a projection and linking service for semantic web applications. In *Proceedings of the 17th International Conference on Database and Expert Systems Applications (DEXA 2006), Krakow, Poland.* IEEE, September 4–8 2006.
59. K. Viljanen, J. Tuominen, T. Känsälä, and E. Hyvönen. Distributed semantic content creation and publication for cultural heritage legacy systems. In *Proceedings of the 2008 IEEE International Conference on Distibuted Human-Machine Systems, Athens, Greece.* IEEE, 2008.

Ontology-Based Recommender Systems

Stuart E. Middleton[1], David De Roure[2], and Nigel R. Shadbolt[2]

[1] IT Innovation Centre, University of Southampton, Southampton SO16 7NP, UK, sem@it-innovation.soton.ac.uk

[2] Intelligence, Agents, Multimedia Group, Department of Electronics and Computer Science, University of Southampton, Southampton SO17 1BJ, UK, dder@ecs.soton.ac.uk, nrs@ecs.soton.ac.uk

Summary. We present an overview of the latest approaches to using ontologies in recommender systems and our work on the problem of recommending on-line academic research papers. Our two experimental systems, Quickstep and Foxtrot, create user profiles from unobtrusively monitored behaviour and relevance feedback, representing the profiles in terms of a research paper topic ontology. A novel profile visualization approach is taken to acquire profile feedback. Research papers are classified using ontological classes and collaborative recommendation algorithms used to recommend papers seen by similar people on their current topics of interest. Ontological inference is shown to improve user profiling, external ontological knowledge used to successfully bootstrap a recommender system and profile visualization employed to improve profiling accuracy.

In a specific case study we report results from two small-scale experiments, with 24 subjects over 3 months, and a large-scale experiment, with 260 subjects over an academic year, are conducted to evaluate different aspects of our approach. The overall performance of our ontological recommender systems are favourably compared to other systems in the literature.

1 Introduction

The mass of content available on the World-Wide Web raises important questions over its effective use. Information on the web is largely unstructured, with web pages authored by many people on a diverse range of topics. This often makes simple browsing too time consuming to be practical. The emergence of e-commerce sites means many vendors are offering potentially great deals on very similar products. Web information filtering has thus become necessary for most web users in order to find the things they really need.

Recommender systems have emerged as one successful approach that can help tackle the problem of information overload. They exploit patterns in item metadata and reviews posted by groups of people to find new items that might

S. Staab and R. Studer (eds.), *Handbook on Ontologies*, International Handbooks on Information Systems, DOI 10.1007/978-3-540-92673-3,

be of interest to a user. Ontologies are increasingly being used within the field of recommender systems, allowing knowledge-based techniques to supplement classical machine learning and statistical approaches.

1.1 Recommender Systems

People find articulating exactly what they want difficult, but they are good at recognizing it when they see it. This insight has led to the utilization of relevance feedback, where people rate items as interesting or not interesting and the system tries to find items that match the "interesting", positive examples and do not match the "not interesting", negative examples. With sufficient positive and negative examples, modern machine learning techniques can classify new pages with impressive accuracy. Recommender systems can recommend many types of item, including web pages, new articles, music CDs and books.

Unobtrusive monitoring provides positive examples of what the user is looking for, without interfering with the user's normal work activity. Heuristics can also be applied to infer negative examples from observed behaviour, although generally with less confidence. This idea has led to content-based recommender systems, which unobtrusively watch user behaviour and recommend new items that correlate with a user's profile.

Another way to recommend items is based on the ratings provided by other people who have liked the item before. Collaborative recommender systems do this by asking people to rate items explicitly and then recommend new items that similar users have rated highly. An issue with collaborative filtering is that there is no direct reward for providing examples since they only help other people. This leads to initial difficulties in obtaining a sufficient number of ratings for the system to be useful, a problem known as the cold-start problem [15].

Hybrid systems, attempting to combine the advantages of content-based and collaborative recommender systems, have also proved popular to-date. The feedback required for content-based recommendation is shared, allowing collaborative recommendation as well.

1.2 User Profiling

User profiling is typically either knowledge-based or behaviour-based. Knowledge-based approaches use static models of users and dynamically match users to the closest model. Questionnaires and interviews are often employed to obtain this user knowledge. Once a model is selected for a user, specialist domain knowledge for that user type can be applied to help the user. Behaviour-based approaches use the user's behaviour as a model, commonly using machine-learning techniques to discover useful patterns in the behaviour. Behavioural logging is employed to obtain the data necessary from which to extract patterns. Kobsa [9] provides a good survey of user modelling techniques.

The user profiling approach used by most recommender systems is behavioural-based, commonly using a binary class model to represent what users find interesting and not interesting. Machine-learning techniques are then used to find potential items of interest in respect to the binary model, recommending items that match the positive examples and do not match the negative examples. There are a lot of effective machine learning algorithms based on two classes. A binary profile does not, however, lend itself to sharing examples of interest or integrating any domain knowledge that might be available. Sebastiani [19] provides a good survey of current machine learning techniques.

1.3 Ontologies

An ontology is a conceptualisation of a domain into a human-understandable, but machine-readable format consisting of entities, attributes, relationships, and axioms [8]. Ontologies can provide a rich conceptualisation of the working domain of an organisation, representing the main concepts and relationships of the work activities. These relationships could represent isolated information such as an employee's home phone number, or they could represent an activity such as authoring a document, or attending a conference. Part III contains examples of the types of ontology that are in use today, such as chapter "COMM: A Core Ontology for Multimedia Annotation".

Ontologies help extend recommender systems to a multi-class environment, allowing knowledge-based approaches to be used alongside classical machine learning algorithms. Section 2 provides an in-depth overview of how ontologies are integrated into the techniques used for recommendation. Part IV of this book contains details on the current best practice for supporting infrastructures and for ontologies, especially chapters "Ontology Repositories" and "Ontology Mapping".

1.4 Chapter Structure

In this chapter we show how ontologies are used in recommender systems today, providing an overview of the technology space and some further reading on specific approaches. We then examine in some depth a case study of two recommender systems that were among the first to adopt ontological techniques. In these case studies the problem domain, algorithms and results are detailed along with a discussion that highlights some of the practical difficulties experienced running a recommender system for real.

2 Ontology Use in Recommender Systems

Ontologies are now used routinely in recommender systems in combination with machine learning, statistical correlations, user profiling and domain specific heuristics. Commercial recommender systems generally either maintain simple product ontologies (e.g. books) that they can then utilize via heuristics or have a large community of users actively rating content (e.g. movies) suitable for collaborative filtering. More research oriented recommender systems use a much wider variety of techniques that offer advantages such as improved accuracy coupled with constraints such as requiring explicit relevance feedback or intrusive monitoring of user behaviour over prolonged periods of time.

Recommendation of new items to users can be performed by looking at item to item similarity (content-based filtering), item reviews within a community of users (collaborative filtering), semantic relationships between items (heuristic-based recommendation) or a hybrid approach. In many cases the type of approach adopted will depend heavily on how much metadata is available about the items and how much user feedback is available, both implicit and explicit. Content-based techniques work well if training data is available in advance. Collaborative techniques work well when a system has a large community of users. There are, however, no definitive rules to decide on an approach and normally experience and expertise is required to pick the best approach for a given problem domain.

2.1 Content-Based Recommendation

Early recommender systems used content-based binary classification approaches looking at training sets of what was, and what was not interesting to a specific user. Machine learning techniques were employed to perform supervised learning based on sets of observed training examples that a user labelled either as "good" or "bad". A classic example of a content-based recommender system is Fab [1], which uses a binary class k-Nearest Neighbour classifier. Other binary class examples include personal assistant agents such as NewsDude [2], using a naive Bayes classifier, and NewsWeeder [11], using a TF-IDF based classifier, which profile individual user interests and try to find items of interest.

To enhance binary classification domain ontologies were introduced allowing multi-class classification and hence multi-class recommendation. Typically the classes in a domain ontology, such as a product ontology defining all the products of an e-commerce website, would be used to classify the previously observed products / web pages a user had purchased / viewed. A good example of multi-class recommendation is RAAP [4], which uses a simple set of categories to represent individual user profiles.

Once a domain has been classified in terms of ontological concepts the relationships defined by the domain ontology can be used to infer interest and relevance of one concept from observed interest in another. A knowledge-based system can use expert system rules to infer probabilistic interest in classes of item with a semantic connection to an observed item of interest. Typically the semantic distance (number of relationships away one topic is from another) is used to calculate semantic similarity, and this is used to weight likely interest. Entre [3] is a restaurant recommender system that uses a knowledge-base and heuristic rules for recommendation. Where users articulate queries via a web interface the query criteria can drive a knowledge-based decision tree for advanced query refinement. The CWAdvisor [5] system is an example of such an approach where a finite state model is used to refine queries for available financial service products that match the user's stated requirements.

2.2 Clustering and Topic Diversification

Some domains do not have well identified classes of item from which content can be classified. In these cases recommender systems have employed clustering techniques to identify within groups of items potentially similar classes. Hierarchical clustering has been used to categorize document collections for recommender algorithms [18] and sub-divides into either distance-based clustering or concept-based clustering.

Distance-based clustering [21] takes either a top-down (partitioning) or bottom-up (agglomerative) approach to building a hierarchical class tree. A distance function is defined to compute similarity between documents, often based on the similarity of frequency of the words within the document. The clustering algorithm iterates, either dividing super-clusters or merging small clusters into larger ones, until the final concept tree is formed.

Concept-based clustering takes items represented as attribute-pairs and builds relationships based on the probability of occurrence of attribute-pairs within nodes. An early example of concept-based clustering is the COBWEB [6] algorithm. Nodes are created in a top-down approach where nodes are split or merged according to a category utility value; category utility is a measure of differentiation power of that node.

Often recommender systems will recommend clusters of items that are very similar, or variants of the same item (e.g. different formats of the film/DVD). To avoid this topic diversification [22] can be employed to ensure each recommendation is on a well defined concept, hopefully increasing the usefulness of a set of recommendations to the user. Algorithms to perform topic diversification will compute a dissimilarity ranking and merge this with the recommendation ranking. Semantic distance and super-class relationships can be used to compute dissimilarity between item sets.

2.3 Collaborative Filtering

Collaborative filtering works by using the ratings provided by a community of users to recommend items for a specific user. There are two complementary approaches available, user-based or item-based collaborative filtering. User-based collaborative filtering is where similar users are found and items recommended that these similar users also liked. Item-based collaborative filtering is where items are grouped if people rate them similarly.

In order to perform collaborative filtering a user profile must be created from the available historical records of what items people have reviewed and rated. Often a 5-point scale is used for ratings (very good to very bad). A common user profile representation is a weighted vector if interest with as many dimensions as the domain has classes. Vectors can also be used for item to item similarity. Domains where item metadata is not accessible as ontological terms will usually apply pre-processing techniques to compute word/document/metadata term frequencies, remove common words and merge similar words using a thesaurus like WordNet.

User-based collaborative filtering is the most popular recommendation algorithm due to its simplicity and excellent quality of recommendation. First neighbourhoods are formed using a similarity metric, such as a statistical correlation metric like Pearson-r correlation. Second a set of rating predictions are created using profiles that are within the same neighbourhood as the user's own profile. Recommendations are created from the top-N items. The GroupLens project [10] is an early exploiter of user-based collaborative filtering.

Item-based collaborative filtering has become popular in the last 5 years since it decouples the model computation from the prediction process; Amazon [13] have used this technique successfully. Just as in user-based similarity items are compared on the basis of how many users rank them similarly. The neighbourhoods computed are therefore collections of items that are similar. This technique scales well since new items will be added to neighbourhoods as users rate them without the need for explicit ontology maintenance.

Sometimes a recommender system will have to compare items from different domain ontologies, such as two product lists. In these cases an ontology can be created for both domains in a common language (such as OWL) and the mapping between them formulated, either manually or using a automated technique [12] such as a Bayesian belief network. Once concepts are successfully mapped the normal approaches for recommendation can be applied.

2.4 Use of the Semantic Web and Web 2.0 Approaches

Recent work has also used some of the emerging Web 2.0 resources from the Semantic Web to help identify classes of item. One such system [20] has used an internet movie database that contains extensive information about actors, movies, etc., and mapped this semantic information to user behaviour on a movie recommendation website. Tag clouds are created based on the keyword frequencies behind the items they have rated. Data mining techniques [4] can also be coupled with ontological knowledge to improve similarity matching and recommendation within historical usage data.

3 Case Study: Two Ontological Recommender Systems

For a case study two experimental recommender systems are presented, Quickstep and Foxtrot, that explored the novel idea of using an ontological approach to user profiling in the context of recommender systems. Representing user interests in ontological terms involves losing some of the fine grained information held in the raw examples of interest, but in turn allows inference to assist user profiling, communication with other external ontologies and visualization of the profiles using ontological terms understandable to users. Figure 1 shows the general approach taken by both our recommender systems. Quickstep implements only the basic recommendation interface, while Foxtrot implements all the shown features.

A research paper topic ontology is shared between all system processes, allowing both classifications and user profiles to use a common terminology. The ontology itself contains is-a relationships between appropriate topic classes; a section from the topic ontology is shown in Fig. 2. The Quickstep ontology was

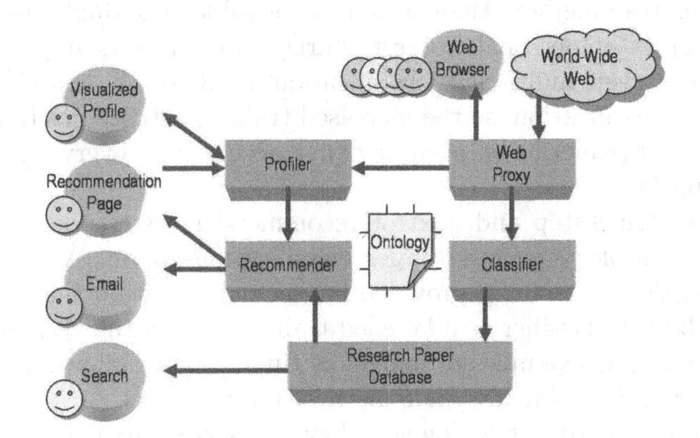

Fig. 1. Quickstep and Foxtrot recommender system data flow

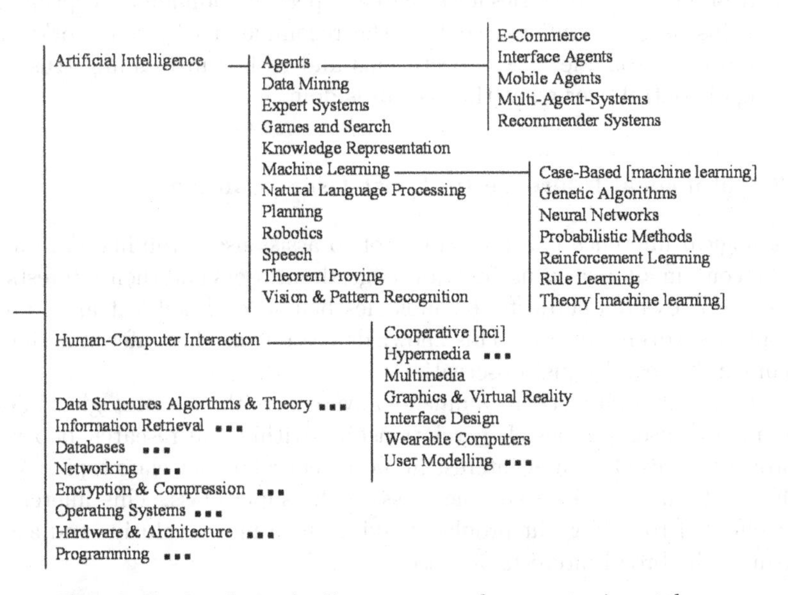

Fig. 2. Section from the Foxtrot research paper topic ontology

based on the open directory project's [7] computer science topic classification, while the Foxtrot ontology was based on the CORA [14] digital library paper classification; manual enhancements were made to each ontology to better reflect some of the more specialist sub-topics researchers required. Reusing existing classifications saves time and provides a source for training examples, especially with the CORA digital library, which contained many pre-classified research papers.

3.1 Classification Using a Research Paper Topic Ontology

Sharing training examples, within the structure of an ontology, allows for much larger training sets than would be possible if a single user just provided examples of personal interest. Larger training sets improve classifier accuracy. However, multi-class classification is inherently less accurate than binary class classification, so the increased training set size has to be weighed along with the reduction in accuracy that occurs with every extra class the system supports.

Both the Quickstep and Foxtrot recommender systems use the research paper topic ontology to base paper classifications upon. A set of labelled example papers is manually provided for each class within the ontology, and then used by the classifier as a labelled training set. In the Quickstep system users can add new examples of papers as time goes by, allowing the training set to reflect the continually changing needs of the users.

In addition to larger training sets, having users share a common ontology enforces a consistent conceptual model, which removes some of the subjective nature of selecting categories for research papers. A common conceptual model also helps users to understand how the recommender system works, which helps form reasonable user expectations and assists in building trust and a feeling of control over what the system is doing.

3.2 Ontological Inference to Assist User Profiling

Ontological inference is a powerful tool to assist user profiling. An ontology could contain all sorts of useful knowledge about users and their interests, such as related research subjects, technologies behind each subject area, projects people are working on, etc. This knowledge can be used to infer more interests than can be seen by just observation.

Our two experimental recommender systems both use ontological inference to enhance user profiles. Is-a relationships within the research paper topic ontology are used to infer interest in more general, super-class topics. We add 50% of the interest in a specific class to the super-class. This inference has the effect of rounding out profiles, making them more inclusive and attuning them to the broad interests of a user.

$$\text{Topic interest} = \sum_{1..\text{no of instances}}^{n} \text{Interest value(n) / days old(n)}$$

Event interest values	Paper browsed = 1
	Recommendation followed = 2
	Topic rated interesting = 10
	Topic rated not interesting = −10

Interest value for super-class per instance = 50% of sub-class

Fig. 3. Profiling algorithm

The profiling algorithm used is shown in Fig. 3. A time-decay function is applied to the observed behaviour events to form the basic profile. Inference is then used to enhance the interest profile, with the 50% inference rule applied to all ontological is-a relationships, up to the root class, for each observed event.

The event interest values were chosen to balance the feedback in favour of explicitly provided feedback, which is likely to be the most reliable. The 50% inference value was chosen to reflect the reduction in certainty you get the further away from the observed behaviour you move. Determining optimal values for these parameters would require further empirical evaluation.

3.3 Bootstrapping with an External Ontology

Recommender systems suffer from the cold-start problem [15], where the lack of initial behavioural information significantly reduces the accuracy of user profiles, and hence recommendations. This poor performance can deter users from adopting the system, which of course prevents the system from acquiring more behaviour data; it is possible that a recommender system will never be used enough to overcome its cold-start.

In one of our experiments we take an external ontology containing publication and personnel data about academic researchers and integrate it with the Quickstep recommender system. The knowledge held within the external ontology is used to bootstrap initial user profiles, with the aim of reducing the cold-start effect. The external ontology uses the same research topic ontology as the Quickstep system, providing a firm basis for communication. The external ontology contains publications and authorship relationships, projects and project membership, staff and their roles and other such knowledge. Knowledge of publications held within the external ontology is used to infer historical interests for new users, and network analysis of ontological relationships is used to discover similar users whose own interests might be used to bootstrap a new user's profile.

The two bootstrapping algorithms used in our experiment are shown in Figs. 4 and 5. The new-system initial profile algorithm takes all the publications of a user and creates a profile of historical interests. The assumption is that a user's previous publications indicate that user's interests. The new-user initial profile algorithm takes a set of similar users, obtained via network analysis of the external ontologies project membership and inter-staff relationships, and includes these users interests into the bootstrap profile. Historical publication interests from the new user are also added as before. The parametric values shown in Figs. 4 and 5 were empirically determined after several experimental runs using test data.

In addition to using the ontology to bootstrap the recommender system, our experiment uses the interest profiles held within the recommender system to continually update the external ontology. Interest acquisition is a problematic task for ontologies that are based on static knowledge sources, and this synergistic relationship provides a useful source of personal knowledge about individual researchers.

$$\text{topic interest}(t) = \sum_{\substack{1..\text{ publications} \\ \text{belonging to class } t}}^{n} 1 / \text{publication age}(n)$$

new-system initial profile = (t, topic interest(t))*

t = <research paper topic>

Interest value for super-class per topic = 50% of sub-class

Fig. 4. New-system initial profile algorithm

$$\text{topic interest}(t) = \frac{\gamma}{N_{similar}} \sum_{1..N_{similar}}^{u} \text{profile interest}(u,t) + \sum_{1..N_{pubs\,t}}^{n} 1 / \text{publication age}(n)$$

profile interest(u,t) = interest of user u in topic t * confidence
new-user initial profile = (t, topic interest(t))*

t = research paper topic
u = user
γ = weighting constant >= 0
$N_{similar}$ = number of similar users
$N_{pubs\,t}$ = number of publications belonging to class t
confidence = confidence in similarity of user

Interest value for super-class per topic = 50% of sub-class

Fig. 5. New-user initial profile algorithm

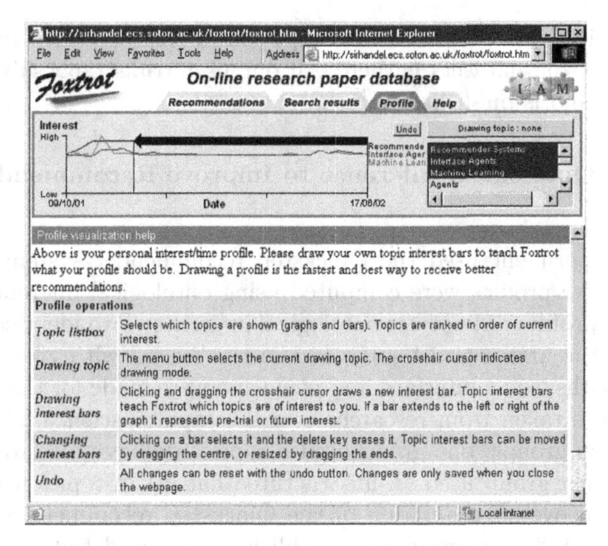

Fig. 6. Foxtrot profile visualization interface

3.4 Profile Visualization Using Ontological Concepts

Since users can understand the topics held within the ontology, the user profiles can be visualized. These visualizations allow users to see what the system thinks they are interested in and hence allow them to gain an insight into how the system works. Profile visualization thus provides users with a conceptual model of how the profiling algorithm works, allowing users to gain trust in the system and providing users with a feeling of control over what's going on. With a better conceptual model user expectations should be more realistic.

The Foxtrot recommender system visualizes profiles using a time/interest graph. In addition to simply visualizing profiles, a drawing package metaphor is used to allow users to draw interest bars directly onto the time/interest graph. This allows the system to acquire direct profile feedback, which can be used by the profiler to improve profile accuracy and hence recommendation accuracy. Figure 6 shows the profile visualization interface.

4 Case Study: Experimentation Results

We have conducted three experiments with our two recommender systems. The Quickstep recommender system is used to measure the performance gain seen when using profile inference, and the reduction in the cold-start seen when an external ontology is used for bootstrapping. The Foxtrot recommender system is used to measure the effect profile visualization has on profile accuracy, and to perform a large-scale assessment of our overall ontological approach to recommender systems.

A more in-depth statistical investigation of this approach has been performed using the datasets gathered in our user trials (260 subjects, 15,792 documents) and is published in [17].

4.1 Using Ontological Inference to Improve Recommendation Accuracy

Our first experiment used the Quickstep recommender system to compare subjects whose profiles were computed using ontological inference with subjects whose profiles did not use ontological inference. Two identical trials were conducted, the first with 14 subjects and the second with 24 subjects, both over 1.5 months; some interface improvements were made for the second trial. Subjects were taken from researchers in a computer science laboratory and split into two groups; one group used a topic ontology and profile inference while the other group used an unstructured flat list of topics with no profile inference. An overall evaluation of the Quickstep recommender system was also performed. This experiment is published in more detail in [16].

This experiment found that ontological profile users provided more favourable feedback and had superior recommendation accuracy. Figures 7 and 8 shows these results. Users provide feedback on their individual recommendations, rating them as "interesting", "uninteresting" or "no comment". Good topics are defined as those not rated as "uninteresting" by users. A jump is where the user jumps to a recommended paper by opening it via the web browser. Jumps are correlated with topic interest feedback, so a good jump is a jump to a paper on a good topic. Recommendation accuracy is the

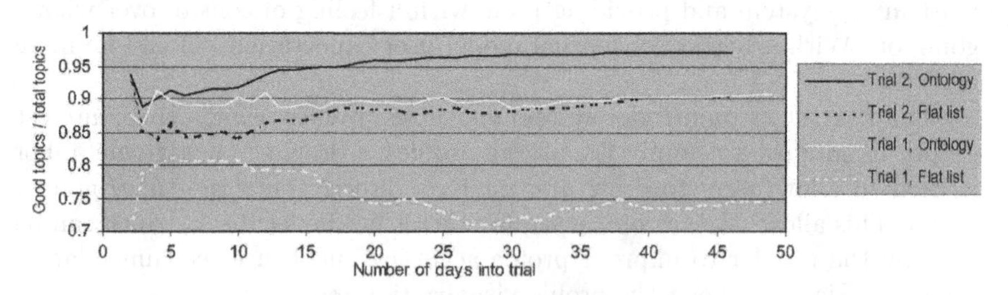

Fig. 7. Good topics to total topics ratio

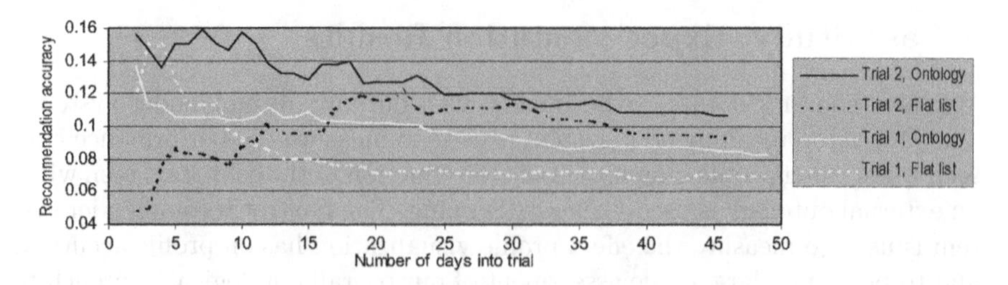

Fig. 8. Recommendation accuracy

ratio of good jumps to recommendations, and is an indication of the quality of the recommendations being made as well as the accuracy of the profile.

The ontology groups from the two trials have a 7% and 15% higher topic acceptance. In addition to this trend, the first trial ratios are about 10% lower than the second trial ratios, probably as a result of the interface improvements that made the feedback options less confusing. There is a small 1% improvement in recommendation accuracy by the ontology group. Both trials show between 7% and 11% recommendation accuracy.

Since 10 recommendations were provided at a time, a recommendation accuracy of 10% means that on average there was one good recommendation in each set presented to the user. We regard providing one good recommendation upon each visit to the recommendation web site as demonstrating significant utility.

While not statistically significant due to sample size, the results suggest how using ontological inference in the profiling process results in superior performance over using a flat list of unstructured topics. The ontology users tended to have more "rounder" profiles, including topics of interest that were not directly browsed. This increased the accuracy of the profiles, and hence usefulness of the recommendations.

4.2 Ontological Bootstrapping to Reduce the Cold-Start Problem

Our second experiment integrated the Quickstep recommender system with an external ontology to evaluate how using ontological knowledge could reduce the cold-start problem . The external ontology used was based on a publication database and personnel database, coupled with a tool for performing network analysis of ontological relationships to discover similar users. The behavioural log data from the previous experiment was used to simulate the bootstrapping effect both the new-system and new-user initial profiling algorithms would have. This experiment is published in more detail in [15].

Subjects were selected from those in the previous experiment who had entries within the external ontology. We selected nine subjects in total and their URL browsing logs were broken up into weekly log entries. Seven weeks of browsing behaviour were taken from the start of the Quickstep trials, and an empty log created to simulate the very start of the trial where no behaviour has yet been recorded.

Two bootstrapping algorithms were tested, the new-system and new-user initial profile algorithms described earlier. As the new-system algorithm bootstraps a completely cold-start we tested from week 0 to week 7. The new-user algorithm requires the system to have been running for a while, so we added the new user on week 7, after the new-system cold-start was over.

Two measurements were made to measure the reduction in the cold-start. The first, profile precision, measures how many topics were mentioned in both the bootstrapped profile and benchmark profile. Profile precision is an indication of how quickly the profile is converging to the final state, and thus

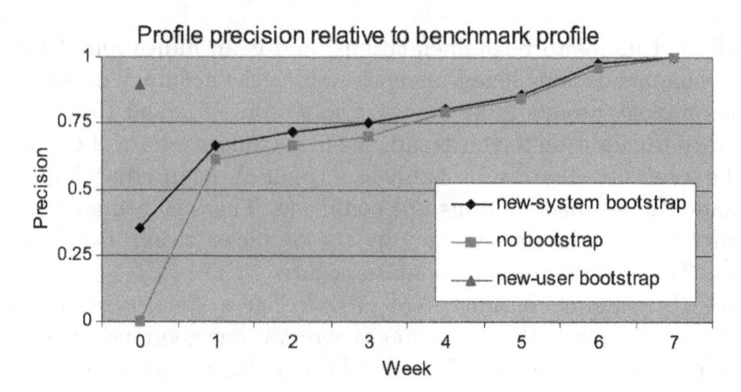

Fig. 9. Bootstrapping algorithm performance

how quickly the effects of the cold-start are overcome. The second, profile error rate, measures how many topics appeared in the bootstrapped profile that did not appear within the benchmark profile. Profile error rate is an indication of the errors introduced by the two bootstrapping algorithms. Figure 9 shows the precision results. The new-user result appears on week 0 to indicate the first week for the new-user, even though the system itself had been running for 7 weeks.

The new-system algorithm produced profiles with a low error rate of 0.06 and a reasonable precision of 0.35. This reflects that previous publications are a good indication of users current interests, and so can produce a good starting point for a bootstrap profile. The new-user algorithm achieved good precision of 0.84 at the expense of a significant 0.55 error rate.

This experiment suggests that using an ontology to bootstrap user profiles can significantly reduce the impact of the recommender system cold-start problem. It is particularly useful for the new-system cold-start problem, where the alternative is to start with no information at all and hence a profile precision of zero.

4.3 Visualizing Profiles to Improve Profile Accuracy

Our third experiment used the Foxtrot recommender system to compare subjects who could visualize their profiles and provide profile feedback with subjects who could only use traditional relevance feedback. An overall evaluation of the Foxtrot recommender system was also performed.

This experimental trial took place over the academic year 2002, starting in November and ending in July. Of the 260 subjects registered to use the system, 103 used the web page, and of these 37 subjects used the system three or more times. All 260 subjects used the web proxy and hence their browsing was recorded and daily profiles built. By the end of the trial the research paper database had grown from 6,000 to 15,792 documents as a result of subject web browsing.

Subjects were divided into two groups. The first "profile feedback" group had full access to the system and its profile visualization and profile feedback options; the second "relevance feedback" group were denied access to the profile interface. A total of nine subjects provided profile feedback.

Towards the end of the trial an additional email feature was added to the recommender system. This email feature sent out weekly emails to all users who had used the system at least once, detailing the top three papers in their current recommendation set. Email notification was started in May and ran for the remaining 3 months of the trial.

Recommendation accuracy, profile accuracy and profile predictive accuracy was measured. Profile accuracy measures the number of papers jumped to or browsed that match the top three profile topics each day. This is a good measure of the accuracy of the current interests within a profile at any given time. Profile predictive accuracy measures the number of papers jumped to or browsed that match the top three profile topics in a 4-week period after the day the profile was created. This measures the ability of a profile to predict subject interests. Figures 10 and 11 show these results.

The "profile feedback" group outperformed the "relevance feedback" group for most of the metrics, and the experimental data revealed several trends. Email recommendation appeared to be preferred by the "relevance feedback" group, and especially by those users who did not regularly check their web page recommendations. A reason for this could be that since the "profile feedback" group used the web page recommendations more, they needed to use the email recommendations less. There is certainly a limit to how many recommendations any user needs over a given time period; in our case nobody regularly checked for recommendations more than once a week. The overall recommendation accuracy was about 1%, or 2–5% for the profile feedback group.

This third experiment shows that both profile visualization and profile feedback can significantly improve the profiling accuracy and the

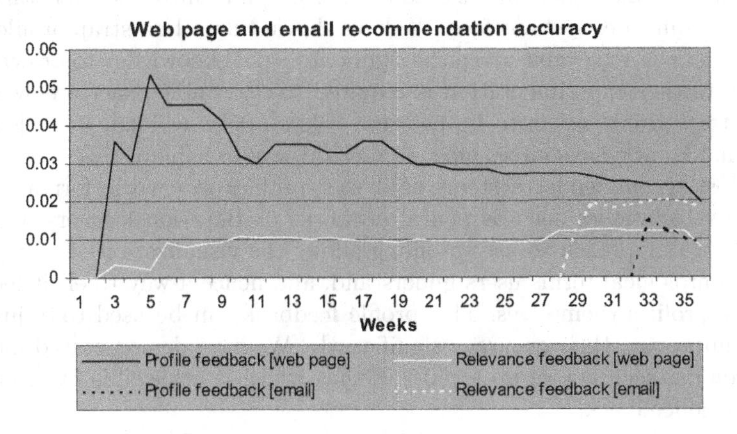

Fig. 10. Recommendation accuracy

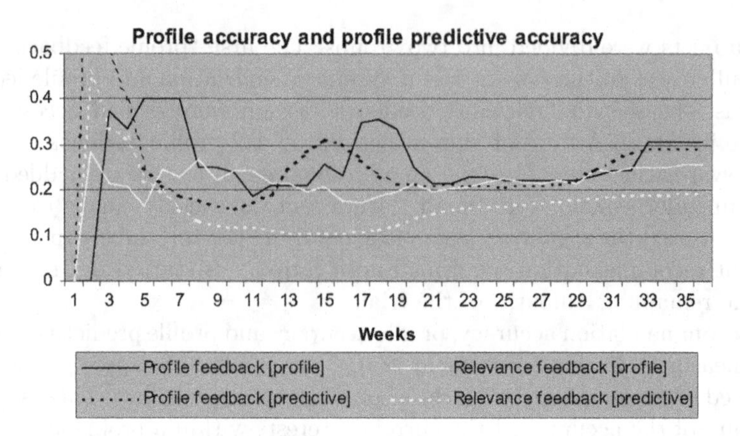

Fig. 11. Profile accuracy and predictive profile accuracy

recommendation process. Our ontological approach makes this possible because user profiles are represented in terms the users can understand.

5 Case Study: Conclusions

Through our three experiments we have demonstrated that using an ontological approach to user profiling offers significant benefits to recommender systems.

Ontological inference, even simple inference such as using is-a relationships to infer general interests, can improve profiling process and hence the recommendation accuracy of a recommender system. We achieve a 7–15% increase recommendation accuracy using just is-a relationships, and we feel it is clear that a more complete domain ontology, with more informative relationships, could perform significantly better.

External ontologies can be used to reduce significantly the cold-start problem recommender systems face. We have shown that a bootstrap profile precision of 35% is achievable given the right ontological knowledge to drawn upon. While further experimentation is required to determine exactly how good a bootstrap profile needs to be before a cold-start is avoided, it is clear that external knowledge sources offer a practical way to achieve this.

Most recommender systems hold user profiles in cryptic formats generated by techniques such as neural networks or Bayesian learners. Using an ontological approach to user profiling allows the visualization of user profiles using ontological terms users understand, and hence a way to elicit feedback on the profiles themselves. This profile feedback can be used to adjust profiles, improving their accuracy significantly. We have demonstrated increases in profiling accuracy of up to 50% of that which is achievable by traditional relevance feedback.

These three features are implemented in our two experimental recommender systems. Overall recommendation accuracy, for individual recommendations, of 7–11% for a laboratory based subject group and 2–5% recommendation accuracy for a larger department based group is demonstrated. This gives an average of one good recommendation per set of recommendations provided for the small group of about 20 users, and one every other set for the larger group of about 200 users. Both these systems compare favourably with other systems in the literature when the problem domains are taken into account.

Acknowledgements

This work was funded by EPSRC studentship award number 99308831 and the Interdisciplinary Research Collaboration In Advanced Knowledge Technologies (AKT) project GR/N15764/01.

References

1. Balabanovic M, Shoham Y (1997) Fab: Content-based, collaborative recommendation. Communications of the ACM 40(3):67–72.
2. Billsus D, Pazzani MJ (1998) A personal news agent that talks, learns and explains. In Autonomous Agents 98, Minneapolis MN, USA.
3. Burke R (2000) Knowledge-based Recommender Systems. In: Kent A (ed.) Encyclopedia of Library and Information Systems, vol. 69, supplement 32. Marcel Dekker, New York.
4. Eirinaki M, Lampos C, Paulakis S, Vazirgiannis M (2004) Web personalization integrating content semantics and navigational patterns. In Proceedings of the 6th annual ACM international workshop on Web information and data management, Washington DC, USA.
5. Felfernig A, Friedrich G, Jannach D, Zanker M (2006) An integrated environment for the development of knowledge-based recommender applications. International Journal of Electronic Commerce 11(2):11–34.
6. Fisher DH (1987) Knowledge acquisition via incremental concept clustering. Machine Learning 2(2):139–172.
7. Gerhart A (2002) Open directory project search results and ODP status. Search Engine Guide.
8. Guarino N, Giaretta P (1995) Ontologies and knowledge bases: towards a terminological clarification. In Mars N (ed.) Towards Very Large Knowledge Bases: Knowledge Building and Knowledge Sharing. IOS Press, Amsterdam, pp. 25–32.
9. Kobsa A (1993) User modeling: recent work, prospects and hazards. In Schneider-Hufschmidt M, Khme T, Malinowski U (ed.) Adaptive User Interfaces: Principles and Practice. Elsevier Amsterdam.
10. Konstan JA, Miller BN, Maltz D, Herlocker JL, Gordon LR, Riedl J (1997) GroupLens: applying collaborative filtering to usenet news. Communications of the ACM 40(3):77–87.
11. Lang K (1995) NewsWeeder: learning to filter NetNews. In ICML95 Conference Proceedings, pp. 331–339.

12. Lee T, Chun J, Shim J, Lee S (2006) An ontology-based product recommender system for B2B marketplaces. International Journal of Electronic Commerce 11(2):125–155.
13. Linden G, Smith B, York J (2003) Amazon.com recommendations: Item-to-Item collaborative filtering. IEEE Internet Computing 7(1):76–80.
14. Mccallum AK, Nigam K, Rennie J, Seymore K (2000) Automating the construction of internet portals with machine learning. Information Retrieval 3(2):127–163.
15. Middleton SE, Alani H, Shadbolt NR, De Roure DC (2002) Exploiting synergy between ontologies and recommender systems. In International Workshop on the Semantic Web, Proceedings of the 11th International World Wide Web Conference WWW-2002, Hawaii, USA.
16. Middleton SE, De Roure DC, Shadbolt NR (2001) Capturing knowledge of user preferences: ontologies on recommender systems. In Proceedings of the First International Conference on Knowledge Capture K-CAP 2001, Victoria, BC, Canada.
17. Middleton SE, Shadbolt NR, De Roure DC (2004) Ontological user profiling in recommender systems. ACM Transactions on Information Systems (TOIS) 22(1):54–88, ACM Press, New York.
18. Schickel-Zuber V, Faltings B (2007) Using hierarchical clustering for learning the ontologies used in recommendation systems. In KDD 2007, California, USA.
19. Sebastiani F (2002) Machine learning in automated text categorization. ACM Computing Surveys 34(1):1–47.
20. Szomszor M, Cattuto C, Alani H, O'Hara K, Baldassarri A, Loreto V, Servedio VDP (2007) Folksonomies, the Semantic Web, and Movie Recommendation. In Proceedings of 4th European Semantic Web Conference, Bridging the Gap between Semantic Web and Web 2.0 (in press), Innsbruck, Austria.
21. Zhao Y, Karypis G (2005) Hierarchical clustering algorithms for document datasets. Data Mining and Knowledge Discovery 10:141–168.
22. Ziegler C, McNee SM, Konstan JA, Lausen G (2005) Improving recommendation lists through topic diversification. In Proceedings of the 14th international conference on World Wide Web, Chiba, Japan.

Author Index

Subject Index

Printed in the United States
By Bookmasters